Interdisciplinary Applied Mathematics

Volume 29

Problems in engineering, computational science, and the physical and biological sciences are using increasingly sophisticated mathematical techniques. Thus, the bridge between the mathematical sciences and other disciplines is heavily traveled. The correspondingly increased dialog between the disciplines has led to the establishment of the series: *Interdisciplinary Applied Mathematics.*

The purpose of this series is to meet the current and future needs for the interaction between various science and technology areas on the one hand and mathematics on the other. This is done, firstly, by encouraging the ways that mathematics may be applied in traditional areas, as well as point towards new and innovative areas of applications; and, secondly, by encouraging other scientific disciplines to engage in a dialog with mathematicians outlining their problems to both access new methods and suggest innovative developments within mathematics itself.

The series will consist of monographs and high-level texts from researchers working on the interplay between mathematics and other fields of science and technology.

Interdisciplinary Applied Mathematics

Volumes published are listed at the end of the book.

George Karniadakis
Ali Beskok
Narayan Aluru

Microflows and Nanoflows

Fundamentals and Simulation

Foreword by Chih-Ming Ho

Springer

George Karniadakis
Center for Fluid Mechanics
Brown University
Providence, RI 02912
USA

Ali Beskok
Department of Mechanical Engineering
Texas A & M University
College Station, TX 77843
USA

Narayan Aluru
Beckmann Institute for the Advancement of Science and Technology
University of Illinois at Urbana-Champaign
Urbana, IL 61801
USA

Editors

S.S. Antman
Department of Mathematics and Institute for Physical Science and Technology
University of Maryland
College Park, MD 20742
USA
ssa@math.umd.edu

J.E. Marsden
Control and Dynamical Systems
Mail Code 107-81
California Institute for Technology
Pasadena, CA 91125
USA
marsden@cds.caltech.edu

L. Sirovich
Division of Applied Mathematics
Brown University
Providence, RI 02912
USA
chico@camelot.mssm.edu

Mathematics Subject Classification (2000): 70-01, 70-02, 70-08, 76-02, 76D05, 76D06, 76D07, 76D08, 76D45

Library of Congress Control Number: 2005923507

ISBN-10: 0-387-22197-2
ISBN-13: 978-0387-22197-7

Printed on acid-free paper.

Printed in the United States of America.

9 8 7 6 5 4 3 2

springer.com

Foreword by Chih-Ming Ho

Fluid flow through small channels has become a popular research topic due to the emergence of biochemical lab-on-the-chip systems and micro electromechanical system fabrication technologies, which began in the late 1980s. This book provides a comprehensive summary of using computational tools (Chapters 14–18) to describe fluid flow in micro and nano configurations. Although many fundamental issues that are not observed in macro flows are prominent in microscale fluid dynamics, the flow length scale is still much larger than the molecular length scale, allowing for the continuum hypothesis to still hold in most cases (Chapter 1). However, the typical Reynolds number is much less than unity, due to the small transverse length scale, which results in a high-velocity gradient. For example, a $10^5\ \text{sec}^{-1}$ shear rate is not an uncommon operating condition, and thus high viscous forces are prevalent, resulting in hundreds or thousands of ψ hydrodynamic pressure drops across a single fluidic network. Consequently, it is not a trivial task to design micropumps that are able to deliver the required pressure head without suffering debilitating leakage. Electrokinetic and surface tension forces (Chapters 7 and 8) are used as alternatives to move the embedded particles and/or bulk fluid. The high viscous damping also removes any chance for hydrodynamic instabilities, which are essential for effective mixing. Mixing in micro devices is often critical to the overall system's viability (Chapter 9). Using electrokinetic force to reach chaotic mixing is an interesting research topic. In these cases, the electrical properties, e.g., dielectric constants, rather than the viscosity determine the efficiency of transport.

The National Nano Initiative, established first in the USA (www.nano.gov)

and subsequently in many other countries, has pushed the length scale range of interest from microns down to nanometers. Flows in these regimes start to challenge the fundamental assumptions of continuum mechanics (Chapter 1). The effects of the molecules in the bulk of the fluid versus those molecules in proximity to a solid boundary become differentiated (Chapter 10). These are extremely intriguing aspects to be investigated for flows in small configurations. The demarcation between the continuum and the noncontinuum boundary has yet to be determined and inevitably will have a tremendous influence on the understanding of small-scale fluid behavior as well as system design.

The ratio between the size of the channel and that of the molecule is not the only parameter that validates the continuum assumption. In biological applications, for example, molecules with large conformation changes, electrical charges, and polar structures are frequently encountered. These variables make it impossible to determine whether a flow can be considered a continuum based only on a ratio of sizes (Chapter 11). When a continuum flow of a Newtonian fluid is assumed, molecular effects are defined by the governing equations of traditional fluid mechanics. Interactions among fluid molecules are expressed by a physical constant, which is viscosity. The no-slip condition represents the interactions between the fluid and the solid surface molecules. Both viscosity and the no-slip condition are concepts developed under the framework of continuum. Deviations from the bulk viscosity and the no-slip condition can lead to other results due to the breakdown of the continuum assumption (Chapters 2 and 10).

In the nanoflow regime, not many molecules are situated far away from the channel wall. Therefore, the motion of the bulk fluid is significantly affected by the potential fields generated by the molecules near the solid wall. Near the surface, the fluid molecules do not flow freely. At a distance of a few fluid molecule layers above the surface, the flow has very different physical constants from the bulk flow. The surface effects are strong not only in nano configurations (Chapter 10); even in microfluidic devices, the performance, e.g., surface fouling, is dependent on the surface property. We frequently spend more time on modifying the surface properties than on designing and fabricating devices. As a result of our limited understanding of fluidic behavior within nanoscale channels (Chapters 10 through 13), many vital systematic processes of today's technology are arduously, yet imperfectly, designed. Delivering and stopping a picoliter volume of fluid to a precise location with high accuracy as well as the separation and mixing of nano/micro particles in a fluid medium of high ionic concentration remains a challenging task. By furthering the understanding of fluid interactions in the nano world, many of the interesting mysteries and challenges that have puzzled scientists will be revealed.

June 2004, Los Angeles, California, USA Chih-Ming Ho

Preface

In the early 1990s, microchannel flow experiments at the University of Pennsylvania by the groups of H. Bau and J. Zemel revealed intriguing results for both liquids and gases that sparked excitement and new interest in the study of low Reynolds number flows in microscales. Another influential development at about the same time was the fabrication of the first microchannel with integrated pressure sensors by the groups of C.M. Ho (UCLA) and Y.C. Tai (Caltech). While the experimental results obtained at the University of Pennsylvania indicated global deviations of microflows from canonical flows, pointwise measurements for gas flows with pressure sensors, and later with temperature sensors, revealed a new flow behavior at microscales not captured by the familiar continuum theory. In microgeometries the flow is *granular* for liquids and *rarefied* for gases, and the walls "move." In addition, other phenomena such as thermal creep, electrokinetics, viscous heating, anomalous diffusion, and even quantum and chemical effects may become important. Most important, the material of the wall and the quality of its surface play a very important role in the momentum and energy exchange. One could argue that at least for gases the situation is similar to low-pressure high-altitude aeronautical flows, which were studied extensively more than 40 years ago. Indeed, there is a similarity in a certain regime of the Knudsen number. However, most gas microflows correspond to a low Reynolds number and low Mach number, in contrast to their aeronautical counterparts. Moreover, the typical microgeometries are of very large aspect ratio, and this poses more challenges for numerical modeling, but also creates opportunities for obtaining semianalytical results. For liquids no such analogy exists and their dynamics in confined microgeometries, especially at the submicron range, is much more complex.

The main differences between fluid mechanics at microscales and in the macrodomain can be broadly classified into four areas:

- Noncontinuum effects,
- surface-dominated effects,
- low Reynolds number effects, and
- multiscale and multiphysics effects.

Some of these effects can be simulated with relatively simple modifications of the standard numerical procedures of computational fluid dynamics. However, others require new simulation approaches not used typically in the macrodomain, based on multiscale algorithms. For gas microflows, compressibility effects are very important because of relatively large density gradients, although the Mach number is typically low. Depending on the degree of rarefaction, corrections at the boundary or everywhere in the domain need to be incorporated. Increased rarefaction effects may make the constitutive models for the stress tensor and the heat flux vector in the Navier–Stokes equations invalid. On the other hand, working with the Boltzmann equation or with molecular dynamics implementation of Newton's law directly is computationally prohibitive for complex microgeometries. The same is true for liquids, since atomistic simulation based on Newton's law for individual atoms is restricted to extremely small volumes. Therefore, mesoscopic and hybrid atomistic–continuum methods need to be employed for both gas and liquid microflows to deal effectively with deviations from the continuum and to provide a link with the large domain sizes. Most important, microflows occur in devices that involve simultaneous action in the flow, electrical, mechanical, thermal, and other domains. This, in turn, implies that fast and flexible algorithms and low-dimensional modeling are required to make full-system simulation feasible, similar to the achievements of the 1980s in VLSI simulation.

There has been significant progress in the development of microfluidics and nanofluidics at the application as well as at the fundamental and simulation levels since the publication of an earlier volume of this book (2001). We have, therefore, undertaken the "nontrivial" task of updating the book in order to include these new developments. The current book covers length scales from angstroms to microns (and beyond), while the first volume covered scales from one hundred nanometers to microns (and beyond). We have maintained the emphasis on fundamental concepts with a mix of semi-analytical, experimental, and numerical results, and have outlined their relevance to modeling and analyzing functional devices. The first two co-authors (GK and AB) are very pleased to have a new co-author, Prof. N.R. Aluru, whose unique contributions have made this new volume possible. We are also grateful to Springer, and in particular to Senior Editor in Mathematics Dr. Achi Dosanjh, who gave us this opportunity.

The majority of the new developments are in Chapters 7 through 18, most of which contain totally new material. In addition, all other Chapters (1 through 6) have been modified, and in some cases new material has also been added. We have divided the material into three main categories by subject:

1. Gas Flows (Chapters 2–6).
2. Liquid Flows (Chapters 7–13)
3. Simulation Techniques (Chapters 14–18)

The last category also contains two Chapters (17 and 18) on low-dimensional modeling and simulation, in addition to chapters on multiscale modeling of gas and liquid flows. The entire material can be used in a two-semester first- or second-year graduate course. Also, selected chapters can be used for a short course or an undergraduate-level course.

In the following we present a brief overview of the material covered in each chapter.

In Chapter 1 we provide highlights of the many concepts and devices that we will discuss in detail in the subsequent chapters. For historic reasons, we start with some prototype Micro-Electro-Mechanical-Systems (MEMS) devices and discuss such fundamental concepts as breakdown of constitutive laws, new flow regimes, and modeling issues encountered in microfluidic and nanofluidic systems. We also address the question of full-system simulation of microsystems and introduce the concept of macromodeling.

In Chapter 2 we first present the basic equations of fluid dynamics for both incompressible and compressible flows, and discuss appropriate nondimensionalizations. Subsequently, we consider the compressible Navier–Stokes equations and develop a general boundary condition for velocity slip. The validity of this model is assessed in subsequent chapters.

In Chapter 3 we consider shear-driven gas flows with the objective of modeling several microsystem components. In order to circumvent the difficulty of understanding the flow physics for complex engineering geometries, we concentrate on prototype flows such as the linear and oscillatory Couette flows in the slip, transition, and free-molecular flow regimes, and flow in shear-driven microcavities and microgrooves.

In Chapter 4 we present pressure-driven gas flows in the slip, transition and free molecular flow regimes. In the slip flow regime, we first validate simulation results based on compressible Navier–Stokes solutions employing various slip models introduced in Chapter 2. In addition, we examine the accuracy of the one-dimensional Fanno theory for microchannel flows, and we study inlet flows and effects of roughness. In the transition and free-molecular regime we develop a unified model for predicting the velocity profile and mass flowrate for pipe and duct flows.

In Chapter 5 we consider heat transfer in gas microflows. In the first section we concentrate on the thermal creep (transpiration) effects that may be important in channels with tangential temperature gradients on their surfaces. We also study other temperature-induced flows and investigate the validity of the heat conduction equation in the limit of zero Knudsen number. In the second and third sections we investigate the combined effects of thermal creep, heat conduction, and convection in pressure-, force-, and shear-driven channel flows.

In Chapter 6 we consider rarefied gas flows encountered in applications other than simple microchannels. In the first section, we present the lubrication theory and its application to the slider bearing and squeezed film problems. In the second and third sections, we consider separated flows in internal and external geometries in the *slip flow* regime in order to investigate the validity of continuum-based slip models under flow separation. In the fourth section, we present theoretical and numerical results for Stokes flow past a sphere including rarefaction effects. In the fifth section we summarize important results on gas flows through microfilters used for capturing and detecting airborne biological and chemical particles. In the last section, we consider high-speed rarefied flows in micronozzles, which are used for controlling the motion of microsatellites.

In Chapter 7 we present basic concepts and a mathematical formulation of microflow control and pumping using electrokinetic effects, which do not require any moving components. We cover electroosmotic and electrophoretic transport in detail both for steady and time-periodic flows, and we discuss simple models for the near-wall flow. We also present dielectrophoresis, which enables separation and detection of similar size particles based on their polarizability.

In Chapter 8 we consider surface tension-driven flows and capillary phenomena involving wetting and spreading of liquid thin films and droplets. For microfluidic delivery on open surfaces, electrowetting and thermocapillary along with dielectrophoresis have been employed to move continuous and discrete streams of fluid. A new method of actuation exploits optical beams and photoconductor materials in conjunction with electrowetting. Such electrically or chemically defined paths can be reconfigured dynamically using electronically addressable arrays that respond to electric potential, temperature, or laser beams and control the direction, timing, and speed of fluid droplets. In addition to the above themes, we also study bubble transport in capillaries including both classical theoretical results and more recent theoretical and experimental results for electrokinetic flows.

In Chapter 9 we consider micromixers and chaotic advection. In microchannels the flow is laminar and steady, so diffusion is controlled solely by the diffusivity coefficient of the medium, thus requiring excessive amounts of time for complete mixing. To this end, chaotic advection has been exploited in applications to accelerate mixing at very low speeds. Here, we present the basic ideas behind chaotic advection, and discuss examples of passive and active mixers that have been used in microfluidic applications. We also provide effective quantitative measures of characterizing mixing.

In Chapter 10 we consider simple liquids in nanochannels described by standard Lennard–Jones potentials. A key difference between the simulation of the fluidic transport in confined nanochannels and at macroscopic scales is that the well-established continuum theories based on Navier–Stokes equations may not be valid in confined nanochannels. Therefore, atomistic scale simulations are required to shed fundamental insight on fluid transport. Here we discuss density distribution, diffusion transport, and validity of the Navier–Stokes equations. In the last section we discuss in detail the slip condition at solid–liquid interfaces, and present experimental and computational results as well as conceptual models of slip. We also revisit the lubrication problem and present the Reynolds–Vinogradova theory for hydrophobic surfaces.

In Chapter 11 we focus on water and its properties in various forms; this is one of the most actively investigated areas because of its importance in nature. The anomalies that exist in the bulk properties of water make it very interesting and challenging for research, and a vast deal of literature is already available. Even though water has been studied for more than 100 years now, its properties are far from understood. With the advances in fabrication of nanochannels that are only a few molecular diameters in critical dimension, the properties of water in confined nanochannels have recently received a great deal of attention. In this chapter, after introducing some definitions and atomistic models for water, we present the static and dynamic behavior of water in confined nanochannels.

In Chapter 12 we discuss the fundamentals and simulation of electroosmotic flow in nanochannels. The basic theory was covered in Chapter 7, so here the limitations of the continuum theory for electroosmotic flow in nanochannels are identified by presenting a detailed comparison between continuum and MD simulations. Specifically, the significance of the finite size of the ions and the discrete nature of the solvent molecules are highlighted. A slip boundary condition that can be used in the hydrodynamic theory for nanochannel electroosmotic flows is presented. Finally, the physical mechanisms that lead to the charge inversion and flow reversal phenomena in nanochannel electroosmotic flows are discussed.

In Chapter 13 we focus on functional fluids and on functionalized devices, specifically nanotubes. The possibility to target and precisely control the electrooptical as well as the mechanical properties of microstructures in a dynamic way using external fields has opened new horizons in microfluidics research including new concepts and protocols for micro- and nanofabrication. On the more fundamental level, systematic studies of paramagnetic particles or charged particles and their dynamics offer insight into the role of Brownian noise in microsystems as well as conceptual differences between deterministic and stochastic modeling. This is studied in the first part of this chapter. In the second part of the chapter we study carbon nanotubes and their properties. Carbon nanotubes with diameters as small as 5–10 Å are comparable to the diameters encountered in biological ion channels. By *functionalizing* carbon nanotubes, it is possible to tune the surface properties of carbon nanotubes to investigate the function of a variety of ion channels. To enable such advances, it is important to understand how water, ions, and various electrolytes interact with carbon nanotubes and functionalized nanotubes.

In Chapter 14 we discuss representative numerical methods for continuum-based simulations. The significant geometric complexity of flows in microsystems suggests that finite elements are more suitable than finite differences, while high-order accuracy is required for efficient discretization. To this end, we focus on spectral element and meshless methods in stationary and moving domains. We also discuss methods for modeling particulate microflows and focus on the force coupling method, a particularly fast approach suitable for three-dimensional simulations. These methods represent three different classes of discretization philosophies and have been used with success in diverse applications of microsystems.

In Chapter 15 we discuss theory and numerical methodologies for simulating gas flows at the mesoscopic and atomistic levels. Such a description is necessary for gases in the transition and free-molecular regimes. First, we present the Direct Simulation Monte Carlo (DSMC) method, a stochastic approach suitable for gases. We discuss limitations and errors in the steady version of DSMC and subsequently present a similar analysis for the unsteady DSMC. In order to bridge scales between the continuum and atomistic scales we present the Schwarz iterative coupling algorithm and apply it to modeling microfilters. We then give an overview of the Boltzmann equation, describing in some detail gas–surface interactions, and include benchmark solutions for validation of numerical codes and of macromodels. A main result relevant to accurately bridging microdynamics and macrodynamics is the Boltzmann inequality, which we also discuss in the last section on lattice Boltzmann methods (LBM). These methods represent a "minimal" discrete form of the Boltzmann equation, and they are applicable to both compressible and incompressible flows; in fact, the majority of LBM applications focuses on incompressible flows.

In Chapter 16 we discuss theory and numerical methodologies for simulating liquid flows at the atomistic and mesoscopic levels. The atomistic description is necessary for liquids contained in domains with dimension of fewer than ten molecules. First, we present the Molecular Dynamics (MD) method, a deterministic approach suitable for liquids. We explain details of the algorithm and focus on the various potentials and thermostats that can be used. This selection is crucial for reliable simulations of liquids at the nanoscale. In the next section we consider various approaches in coupling atomistic with mesoscopic and continuum level. Such coupling is quite difficult, and no fully satisfactory coupling algorithms have been developed yet, although significant progress has been made. An alternative method is to embed an MD simulation in a continuum simulation, which we demonstrate in the context of electroosmotic flow in a nanochannel. In the last section we discuss a new method, developed in the late 1990s primarily in Europe: the dissipative particle dynamics (DPD) method. It has features of both LBM and MD algorithms and can be thought of as a coarse-grained version of MD.

In Chapter 17 we turn our attention to simulating full systems across heterogeneous domains, i.e., fluid, thermal, electrical, structural, chemical, etc. To this end, we introduce several reduced-order modeling techniques for analyzing microsystems. Specifically, techniques such as generalized Kirchhoff networks, black box models, and Galerkin methods are described in detail. In black box models, detailed results from simulations are used to construct simplified and more abstract models. Methods such as nonlinear static models and linear and nonlinear dynamic models are described under the framework of black box models. Finally, Galerkin methods, where the basic idea is to create a set of coupled ordinary differential equations, are described. The advantages and limitations of the various techniques are highlighted.

Finally, in Chapter 18 we discuss the application of these techniques to several examples in microflows. First, we present circuit and device models and their application to lab-on-a-chip systems. Then, we discuss reduced-order modeling of squeezed film damping by applying equivalent circuit, Galerkin, mixed-level, and black box models. Next, we present a compact model for electrowetting. Finally, we summarize some of the software packages that are available for reduced-order simulation.

We are very grateful to Prof. Chih-Ming Ho who agreed to provide a foreword to our book. We would like to thank all our colleagues from many different countries who have allowed us to use their work in the previous and this new and expanded edition of the book. We also want to thank Ms. Madeline Brewster at Brown University for her assistance with all aspects of this book, and our students who helped with formatting the figures, especially Vasileios Symeonidis, Pradipkumar Bahukudumbi, and Aveek Chatterjee. AB would like to thank his students I. Ahmed, P. Bahukudumbi, Prof. P. Dutta, Dr. J. Hahm, H.J. Kim, S. Kumar, Dr. J.H. Park, and Prof. C. Sert. The last author (NRA) would like to acknowledge the help of all his students, especially Chatterjee, De, Joseph, and Qiao for letting him use some of the results from their thesis work. NRA is very grateful to Profs. Karniadakis and Beskok for the opportunity to co-author this book with them. NRA would like to thank Profs. Dutton (Stanford), Hess (UIUC), Karniadakis (Brown), Law (Stanford), Pinsky (Stanford), Senturia (MIT), and White (MIT) for mentoring his career.

The first author (GK) would like to thank all members of his family for their support during the course of this effort. The second author (AB) would like to thank Carolyn, Sarah and Sinan for their continuous love, support and patience. In addition, AB would like to dedicate his work to the memory of his parents, Güngör and Çetin Beşkök. Finally, NRA is deeply indebted to all his family members, especially his parents, Subhas and Krishna Aluru, his brother, Ravi, his wife, Radhika, and his daughter, Neha, for their love, encouragement, and support.

Providence, Rhode Island, USA — George Em Karniadakis
College Station, Texas, USA — Ali Beskok
Urbana, Illinois, USA — Narayan R. Aluru

Contents

1

Basic Concepts and Technologies

In this chapter we highlight some of the concepts, devices, and modeling approaches that we shall discuss in more detail in all subsequent chapters. We have included a section on the pioneers of the field, and we present some of the key results that have had great impact on the development and the rapid growth of microfluidics and nanofluidics. Our emphasis is on fundamental concepts such as breakdown of constitutive laws, new flow regimes, and modeling issues encountered in flow microsystems. We also discuss fluid–surface interactions for liquids, such as electrokinetic effects and wetting, which become very important at very small scales. Finally, we address the question of full-system simulation of micro-electro-mechanical systems (MEMS) and introduce the concept of macromodeling.

1.1 New Flow Regimes in Microsystems

Micron- and submicron-size mechanical and biochemical devices are becoming more prevalent both in commercial applications and in scientific inquiry. Small accelerometers with dimensions measured in microns are being used to deploy air bag systems in automobiles. Tiny pressure sensors for the tip of a catheter are smaller than the head of a pin. Microactuators are moving scanning electron microscope tips to image single atoms. Novel bioassays consisting of microfluidic networks are designed for patterned drug delivery. New fabrication techniques, such as surface silicon micromachining, bulk silicon micromachining, LIGA (Lithographie Galvanoformung Abformung), and EDM (Electro Discharge Machining) have been successfully applied to microfabrication in recent years, making these microdevices possible. The capability to batch fabricate and automate these fabrication technologies makes such microdevices inexpensive (Howe et al., 1990; Bryzek et al., 1994; Reed, 1993; Trimmer, 1997). New nanofabrication techniques have emerged exploiting the concept of self-assembly for submicron-size objects (Whitesides and Grzybowski, 2002; Doyle et al., 2002).

Inherent in these new technologies is the need to develop the fundamental science and engineering of small devices. Microdevices tend to behave differently from the objects we are used to handling in our daily life (Gad-el-Hak, 1999; Ho and Tai, 1998). The inertial forces, for example, tend to be quite small, and surface effects tend to dominate the behavior of these small systems. Friction, electrostatic forces, and viscous effects due to the surrounding air or liquid become increasingly important as the devices become smaller. In general, properties (p) that are a function of the area of interaction (A) decrease more slowly than properties that depend on the volume (V), as expressed by the "square-cube" law:

$$\frac{p_1(A)}{p_2(V)} \propto \frac{L^2}{L^3} \propto \frac{1}{L}, \tag{1.1}$$

where L is the characteristic dimension of the microdevice; a typical order of magnitude is 10^6 $\mathrm{m}^2/\mathrm{m}^3$. Surface tension effects are dominant at these scales, and micropumps and microvalves have been fabricated taking advantage of this principle (Evans et al., 1997).

Typical early applications can be found in the micro- and nanoscale design of computer components such as the Winchester-type hard disk drive mechanism, where the read/write head floats 50 nm above the surface of the spinning platter (Tagawa, 1993). The head and platter together with the air layer in between form a slider bearing. The typical operating conditions correspond to low values of both Reynolds and Mach number, e.g., less than 0.6 and 0.3, respectively. The corresponding Knudsen number, which expresses the relative size of the mean free path to the size of the microflow domain, is relatively large. It is expected to increase further for the next generation of devices, since the smaller the gap between the spinning platter and the read/write head, the greater the recording capacity.

Turning now to micro-electro-mechanical systems (MEMS), one of the first microfabricated products is a polysilicon, surface micromachined side-driven motor; its fabrication, operation, and performance have been studied extensively in (Mehregany et al., 1990; Tai et al., 1989; Trimmer, 1997). A diagram of such a motor is shown in the sketch of Figure 1.1 along with the characteristic dimensions. It creates a variable capacitance by means of salient poles distributed along the periphery; a typical design may employ 12 stator poles and eight rotor poles. During operation, the shield (lower surface), bearing, and rotor are electrically grounded, and only one stator phase is excited at a given time. The motive torque has been determined through step transient measurements and it is of order 10 PN $\cdot$ m. The axial

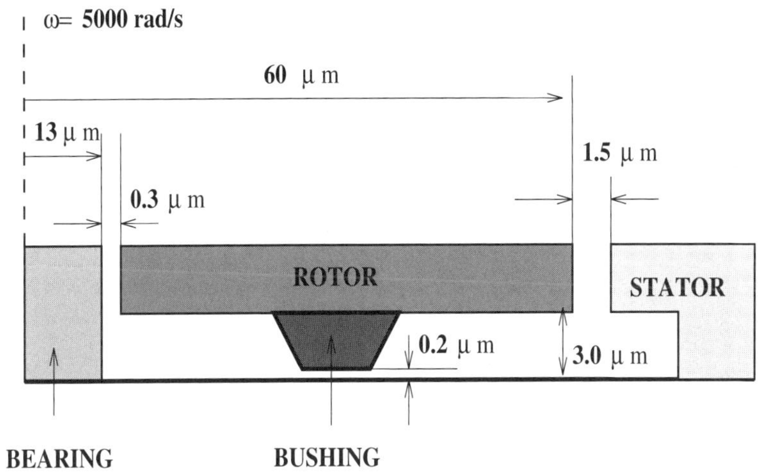

FIGURE 1.1. Diagram of an electrostatically side-driven micromotor with typical dimensions. The motor consists of a set of stator poles and rotor poles which are excited sequentially, creating an electrostatic motive torque.

force is of order 10^{-7} N, which is much larger than the typical weight of the motor, of order 10^{-10} N. Typical operating conditions for an angular speed of $\omega = 5000$ rad/sec show that the Reynolds number Re ≈ 1 based on the gap between the base and the rotor ($3\,\mu$m) and that the Mach number M is less than 0.1 based on the rotor tip speed (rotor radius is $60\,\mu m$). Under macroscale continuum conditions it is possible to have creeping flows that result in small Reynolds number Re and Mach number M. However, in the case of microflows, the Reynolds number Re is small due to the small length scales of the microdevice rather than very small velocities. Therefore, higher Mach numbers M could be achieved in microflows compared to the creeping (i.e., very slow) continuum flows. Results from a steady-state axisymmetric analysis reported in (Omar et al., 1992) showed that 75% of the viscous drag is caused by the lower surface. However, that analysis did not include slip effects, which may modify the viscous drag contribution.

A similar type of gas microflow occurs in another classic MEMS device, the electrostatic comb microdrive (Tang et al., 1989), which is shown in Figure 1.2. Electrostatic comb-drives are excellent resonant actuators that produce large motions at low drive voltage. For typical operating conditions with a resonance frequency of 75 kHz, the dimensions shown in the figure, and with the gap between the stationary and movable comb arms of order $1\,\mu$m, we calculate that Re $= 0.74$ and $M = 0.014$. Both the micromotor and the comb-drive flows are sustained due to the motion of a thin layer of polysilicon across the silicon substrate. In the simplest form, these flows can be modeled by a shear-driven flow (see Chapter 3).

An electrostatic comb-drive is one of the most important first-generation

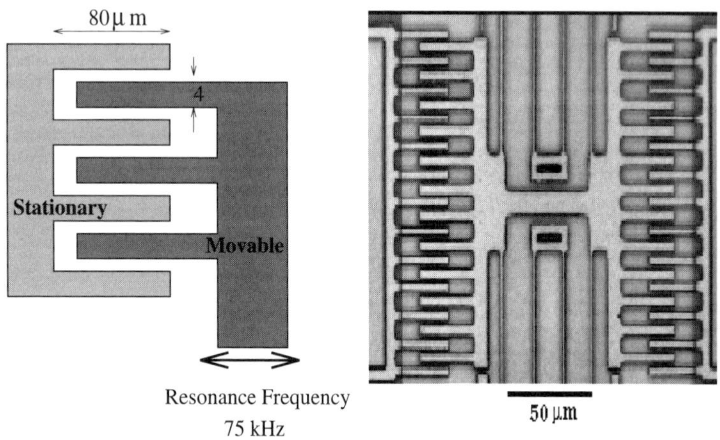

FIGURE 1.2. Left: Diagram of a typical electrostatic comb-drive mechanism with typical dimensions. Right: Actual electrostatic comb-drive. The resonator consists of a central shuttle that is suspended by a cantilever beam so that it can move horizontally. (Courtesy of D. Freeman.)

MEMS devices. It is driven by interdigitated capacitors, the electrostatic combs. In a standard comb-drive, the capacitance varies linearly with the displacement, resulting in an electrostatic force that is independent of the position of the moving fingers, except at the end of the travel, where the force becomes large (full overlap). An approximate equation for the driving electrostatic force F_e on each of the moving fingers is

$$F_e = \frac{\epsilon h V^2}{d},$$

where V is the voltage, h is the height of the fingers (direction perpendicular to the page in Figure 1.2), d is the gap, and $\epsilon = 8.85 \times 10^{-12}$ C^2/Nm^2. This formula does not include the effect of the width of the fingers, but accurate simulations performed in (Shi, 1995) show that the variation is almost linear. The magnitude of this force is very small; for example, for $h = 1$ μm; $d = 2.5\,\mu$m, and $V = 40$ V the above formula gives $F_e = 5.7$ nN, which is smaller than the more accurate value 6.3 nN obtained with a boundary element simulation in (Shi, 1995).

We now turn to the flow analysis of the comb-drive shown in Figure 1.2, for which detailed measurements were obtained by Freeman using computer microvision (Freeman et al., 1998). Specifically, stroboscopic illumination was used to obtain images at evenly spaced phases during a sinusoidal excitation. The displacements between the images were obtained using algo-

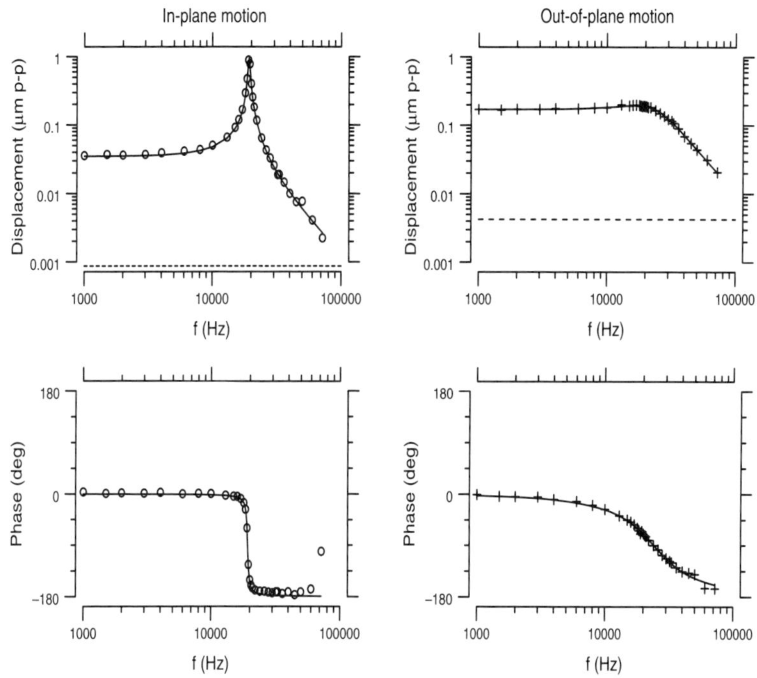

FIGURE 1.3. Magnitude and phase measurements of the comb-drive using computer microvision techniques. (Courtesy of D. Freeman.)

rithms originally developed for machine vision, and they were subsequently integrated to produce a time series for which the magnitude and phase of the motion was determined (see Figure 1.3). The frequency response was fit by a second-order system with mass m, dashpot damping coefficient C, and spring stiffness k. The quality factor defined as

$$Q = \frac{\sqrt{km}}{C}$$

was found to be $Q = 27$, and the best frequency for resonance was $f_0 = 19.2$ kHz. The mass parameter was derived from the geometry with density 2300 kg/m^3 resulting in $m = 50.06$ ng. The stiffness was obtained from the mass and measured resonant frequency resulting in $k = 0.729$ N/m.

In order for this system to be simulated accurately, the damping forces, which are primarily due to fluid motion, i.e., the viscous drag forces, should be computed accurately. A full three-dimensional simulation of this system was performed for the first time by (Ye et al., 1999) using the *FastStokes* program, which is based on boundary element methods and precorrected FFTs. A total of 23,424 panels were employed in their simulation, as shown in Figure 1.4. For kinematic viscosity of $\nu = 0.145$ cm^2/sec and density $\rho = 1.225$ kg/m^3, *FastStokes* predicted a drag force of 207.58 nN and cor-

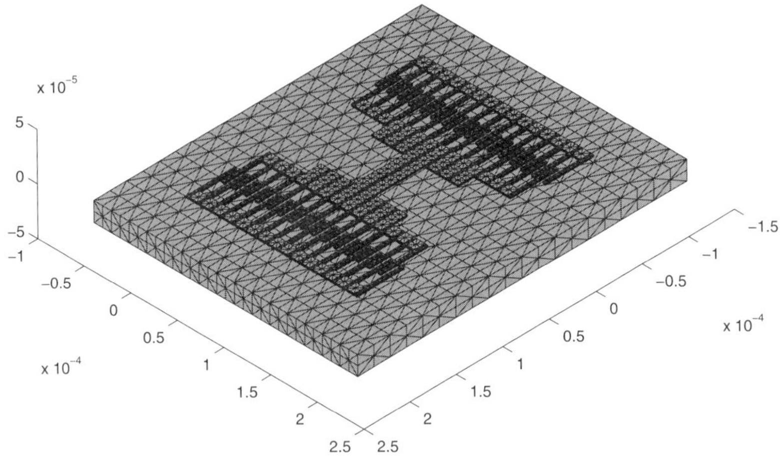

FIGURE 1.4. Dimensions and boundary element discretization employed in the comb-drive three-dimensional simulation using the *FastStokes program*. (Courtesy of W. Ye and J. White.)

responding quality factor $Q = 29.1$. These predictions are very close to the experimental values. Simple steady or unsteady models based on Couette flow (see Chapter 3) overpredict the Q factor by almost 100%. The estimated Knudsen number in this case was $\mathrm{Kn} \approx 0.03$, and the Reynolds number was $\mathrm{Re} \approx 0.02$, so rarefaction and nonlinearities were apparently second-order effects compared to very strong three-dimensional effects. The complete simulation of this problem requires models for the electrostatic driving force, flow models as above, and also mechanical models to investigate possible vibrational effects because of the very small width of the moving fingers. This *mixed domain* simulation requirement for the comb-drive is representative in the field of MEMS (see Section 1.7 and Chapters 17 and 18 for issues in full-system simulation of MEMS devices).

Another application area is microdevices that involve *particulate flows* for sorting, analysis, and removal of particles or cells from a sample, with both liquid and gas microflows (Ho, 2001; Green and Morgan, 1998; Telleman et al., 1998; Wolff et al., 1998; Yager et al., 1998). Examples of two different devices for cell sorting are shown in Figure 1.5 (Telleman et al., 1998). The device on the left is based on microfluorescent activated cell sorting (μFACS), while the device on the right is based on micromagnetic activated cell sorting (μMACS). In the former, the targeted cells are labeled with fluorescent antibodies, and as they pass through an optical sensor a valve is activated, letting the desired cells collected at one outlet. However, there is

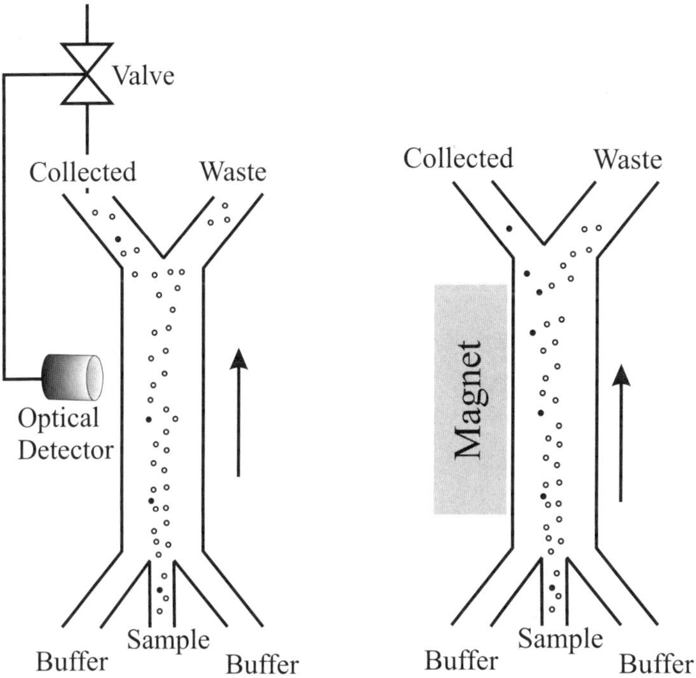

FIGURE 1.5. Diagrams of particle separators in microflows based on microfluorescent activated cell sorting (μFACS; left) and micromagnetic activated cell sorting (μMACS; right). (Courtesy of S. Lomholt.)

always a residual amount of undesired cells, and thus the process should be repeated using multiple μFACS devices. In the second device, the targeted cells are labeled with paramagnetic antibodies, and only the desired cells reach the collection outlet. The typical size of the channels is 100 μm, and the cells are about 5 to 10 μm (Lomholt, 2000). The carrying fluid and the buffers are neutral liquids for living cells. Similar devices exist for removing particles from gases, e.g., an airstream for environmental applications. In a device presented in (Yager et al., 1998), multisized particles of up to 10 μm were removed at various stages. Such particulate microflows require special numerical modeling to deal efficiently with the multiple moving surfaces, i.e., cells or particles present in the domain (see Section 14.3.2).

Microparticles, from 20 nm to about 3 μm can also be used to fabricate microdevices, such as pumps and valves, which in turn can be used for microfluidic control. Several studies have focused on fabricating self-assembled structures using paramagnetic particles carried by liquids in microchannels (Hayes et al., 2001; Doyle et al., 2002). The ability to form supraparticle structures and precisely control their arrangement and motion externally by magnetic fields could lead to many novel applications such as micro-optical filters and gratings, but also to new materials and new micro- and

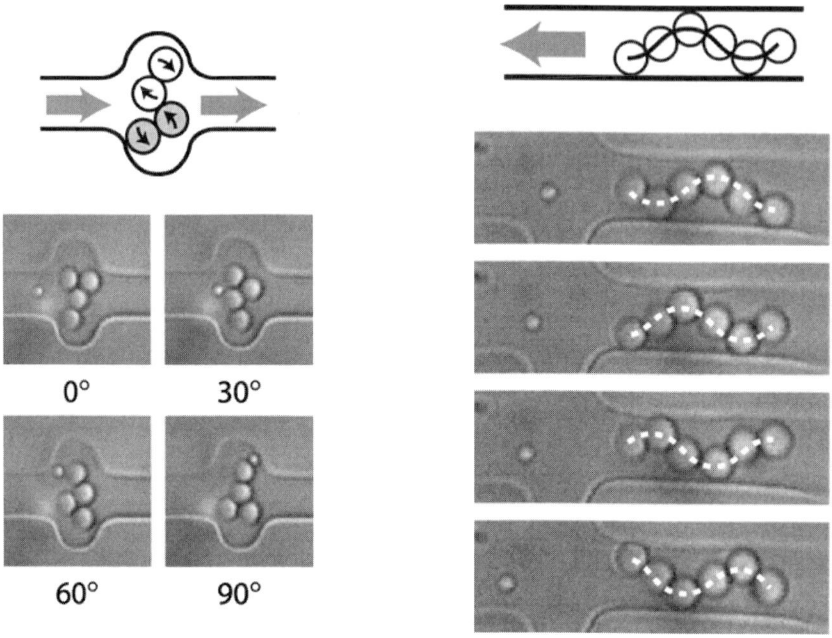

FIGURE 1.6. Colloidal micropumps using 3-micron silica microspheres. (a) Lobe movement of a gear pump. (b) Peristaltic pump. The channel is 6 microns, and the motion is induced by optical traps. (Courtesy of D. Marr.)

nanofabrication protocols (Furst et al., 1998; Hayes et al., 2001; Whitesides and Grzybowski, 2002).

Colloidal micropumps and colloidal microvalves are already in existence and have been used for active microfluidic control. For example, in (Terray et al., 2002), latex microspheres were manipulated by optical traps to pump fluids. These devices are about the size of a human red blood cell; see Figure 1.6. These colloidal micropumps are based on positive-displacement pumping techniques and operate by imparting forward motion to small volumes of fluid. The two micropumps shown in Figure 1.6 induce motions of 2 to 4 μm/s with corresponding flow rate of 0.25 nl/hour; see (Terray et al., 2002) for details.

1.2 The Continuum Hypothesis

Important details of the operation of micromachines involve complex dynamical processes and unfamiliar physics. The dynamics of fluids and their interaction with surfaces in microsystems are very different from those in large systems. In microsystems the flow is *granular* for liquids and *rarefied* for gases, and the walls "move." In addition, other phenomena such as thermal creep, electrokinetics, viscous heating, anomalous diffusion, and even quantum and chemical effects may be important (Chan et al., 2001). In particular, the material of the wall is very important in the dynamics; for example, a simple graphite submicron bearing exhibits complex vibrational modes and interacts differently with the fluid than does a diamondoid submicron bearing. Similarly, for gas microflows the material surface, i.e., its type and roughness, determines the fluid-wall interactions, which lead to definition of thermal and momentum accommodation coefficients (see Section 2.2.2).

Such interaction with the wall material can be studied with molecular dynamics (MD) simulations; see Section 16.1. In a typical molecular dynamics simulation, a set of molecules is introduced initially with a random velocity for each molecule corresponding to a Boltzmann distribution at the temperature of interest. The interaction of the molecules is prescribed in the form of a potential energy, and the time evolution of the molecular positions is obtained by integrating Newton's equations of motion. Realistic intermolecular potentials are constructed by modeling the atom–atom interaction potential using relatively simple equations, such as the Lennard–Jones potential

$$V(r) = 4\epsilon \left[\left(\frac{r}{\sigma} \right)^{-12} - \left(\frac{r}{\sigma} \right)^{-6} \right],$$

written for a pair of two atoms separated by distance r. The Lennard–Jones potential incorporates the shape effects by an anisotropic repulsive core and anisotropic dispersion interactions. For an appropriate choice of these parameters a reasonable description of real liquids is possible. For example, using $\epsilon/k_{\rm B} \approx 120$ K, where $k_{\rm B}$ is Boltzmann's constant and $\sigma \approx 0.34$ nm, a reasonable description of *liquid argon* can be obtained; see Section 16.1 for more details.

In Figure 1.7 we plot results from a molecular dynamics simulation of (Koplik et al., 1989) that shows a large density fluctuation very close to the wall. Specifically, the fluid is governed by a Lennard–Jones potential, and there is no net flow. The total number of atoms in this simulation is 27,000. The geometry is a three-dimensional periodic channel made of two atomic walls with 2,592 atoms each of FCC lattice type. The size of the channel is 51.30×29.7 (in the plane) $\times\, 25.65$ (out of plane), with all dimensions in molecular units; i.e., the atom diameter is 1.0. The density profile is obtained by binning the atomic positions in 170 slabs parallel to the walls, and the overall density is 0.8 units, while the temperature is kept at 1.0 units. In Figure 1.8 we show a snapshot of the flow in the near-wall region to demonstrate the *layering phenomenon*, where the fluid atoms are organized in horizontal layers parallel to the wall atomic layers. This layering is responsible for the large density fluctuations very near to

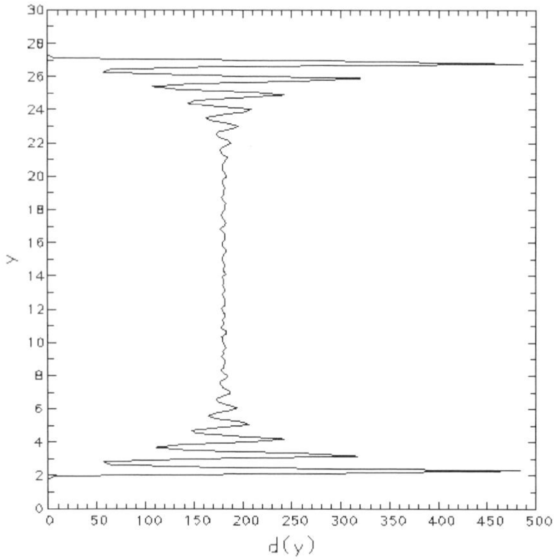

FIGURE 1.7. Density profile of Lennard–Jones fluid in a channel made of two atomic walls. The length dimensions are in molecular units (diameter of a molecule is 1.0). (Courtesy of J. Koplik and J. Banavar.)

the wall. While in liquids this effect extends only a few atom diameters from the wall, in gases the wall–fluid interaction extends over much greater length, and this has to be accounted for explicitly.

The amount of slip revealed in the above MD simulations depends strongly on the wall type and its modeling, which is determined by the strength of the liquid–solid coupling and the wall–liquid density ratio, among others. A shear-driven (Couette) microflow was simulated by (Thompson and Troian, 1997) in a channel with height $h = 24.57\sigma$ using a truncated Lennard–Jones potential, which was set to zero for $r > r_c = 2.2\sigma$. The wall–liquid interaction was also modeled with a Lennard–Jones potential but with different energy and length scales ϵ_w and σ_w, respectively. The liquid density was described by $\rho = 0.81\sigma^{-3}$, and its temperature was maintained at $T = 1.1k_B/\epsilon$. A very wide range of values of shear rate $\dot{\gamma}$ was investigated in (Thompson and Troian, 1997), leading to both linear and nonlinear responses. The shear rate was scaled with the characteristic time of the Lennard–Jones potential

$$\tau = \sqrt{\frac{m\sigma^2}{\epsilon}},$$

where m is the mass of the molecule. A linear velocity profile was obtained in the bulk of the flow, in accordance with Navier–Stokes solutions, suggesting that the dynamic viscosity was constant (Newtonian fluid).

Results from these MD simulations showed an intriguing response. In

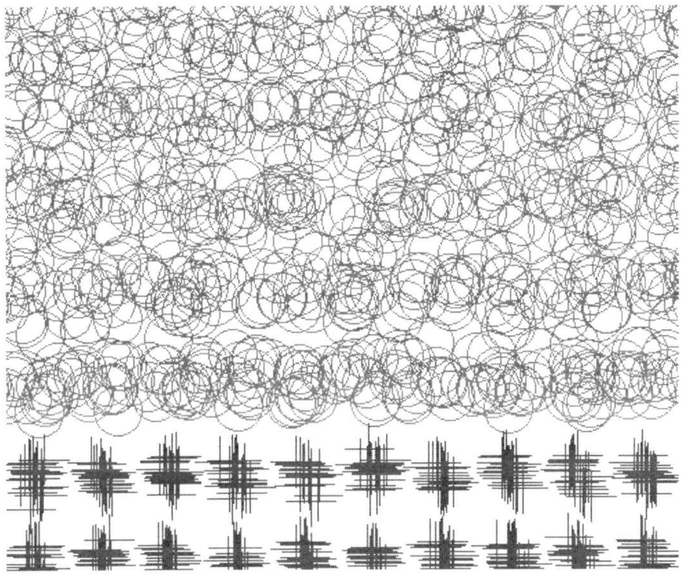

FIGURE 1.8. Snapshot of the Lennard–Jones fluid near a wall. The wall atoms are denoted by crosses, and fluid atoms by circles. This layered structure of the fluid molecules in close proximity with the wall is responsible for the density fluctuations shown in the previous figure. (Courtesy of J. Koplik and J. Banavar.)

particular, at low values of the shear rate, a slip velocity proportional to $\dot{\gamma}$ was obtained with corresponding values of the *slip length* b ranging from 0 to about 17σ. The slip length increases as the wall energy ϵ_w decreases or the wall density ρ_w increases. This is the linear response, and it is consistent with previous investigations. However, beyond 17σ a strongly nonlinear response was observed with the slip length b diverging beyond a critical value of the shear rate $\dot{\gamma}_c$. The results of MD simulations of (Thompson and Troian, 1997) are summarized in Figure 1.9 in a normalized form and for various conditions. The dashed line represents a best fit to the data in the form

$$\frac{b_s}{b_s^0} = \left[1 - \frac{\dot{\gamma}}{\dot{\gamma}_c}\right]^\alpha , \tag{1.2}$$

where the exponent $\alpha = -1/2$ is specific for the conditions that were tested in (Thompson and Troian, 1997), but may be different for other conditions. Such results suggest that at high shear rates and even for Newtonian fluids the liquid behavior in the near-wall vicinity is non-Newtonian; see also equation (10.7). At values of shear rate close to a critical value, such non-Newtonian behavior may propagate into the flow, and in that case even small variations in the wall surface may have a significant effect. It is not clear whether the conditions employed in MD simulations can match the experimental conditions. Experiments in submicron channels and gaps us-

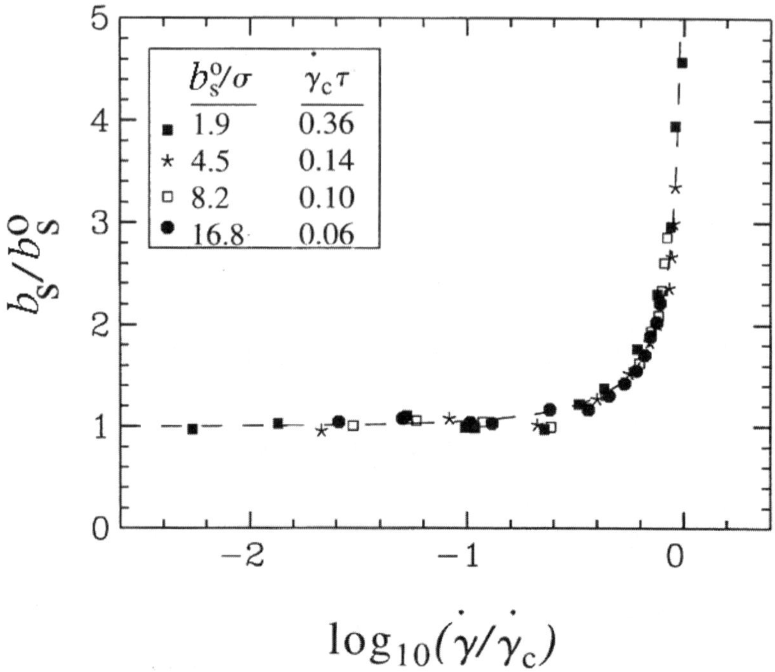

FIGURE 1.9. Summary of results from MD simulations reported in (Thompson and Troian, 1997). The normalized slip length is plotted against the normalized shear rate. All data collapse into a universal curve as shown; τ is a relaxation time scale. (Courtesy of S. Troian.)

ing the surface force apparatus (see Section 10.5) have revealed a slip length much larger than what is predicted by MD simulations, often by an order of magnitude!

The question of validity of the continuum approach arises also in *particulate* flows, especially in applications involving nanoparticles. A systematic MD study was undertaken by (Drazer et al., 2002) for a colloidal spherical particle through a nanotube containing a partially wetting fluid. They used a *generalized* Lennard–Jones liquid of the form

$$V_{FS}(r) = 4\epsilon\left[\left(\frac{r}{\sigma}\right)^{-12} - c_{FS}\left(\frac{r}{\sigma}\right)^{-6}\right],$$

where c_{FS} is an attractive strength that controls the wetting properties of the fluid–wall system. Full wetting corresponds to $c_{FS} = 1$, while poor wetting corresponds to $c_{FS} \ll 1$. Drazer et al. demonstrated that the MD simulations are in good agreement with the continuum simulations of (Bungay and Brenner, 1973) despite the large thermal fluctuations present in the system. This is true even for very small particles of order 2 nm. In Figure 1.10 we plot comparisons of MD and continuum simulations, first reported in (Drazer et al., 2002), for different values of the nanotube radius

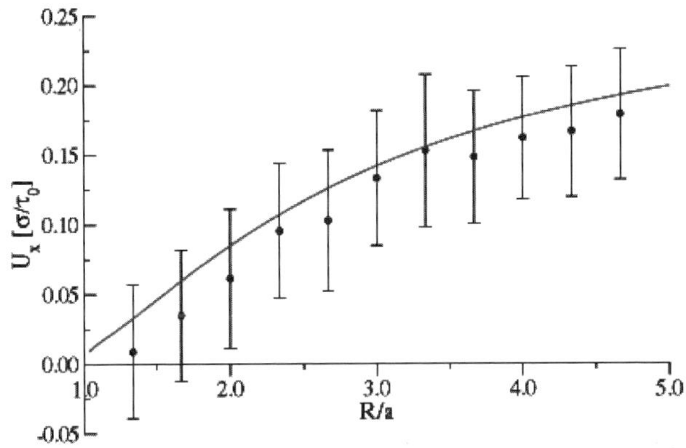

FIGURE 1.10. Mean sphere velocity in a microtube as a function of the radii ratio. Points correspond to MD simulations of Drazer et al. (2002) and the solid line the continuum results of Bungay & Brenner (1973). The error bars denote temporal fluctuations. (Courtesy of J. Koplik.)

R and particle radius a and $c_{\rm FS} = 1$. The fact that the continuum slightly overpredicts the mean particle velocity was attributed by Drazer et al. to the transverse random motion of the particles in the MD simulation. At later times it is possible for the particle to execute an intermittent *stick-slip* motion, especially for poorly wetting fluid–wall systems. For $c_{\rm FS} \leq 0.7$ the particle is eventually adsorbed to the tube wall, and in the stick regime almost all the fluid atoms between the particle and the wall have been squeezed out. This total depletion of fluid atoms would require an infinite force in the continuum limit. This phenomenon is also encountered in *capillary drying*, in which liquid is suddenly ejected from the gap formed between two hydrophobic surfaces when the width falls below a critical value (Lum et al., 1999). The robustness of continuum calculations in this context has been demonstrated also in (Israelachvili, 1992a; Vergeles et al., 1996) for spheres approaching a plane wall; see also Section 10.5.

1.2.1 Molecular Magnitudes

In this section we present relationships for the number density of molecules n, mean molecular spacing δ, molecular diameter d, mean free path λ, mean collision time t_c, and mean-square molecular speed $\bar{c}$ for gases.

The number of molecules in one mole of gas is a constant known as **Avogadro's number** 6.02252×10^{23}/mole, and the volume occupied by a mole of gas at a given temperature and pressure is constant irrespective of the composition of the gas (Vincenti and Kruger, 1977). This leads to the **perfect gas** relationship given by

$$p = nk_{\rm B}, T \tag{1.3}$$

where p is the pressure, T is the temperature, n is the number density of the gas, and k_B is the Boltzmann constant ($k_B = 1.3805 \times 10^{-23}$ J/K). This ideal gas law is valid for dilute gases at any pressure (above the saturation pressure and below the critical point). Therefore, for most of the microscale gas flow applications we can predict the number density of the molecules at a given temperature and pressure using equation (1.3). At atmospheric pressure and 0^oC the number density is $n \approx 2.69 \times 10^{25}$ m^{-3}. If we assume that all these molecules are packed uniformly, we obtain the **mean molecular spacing** as

$$\delta \propto n^{-1/3}. \tag{1.4}$$

Under standard conditions the mean molecular spacing is $\delta \approx 3.3 \times 10^{-9}$ m.

The mean molecular diameter of typical gases, based on the measured coefficient of viscosity and the Chapman–Enskong theory of transport properties for hard sphere molecules (Chapman and Cowling, 1970), is of order 10^{-10} m (see Table 1.1 for various thermophysical properties of common gases). For air under standard conditions, $d \approx 3.7 \times 10^{-10}$ m (Bird, 1994). Comparison of the mean molecular spacing δ and the typical molecular diameter d shows an order of magnitude difference. This leads us to the concept of *dilute gas* where $\delta/d \gg 1$. For dilute gases, binary intermolecular collisions are more likely than simultaneous multiple collisions. On the other hand, dense gases and liquids go through multiple collisions at a given instant, making the treatment of intermolecular collision processes more difficult. The dilute gas approximation, along with the molecular chaos and equipartition of energy principles, leads us to the well-established kinetic theory of gases and formulation of the Boltzmann transport equation starting from the Liouville equation. The assumptions and simplifications of this derivation are given in (Sone, 2002; Cercignani, 1988; Bird, 1994). In Section 15.4 we present an overview of the Boltzmann equation and some benchmark solutions appropriate for microflows, and in Section 15.5 we explain the BBGKY hierarchy that leads from the atomistic to the continuum description.

Momentum and energy transport in a fluid and convergence to a thermodynamic equilibrium state occur due to intermolecular collisions. Hence, the time and length scales associated with the intermolecular collisions are important parameters for many applications. The distance traveled by the molecules between collisions is known as the **mean free path** λ. For a simple gas of hard spherical molecules in thermodynamic equilibrium the mean free path is given in the following form (Bird, 1994):

$$\lambda = (2^{\frac{1}{2}} \pi d^2 n)^{-1}. \tag{1.5}$$

For example, for air at standard conditions, $\lambda \approx 6.5 \times 10^{-8}$ m.

TABLE 1.1. Thermophysical properties of typical gases used in microdomain applications at atmospheric conditions (298 K and 1 atm).

Gas	Density [kg/m^3]	Dynamic Viscosity [kg/(m s)]	Thermal Conductivity [W/(m K)]	Thermal Diffusivity [m^2/s]	Specific Heat [J/(kg K)]	Mean Free Path [m]
Air	1.293	1.85E-5	0.0261	2.01E-5	1004.5	6.111E-8
N_2	1.251	1.80E-5	0.0260	2.00E-5	1038.3	6.044E-8
CO_2	1.965	1.50E-5	0.0166	1.00E-5	845.7	4.019E-8
O_2	1.429	2.07E-5	0.0267	2.04E-5	916.9	6.503E-8
He	0.179	1.99E-5	0.150	1.60E-4	5233.5	17.651E-8
Ar	1.783	2.29E-5	0.0177	1.93E-5	515.0	6.441E-8

The gas molecules are traveling with speeds proportional to the speed of sound. The **mean-square molecular speed** of gas molecules is given in (Vincenti and Kruger, 1977):

$$\bar{c} = \sqrt{\frac{3p}{\rho}} = \sqrt{3RT}, \tag{1.6}$$

where R is the specific gas constant. For air under standard conditions this corresponds to 486 m/s. This value is about three to five orders of magnitude greater than the typical average speed in microscale gas flows.

With regard to the time scales of intermolecular collisions, we can obtain an average value by taking the ratio of the mean free path to the mean-square molecular speed. This results in $t_c \approx 10^{-10}$ seconds for air under standard conditions. This time scale should be compared to a typical scale in the microdomain to determine the validity of thermodynamic equilibrium.

In engineering practice, it is convenient to lump all the molecular effects to space-averaged macroscopic or continuum-based quantities, such as the fluid density, temperature, and velocity. It is important, however, to determine the limitations of these continuum-based descriptions. Specifically, we ask:

- How small should a sample size be so that we can assign it *mean* properties?

- At what scales will the statistical fluctuations be significant?

It turns out that sampling a volume that contains 10,000 molecules results in 1% statistical fluctuations in the averaged quantities. Based on that, for air at standard conditions the smallest sample volume that will result in 1% statistical variations is about 3.7×10^{-22} m^3. If we try to measure the macroscopic gradients (like velocity, density, and temperature) in three-dimensional space, one side of our sampling volume will be about 65nm.

A key nondimensional parameter for gas microflows is the **Knudsen number**, which is defined as the ratio of the mean free path over a characteristic geometric length or a length over which very large variations of a macroscopic quantity may take place. The Knudsen number is related to the Reynolds and Mach numbers as follows:

$$\mathrm{Kn} \equiv \frac{\lambda}{L} = \sqrt{\frac{\gamma\pi}{2}} \frac{M}{\mathrm{Re}}. \tag{1.7}$$

In complex microgeometries where three-dimensional spatial gradients are expected, definitions of instantaneous macroscopic values and their gradients become problematic for flows with Kn > 1 as the concept of macroscopic property distribution breaks down. However, for microchannels with large aspect ratio (width to height), we can perform spanwise space averaging to define an averaged velocity profile, and thus define the equivalent macroscopic quantities.

Rarefaction effects become more important as the Knudsen number increases and thus pressure drop, shear stress, heat flux, and corresponding mass flowrate cannot be predicted from flow and heat transfer models based on the continuum hypothesis. On the other hand, models based on kinetic gas theory concepts are not appropriate either, except in the very high Knudsen number regime corresponding to near vacuum conditions or very small clearances. The appropriate flow and heat transfer models depend on the range of the Knudsen number. A classification of the different flow regimes is given in (Schaaf and Chambre, 1961):

- for $\mathrm{Kn} \leq 10^{-2}$ the fluid can be considered as a continuum, while
- for $\mathrm{Kn} \geq \mathcal{O}(10)$ it is considered a free-molecular flow.

A rarefied gas can be considered neither an absolutely continuous medium nor a free-molecular flow in the Knudsen number range between 10^{-2} and 10. In that region, a further classification is needed, i.e.,

- slip flow ($10^{-2} <$ Kn < 0.1), and
- transition flow ($0.1 <$ Kn < 10).

This classification is based on empirical information and thus the limits between the different flow regimes may depend on the problem geometry.

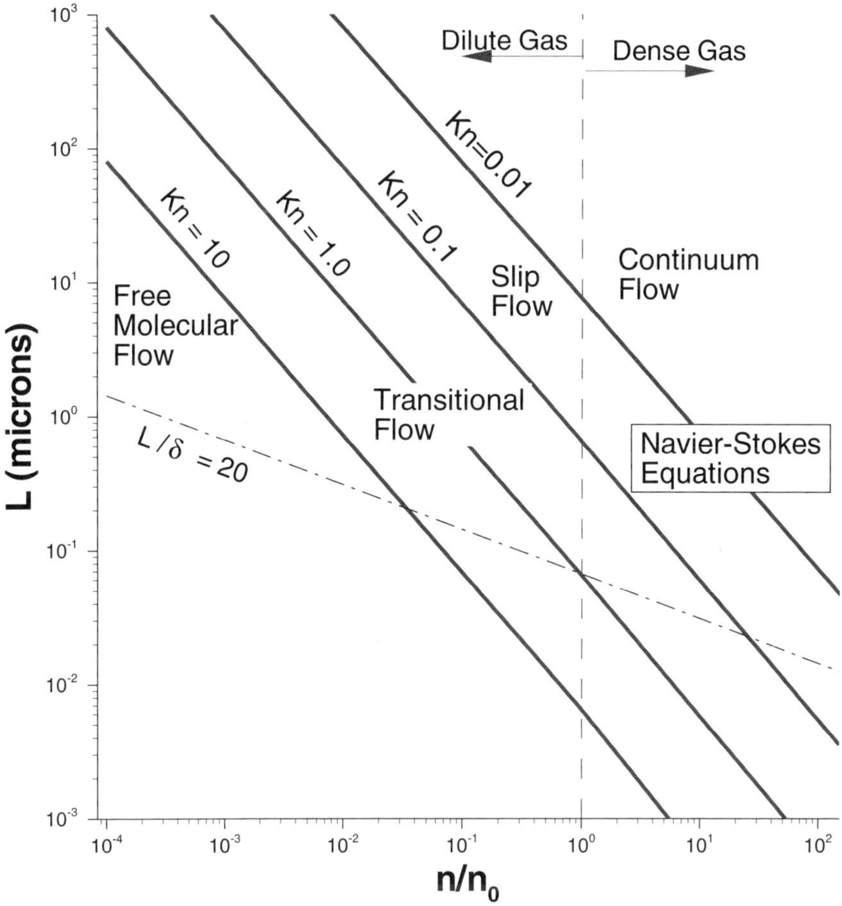

FIGURE 1.11. Limits of approximations in modeling gas microflows. L (vertical axis) corresponds to the characteristic length and n/n_0 is the number density normalized with corresponding atmospheric conditions. The lines that define the various Knudsen number regimes are based on air at *isothermal* conditions at $T = 273$ K. Statistical fluctuations are significant below the line $L/\delta = 20$.

This separation in different regimes is plotted in Figure 1.11, where we define the various flow regions as a function of the characteristic length scale L in microns, and also the number density. In addition, we have included a line that corresponds to $L/\delta = 20$, below which statistical fluctuations are present; this line corresponds to 1% fluctuations in macroscopic measurements.

In many fluid-mechanical applications an analogy between different geometric scales and dynamic conditions can be obtained by invoking the concept of **dynamic similarity**. This enables us to determine the performance of a fluidic device by experimenting on a scaled prototype under similar physical conditions, characterized by a set of nondimensional parameters, such as the Reynolds, Mach, Prandtl, and Knudsen numbers. It is therefore appropriate to pose this question for *gas* microflows:

- Are the low-pressure rarefied gas flows *dynamically similar* to the gas microflows?

The answer to this question depends on the onset state of statistical fluctuations and also on wall surface effects. For example, at standard conditions for air, the value of the Knudsen number Kn $= 1$ is obtained at about a 65 nm length scale. For smaller length scales, corresponding to higher Knudsen number regimes, the average macroscopic quantities cannot be defined. However, for low-pressure flows, for example, at 100 Pa and 270 K, the 1% statistical scatter limit sets in at about $L \approx 0.65\mu$m, since $\delta \sim p^{-1/3}$. However, at this low-pressure condition, Kn $= 1$ corresponds to the characteristic length of about 65μm. This length scale is two orders of magnitude larger than the one at the onset of the statistical scatter at these conditions. Therefore, macroscopic property distributions can be defined without any significant statistical fluctuations. Hence, for dynamic similarity approaches for gas microflows to be valid, the onset of statistical scatter should be carefully considered. Also, Figure 1.11 shows a dense gas region where the $Kn = 0.1$ line crosses the 1% statistical scatter line. For dense gas flows in this region, the Navier–Stokes equations are valid, but the results show large statistical deviations due to the onset of the Brownian motion.

1.2.2 Mixed Flow Regimes

In the examples of Section 1.1, the gas flow cannot be modeled based on the continuum hypothesis. The mean free path of air, which at standard atmospheric conditions is about 65 nm, is comparable to the characteristic geometric scale, and therefore microscopic effects are important. For example, in the case of computer hard drives, the load capacity predicted by the continuum Reynolds equations without slip is in error by more than 30% (Fukui and Kaneko, 1988; Alexander et al., 1994). This deviation of the state of the gas from continuum is measured by the Knudsen number Kn. For the micromotor, using a length scale of $L = 3\mu$m (the gap between rotor and the base) and assuming that the operation conditions are atmospheric, we obtain the value Kn $= 0.022$. For the magnetic disk drive (slider bearing) the Knudsen number is Kn $= 1.3$, and in ultralow clearances corresponding to increased recording capacity, the Knudsen number is well above unity. Also, in other capillary flows, such as in helium leak

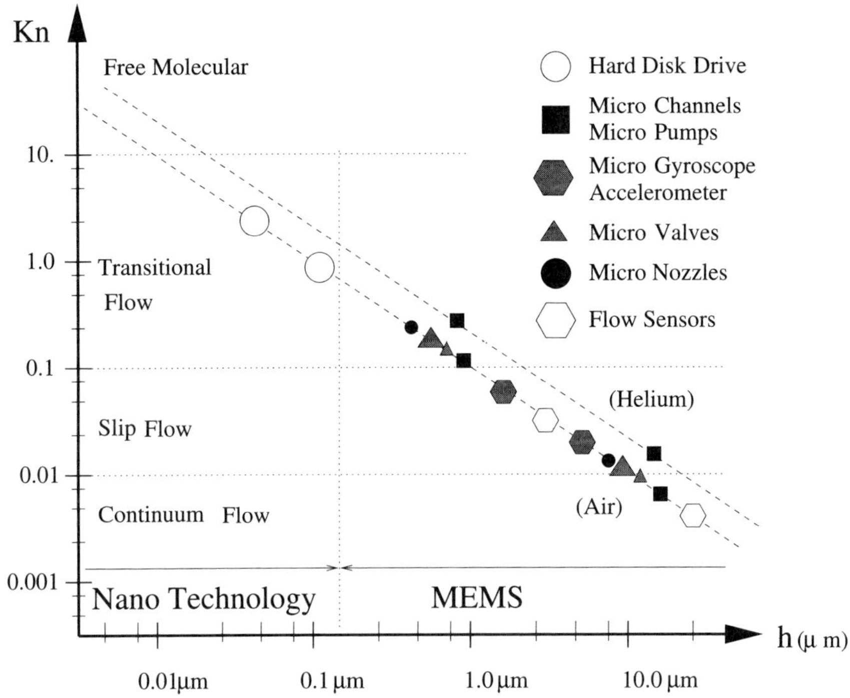

FIGURE 1.12. Typical MEMS and nanotechnology applications in standard atmospheric conditions span the entire Knudsen regime (Continuum, slip, transition, and free-molecular flow). Here h denotes a characteristic length scale for the microflow.

detection microdevices and mass spectrometers, the Knudsen number may achieve values up to 200 (Tison, 1993).

The operation regimes of typical microsystems at standard temperature and pressure are shown in Figure 1.12. MEMS devices operate in a wide range of flow regimes covering the continuum, slip, and transition flow. Further miniaturization of MEMS device components and applications in the emerging field of nano-electro-mechanical systems (NEMS) (Craighead, 2000; Ho, 2001) would result in higher Knudsen numbers, making it necessary to study mass, momentum, and energy transport in the entire Knudsen regime.

1.2.3 Experimental Evidence

An experimental illustration of the taxonomy described in Figure 1.11 is provided in Figure 1.13, where we plot data obtained by S. Tison at the National Institute of Standards (NIST) (Tison, 1995) at very low pressures in a pipe of diameter $2a = 2$ mm (a is the radius) and length $L = 200$ mm. Both inlet and outlet pressures were varied in the experiment, with

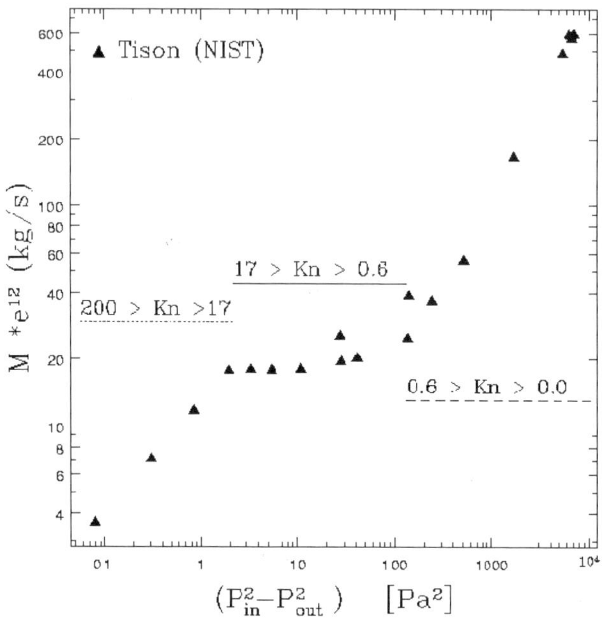

FIGURE 1.13. Variation of mass flowrate as a function of $(P_{\text{in}}^2 - P_{\text{out}}^2)$. The data are for rarefied gas flow experiments conducted by S. Tison at NIST (Kn is based on the exit pressure).

the corresponding Knudsen number varied from almost 0 to 200. In this log-log plot, we can easily identify *three distinct flow regimes*, although the corresponding values at the boundaries between the different flow regimes, are somewhat different from the aforementioned ones (Schaaf and Chambre, 1961). In particular, the *slip flow* regime extends up to Kn = 0.6 and the *transition regime* up to Kn = 17. It is interesting to notice the very slow variation of mass flowrate in terms of the pressure difference in the transition regime. The form of the plot in Figure 1.13 also suggests that a nonlinear pressure drop exists in this rarefied pipe flow. This was also verified in the slip flow experiments of (Sreekanth, 1969) in pipes with a diameter of 2 inches at low pressures.

The experimental data of Figure 1.13 are for a relatively large pipe, but at reduced variable pressure, so that a wide range of Knudsen numbers is covered. However, similar trends have been observed in microchannels. Specifically, the pressure distribution along the microchannel was measured by using a surface micromachined system with a number of sensors as part of the surface (Liu et al., 1993; Pong et al., 1994). A nonlinear pressure distribution was clearly demonstrated in these experiments. Scaling of the mass flowrate with the difference of pressure squares is characteristic of

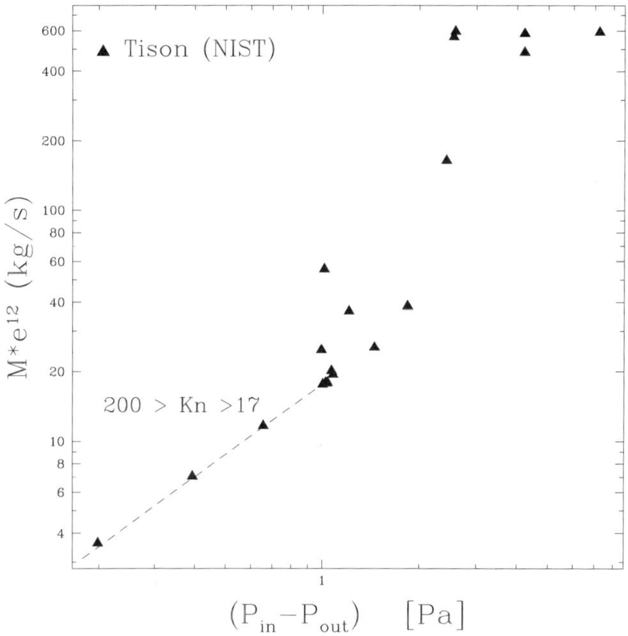

FIGURE 1.14. Variation of mass flowrate as a function of $(P_{\mathrm{in}} - P_{\mathrm{out}})$. Linear pressure drop is obtained in the high Knudsen number flow regime (i.e., $\mathrm{Kn} > 17$).

compressible low Reynolds number flows in long channels. To demonstrate this more clearly we replot the same data as mass flowrate versus linear pressure drop, which would be more appropriate for an incompressible pipe flow. This is shown in Figure 1.14, where we see that in this form there are no particular trends or correlations, except in the free-molecular flow regime, where the pressure drop is linear, in accordance with the free-molecular flow theory, see (Kennard, 1938), p. 304.

Systematic experiments that show significant deviations from continuum behavior in microchannel flows were performed (starting in 1988) at the University of Pennsylvania (Pfahler et al., 1991). Figure 1.15 shows results from these first experiments. In particular, the ratio of Poiseuille number obtained experimentally over the Poiseuille number obtained theoretically for two different theoretical approaches is plotted. The ratio is defined as

$$C^* = \frac{(\mathrm{Po})_{\mathrm{exp}}}{(\mathrm{Po})_{\mathrm{th}}}, \tag{1.8}$$

where the Poiseuille number is

$$\mathrm{Po} = C_{\mathrm{f}}\mathrm{Re}$$

with C_{f} the friction factor and Re the Reynolds number. For flow in a circular pipe the theoretical value is $(\mathrm{Po})_{\mathrm{th}} = 64$, and for a two-dimensional

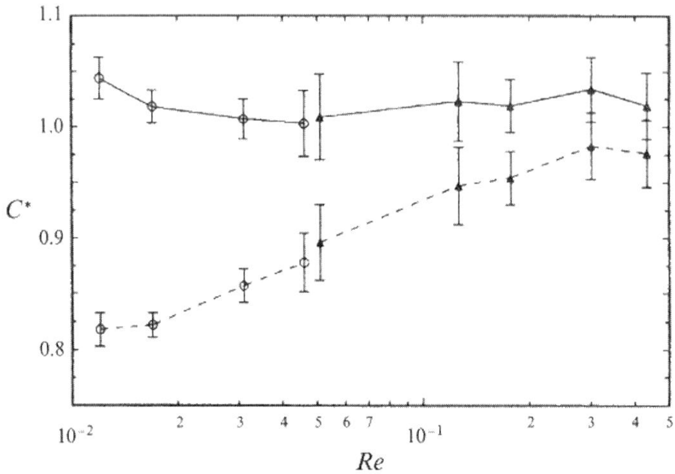

FIGURE 1.15. Ratio of the Poiseuille number obtained experimentally normalized by the Poiseuille number obtained theoretically (equation (1.8)). The dashed line corresponds to normalization with no-slip theory and the solid line to slip theory. Here triangles and circles correspond to the nitrogen and helium flow in a 0.51μm microchannel, respectively. (Courtesy of H. Bau.)

channel $(\mathrm{Po})_{\mathrm{th}} = 96$. The data are from the experiments of (Harley et al., 1995) for a microchannel of 0.51μm deep; this is case (JH6) studied in Section 4.1.2. The inlet pressure varied from 1.1 MPa to 3.5 MPa, and the exit pressure was atmospheric. The Mach number was kept below 0.02 for all experiments and the Knudsen number ranged from 0.004 to 0.373. It is interesting to see that the velocity slip effect is more pronounced in the low Reynolds number regime. This is consistent with the fact that the Knudsen number scales inversely proportionally to the Reynolds number (see equation (1.7)). For example, for $\mathrm{Re} = 0.012$ the inlet Knudsen number was 0.025 and the outlet Knudsen number was 0.373.

For **liquids** the noncontinuum behavior is more difficult to detect. It is manifested as *anomalous diffusion*, i.e., different diffusion in the near-wall region than in the bulk, and is associated with the rheology of the liquid. Early studies by (Debye and Cleland, 1959) with paraffin in porous Vycor glass indicated the presence of a slipping adsorbed thin layer. This could either increase or decrease the pressure drop compared to the continuum behavior; both $C^* < 1$ and $C^* > 1$ were measured. A possibility of permanent structural modification of water in a very thin film was reported in (Derjaguin, 1970), but such results have been largely rejected. In a follow-up study (Derjaguin et al., 1983), it was demonstrated that $C^* \approx 2.5$ for sebacine-amyl ester within a thickness of 12 nm, and 100 nm for nitrobenzene. This high viscosity was attributed to an ordered boundary phase. However, careful experiments by (Israelachvili, 1992b) using the

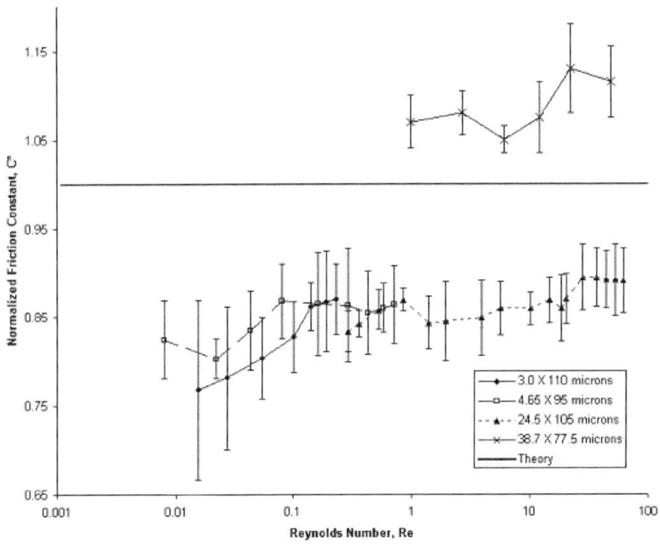

FIGURE 1.16. Results for silicone oil. Ratio of the Poiseuille number obtained experimentally over the Poiseuille number obtained theoretically (equation (1.8)). Channels with different sizes (cross-sections) are tested as shown in the legend. (Courtesy of H. Bau.)

atomic force microscope (AFM) showed that water films as thin as $2\,nm$ (corresponding to approximately ten molecular diameters) have continuum-like diffusion characteristics. We have already seen in Figure 1.7 that large density fluctuations occur below ten molecular diameters, and thus in that region it is possible to have $C^* \gg 1$. Water is a seemingly simple fluid, but its numerical modeling is quite difficult; its static and dynamic behavior deserves a separate study, as we present in Chapter 11.

The work of Bau and collaborators (see (Bau and Pfahler, 2001), and references therein) included experiments with different types of liquids in microducts with diameter from 0.48 to $40\,\mu m$. Specifically, systematic experiments with isopropyl alcohol, silicone oil, and distilled water were conducted. A typical result for silicone oil is shown in Figure 1.16, where C^* is plotted against the Reynolds number. For the small channels and for low Reynolds number, $C^* < 1$ is observed, but for larger channels, $C^* > 1$ is observed. In general, the deviation of C^* from the theoretical value $C^* = 1$ (using continuum theory) is of order 20% for all cases tested. A history of the research efforts in establishing the no-slip condition as well as conceptual models of slip in liquids are discussed in Section 10.5.

1.3 The Pioneers

A systematic research effort in micromechanics in the context of MEMS devices, i.e., fabrication and operation, began in the late 1980s. Richard Feynman, in his prophetic lecture in 1959 "There's plenty of room at the bottom," described new vistas and novel new applications in microscale science and engineering. In a follow-up lecture in 1983 at the Jet Propulsion Laboratory, Feynman revisited this subject, anticipating some of today's standard MEMS technologies such as the sacrificial-layer method of making silicon micromotors, the use of electrostatic actuation, and the importance of friction and contact sticking in such devices. Both lectures are included in (Trimmer, 1997), where other classical and seminal papers in MEMS and micromechanics up to 1990 can be found.

Fundamental work in microflows started much earlier. In 1846, Poiseuille published the first paper describing flow in tubes with diameters ranging from 30 μm to 150 μm (Poiseuille, 1846). His studies with liquids led to the well-known relationship between flowrate, pressure drop, and tube geometry, although he seemed to be unaware of the viscosity concept at that time. In 1909, Knudsen studied gas flows through glass capillary tubes in the transition and free molecular flow regimes (Knudsen, 1909). In these experiments, the volumetric flowrate normalized with the inlet to exit pressure difference ($\dot{Q}/(P_{\rm i} - P_{\rm o})$ showed a minimum at Kn ≈ 1, when plotted against the average pressure in the capillary. This counter intuitive behavior is known as the **Knudsen's paradox** or Knudsen's minimum. The first known experiment of flow in a microchannel was performed by Gaede in 1913, who placed two parallel plates 4 μm apart (Gaede, 1913). Gaede found that the flowrate of hydrogen decreases about 50% from the free-molecular value while passing through a minimum and then rising with increasing pressure levels. In long capillaries, the difference between the minimum and the free-molecular value is only 5 to 8%, as also seen in Figure 1.17. Although Knudsen's minimum is measured in smooth capillaries, crimped tubes do not exhibit this behavior (Tison, 1993). Therefore, rarefied flows behave differently in pipes and channels, and the surface conditions can be important in gas transport in the transitional and free-molecular flow regimes.

Driven by the growing number of microsystem applications, a number of experimental studies with microchannels were conducted in the late 1980s both for gases and liquids. The first recent experimental study of microflows using micromachined channels was reported in (Pfahler et al., 1991; Harley et al., 1995), in the Reynolds number range $0.50 \leq$ Re ≤ 20. For gases, the Knudsen number was $0.001 \leq$ Kn ≤ 0.363 for hydraulic channel diameter $D_{\rm H} = 8\,\mu$m and channel length 11 mm. The corresponding Mach number at the inlet was very small, but at the exit it reached in some cases the value $M = 0.7$, corresponding to substantial pressure drop. The reported skin friction reduction due to apparent slip of the flow was confirmed in other

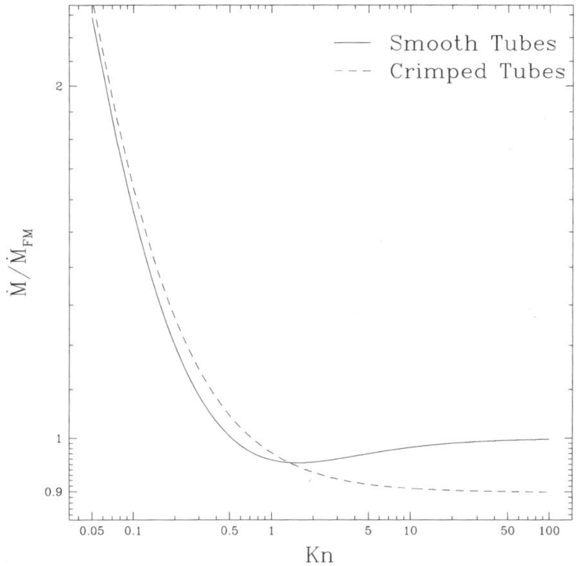

FIGURE 1.17. Mass flowrate normalized with its free-molecular value versus Knudsen number. Gas flow in smooth capillary tubes exhibits Knudsen's minimum around Kn $= 1$, but crimped tubes do not exhibit a minimum (Tison, 1993).

similar experimental studies with gases (Liu et al., 1993; Pong et al., 1994; Arkilic et al., 1994) using different microfabrication techniques to make the microchannels. In particular, the paper in (Liu et al., 1993), describes the first-ever fabricated microchannel with several in situ pressure sensors as an integral part of the microchannel.

One particular set of highly accurate data, which can be used for validation of theory and simulations in microflows, was obtained by Breuer and collaborators (Arkilic and Breuer, 1993; Arkilic et al., 1994; Arkilic, 1997). A schematic of the microchannel used in the measurements is shown in Figure 1.18. It has length $L = 7.5$ mm, width $W = 52.25\,\mu$m, and height $H = 1.33\,\mu$m. Examination of scanning electron microscope (SEM) images showed that the height was uniform and that the roughness was below 65 nm. The pressures at the inlet and outlet are measured, as well as the temperature of a small tank where the mass is accumulated. One set of measurements for argon flow is shown in Figure 1.19; plotted is the mass flowrate versus the inlet-to-outlet pressure ratio. The exhaust pressure was kept at 101 kPa. Theoretical predictions based on no-slip and slip theory are also included in the plot, and they reveal a clear rarefaction effect. These measurements have been used by Breuer and collaborators (Arkilic et al., 1994) to obtain values of tangential momentum accommodation coefficients

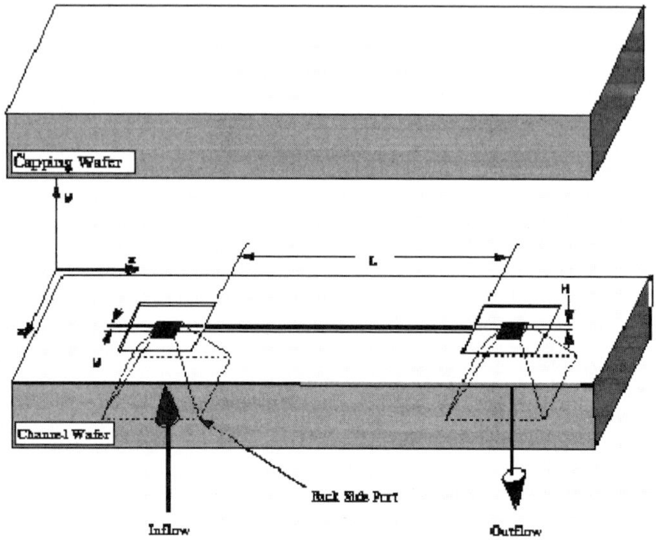

FIGURE 1.18. Schematic of the channel used in high resolution measurements of mass flowrate (Arkilic, 1997). The channel height is H and the channel length is L. This is a dual-tank accumulation technique, coupled to high precision control of temperature and pressure fluctuations in the system. (Courtesy of K. Breuer.)

in silicon microchannels (see Section 2.2.2 for the details). The resolution of this mass flow system is down to 10^{-12} kg/sec. The system has also good rejection of common-mode noise (usually due to microthermal fluctuations in the testing environment) and can be adapted to measure mass flows at any range with any working fluid, including liquids.

An important and exciting result on the experimental side has been the development of techniques that can measure velocity profiles accurately in microchannels, e.g., using micro-particle-image velocimetry (microPIV) (Meinhart et al., 1999; Santiago et al., 1998). In Figure 1.20 we plot such measurements in a 30 μm $\times$300 μm channel for Stokes flow (Meinhart et al., 1999). The two-dimensional velocity field is measured using micron resolution PIV. The spatial resolution, defined by the size of the first interrogation window, is 13.8 μm $\times$0.9 μm. A 50% overlap between the interrogation spots yields a velocity vector spacing of 450 nm in the wall-normal direction near the wall. This high spatial resolution is made possible by using relatively low particle concentrations in the flow, and by incorporating a specialized interrogation algorithm to increase the signal to noise ratio. This algorithm averages in time a series of particle image correlation functions, and searches the time-averaged correlation function to determine the location of the displacement signal peak (Meinhart et al., 2000).

A three-dimensional particle-tracking technique was developed by (Kihm

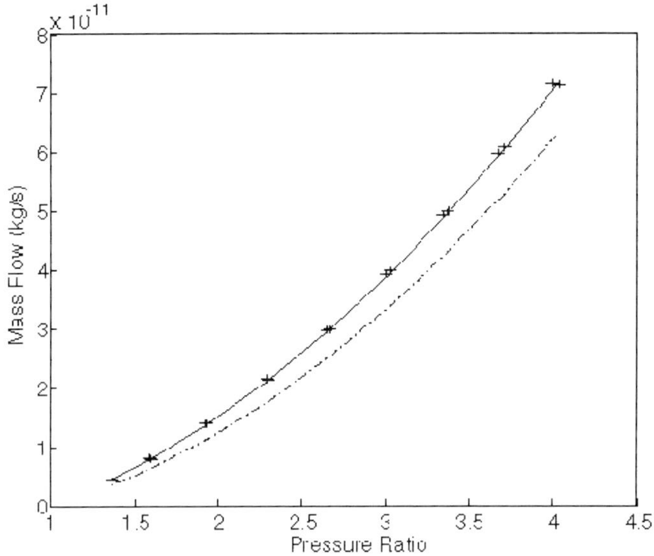

FIGURE 1.19. Mass flowrate versus pressure ratio for argon microflow. The solid line denotes slip-flow theory, the dash-dot no-slip theory, and the crosses the measurements. The flow is accumulated in a chamber attached to the outlet as shown in Figure 1.18. (Courtesy of K. Breuer.)

et al., 2004) to investigate nanoparticle motion near a solid surface using the Ratiometric Total Internal Reflection Microscopy (R-TIRFM) technique. This allowed three-dimensional reconstruction of particle trajectories by combining two-dimensional lateral tracking techniques with R-TIRFM for tracking vertical particle motions. Using R-TIRFM, they were able to measure motions of fluorescence-coated polystyrene spheres of 200 ± 20 nm diameter (specific gravity 1.05). Figure 1.21 shows Brownian motion of a single particle suspended in water at 293 K. Figure 1.21 (upper) shows the time history of three-dimensional particle locations over 67 imaging frames, recorded for a duration of 2.23 seconds. The symbols in the upper plot show particle–surface locations closest to the bottom glass surface. Figure 1.21 (lower) shows lateral displacement of particles in the x-y plane and vertical displacements in the z-direction. The results indicate horizontal and vertical resolutions of approximately 200 nm and 50 nm, respectively.

Biomimetic concepts can play an important role in designing functionalized devices at microscales; see Chapter 13. Self-assembly is a field that originated in organic chemistry, but it has had a great impact on micro- and nanofabrication of complex structures thanks to the pioneering studies of Whitesides and his collaborators (Whitesides and Grzybowski, 2002). Self-

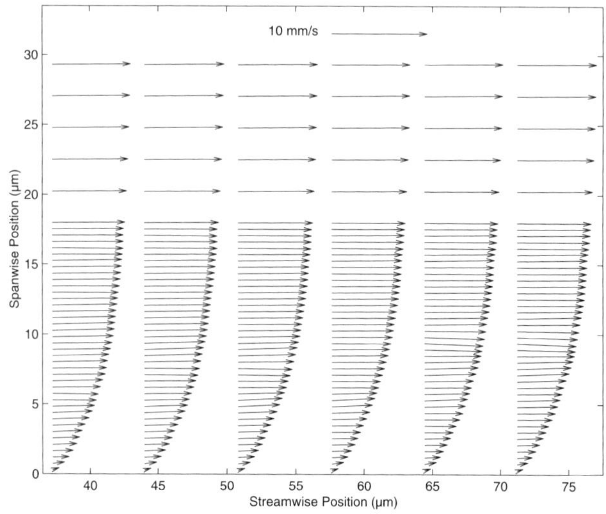

FIGURE 1.20. MicroPIV measurements. Shown is a time-averaged velocity vector field of Stokes' flow through a $30 \times 300\,\mu$m channel. (Courtesy of C.D. Meinhart.)

assembly is one of the few practical methods for making nanostructures; it is simple and can use a wide range of materials, in contrast to other techniques, e.g., micromachining, stereolithography, three-dimensional printing, or holographic lithography.

Self-assembly is the autonomous organization of components into different structures without human intervention. It is classified as *static self-assembly* or *dynamic self-assembly.* In the former type, systems are at equilibrium and do not dissipate energy, e.g., folded proteins and molecular crystals. In the latter type, interactions between the members of the assembly occur only if energy is dissipated, e.g., oscillating and reaction–diffusion reactions and bacteria swarms. (Whitesides and Grzybowski, 2002). Self-assembly requires that the components be mobile, and therefore a fluidic environment can accommodate this requirement. Correspondingly, self-assembly is typically driven by van der Waals, electrostatic, magnetic, capillary, and entropic interactions.

An example of fabrication of a functionalized microdevice using self-assembly exploiting capillary interactions is shown in Figure 1.22, taken from (Jacobs et al., 2002). Specifically, unpackaged GaAs/GaAlAs LEDs with a chip size of 280 μm $\times 200$ μm are used as components, which are assembled on a cylindrical solder substrate. To induce the LEDs to assemble into a well-defined array, 113 solder-based receptors were fabricated on the substrate. The surface of the liquid solder wets and adheres to the back side of the LEDs. The components were suspended in water and agitated

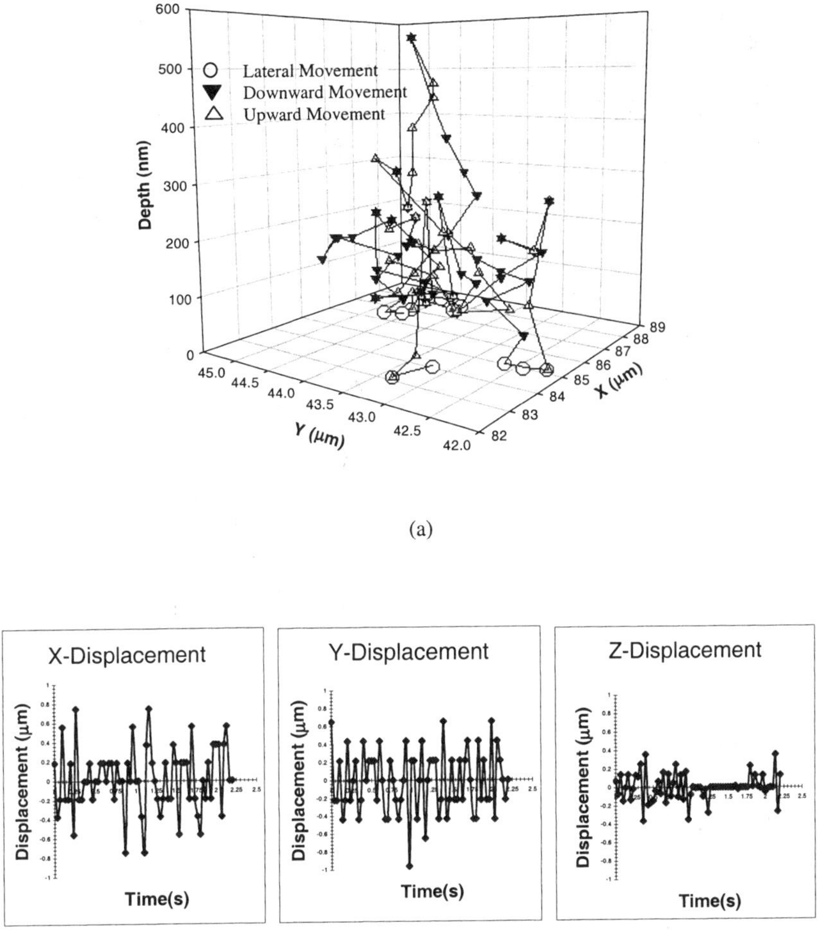

(a)

(b)

FIGURE 1.21. Manual tracking of Brownian motion of a single 200 nm particle suspended in water at 293 K. Upper: History of three-dimensional locations over 67 imaging frames recorded for the duration of 2.23 s; and Lower: History of its x-y-z directional displacements. (Courtesy of K.D. Kihm.)

gently initially. The driving force for the assembly is due to the minimization of free energy of the solder–water interface. Figure 1.22 shows a fully addressable cylindrical display, but other planar arrays were also fabricated with this approach on a much larger area. The one shown in the figure has eight columns of eight receptors interleaved with seven columns of seven receptors on the bottom electrode that connect to 15 rows of crossing copper wires on the top electrode. To test the functionality of the display with all its interconnects, a potential of 2 V was applied between the top and substrate electrodes. Solder-based assembly is appropriate for connecting components electrically and also provides mechanical strength.

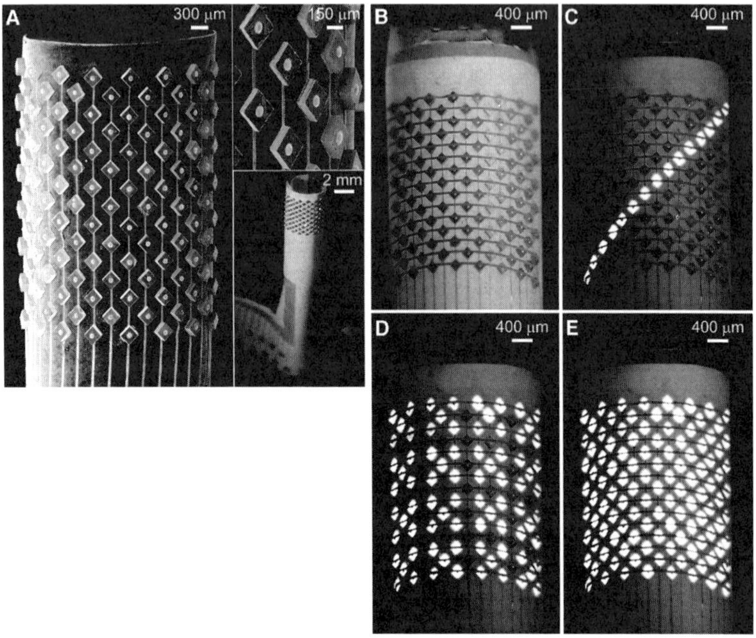

FIGURE 1.22. Scanning electron micrographs (SEM) of a cylindrical display. (A) Images of LEDs that assembled into a regular array. (B) Display after self-alignment of the top electrode. (C to E) Photographs of the operating display performing different functions. (Courtesy of G.M. Whitesides.)

1.4 Modeling of Microflows

We consider in this book *gas*, *liquid*, and *particulate* microflows and present governing equations, simplified models, and efficient numerical methods that can be used in simulations. Specifically, modeling of microflows is discussed in Chapters 2 to 9 based on continuum or quasi-continuum approximations. Because of the unfamiliar physics involved, detailed simulations are required, and this necessitates either using atomistic (particle-based) simulations or adding correction terms to the macroscopic full simulation methods. For example, in the slip flow regime it is reasonable to employ the Navier–Stokes equations modified at the wall surface with appropriate velocity slip and temperature jump conditions.

In Chapter 2 we present continuum-based velocity slip and temperature jump models for gas microflows, in conjunction with the appropriate governing equations for the slip and transitional flow regimes.

> *The fundamental assumption in this analysis is the dynamic similarity of microflows with rarefied flows encountered in a low-pressure environment.*

This assumption is justified theoretically based on the analysis of the Boltzmann equation for small Reynolds number and Knudsen number (Sone, 2002). Based on this assumption, gas microflows are simulated subject to slip boundary conditions. In order to demonstrate the effects of rarefaction and slip in microgeometries we compare these findings with the continuum no-slip solutions whenever possible. In **gas** microflows we encounter four important effects:

- rarefaction,
- compressibility,
- viscous heating, and
- thermal creep.

In particular, we investigate the *competing effects* of compressibility and rarefaction, which result in a nonlinear pressure distribution in microchannels in the slip flow regime. Curvature in pressure distribution is due to the *compressibility* effects, and it increases with increased inlet to outlet pressure ratios across the channels. The effect of *rarefaction* is to reduce the curvature in the pressure distribution. This finding is consistent with the fact that the pressure distribution becomes linear as the free-molecular flow regime is reached (Kennard, 1938). The *viscous heating* effects are due to the work done by viscous stresses (dissipation), and they are important for microflows, especially in creating temperature gradients within the domain even for isothermal surfaces. The *thermal creep* (transpiration) phenomenon is a rarefaction effect. For a rarefied gas flow it is possible to start the flow with tangential temperature gradients along the channel surface. In such a case the gas molecules start creeping from cold toward hot (Kennard, 1938; Kruger et al., 1970). Thermal creep effects can be important in causing variation of pressure along microchannels in the presence of tangential temperature gradients (Fukui and Kaneko, 1988). This mechanism is also significant for transport through porous media in atmospheric conditions (Loeb, 1961; Vargo and Muntz, 1996). Other temperature-induced flows are studied in Section 5.2.

Modeling of **liquids** in microdomains, see Chapters 7–9, requires a different approach. In mesoscopic scales a continuum description suffices (see Chapter 14), whereas in submicron dimensions atomistic modeling is required (see Chapter 16). We have already discussed slip phenomena in liquids in Section 1.2; however, other phenomena may be present, for example:

- wetting,
- adsorption, and
- electrokinetics.

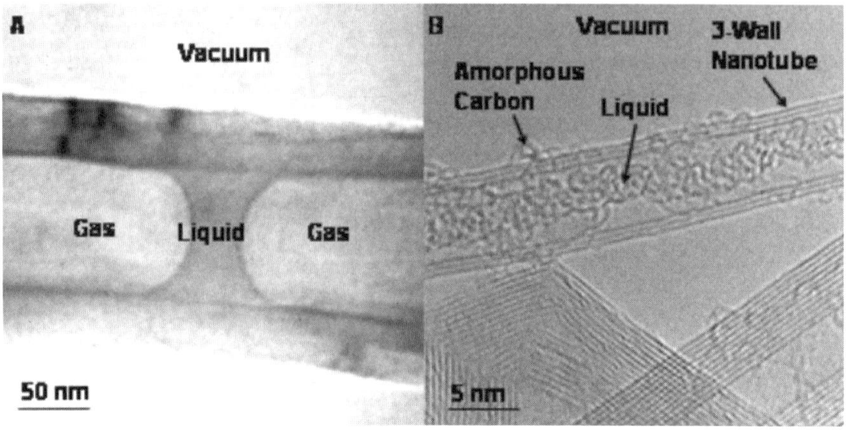

FIGURE 1.23. Transmission electron micrographs showing continuum behavior in a nanotube of large-diameter (left) and non-continuum behavior in a nanotube of diameter about ten times smaller (right). These two images demonstrate a dramatically different degree of interaction between fluid/wall for the two cases. (Courtesy of C.M. Megaridis, A.G. Yazicioglu and Y. Gogotsi.)

The *wetting* of the solid surfaces by liquids can be exploited in microfluidics to determine precisely defined routes based on surface tension gradients and different types of surfaces, i.e., hydrophobic or hydrophilic. Control can be accomplished via active contaminants or thermally; see Chapter 8. Wetting may also affect flow and performance of a MEMS device by altering its mechanical response or even blocking flow channels. While there exist macroscopic descriptions of wetting, they do not fully incorporate effects such as hysteresis (Dussan, 1979; deGennes, 1985), layering, and monolayer spreading (Heslot et al., 1989; Jin et al., 1997). Hysteretic effects can have a dramatic effect on the value of the contact angle and the strength of surface tension. In modeling wetting phenomena it is important to incorporate the possible deformation of the solid in wetting processes, since local regions of high stress can produce distortions of the elastic solid (Shanahan, 1988), e.g., the elastomeric walls used in some patterned drug delivery applications.

Adsorption is important in interactions of liquids with nanoporous materials such as glasses. The boundary conditions for fluid flow are very sensitive to the type and amount of adsorption on the walls of a MEMS device. Liquids confined in microgeometries exhibit a supercooling of the liquid–solid phase transition, which can be quite substantial and depends strongly on the geometry of the pores (Tell and Maris, 1983). Anomalous diffusion has been observed experimentally, and it is manifested as either suppression of the diffusion coefficient by two orders of magnitude at the boundary (Dozier et al., 1986), or an enhancement by an order of magnitude over the bulk (D'Orazio et al., 1989). The key parameter in such surface–bulk interactions is the degree of pore filling.

An intriguing flow regime has been discovered by Megaridis and his collaborators, see (Megaridis et al., 2002; Ye et al., 2004; Naguib et al., 2004), in nanotubes containing multiphase flow at high pressure, see Figure 1.23. The left figure shows a transmission electron microscope image of a fluid-filled multi-walled closed carbon nanotube synthesized using a hydrothermal procedure. The nanotube contains a stable liquid membrane bordered by two gas bubbles. Note the well-defined menisci separating the gas and liquid phases. On the right a much smaller nanotube is shown, about ten time less in diameter. This three-walled carbon nanotube, which was produced using arc evaporation, has been subsequently filled with water via high-temperature high-pressure treatment in an autoclave. The fluid interfaces are no longer smooth, as in the larger nanotube. The liquid molecules appear to form a long chain, with dry areas visible close to the wall. Amorphous carbon is present on the nanotube wall exterior. The graphene sheets seen in the lower portion of the micrograph belong to larger carbon nanotubes, which, however, are void of fluids. The walls of the nanotubes are made of carbon (graphene) layers, but the gaps in between the carbon atoms do not allow water to seep through. Only hydrogen can escape through these very fine pores. This explains why these multilayered walls can contain the fluid contents at such high pressures for quite some time. For the larger nanotube, the volume of the liquid inclusion is about 10^{-18} liters, and the pressure difference across the liquid is in the range of 1–3 MPa. The double meniscus bounding the liquid in the larger nanotube shows good wettability of the inner carbon walls by the water-based fluid; it can be possibly described by continuum-based simulations, see Chapter 14. However, on the smaller nanotube the noncontinuum behavior requires atomistic simulations, see Chapter 16. The closed-end tube is typical of samples made through a hydrothermal process (Libera and Gogotsi, 2001; Ye et al., 2004). These nanotube samples were frequently found to contain a high-pressure multicomponent fluid showing curved interfaces separating the liquid from the gas phase. The heating experiments performed in (Megaridis et al., 2002), reported an array of dynamic interface phenomena visualized in real time in the nanotube interior.

Another very interesting multiphase flow at nanoscales is shown in Figure 1.24. The cross-flow device is a T-junction consisting of two nanochannels, one containing water and the other hexadecane plus 2% Span80. Because the two fluids are immiscible, an instability is formed. The pictures show the resulting patterns of water in oil emulsions in the outlet channel. The complexity varies dramatically as a function of the pressure applied to the input channels, as indicated in Figure 1.24. Each of these streams is generated spontaneously and is continuously moving as more droplets are formed. This regime too can be described by quasi-continuum approximation.

Electrokinetic effects are important in microfluidics; they provide a means of maintaining a certain level of flowrate with practically uniform profiles; see Chapter 7. They were first discovered in experiments with porous clay (Reuss, 1809; Wiedmann, 1852). The basic mechanism is based on the interaction of an ionized solution with static charges on dielectric surfaces such as glass. A high concentration of ions is present next to the wall with strong electrostatic forces. If a voltage difference is applied along a microchannel, flow is initiated very close to the wall within a distance of less than 100 nm. This situation can be modeled with a slip velocity proportional to the electric field, and a continuum approach suffices to describe this flow.

1.5 Modeling of Nanoflows

Fluid flows in nanometer scale channels and pores, referred to as nanoflows, play an important role in determining the functional characteristics of many biological and engineering devices and systems. In this section, we first introduce a few applications in which nanoflows play an important role and then discuss issues in modeling nanoflows. In Chapters 10–13 we present fundamental aspects of nanoflows.

Ionic channels are naturally occurring nanotubes found in the membranes of all biological cells (Hille, 2001). They are defined as a class of proteins, and each channel consists of a chain of amino acids folded in such a way that the protein forms a nanoscopic water-filled pore controlling the transport of ions in and out of the cell, and in and out of compartments inside cells like mitochondria and nuclei, thereby maintaining the correct internal ion composition that is crucial for cell survival and function. Each channel carries a strong and steeply varying distribution of permanent charge, which depends on the particular combination of channel and prevalent physiological conditions. The narrowest diameter of an ion channel can vary from a few angstroms to tens of angstroms. For example, shown in Figure 1.25 is a gramicidin ion channel, which is one of the smallest ion channels, with a critical diameter of 4 Å and a length of 25 Å. Many ion channels have the ability to selectively transmit or block a particular ion species, and most exhibit switching properties similar to electronic devices. Malfunctioning channels cause or associate with many diseases, and a large number of drugs act directly or indirectly on ion channels. The possibility to incorporate ion channel structures in electronic circuits as sensing, memory, or even as computational elements opens up exciting new opportunities and great challenges. Given the physiological importance of ion channels, it is important to understand the flow of water and electrolytes in naturally existing nanoscopic pores in the presence of a strong permanent charge; see Chapters 12 and Section 13.2.

Another application in which nanoflows are gaining considerable attention is the translocation of deoxyribonucleic acid (DNA) through a nanopore. Functional analysis of the genome will require sequencing the DNA of many organisms, and nanometer–scale pores are being explored for DNA sequencing and analysis. The characteristic that makes nanopores useful for analysis of DNA or other individual macromolecules is that the scale of the pores is the same as that of the molecules of interest. For example, the diameter of single-stranded DNA (ssDNA) is approximately 1.3 nm, while the diameter of the narrowest restriction in α-haemolysin protein, the most commonly used organic pore for DNA analysis, is approximately 1.5 nm (Nakane et al., 2003; Kasianowicz et al., 1996). Although natural ion channels have many desired properties for sensing and analysis of macromolecules, they usually work only in carefully controlled conditions and are difficult to integrate with other components of a device. As a result, there has been a significant amount of work devoted to building robust and easy-to-integrate sensing and analysis devices based on synthetic nanopores (Li et al., 2001). Figure 1.26 shows the schematic of a nanopore-based DNA sequencing system. The basic idea is that as a DNA molecule is pulled through the nanopore immersed in an electrolyte solution by an external electric field, the partial charges on the DNA, which are different for different DNA sequences, will change the electric current through the pore. The observed current variation will serve as a signature to differentiate various DNA sequences. The diameter of the nanopore must be sufficiently small (typically less than 5 nanometers) so that the electric current can be sensitive to the DNA molecule passage. Since a nanopore device needs only a few DNA molecules to obtain reliable results, it can be much cheaper and faster compared to the traditional fluorescence-based detection schemes, in which much of the time and cost are devoted to making copies and purifying the DNA molecules so as to obtain a reasonable signal-to-noise ratio.

Another emerging application in which nanoflows are gaining importance is that of molecular gates (Kuo et al., 2003a; Kuo et al., 2003b). Nanoporous membranes can be used to interface vertically separated microfluidic channels to create a truly three-dimensional fluidic architecture, as shown in Figure 1.27. Such hybrid three-dimensional architectures can be used for gateable transfer of selected solution components between vertically separated microfluidic channels. For example, by adjusting the voltages applied at the terminals, and by controlling the polarity and density of the surface charges of the nanopores, vertical transport of analyte through the nanopores can be controlled precisely. Integration of such fluidic circuits into a single chip can thus enable very complicated fluidic and chemical manipulations. Bohn and colleagues (Kuo et al., 2003b) have also shown that even when operated in a passive mode, without external manipulation, the nanochannels can separate solutions in different layers and inhibit mixing while the fluidic manipulations are performed in each of the microchannels. The diameter of the cylindrical pores is usually on the order of 10 to 100 nm. In nanopores, flow occurs in structures of the same size as physical parameters that govern the flow. For example, in molecular gates, the Debye length (see Figure 1.27), which characterizes the length scale of ionic interactions in a solution, can be comparable to the diameter of the pore (Kuo et al., 2001). In microflows, the Debye length is negligible compared to the channel diameter. As a result, the flow characteristics in nanochannels can be different compared to the flows in microchannels; see Chapter 12.

Even though we have discussed only three examples in which nanoflows play an important role in determining the characteristics of the system, there are a number of other applications such as fuel cell devices, drug delivery systems, chemical and biological sensing and energy conversion devices, and a number of other nanodevices in which nanoflows are important and need to be understood in great detail. With advances in nanofabrication techniques, it is now possible to fabricate devices with diameters ranging from a few angstroms to few hundred nanometers. Such advances make it possible to understand fundamental physical mechanisms in nanoflows through a detailed comparison between experimental and theoretical studies.

The study of nanoflows has attracted significant attention in recent years, since the fundamental issues encountered in nanoflows can be different from those of microflows and macroflows for of the following reasons:

1. The surface-to-volume ratio is very high in nanofluidic systems.

2. The critical channel dimension in nanoflows can be comparable to the size of the fluid molecules under investigation.

3. Density fluctuations over interatomic distances can be important in nanoflows, while they can be largely neglected at larger scales.

4. Transport properties such as the diffusion coefficient and viscosity can be different in confined nanoflows.

5. The interaction of the fluid with the surface (e.g., hydrophilic versus hydrophobic) can have a profound influence on the flow characteristics in nanochannels.

6. The validity of the continuum theory can be questionable for confined nanoflows.

7. The issue of boundary conditions at solid–liquid interfaces at nanoscales is not very well understood.

8. Anomalous behavior has been observed in nanoflows.

In this book, we discuss a number of fundamental issues encountered in nanoflows. Much of the new physics encountered in nanoflows can be understood by studying simple fluids (such as Lennard–Jones liquids) in confined nanochannels. In Chapter 10, we discuss simple fluids in nanochannels. In Chapter 11, we discuss the static and dynamic properties of water confined in nanochannels. Finally, in Chapter 12, we discuss electroosmotic flow in confined nanochannels.

1.6 Numerical Simulation at All Scales

While the first part (Chapters 2–6) and second part (Chapters 7–13) address physical modeling issues of gas flows and liquid flows, respectively, at all scales, the third part of the book is devoted to descriptions of different numerical methods. A summary of possible simulation approaches at the atomistic and continuum levels for both gas and liquid microflows is shown in Figure 1.28. In this book we present in some detail the atomistic approaches (DSMC, MD, DPD, Lattice Boltzmann) and representative discretizations for the continuum approaches (spectral elements and meshless methods). Specifically, Chapter 14 describes continuum-based approaches, while Chapters 15 and 16 describe atomistic-based approaches.

Some of these algorithms, even for the continuum, have been used extensively in simulating microflows, such as the direct simulation Monte Carlo method (DSMC), while others have only recently been applied to this field, e.g., meshless methods. The force coupling method for particulate microflows that we present in Section 14.3.2 is an example of an efficient method for a very difficult problem involving many moving interfaces. We selected these methods carefully, taking into account the interdisciplinary character of microsystems applications, where full-system simulation requires simultaneous solutions of fluidic, electrical, mechanical, and thermal domains. Therefore, suitable methods for microsystems should be very efficient but also very accurate, since new and unfamiliar physics in the microdomain requires resolution-independent simulation studies. They should also be very flexible, especially in the context of moving domains, and therefore meshless methods seem a good candidate.

In Chapter 15 we present atomistic methods for *gas* microflows and nanoflows. We start with the most popular approach, direct simulation Monte Carlo (DSMC), for steady and unsteady flows, and summarize practical guidelines that need to be followed in simulations of rarefied gases. We then discuss techniques of coupling DSMC with continuum descriptions, and we analyze microfilters as an example. We also include an overview of the Boltzmann equation as well as benchmark solutions in the slip flow and transitional regimes, that could be useful to the reader for validation of experimental and numerical results. Finally, we present the lattice Boltzmann method, which solves the Boltzmann equation fast but in a constrained subspace. This method too could be a very effective tool for simulating flows in complex microgeometries.

In Chapter 16 we present atomistic methods for *liquid* microflows and nanoflows. We provide the general formulation of the molecular dynamics (MD) method as well as specific subsections on different types of potentials, thermostats, and practical guidelines for data analysis as well as sources for available software. We then discuss multiscale modeling, i.e., coupling of atomistic and continuum descriptions, and we apply an embedding multiscale method to electroomostic flows at nanoscales. A new method, dissipative particle dynamics (DPD), is presented last; it is a stochastic MD approach for modeling liquid flows at the mesoscopic level, and it is a potentially very effective tool for multiscale modeling of liquids.

1.7 Full-System Simulation of Microsystems

MEMS and microfluidic systems are characterized by the presence of multiple energy domains like mechanical, electrical, magnetic, and thermal, as shown in Figure 1.29. These devices are often integrated with components such as electronics, optics, and other micromechanical components to create integrated systems, thereby adding more energy domains and complexity to the system. A generic example of a largely simplified microfluidic system is shown schematically in Figure 1.30. This system is made up of a micropump, a microflow sensor, and an electronic control circuit. The electronic circuit may be used to adjust the pump flowrate so that a constant flow is maintained in a microchannel. The simulation of the complete system requires models for the micropump, the microflow sensor, and the electronic components associated with the control circuit. Another example of an integrated system is the large-scale integration of microfluidic channels on a chip. Quake and colleagues (Thorsen et al., 2002) designed and fabricated high-density microfluidic chips that contain plumbing networks with thousands of micromechanical valves and hundreds of individually addressable chambers (see Figure 1.31). These fluidic devices are analogous to electronic integrated circuits fabricated using large-scale integration. These integrated microfluidic networks are used to construct the microfluidic analogue of a comparator array and a microfluidic memory storage device whose behavior resembles random-access memory.

The design complexity and functionality complexity of microsystems and nanosystems can exceed the complexity of VLSI systems. More than three decades ago, VLSI simulation was considered a formidable task, but today VLSI systems are simulated routinely thanks to the many advances in CAD and simulation tools achieved over that period. It is clear that similar and even greater advances are required in the microsystems field in order to make *full-system simulation* of microsystems a reality in the near future. This will enable the microsystems community to explore new pathways of discovery and expand the role and influence of microsystems at a rapid rate.

There are broadly two levels of simulation in the field of microsystems.

1. **Physical Level:** At this level, the full behavior of the continuum is captured by means of highly meshed 3D simulations. These simulations involve many degrees of freedom. This level is primarily divided into process, material, and structural modeling. Multiple-energy domains (thermal, chemical, electrical, mechanical, etc.), large amplitude motions, and inherent nonlinearities (e.g., forces on capacitor plates; squeeze film damping) make physical-level simulation very complicated. Physical-level models, although very accurate, are computationally very expensive, and it is impractical to use them in the iterative microsystem design process.

2. **System Level:** In practice, every microdevice is connected to a full system (see, e.g., Figure 1.30). The use of highly comprehensive physical models at this level makes the simulations slow and computationally expensive. System level modeling requires low-order behavioral models of the various components of the system, which are electronic, micromechanical, fluidic, optical, chemical, biological, etc. In the general case, these system-level models are represented as a network of lumped elements analogous to the electric circuit components. More generally, they are represented as a small set of coupled ordinary differential equations that can be easily integrated in time. The results of detailed numerical simulations (with their enormous number of degrees of freedom) have to be projected onto spaces spanned by a very small number of appropriately chosen dynamical variables used in the system-level simulation.

In order to develop a system-level simulation framework, that is sufficiently simple, accurate, and robust, *all processes* involved need to be simulated at a comparable degree of accuracy and integrated seamlessly. That is, circuits, semiconductors, springs and masses, beams and membranes, as well as the flow field need to be simulated in a consistent and compatible way and in reasonable computational time! There are two well-known approaches for system-level simulation of microsystems. In the first approach, reduced-order models or macromodels for microdevices are combined with circuit simulation tools. In the second approach, the full-physics-based simulation tools for microdevices are directly combined with the circuit simulation tools. Since circuit simulation tools play an important role in system-level modeling, a brief description of circuit simulation tools is provided before we discuss the two approaches in more detail.

Circuit simulation is today a mature subject, and various software packages exist for design purposes. The program SPICE, which is an acronym for **S**imulation **P**rogram with **I**ntegrated **C**ircuit **E**mphasis, was developed in the 1970s at UC Berkeley (Nagel and Pederson, 1973), and since then it has become the unofficial industry standard among integrated circuit (IC) designers. SPICE is a general-purpose simulation program for circuits that may contain resistors, capacitors, inductors, switches, transmission lines, etc., as well as the five most common semiconductor devices: Diodes, BJTs, JFETs, MESFETs, and MOSFETs. SPICE has built-in models for the semiconductor devices, and the user specifies only the pertinent model parameter values. However, these devices are typically simple and can be described by lumped models, i.e., combinations of ordinary differential equations and algebraic equations. In some cases, such as in submicron devices, even usual semiconductor devices, e.g., MOSFET, simple modeling is not straightforward, and it is rather art than science to transfer from basic PDEs to approximated ODEs and algebraic equations.

1.7.1 Reduced-Order Modeling

Reduced-order models or macromodels are often considered as a link between physical and systems level-modeling, as shown in Figure 1.32. In this level of modeling, the prime focus is model-order reduction, i.e., to reduce the number of degrees of freedom present in the physical level simulation to many fewer degrees of freedom. Reduced-order modeling essentially acts as a link between physical and system levels by projecting the results of detailed numerical simulations (physical level) onto spaces spanned by a very small number of appropriately chosen dynamical variables used in the system-level simulation. Reduced-order models are generally characterized by the following attributes (Senturia et al., 1997):

1. They are generally analytical, rather than numerical, permitting the designer to reason about the effects of design changes (dimensions and material properties) without having to resimulate at the physical level.
2. They have a minimal number of degrees of freedom.
3. Both large- and small-amplitude excitations are represented.
4. They incorporate correct dependencies on device geometry and constitutive properties.
5. They should accurately capture both quasi-static and dynamic behavior.
6. They should be expressible in a simple-to-use form, i.e., an equation, a network analogy, or a small set of coupled ODEs.
7. They should be easy to connect to system–level simulators.
8. They should be sufficiently accurate when compared with experiments on suitable test devices and with fully meshed 3D simulations.
9. They should account for correct explicit energy conservation and dissipation behavior.

To clarify the concept of a macromodel or a reduced model, we give a specific example, taken from (Senturia et al., 1997), for a suspended membrane of thickness b deflected at its center by an amplitude d under the action of a uniform pressure force p. Let us also denote by $2a$ the length of the membrane, by E the Young's modulus, and by ν_p the Poisson ratio. The form of the pressure-deflection relation can be obtained analytically, for example, by employing power series assuming a circular thin membrane. This can be extended to more general shapes and nonlinear responses, for example,

$$p = \frac{C_1 b}{a^2} + \frac{C_2 f(\nu_p)}{a^4} \frac{E}{1-\nu_p} d^3,$$

where C_1 and C_2 are dimensionless constants that depend on the shape of the membrane, and $f(\nu_p)$ is a slowly varying function of the Poisson ratio. This function is determined from detailed finite-element simulations over a range of length a, thickness b, and material properties ν_p and E. Such "best-fits" are tabulated and are used in the simulation according to the specific structure considered, without the need for solving the partial differential equations governing the dynamics of the structure.

One of the simplest approaches for reduced-order modeling is to recast mechanical systems into electrical systems, so that mechanical systems can be integrated into SPICE. This can be understood more clearly by considering the analogy of a mass–spring–damper system driven by an external force with a parallel-connected RLC circuit with a current source. In this example, mass corresponds to capacitance, dampers to resistors, springs to inductive elements, and forces to currents. An effort has already been made to produce equivalent RLC circuits for microflows such as the squeezed gas film in silicon accelerometers, which we discuss in detail in Chapter 18. A variety of other sophisticated and complicated approaches, such as generalized Kirchhoffian networks, black box models, and Galerkin techniques, can be used for generation of reduced-order models and these are discussed in detail in Chapter 17. The application of various reduced-order methods for simulation of microflows is presented in Chapter 18.

1.7.2 Coupled Circuit/Device Modeling

For highly nonlinear behavior of microsystems, the development of simple, easy–to–use, and accurate reduced-order models can be challenging. In such cases, simulation based on full-physics models of all the energy domains contained in the microsystem may become necessary. Coupled circuit/device modeling refers to the simultaneous simulation of different functional units of the microsystem. The concept of a *coupled* circuit and device simulator has proved to be extremely beneficial in the domain of integrated circuits. Since the first of such simulators, MEDUSA (Engl et al., 1982), became available in the early 1980s, there has been significant work addressing coupled simulation. These activities have focused on improved algorithms, faster execution speeds, and applications. Today, commercial CAD vendors also support a mixed circuit–device simulation capability (Technology Modeling Associates, 1997; Silvaco International, 1995).

An effective *hierarchical* strategy for a full-system simulation employing the coupled circuit–device simulator for microfluidic applications is illustrated in Figure 1.33. This simulator supports compact models for the electronic components and available macromodels for microfluidic devices. In addition, full-physics models are available for the microfluidic components that can be utilized when detailed and accurate modeling is required. As an example, specific components such as microvalves, pumps, and flow sensors are shown in Figure 1.33. However, the list for the flow domain is much broader and could include networks of microchannels, micronozzles, as well as more complex flow systems such as a gas microturbine. The coupling of the circuit and microfluidic components is handled by imposing suitable boundary conditions on the fluid solver. This simulator allows the simulation of a complete microfluidic system including the associated control electronics. A coupled circuit/device modeling tool, CODECS (acronym for **C**oupled **D**evice and **C**ircuit **S**imulator), provides a truly mixed-level description of both circuits and devices. This program was developed at UC Berkeley (Mayaram and Pederson, 1987), and it employs combinations of both ODEs and PDEs with algebraic equations. CODECS incorporates SPICE3, the latest version of SPICE written in C (Quarles, 1989), for the circuit simulation capability. The multirate dynamics introduced by combinations of devices and circuits is handled efficiently by a multilevel Newton method or a full-Newton method for transient analysis, unlike the standard Newton method employed in SPICE. CODECS is appropriate for one-dimensional and two–dimensional devices, but other developments have produced efficient algorithms for three-dimensional devices as well (Mayaram et al., 1993).

In coupled-domain problems, such as flow–structure, structure–electric, or a combination of both, there are significant *disparities* in temporal and spatial scales. This, in turn, implies that multiple grids and heterogeneous time-stepping algorithms may be needed for discretization, leading to very complicated and consequently computationally prohibitive simulation algorithms. The main disadvantage of a full-system simulation approach is the high computational cost involved. The principal cost comes from solving the three–dimensional time–dependent flow equations in complex geometric domains, in transition regimes, and with unfamiliar physics. It is therefore important to obtain a fundamental understanding of microflows and nanoflows first in order to construct low–dimensional models similar to what has been done in flows at large scales (see, for example, (Berkooz et al., 1993; Deane et al., 1991; Ball et al., 1991)), and second to explore new design concepts based on new physics.

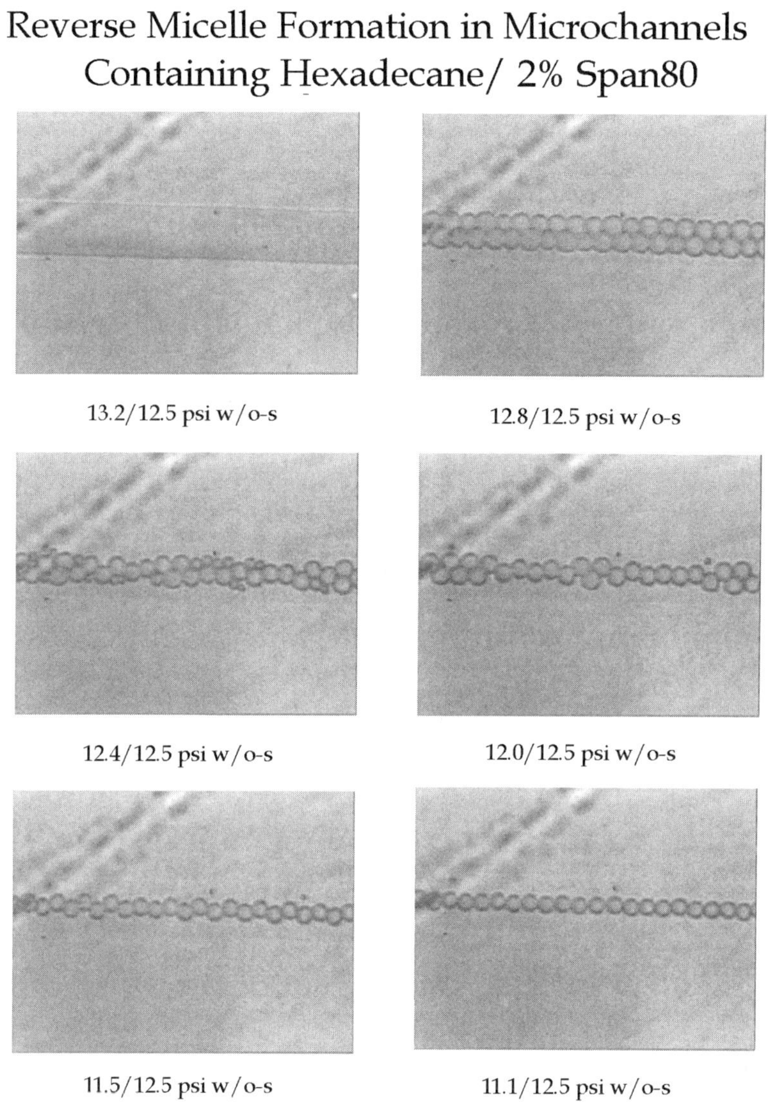

FIGURE 1.24. T-junction flow of water and hexadecane (2% Span80) in submicron channels. Depending on the relative pressure in the channels, different patterns of droplets are formed due to an instability. (Courtesy of S. Quake and T. Thorsen.)

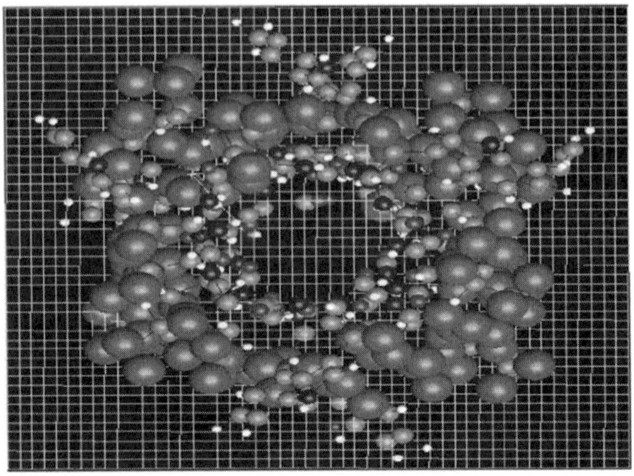

FIGURE 1.25. Front view of a gramicidin ion channel. Only the protein with the nanoscopic pore is shown. The diameter of the pore is 4 Å, and the length is 25 Å.

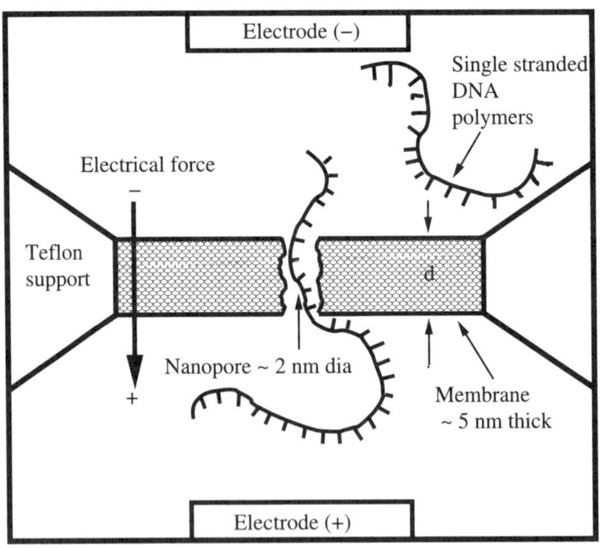

FIGURE 1.26. A schematic diagram of a nanopore–based DNA sequencing device.

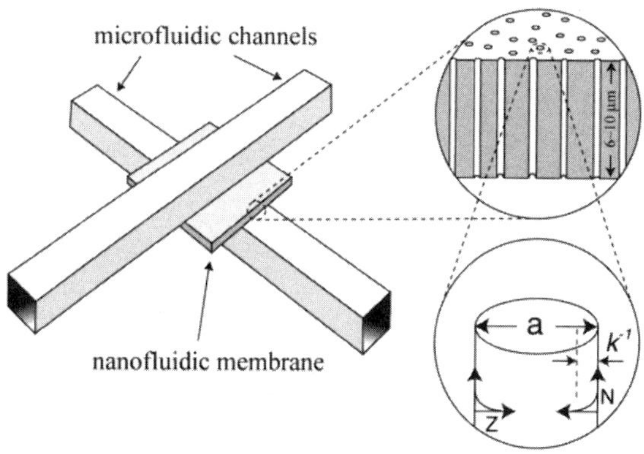

FIGURE 1.27. A schematic diagram of the crossed microfluidic channels separated by a nanometer-diameter capillary array interconnect. Upper right: the cross-sectional schematic of the nanocapillary. Lower right: schematic diagram showing the relative sizes of the channel diameter (denoted by a) and the Debye length (denoted by k^{-1}) of the electrolyte solution. (Courtesy of P.W. Bohn.)

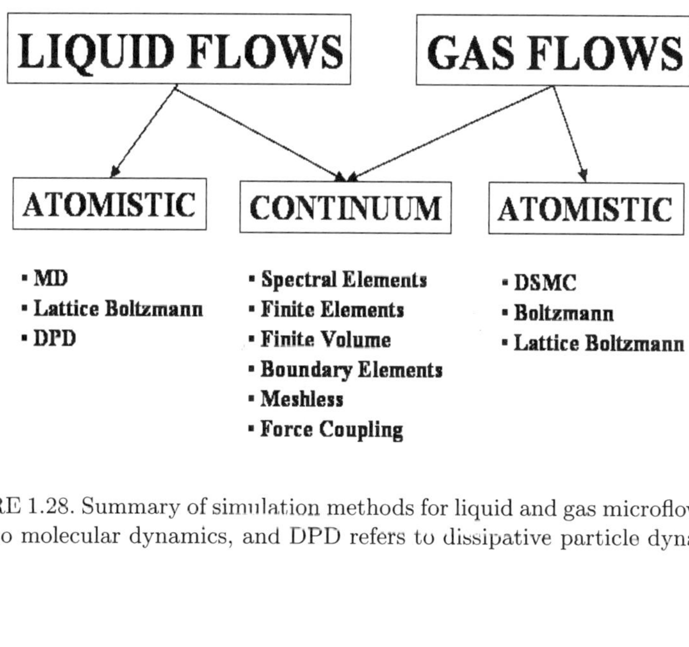

FIGURE 1.28. Summary of simulation methods for liquid and gas microflows. MD refers to molecular dynamics, and DPD refers to dissipative particle dynamics

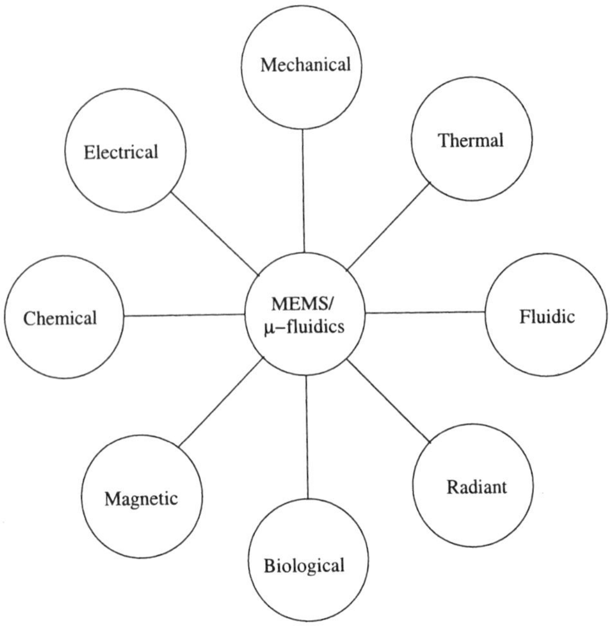

FIGURE 1.29. Energy domains in MEMS/microfluidics.

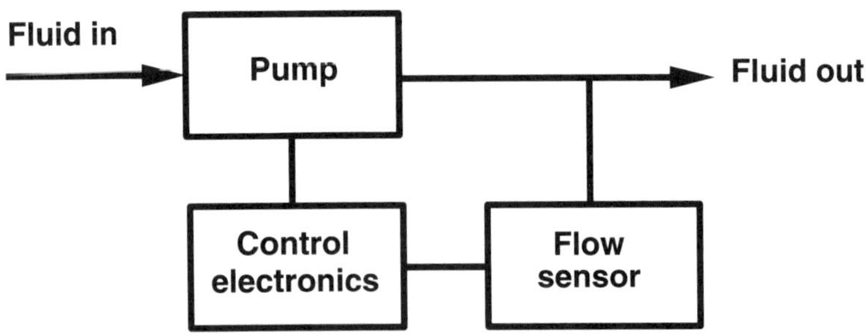

FIGURE 1.30. Block diagram of a generic microfluidic system. The flow sensor senses the flowrate which is controlled by the electronic circuit controlling the pump.

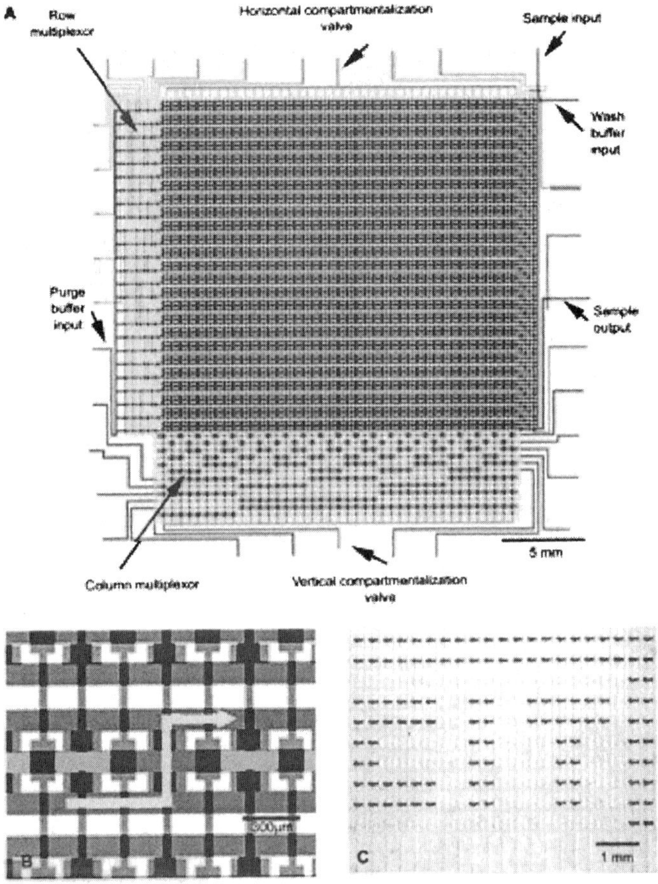

FIGURE 1.31. (A) Mask design for the microfluidic memory storage device. The chip contains an array of 25 × 40 chambers, each of which has a volume of 250 pl. Each chamber can be individually addressed using the column and row multiplexors. The contents of each memory location can be selectively programmed to be either blue dye (sample input) or water (wash buffer input). (B) Purging mechanics for a single chamber within a selected row of the chip. Each row contains three parallel microchannels. A specific chamber is purged as follows: (i) Pressurized fluid is introduced into the purge buffer input. (ii) The row multiplexor directs the fluid to the lower channel of the selected row. (iii) The column multiplexor releases the vertical valves of the chamber, allowing the pressurized fluid to flow through the chamber and purge its contents. (C) Demonstration of microfluidic memory display: Individual chambers are selectively purged to spell out "C I T". (Courtesy of S. Quake.)

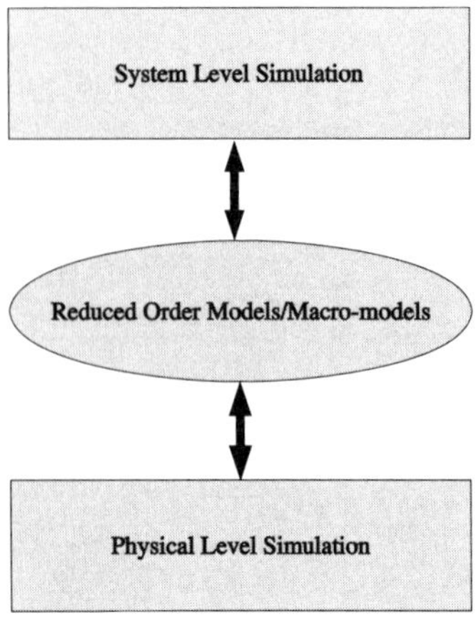

FIGURE 1.32. Simulation levels in microsystem design.

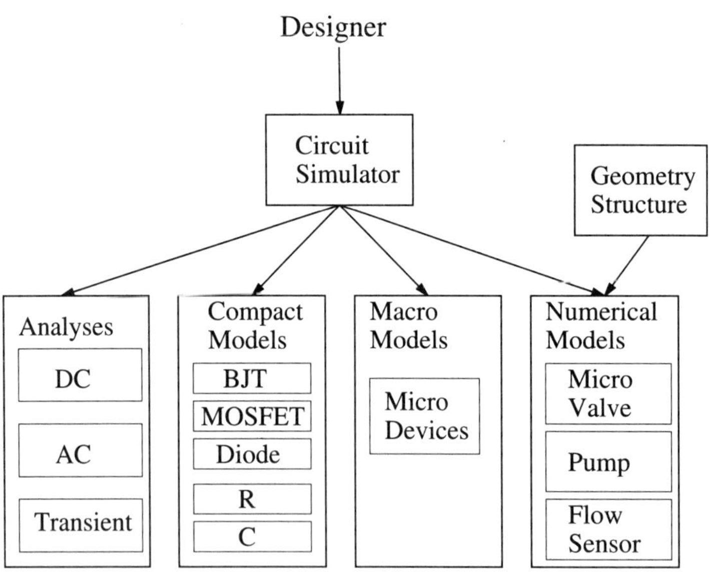

FIGURE 1.33. The coupled circuit–fluidic device simulator. Microfluidic systems, including the control electronics, can be simulated using accurate numerical models for all microcomponents.

Part I:
Gas Flows

2

Governing Equations and Slip Models

In this chapter we first present the basic equations of fluid dynamics both for incompressible and compressible flows, and discuss appropriate nondimensionalizations for low-speed and high-speed flows. Although most of the flows encountered in microsystems applications are typically of low speed, micropropulsion applications may involve high-speed supersonic flows (see Section 6.6). Subsequently, we consider the compressible Navier–Stokes equations and develop a general boundary condition for velocity slip. This applies to a regime for which $\mathrm{Kn} < 1$, and it corresponds to a second-order correction in Knudsen number. It improves Maxwell's original first-order formula, which is limited to $\mathrm{Kn} \leq 0.1$. The validity of this model is assessed in Chapter 4 with DSMC data, linearized Boltzmann equation solutions, as well as with experimental results. A more rigorous derivation of the governing equations from the Boltzmann equation is given in Section 15.4.2.

2.1 The Basic Equations of Fluid Dynamics

Consider fluid flow in the nondeformable control volume Ω bounded by the control surface $\partial\Omega$ with $\boldsymbol{n}$ the unit outward normal. The equations of motion can then be derived in an absolute reference frame by applying the principles of mechanics and thermodynamics (Batchelor, 1998). They can be formulated in integral form for mass, momentum, and total energy, respectively, as

$$\frac{d}{dt}\int_{\Omega} \rho d\Omega + \int_{\partial\Omega} \rho \boldsymbol{v} \cdot \boldsymbol{n}\, dS = 0, \tag{2.1a}$$

$$\frac{d}{dt}\int_{\Omega} \rho \boldsymbol{v} d\Omega + \int_{\partial\Omega} [\rho \boldsymbol{v}(\boldsymbol{v} \cdot \boldsymbol{n}) - \boldsymbol{n}\sigma]\, dS = \int_{\Omega} \boldsymbol{f} d\Omega, \tag{2.1b}$$

$$\frac{d}{dt}\int_{\Omega} E d\Omega + \int_{\partial\Omega} [E\boldsymbol{v} - \sigma\boldsymbol{v} + \boldsymbol{q}] \cdot \boldsymbol{n}\, dS = \int_{\Omega} \boldsymbol{f} \cdot \boldsymbol{v} d\Omega. \tag{2.1c}$$

Here $\boldsymbol{v}(\boldsymbol{x}, t) = (u, v, w)$ is the velocity field, ρ is the density, and $E = \rho(e + 1/2\boldsymbol{v} \cdot \boldsymbol{v})$ is the total energy, where e represents the internal specific energy. Also, σ is the stress tensor, $\boldsymbol{q}$ is the heat flux vector, and $\boldsymbol{f}$ represents all external forces acting on this control volume. For Newtonian fluids, the stress tensor, which consists of the normal components (p for pressure) and the viscous stress tensor τ, is a *linear* function of the velocity gradient, that is,

$$\sigma = -p\mathbf{I} + \tau, \tag{2.2a}$$

$$\tau = \mu[\nabla \boldsymbol{v} + (\nabla \boldsymbol{v})^T] + \zeta(\nabla \cdot \boldsymbol{v})\mathbf{I}, \tag{2.2b}$$

where $\mathbf{I}$ is the unit tensor, and μ and ζ are the first (absolute) and second (bulk) coefficients of viscosity, respectively. They are related by the Stokes hypothesis, that is, $2\mu + 3\zeta = 0$, which expresses local thermodynamic equilibrium. (We note that the Stokes hypothesis is valid for monoatomic gases but it may not be true in general.) The heat flux vector is related to temperature gradients via the Fourier law of heat conduction, that is,

$$\boldsymbol{q} = -k\nabla T, \tag{2.3}$$

where k is the thermal conductivity, which may be a function of temperature T.

In the case of a *deformable control volume*, the velocity in the flux term should be recognized as in a frame of reference relative to the control surface, and the appropriate time rate of change term should be used. Considering, for example, the mass conservation equation, we have the form

$$\frac{d}{dt}\int_{\Omega} \rho d\Omega + \int_{\partial\Omega} \rho \boldsymbol{v}_r \cdot \boldsymbol{n}\, dS = 0,$$

or

$$\int_{\Omega} \frac{\partial \rho}{\partial t}\, d\Omega + \int_{\partial\Omega} \rho \boldsymbol{v}_r \cdot \boldsymbol{n}\, dS + \int_{\partial\Omega} \rho \boldsymbol{v}_{\mathrm{cs}} \cdot \boldsymbol{n}\, dS = 0,$$

where $\boldsymbol{v}_{\mathrm{cs}}$ is the velocity of the control surface, $\boldsymbol{v}_r$ is the velocity of the fluid with respect to the control surface, and the total velocity of the fluid with respect to the chosen frame is $\boldsymbol{v} = \boldsymbol{v}_r + \boldsymbol{v}_{cs}$. The above forms are equivalent, but the first expression may be more useful in applications in which the time history of the volume is of interest.

Equations (2.1a) through (2.1c) can be transformed into an equivalent set of partial differential equations by applying Gauss's theorem (assuming that sufficient conditions of differentiability exist), that is,

$$\frac{\partial \rho}{\partial t} + \nabla \cdot (\rho \boldsymbol{v}) = 0, \tag{2.4a}$$

$$\frac{\partial}{\partial t}(\rho \boldsymbol{v}) + \nabla \cdot [\rho \boldsymbol{v}\boldsymbol{v} - \sigma] = \boldsymbol{f}, \tag{2.4b}$$

$$\frac{\partial}{\partial t} E + \nabla \cdot [E\boldsymbol{v} - \sigma \boldsymbol{v} + \boldsymbol{q}] = \boldsymbol{f} \cdot \boldsymbol{v}. \tag{2.4c}$$

The momentum and energy equations can be rewritten in the following form by using the continuity equation (2.4a) and the constitutive equations (2.2a), (2.2b):

$$\rho \frac{D\boldsymbol{v}}{Dt} = -\nabla p + \nabla \cdot \tau + \boldsymbol{f}, \tag{2.5a}$$

$$\rho \frac{De}{Dt} = -p \nabla \cdot \boldsymbol{v} - \nabla \cdot \boldsymbol{q} + \Phi, \tag{2.5b}$$

where $\Phi = \tau \cdot \nabla \boldsymbol{v}$ is the dissipation function and $D/Dt = \partial/\partial t + \boldsymbol{v} \cdot \nabla$ is the material derivative.

In addition to the governing conservation laws, an equation of state is required. For ideal gases, it has the simple form

$$p = \rho RT, \tag{2.6}$$

where R is the ideal gas constant defined as the difference of the constant specific heats; that is, $R = C_p - C_v$, where $C_v = \frac{\partial e}{\partial T}|_\rho$ and $C_p = \gamma C_v$ with γ the adiabatic index. For ideal gases, the energy equation can be rewritten in terms of the temperature, since $e = p/(\rho(\gamma - 1)) = C_v T$, and so equation (2.5b) becomes

$$\rho C_v \frac{DT}{Dt} = -p \nabla \cdot \boldsymbol{v} + \nabla \cdot [k \nabla T] + \Phi. \tag{2.7}$$

The system of equations (2.4a; 2.5a), (2.6), and (2.7) is called *compressible Navier–Stokes equations*, contains six unknown variables $(\rho, \boldsymbol{v}, p, T)$ with six scalar equations. Mathematically, it is an *incomplete parabolic* system, since there are no second-order derivative terms in the continuity equation.

A hyperbolic system arises in the case of inviscid flow, that is, $\mu = 0$ (assuming that we also neglect heat losses by thermal diffusion, that is, $k = 0$). In that case we obtain the *Euler equations*, which in the absence of external forces or heat sources have the form

$$\frac{\partial \rho}{\partial t} + \nabla \cdot (\rho \boldsymbol{v}) = 0, \tag{2.8a}$$

$$\frac{\partial (\rho \boldsymbol{v})}{\partial t} + \nabla \cdot (\rho \boldsymbol{v}\boldsymbol{v}) = -\nabla p, \tag{2.8b}$$

$$\frac{\partial E}{\partial t} + \nabla \cdot [(E + p)\boldsymbol{v}] = 0. \tag{2.8c}$$

This system admits discontinuous solutions, and it can also describe the transition from a supersonic flow (where $|\boldsymbol{v}| > c_s$) to subsonic flow (where $|\boldsymbol{v}| < c_s$), where $c_s = (\gamma RT)^{1/2}$ is the speed of sound. Typically, the transition is obtained through a shock wave, which represents a discontinuity in flow variables. In such a region the integral form of the equations should be used by analogy with equations (2.1a)–(2.1c).

2.1.1 Incompressible Flow

For an incompressible fluid, where $D\rho/Dt = 0$, the mass conservation (or continuity) equation simplifies to

$$\nabla\cdot\boldsymbol{v} = 0. \tag{2.9}$$

Typically, when we refer to an incompressible fluid we mean that $\rho =$ constant, but this is not necessary for a divergence-free flow; for example, in thermal convection the density varies with temperature variations. The corresponding momentum equation has the form:

$$\rho\frac{D\boldsymbol{v}}{Dt} = -\nabla p + \nabla\cdot\left[\mu[\nabla\boldsymbol{v} + (\nabla\boldsymbol{v})^T]\right] + \boldsymbol{f}, \tag{2.10}$$

where the viscosity $\mu(\boldsymbol{x}, t)$ may vary in space and time. The pressure $p(\boldsymbol{x}, t)$ is not a thermodynamic quantity but can be thought of as a constraint that projects the solution $\boldsymbol{v}(\boldsymbol{x}, t)$ onto a divergence-free space. In other words, an equation of state is no longer valid, since it will make the incompressible Navier–Stokes system overdetermined.

The acceleration terms can be written in various equivalent ways, so that in their discrete form, they conserve total linear momentum $\int_\Omega \rho\boldsymbol{v}\, d\Omega$ and total kinetic energy $\int_\Omega \rho\boldsymbol{v}\cdot\boldsymbol{v}\, d\Omega$ in the absence of viscosity and external forces. In particular, the following forms are often used:

- Convective form: $D\boldsymbol{v}/Dt = \partial\boldsymbol{v}/\partial t + (\boldsymbol{v}\cdot\nabla)\boldsymbol{v}$,
- Conservative (flux) form: $D\boldsymbol{v}/Dt = \partial\boldsymbol{v}/\partial t + \nabla\cdot(\boldsymbol{v}\boldsymbol{v})$,
- Rotational form: $D\boldsymbol{v}/Dt = \partial\boldsymbol{v}/\partial t - \boldsymbol{v}\times(\nabla\times\boldsymbol{v}) + 1/2\nabla(\boldsymbol{v}\cdot\boldsymbol{v})$,
- Skew-symmetric form: $D\boldsymbol{v}/Dt = \partial\boldsymbol{v}/\partial t + 1/2[(\boldsymbol{v}\cdot\nabla)\boldsymbol{v} + \nabla\cdot(\boldsymbol{v}\boldsymbol{v})]$.

The incompressible Navier–Stokes equations (2.9), (2.10) are written in terms of the primitive variables $(\boldsymbol{v}, p)$. An alternative form is to rewrite these equations in terms of the velocity $\boldsymbol{v}$ and vorticity $\omega = \nabla\times\boldsymbol{v}$. This is a more general formulation than the standard vorticity-streamfunction, which is limited to two dimensions. The following system is equivalent to equations (2.10) and (2.9), assuming that ρ, μ are constant:

$$\rho \frac{D\omega}{Dt} = (\omega \cdot \nabla)\boldsymbol{v} + \mu \nabla^2 \omega \quad \text{in} \quad \Omega, \tag{2.11a}$$

$$\nabla^2 \boldsymbol{v} = -\nabla \times \omega \quad \text{in} \quad \Omega, \tag{2.11b}$$

$$\nabla \cdot \boldsymbol{v} = 0 \quad \text{in} \quad \Omega, \tag{2.11c}$$

$$\omega = \nabla \times \boldsymbol{v} \quad \text{in} \quad \Omega, \tag{2.11d}$$

where the elliptic equation for the velocity $\boldsymbol{v}$ is obtained using a vector identity and the divergence-free constraint. We also assume here that the domain Ω is simply connected. An equivalent system in terms of velocity and vorticity is studied in (Karniadakis and Sherwin, 1999). The problem with the lack of direct boundary conditions for the vorticity also exists in the more often used vorticity-streamfunction formulation in two dimensions.

Finally, a note regarding **nondimensionalization**. Consider the free-stream flow U_0 past a body of characteristic size D in a medium of dynamic viscosity μ as shown in Figure 2.1. There are two characteristic time scales in the problem, the first one representing the *convective time scale* $t_c = D/U_0$, and the second one representing the *diffusive time scale* $t_d = D^2/\nu$, where $\nu = \mu/\rho$ is the kinematic viscosity. If we nondimensionalize all lengths with D, the velocity field with U_0, and the vorticity field with U_0/D, we obtain two different nondimensional equations corresponding to the choice of the time nondimensionalization:

Incompressible High-Speed Flows:

$$\frac{\partial \omega}{\partial t_c^*} + \nabla \cdot (\boldsymbol{v}\omega) = (\omega \cdot \nabla)\boldsymbol{v} + Re^{-1} \nabla^2 \omega,$$

Incompressible Low-Speed Flows:

$$\frac{\partial \omega}{\partial t_d *} + \mathrm{Re} \nabla \cdot (\boldsymbol{v}\omega) = \mathrm{Re}(\omega \cdot \nabla)\boldsymbol{v} + \nabla^2 \omega,$$

where t_c^* and t_d^* are the nondimensionalized time variables with respect to t_c and t_d, respectively, and $\mathrm{Re} = U_0 D/\nu$ is the Reynolds number. Both forms are useful in simulations, the first in high Reynolds number simulations (e.g., micronozzles, Section 6.6), and the second in low Reynolds number flows (e.g., microchannels).

In many microflow problems the nonlinear terms can be neglected, and in such cases the governing equations are the **Stokes equations**, which we can cast in the form

$$-\nu \nabla^2 \boldsymbol{v} + \nabla p/\rho = \boldsymbol{f} \quad \text{in} \quad \Omega, \tag{2.12a}$$

$$\nabla \cdot \boldsymbol{v} = 0 \quad \text{in} \quad \Omega, \tag{2.12b}$$

along with appropriate boundary conditions for $\boldsymbol{v}$.

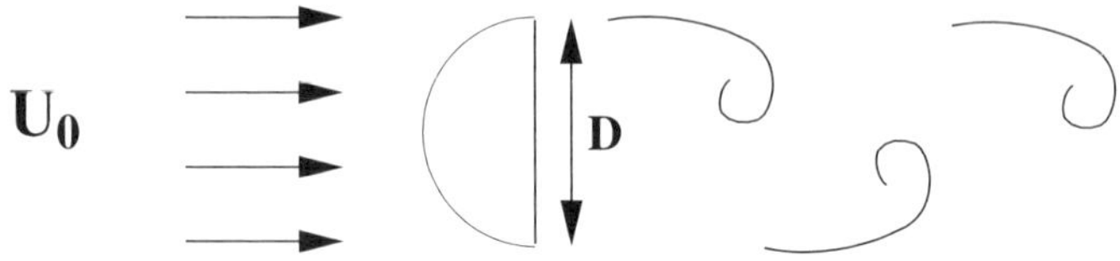

FIGURE 2.1. A schematic of free-stream flow past a microprobe in a viscous fluid.

2.1.2 Reduced Models

The mathematical nature of the Navier–Stokes equations varies depending on the flow that we model and the corresponding terms that dominate in the equations. For example, for an inviscid compressible flow, we obtain the Euler equations, which are of hyperbolic nature, whereas the incompressible Euler equations are of hybrid type corresponding to both real and imaginary eigenvalues. The unsteady incompressible Navier–Stokes equations are of mixed parabolic/hyperbolic nature, but the steady incompressible Navier–Stokes equations are of elliptic/parabolic type. It is instructive, especially for a reader with not much experience in fluid mechanics, to follow a hierarchical approach in reducing the Navier–Stokes equations to simpler equations so that each introduces one new concept.

Taking as an example the incompressible Navier–Stokes equations (2.9), (2.10), a simpler model is the *unsteady Stokes* system. This retains all the complexity but not the nonlinear terms; that is,

$$\frac{\partial \boldsymbol{v}}{\partial t} = -\nabla p/\rho + \nu \nabla^2 \boldsymbol{v} + \boldsymbol{f}$$
$$\nabla \cdot \boldsymbol{v} = 0.$$

The Stokes system [equations (2.12a) and (2.12b)] is recovered by dropping the time derivative. Alternatively, we can drop the divergence-free constraint and study the purely parabolic scalar equation for a variable u, that is,

$$\frac{\partial u}{\partial t} = \nu \nabla^2 u + f. \tag{2.13}$$

This equation expresses unsteady diffusion and includes volumetric source terms. If we instead drop all terms on the right-hand side of (2.10), as well as the divergence-free constraint, we obtain a nonlinear advection equation. Finally, by dropping the time derivative in the parabolic equation (2.13), we obtain the **Poisson equation**,

$$-\nu \nabla^2 u = f,$$

which is encountered often in MEMS (micro electro mechanical systems), e.g., in electrostatics.

2.2 Compressible Flow

The flow regime for Kn < 0.01 is known as the *continuum regime*, where the Navier–Stokes equations with no-slip boundary conditions govern the flow. In the *slip flow regime* ($0.01 \leq$ Kn ≤ 0.1) the often-assumed no-slip boundary conditions seem to fail, and a sublayer on the order of one mean free path, known as the Knudsen layer, starts to become dominant between the bulk of the fluid and the wall surface. The flow, in the Knudsen layer cannot be analyzed with the Navier–Stokes equations, and it requires special solutions of the Boltzmann equation (see Section 15.4 and also (Sone, 2002)). However, for Kn ≤ 0.1, the Knudsen layer covers less than 10% of the channel height (or the boundary layer thickness for external flows), and this layer can be neglected by extrapolating the bulk gas flow towards the walls. This results in a finite velocity slip value at the wall, and the corresponding flow regime is known as the *slip flow regime* (i.e., $0.01 \leq$ Kn ≤ 0.1). In the slip flow regime the flow is governed by the Navier–Stokes equations, and rarefaction effects are modeled through the partial slip at the wall using Maxwell's velocity slip and von Smoluchowski's temperature jump boundary conditions (Kennard, 1938).

> For example, it may, however, be misleading to identify the flow regimes as "slip" and "continuum," since the "no-slip" boundary condition is just an empirical finding, and the Navier-Stokes equations are valid for both the slip and the continuum flow regimes. Nevertheless, this identification was first made for rarefied gas flow research almost a century ago, and we will follow this terminology throughout this book.

In the *transition regime* (Kn ≥ 0.1) the constitutive laws that define the stress tensor and the heat flux vector break down (Chapman and Cowling, 1970), requiring higher-order corrections to the constitutive laws, resulting in the Burnett or Woods equations (Woods, 1993). It is also possible to use the Boltzmann equation directly, which is valid at the microscopic level (see Section 15.4). The Burnett and Woods equations are derived from the Boltzmann equation based on the Chapman–Enskog expansion of the velocity distribution function f, including terms up to Kn^2 in the following form:

$$f = f_0(1 + a\,\mathrm{Kn} + b\,\mathrm{Kn}^2), \tag{2.14}$$

where a and b are functions of gas density, temperature, and macroscopic velocity vector, and f_o is the equilibrium (Maxwellian) distribution function (Chapman and Cowling, 1970):

$$f_0 = \left(\frac{m}{2\pi k_B T_0}\right)^{3/2} \exp\left(-\frac{mv^2}{2k_B T_0}\right), \tag{2.15}$$

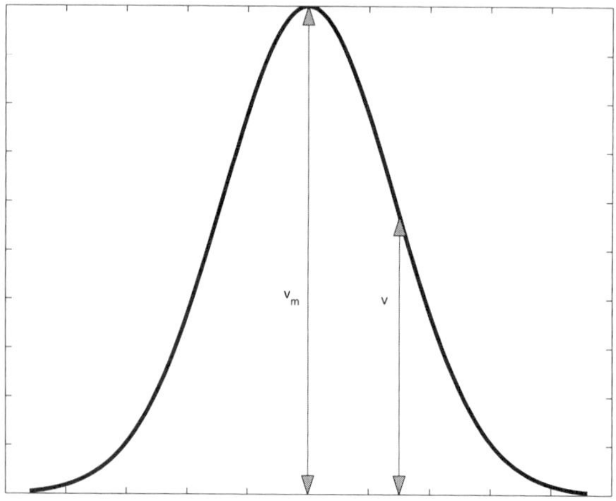

FIGURE 2.2. A plot of the Maxwellian distribution showing the most probable velocity and the mean thermal velocity, equation (2.15).

which is plotted in Figure 2.2. Here m is the molecular mass, k_B is the Boltzmann constant, T_0 is the temperature, and v is the mean thermal velocity of the molecules. The zeroth–order solution of equation (2.14) is the equilibrium solution, where flow gradients vanish; i.e., the viscous stress tensor and the heat flux vector vanish, giving the Euler equations (Chapman and Cowling, 1970). Therefore, $\mathrm{Kn} \equiv 0$ corresponds to the Euler equations. This is a *singular limit* in transition from the Navier–Stokes equations to the Euler equations, where the infinitesimally small viscosity (or heat conduction coefficient) vanishes.

Remark: In this book $\mathrm{Kn} = 0$ is commonly used to indicate the *no-slip* flow limit, and hence in the rest of this work $\mathrm{Kn} = 0$ indicates a limit that $\mathrm{Kn} \to 0$, but never $\mathrm{Kn} \equiv 0$!

The first-order solution in Kn yields the Navier–Stokes equations, and the second-order solution in Kn yields the Burnett equations. The Woods equations have a different form in the high-order corrections of the stress tensor and heat flux terms (Woods, 1993; Welder et al., 1993).

We rewrite here equations (2.4b), (2.4c) for compressible flows in two dimensions:

$$\frac{\partial}{\partial t}\begin{pmatrix}\rho\\ \rho u_1\\ \rho u_2\\ E\end{pmatrix}+\frac{\partial}{\partial x_1}\begin{pmatrix}\rho u_1\\ \rho u_1^2+p+\sigma_{11}\\ \rho u_1 u_2+\sigma_{12}\\ (E+p+\sigma_{11})\cdot u_1+\sigma_{12}\cdot u_2+q_1\end{pmatrix}$$
$$+\frac{\partial}{\partial x_2}\begin{pmatrix}\rho u_2\\ \rho u_1 u_2+\sigma_{21}\\ \rho u_2^2+p+\sigma_{22}\\ (E+p+\sigma_{22})\cdot u_2+\sigma_{21}\cdot u_1+q_2\end{pmatrix}=0, \quad (2.16)$$

where the two velocity components are denoted by $(u_1, u_2) \equiv (u, v)$ in the Cartesian coordinate system $(x_1, x_2) \equiv (x, y)$.

Remark: The conservation equations (2.16) are valid for continuum as well as for rarefied flows. However, the viscous stresses (σ_{ij}) and the heat flux (q_i) have to be determined differently for different flow regimes (see ection 15.4.2). Specifically, the *thermal stresses*

$$\frac{\partial^2 T}{\partial x_i \partial x_j} - \frac{1}{3}\frac{\partial^2 T}{\partial x_k^2}\delta_{ij}$$

in the momentum equation (derived from the Boltzmann equation) are not included in the Newtonian law for fluids. Similarly, the term in the energy equation

$$\frac{\partial^2 u_i}{\partial x_j^2}$$

is not present in the Fourier law. These terms are derived in the asymptotic analysis of the Boltzmann equation in the limit of small deviation from equilibrium (Sone, 2002). For small Knudsen number flows and with $\mathcal{O}(M) \sim \mathcal{O}(\mathrm{Kn})$, the thermal stress in the momentum equation can be absorbed in the pressure term. However, if the Reynolds number of the system is large or the temperature variation is not small, then the thermal stress cannot be included in the pressure term. In this case, these extra terms have to be included explicitly in the governing equations, which are different from the above compressible Navier–Stokes equations (Sone, 2002). To this end, also the work of (Myong, 1998) may be consulted. He derived thermodynamically consistent hydrodynamic models for high Knudsen number gas flows, valid uniformly for all Mach number flows and satisfying the second law of thermodynamics.

2.2.1 First-Order Models

By first-order models we refer to the approximation of the Boltzmann equation up to $\mathcal{O}(Kn)$, i.e., the compressible Navier-Stokes equations. The constitutive laws from equations (2.2a) and (2.2b) are

$$\sigma_{ij}^{NS} = -\mu \left(\frac{\partial u_j}{\partial x_i} + \frac{\partial u_i}{\partial x_j} \right) + \mu \frac{2}{3} \frac{\partial u_m}{\partial x_m} \delta_{ij} - \zeta \frac{\partial u_m}{\partial x_m} \delta_{ij}, \tag{2.17}$$

where μ and ζ are the dynamic (first coefficient)) and bulk (second coefficient) viscosities of the fluid, and δ_{ij} is the Kronecker delta. The heat flux is determined from Fourier's law (equation (2.3)). This level of conservation equations defines the *compressible Navier–Stokes equations.*

In the *slip flow* regime, the Navier–Stokes equations (2.16), (2.17) are solved subject to the velocity slip and temperature jump boundary conditions given by

$$u_s - u_w = \frac{2-\sigma_v}{\sigma_v} \frac{1}{\rho(2RT_w/\pi)^{1/2}} \tau_s + \frac{3}{4} \frac{\Pr(\gamma-1)}{\gamma \rho R T_w}(-q_s), \tag{2.18}$$

$$T_s - T_w = \frac{2-\sigma_T}{\sigma_T} \left[\frac{2(\gamma-1)}{\gamma+1} \right] \frac{1}{R\rho(2RT_w/\pi)^{1/2}}(-q_n), \tag{2.19}$$

where q_n, q_s are the normal and tangential heat flux components to the wall. Also, τ_s is the viscous stress component corresponding to the skin friction, γ is the ratio of specific heats, u_w and T_w are the reference wall velocity and temperature, respectively. Here Pr is the Prandtl number

$$\Pr = \frac{C_p \mu}{k}.$$

Equation (2.19) was proposed by Maxwell in 1879. The second term in (2.19) is associated with the thermal creep (transpiration) phenomenon, which can be important in causing pressure variation along channels in the presence of tangential temperature gradients (see Section 5.1). Since the fluid motion in a rarefied gas can be started with tangential temperature variations along the surface, the momentum and energy equations are coupled through the thermal creep effects. In addition, there are other thermal stress terms that are omitted in classical gas dynamics, but they may be present in rarefied microflows, as we discuss in Section 5.1. Equation (2.19) is due to von Smoluchowski (Kennard, 1938); it models temperature jump effects. Here σ_v, σ_T **are the tangential momentum and energy accommodation coefficients**, respectively (see Section 2.2.2). After nondimensionalization with a reference velocity and temperature, the slip conditions are written as follows:

$$U_s - U_w = \frac{2-\sigma_v}{\sigma_v} \mathrm{Kn} \frac{\partial U_s}{\partial n} + \frac{3}{2\pi} \frac{(\gamma-1)}{\gamma} \frac{\mathrm{Kn}^2 \mathrm{Re}}{\mathrm{Ec}} \frac{\partial T}{\partial s}, \tag{2.20a}$$

$$T_s - T_w = \frac{2-\sigma_T}{\sigma_T} \left[\frac{2\gamma}{\gamma+1} \right] \frac{\mathrm{Kn}}{\Pr} \frac{\partial T}{\partial n}, \tag{2.20b}$$

where the capital letters are used to indicate nondimensional quantities. Also, **n** and **s** denote the outward normal (unit) vector and the tangential (unit) vector.

Remark: Note that while the second term on the right-hand side of equation (2.20b) (thermal creep effect) appears to be $\mathcal{O}(\mathrm{Kn}^2)$, it actually corresponds to a first-order expansion (in Kn) of the Boltzmann equation. So both velocity jump and thermal creep are derived from an $\mathcal{O}(\mathrm{Kn})$ asymptotic expansion of the Boltzmann equation (Sone, 2002).

To determine fully the momentum and energy transport in microdomains, we need the following nondimensional numbers:

- *Reynolds number:* $\mathrm{Re} = \frac{\rho u h}{\mu}$,
- *Eckert number:* $\mathrm{Ec} = \frac{u^2}{C_p \Delta T}$, and
- *Knudsen number:* $\mathrm{Kn} = \frac{\lambda}{h}$.

However, it is possible to introduce a functional relation for Knudsen number and Eckert number in terms of the Mach number

$$M = \frac{u}{\sqrt{\gamma R T_0}}.$$

The Knudsen number can be written in terms of the Mach number and Reynolds number as

$$\mathrm{Kn} = \frac{\mu}{h\rho(2RT_w/\pi)^{1/2}} = \sqrt{\pi\gamma/2}\frac{M}{\mathrm{Re}}, \tag{2.21}$$

while the Eckert number can be written as

$$\mathrm{Ec} = (\gamma - 1)\frac{T_0}{\Delta T}M^2, \tag{2.22}$$

where ΔT is a specified temperature difference in the domain, and T_0 is the reference temperature used to define the Mach number. Using these relations for Ec and M, the independent parameters of the problem are reduced to three:

- **Prandtl number Pr, Reynolds number Re, and Knudsen number Kn.**

2.2.2 The Role of the Accommodation Coefficients

Momentum and energy transfer between the gas molecules and the surface requires specification of interactions between the impinging gas molecules and the surface. A detailed analysis of this is quite complicated and requires complete knowledge of the scattering kernels (see Section 15.4). From the macroscopic viewpoint, it is sufficient to know some average parameters in terms of the so-called momentum and thermal accommodation coefficients in order to describe gas–wall interactions. The **thermal accommodation coefficient** (σ_T) is defined by

$$\sigma_T = \frac{dE_i - dE_r}{dE_i - dE_w}, \tag{2.23}$$

where dE_i and dE_r denote the energy fluxes of incoming and reflected molecules per unit time, respectively, and dE_w denotes the energy flux if all the incoming molecules had been reemitted with the energy flux corresponding to the surface temperature T_w. The perfect energy exchange case corresponds to $\sigma_T = 1$. A separate thermal accommodation coefficient can be defined for the effects of gas–surface interactions on transitional, rotational, and vibrational energies of the molecules. Experimental evidence indicates that under such interactions the transitional and rotational energy components are more affected compared to the vibrational energy of the molecules (Schaaf and Chambre, 1961). However, such refinements cannot be applied to macroscopic models, since the rarefaction effects are treated by solving the continuum energy equation with the temperature jump boundary condition. DSMC models (see Section 15.1) can be more flexible in employing various molecule–wall collision models for different modes of energy transfer, as we show in Section 15.4.

The **tangential momentum accommodation coefficient** (σ_v) can be defined for tangential momentum exchange of gas molecules with surfaces

$$\sigma_v = \frac{\tau_i - \tau_r}{\tau_i - \tau_w}, \tag{2.24}$$

where τ_i and τ_r show the tangential momentum of incoming and reflected molecules, respectively, and τ_w is the tangential momentum of reemitted molecules, corresponding to that of the surface ($\tau_w = 0$ for stationary surfaces).

- The case of $\sigma_v = 0$ is called **specular reflection**,

where the tangential velocity of the molecules reflected from the walls is unchanged, but the normal velocity of the molecules is reversed due to the normal momentum transfer to the wall. In this case there is no tangential momentum exchange of fluid with the wall, resulting in zero skin friction. This is a limit of *inviscid flow*, where viscous stresses are zero. Hence

$$\frac{\partial u_s}{\partial n} \to 0 \quad \text{as} \quad \sigma_v \to 0,$$

and equation (2.20b) becomes obsolete, since the Euler equations require only the no-penetration boundary condition in this limit.

- The case of $\sigma_v = 1$ is called **diffuse reflection**.

TABLE 2.1. Thermal and tangential momentum accommodation coefficients for typical gases and surfaces (Seidl and Steinheil, 1974; Lord, 1976).

Gas	Surface	σ_T	σ_v
Air	Al	0.87–0.97	0.87–0.97
He	Al	0.073	
Air	Iron	0.87–0.96	0.87–0.93
H_2	Iron	0.31–0.55	
Air	Bronze		0.88–0.95

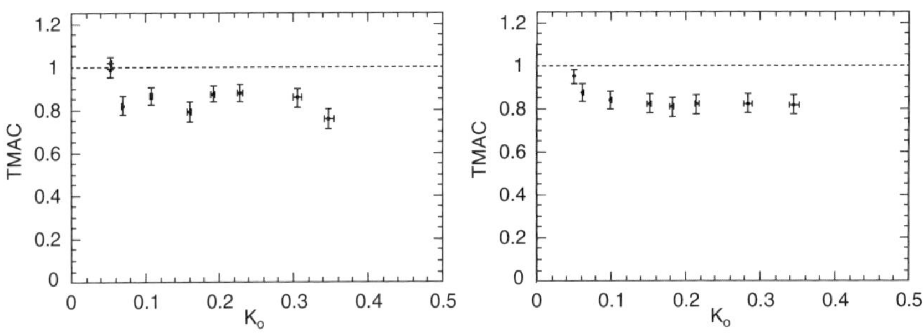

FIGURE 2.3. Tangential momentum accommodation coefficient σ_v (TMAC) versus Knudsen number obtained from mass flowrate measurements for argon (left) and for nitrogen (right). (Courtesy of K. Breuer.)

In this case the molecules are reflected from the walls with zero average tangential velocity. Therefore, the diffuse reflection is an important case for tangential momentum exchange (and thus friction) of the gas with the walls.

The tangential momentum and thermal accommodation coefficients depend on the gas and surface temperatures, local pressure, and possibly the velocity and the mean direction of the local flow. They are usually tabulated for some common gases and surfaces; see Table 2.1 and for details (Seidl and Steinheil, 1974; Lord, 1976). Diffuse reflection is likely to occur for rough surfaces. The values of σ_v and σ_T are not necessarily equal, as shown in Table 2.1. Typically, it takes a few surface collisions for a molecule to adopt the average tangential momentum of the surface, but it takes more surface collisions to obtain the energy level of the surface. Under laboratory conditions, values as low as 0.2 have been observed (Lord, 1976). Very low values of σ_v will increase the slip on the walls considerably even for small Knudsen number flows due to the $(2 - \sigma)/\sigma$ factor multiplying the velocity slip and temperature jump equations.

Measurements or direct computation of accommodation coefficients are very difficult to obtain. The accommodation coefficients for microchannel flows were measured indirectly using the first-order (Arkilic et al., 2001) and second-order slip flow theories (Maurer et al., 2003; Colin et al., 2004). Measurements of accommodation coefficients in (Arkilic et al., 2001) were obtained in the microchannel described in Chapter 1 (see Figure 1.18). Using high-resolution measurements for the mass flowrate and plotting it against the inverse pressure, the slope was computed, and based on the slip theory equations (see Chapter 4), the tangential momentum accommodation coefficient (TMAC) was obtained. The results of such measurements for argon and nitrogen are plotted in Figure 2.3 as a function of the Knudsen number. The measured value is $\sigma_v \approx 0.80$ for nitrogen or argon or carbon dioxide in contact with prime silicon crystal in the slip and early transitional flow regime ($0 < \mathrm{Kn} \leq 0.4$). It is observed that σ_v is less than unity, and independent of Kn in that range (Arkilic, 1997; Arkilic et al., 2001). Lower accommodation coefficients are possible due to the low surface roughness of prime silicon crystal.

(Maurer et al., 2003) presented experimental results for helium and nitrogen flow in 1.14 μm deep 200 μm wide glass channel covered by an atomically flat silicon surface. Flow behavior in the slip and early transition regimes was investigated for channel-averaged Knudsen numbers of 0.8 and 0.6 for helium and nitrogen flows, respectively. Using the flowrate data and a second-order slip model represented by equation (2.42), TMAC values of 0.91 ± 0.03 for helium, and 0.87 ± 0.06 for nitrogen were obtained. The authors also estimated the upper limit of slip flow regime as $\mathrm{Kn} = 0.3 \pm 0.1$, where Kn is based on the channel height. In a separate study, (Colin et al., 2004) presented experimental results for nitrogen and helium flow in a series of silicon microchannels fabricated using deep reactive ion etching (DRIE). Using mass flowrate and the corresponding pressure drop data, and the second-order slip model by (Deissler, 1964), they reported TMAC values of 0.93 for both helium and nitrogen. These authors reported breakdown of the first-order slip model for $\mathrm{Kn} \geq 0.05$, and of the second-order theory of Deissler for $\mathrm{Kn} \geq 0.25$, where Kn was based on the channel depth. These limits are unusually low compared to the values commonly accepted in the literature. A comparison between the experiments of (Colin et al., 2004) and (Maurer et al., 2003) shows that the uncertainty in the channel depth was ± 0.1 μm for Colin's channels, where it was 0.02 μm for Maurer's channels. In addition, the fabrication methods and the channel aspect ratios in these studies were different. These are certainly some of the reasons for the differences between the measurements of TMAC values by these two groups.

Finally, we note that

- *It is possible to predict the (pressure–driven) channel flowrate in the early transition flow regime, using a second–order slip solution of the Navier–Stokes equations. However, this procedure may create erroneous velocity profiles*, as shown in Figures 4.11 and 4.17.

We caution the reader about these limitations of the second-order theory

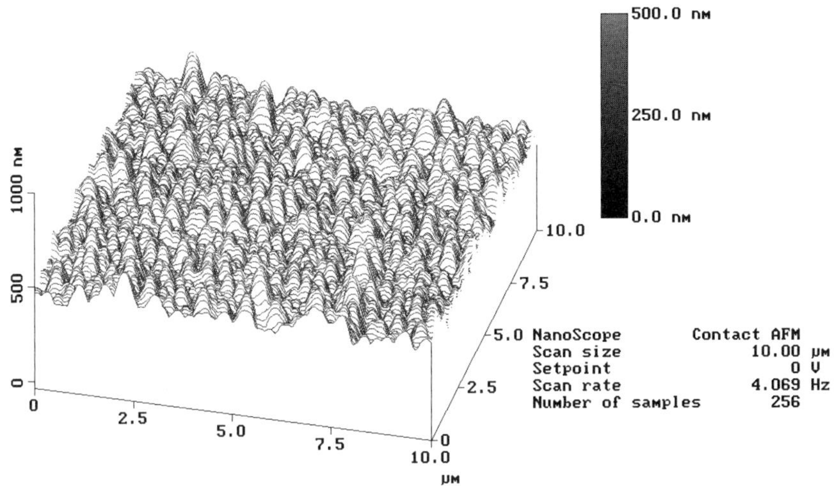

FIGURE 2.4. Profile of a polysilicon mirror surface; scan area 10×10 μm. (Courtesy of C. Liu.)

in the context of validation efforts using integral (e.g., mass flowrate versus pressure drop) measurements.

Roughness plays a very important role in microscales, but it is difficult to quantify its effect. In practice, it can be characterized using an atomic force microscope (AFM) for nonconductive surfaces, scanning tunnelling microscope (STM) for conductive surfaces, WYCO interferometer for optical nondestructive evaluation, and scanning electron microscope (SEM). Micron–scale roughness can be produced by wet chemical etching. A typical surface profile for a polysilicon mirror surface is shown in Figure 2.4.

The effects of roughness are difficult to analyze theoretically or numerically, but some progress has been made. (Richardson, 1973) considered a periodically modulated (rough) wall and applied a shear stress-free boundary condition. He showed analytically that the no-slip boundary condition is actually a consequence of surface roughness. In a systematic molecular dynamics study, (Mo and Rosenberger, 1990) investigated the effects of both periodic and random roughness with amplitude A. They found that as the roughness height (amplitude) A increases compared to the mean free path λ, the velocity slip at the wall decreases. Specifically, they proposed a criterion for the no-slip condition to be valid based on the ratio λ/A. If this ratio is of order unity, that is, if the roughness height is smaller but comparable to the mean free path, then the no-slip condition is satisfied. Otherwise, significant slip at the wall is present, which for atomically smooth walls occurs if the global Knudsen number, i.e., the ratio λ/h (with h the channel height), is finite. In summary, it was concluded that:

- For a microchannel flow with atomically **smooth** walls, if the global Knudsen number $\mathrm{Kn}_g = \lambda/h$ is less than 0.01, then the no-slip condition at the walls is valid (h is the channel height).
- For a microchannel flow with atomically **rough** walls, if the local Knudsen number $\mathrm{Kn}_l = \lambda/A$ is of order unity, then the no-slip condition at the walls is valid (A is the roughness height).
- Otherwise, in both smooth or rough walls, there is significant velocity slip at the walls.

In another study, (Li et al., 2002) considered surface roughness effects on gas flows through long microtubes. They treated the rough surface as a porous film covering an impermeable surface. In the porous film region they used the Brinkman-extended Darcy model, and they employed a high-order slip model in the core region of the microtubes. Solutions in these two different regions of the tube were combined by matching the velocity slip and the shear stress at the porous-core flow interface. This enabled derivation of expressions for the pressure distribution in microtubes, including the slip effects.

2.3 High-Order Models

The conservation equations (2.16) are still valid for larger deviations from the equilibrium conditions; however, the stress tensor (and heat flux vector) have to be corrected for high-order rarefaction effects. The general tensor expression of the Burnett level stress tensor is

$$\begin{aligned}\sigma_{ij}^B = -2\mu\overline{\frac{\partial u_i}{\partial x_j}} + \frac{\mu^2}{p}\Bigg[\omega_1 \frac{\partial u_k}{\partial x_k}\overline{\frac{\partial u_i}{\partial x_j}} + \omega_2\left(\overline{\frac{D}{Dt}\frac{\partial u_i}{\partial x_j}} - 2\overline{\overline{\frac{\partial u_i}{\partial x_k}}\frac{\partial u_k}{\partial x_j}}\right) \\ + \omega_3 R\overline{\frac{\partial^2 T}{\partial x_i \partial x_j}} + \omega_4 \overline{\frac{1}{\rho T}\frac{\partial p}{\partial x_i}\frac{\partial T}{\partial x_j}} + \omega_5 \overline{\frac{R}{T}\frac{\partial T}{\partial x_i}\frac{\partial T}{\partial x_j}} \\ + \omega_6 \overline{\overline{\frac{\partial u_i}{\partial x_k}}\frac{\partial u_k}{\partial x_j}}\Bigg], \end{aligned} \tag{2.25}$$

where a bar over a tensor designates a nondivergent symmetric tensor, i.e.,

$$\overline{f_{ij}} = (f_{ij} + f_{ji})/2 - \delta_{ij}/3 f_{mm}.$$

Similar expressions are valid for the heat flux q_i^B (Zhong, 1993). The coefficients ω_i depend on the gas model and have been tabulated for hard spheres and Maxwellian gas models (Schamberg, 1947; Zhong, 1993). Since the Burnett equations are of second order in Kn, they are valid in the early transition flow regime. However, fine-grid numerical solutions of certain

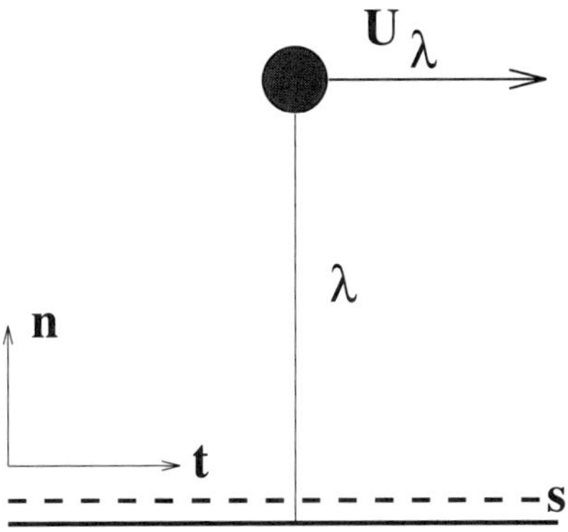

FIGURE 2.5. Control surface for tangential momentum flux near an isothermal wall moving at velocity U_w.

versions of the Burnett equations result in small wavelength instabilities. The cause of this instability has been traced to violation of the second law of thermodynamics (Balakrishnan, 2004). Using the Chapman–Enskog expansion and the Bhatnagar–Gross–Krook model of the collision integral, Balakrishnan (2004) derived the BGK–Burnett equations, and reported that the entropy-consistent behavior of the BGK–Burnett equations depends on the moment closure coefficients and approximations of the total derivative terms $(\frac{D}{Dt})$ in equation (2.25). In the following we use the exact definition of the total derivative instead of the Euler approximation most commonly used in hypersonic rarefied flows (Zhong, 1993). Numerical solutions of the Burnett equations for sevearal gas microflows can be found in (Agarwal et al., 2001; Agarwal and Yun, 2002; Xu, 2003; Lockerby and Reese, 2003; Xue et al., 2003).

Since the Burnett equations are obtained by a second-order Chapman–Enskog expansion in Kn, they require second-order slip boundary conditions. Such boundary conditions were derived by (Schamberg, 1947); however, numerical experiments with aerodynamic rarefied flows (Zhong, 1993) showed that Schamberg's boundary conditions are inaccurate for Kn > 0.2. Similar second-order slip boundary conditions have also been proposed in (Deissler, 1964) and (Sreekanth, 1969). Detailed discussions of performance of these second-order slip models will be presented in Sections 4.1.3 and 4.2, with comparisons of the DSMC and the linearized Boltzmann results against the analytical predictions for the velocity profile.

2.3.1 Derivation of High-Order Slip Models

Maxwell's derivation of equation (2.19) is based on kinetic theory. A similar boundary condition can be derived by an approximate analysis of the motion of gas in isothermal conditions. We write the tangential momentum flux on a surface s located near the wall (see Figure 2.5) as

$$\frac{1}{4} n_s m \bar{v} u_s,$$

where n_s is the number density of the molecules crossing surface s, m is the molecular mass, $\bar{v}$ is the mean thermal speed defined as

$$\bar{v} = (8/\pi RT)^{0.5},$$

and u_s is the tangential (slip) velocity of the gas on this surface. If we assume that approximately half of the molecules passing through s are coming from a layer of gas at a distance proportional to one mean free path ($\lambda = [\mu (RT\pi/2)^{\frac{1}{2}}/p]$) away from the surface, the tangential momentum flux of these incoming molecules is written as

$$\frac{1}{4} n_\lambda m \bar{v}_\lambda u_\lambda,$$

where the subscript λ indicates quantities evaluated one mean free path away from the surface. Since we have assumed that half of the molecules passing through s are coming from λ away from this surface $n_\lambda = \frac{1}{2} n_s$, the other half of the molecules passing through s are reflected from the wall (see Figure 2.5), and they bring to surface s a tangential momentum flux of

$$\frac{1}{4} n_w m \bar{v}_w u_r,$$

where the subscript w indicates wall conditions and the number density n_w is equal to $\frac{1}{2} n_s$. The average tangential velocity of the molecules reflected from the wall is shown by u_r. For determination of u_r we will use the definition of tangential momentum accommodation coefficient σ_v. Assuming that σ_v (in percentage) of the molecules are reflected from the wall *diffusely* (i.e., with average tangential velocity corresponding to that of the wall u_w), and $(1 - \sigma_v)$ (in percentage) of the molecules are reflected from the wall *specularly* (i.e., conserving their average incoming tangential velocity u_λ), we have

$$u_r = (1 - \sigma_v) u_\lambda + \sigma_v u_w.$$

Therefore, the total tangential momentum flux on surface s is written as

$$\frac{1}{4} n_s m \bar{v} u_s = \frac{1}{4} n_\lambda m \bar{v}_\lambda u_\lambda + \frac{1}{4} n_w m \bar{v}_w \left[(1 - \sigma_v) u_\lambda + \sigma_v u_w \right].$$

Since we have assumed that the temperatures of the fluid and the surface are the same, the mean thermal speeds are identical (i.e., $\bar{v}_s = \bar{v}_\lambda = \bar{v}_w$); this is a rather strong assumption in our derivation. The number density

n_s of molecules passing through the surface is composed of n_λ and n_w. We have assumed that $n_\lambda = n_w = \frac{1}{2}n_s$, which is approximately true if there is no accumulation or condensation of gas on the surface. Using the tangential momentum flux relation, the mean tangential velocity of the gas molecules on the surface, called **slip velocity**, is

$$u_s = \frac{1}{2}[u_\lambda + (1 - \sigma_v)u_\lambda + \sigma_v u_w]. \tag{2.26}$$

Schaaf and Chambre (1961) have written this expression as an average tangential velocity on a surface adjacent to an isothermal wall. Our derivation results in the same relation with approximately similar assumptions. Notice that instead of obtaining the slip information u_λ one mean free path away from the wall, a fraction of λ may be used; see (Thompson and Owens, 1975). Using a Taylor series expansion for u_λ about u_s, we obtain

$$\begin{aligned} u_s = & \frac{1}{2}\left[u_s + \lambda\left(\frac{\partial u}{\partial n}\right)_s + \frac{\lambda^2}{2}\left(\frac{\partial^2 u}{\partial n^2}\right)_s + \cdots\right] \\ & + \frac{1}{2}\left\{(1 - \sigma_v)\left[u_s + \lambda\left(\frac{\partial u}{\partial n}\right)_s + \frac{\lambda^2}{2}\left(\frac{\partial^2 u}{\partial n^2}\right)_s + \cdots\right] + \sigma_v \cdot u_w\right\}, \end{aligned}$$

where the normal coordinate to the wall is denoted by n. This expansion results in the following slip relation on the boundaries:

$$u_s - u_w = \frac{2 - \sigma_v}{\sigma_v}\left[\lambda\left(\frac{\partial u}{\partial n}\right)_s + \frac{\lambda^2}{2}\left(\frac{\partial^2 u}{\partial n^2}\right)_s + \cdots\right]. \tag{2.27}$$

After nondimensionalization with a reference length and velocity scale (such as free-stream velocity), we obtain

$$U_s - U_w = \frac{2 - \sigma_v}{\sigma_v}\left[\mathrm{Kn}\left(\frac{\partial U}{\partial n}\right)_s + \frac{\mathrm{Kn}^2}{2}\left(\frac{\partial^2 U}{\partial n^2}\right)_s + \cdots\right], \tag{2.28}$$

where we have denoted the nondimensional quantities with capital letters. By neglecting the higher-order terms in the above equation we recover Maxwell's first-order slip boundary condition (2.19) in nondimensional form. Similarly, if we truncate the above equation to include only up to second-order terms in Kn, we obtain

$$U_s - U_w = \frac{2 - \sigma_v}{\sigma_v}\left[\mathrm{Kn}\left(\frac{\partial U}{\partial n}\right)_s + \frac{\mathrm{Kn}^2}{2}\left(\frac{\partial^2 U}{\partial n^2}\right)_s\right]. \tag{2.29}$$

We will use this equation for comparison of various slip models in Section 2.3.3 and in Section 4.2 .

Equation (2.26) excludes the thermal creep terms of equation (2.19), since isothermal conditions are assumed in its derivation. For nonisothermal flows, the thermal creep effects are included to equation (2.26) separately, resulting in the following relation:

$$u_s = \frac{1}{2}[u_\lambda + (1-\sigma_v)u_\lambda + \sigma_v u_w] + \frac{3}{4}\frac{\mathrm{Pr}(\gamma-1)}{\gamma\rho R T_w}(-q_s).$$

For the temperature jump boundary condition, a derivation based on the kinetic theory of gases is given in (Kennard, 1938). We propose the following form for the high-order temperature jump condition by analogy with equation (2.28):

$$T_s - T_w = \frac{2-\sigma_T}{\sigma_T}\left[\frac{2\gamma}{\gamma+1}\right]\frac{1}{Pr}\left[Kn\left(\frac{\partial T}{\partial n}\right)_s + \frac{Kn^2}{2}\left(\frac{\partial^2 T}{\partial n^2}\right)_s \right.$$
$$\left. + \frac{\mathrm{Kn}^3}{6}\left(\frac{\partial^3 T}{\partial n^3}\right)_s + \cdots\right], \qquad (2.30)$$

which can be rearranged by recognizing the Taylor series expansion of T_λ about T_s to give a bf temperature jump boundary condition similar to equation (2.26) as

$$T_s = \left(\frac{(2-\sigma_T)}{\mathrm{Pr}}\frac{2\gamma}{(\gamma+1)}T_\lambda + \sigma_T T_w\right) \Big/ \left(\sigma_T + \frac{2\gamma}{(\gamma+1)}\frac{(2-\sigma_T)}{\mathrm{Pr}}\right). \qquad (2.31)$$

Here T_λ is the temperature at the edge of the Knudsen layer, i.e., one mean free path (λ) away from the wall.

2.3.2 General Slip Condition

The expansion originally given in (Schaaf and Chambre, 1961) is of first order in Kn. However, for higher Knudsen numbers, second-order corrections to these boundary conditions may become necessary. The velocity slip near the wall is coupled with the first and second variations of the tangential velocity in the normal direction to the wall. Numerical implementation of the slip formula in this form is computationally difficult. Therefore, further simplification of (2.28) without changing the second-order dependence on Kn is desired. For this purpose we assume that the transition from no-slip flow to slip flow occurs smoothly. Thus, a regular perturbation expansion of the velocity field in terms of Kn is defined in equation (2.32) below, where the no-slip Navier–Stokes velocity field is denoted by $U_0(\mathbf{x}, \mathbf{t})$, and corrections to the velocity field due to different orders of Kn dependence are denoted by $U_i(\mathbf{x}, \mathbf{t})$ $(i = 1, 2, 3 \ldots)$. We then have

$$U = U_0 + \mathrm{Kn}\, U_1 + \mathrm{Kn}^2\, U_2 + \mathrm{Kn}^3\, U_3 + \mathcal{O}(\mathrm{Kn}^4). \qquad (2.32)$$

This substitution enables us to rewrite the Navier–Stokes equations for different orders of Kn dependence in the following form:

$$
\begin{aligned}
\mathcal{O}(1): \quad & \frac{\partial \mathbf{U}_0}{\partial t} + (\mathbf{U}_0 \cdot \nabla)\mathbf{U}_0 = -\nabla P_0 + \mathrm{Re}^{-1}\nabla^2 \mathbf{U}_0; \qquad (2.33)\\
\mathcal{O}(\mathrm{Kn}): \quad & \frac{\partial \mathbf{U}_1}{\partial t} + (\mathbf{U}_1 \cdot \nabla)\mathbf{U}_0 + (\mathbf{U}_0 \cdot \nabla)\mathbf{U}_1 = -\nabla P_1 + \mathrm{Re}^{-1}\nabla^2 \mathbf{U}_1;\\
\mathcal{O}(\mathrm{Kn}^2): \quad & \frac{\partial \mathbf{U}_2}{\partial t} + (\mathbf{U}_0 \cdot \nabla)\mathbf{U}_2 + (\mathbf{U}_2 \cdot \nabla)\mathbf{U}_0 + (\mathbf{U}_1 \cdot \nabla \mathbf{U}_1)\\
& = -\nabla P_2 + Re^{-1}\nabla^2 \mathbf{U}_2;\\
\mathcal{O}(Kn^3): \quad & \frac{\partial \mathbf{U}_3}{\partial t} + (\mathbf{U}_0 \cdot \nabla)\mathbf{U}_3 + (\mathbf{U}_3 \cdot \nabla)\mathbf{U}_0 + (\mathbf{U}_2 \cdot \nabla)\mathbf{U}_1\\
& + (\mathbf{U}_1 \cdot \nabla)\mathbf{U}_2 = -\nabla P_3 + \mathrm{Re}^{-1}\nabla^2 \mathbf{U}_3.
\end{aligned}
$$

The boundary conditions for these equations are obtained similarly by substitution of the asymptotic expansion into the slip boundary condition formula:

$$
\begin{aligned}
\mathcal{O}(1): \quad & U_0|_s = U_w, \qquad (2.34)\\
\mathcal{O}(Kn): \quad & U_1|_s = \frac{2-\sigma}{\sigma}(U_0')\Big|_s,\\
\mathcal{O}(Kn^2): \quad & U_2|_s = \frac{2-\sigma}{\sigma}\left(\frac{1}{2}U_0'' + U_1'\right)|_s,\\
\mathcal{O}(Kn^3): \quad & U_3|_s = \frac{2-\sigma}{\sigma}\left(U_2' + \frac{1}{2}U_1'' + \frac{1}{6}U_0'''\right)\Big|_s,
\end{aligned}
$$

where U_i', U_i'', and U_i''' denote first, second, and third derivatives of the ith-order tangential velocity field along the normal direction to the surface.

A possible solution methodology for slip flow with high-order boundary conditions can be the solution of the Navier–Stokes equations ordee by order. However, this approach is computationally expensive, and there are numerical difficulties associated with accurate calculation of higher-order derivatives of velocity near walls with an arbitrary surface curvature.

We propose a formulation where the governing equations are directly solved without an asymptotic expansion in velocity, as mentioned above. The objective is to establish a methodology to develop slip boundary conditions accurate up to the second-order terms in Kn. First, we introduce a **new slip boundary condition**

$$
U_s - U_w = \frac{2-\sigma_v}{\sigma_v}\frac{\mathrm{Kn}}{1 - B(\mathrm{Kn})\,\mathrm{Kn}}\left(\frac{\partial U}{\partial n}\right), \qquad (2.35)
$$

where $B(\mathrm{Kn})$ is an empirical parameter to be determined. For a general choice of $B(\mathrm{Kn})$, equation (2.35) is first-order accurate in Kn, provided that $|B(\mathrm{Kn})| < 1$. However, for the continuum flow regime ($\mathrm{Kn} \to 0.0$)

the parameter $B(\text{Kn})$ has a definite value. This value can be used to make equation (2.35) second-order accurate in Kn for finite Kn. For the rest of the Kn values, $B(\text{Kn})$ can be curve-fitted accurately using the solutions of corresponding numerical models (i.e., Navier–Stokes and DSMC models) for the entire Kn range ($0.0 < \text{Kn} < \infty$). Equation (2.35) suggests finite corrections for slip effects for the entire Kn range, provided that $B(\text{Kn}) \leq 0$. It is possible to obtain the value of the parameter $B(\text{Kn})$ for small Kn, especially for the slip flow regime, by Taylor series expansion of $B(\text{Kn})$ about $\text{Kn} = 0$. We thus obtain

$$B(\text{Kn}) = B|_0 + \left.\frac{dB}{d\,\text{Kn}}\right|_0 \text{Kn} + \cdots = b + \text{Kn}\; c + \cdots. \tag{2.36}$$

Assuming that $|B(\text{Kn})| < 1$, we expand equation (2.35) in geometric series, including also the expansion given in equation (2.36) for $B(\text{Kn})$. This results in

$$U_s - U_w = \frac{2-\sigma_v}{\sigma_v}\,\text{Kn}\,\frac{\partial U}{\partial n}[1 + b\,\text{Kn} + (b^2 + c)\,\text{Kn}^2 + \cdots]. \tag{2.37}$$

Next, we substitute the asymptotic expansion for the velocity field (equation (2.32)) to the general slip condition given above, and rearrange the terms as a function of their Knudsen number order. This results in

$$\begin{aligned} \mathcal{O}(1): \quad & U_0|_s = U_w; \\ \mathcal{O}(\text{Kn}): \quad & U_1|_s = \frac{2-\sigma_v}{\sigma_v}(U_0')|_s; \\ \mathcal{O}(\text{Kn}^2): \quad & U_2|_s = \frac{2-\sigma_v}{\sigma_v}(bU_0' + U_1')|_s; \\ \mathcal{O}(\text{Kn}^3): \quad & U_3|_s = \frac{2-\sigma_v}{\sigma_v}(U_2' + bU_1' + (b^2 + c)U_0')|_s. \end{aligned} \tag{2.38}$$

Comparing these equations with the conditions obtained from the Taylor series expansion in equation (2.35) order by order, we obtain that the two representations are identical up to the first-order terms in Kn. To match the second-order terms we must choose the parameter b as

$$b = \left(\frac{1}{2}\frac{U_0''}{U_0'}\right)_s = \frac{1}{2}\left[\frac{\left(\frac{\partial \omega}{\partial n}\right)_0}{\omega_0}\right]_s. \tag{2.39}$$

The quantities U_0' and U_0'' for an arbitrary curved surface denote first and second derivatives of the tangential component of the velocity vector along the normal direction to the surface, corresponding to a *no-slip* solution.

- *The parameter b in equation (2.39) is the ratio of the vorticity flux to the wall vorticity, obtained in no-slip flow conditions. The value of b for simple flows can be found analytically.*

Similarly, third-order terms in Kn can be matched if c is chosen as

$$c = \frac{1}{U_0'}\left(\frac{1}{2}U_1'' + \frac{1}{6}U_0''' - b^2 - bU_1'\right). \tag{2.40}$$

However, the third-order-accurate slip formula is computationally more expensive, since it requires the solutions for the U_1 field. We can obtain a second-order-accurate slip formula by approximating equation (2.35) as

$$U_s - U_w = \frac{2-\sigma_v}{\sigma_v}\frac{\mathrm{Kn}}{1-B\,\mathrm{Kn}}\frac{\partial U}{\partial n} = \frac{2-\sigma_v}{\sigma_v}\frac{\mathrm{Kn}}{1-b\,\mathrm{Kn}}\frac{\partial U}{\partial n} + \mathcal{O}(\mathrm{Kn}^3), \tag{2.41}$$

where b is the high-order slip coefficient given in equation (2.39). The error for equation (2.41) is $\mathcal{O}(\mathrm{Kn}^3)$, i.e.,

$$\text{Error} = c\,U_0'\,\mathrm{Kn}^3\,.$$

Truncated geometric series containing only $\mathcal{O}(\mathrm{Kn}^2)$ terms could have also been used to implement the new second-order slip-boundary condition (see equation (2.37)). The error in this case is also $\mathcal{O}(\mathrm{Kn}^3)$, and is given as

$$\text{Error}_{g.s.} = [U_2' + bU_1' + (b^2 + c]U_0')\,\mathrm{Kn}^3\,.$$

Since we do not know the magnitude of the U_1' and U_2' terms, it is difficult to decide which approach is better. However, we believe that using equation (2.41) is better, since this equation keeps the original form suggested in (2.35). Also for separated flows, equation (2.41) gives no slip at the separation or reattachment points (as predicted from the first-order slip formula), since the shear stress (therefore $\frac{\partial U}{\partial n} = 0$) is zero at these points. However, the truncated geometric series (equation (2.37)) will give multiplication of infinitesimally small wall shear stress ($\tau_{\text{wall}} = \mu\frac{\partial U}{\partial y} \to 0.0$) with large b ($b = \frac{U_o''}{2U_o'} \to \infty$, since $U_o' \to 0$). This may result in a velocity slip at the separation point based on some numerical truncation error in the calculations.

In this section we have developed various second- and higher-order slip conditions for gas microflows. We note that the Navier–Stokes equations require only the first-order slip conditions, and the second-order slip models should be used strictly for the second-order equations, such as the Burnett or Woods equations. Throughout this book we will utilize the second-order slip conditions routinely for the Navier–Stokes equations. This can be justified by the following arguments:

TABLE 2.2. Coefficients for first- and second-order slip models.

Author	C_1	C_2
Cercignani (Cercignani and Daneri, 1963)	1.1466	0.9756
Cercignani (Hadjiconstantinou, 2003a)	1.1466	0.647
Deissler (Deissler, 1964)	1.0	9/8
Schamberg (Schamberg, 1947)	1.0	$5\pi/12$
Hsia and Domoto (Hsia and Domoto, 1983)	1.0	0.5
Maxwell (Kennard, 1938)	1.0	0.0
Equation (2.29)	1.0	−0.5

- In the small Reynolds number limit, i.e., $\text{Re} \ll \text{Kn} \ll 1$, asymptotic analysis of the Boltzmann equation shows that a consistent set of governing equations and boundary conditions up to $\mathcal{O}(\text{Kn}^2)$ is the Stokes system with second-order slip boundary conditions; see Section 15.4.2 and for details (Sone, 2002; Aoki, 2001).

- Rarefaction effects both in the aforementioned limit as well as in the limit of $\text{Re} \sim \mathcal{O}(1) \rightarrow M \sim \mathcal{O}(Kn)$ come in only through the boundary condition. This has been proven rigorously using the Boltzmann equation in (Sone, 2002).

- The high-order boundary conditions proposed include Maxwell's first-order slip conditions (2.19), (2.19) as the leading-order term. Hence, these results are correct up to $\mathcal{O}(\text{Kn})$ in the slip flow regime, irrespective of the formal order of the utilized slip conditions.

- The general boundary condition for slip (equation (2.43)) converges to a finite value for large Kn, unlike the first-order Maxwell's boundary condition.

2.3.3 *Comparison of Slip Models*

For isothermal flows with tangential momentum accommodation coefficient $\sigma_v = 1$, the general second-order slip condition has the *nondimensional* form

$$U_s - U_w = C_1 \,\text{Kn} \left(\frac{\partial U}{\partial n}\right)_s - C_2 \,\text{Kn}^2 \left(\frac{\partial^2 U}{\partial n^2}\right)_s, \tag{2.42}$$

where $(\partial/\partial n)$ denotes gradients normal to the wall surface. The coefficients C_1 and C_2 are the slip coefficients. Typical values of the slip coefficients developed by different investigators are shown in Table 2.2.

We will apply the second-order slip boundary conditions given above for channel flows in Chapter 4 to examine their accuracy in representing the flow profile, including the velocity slip predictions. According to Sreekanth

(Sreekanth, 1969), Cercignani's second-order boundary conditions should be used only for evaluating the flow states *far from the wall*, and these conditions should not be used to evaluate space integrals in regions extending close to the walls. Sreekanth reports good agreement of second-order slip boundary conditions with his experimental results for Kn as high as Kn = 1.5 (Sreekanth, 1969). However, Sreekanth used a different second-order slip coefficient ($C_2 = 0.14$) than the original ones shown in Table 2.2. He also reports a change of the first slip coefficient (C_1) from 1.00 to 1.1466 as the Knudsen number is increased. First-order boundary conditions cease to be accurate, according to Sreekanth's study, above Kn > 0.13. More recent studies also show that Maxwell's slip boundary condition breaks down around Kn = 0.15 (Piekos and Breuer, 1995).

Implementation of second-order slip boundary conditions using equation (2.29) requires obtaining the second derivative of the tangential velocity in the normal direction to the surface ($\partial^2 U/\partial n^2$), which may lead to computational difficulties, especially in complex geometric configurations. To circumvent this difficulty we have proposed in the previous section the following *general* velocity slip boundary condition.

$$U_s - U_w = \frac{2-\sigma_v}{\sigma_v}\left[\frac{\mathrm{Kn}}{1-b\,\mathrm{Kn}}\left(\frac{\partial U}{\partial n}\right)_s\right], \tag{2.43}$$

where b is a general slip coefficient. Notice that the value of b can be determined such that for $|b\,\mathrm{Kn}\,| < 1$ the geometric series obtained from the boundary condition of equation (2.43) matches exactly the second-order equation (2.29), and thus for slip flow the above boundary condition is *second-order accurate* in the Knudsen number.

An alternative way of implementing the slip boundary condition is to use equation (2.26) derived directly from the tangential momentum flux analysis. Such a boundary condition has not been tested before, so in Section 4.1.3 we will determine the region of its validity, and in particular at what distance from the wall it should be applied, i.e., λ or $C\lambda$, where $C \neq 1$ (see Figure 2.5 and (Thompson and Owens, 1975)).

As regards the accuracy of two velocity slip boundary conditions, i.e., equation (2.26) versus equation (2.43), we can analyze the differences for the two-dimensional pressure-driven incompressible flow between parallel plates separated by a distance h in the slip-flow regime. Assuming isothermal conditions and that the slip is given by equation (2.26), the corresponding velocity distribution is

$$U(y) = \frac{h^2}{2\mu}\frac{dP}{dx}\left[\frac{y^2}{h^2} - \frac{y}{h} - \frac{2-\sigma_v}{\sigma_v}(\mathrm{Kn} - \mathrm{Kn}^2)\right]. \tag{2.44}$$

This is identical to the results obtained using equation (2.43) up to second-order terms in Kn, given below:

$$U(y) = \frac{h^2}{2\mu}\frac{dP}{dx}\left[\frac{y^2}{h^2} - \frac{y}{h} - \frac{2-\sigma_v}{\sigma_v}\frac{\mathrm{Kn}}{1+\mathrm{Kn}}\right]. \tag{2.45}$$

This equivalence can be seen by expanding the last term in equation (2.45) as a geometric series expansion in terms of powers of Kn. The leading error in equation (2.45) is therefore proportional to

$$\frac{h^2}{2\mu}\left|\frac{\partial P}{\partial x}\right| \mathrm{Kn}^3,$$

where h is the microchannel height.

Remarks: We summarize here observations that will aid in evaluating the proper application and limitations of the slip boundary conditions given by equations (2.42) and (2.43).

1. The first-order slip boundary condition should be used for $\mathrm{Kn} \leq 0.1$ flows. Since rarefaction effects gradually become important with increased Kn (regular perturbation problem), inclusion of second- and higher-order slip effects into a Navier–Stokes–based numerical model is neither mathematically nor physically inconsistent.

2. Using the high-order slip boundary conditions with the Navier–Stokes equations can lead to some physical insight. For example, using equation (2.42) for pressure-driven flows with various slip coefficients from Table 2.2 results in different velocity profile and flowrate trends. All the models in Table 2.2, with the exception of equation (2.29), result in *increased flowrate* due to the second-order slip terms. Although this is a correct trend for flowrate, the velocity distribution predicted by these models become erroneous with increased Kn, as shown in Figures 4.11 and 4.17. This indicates that *solely using the high-order slip correction in the transition flow regime is insufficient to predict the velocity profile and the flowrate simultaneously.* In Section 4.2, we address this problem by introducing a *rarefaction correction parameter* that leads to a unified flow model for pressure-driven channel and pipe flows, when combined with the *general* slip condition (equation (2.43)). The unified model predicts the correct velocity profile, flowrate, and pressure distribution in the *entire Knudsen regime* (see Section 4.2 for details).

3. Steady plane Couette flows have linear velocity profiles, which result in $\partial^2 U/\partial n^2 = 0$. Therefore, the high-order slip effects in equations (2.42) and (2.43) diminish for plane Couette flows. In Section 3.2, we demonstrate a generalized slip model for linear Couette flows that is valid for $\mathrm{Kn} \leq 12$.

4. As a final remark, interfacial interactions between the gas and surface molecules may result in inelastic reflections, due to the long-range interaction forces between the gas and surface molecules. Consequently, the gas molecules may condense and then evaporate after a certain time. This results in deposition of a thin layer of gas molecules on the surface. Using Langmuir's theory of adsorption, Myong (2004) explained the accommodation coefficient concept, and studied velocity slip for both monatomic and diatomic molecules. He has shown that the Langmuir model recovers Maxwell's first-order slip conditions, and he also described equation (2.43) in the context of the Langmuir model (Myong, 2004).

3

Shear-Driven Flows

In this chapter we consider shear-driven gas microflows with the objective of modeling a certain class of flows encountered in microsystems. For example, the flow between the rotor and base plate of a micromotor and the flow between the stationary and movable arms of a comb-drive mechanism are shear-driven microflows (see Figures 1.1 and 1.2, respectively). In order to circumvent the difficulty of understanding the flow physics for complex engineering geometries, we concentrate on prototype flows such as the linear Couette flow, and flow in shear-driven microcavities and grooved microchannels. First, we present analytical and numerical results for steady Couette flow in the slip flow regime. This is followed in Section 3.2 by the development and validation of an empirical model for steady Couette flow in the transition and free-molecular flow regimes. In Section 3.3, we present simulation results and analysis for oscillatory shear-driven flows in the entire Knudsen regime. The last two sections include flows in prototype complex geometries, such as the microcavity, and grooved microchannel flows.

3.1 Couette Flow: Slip Flow Regime

Shear-driven flows are encountered in micromotors, comb mechanisms, and microbearings. In the simplest form, the linear Couette flow can be used as a prototype flow to model such flows driven by a moving plate. Since the Couette flow is shear-driven, the pressure does not change in the streamwise direction. Hence, the compressibility effects become important for large temperature fluctuations or at high speeds. Considering this, we first present an analysis of incompressible Couette flow with slip.

The velocity profile for incompressible Couette flows with slip can be obtained by considering a two-dimensional channel extending between $y = 0$ to $y = h$, with the top surface moving with a prescribed velocity U_∞. By integrating the momentum equation (2.10) assuming no dependence on the streamwise direction, the following velocity profile is obtained in nondimensional form,

$$\frac{U}{U_\infty} = \frac{y/h + \frac{2-\sigma_v}{\sigma_v}\,\mathrm{Kn}}{1 + 2\frac{2-\sigma_v}{\sigma_v}\,\mathrm{Kn}} + \frac{3}{2\pi}\frac{(\gamma-1)}{\gamma}\frac{\mathrm{Kn}^2\,\mathrm{Re}}{\mathrm{Ec}}\frac{\partial T_s}{\partial x}, \tag{3.1}$$

where $\partial T_s/\partial x$ is the tangential temperature gradient along the surfaces due to the thermal creep effects, and Ec is the Eckert number defined as $(\mathrm{Ec} = \mathrm{M}\frac{u^2}{C_p \Delta T})$. The linear velocity profile of the Couette flow makes it *impossible* to incorporate high-order slip effects using equation (2.42), since $\partial^2 U/\partial n^2 = 0$. Therefore, high-order rarefaction effects cannot be captured unless the first-order slip coefficient C_1 is modified as a function of Kn. In Section 3.2, we present a high-order slip model for Couette flows that is valid in the transition and free molecular flow regimes. In deriving the velocity profile in equation (3.1), we utilized $C_1 = 1.0$. However, more accurate values for the slip coefficient have been determined using the Boltzmann equation, DSMC, and molecular dynamics simulations. For example, the theoretical model derived in (Ohwada et al., 1989b), using linearized Boltzmann equation for hard-sphere molecules predicts $C_1 = 1.111$, for $\mathrm{Kn} \leq 0.1$ for linear Couette flows.

The volumetric flowrate per channel width can be found by integrating equation (3.1)

$$\frac{\dot{Q}}{U_\infty h} = \frac{1}{2} + \frac{3}{2\pi}\frac{(\gamma-1)}{\gamma}\frac{\mathrm{Kn}^2\,\mathrm{Re}}{\mathrm{Ec}}\frac{\partial T_s}{\partial x}. \tag{3.2}$$

The first term is independent of the Knudsen number, while the second term is due to the thermal creep effects, which can result in change of the flowrate in the channel. In Section 5.1 we will present heat transfer analysis of shear-driven microflows, including the thermal creep effects corresponding to various prescribed heat flux conditions.

Once the analytical solution for the velocity distribution is known, the ratio of the skin friction coefficient for shear-driven slip flows and no-slip flows is given by

$$\frac{C_f}{C_{f_0}} = \frac{1}{1 + 2\frac{2-\sigma_v}{\sigma_v}\,\mathrm{Kn}}, \tag{3.3}$$

where $C_f \equiv \tau_w/(\frac{1}{2}\rho U_\infty^2)$, with τ_w the wall shear stress. The above equation is obtained for constant mass flow rate in the channel; hence the contribution of thermal creep is neglected. If the thermal creep effects were considered, the driving velocity U_∞ should have been modified in order to keep the volumetric flowrate of slip and no-slip flows the same.

For *compressible* shear-driven flows, an analytical solution can also be obtained given the simplicity of the geometry. In order to include the compressibility effects we must also specify the thermal boundary conditions. For the following analysis, we assumed that the upper plate temperature is specified to be T_∞, while the bottom plate is adiabatic. Also for simplification, viscosity and thermal conductivity are assumed to vary linearly with temperature (i.e., $k \sim \mu \sim T$), and the Prandtl number is fixed. In this case it is possible to obtain the friction coefficient (see page 313 in Liepmann and Roshko, 1957):

$$C_{f_0} = 2\frac{1 + \mathrm{Pr}\frac{\gamma-1}{3}{\mathrm{M}_\infty}^2}{\mathrm{Re}}, \tag{3.4}$$

where M_∞ is the Mach number based on the upper plate velocity and temperature, and Re is the Reynolds number based on the channel height.

We performed a series of simulations with the program μ**Flow**, which is based on high-order discretization (see Section 14.1 for details) to explore the compressibility effects in shear-driven flows. One set of simulation results corresponds to top plate temperature $T_\infty = 300\,\mathrm{K}$ and Reynolds number Re $= 5$. The simulations are performed using nine spectral elements with sixth-order polynomial expansions per direction in each element. The Mach number M_∞ is specified by varying the driving velocity of the top plate U_∞. Correspondingly, rarefaction effects are specified through the Knudsen number, since

$$\mathrm{Kn}_\infty = \sqrt{\pi\frac{\gamma}{2}}\frac{\mathrm{M}_\infty}{\mathrm{Re}}. \tag{3.5}$$

The variation of friction coefficient as a function of Mach number and corresponding Knudsen number is shown in Figure 3.1. The friction coefficient of no-slip compressible flow increases quadratically, in agreement with equation (3.4), well above the constant value of the corresponding incompressible flow. The no-slip compressible flow simulations match the theoretical results exactly. For rarefied flows, slip effects change the friction coefficient significantly. Compressible slip flow results are denoted by open circles in Figure 3.1. It is seen that the compressible slip flow results correspond to small deviations from the incompressible slip flow results obtained from

$$C_f = \frac{2}{\mathrm{Re}}\left[\frac{1}{1 + 2\frac{2-\sigma_v}{\sigma_v}\mathrm{Kn}}\right]. \tag{3.6}$$

In linear Couette flow the pressure is constant, and therefore compressibility effects are due to the temperature changes only. As M_∞ increases, the temperature difference between the two plates gets larger due to viscous heating. Thus, compressibility effects become significant. It is seen in

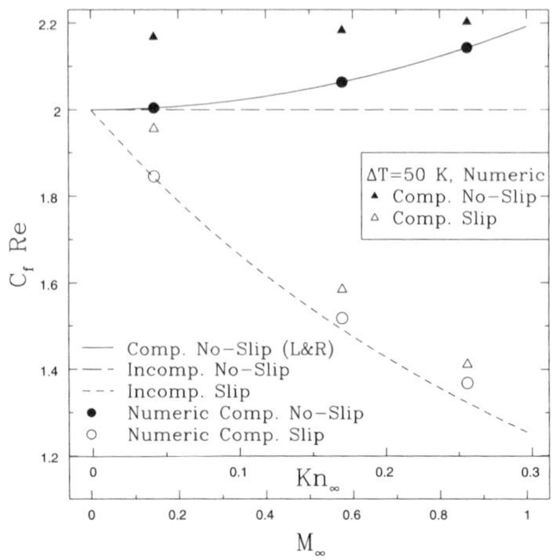

FIGURE 3.1. Variation of skin friction as a function of M and Kn for Couette flow. (Re $= 5, T_\infty = 300\,K$). Lines and symbols denote analytical and numerical results, respectively. Compressible no-slip results denoted by L&R are from (Liepmann and Roshko, 1957).

Figure 3.1 that significant deviations from incompressible flows (with either slip or no-slip) are obtained for $\mathrm{M}_\infty > 0.3$. In particular, we investigated a case in which the bottom plate is kept at $T_w = 350\,\mathrm{K}$ while the top plate is kept at $T_\infty = 300\,K$. The friction coefficient of this case is also given in Figure 3.1. The results are shown by solid and open triangles for the no-slip and the slip cases, respectively. The trend is different from the adiabatic bottom plate case. The no-slip results show small variation of C_f as a function of M, while for slip flows C_f is reduced significantly as Kn is increased.

The density variation across the channel for compressible no-slip as well as slip flows is shown in Figure 3.2 (left) for the case with adiabatic bottom wall. Here, we normalized the density variation by the top plate density of the no-slip case (solid line). The no-slip cases exhibit large density variations for relatively large values of M_∞. Since the pressure is constant, density variation across the channel is due to the drastic change in temperature, which is attributed to viscous heating. However, density variations are reduced in slip flows. There are two reasons for this behavior:

1. The shear stress is reduced due to slip, reducing the viscous heating effects, i.e., work done by viscous stresses in the energy equation.

2. A temperature jump exists at the driving (top) plate; there is no temperature jump on the bottom plate, since it is adiabatic.

We also included the variation of velocity in the channel (normalized with the driving plate velocity) in Figure 3.2 (right). It is seen that the lin-

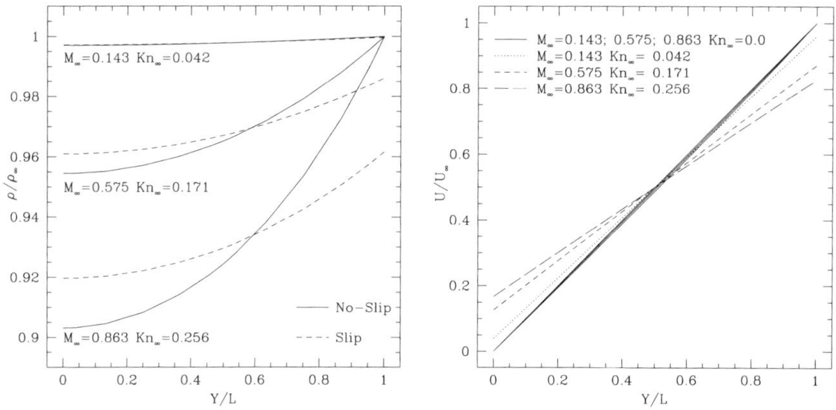

FIGURE 3.2. Couette flow: Density variation (left) and velocity variation (right) across the channel for various values of Mach number M and Knudsen number Kn; $L = h$ is the height of the channel (Re $= 5$, $T_\infty = 300\,\mathrm{K}$).

ear velocity profile for the incompressible flows is modified for compressible no-slip flows (solid lines); for the range of simulations performed here the deviations from linear profile is small. This is expected, because the compressible no-slip flow friction coefficient for $M_\infty = 0.863$ is only 10% larger than the corresponding incompressible flow. However, the velocity profiles for the compressible slip flow given in Figure 3.2 (right) show significant reduction in the slope of velocity, which explains the large reduction in the friction coefficient.

3.2 Couette Flow: Transition and Free-Molecular Flow Regimes

In this section, we analyze plane Couette flows in the transition and free-molecular flow regimes. Following (Bahukudumbi et al., 2003), we present empirical models for velocity distribution and shear stress, developed using the linearized Boltzmann equation solutions and extensive direct simulation Monte Carlo (DSMC) results.

3.2.1 Velocity Model

We consider rarefied gas flow between two infinite parallel plates, separated by a distance L. The plates are maintained at the same uniform temperature T_w, and they are moving with a uniform velocity of $+U$. We investigate steady one-dimensional plane Couette flow induced between the plates subject to the following assumptions:

- The gas molecules are hard spheres of uniform size, and they undergo diffuse reflections with the surfaces ($\sigma_v = 1$).
- The plate velocity is small compared to the mean thermal velocity. Hence the compressibility effects, temperature fluctuations, and viscous heating effects are negligible.

Based on these, the velocity field in the slip flow regime is given as

$$u_c(Y) = \frac{2UY}{1 + 2\frac{2-\sigma_v}{\sigma_v} C_1 \,\mathrm{Kn}}, \tag{3.7}$$

where $\mathrm{Kn} = \lambda/L$, and $Y = y/L$. Utilizing this solution with $C_1 = 1.111$, it has been shown that equation (3.7) is valid for $Kn \leq 0.25$ (Marques et al., 2000).

Steady flow calculations in (Sone et al., 1990) using the linearized Boltzmann equation indicate that the bulk flow velocity profile is essentially linear for all Knudsen numbers. However, a kinetic boundary layer (Knudsen layer) on the order of a mean free path starts to become dominant between the bulk flow and solid surfaces in the transition flow regime. The velocity distribution and other physical variables are subject to appreciable changes within the Knudsen layer, which can be predicted only by solution of the Boltzmann equation. In order to validate Sone's linearized Boltzmann solutions, (Bahukudumbi et al., 2003) performed a series of hard-sphere DSMC simulations at various Knudsen numbers and wall speeds. In these simulations, solid surfaces were maintained at 273 K, and they were assumed to be fully accommodating ($\sigma_v = 1$). Argon with molecular mass $m = 6.63 \times 10^{-26}$ kg and hard-sphere diameter $d_{\mathrm{hs}} = 3.66 \times 10^{-10}$ m was simulated. Since the collision diameter for hard-sphere molecules is independent of the relative velocity of the colliding molecules, the viscosity dependence on temperature is in the form of $\mu_0 \propto T_0^{0.5}$. This trend is slightly different from a more reliable variable hard-sphere model (VHS), which has viscosity dependence on temperature in the form of $\mu_0 \propto T_0^{0.81}$ for argon molecules. However, it is noted that DSMC simulations for hard-sphere and variable hard-sphere molecules yield similar results when the flow is isothermal. Simulations were performed for a wide range of Knudsen and Mach numbers. A minimum of 20 cells across the channel width was used in the simulations, and the domain discretization always ensured a minimum of 3 cells per mean free path. With this discretization, 10,000 hard-sphere particles were utilized, and the results were sampled for 1.0×10^6 time steps. The simulation time step was one-fifth of the mean collision time $(\lambda/\sqrt{2RT_0})$. These sets of parameters are sufficient to obtain accurate DSMC results (Chen and Boyd, 1996).

In Figure 3.3, velocity profiles for linear Couette flow in the upper half of the channel at three different Knudsen numbers are presented, where $k = (\sqrt{\pi}/2)\,\mathrm{Kn}$ is the rescaled Knudsen number. The linearized Boltzmann

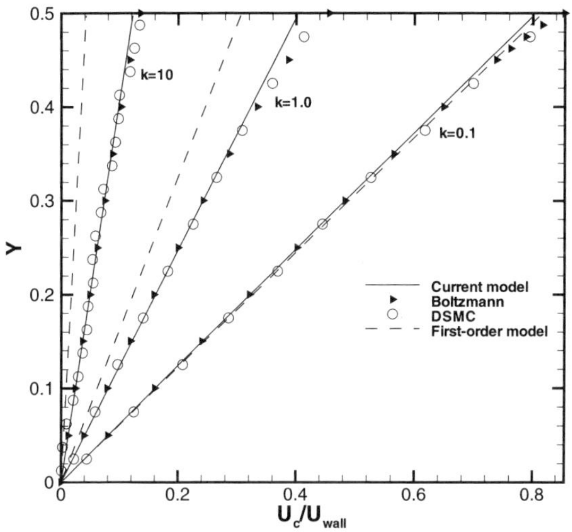

FIGURE 3.3. Velocity profiles for linear Couette flow in the upper half of the channel at rescaled Knudsen numbers $k = 0.1, 1.0$, and 10.0. The wall speed corresponds to $M = 0.05$.

solutions (triangles) and DSMC (circles) agree quite well, and they both predict essentially *linear velocity distribution in the bulk flow region* with significant slip effects for increased Kn. Knudsen layers are also visible in this plot, and they become dominant especially in the transition flow regime. The nondimensional results are valid for any monoatomic hard-sphere dilute gas. Figure 3.3 also shows the velocity distribution predicted using Maxwell's first-order slip model (dashed lines). These analytical results are obtained using Equation (3.7) with $C_1 = 1.111$, and the first-order slip solution is reasonable for $\mathrm{Kn} \leq 0.1$.

- *The velocity profile for incompressible Couette flow remains essentially linear in the entire Knudsen regime. Therefore, high-order slip conditions that utilize the second- or higher-order derivatives of the velocity field cannot predict the desired slip corrections.*

In order to develop an engineering model for these flows, (Bahukudumbi et al., 2003) introduced a modified slip coefficient (C_1) for equation (2.42), and similarly for equation (3.7) in the following form:

$$C_1 = \beta_0 + \beta_1 \tan^{-1}(\beta_2 \,\mathrm{Kn}^{\beta_3}), \tag{3.8}$$

where $\beta_i\ (i = 0, 1, 2, 3)$ are empirical constants that are obtained by comparing the *slope of the velocity profile*, obtained by the linearized Boltzmann solution in (Sone et al., 1990), with that obtained from equation (3.7), using

C_1. This velocity slip coefficient (C_1) is in essence a correction term applied to extend the validity of the original first-order slip boundary condition. In this model, the values of β_i were determined as $\beta_0 = 1.2977$, $\beta_1 = 0.7185$, $\beta_2 = -1.1749$, and $\beta_3 = 0.5864$, using a least square fit to the linearized Boltzmann solutions presented in (Sone et al., 1990).

Figure 3.3 also presents the predictions obtained by the new model (solid lines). Unlike the first-order model, the new model accurately matches the velocity profile in the bulk flow region for a wide range of Knudsen numbers. It must be noted that the new model fails to predict the velocity distribution in the Knudsen layer. This is expected, since the model is based on the Navier–Stokes equations, and the velocity profiles cannot be different from a linear profile for Couette flows. For Kn < 0.1, presence of the Knudsen layers can be neglected by extrapolating the bulk flow toward the wall. When Kn > 1, the Knudsen layer occupies the entire channel.

The results presented in Figure 3.3 are obtained for $M = 0.05$, and they exhibit rarefaction effects alone. Although the compressibility and viscous heating effects are insignificant for low-speed flows, they become important for nonisothermal as well as high-speed flows, where viscous heating creates temperature variations in the domain, as discussed in Section 3.1. Figure 3.4 shows the compressibility effects on Couette flow velocity distribution. The results are obtained using hard-sphere DSMC at various wall speeds. The new velocity model and the linearized Boltzmann solution in (Sone et al., 1990); exhibit significant deviations from the hard-sphere DSMC simulation results when $M > 0.3$. In order to extend the velocity model for $M > 0.3$, a slip boundary condition that couples the velocity and temperature fields is necessary, as presented in (Schamberg, 1947). For the range of simulations performed here, the maximum deviation of the new velocity model from the DSMC results is around 7% when $M = 0.5$, and the deviation is 14% when $M = 1.0$.

3.2.2 *Shear Stress Model*

The shear stress for Couette flows exhibits two distinct behaviors in the continuum and free-molecular flow regimes. Using the classical constitutive laws employed in the Navier–Stokes equations, the shear stress for plane Couette flow is given by

$$\tau_{\text{cont}} = \mu \frac{dU}{dy} = -\mu \frac{2U}{L}, \tag{3.9}$$

where the viscosity $\mu = (2RT_w/\pi)^{1/2}\rho_0\lambda$ does not depend on pressure, and the minus sign is due to direction of the shear stress on the fluid. So, in the hydrodynamic approximation the shear stress is proportional to the velocity gradient, and this representation is also valid in the slip flow regime with the appropriate velocity slip corrections. In the free-molecular flow regime,

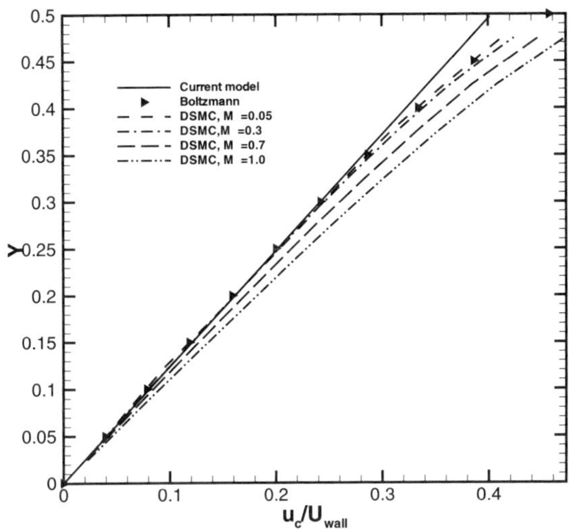

FIGURE 3.4. Couette flow velocity distribution in the upper half of the channel at $k = 1.0$, at various wall speeds.

the shear stress is proportional to the density and relative velocity of the plates, and it is given by (Kogan, 1969)

$$\tau_\infty = -\rho_0 U \sqrt{\frac{2RT_w}{\pi}}. \tag{3.10}$$

In the free-molecular flow regime, the shear stress is due to the tangential momentum exchange between the two plates that interact via the impinging and reflecting molecules. We note that this behavior is *independent* of $(\partial u/\partial y)$; hence, a free-molecular shear stress exists even if $\partial u/\partial y \to 0$.

An analytical expression was derived by (Cercignani, 1963) for the shear stress using different molecular interaction models, i.e.,

$$\pi_{xy} = \frac{\tau}{\tau_\infty} = -\frac{a_1 \,\mathrm{Kn}^2 + a_2 \,\mathrm{Kn}^2}{a_1 \,\mathrm{Kn}^2 + a_3 \,\mathrm{Kn} + a_4}, \tag{3.11}$$

$$a_1 = 1.3056, \qquad a_2 = 2\pi, \qquad a_3 = 7.5939, \qquad a_4 = \pi,$$

where π_{xy} is the shear stress normalized with the free molecular shear stress. The values of the coefficients a_1, a_2, a_3, and a_4 in equation (3.11) depend on the molecular interaction model used, and here we present the coefficients corresponding to the hard-sphere molecules. (Sone et al., 1990) have obtained accurate numerical solutions for shear stress in the entire Knudsen regime. They also developed the following analytical solution valid for $\mathrm{Kn} \to 0$, using perturbation expansions

$$\pi_{xy} = -\frac{\gamma_1 \pi \,\mathrm{Kn}}{2(1+2C_1\,\mathrm{Kn})}, \quad \gamma_1 = 1.270042, \qquad C_1 = 1.111.$$

(Bahukudumbi and Beskok, 2003) presented a shear stress model, similar in form to the result of (Cercignani, 1963), as follows:

$$\pi_{xy} = -\frac{a\,\mathrm{Kn}^2 + 2b\,\mathrm{Kn}}{a\,\mathrm{Kn}^2 + c\,\mathrm{Kn} + b}, \qquad a = 0.5297, \quad b = 0.6030, \quad c = 1.6277, \tag{3.12}$$

where the coefficients a, b, and c are obtained by a least squares fit to the linearized Boltzmann solution of (Sone et al., 1990). It is important to test the model for uniform convergence to the correct continuum ($\mathrm{Kn} \to 0$) and free-molecular ($\mathrm{Kn} \to \infty$) limits. A Taylor series expansion of the new model in the above-mentioned limits is given by

$$\frac{\tau}{\tau_\infty} = \pi_{xy} = 1 + \frac{2b-c}{a}\frac{1}{\mathrm{Kn}} + \frac{-b - (\frac{2b-c}{a})c}{a}\frac{1}{\mathrm{Kn}^2} + O(Kn^{-3}) \quad \text{as } \mathrm{Kn} \to \infty, \tag{3.13}$$

$$\frac{\tau}{\tau_{\mathrm{cont}}} = \pi_{xy} = 1 + \frac{a-2c}{2b}\,\mathrm{Kn} + \frac{(\frac{2c-a}{b})c - 2a}{2b}\,\mathrm{Kn}^2 + O(\mathrm{Kn}^3) \quad \text{as } \mathrm{Kn} \to 0, \tag{3.14}$$

where the coefficients of the $O(\mathrm{Kn}^i)$ terms ($i = \ldots, -2, -1, 1, 2, \ldots$) are corrections to the shear stress due to different orders of Kn dependence. It can be seen that the coefficient of the $O(\mathrm{Kn})$ term in the expansion for Π is $\frac{a-2c}{2b} = -2.2601 \approx -2C_1 = 2.222$. Comparing equation (3.14) with the asymptotic theory in (Sone et al., 1990), for the continuum limit, we obtain that the two representations are similar up to second-order terms in Kn.

In Figure 3.5, we compare the normalized shear stress π_{xy} predicted by the new model with the hard-sphere DSMC results, linearized Boltzmann solution of (Sone et al., 1990), and 4-moment solution of (Gross and Ziering, 1958). All solutions uniformly approach to the free-molecular flow shear stress limit as $\mathrm{Kn} \to \infty$. However, the asymptotic solution of Sone (derived for $\mathrm{Kn} \to 0$) systematically deviates from the linearized Boltzmann and DSMC results for $\mathrm{Kn} > 1.0$. Differences between the current hard-sphere DSMC results and the linearized Boltzmann solution are almost invisible in the plot, and any discrepancy in the numerical values can be attributed to the statistical nature of the DSMC. The new model has a maximum

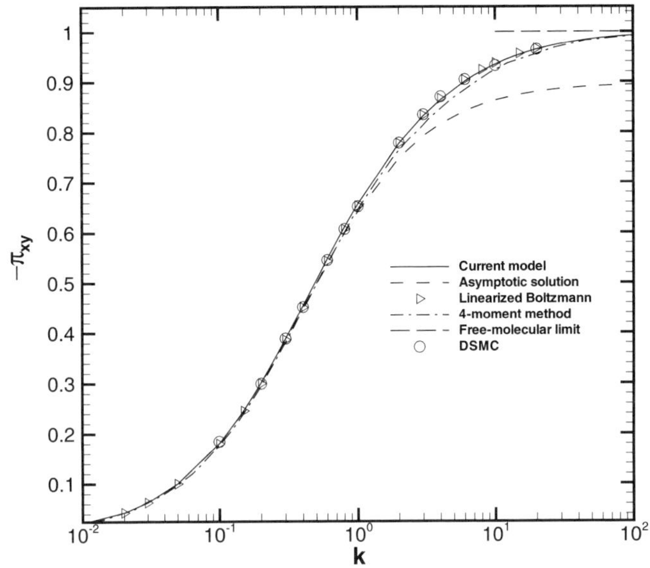

FIGURE 3.5. Variation of normalized shear stress with the rescaled Knudsen number k.

deviation of 0.3% from the linearized Boltzmann solution, while the disagreement is about 1% for the solution of (Cercignani, 1963) and 0.7% for the empirical model in (Veijola and Turowski, 2001).

Results presented in Figure 3.5 were obtained for a Mach number $M = 0.05$ flow. The shear stress variation as a function of the wall speed using hard-sphere DSMC simulations results in deviations from this model. The maximum increase observed in the magnitude of the shear stress is about 6% as the Mach number is increased from 0.05 to 1.0 (Bahukudumbi et al., 2003).

In another study, (Lockerby and Reese, 2003) presented numerical solutions of Burnett equations for linear Couette flows using the no-slip, first-order slip (2.19) and second-order slip (2.42) boundary conditions. Their shear stress results for high-speed Couette flows at $M = 3$ are shown in Figure 3.6, where the Burnett predictions using various boundary conditions are compared with the DSMC computations of (Nanbu, 1983). It is clearly seen that the first-order slip condition is valid up to $\mathrm{Kn} = 0.1$, while the second-order slip condition (2.42) is ineffective for this flow (see Remark 3 at the end of Section 2.3.3).

Remark: We presented empirical models for velocity distribution and shear stress for engineering analysis of shear-driven gas flows encountered

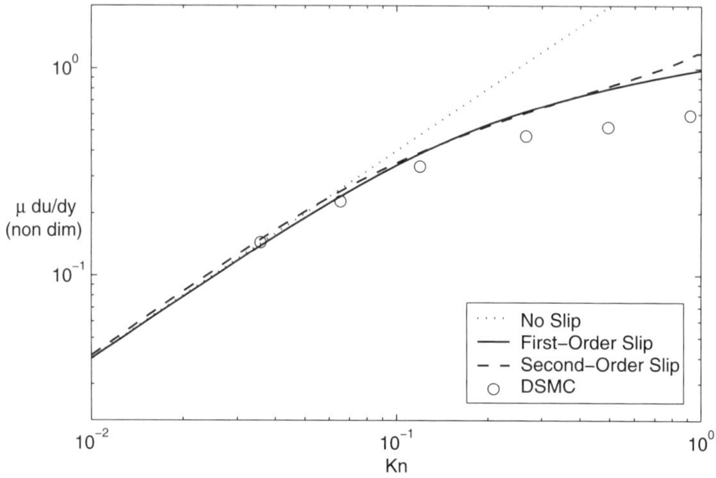

FIGURE 3.6. Variation of normalized shear stress with Kn for $M = 3$ flow. Comparison of DSMC results with Burnett solutions using various boundary conditions. (Courtesy of J.M. Reese.)

in microsystems. The new velocity model is based on a generalized slip coefficient, and it uniformly converges to the first-order slip model for Kn ≤ 0.1 flows, while it accurately matches the DSMC and the linearized Boltzmann results in the transition and the free-molecular flow regimes. Validity of the velocity model was established for Kn ≤ 12. The new shear stress model is valid in the entire Knudsen regime, and it is asymptotically consistent in the continuum (Kn $\to 0$) and free-molecular flow (Kn $\to \infty$) regimes. These models are valid for monoatomic dilute hard-sphere gases, and are restricted to low subsonic $M \leq 0.3$, nearly isothermal flows, since extreme compressibility effects and viscous heating are neglected in their derivation.

3.3 Oscillatory Couette Flow

Oscillatory Couette flow is the simplest approximation for time-periodic shear-driven gas flows encountered in several microsystem applications, such as microaccelerometers, inertial sensors, and resonant filters. Oscillatory Couette flow can also be interpreted as a variation of Stokes's second problem (Batchelor, 1998). This classical problem has been investigated extensively using continuum-based flow models for Kn ≤ 0.1 (Veijola and Turowski, 2001); however, rarefaction effects in the transition and free-molecular flow regimes have not been studied in detail.

Let us consider rarefied gas flow between two infinite parallel plates that are at a distance L apart, where the top plate oscillates harmonically with frequency (ω) in the lateral (flow) direction, and the bottom plate is stationary. The two plates are maintained at the same temperature $T_w = 273$ K. The gas is initially at rest; it has an equilibrium number density n_0 and equilibrium temperature T_e that is equal to the wall temperature T_w. For a given set of parameters, the oscillatory rarefied Couette flow can be characterized by the Knudsen (Kn), and Stokes (β) numbers. The Stokes number represents balance between the unsteady and viscous effects. It can also be interpreted as the ratio of the diffusion and oscillation characteristic time scales, and it is defined as

$$\beta = \sqrt{\frac{\omega L^2}{\nu}} = \left(\frac{L^2/\nu}{1/\omega}\right)^{1/2}, \tag{3.15}$$

where ν is the kinematic viscosity.

3.3.1 Quasi-Steady Flows

In this Section, we present extensions of the steady flow velocity and shear stress models in Section 3.2 to include oscillatory Couette flows. This particular approach will be valid for

- any Stokes number flow in the slip flow regime (Kn < 0.1),
- quasi-steady flows for Kn < 12.

We define the *quasi-steady* conditions as the flow, where the velocity amplitude distribution always passes through $(y/L, u/u_0) = (0.5, 0.5)$, resulting in a linear velocity distribution with *equal* amounts of *slip* on the oscillating and stationary walls. Our observations have shown that such conditions are typically achieved when $\beta \leq 0.25$.

Velocity Model

For oscillatory Couette flows, the momentum equation reduces to the following form:

$$\frac{\partial u(y,t)}{\partial t} = \nu \frac{\partial^2 u(y,t)}{\partial y^2}. \tag{3.16}$$

An analytical solution of the above equation can be obtained for oscillatory flows with a specified frequency ω and amplitude U_0, as shown in (Sherman, 1990). For a sinusoidal velocity excitation, a velocity response of the form $u(y,t) = \Im\{V(y)\ \exp(j\omega t)\}$ is expected, where the symbol $\Im$ denotes the imaginary part of a complex expression, and $V(y)$ is the amplitude governed by

$$j\omega V(y) = \nu \frac{\partial^2 V(y)}{\partial y^2}. \tag{3.17}$$

Alternatively, we can write this as

$$\frac{\partial^2 V(y)}{\partial y^2} - \psi^2 V(y) = 0, \tag{3.18}$$

where $\psi = \sqrt{\frac{j\omega}{\nu}}$ is the complex frequency variable, and $j = \sqrt{-1}$. The generalized solution of the above equation can be written in the following form,

$$V(y) = A_1 \sinh(\psi y) + A_2 \cosh(\psi y), \tag{3.19}$$

where the constants A_1 and A_2 are determined by applying the boundary conditions. The complex variable ψ can be expressed in terms of the Stokes number, since

$$\psi = \sqrt{\frac{j\beta^2}{L^2}}.$$

Utilizing the slip boundary condition (2.42) with $C_2 = 0$, and the generalized slip coefficient C_1 from equation (3.8), we obtain the following time-dependent velocity distribution:

$$u(y,t) \tag{3.20}$$

$$= \Im\left[\left(U_0 \frac{\sinh(\sqrt{j}\beta Y) + \sqrt{j}\beta C_1 \,\mathrm{Kn}\, \cosh(\sqrt{j}\beta Y)}{(1 + j\beta^2 C_1^2 \,\mathrm{Kn}^2)\sinh(\sqrt{j}\beta) + 2\sqrt{j}\beta C_1 \,\mathrm{Kn}\, \cosh(\sqrt{j}\beta)}\right)\exp(j\omega t)\right],$$

where $Y = y/L$ and $\mathrm{Kn} = \lambda/L$. This is a *general* solution for the velocity profile that captures the no-slip solution simply by setting $\mathrm{Kn} = 0$, and the first-order slip solution by setting $C_1 = 1.111$ in equation (3.20). Therefore, the above equation is expected to be valid for *any Stokes number flow in the continuum and slip flow regimes*, since it uniformly captures these well-explored limits.

Figure 3.7 shows variation of the normalized velocity amplitude between the two surfaces. We compare the DSMC results with the predictions of the extended slip model for (a) quasi-steady flows in the entire Knudsen regime, and (b) slip flows for a wide range of Stokes numbers ($\beta \leq 7.5$). The velocity amplitudes are obtained by measuring the magnitude of the maximum velocity at different cross-flow ($Y = y/L$) locations. Note that the generalized velocity model given by equation (3.20) converges to the first-order slip model for $\mathrm{Kn} < 0.1$. Hence, only the predictions of the extended slip model are shown in the figure. For quasi-steady flows, the velocity amplitude distribution always passes through $(y/L, u/u_0) = (0.5, 0.5)$ resulting in a linear velocity distribution with equal amounts of slip on the oscillating and stationary walls. The extended slip model accurately matches the DSMC velocity profile for a wide range of Knudsen numbers ($\mathrm{Kn} \leq 12$). However, it fails to predict the Knudsen layers that are captured by the DSMC results, as expected. The extended slip model is also valid for high Stokes number flows in the continuum and slip flow regimes due to the use of the Navier–Stokes equations in its derivation. Figure 3.7(b) shows

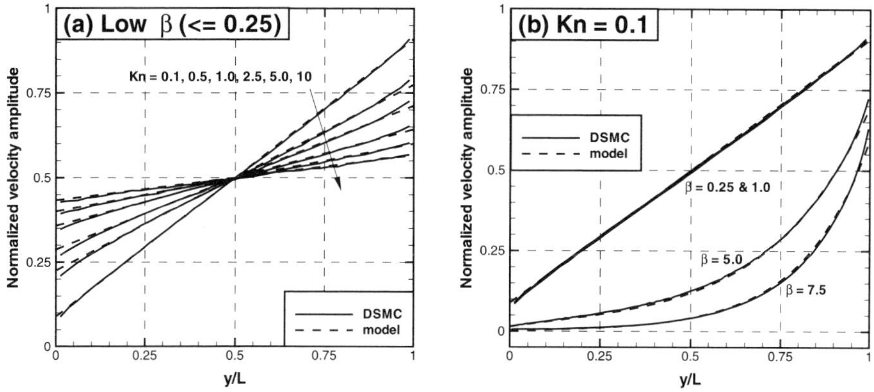

FIGURE 3.7. Velocity amplitudes for (a) quasi-steady and (b) low Kn cases.

that the DSMC results accurately capture the slip-flow limit even for large Stokes numbers. For high Stokes number flows there are deviations from the linear velocity profile, and the velocity amplitude distribution loses its symmetry beyond $\beta = 1.0$ for Kn $= 0.1$ flow. High Stokes number cases ($\beta \geq 5$) result in bounded Stokes layers, where the flow is confined to a near wall region. Significant velocity slip can be observed with increased Stokes number beyond the quasi-steady flow limit, while the slip velocity for quasi-steady flows is independent of the Stokes number, as can be deduced by comparing the $\beta \leq 1.0$ cases with the $\beta \geq 5$ cases in Figure 3.7(b).

Figure 3.8 shows the dynamic response characteristics for Kn $= 0.1$ and $\beta = 5.0$ flow. Snapshots of velocity distribution at different times are shown in Figure 3.8(a). With the exception of the velocity slip, the dynamics are similar to those of no-slip continuum flows. The velocity distribution predicted by the extended slip model and the DSMC simulations are in good agreement, despite a slight phase difference between the DSMC and the model. In the context of this work,

- *the phase angle is defined as the fraction of the time period by which the solution felt at any streamwise position lags or leads the reference velocity solution imposed on the oscillating wall.*

A general representation of the velocity solution at an arbitrary location Y is given by

$$u(Y,t) = u_0 \sin(\omega t + \Psi), \tag{3.21}$$

where Ψ is the phase angle. Expanding and rearranging equation (3.21), we obtain

$$u(Y,t) = u_0 \left[\sin(\omega t)\cos\Psi + \cos(\omega t)\sin\Psi\right] \\ = A(Y)\sin(\omega t) + B(Y)\cos(\omega t), \tag{3.22}$$

where

$$A(Y) = u_0 \sin\Psi, \quad B(Y) = u_0\cos\Psi. \tag{3.23}$$

The phase angle can then be determined from equation (3.23) as

$$\Psi(Y) = \tan^{-1}\left[\frac{B(Y)}{A(Y)}\right], \tag{3.24}$$

where the coefficients A and B are determined from the DSMC results, using a χ-square fit. In addition, a theoretical expression for the phase angle of the extended slip model of (Bahukudumbi and Beskok, 2003) is also presented in (Park et al., 2004).

Figure 3.8(b) shows the velocity time history at various streamwise locations (y/L) in the flow domain. The velocity solutions at different y/L locations exhibit reduced amplitudes and different phase angles. Note that the peak values of the velocity solution in Figure 3.8(b) correspond to the velocity amplitudes in Figure 3.7. In Figure 3.8(c), the phase angle predicted by the extended slip model and DSMC show similar trends. However, the initial deviation at $y/L \geq 0.9$, due to the Knudsen layer effects, offsets the DSMC results from the model solution. The phase angle variation is essentially linear in most of the domain, except within the Knudsen layers near the walls. The wave propagation speed c' (phase speed) can be computed from the phase angle variation using the relation

$$c' = \omega \bigg/ \left(\frac{\Delta\Psi}{\Delta Y}\right). \tag{3.25}$$

Consequently, the phase speed, computed using the above definition, is constant in the region of linearly varying phase angle. The extended slip model predicts a wave propagation speed of $c' = 1.770$, which is in good agreement with the corresponding DSMC prediction of $c' = 1.790$. The classical Stokes second problem without the stationary wall also predicts a very similar wave propagation speed, $c' = 1.777$. The phase speed is *not* constant near the walls due to the presence of Knudsen layers. In addition, the phase speed decays near the stationary wall, due to the interference between the incident and reflected solutions. The normalized velocity amplitude, plotted in log-scale in Figure 3.8(d), shows exponential decay in the amplitude with small alterations when $y/L \leq 0.1$, due to the presence of the stationary wall. It can be seen that the slip model results and DSMC solution are consistent.

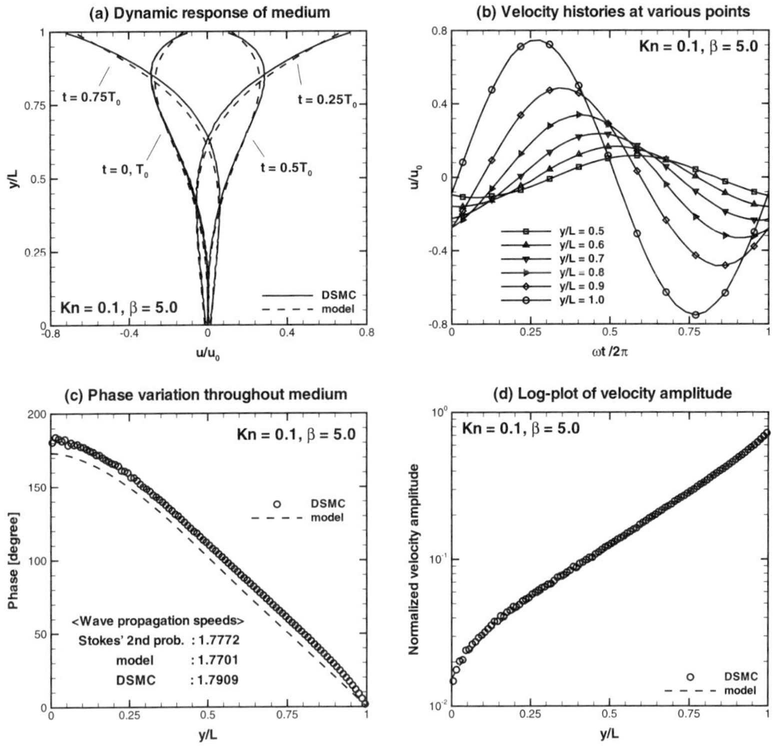

FIGURE 3.8. Details of the flow dynamics for Kn = 0.1 and $\beta = 5.0$.

Shear Stress Model

In this section, we extend the shear stress model developed for plane Couette flows in Section 3.2.2 to predict the shear stress on the *oscillating wall*. Our main assumption is that quasi-steady oscillatory flows should behave more or less like the steady Couette flow. Hence, an *effective viscosity* derived from the steady Couette flow results can be used as the viscosity coefficient for the quasi-steady oscillatory flows, an assumption that we validate later using the DSMC data. Therefore, using the new velocity model for steady flows and the shear stress model given by equation (3.12), we define an "effective viscosity" as

$$\mu_e = \frac{\tau_c}{\frac{du_c}{dy}} = \frac{\mu_0}{2} \frac{a\,\mathrm{Kn} + 2b}{a\,\mathrm{Kn}^2 + c\,\mathrm{Kn} + b} (1 + 2C_1\,\mathrm{Kn}), \tag{3.26}$$

where the subscript c indicates plane Couette flow, and u_c is the plane Couette flow velocity profile given by equation (3.7) with the slip coefficient C_1. The coefficients a, b, and c are due to the shear stress model given in (3.12). For quasi-steady oscillatory flows, we employ this "effective

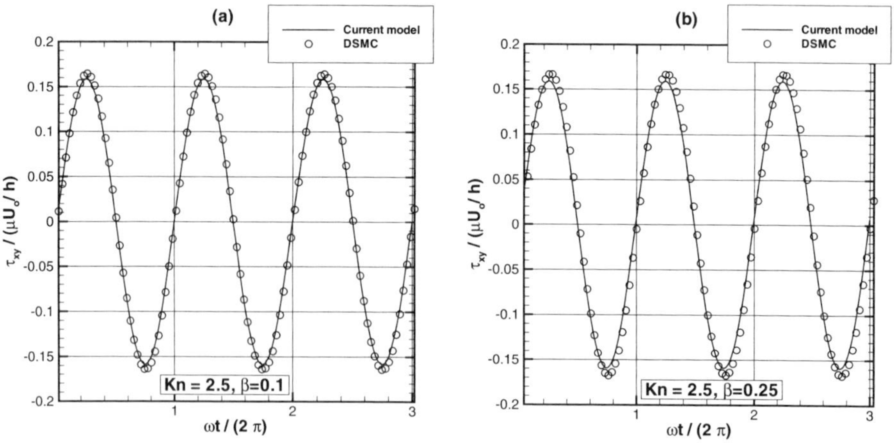

FIGURE 3.9. Time history of the normalized shear stress at the laterally oscillating wall, predicted by the model of equation (3.27) and the *unsteady* DSMC simulations at (a) Kn = 2.5, $\beta = 0.1$ and (b) Kn = 2.5, $\beta = 0.25$.

viscosity" coefficient, and define the shear stress at the oscillating plate as

$$(\tau)_{\rm qs} = \mu_e \left.\frac{du(y,t)}{dy}\right|_{y=L} = \frac{\mu_0 U_0}{2h}\frac{a\,\mathrm{Kn}+2b}{aKn^2+c\,\mathrm{Kn}+b}\,(1+2C_1\,\mathrm{Kn}) \qquad (3.27)$$

$$\times\Im\left[\left(\sqrt{j}\beta(\frac{\cosh(\sqrt{j}\beta)+\sqrt{j}\beta C_m Kn\ \sinh(\sqrt{j}\beta)}{(1+j\beta^2C_m^2\,\mathrm{Kn}^2)\sinh(\sqrt{j}\beta)+2\sqrt{j}\beta C_m\,\mathrm{Kn}\ \cosh(\sqrt{j}\beta)}\right)\exp(j\omega t)\right].$$

Figure 3.9 shows the shear stress time history at the oscillating wall, normalized with its continuum value. Predictions of the current model and DSMC results are also presented for Kn = 2.5 flow at $\beta = 0.1$ and 0.25 conditions. There is a difference of very small magnitude in the shear stress between the $\beta = 0.1$ and 0.25 cases at Kn = 2.5. The above model accurately predicts the shear stress magnitude on the oscillating wall, and it also matches the DSMC results for a wide range of Kn values under quasi-steady flow conditions ($\beta \leq 0.25$).

3.3.2 Unsteady Flows

In this section we present time-periodic oscillatory Couette flows in the transition and free-molecular flow regimes at relatively large Stokes numbers. Therefore, the quasi-steady approximation is no longer valid, and "bounded Stokes layer"–type behaviors are observed. We also discuss the physical aspects, such as the penetration depth, wall shear stress, and slip

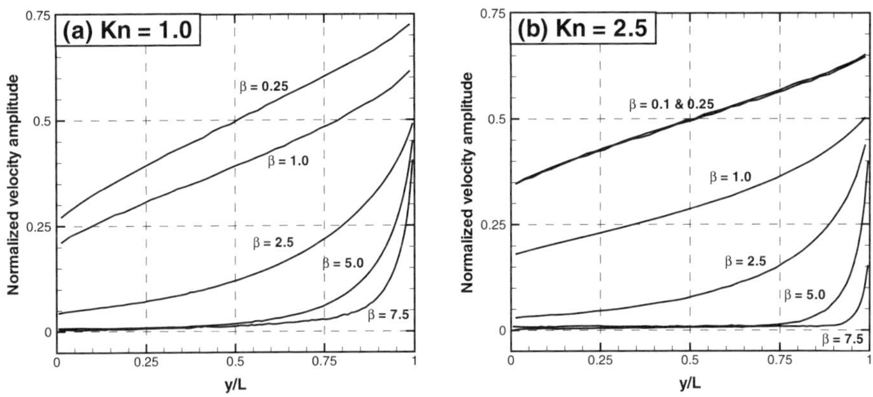

FIGURE 3.10. Effect of β in transition flow regime.

velocity variations, as well as the energy dissipation and damping characteristics of oscillatory Couette flows. Computational details about the unsteady DSMC method can be found in Section 15.1.2.

Transition Flow Regime

Figure 3.10 shows the effect of Stokes number on the velocity amplitude in the transition flow regime. At fixed Kn, the slip velocity increases with increasing β. For Kn $= 1.0$, it can be seen that beyond $\beta = 0.25$ the quasi-steady approximation breaks down. We observe a "bounded Stokes layer" type of behavior for $\beta \geq 5$ in both figures. Comparing the Kn $= 1.0$ and Kn $= 2.5$ cases, we find that the slip velocity increases with increasing Kn at constant β. For a fixed Kn, the Stokes layer thickness decreases with increasing β, as expected.

The effect of Kn on the velocity amplitude for moderate Stokes number conditions is shown in Figure 3.11. It can be seen that the slip velocity magnitude on the oscillating wall increases with increasing Kn for a fixed Stokes number. For $\beta = 1.0$, quasi-steady flow behavior is observed for Kn $= 0.1$, since the velocity amplitude distribution is linear and passes through $(y/L, u/u_0) = (0.5, 0.5)$. Hence, the quasi-steady flow approximation also depends on Kn, as can be deduced by comparison of Figures 3.10 and 3.11. The most interesting observation in Figure 3.11 is the emergence of a "*bounded rarefaction layer*" with increasing Kn. By this term we emphasize that this behavior is due to the rarefaction effects alone, and not to the influence of the Stokes number, which is kept constant. Transition to this "bounded rarefaction layer" occurs even at moderate Stokes number flows (see Figure 3.11(a)) by increasing Kn. However, these effects are more pronounced when the Stokes number increases, as can be deduced by comparing Figures 3.11 (a) and (b).

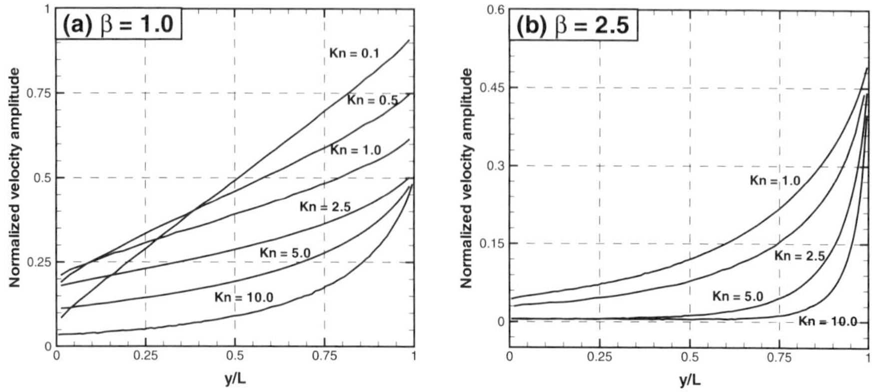

FIGURE 3.11. Effect of Knudsen number Kn for moderate values of the Stokes number β.

Figures 3.12, 3.13, and 3.14 show the dynamic response of the system for moderate and high Stokes number flows in the transition flow regime (Kn $= 1.0, \beta = 2.5$; Kn $= 1.0, \beta = 5.0$, and Kn $= 5.0, \beta = 2.5$). Here, we will not present detailed discussions of the dynamic system response for the individual cases, since the behavior is qualitatively similar to that of Figure 3.8. Comparing Figures 3.12(a) and 3.13(a), we observe that a more pronounced Stokes layer forms by increasing the Stokes number. Alternatively, comparing Figures 3.12(a) and 3.14(a), we observe a more pronounced "bounded rarefaction layer" when Kn is increased. In all three cases, reduced velocity amplitudes and different phase angles are observed at different streamwise locations. Note that while the phase angle reaches $210°$ at $y/L = 0$ in Figure 3.12(c), the same value is reached at $y/L \approx 0.75$ in Figure 3.13(c), and at $y/L \approx 0.8$ in Figure 3.14(c). This indicates that the phase speed, defined in equation (3.25), increases with increasing β and Kn. It is also worthwhile to compare the level of statistical scatter between these three results. Statistical scatter in Figure 3.12(d) is insignificant, since the normalized velocity amplitude does not drop below 1% of the solution. However, with increasing β and Kn, the normalized velocity amplitude drops below 1% outside the "bounded layers," and the statistical scatter becomes important, as can be observed in Figures 3.13(d) and 3.14(d).

Free-Molecular Regime

In this section, we follow (Park et al., 2004), to derive the velocity distribution and shear stress for oscillatory Couette flows in the free-molecular flow limit (Kn ≥ 10). Our objectives are to provide a theoretical solution to compare and validate the DSMC results, and enhance our understanding of flow physics in this regime. As the Knudsen number is increased, inter-

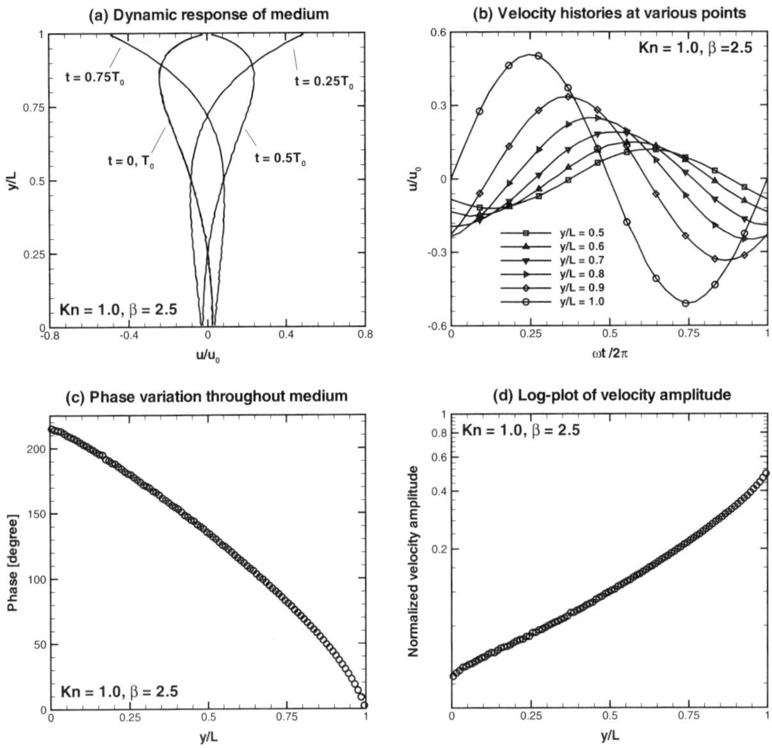

FIGURE 3.12. Details of the flow dynamics for Kn = 1.0 and $\beta = 2.5$.

molecular collisions become negligible compared to the molecule/surface collisions. Therefore, the flow can be modeled using the collisionless Boltzmann equation

$$\frac{\partial f}{\partial t} + \eta \frac{\partial f}{\partial y} = 0, \tag{3.28}$$

where f is the velocity distribution function and η is the cross-flow (y) component of the molecular velocity. Due to the simple geometry, f changes only in the cross-flow direction, and there are no external force fields. We assume that both top and bottom walls are fully diffusive, and a sinusoidal excitation is exerted on the top wall ($y = L$). The boundary conditions for equation (3.28) are

$$f(y=0) = f_0 = \frac{\kappa^3}{\pi^{3/2}} \exp\left[-\kappa^2\left(\xi^2 + \eta^2 + \zeta^2\right)\right]; \quad \eta > 0, \tag{3.29a}$$

$$f(y=L\,;\,u_w) = \frac{\kappa^3}{\pi^{3/2}} \exp\left[-\kappa^2\left(\xi^2 + (\eta - u_w)^2 + \zeta^2\right)\right]; \; \eta < 0, \tag{3.30a}$$

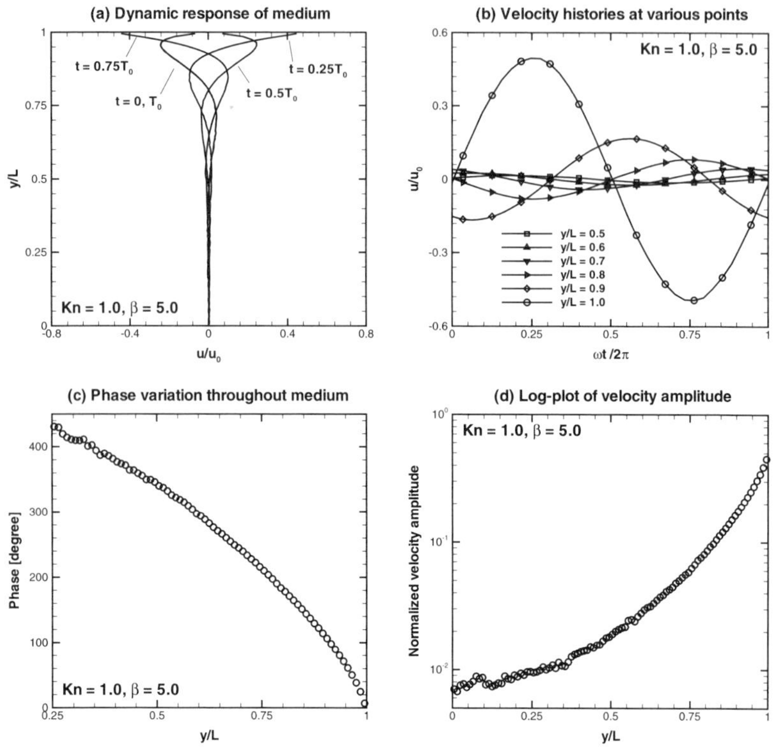

FIGURE 3.13. Details of the flow dynamics for Kn $= 1.0$ and $\beta = 5.0$.

where $u_w = u_0 \sin(\omega\, t)$ and $\kappa = \sqrt{\frac{m}{2\, k_B\, T_e}}$. Here T_e is the initial equilibrium temperature, k_B is the Boltzmann constant, and ξ and ζ are the streamwise and spanwise components of the molecular velocity, respectively. Diffuse reflections of gas molecules from the surfaces require that the reflected molecules have a Maxwellian distribution f_0, characterized by the velocity and temperature of the plates. Here, we assume that the amplitude of velocity oscillations is less than the speed of sound. This enables linearization of the collisionless Boltzmann equation and the boundary conditions. Following the work of Sone, the velocity distribution function can be decomposed into its equilibrium and fluctuating components as follows (Sone, 1964; Sone, 1965):

$$f = f_0\,(1 + \phi)\,, \tag{3.31}$$

where ϕ is the normalized fluctuation. We can obtain the linearized forms of the Boltzmann equation and the boundary conditions by substituting equation (3.31) into equations (3.28), (3.29a), and (3.30a) and neglecting

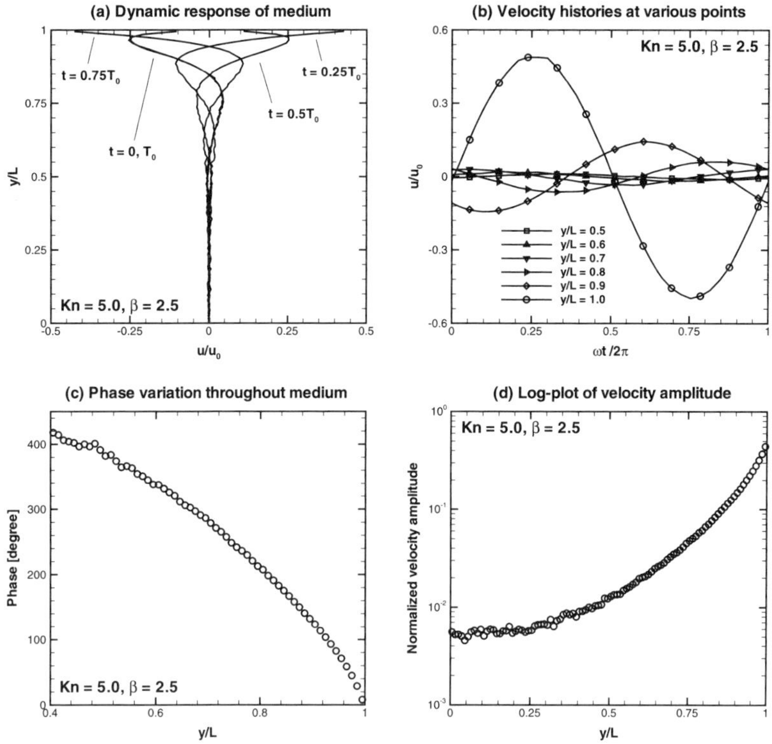

FIGURE 3.14. Details of the flow dynamics for Kn = 5.0 and $\beta = 2.5$.

all the higher order terms in ϕ, i.e.,

$$\frac{\partial \phi}{\partial t} + \eta \frac{\partial \phi}{\partial y} = 0, \tag{3.32a}$$

$$\phi_0 = \phi\,(y = 0) = 0; \quad \eta > 0, \tag{3.32b}$$

$$\phi_L = \phi\,(y = L) = 2\,\kappa^2 \xi\, u_w; \quad \eta < 0. \tag{3.32c}$$

The streamwise component of the velocity u and the shear stress τ are defined as follows:

$$u = \int \xi\, \phi\, f_0\, d\xi\, d\eta\, d\varsigma, \tag{3.33a}$$

$$\tau = \rho_0 \int \xi\eta\, \phi\, f_0\, d\xi\, d\eta\, d\varsigma, \tag{3.33b}$$

where $\int (\cdots)\, d\xi\, d\eta\, d\zeta$ shows integration over the velocity space, and ρ_0 is the mean density given by $\rho_0 = n_0 m$, with n_0 being the equilibrium number density and m being the molecular mass. By taking the Laplace transform of equation (3.32a) and the boundary conditions (3.32b, 3.32c), we calculate the integral formulations of the velocity and shear stress given by equations (3.33a) and (3.33b). The Laplace transformed variables $\hat{\phi}$, $\hat{u}$, and $\hat{\tau}$ are given as

$$\hat{\phi} = \hat{\phi}_L \exp\left[-\frac{s}{\eta}(y-L)\right] = 2\kappa^2 \xi \hat{u}_w \exp\left[-\frac{s}{\eta}(y-L)\right]; \quad \eta < 0, \tag{3.34a}$$

$$\hat{u} = \int \xi \hat{\phi} f_0 \, d\xi \, d\eta \, d\zeta, \tag{3.34b}$$

$$\hat{\tau} = \rho_0 \int \xi \eta \hat{\phi} f_0 \, d\xi \, d\eta \, d\zeta, \tag{3.34c}$$

where s is the Laplace transformation variable. After eliminating $\hat{\phi}$ from equations (3.34a), (3.34b), and (3.34c), we obtain integral formulations for $\hat{u}$ and $\hat{\tau}$ as follows:

$$\hat{u} = \frac{\kappa}{\sqrt{\pi}} \int_0^\infty \hat{u}_w \exp\left(-\frac{s}{\eta''}(L-y) - \kappa^2 \eta''^2\right) d\eta'', \tag{3.35a}$$

$$\hat{\tau} = \rho_0 \frac{\kappa}{\sqrt{\pi}} \int_0^\infty \hat{u}_w \, \eta'' \exp\left(-\frac{s}{\eta''}(L-y) - \kappa^2 \eta''^2\right) d\eta'', \tag{3.35b}$$

where $\hat{u}_w$ is the transformed function of u_w. Finally, the inverse Laplace transform provides u and τ as functions of y and t:

$$u(y,t) = \frac{u_0}{\sqrt{\pi}} \int_{\frac{\kappa(L-y)}{t}}^\infty \sin\left[t - \frac{\kappa(L-y)}{\eta'}\right] \exp(-\eta'^2) \, d\eta', \tag{3.36a}$$

$$\tau(y,t) = \frac{\rho_0 u_0}{\sqrt{\pi}} \int_{\frac{\kappa(L-y)}{t}}^\infty \eta' \sin\left[t - \frac{\kappa(L-y)}{\eta'}\right] \exp(-\eta'^2) \, d\eta'. \tag{3.36b}$$

The gas velocity $\tilde{u}_{w,g}(t)$ and shear stress on the oscillating (top) wall are calculated as

$$\begin{aligned} \tilde{u}_{w,g}(t) = u(L,t) &= \frac{u_0}{\sqrt{\pi}} \int_0^\infty \sin(\omega t) \exp(-\kappa^2 \eta''^2) \, d\eta'' \\ &= \frac{u_0}{2} \sin(\omega t) = u_{w,g} \sin(\omega t), \end{aligned} \tag{3.37a}$$

$$\begin{aligned} \tilde{\tau}_w = \tau(L,t) &= \frac{\rho_0 u_0}{\sqrt{\pi}} \int_0^\infty \eta'' \sin(\omega t) \exp(-\kappa^2 \eta''^2) \, d\eta'' \\ &= \frac{1}{2\sqrt{\pi}} \frac{\rho_0 u_0}{\kappa} \sin(\omega t) = \tau_w \sin(\omega t). \end{aligned} \tag{3.37b}$$

The magnitude of gas velocity on the oscillating plate and the corresponding shear stress are $u_w = \frac{u_0}{2}$ and $\tau_w = \frac{1}{2}\rho_0 u_0 \sqrt{\frac{2k_B T_w}{\pi}}$, respectively. This shows that

- *for oscillatory Couette flows, the magnitude of the gas velocity and shear stress on the oscillating surface reach the same asymptotic limit as their steady counterpart when* $\mathrm{Kn} \to \infty$.

Validation of the DSMC results in the free-molecular flow regime is presented in Figure 3.15. We compare the normalized velocity amplitudes obtained from the DSMC with the solution of the linearized collisionless Boltzmann equation at different Stokes numbers. The free-molecular solution plotted in this figure is obtained from equation (3.36a). Overall, a very good agreement between the DSMC results and the free-molecular solution is obtained. The DSMC results in Figure 3.15(a) show statistical scatter associated with a high Knudsen number simulation. However, agreement between the theory and simulations is remarkable in Figures 3.15(b–d).

In Figures 3.16 and 3.17, we compare the dynamic response of the flow obtained from the DSMC results and the collisionless Boltzmann equation solutions for the $\mathrm{Kn} = 10, \beta = 1.0$, and $\mathrm{Kn} = 10, \beta = 2.5$ cases. Predictions of the velocity profiles, phase angles, and the slip velocities are presented in the figures. As observed in Figure 3.17(a), both methods capture the "bounded rarefaction layer" equally well. Due to the onset of statistical scatter outside this layer, we plotted the DSMC phase angle only for $y/L \geq 0.65$ in Figure 3.17(c). Nevertheless, the DSMC and Boltzmann solutions match remarkably well within the bounded layer, confirming the accuracy of the DSMC results.

Basic Characteristics

In the following we present results for the penetration depth, shear stress, slip velocity, and viscous dissipation as functions of the Knudsen and Stokes numbers.

Penetration Depth Variation

It is important to note that the bounded Stokes and rarefaction layers observed in the results create a new length scale in the problem. This new length scale is related to the thickness of the Stokes/rarefaction layers, and becomes particularly important for high values of Kn or β. The Stokes layer thickness ($\delta \approx \sqrt{\nu/\omega}$), also referred to as the "penetration depth," is defined as the distance from the moving wall where the velocity amplitude decays to 1% of its excitation value ($\frac{u}{u_0} = 0.01$). Most of the flow is confined within this layer, and the moving wall no longer interacts with the stationary wall. For these cases, the characteristic length scale of the problem should be based on the penetration depth δ, rather than the separation distance between the two plates. This would require redefinition of the nondimensional parameters Kn and β, based on the penetration depth ($\mathrm{Kn}' = \frac{\lambda}{\delta}, \beta' = \sqrt{\frac{\omega\delta^2}{\nu}}$). However, there are no relations for variation of δ as a function of Kn and β. Hence, a priori estimation of the penetration

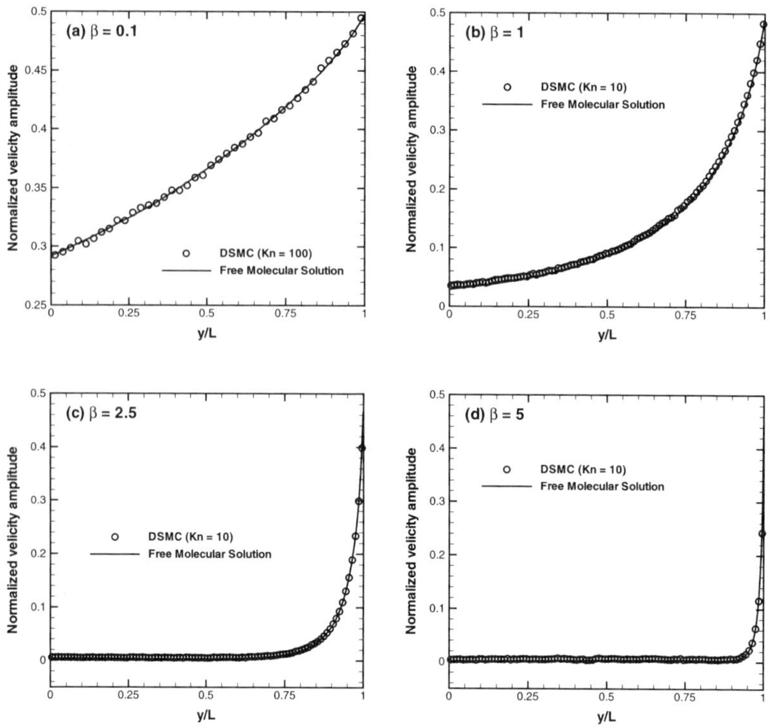

FIGURE 3.15. Velocity amplitudes for the free-molecular flow regime.

depth is not possible. For the sake of consistency, Kn and β are defined using the plate separation distance throughout this work. Hence, no switch is made in the characteristic length scale. However, change in the characteristic length scale has physical implications. For example, the actual Knudsen number for these cases can be found by $\mathrm{Kn}' = \mathrm{Kn}\, L/\delta$. Figure 3.18 shows variation of the normalized penetration depth (δ/L) with Kn and β. For the cases not shown in this figure, the solution does not attenuate enough to observe a "bounded layer." The penetration depth decreases with increasing β, as expected. The penetration depth approaches different values in the free-molecular limit for different Stokes numbers. For fixed β, the penetration depth decreases by increasing Kn, reflecting the "bounded rarefaction layer" concept introduced in (Park et al., 2004). It can be seen from Figure 3.18 that $\frac{\delta}{L} \propto \frac{1}{\mathrm{Kn}}$ for a given β. This figure also clarifies the need for redefinition of the characteristic length scale for high β and Kn flows.

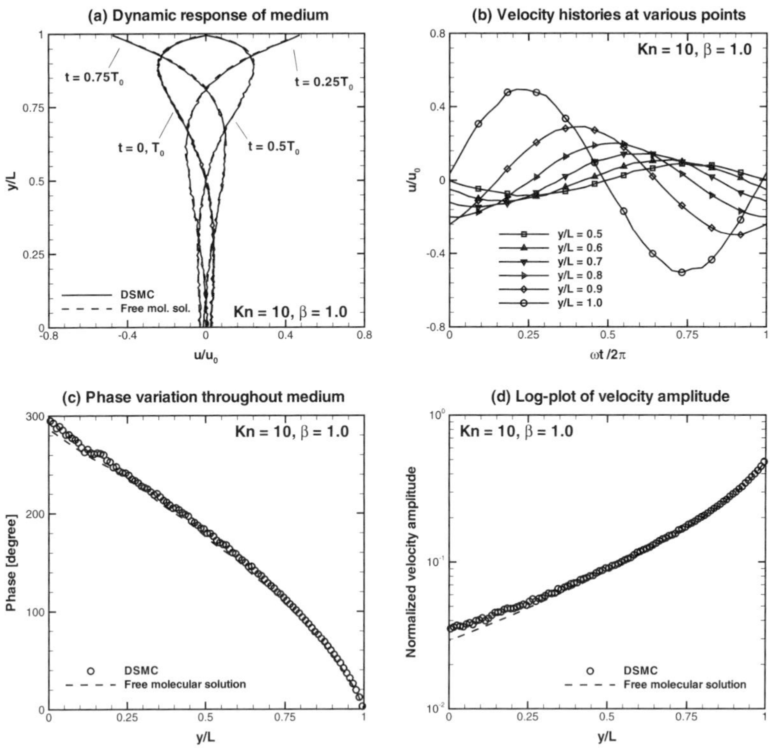

FIGURE 3.16. Details of the flow dynamics for $\text{Kn} = 10$ and $\beta = 1.0$.

Shear Stress Variation

In Figure 3.19, we present the effects of Kn and β on the wall shear stress using the DSMC results. We plot the shear stress normalized with the free-molecular and continuum shear stress values, to show that the DSMC results uniformly approach the correct asymptotic limits. We also compare the DSMC results with our empirical model for quasi-steady oscillatory flows given by equation (3.27). Good agreement between the empirical model and the DSMC results are observed for quasi-steady flows ($\beta \leq 0.25$). Beyond the quasi-steady flow regime there is a significant increase in the shear stress magnitude, especially for low Kn values. This is expected, since the shear stress is proportional to the velocity gradient, which increases with β, especially due to the formation of bounded Stokes layers. In the free-molecular flow limit, the shear stress reaches the same asymptotic limit of the steady plane Couette flow regardless of the Stokes number, as shown in equation (3.37b). In Figure 3.19, we observed similar behavior in the DSMC results. Interestingly, the DSMC data reached the asymptotic shear stress value in the transition flow regime for large Stokes number cases. This behavior is a manifestation of our definition of

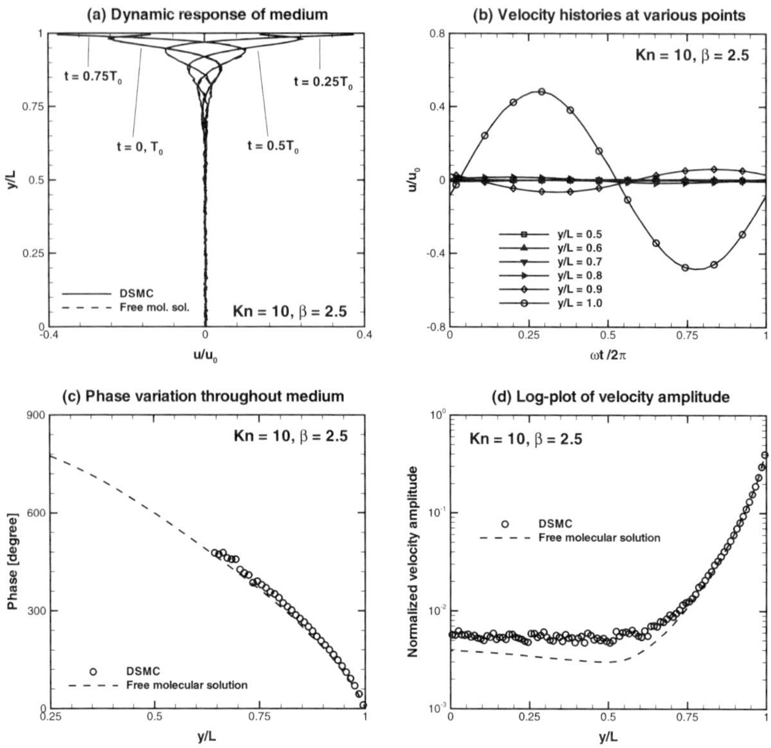

FIGURE 3.17. Details of the flow dynamics for $\mathrm{Kn} = 10$ and $\beta = 2.5$.

Kn, which is constant irrespective of β. If we consider changing the characteristic length scale from the distance between the two plates (L) to the penetration length (δ), we would observe that the effective Knudsen numbers for such cases are indeed in the free-molecular flow regime. For example, for the $\mathrm{Kn} = 2.5, \beta = 7.5$ case, the penetration depth δ is equal to $0.15L$ (see Figure 3.18). Hence, the effective Knudsen number for this case is about $\mathrm{Kn}' \approx 17$, which corresponds to a free-molecular regime.

Slip Velocity Variation

Here we present the slip velocity variation as a function of the Knudsen and Stokes numbers. We simply present the slip amplitude variations, since the flow is time-periodic and there are no phase differences between the wall oscillation frequency and the slip velocity response. Figure 3.20 shows Kn and β effects on the slip amount on the moving wall. We had to determine the slip velocity by extrapolation of the DSMC cell-center velocities onto the oscillating wall. For low Knudsen flows, slip velocity increases with increasing β. This is due to the increased velocity gradients near the wall. For a fixed β, the slip amount increases by increasing Kn. Noticeably, all

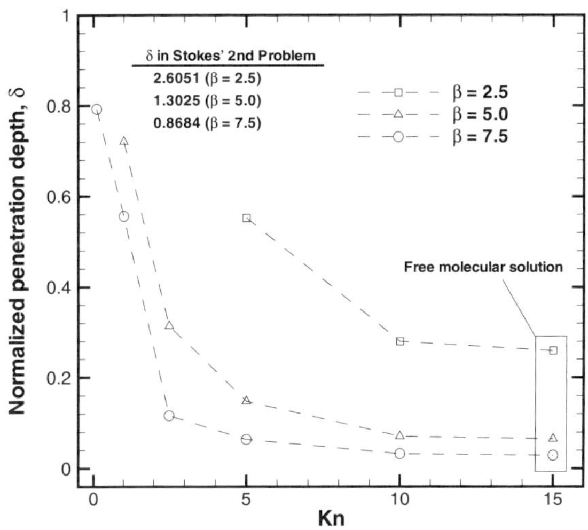

FIGURE 3.18. Effects of Kn and β on penetration depth.

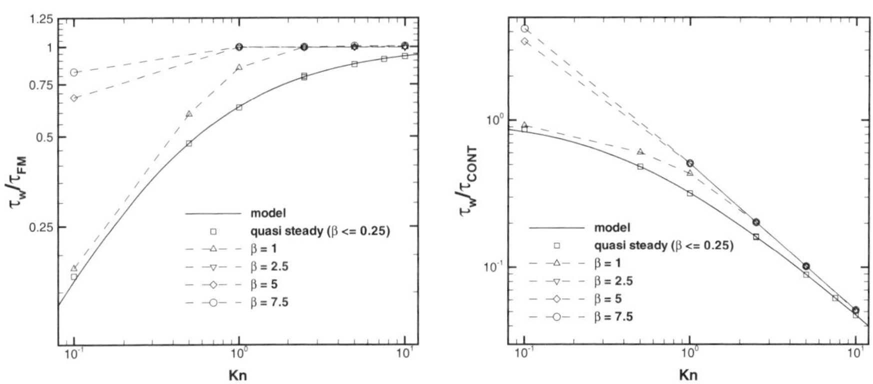

FIGURE 3.19. Effects of Kn and β on wall shear stress.

slip values reach the limit of $0.5u_0$ for high Kn values, as theoretically predicted in Section 3.3.2.

Energy Dissipation

Characterization of energy dissipation in laterally oscillating microstructures is important for designing several microsystem components, such as the microcomb drive mechanisms. The MEMS community utilizes the "quality (Q) factor" concept, which is the ratio of the total energy of the

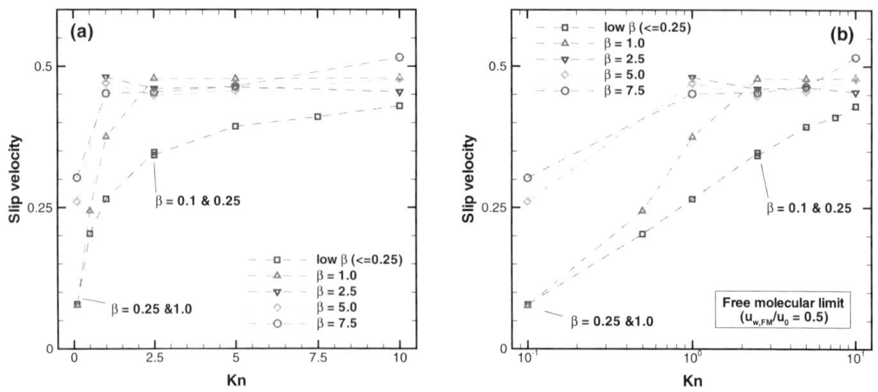

FIGURE 3.20. Rarefaction and Stokes number effects on the slip velocity at the moving wall.

system to the dissipated energy per cycle. In the laterally oscillating plate problem, the energy dissipation per cycle is due to the viscous dissipation (work done by the viscous forces), which is given as

$$D = \frac{1}{\omega} \int_0^{2\pi} \tau_w(\omega t) u_{w,g}(\omega t) d(\omega t). \tag{3.38}$$

Therefore, accurate characterization of the shear stress (τ_w) and gas velocity ($u_{w,g}$) on the wall as functions of the Knudsen and Stokes numbers is required to estimate the energy dissipation. For *quasi-steady* oscillatory Couette flow in the *continuum* flow regime, the energy dissipation per cycle is

$$D_c = \frac{\pi}{\omega} U^2 \left(\frac{\mu}{L}\right). \tag{3.39}$$

Using this value as a reference, we define

$$D^* = D_c / D(\text{Kn}, \beta), \tag{3.40}$$

where $D(\text{Kn}, \beta)$ is the energy dissipation per cycle obtained using the DSMC results as a function of β and Kn. Figure 3.21 displays the effects of Kn and β on D^*. Owing to the smaller viscous dissipation due to the rarefaction effects, D^* increases as Kn is increased. We also observe that D^* decreases with increasing β in the slip flow regime. This is due to the enhanced shear stress with increased β. However, such influence of the Stokes number is drastically reduced with increasing Kn, since both the gas velocity and the shear stress on the wall are reduced by increasing Kn (see Figures 3.19 and 3.20).

Assuming that the viscous dissipation for free molecular flows can be obtained using eqution (3.38), it is possible to predict the behavior of D^* in

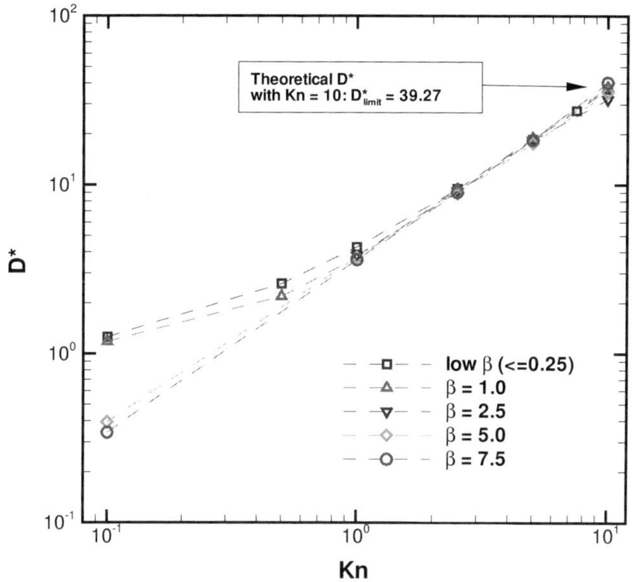

FIGURE 3.21. Rarefaction and Stokes number effects on the normalized dissipation parameter D^*.

the free-molecular flow regime using the analytical solution of the linearized Boltzmann equation. Since, $\tau_{\text{cont}}/\tau_{\infty} = \frac{5\pi}{8}\,\text{Kn}$, and $u_{w,g} = 0.5u_0$, we obtain

$$D^* = \frac{5\pi}{16}\,\text{Kn}, \quad \text{as Kn} \rightarrow \infty. \tag{3.41}$$

Therefore, D^* continuously increases with increasing Kn. For example, Kn $= 10$ flow in Figure 3.21 reaches $D^* = 39.27$ regardless of the Stokes number. The figure shows that the Stokes number dependence of D^* is important until Kn ≈ 1. For higher Kn values, D^* can be predicted using equation (3.41).

3.3.3 *Summary*

In this section, time-periodic Couette flows are studied systematically in the entire Knudsen regime and a wide range of Stokes numbers using the unsteady DSMC method. Simulations show that the quasi-steady flow conditions, which result in linear velocity distribution with equal velocity slip on the oscillating and stationary surfaces, diminish beyond a certain Stokes number. Although this limit also depends on Kn, $\beta \leq 0.25$ can be taken as the limit for quasi-steady flows. The empirical models presented in Section 3.3.1 are valid in this regime for Kn ≤ 12, and they can be easily substituted in place of the DSMC simulations. For moderate Stokes number flows, the aforementioned empirical model is valid only in the slip flow regime.

For high Stokes number flows, we observed formation of "bounded Stokes layers," where the stationary wall no longer interacts with the flow. In the slip flow regime, this results in the classical Stokes's second problem with velocity slip. Once again, the empirical model in Section 3.3.1 is valid in this regime, where the wave speed is constant outside the Knudsen layer, and the velocity amplitude decays exponentially as a function of the distance from the oscillating surface. However, there are small deviations from this behavior within the Knudsen layer. Such deviations are captured by the DSMC, but they cannot be modeled using continuum-based approaches. In the transition and free-molecular flow regimes we observed that the solution decay is not exactly exponential and the wave speed is no longer constant. These are interesting deviations, which are also validated using the analytical solution of the linearized collisionless Boltzmann equation in the free-molecular flow limit. In all simulations, the results have consistently shown that the slip velocity and wave propagation speed increase with increased Kn and β.

An interesting behavior is observed when Kn is increased while the Stokes number is fixed. For such cases, the slip velocity increases, and a bounded layer with a finite penetration depth is formed after a certain value of Kn. This is named the "bounded rarefaction layer" (Park et al., 2004). The penetration depth for this layer is a function of both Kn and β, and it becomes a new length scale in the problem. For such cases, it is necessary to redefine the Knudsen number based on the penetration depth, rather than the separation distance between the two plates. However, without a priori knowledge of the penetration depth it is not possible to predetermine Kn in the simulations. In order to remain consistent, we kept the characteristic length scale of the problem as the plate separation distance. However, the reader can use Figure 3.18 to estimate the actual Knudsen number based on the penetration depth. Due to this switch in the length scales, we observed that the shear stress on the oscillating wall reaches the asymptotic free-molecular limit at earlier Kn values. Solution of the linearized collisionless Boltzmann equation in the free-molecular flow limit indicates that the shear stress and the slip velocity amplitude for oscillatory Couette flows are *identical* to those of the steady plane Couette flows. This interesting finding is also confirmed by DSMC results (Park et al., 2004).

3.4 Cavity Flow

We now address rarefaction effects in a microcavity, where the flow is driven by the moving upper wall. The results presented here were obtained by Nie, Doolen, and Chen (1998) using the lattice Boltzmann method (see Section 15.5). The cavity size is 40×40 in lattice units. The upper wall moves with a constant velocity U_0 from left to right, while the other three walls are at rest. The Reynolds number and the Mach numbers are both very small

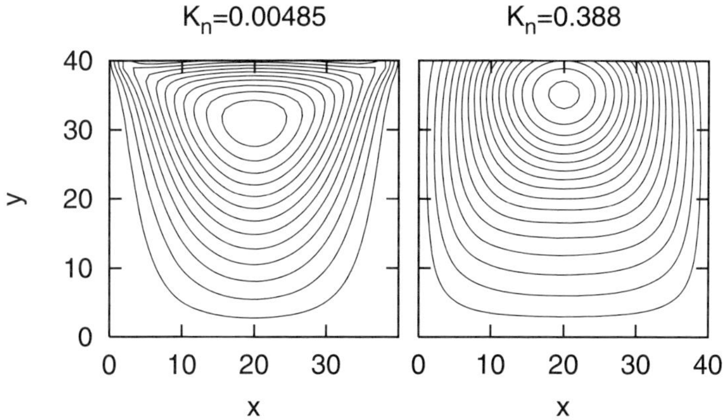

FIGURE 3.22. Streamlines for the microcavity flow in the slip flow regime (left) and transition regime (right). (Courtesy of X. Nie, G.D. Doolen, and S. Chen.)

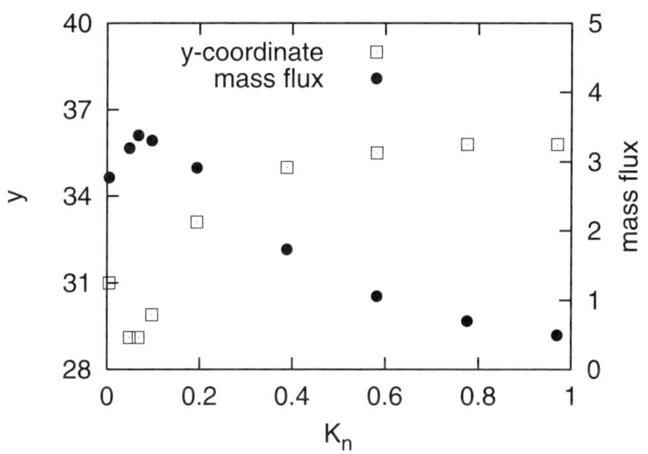

FIGURE 3.23. Distance of the vortex center (y-coordinate) from the lower wall and corresponding mass flux as functions of Knudsen number. (Courtesy of X. Nie, G.D. Doolen, and S. Chen.)

and less than 10^{-3}. A similar study using the (entropic) lattice Boltzmann method has been performed by (Ansumali, 2004); see also Section 15.5.

In Figure 3.22 we plot the streamlines for a small Knudsen number Kn = 0.00485 and a larger Knudsen number Kn = 0.388. We see that the effect of rarefaction is to push the center of the vortex towards the moving wall. This was studied by (Nie et al., 1998) more systematically for a range of values of

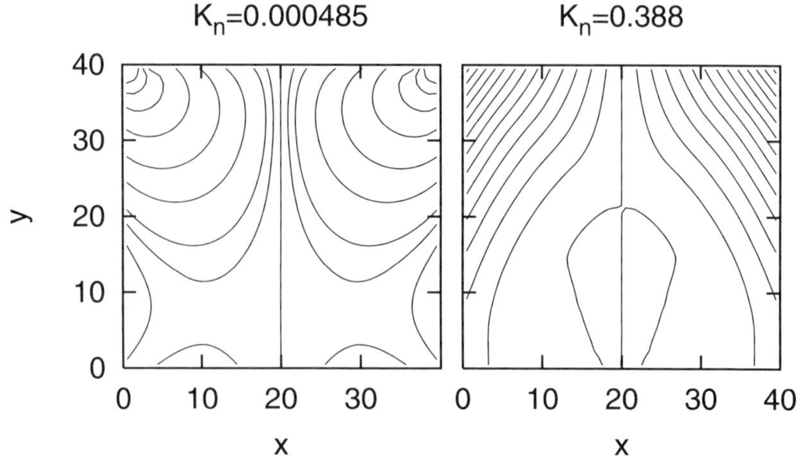

FIGURE 3.24. Pressure contours for the microcavity flow in the slip flow regime (left) and transition regime (right). (Courtesy of X. Nie, G.D. Doolen, and S. Chen.)

Knudsen number, and their results are shown in Figure 3.23. The distance of the vortex center from the lower wall along with the mass flux between the lower wall and the vortex center are plotted. We see that in the slip flow regime the vortex center moves upward rapidly, but then it stabilizes for Kn ≥ 0.3. The corresponding mass flux has a nonmonotonic trend. Initially, it increases due to larger area, but then rarefaction effects are strong and dominate, so the circulation motion is not as intense and the mass flux drops significantly. It is also interesting to examine the pressure contours in the microcavity. This is shown in Figure 3.24, obtained by Nie et al. In the slip flow regime the pressure contours are circular arcs centered on the corners of the moving lid. However, in the transition regime the pressure contours become almost vertical lines. The symmetry around the $x = 20$ plane is due to the low Reynolds number. Other work reported in (Hou et al., 1995) shows that for low Knudsen numbers, as the Reynolds number increases the vortex actually moves toward the lower wall, in obvious contrast with the low Reynolds number limit.

3.5 Grooved Channel Flow

Flow in a micromotor or a microbearing is more complicated than the linear Couette flow. We consider a two-dimensional shear-driven grooved channel flow (Figure 3.25) in order to model the geometric complexity of these microdevices. The presence of grooves complicates the geometry, and an analytical solution for this flow does not exist. The flow separates and

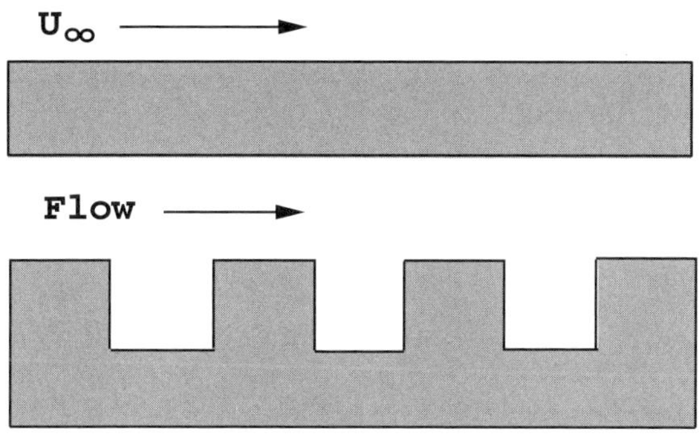

FIGURE 3.25. Schematic of a grooved channel. This geometry is a prototype for modeling geometric complexity and flow reversal encountered in complex microgeometries.

starts to recirculate in the grooves even for small Reynolds number flows. In the numerical model we assumed that the top wall is moving with a speed U_∞, and both surfaces are kept at the same temperature (300 K). We also assumed that the geometry repeats itself along the flow direction. Therefore, the flow is periodic, and only a section of the channel is simulated. In the simulations the Reynolds number is fixed (Re $= 5.0$), and the Knudsen number is increased by decreasing the channel gap. Therefore, the top wall speed U_∞ is increased to keep the Reynolds number constant, resulting in an increased flow Mach number according to equation (3.5). The domain is discretized with 12 elements of sixth-order polynomial expansions in each direction, per element.

The streamwise velocity variation across the middle of the channel, normalized by the top wall velocity (U_∞), is given in Figure 3.26. The flow separates due to the presence of the groove and starts to recirculate. This recirculation zone is seen as negative streamwise velocity (i.e., $\frac{U}{U_\infty} \leq 0$). Figure 3.26 shows a reduction in the separation zone for slip flows compared to the no-slip case ($Kn = 0$). The slope of the velocity profile at the top wall is decreased due to the velocity slip effects, resulting in reduction of the shear stress on the top wall. We also see that the net volumetric flowrate, which is proportional to ($\int_0^h U dy$), is decreased as the Kn is increased.

In order to identify the limitations and accuracy of an incompressible model, we compare the results of the incompressible model, with the compressible model for the same Reynolds and Knudsen numbers. We present the variation of mass flowrate in the channel versus the drag on the top channel wall, both normalized with their no-slip incompressible counterparts in Figure 3.27. Drag reduction due to rarefaction effects is seen. For example, for Kn $= 0.128$ flow, about 30% drag reduction is observed compared to the no-slip case. Both models predict a reduction in mass flowrate for slip flow. This is due to the reduction in the volumetric flowrate as seen in Figure 3.26. The mass flowrate predicted by the incompressible model

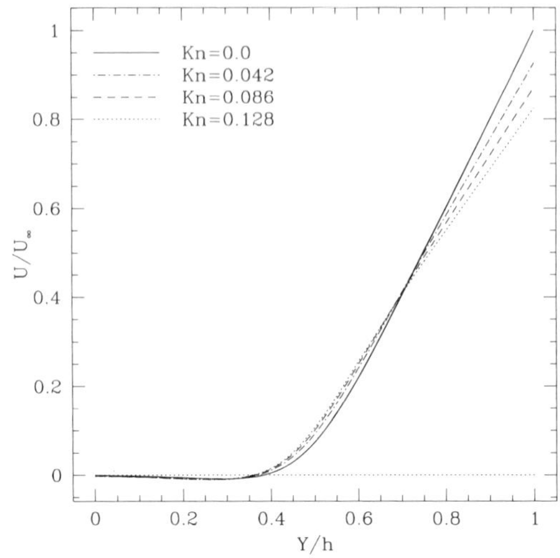

FIGURE 3.26. Streamwise velocity distribution along the center of the grooved channel. $y = 0$ corresponds to the lower boundary of the grooves, and $y = h$ corresponds to the top wall.

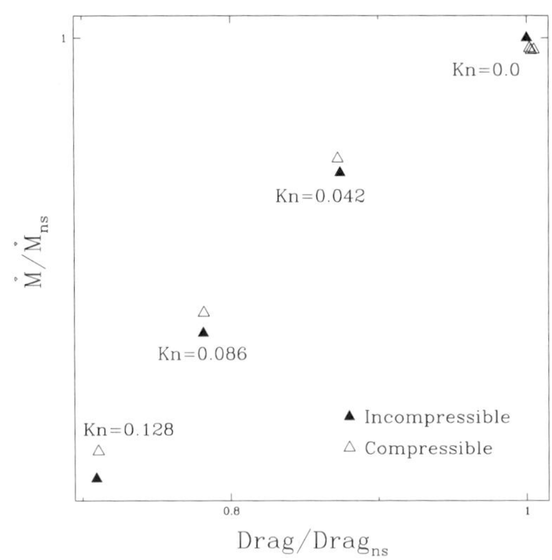

FIGURE 3.27. Variation of mass flowrate versus drag force for the grooved channel, normalized with values of the corresponding incompressible no-slip model.

is less than the mass flowrate predicted by the compressible model. This is due to the inability of the incompressible model to predict the variations in fluid density.

The temperature contours for no-slip and slip (Kn $=$ 0.086) flows are

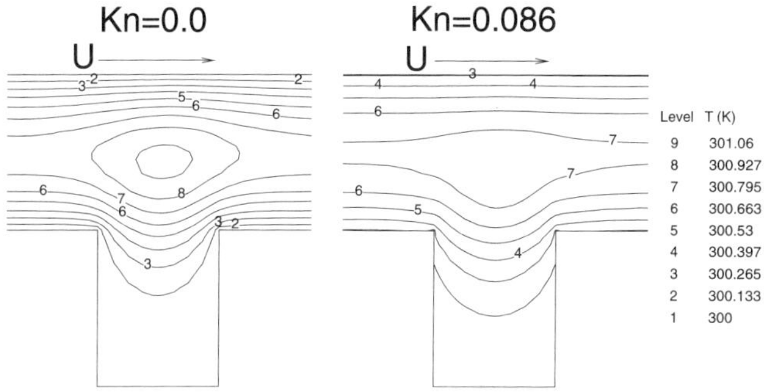

FIGURE 3.28. Temperature contours in no-slip and slip grooved channels.

given in Figure 3.28. The increase in temperature in the middle of the channel is due to viscous heating resulting from large shear stresses in this low Reynolds number flow. The viscous heating effect for slip flow is less than that of no-slip flow due to the reduction in the shear stresses. The temperature of the gas at the wall is different from the prescribed wall temperature. Since the temperature of the fluid is higher in the middle of the channel, the channel loses heat. Therefore, the gas temperature is higher than the surface temperature due to the temperature jump. This may create a problem for gas microflow temperature measurements. Although the change in the temperature due to the viscous heating seems to be small in magnitude, the *gradients* in temperature (as seen by the contour density in Figure 3.28) can be quite large due to the small length scales associated with the microflows.

4

Pressure-Driven Flows

In this Chapter we present models for pressure-driven gas flows in the slip, transition, and free-molecular flow regimes. We are particularly interested in microchannel, pipe, and duct flows, since such flows have primary engineering importance, and they also allow analytical solutions due to their simple geometry. In the first section, we present analysis in the slip flow regime. This is followed by an analysis in the transition and free-molecular flow regimes. In particular, in Section 4.2.2 we develop a *unified* flow model that can accurately predict the volumetric flowrate, velocity profile, and pressure distribution in the *entire* Knudsen regime for pipes and ducts, as well as the Knudsen minimum.

4.1 Slip Flow Regime

In this section, we present pressure-driven isothermal adiabatic flows. While the adiabatic flows are treated using Fanno flow theory, the isothermal flows are studied using compressible Navier–Stokes equations subject to the slip models presented in Chapter 2. We give comparisons of the numerical results with the available experimental data. Due to some scatter in experimental measurements, we further validate the continuum-based slip flow results, using the DSMC method and solutions of the linearized Boltzmann equation (see Chapter 15). This approach enables pointwise comparisons between the fundamentally different atomistic and continuum simulation models. Finally, we discuss the surface roughness effects and present results of inlet flow simulations.

4.1.1 Isothermal Compressible Flows

We consider two-dimensional isothermal flow between two parallel plates of length L, separated a distance h apart, where $L/h \gg 1$. The flow is sustained by a pressure drop from inlet (i) to outlet (o) of the channel ($\Delta P = P_\mathrm{i} - P_\mathrm{o}$). Since the channel thickness h is fixed, the equation of conservation of mass can be simplified as

$$\rho_\mathrm{i} u_\mathrm{i} = \rho_\mathrm{o} u_\mathrm{o},$$

where ρ and u denote the channel averaged density and velocity, respectively. The momentum equation in the streamwise direction results in

$$(P_\mathrm{i} - P_\mathrm{o})h - 2L\tau = \dot{M}(u_\mathrm{o} - u_\mathrm{i}). \tag{4.1}$$

The pressure, density, and velocity are averaged across the channel at respective streamwise locations. The shear stress (denoted by τ) is also averaged, but along the streamwise direction. The mass flowrate (per unit channel width) is denoted by $\dot{M}$.

If we divide the momentum equation by (hP_o), we obtain

$$\frac{(P_\mathrm{i} - P_\mathrm{o})}{P_\mathrm{o}} = \frac{\Delta P}{P_\mathrm{o}} = 2\frac{L}{h}\frac{\tau}{P_\mathrm{o}} + \frac{\dot{M}(u_\mathrm{o} - u_\mathrm{i})}{hP_\mathrm{o}},$$

where $\Delta P/P_\mathrm{o}$ represents the nondimensional pressure drop. Concentrating on the term $\dot{M}/hP_\mathrm{o} = (\rho_\mathrm{o} u_\mathrm{o})/P_\mathrm{o}$ and using the continuity equation ($u_\mathrm{o} = u_\mathrm{i}\rho_\mathrm{i}/\rho_\mathrm{o}$) and the equation of state for an ideal gas ($\rho_\mathrm{i}/\rho_\mathrm{o} = P_\mathrm{i}/P_\mathrm{o}$, assuming isothermal conditions), we obtain

$$\frac{\Delta P}{P_\mathrm{o}} = 2\frac{L}{h}\frac{\tau}{P_\mathrm{o}} + \frac{\rho_\mathrm{o} u_\mathrm{o} u_\mathrm{i}}{P_\mathrm{o}}\left(\frac{P_\mathrm{i}}{P_\mathrm{o}} - 1\right).$$

Since $P_\mathrm{o} = \rho_\mathrm{o} RT$, and $c_\mathrm{s}^2 = \gamma RT$, where c_s is the speed of sound, the above equation can be simplified as

$$\frac{\Delta P}{P_\mathrm{o}} = 2\frac{L}{h}\frac{\tau}{P_\mathrm{o}} + \gamma M_\mathrm{o} M_\mathrm{i}\frac{\Delta P}{P_\mathrm{o}}, \tag{4.2}$$

where M denotes the Mach number at respective locations. Rearranging, we obtain

$$\frac{\Delta P}{P_\mathrm{o}}(1 - \gamma M_\mathrm{o} M_\mathrm{i}) = 2\frac{L}{h}\frac{\tau}{P_\mathrm{o}}.$$

Without further simplification we see that the inertial terms in the momentum equation (right-hand side of equation (4.1)) can be neglected if $\gamma M_\mathrm{o} M_\mathrm{i} \ll 1$. To this end, we note that:

1. In microchannels with $\frac{L}{h} \approx 10^3$ to 10^4, relatively large pressure drops can be sustained for small Mach number flows.

2. Since the Mach number in microflows is usually small, the inertial effects are small. Therefore, we expect semianalytic formulas based on balancing the pressure drop with drag on the channel walls to work reasonably well. (This is not true for micronozzles; see Section 6.6.)

3. If the diffusion term is simplified by approximating the wall shear stress as τ to $\mu u/h$ and recognizing μ/P_o to λ/c_s, we obtain

$$\frac{\Delta P}{P_\mathrm{o}}(1 - \gamma M_\mathrm{o} M_\mathrm{i}) \approx 2\frac{L}{h} M_\mathrm{o}\,\mathrm{Kn}_\mathrm{o}\,. \qquad (4.3)$$

   The above relation indicates the relative importance of compressibility effects in the slip flow regime.

In order to identify the relative importance of inertial terms in the momentum equation compared to the diffusion terms, we compare their respective magnitudes:

$$\frac{\rho u \frac{\partial u}{\partial x}}{\mu \frac{\partial^2 u}{\partial y^2}} \approx \frac{\rho u^2/L}{\mu u/h^2} = \frac{\rho u h}{\mu}\left(\frac{h}{L}\right) = \mathrm{Re}\left(\frac{h}{L}\right).$$

A similar estimate can be obtained by taking the ratio of inertial terms to diffusion terms in equation (4.3) as

$$\frac{\frac{\Delta P}{P_\mathrm{o}}(\gamma M_\mathrm{o} M_\mathrm{i})}{\frac{L}{h} M_\mathrm{o}\,\mathrm{Kn}_\mathrm{o}} \approx \frac{M_\mathrm{i}}{\mathrm{Kn}_\mathrm{o}}\frac{h}{L}\frac{\Delta P}{P_\mathrm{o}} \approx \frac{M_\mathrm{i}}{\mathrm{Kn}_\mathrm{i}}\frac{h}{L}\frac{\Delta P}{P_\mathrm{i}} \approx \frac{h}{L}\mathrm{Re}\frac{\Delta P}{P_\mathrm{i}},$$

where we have used

$$\mathrm{Kn}_\mathrm{o} = \frac{P_\mathrm{i}}{P_\mathrm{o}}\mathrm{Kn}_\mathrm{i}, \quad \text{and} \quad \mathrm{Kn} \approx \frac{M}{\mathrm{Re}},$$

in order to arrive at the third and the fourth equations, respectively. Therefore, the above two estimates are similar, with the exception of the term

$$\frac{\Delta P}{P_\mathrm{i}} = \frac{P_\mathrm{i} - P_\mathrm{o}}{P_\mathrm{i}},$$

which is always smaller than unity. This analysis verifies that for relatively low Re flows ($\mathrm{Re} \leq \mathcal{O}(1)$) in large aspect ratio channels ($L/h \gg 1$) the inertial effects in the momentum equation can be neglected. Under such conditions the momentum equation in the streamwise direction is reduced to the familiar form

$$\frac{dP}{dx} = \mu \frac{\partial^2 u}{\partial y^2}. \tag{4.4}$$

This equation results in the following analytical solution for the streamwise velocity profile:

$$U(y) = \frac{h^2}{2\mu}\frac{dP}{dx}\left[\frac{y^2}{h^2} - \frac{y}{h} + \frac{2-\sigma_v}{\sigma_v}(Kn^2 - \mathrm{Kn})\right], \tag{4.5}$$

where we have used the high-order boundary condition (equation (2.26)). Notice that the second-order correction, which is typically omitted in other works, has the opposite sign compared to the first-order term; its contribution may be significant, especially for high Kn flows. We will examine this in Section 4.1.3.

The corresponding mass flowrate is computed from

$$\dot{M} = \rho \int_0^h U(y)dy,$$

where $\rho = P/RT$, assuming we can still treat the rarefied gas as an ideal gas. Expressing the Knudsen number at a location x as a function of the local pressure, i.e., $\mathrm{Kn} = \mathrm{Kn_o}\, P_\mathrm{o}/P$, where the subscript "o" refers to outlet conditions, we obtain

$$\dot{M} = \frac{h^3 P_\mathrm{o}^2}{24\mu RTL}[(\Pi^2 - 1) + 12\frac{2-\sigma_v}{\sigma_v}(\mathrm{Kn_o}(\Pi - 1) - \mathrm{Kn_o^2}\log_e \Pi)], \tag{4.6}$$

where we have defined $\Pi \equiv P_\mathrm{i}/P_\mathrm{o}$ as the pressure ratio between inlet and outlet. The corresponding flowrate without rarefaction effects is given by

$$\dot{M}_\mathrm{ns} = \frac{h^3 P_\mathrm{o}^2}{24\mu RTL}(\Pi^2 - 1). \tag{4.7}$$

Therefore, the increase in mass flowrate due to rarefaction can be expressed as

$$\frac{\dot{M}}{\dot{M}_\mathrm{ns}} = 1 + 12\frac{2-\sigma_v}{\sigma_v}\frac{\mathrm{Kn_o}}{\Pi + 1} - 12\frac{2-\sigma_v}{\sigma_v}\mathrm{Kn_o^2}\frac{\log_e \Pi}{\Pi^2 - 1}. \tag{4.8}$$

It is seen from this formula that the effect of the second-order correction is to reduce the increase in mass flowrate due to first-order slip. *This is in disagreement with the experimental data, since in the transition flow regime, the flow rate increases faster than predictions of the first-order slip theory.* This inconsistency will be addressed in Sections 4.1.3 and 4.2.2.

Having obtained the mass flowrate, we can write the corresponding pressure distribution along the channel as

$$1 - \tilde{P}^2 + 12\frac{1-\sigma_v}{\sigma_v}\mathrm{Kn_o}(1 - \tilde{P}) + 12\frac{2-\sigma_v}{\sigma_v}\mathrm{Kn_o^2}\log_e(\tilde{P}) \\ = B(L - x), \tag{4.9}$$

where B is a constant such that $\tilde{P}(0) = \Pi$. Here we have defined $\tilde{P} = P/P_o$, i.e., the pressure at a station x normalized with the outlet pressure. The above equation provides an implicit relation for $\tilde{P}$; the pressure distribution for a first-order boundary condition is obtained explicitly by neglecting the second-order terms ($\mathcal{O}(\mathrm{Kn}^2)$) in equation (4.9).

The formula for the flowrate has been tested directly using experimental data reported in (Arkilic et al., 1997), as well as simulation results obtained using the program μ**Flow** (see Section 14.1). In Figure 4.1 we plot the normalized flowrate

$$M^* = \frac{\dot{M} 24 \mu R T L}{h^3 P_o^2}$$

versus the pressure ratio Π. Three cases are included, corresponding to *no-slip compressible air flow* (lower curve; equation (4.7)), *rarefied air flow* (middle curve; equation (4.6)), and *rarefied helium flow* (upper curve; equation (4.6)). The open circles correspond to microchannel experimental data in (Arkilic et al., 1997), and solid circles denote the corresponding numerical simulation results of rarefied helium flow. It is seen that there is a significant mass flowrate increase, especially for helium flow due to rarefaction ($\mathrm{Kn} = 0.165$), and that in this range there are no significant deviations between the formula and the simulations; the latter correspond to solutions of full compressible Navier–Stokes equations with rarefaction effects expressed via equation (2.26).

The differences between simulation, experiments, and analysis are evaluated more accurately by computing the ratio of slip mass flowrate to the corresponding no-slip flowrate predicted by equation (4.8) as a function of pressure ratio in Figure 4.2. Microchannel helium flow experiments in (Arkilic et al., 1997), show a maximum of 10% deviation from the first-order theoretical curve. The deviations are more significant especially for low pressure ratio cases. The gain in the mass flowrate due to slip effects is reduced as the pressure ratio is increased. This is a trend expected by equation (4.8) as well. However, due to significant scatter in the experimental data, it is difficult to determine the slope of this decrease. Comparison of the high-order formula with the first-order formula shows about 8% deviations for small pressure ratios; the deviations are reduced for higher pressure ratio cases. The numerical predictions for helium flow are consistent with the high-order formula. Both the rarefied air flow ($\mathrm{Kn} = 0.075$) and the no-slip air flow show deviations from the high-order formula, especially for high pressure ratios. The numerical predictions show less mass flowrate than that predicted by the formula. The reason is the pronounced compressibility effects caused by the inertial terms in the Navier–Stokes equations, which were neglected in the derivation of equation (4.8).

Next we examine the pressure distribution along a channel; the experimental results in (Liu et al., 1993; Pong et al., 1994), show a nonlinear pressure distribution. In Figure 4.3 (left) we plot the pressure distribution for air flow for different values of pressure ratios Π obtained from simulation.

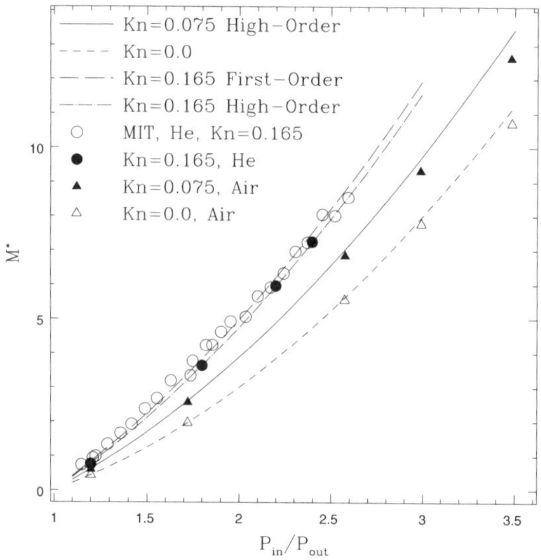

FIGURE 4.1. Variation of normalized mass flowrate as a function of pressure ratio. The experimental data (MIT) are from (Arkilic et al., 1994).

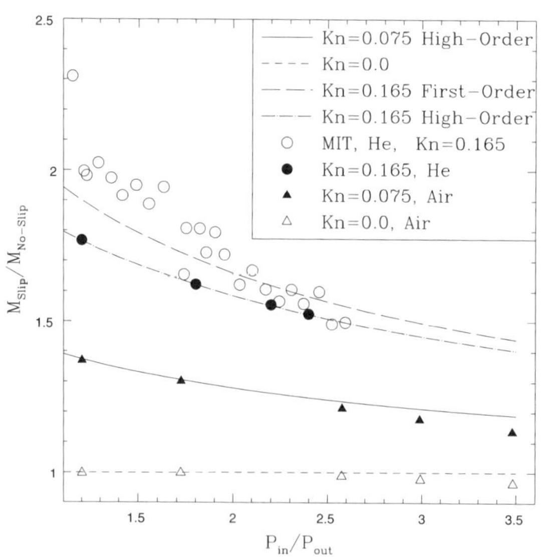

FIGURE 4.2. Variation of mass flowrate normalized with no-slip mass flowrate as a function of pressure ratio. The experimental data (MIT) are from (Arkilic et al., 1994).

TABLE 4.1. Data for air flow simulations for various channel dimensions.

Case	$h(\mu\text{m})$	Re	Kn_o	Π	M_i	M_o
1	0.923	3.367	0.0	2.574	0.091	0.247
	0.923	4.144	0.075	2.576	0.116	0.279
2	0.615	1.522	0.0	2.583	0.064	0.168
	0.615	2.036	0.110	2.585	0.083	0.198
3	0.226	0.208	0.0	2.587	0.026	0.063
	0.226	0.378	0.300	2.586	0.041	0.093

Also included in the plot are the pressure distributions for corresponding compressible no-slip flows. The curvature in pressure distribution is more pronounced for the no-slip compressible flows than for rarefied flows.

- *Therefore, rarefaction effects reduce the curvature in pressure distribution.*

For rarefied air flow (Kn $= 0.3$) presented in Figure 4.3, the curvature in pressure distribution is more pronounced for high pressure ratios. For cases with $\Pi \leq 1.2$, the pressure drop is practically linear, resembling an incompressible flow. To identify the effect of rarefaction in reducing the curvature effects in pressure distribution we plot deviations from the linear pressure drop (i.e., corresponding incompressible flow) for $\Pi = 2.58$ in Figure 4.3 (right). The outlet pressure in these simulations is fixed to be atmospheric. The Knudsen number is varied by changing the channel thickness (h). Corresponding values of h, Re (based on mass flowrate per channel width), Kn_o, and inlet and outlet Mach numbers (M_i and M_o, respectively) are given in Table 4.1. It is clear that the stronger the rarefaction effects, the smaller the deviation from the linear pressure distribution (compare cases 3, 2, and 1). This finding contradicts the conclusions in (Pong et al., 1994), where the deviation from linear distribution for helium microflow is reported to be larger than the corresponding pressure distribution for nitrogen. This was attributed to the increase in the rarefaction effects for helium flow compared to the nitrogen flow. The curvature in pressure distribution for no-slip compressible flow increases with increasing inlet Mach number. For the same pressure ratio, the wider channel (case 1) corresponds to higher mass flowrate than the other cases. It also has larger inlet and outlet Mach numbers at the channel centerline, resulting in larger curvature even for the no-slip case compared to the other cases. The simulations also show a shift of the locus of maximum deviation from linear profile toward the inlet of the channel as the rarefaction effects are increased.

In Figure 4.4 we present the variation of velocity slip along the channel walls normalized with centerline velocity at the channel inlet. Due to the pressure drop along the channel, the local mean free path increases, resulting in an increase in the local Kn. Also, the density along the channel

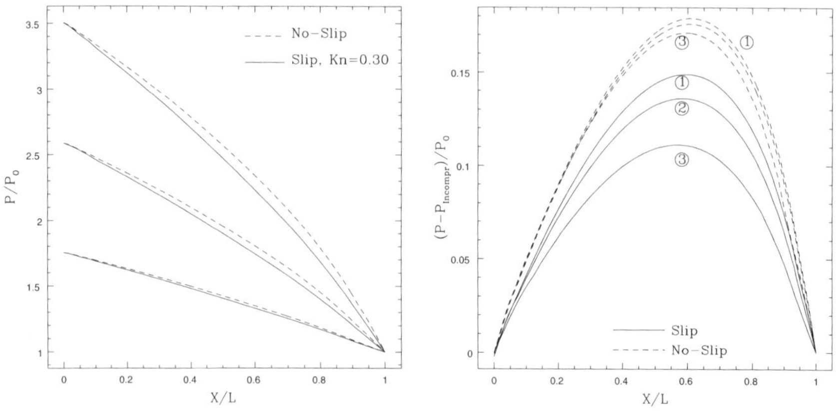

FIGURE 4.3. Left: Pressure distribution for different pressure ratios along the center of microchannels for air flow. Right: Deviation from linear pressure drop for $\Pi = P_{\mathrm{i}}/P_{\mathrm{o}} = 2.58$ air flow. Cases 1, 2, and 3 correspond to different channel thickness given in Table 4.1.

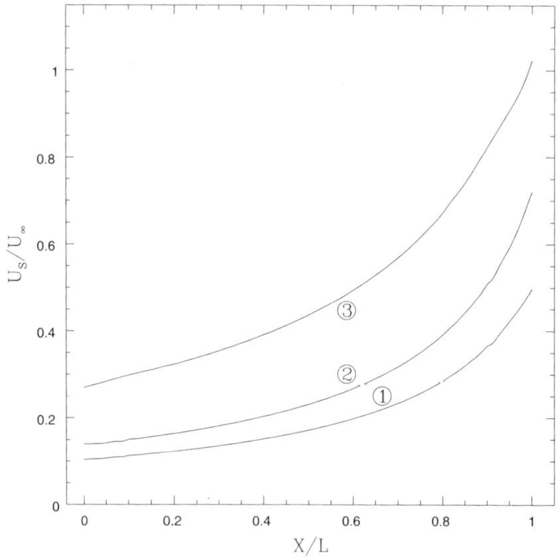

FIGURE 4.4. Velocity slip variation on channel surface for cases described in Table 4.1.

decreases, and thus the average velocity in the channel increases toward downstream to conserve mass. These two effects together increase the slip velocity along the channel walls, as seen in Figure 4.4.

To investigate the compressibility and rarefaction effects further, the de-

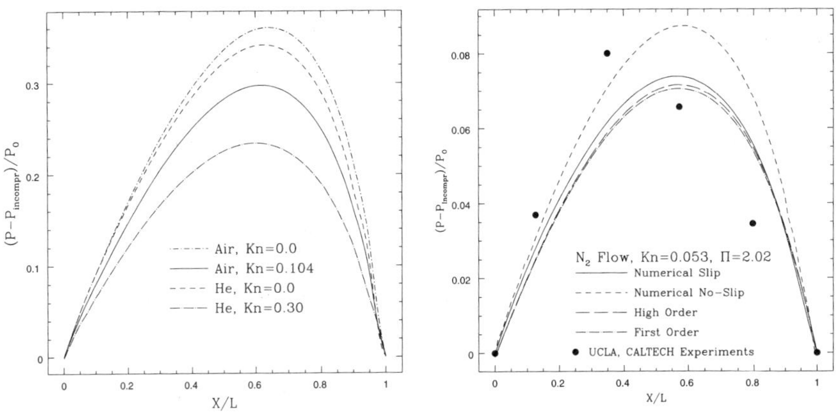

FIGURE 4.5. Left: Deviation from linear pressure drop for air and helium flows ($h = 0.65\,\mu$m, $\Pi = P_i/P_o = 3.5$, $P_o = 1$ atm). Right: Deviation from linear pressure drop for $\Pi = P_i/P_o = 2.02$, nitrogen flow ($h = 1.25\,\mu$m); circles correspond to experimental data of (Pong et al., 1994).

viation from linear pressure distribution for helium and air flows in identical channels with identical inlet and outlet conditions corresponding to pressure ratio $\Pi = 3.5$ is given in Figure 4.5 (left). Here we see that unlike the experimental findings, the curvature in the pressure distribution for helium is less pronounced compared to the air microflow. This trend should be expected, since for the same pressure ratio and outlet pressure, the local Mach number for helium flow is smaller than the Mach number for air flow. Also, the rarefaction effects for helium flow are larger than those of the air flow due to the relatively large mean free path of helium molecules compared to air. The finding from the simulation results that rarefaction causes the opposite effect than compressibility is also evident from the analytical expression (equation (4.9)). This is shown in Figure 4.5 (right), where we plot the analytical predictions taking into account first- and second-order Knudsen number effects. The simulation results for nitrogen flow of Kn $= 0.156$ corresponding to the experiments of (Pong et al., 1994) are also included. We see that the pressure distribution agrees with the high-order curve. The discrepancies are due to the neglected inertial terms in equation (4.9). Also included in the plot are the experimental data for nitrogen taken from (Pong et al., 1994). We see that the trends of experimental data and the simulations are qualitatively the same. They both predict smaller deviations from linear pressure drop than the corresponding no-slip flow. There are quantitative differences such as the maximum deviation location from linear pressure drop, which is at $X/L \approx 0.4$ for the experiments, and $X/L \approx 0.55$ for the simulation.

Remark: Before we end this section, we comment on the *limitations of the* *analytic formulas* given in equations (4.6) and (4.7). First, the second-order corrections to the flowrate and pressure distribution are negligible in the slip flow regime, and equivalent first-order slip formulas can be obtained by simply neglecting the Kn^2 terms in these formulas. Second, derivation of these formulas is based on the additional assumptions that density and pressure *across* the channel at any streamwise location are constant. Thermal effects are also completely neglected. Due to these limitations the analytic formulas can be applied to low Mach number flows (typically $M_o \leq 0.10$). In our simulations we have detected density variations across the channel, especially for large pressure drop cases of nitrogen and air flows. By evaluating the relative importance of inertial terms compared to the cross flow diffusion terms ($\approx \mathrm{Re}\frac{h}{l}$) using the aspect ratio $L/h = 20$, and for the Reynolds number range reported in Table 4.1, the difference between the analytic formula predictions and the solution of full Navier–Stokes equations is approximately 20%. Finally, we also note that entrance effects for pressure-driven flows, e.g., through microfilters and short channels, are addressed in Section 6.5.

4.1.2 Adiabatic Compressible Flows – Fanno Theory

In this section we use the Fanno flow theory to analyze compressible adiabatic gas flows. Most of the microsystem components are fabricated using silicon as the base material. Since silicon is a good heat conductor, the isothermal flow assumption presented in the previous section is more relevant for many silicon-based microsystem components. However, for insulator materials, such as glass, approximate adiabatic flow conditions may be valid. Therefore, study of compressible adiabatic gas flows may be useful in applications that utilize insulator materials.

We will compare the experimental data in (Harley, 1993; Harley et al., 1995), obtained under *almost* isothermal flow conditions, with the *adiabatic* Fanno flow theory. Although the isothermal and adiabatic flows show significant differences for high Mach number flows, Fanno flow analysis can predict the general trends of the experimental results, enabling approximate analysis for many engineering applications.

In the experiments of (Harley et al., 1995), about an order of magnitude pressure drop is imposed through a microchannel of length 10.9 mm. Since the temperature was almost constant, about an order of magnitude change in the inlet to outlet density ratio was observed. This case corresponds to severe compressibility effects, which may eventually result in a *choked flow*. Such large pressure drops and even flow choking may become relevant in analysis of micronozzles for aerospace applications (see Section 6.6).

The approximation is based on one-dimensional adiabatic constant-area flow, i.e., the so-called Fanno flow, conditions. The experimental data reported in (Harley, 1993), are summarized in Table 4.2. The Fanno flow equations can be derived by using a friction factor ($\bar{f} = \frac{1}{L}\int_0^L f dx$) aver-

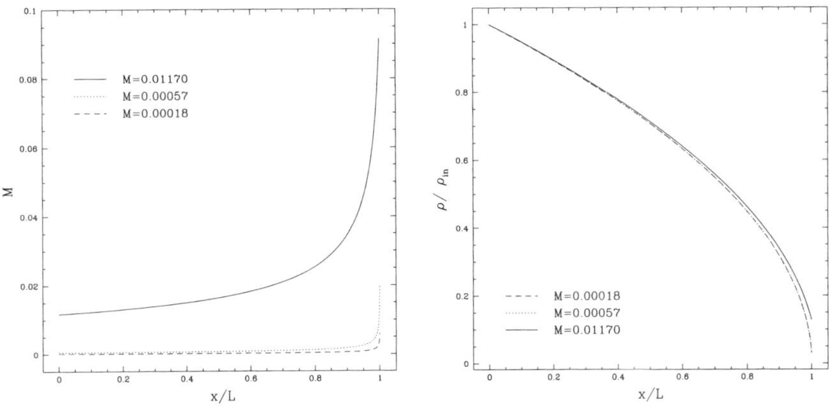

FIGURE 4.6. Mach number (left) and density (right) variation along the channel as a function of $\frac{x}{L}$ for inlet conditions of Table 4.2 (the legend refers to the inlet Mach number).

aged over the channel of length L. The Fanning friction factor at any point in the channel can be evaluated as a function of local conditions at station x,

$$f(x) = \frac{\tau_s(x)}{\frac{1}{2}\rho(x)U(x)^2},$$

where τ_s ($\tau_s = \mu\frac{\partial U}{\partial y}$) is the shear stress on the wall. Since the mass flux (ρU) in the channel is constant, the Fanning friction factor can be assumed to be a function of dynamic viscosity only, i.e., $\mu(T)$, assuming $\frac{\partial U}{\partial y}/U$ is approximately a constant. Since the dynamic viscosity variations can be obtained from Sutherland's law, the Fanning friction factor f essentially becomes a function of temperature. In particular,

$$\frac{\mu}{\mu_o} = \left(\frac{T}{T_o}\right)^{3/2}\frac{T_o + 110}{T + 100},$$

where T and T_o are the local and reference temperatures, respectively, and μ and μ_o are the local and reference dynamic viscosities, respectively. Therefore, for small temperature changes reported in the experiments (see Table 4.2) the friction factor is approximately constant. Similarly, the Reynolds number in the channel ($\text{Re} = \frac{\rho U D}{\mu}$) is a function of temperature, and becomes approximately constant for small temperature variations.

The Mach number variation in the channel for Fanno flow can be obtained for specified friction factor as see (Thompson, 1988), equation (6.34):

$$\frac{dM^2}{dx} = \frac{\gamma M^4(1 + \frac{\gamma-1}{2}M^2)}{1 - M^2}\frac{4f}{D}. \tag{4.10}$$

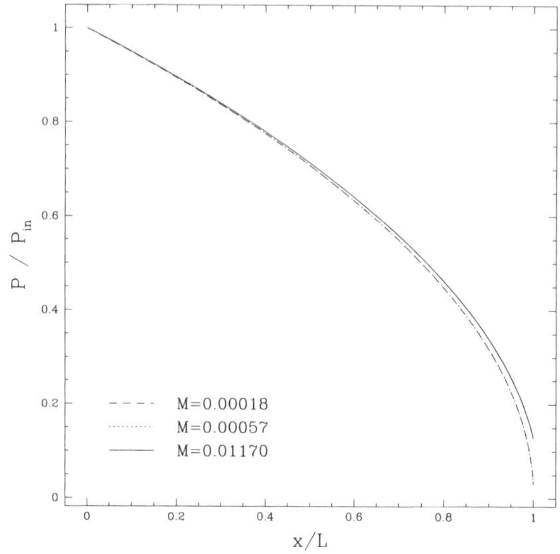

FIGURE 4.7. Pressure ratio ($\frac{P}{P_{\text{in}}}$) variation along the channel as a function of $\frac{x}{L}$ for inlet conditions of Table 4.2.

This equation, integrated from inlet state (M_{in}) to any station x downstream of the channel, gives

$$\frac{4\bar{f}x}{D} = \frac{\gamma+1}{2\gamma}\log_e\left[\frac{M_{\text{in}}^2(1+\frac{\gamma-1}{2}M^2)}{M^2(1+\frac{\gamma-1}{2}M_{\text{in}}^2)}\right] + \frac{(M^2-M_{\text{in}}^2)}{\gamma M^2 M_{\text{in}}^2}. \quad (4.11)$$

The variation of Mach number in the channel as a function of the inlet Mach number and $\frac{4\bar{f}x}{D}$ is found by plotting equation (4.11) for specified $\mathrm{M_{in}}$ values. Once the Mach number variation in the channel is known, the corresponding pressure ratio ($\frac{P}{P_{\text{in}}}$) and the density ratio ($\frac{\rho}{\rho_{\text{in}}}$) variations along the channel can be calculated by using the following relations:

$$\frac{P}{P_{\text{in}}} = \frac{M_{\text{in}}}{M}\left[\frac{(1+\frac{\gamma-1}{2}M_{\text{in}}^2)}{(1+\frac{\gamma-1}{2}M^2)}\right]^{\frac{1}{2}},$$

$$\frac{\rho}{\rho_{\text{in}}} = \frac{M_{\text{in}}}{M}\left[\frac{(1+\frac{\gamma-1}{2}M^2)}{(1+\frac{\gamma-1}{2}M_{\text{in}}^2)}\right]^{\frac{1}{2}}. \quad (4.12)$$

The Mach number variation along the channels is given in Figure 4.6 (left) as a function of $\frac{x}{L}$ for three different inlet conditions reported in (Harley, 1993). Very steep variation of the Mach number is observed toward the exit of the channel. This variation becomes steeper as the inlet Mach

TABLE 4.2. Experimental data in microchannels from (Harley, 1993). D_H is the hydraulic diameter, and all the f and $\bar{f}$ factors denote Fanning friction factors as noted. Also, Po is the Poiseuille number; see Section 1.2.

Case	JP9	JH6	JH6
Width μm	92.40	96.60	96.60
Depth μm	4.44	0.51	0.51
$D_H \mu$m	8.47	1.01	1.01
Length mm	10.90	10.90	10.90
Gas	N_2	N_2	He
P_{in} (kPa)	800	3398	3115
P_{out} (kPa)	102.8	101.2	91.93
ρ_{in} ($\frac{\text{kg}}{\text{m}^3}$)	9.16	38.57	5.05
ρ_{out} ($\frac{\text{kg}}{\text{m}^3}$)	1.18	1.14	0.15
T_{in} (C)	21.11	23.65	23.92
T_{out} (C)	21.2	23.7	23.99
V_{in} ($\frac{m}{s}$)	3.9	0.19	0.18
V_{out} ($\frac{m}{s}$)	30.5	6.55	6.05
Re	17.324	0.433	0.0462
M_{in}	0.0117	0.00057	0.00018
M_{out}	0.0915	0.01965	0.00598
Kn_{in}	0.001	0.002	0.0002
Kn_{out}	0.008	0.067	0.209
$\text{Kn}_{\text{experiment}}$	0.008	0.066	0.214
$\text{Kn}_{\text{average}}$	0.0022	0.0096	0.00096
$\bar{f}_{\text{exp}}$	4.4	215	1820
$\bar{f}_{\text{fanno}}$	4.0	205	1763
$\frac{\text{Po}}{\text{Re}}$	5.18	220	2063
f_{slip}	5.147	214	2057
$\frac{4fL}{D}$	5129	2,215,253	18,501,616
$(\frac{4f}{D}L_{\text{choke}})$	5210	2,217,097	19,043,788
$\frac{L}{D_H}$	1287	10797	10797

number gets smaller. However, the change in the Mach number of the flow is very small at the inlet section of the channel. The density ratio $(\frac{\rho}{\rho_{\text{in}}})$ variation in the channel is given in Figure 4.6 (right). Even if the Mach number variation is small for about 60% of the channel length, a density variation of 20% is obtained in about 35 to 40% of the channel length. The pressure ratio $(\frac{P}{P_{\text{in}}})$ variation is given in Figure 4.7, where the curvature in the pressure distribution can be seen.

The estimated Fanning friction factor $(\bar{f}_{\text{fanno}})$, experimental Fanning friction factor $(\bar{f}_{\text{exp}})$, and the theoretical friction factor for laminar incompressible flows ($\frac{\text{Po}}{\text{Re}}$) are given in Table 4.2. The experimental friction factor ($\bar{f}_{\text{exp}}$) is calculated by using both the adiabatic flow model (equivalent to the Fanno flow approximation), and the isothermal flow model for all three cases. The differences in the two predictions are negligible. For the JP9 case, the friction factor f and the Poiseuille number (Po) are independent of the Reynolds number. However, for cases 2 and 3 there is a steep trend of decreasing Poiseuille number (Po) with decreasing Reynolds number (Harley, 1993). The reason for this trend may be associated with the portion of the channel JP6 being in the early transitional flow regime (Kn > 0.1). The theoretical values are calculated assuming laminar, fully developed, incompressible flows, where the Fanning friction factor is given as

$$f = \frac{\text{Po}}{\text{Re}}.$$

Here the Poiseuille number (Po) is a parameter that depends on the channel's cross sectional geometry, and it is independent of the Reynolds number of the flow. Values of Po are given as a function of the ratio of the channel depth to the channel width (see equation (54) in (Harley, 1993)). The theoretical values of friction factor are based on the incompressibility assumption, because calculations for the friction factor of compressible flows are not available. The theoretical values of the friction factor are greater than the experimental predictions for all of the cases. There are three main reasons for this trend:

- compressibility effects,
- rarefaction effects, and
- two-dimensionality effects.

Here both the predicted and the experimental friction factors are based on one-dimensional arguments.

In order to compensate for compressibility effects, we predicted the average Knudsen number in the channel by numerically integrating the density ratio variation, assuming constant temperature throughout the channel. The average Kn values and the predicted Fanning friction coefficient (f_{slip}) are also given in Table 4.2. It is seen that the predictions overestimate the friction factor compared to the experimental values for case 1 and case 3. However, the predictions for case 2 are quite close to the experimental values. The reason for the discrepancies in the results is due to the use of theoretical (incompressible) Poiseuille number Po for calculating f_{slip}.

The values of parameter $\frac{4\bar{f}L}{D}$ and the corresponding values for the choking length (L_{choke}) are also given in Table 4.2. It is seen that slightly longer channels with the same inlet conditions would have choked the flow. Choking of the microchannels can be a significant disadvantage for applications, limiting the flowrate.

Finally, the one-dimensional theory used in this section predicts only the cross-section averaged quantities. Since the velocity, density, and pressure variations in the cross-flow direction are neglected, the current approach should be used only as a first approximation.

4.1.3 Validation of Slip Models with DSMC

In Section 4.1.1 we compared the continuum-based numerical models with existing experimental data in microflows. Although the experimental data show trends consistent with the slip flow theory, the uncertainties in the experimental measurements are relatively large, and pointwise measurements are not available with the exception of results reported in (Liu et al., 1993; Pong et al., 1994). Therefore, we will use the DSMC method (see Section 15.1) to examine the validity of the slip models proposed in Section 2.3.

The results we present in the next few figures are for a microchannel with aspect ratio ($L/h = 20$) and inlet to exit pressure ratio Π. The Knudsen number at the channel outlet is 0.20. The discretization with the μ**Flow** program involves ten elements in the flow direction and two elements in the cross-flow direction, employing sixth-order polynomial expansions per direction in each element. Convergence is verified by increasing the order of the polynomial expansions while keeping the number of spectral elements fixed. In addition, we monitor the residuals of the global conservation of mass and momentum. The results presented in these runs conserve mass and momentum with 0.01% error. The DSMC results are performed for the case $\Pi = 2.28$ with 24,000 cells, of which 600 cells are in the flow direction and 40 across the channel; a total of 480,000 molecules are simulated. The results are sampled (time-averaged) for 1.0×10^5 time steps. Convergence is also verified by monitoring mass balance; maximum errors are less than 1%.

Diffuse reflection ($\sigma_v = 1.0$) is assumed for the interaction of gas molecules with the channel walls. The value of the accommodation coefficient σ_v is an important factor, since the slip amount is significantly affected by small variations in σ_v due to the factor $(2 - \sigma_v)/\sigma_v$. For example, $\sigma_v = 0.9$ (compared to unity) results in a 22% increase in the slip amount. The DSMC simulations are performed for a nominal value of $\sigma_v = 1.0$. To investigate the effective value of σ_v in the simulation we record the values of the tangential momentum of impinging (τ_{i}) and reflected (τ_{r}) gas molecules at every time step. Based on these values and the tangential momentum of gas corresponding to that of the wall τ_{w}, we compute $\sigma_v = (\tau_{\text{i}} - \tau_{\text{r}})/(\tau_{\text{i}} - \tau_{\text{w}})$. The calculated average value is $\bar{\sigma} = 0.99912$ (with standard deviation of $\sigma_{\text{rms}} = 0.01603$). Thus, we can use these DSMC results to examine the validity of the slip boundary conditions given in equation (2.28) with $\sigma_v = 1$. To this end, we examined several forms where slip information $C\lambda$ away from the surface is utilized (see equation (2.26)). We have determined that the best agreement with the DSMC results is achieved by obtaining the

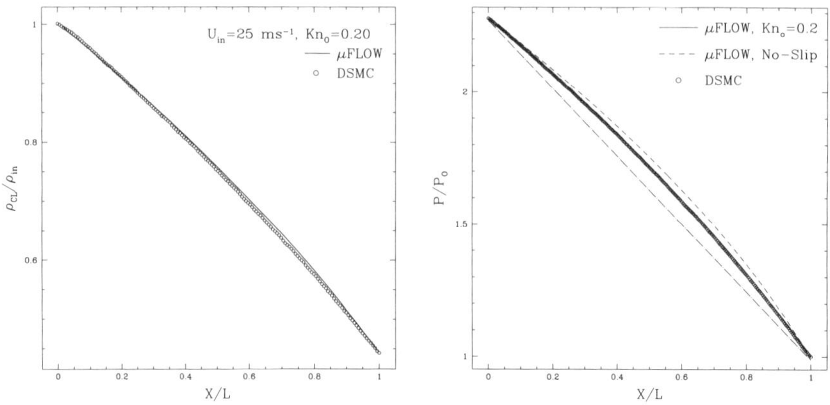

FIGURE 4.8. Comparisons of Navier–Stokes and DSMC predictions. Left: Density variation along the channel. Simulation conditions are for $\Pi = 2.28$ and $M_o = 0.19$. Right: Pressure variation along the channel for pressure ratio $\Pi = 2.28$. The solid line corresponding to μ**Flow** $\text{Kn}_o = 0.2$ prediction is indistinguishable from the DSMC results.

slip information λ away from the surface ($C = 1$).

The density distribution (normalized with the inlet density) along the channel for the case $\Pi = 2.28$ predicted by DSMC and μ**Flow** is shown in Figure 4.8 (left). Good agreement between the microscopic and macroscopic simulations is achieved. The corresponding centerline pressure distribution along the channel is plotted in Figure 4.8 (right). The pressure distribution is nonlinear, as expected for a compressible channel flow. As has been shown earlier, rarefaction and compressibility are competing effects in determining the curvature of the pressure distribution. While compressibility makes the curvature more pronounced, rarefaction makes the variation more linear. This is also verified in Figure 4.8 (right) by comparison of slip-based Navier–Stokes results with the no-slip results.

Velocity profiles for $\Pi = 2.28$, normalized with the reference inlet velocity at three different x/L locations, are shown in Figure 4.9. The DSMC and μ**Flow** results are in good agreement. The velocity profiles obtained by both methods are parabolic. The velocity slip variation along the channel wall is shown in Figure 4.10. Both the DSMC method and μ**Flow** predict similar magnitudes.

To examine the accuracy of the results obtained with DSMC, we compare them with solutions of the *linearized Boltzmann equation* in the slip flow regime obtained in (Ohwada et al., 1989a), for $\text{Kn} = 0.1$. In Figure 4.11 the velocity distribution obtained by both methods is plotted normalized with the local average velocity. The differences of DSMC and linearized Boltzmann predictions are almost indistinguishable. We can then use these profiles to examine the accuracy of the analytic solutions obtained using

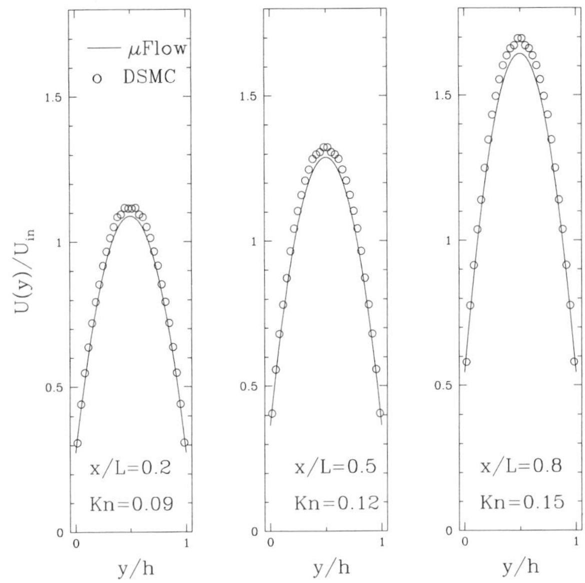

FIGURE 4.9. Velocity profiles predicted by the Navier–Stokes and DSMC simulations at various x/L locations. The inlet is located at $x/L = 0.0$.

various slip boundary conditions. We write all proposed models in the following general form:

$$U_s - U_w = C_1 \,\mathrm{Kn} \left(\frac{\partial U}{\partial n}\right)_s - C_2 \,\mathrm{Kn}^2 \left(\frac{\partial^2 U}{\partial n^2}\right)_s, \tag{4.13}$$

where the first-order (C_1) and second-order (C_2) slip coefficients are given in Table 2.2.

The velocity distribution for an isothermal flow in a long channel ($h/L \ll 1$) of thickness h is predicted by the second-order slip boundary conditions in the following form:

$$U(x,y) = -\frac{dP}{dx}\frac{h^2}{2\mu}\left[-\left(\frac{y}{h}\right)^2 + \left(\frac{y}{h}\right) + C_1 \,\mathrm{Kn} + 2C_2 \,\mathrm{Kn}^2\right],$$

where Kn is the *local* Knudsen number, and it varies in the streamwise direction. Hence, the velocity profile is a function of both x and y. The corresponding volumetric flowrate also increases along the channel as the density is decreased from the inlet to the exit. The local volumetric flowrate is

$$\dot{Q}(x) = -\frac{dP}{dx}\frac{h^3}{2\mu}\left(\frac{1}{6} + C_1 \,\mathrm{Kn} + 2C_2 \,\mathrm{Kn}^2\right). \tag{4.14}$$

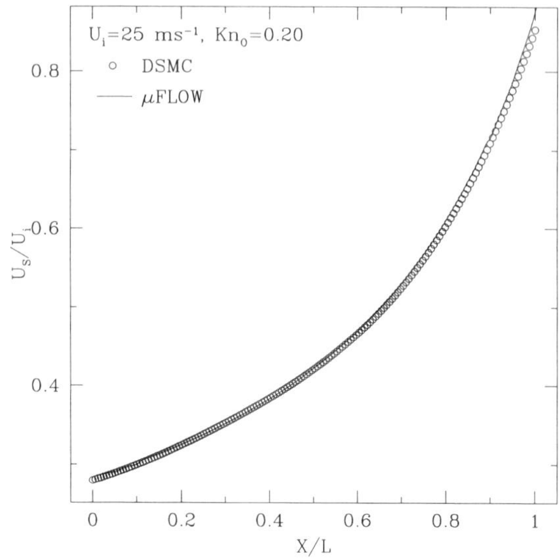

FIGURE 4.10. Slip velocity along the channel wall as predicted by the Navier–Stokes and DSMC simulations.

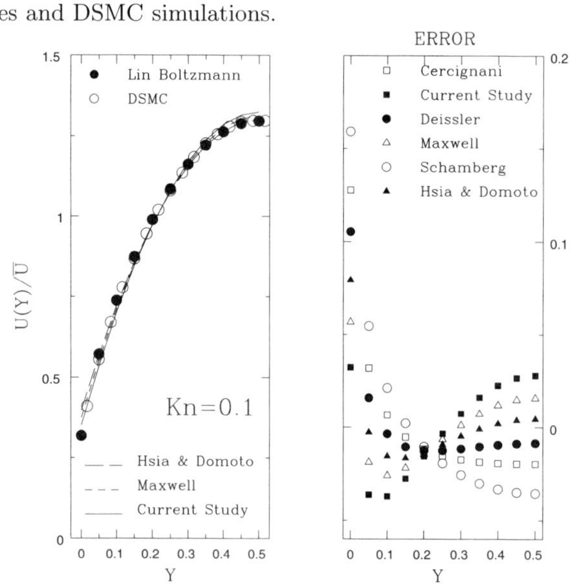

FIGURE 4.11. The velocity distribution normalized with the local average velocity in half of a microchannel (left). The linearized Boltzmann solution is from (Ohwada et al., 1989a). Error in the solution of the Navier–Stokes equations with various slip flow models at Kn = 0.1 (right). For the legend see Table 2.2.

Using this, we can find a local average velocity, which is used to normalize the local velocity distribution. This is the best choice for normalization of the velocity profile, making the magnitude of the nondimensional velocity distribution $\mathcal{O}(1)$ for any Kn value. This normalized velocity distribution is given as

$$U^*(y, \mathrm{Kn}) \equiv \frac{U(x,y)}{\bar{U}(x)} = \frac{\left(-(\frac{y}{h})^2 + \frac{y}{h} + C_1 \mathrm{Kn} + 2C_2 \mathrm{Kn}^2\right)}{\frac{1}{6} + C_1 \mathrm{Kn} + 2C_2 \mathrm{Kn}^2}, \tag{4.15}$$

where $\bar{U}$ is the average (across the channel) velocity. The velocity profiles obtained by the various slip boundary conditions and the corresponding deviations from the linearized Boltzmann solution of (Ohwada et al., 1989a), are shown in Figure 4.11. The error is positive if the model (4.15) overpredicts the velocity at a given point and is negative if the velocity is underpredicted. More specifically, for the slip velocity prediction, the model given by equation (2.29) gives the best agreement with the linearized Boltzmann solution, followed by the first-order model, while Schamberg's boundary condition performs the worst (see Table 2.2). Hsia and Domoto's coefficients give the most accurate description of the centerline velocity, while Schamberg's model once again performs the worst. The maximum errors in Schamberg's model for centerline and velocity slip predictions are about 0.035 and 0.16 units, respectively. The error units are defined as

$$E = U_{\mathrm{Model}} - U_{\mathrm{LB}},$$

and they are not normalized with U_{Model} since it changes as a function of y. The model of equation (2.29) has the smallest overall deviations of at most 0.035 units from the linearized Boltzmann solution U_{LB}.

The second-order slip coefficients cited in Table 2.2 are all positive, with the exception of our model, which corresponds to a Taylor series expansion of the velocity profile including the second-order terms. Therefore, our model uses both the slope of the local velocity profile and the change in the slope in order to predict (extrapolate) the slip velocity near the wall. The volumetric flowrate (equation (4.14)) calculated by other models gives enhancement of the volumetric flowrate compared to the first-order predictions. This is consistent with the experimental data and it is due to the positive contribution of the second-order slip coefficient.

- *In the model of equation (2.29) the second-order slip contribution leads to a reduction in the volumetric flowrate compared to the first-order model.*

- *Other models with second-order slip conditions can predict flowrate accurately, but only at the expense of accuracy in the velocity profile.*

This will be demonstrated further in Section 4.2, where we examine in detail flowrate modeling issues for a wide range of Kn.

Remark: The second-order slip boundary condition given by equation (2.29) is based on equation (2.26), where terms higher than $(\partial^2 U/\partial n^2)$ are neglected. Solutions of Navier–Stokes equations for long channels result in parabolic velocity profiles, with vanishing contribution for derivatives higher than two (see Section 4.2.1). Therefore, the boundary conditions given by equations (2.29) and (2.26) are identical for this problem. Implementation of our second-order model in the slip flow regime was performed by obtaining the necessary slip information at a distance λ away from the surface. For flows with higher Knudsen number in the transition regime we switch our model to the general boundary condition (2.43), the validity of which we investigate in detail in Section 4.2.

4.1.4 Effects of Roughness

Surface roughness in the slip flow regime has been studied in (Li et al., 2002), where the rough surface was represented as a porous film based on the Brinkman-extended Darcy model, and the core region of the flow utilized velocity slip to model the rarefcation effects. Then the velocity solution and the shear stress from the core and the porous regions of the flow were matched at the pore-region/flow interface, enabling derivation of expressions for the pressure distribution in microtubes, including rarefaction effects.

In the following we describe a different approach to modeling surface roughness. The pressure drop in channels with rough walls depends critically on the shape and size of the roughness. For random roughness, the pressure drop should depend on the statistical characteristics of the walls, which are expressed by the correlation function of surface inhomogeneities. A particularly effective method has been developed by Meyerovich and his group on how to extract this dependence at the limit of very large Knudsen numbers (Meyerovich and Stepaniants, 1994; Meyerovich and Stepaniants, 1997). In this description, the transport coefficients, such as mobility, diffusion, effective relaxation time, and mean free path along the walls, are expressed directly via the parameters of the correlation function of surface roughness. The main idea is to perform a nonlinear coordinate transformation, assuming that the boundary can be described by a single-valued function. The transformed equation, e.g., Stokes flow equation, acquires extra random terms, which depend on the boundary roughness. This is demonstrated in Figure 4.12 for a microchannel with rough walls. All variables are transformed in the new coordinate system, including the transport coefficients (here the kinematic viscosity), which are renormalized. In this particular example, L is the channel height, and the inhomogeneities are small and random and are described by $\xi_1(x,z)$ and $\xi_2(x,z)$ for the lower and upper wall, respectively. Also, $\xi_1, \xi_2 \ll L$ have zero mean, i.e., $\langle\xi_1\rangle = \langle\xi_2\rangle = 0$. The transformation essentially shifts the rough boundary to the bulk and makes it flat. It is given by

$$Y = \frac{L[y - 1/2[\xi_2(x, z) - \xi_1(x, z)]]}{L - [\xi_1(x, z) + \xi_2(x, z)]}.$$

The work of Meyerovich has addressed mostly high Knudsen numbers, since the theory for small Knudsen numbers has not been developed yet. This is the most difficult regime, since it requires *simultaneous account for bulk and boundary scattering*, including the interference between bulk and boundary scattering.

Computationally, it is possible to study the additional pressure drop due to inhomogeneities for prototype roughness in microchannels and interwall correlations. For example, it is possible to generate two identical rough walls and shift one with respect to the other. When this shift is zero, the inhomogeneities from different walls are perfectly correlated with each other. With increasing shift, this correlation will gradually decrease and will disappear when the shift becomes much larger than the correlation length. The dependence of the flow parameters on this shift will mimic the dependence on the interwall correlation of inhomogeneities.

We have used the μ**Flow** program as well as the DSMC approach to predict the additional pressure drop due to regularized roughness in long microchannels. We have considered geometries with *in-phase* and *out-of-phase* types of roughness, as shown in Figure 14.1. Typical results are summarized in Figure 4.13. We compare the total pressure as well as the deviation from a linear drop for different roughness types and with a corresponding smooth channel. For all these cases the mass flowrate was maintained constant and the Reynolds number was $\text{Re} = 0.44$, while the Knudsen number at the outlet of the channel was $\text{Kn} = 0.17$. We see that in order to match the pressure drop of the rough channels, the viscosity in the smooth channel should increase by about 18.7%; this case corresponds to $\text{Re} = 0.37$ and $\text{Kn} = 0.2$ at the exit of the channel. Continuum flows do not depend on the details of the roughness, but for rarefied flows, as the Knudsen number increases, the roughness effect becomes more significant. In other words, microflows have "memory" in that they remember their trajectory and specific path, which is geometry-dependent. With respect to the latter, the results we obtained are inconclusive, and new computations are required to support this concept.

4.1.5 Inlet Flows

We consider next the flow between stationary parallel plates at the inlet of a microchannel as a prototype *developing* flow. In this analysis the compressibility effects are neglected, and the effect of rarefaction on developing flows is presented. A uniform velocity profile is specified at the inlet. Gampert analyzed the inlet flows using a vorticity-streamfunction formulation and presented results for different values of the Reynolds and Knudsen

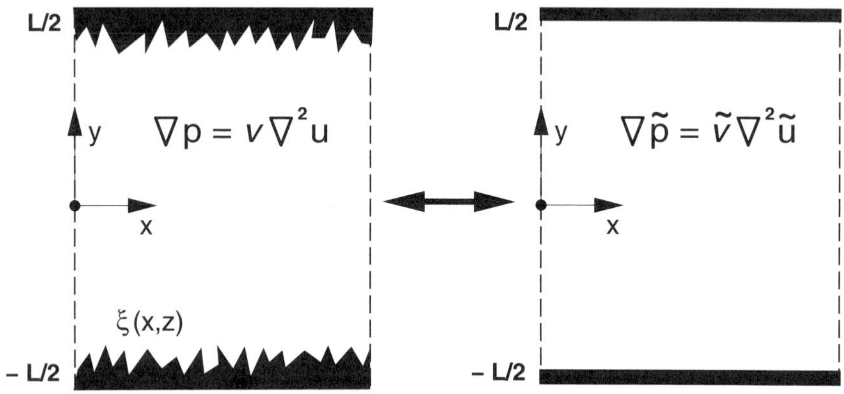

FIGURE 4.12. Sketch illustrating the canonical transformation. As an example, the (linear) Stokes equation in the continuum limit is transformed.

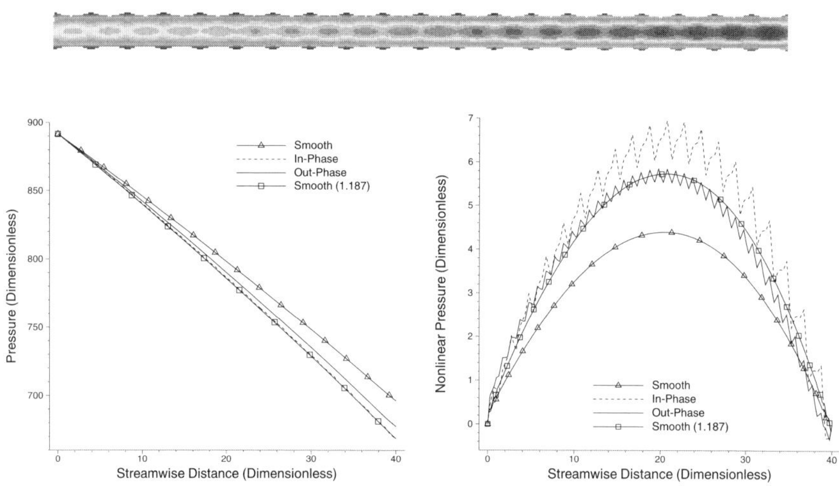

FIGURE 4.13. Upper: Channel with "regularized" wall roughness. Lower: Total pressure distribution (left) and deviations from linear pressure drop (right) for rough and smooth channels. The flowrate is constant for all cases. The case described as "smooth 1.187" corresponds to smooth channel flow with 18.7% enhanced viscosity compared to all other cases.

numbers as well as accommodation coefficients (Gampert, 1976). Here, we present simulation results at $\mathrm{Re} = 20$. In Figure 4.14 (left) we plot the velocity slip along the wall for $\mathrm{Kn} = 0.03$ obtained by using both the first- and high-order slip flow boundary conditions. It is seen that the slip effects are very dominant at the inlet of the channel, and both results predict velocity slip of about 50% of the incoming velocity U_∞. However, these effects are reduced well below 10% in the developed region of the flow. It is evident from the plot that the slip velocity experiences a very steep change from

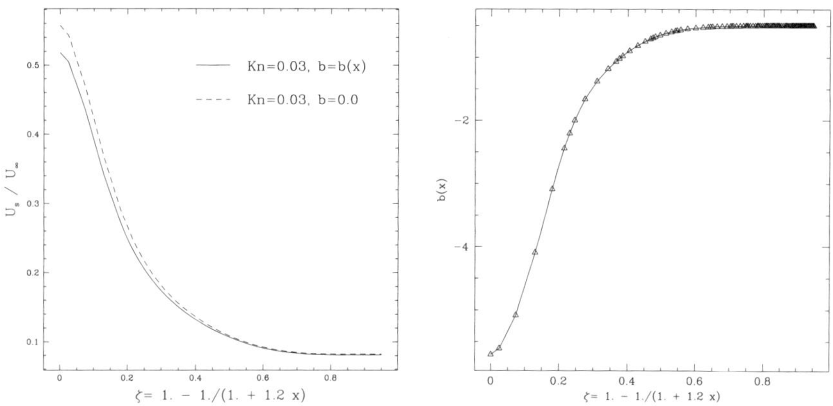

FIGURE 4.14. Developing flow in microchannel: Velocity slip (left) and variation of high-order slip coefficient b as a function of the streamwise location parameter ζ (Kn = 0.03, Re = 20).

the inlet to approximately a distance in the streamwise direction equal to the channel height and achieves an asymptotic value corresponding to the fully developed profile approximately four channel heights downstream (for $Re = 20$ flow). The high-order slip effects reduce the velocity slip compared to the first-order predictions. The value of the high-order slip coefficient (b) is obtained using equation (2.39).

Figure 4.14 (right) shows the variation of high-order slip parameter b as a function of the parameter

$$\zeta = 1 - \frac{1}{1 + 1.2\bar{x}}$$

(where $\bar{x} = x/h$, and h is the half channel width) determining the streamwise location. In Figure 4.15 we plot several streamwise velocity profiles close to inlet for Kn = 0.03 (solid line) and no-slip (dashed line) flows. For various values of ζ close to the inlet ($\zeta \leq 0.75$) the maximum velocity is *off centerline* until a fully developed profile is established. We also see from the plot that the value of the maximum velocity is always smaller in the case of the slip flow. This behavior of maximum velocities off the centerline cannot be obtained if the convective terms are linearized as in (Sparrow et al., 1962). These results agree with the computations reported in (Gampert, 1976); however, in that work the maximum values were underpredicted by more than a factor of two compared to the present results. The reason for this discrepancy may be due to imposition of the incorrect boundary condition (equation (8) in (Gampert, 1976)) close to the inlet. We also verified the grid independence of our results by performing higher-resolution simulations. The flow at the channel entrance is fully two-dimensional. The cross-flow velocity variation as a function of ζ is shown in Figure 4.15. The

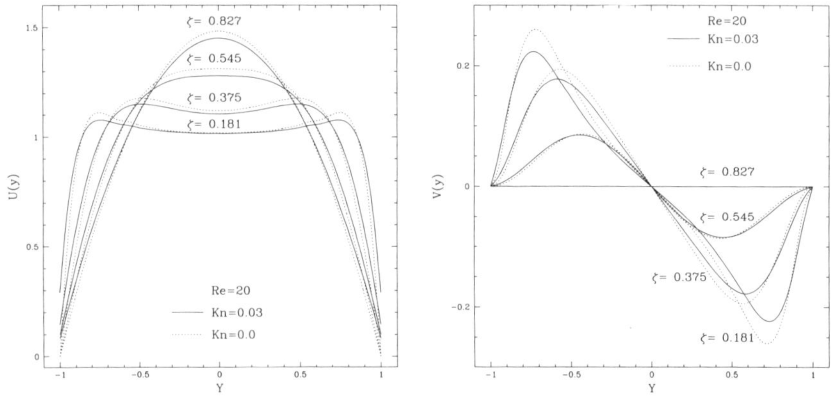

FIGURE 4.15. Streamwise (left) and cross-flow (right) velocity profiles at various streamwise locations specified by ζ. Kn $= 0.03$ (solid lines), and no-slip (dashed lines) cases.

cross-flow component of the velocity decreased for the slip flows compared to the continuum flows. Finally, the pressure distribution along the centerline of the channel is plotted in Figure 4.16. The centerline pressure drop is parabolic very close to the channel inlet and tends to a linear form at about $\frac{x}{h} = 2.5$ ($\zeta = 0.75$) for both slip and no-slip flows. High-order slip effects in pressure distribution are negligible at the centerline of the channel. A comparison of the slopes of the no-slip and slip flow pressure distributions show that a reduction in the pressure drop is obtained in these simulations, where the mass flowrate at the channel inlet is specified.

4.2 Transition and Free-Molecular Regimes

In this section we concentrate on transition and free-molecular flow regimes with the objective of developing *unified* flow models for pipe and duct flows. These models are based on a velocity scaling law, which is valid for a wide range of Knudsen number. Simple no-slip and slip-based descriptions of flowrate in channels and pipes are corrected for effects in transition and free-molecular flow regimes with the introduction of a *rarefaction coefficient*. We follow here the analysis first presented in (Beskok and Karniadakis, 1999).

A number of investigators have considered semianalytical and numerical solutions of the linearized Boltzmann equations for rarefied flow between two parallel plates or a pipe in both transition and free-molecular regimes. Starting with the pioneering work of Cercignani and his associates, the Knudsen's minimum has been rigorously explored (Cercignani and Daneri, 1963). In such studies, simplifications for the collision integral based on the

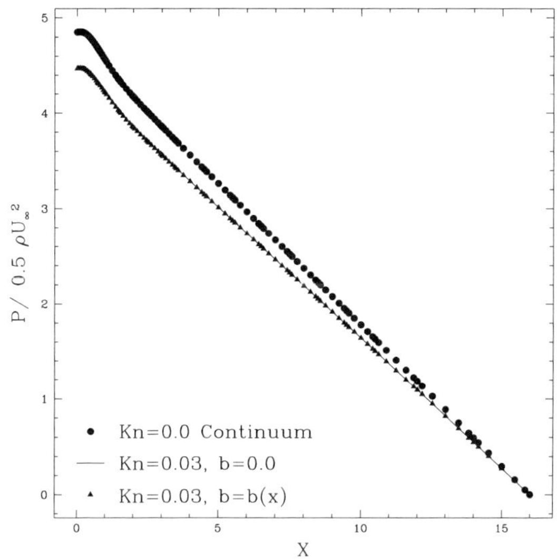

FIGURE 4.16. Pressure distribution along the centerline for slip flow, using first-order ($b = 0$), and high-order ($b = b(x)$) slip boundary conditions, and no-slip flow as a function of $X = \frac{x}{h}$.

BGK model (Bhatnagar et al., 1954) (see also Section 15.4) were extensively used. Other investigators have derived solutions based on the hard-sphere and Maxwellian models for the collision integral (Huang et al., 1997; Sone, 1989; Ohwada et al., 1989a), and have also obtained solutions in cylindrical geometry (Loyalka and Hamoodi, 1990) and ducts with various cross sections (Aoki, 1989). Similar approaches have also been used successfully in modeling gas film lubrication in the transition regime (see Section 6.1 and (Fukui and Kaneko, 1988; Fukui and Kaneko, 1990)).

Before we start developing velocity and flowrate scaling models, we examine the validity of continuum-based slip models in the *transition flow regime*. In Figure 4.17 we present the velocity profiles obtained by the DSMC and linearized Boltzmann solutions at Kn = 0.6. The model based on equation (2.29) is indicated as "Model A," and the one based on equation (2.43) is indicated as "Model B." We have also included models by Cercignani, Deissler, Maxwell, Schamberg, and Hsia and Domoto (see Table 2.2). Model B does very well in the center of the channel but gives slight deviations near the walls. The maximum deviation of the model is observed at the slip location, where Model A is somewhat better. However, Model A gives larger errors than Model B toward the centerline of the channel. The errors in Maxwell's first-order boundary condition and the other second-order models are larger than the errors of either model A or B. The maximum error occurs near the wall with 0.32 units of overestimation us-

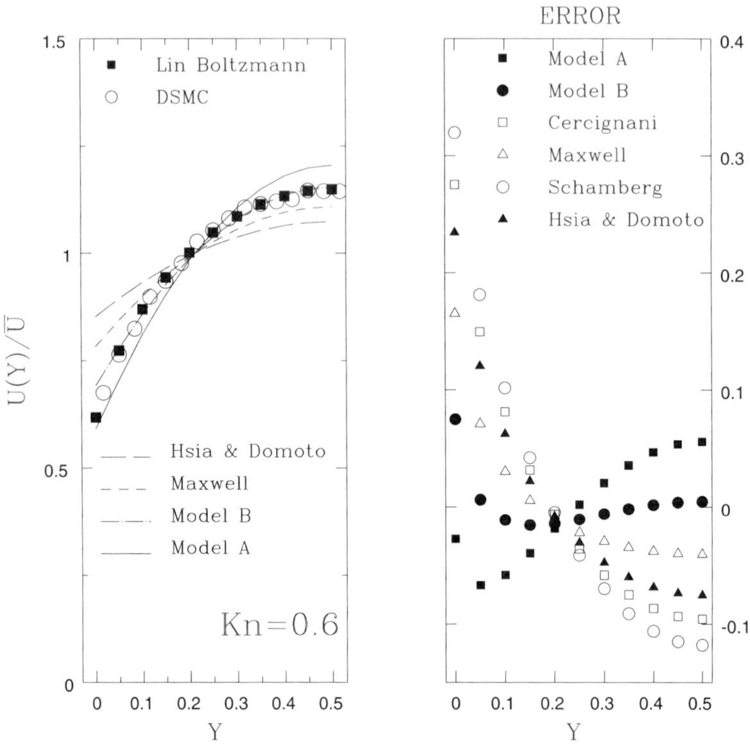

FIGURE 4.17. Nondimensionalized velocity distribution in half of a microchannel (left). The linearized Boltzmann solution is from (Ohwada et al., 1989a). Error in the solution of the Navier–Stokes equations with various slip flow models at Kn = 0.6 (Right).

ing Schamberg's boundary conditions. The models by Cercignani and by Deissler are almost identical for this case, and therefore only one is shown in the figure. The reason for all the models crossing at $Y = (y/h) = 0.2$ in Figure 4.17 is an artifact of nondimensionalization of the velocity profiles with the average velocity $\bar{U}$. Due to equation (4.15), we have $U/\bar{U} = 1$ at $y/h = 0.2$ for every slip model.

We now turn our attention to the **flowrate** in microducts. It is known from Knudsen's and Gaede's experiments in the transition flow regime that there is a *minimum in the flowrate* in pipe and channel flows at about $\mathrm{Kn} \approx 3$ and $\mathrm{Kn} \approx 1$, respectively. This behavior has been investigated by many researchers both theoretically (Cercignani and Daneri, 1963; Cercignani, 1963; Loyalka and Hamoodi, 1990; Ohwada et al., 1989a; Polard and Present, 1948; Kogan, 1969) and experimentally (Tison, 1993). It was first shown by Knudsen that in the free-molecular flow regime in pipes a diffusive transport process proportional to the pressure gradient but independent of

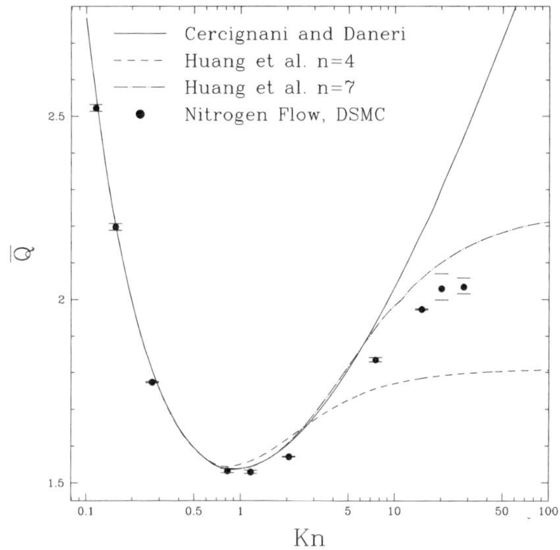

FIGURE 4.18. Variation of normalized flowrate in a channel as a function of Knudsen number. Comparisons are made between DSMC results and various solutions of the linearized Boltzmann equation.

density is observed. Accordingly, the free-molecular mass flowrate in pipes of diameter $2a$ and length L is (Kennard, 1938)

$$\dot{M}_{\mathrm{FM}} = \frac{4}{3}a^3\frac{\Delta P}{L}\sqrt{\frac{2\pi}{RT}}. \tag{4.16}$$

Deviation from this behavior is expected for finite–length pipes (i.e., $a \ll L \ll \lambda$) by a factor $(1 - Ca/L)$ (up to first order in a/L) due to end effects (Polard and Present, 1948), where C is a constant. However, for free-molecular flow in two-dimensional very long channels where $h \ll \lambda \ll L$, the flowrate increases asymptotically to a value proportional to

$$(1/\pi)^{\frac{1}{2}} \log_e(\mathrm{Kn})$$

in the limit $\mathrm{Kn} \to \infty$ (Cercignani, 1963; Huang and Stoy, 1966). This logarithmic behavior is attributed to the degenerate geometry of the two-dimensional channel (Kogan, 1969). For finite-length two-dimensional channels, the flowrate tends toward a finite limit (see the discussion of D.R. Willis at the end of (Cercignani, 1963)). For ducts, the flowrate tends toward a finite limit, resembling the pipe flow behavior. This has been documented in the experiments of (Gaede, 1913) and verified by linearized Boltzmann solutions (Sone and Hasegawa, 1987).

The variation of flowrate in a channel as obtained by DSMC simulations in the transition and early free-molecular flow regimes is shown in Figure

4.18. The volumetric flowrate data are presented at the average Knudsen number in the channel (corresponding to the mean pressure $\overline{P}$ between the inlet and outlet), and it is nondimensionalized in the form

$$\bar{Q} = \frac{\dot{Q}\overline{P}}{\frac{-dP}{dx}h^2(RT_0)^{\frac{1}{2}}},$$

where $\dot{Q}$ is the volumetric flowrate per unit width of the channel, h is the channel height, T_0 is reference temperature, and R is the specific gas constant. The DSMC data is incorporated in the figure by plotting the mass flowrate data as a function of average Kn in the channel, since

$$\dot{M} \sim \dot{Q}\bar{P}.$$

The error bars in the plot correspond to maximum fluctuations in the global mass balance and statistical scatter in pressure gradient (dP/dx), which is used here as an accuracy criterion. Knudsen's minimum is clearly captured by the DSMC results at Kn ≈ 1.0. The DSMC solution is compared with the semianalytic solutions of (Cercignani and Daneri, 1963), where the linearized Boltzmann equations are solved with the BGK model. Also, a comparison with numerical solutions of (Huang et al., 1997) is shown. These solutions were obtained from the linearized Boltzmann equations with the BGK model using the discrete ordinate method; see also (Siewert, 2000). The integrals involved were approximated by various orders (n) of Gauss quadrature. It is seen that Cercignani and Daneri's results are recovered as (n) is increased. The current DSMC results match the Boltzmann solution quite well up to Kn $= 2$. Beyond this value, the DSMC results follow the seventh-order quadrature solution of (Huang et al., 1997), and subsequently become asymptotic to a constant value in the free-molecular flow regime, rather than increasing logarithmically. The reason for deviations of the DSMC data from the theoretical solution for infinitely long channels is the finite length of the channel $(L/h = 20)$ used in these simulations.

4.2.1 Burnett Equations

The governing conservation equations were given in Chapter 2, e.g., equation (2.16). In the transition regime the Burnett equations govern the thermal fluidic transport. The stress tensor for the Burnett equations is given in Section 2.3. To simplify this stress tensor we consider a very long channel of length L and height h, so that

$$\epsilon \equiv h/L \ll 1.$$

We also neglect any temperature gradients in the gas. Under such conditions the Burnett equations can be simplified considerably, resulting in terms including $\mathcal{O}(1)$ up to $\mathcal{O}(\epsilon^5)$. Here we present the results of this expansion including the $\mathcal{O}(\epsilon)$ terms. In the following analysis, Kn_0 and M_0 refer to Knudsen and Mach numbers evaluated at the outlet.

The streamwise component of the Burnett momentum equation in a long channel including $\mathcal{O}(\epsilon)$ terms is

$$
\begin{aligned}
&P_x\left[1-\left(\frac{\omega_2}{3}+\frac{\omega_6}{12}\right)\frac{\gamma\pi}{2}\,\mathrm{Kn}_0^2\,M_0^2\left(\frac{{p_0}^2}{p}\right)(U_y)^2\right]=U_{yy}\\
&+\epsilon\left[\sqrt{\frac{\gamma\pi}{2}}M_0\,\mathrm{Kn}_0\left(\frac{p_0}{p}\right)\left(\left(\frac{\omega_1}{2}-\frac{7\omega_2}{3}+\frac{\omega_6}{3}\right)U_yU_{yx}\right.\right.\\
&\quad\left.\left.+\left(\frac{\omega_1}{2}-\frac{7\omega_2}{2}+\frac{\omega_6}{6}\right)U_xU_{yy}\right)\right]\\
&+\epsilon\left[\sqrt{\frac{\gamma\pi}{2}}M_0\,\mathrm{Kn}_0\left(\frac{p_0}{p}\right)\left(\left(\frac{\omega_1}{2}-\frac{\omega_2}{6}+\frac{\omega_6}{6}\right)U_yV_{yy}+\right.\right.\\
&\quad\left.\left.\left(\frac{\omega_1}{2}+\frac{\omega_2}{2}+\frac{\omega_6}{6}\right)V_yU_{yy}\right)\right]\\
&+\epsilon\left[\sqrt{\frac{\gamma\pi}{2}}M_0\,\mathrm{Kn}_0\left(\frac{p_0}{p}\right)\left(\frac{\omega_2}{2}UU_{yxy}+\frac{\omega_2}{2}VU_{yyy}\right)-\mathrm{Re}(\rho UU)_x-\right.\\
&\quad\left.\mathrm{Re}(\rho UV)_y\right]\\
&-\epsilon\left[\frac{\gamma\pi}{2}\,\mathrm{Kn}_0^2\,M_0^2\left(\frac{p_o}{p}\right)^2\left(\left(\frac{\omega_1}{2}-\frac{7\omega_2}{2}+\frac{\omega_6}{6}\right)P_yU_xU_y+\frac{\omega_2}{2}P_yUU_{yx}+\right.\right.\\
&\quad\left.\left.\frac{\omega_2}{2}P_yVU_{yy}\right)\right].
\end{aligned}
$$

Similarly, the cross-flow Burnett momentum equation including $\mathcal{O}(\epsilon)$ terms is

$$
\begin{aligned}
&P_y\left[1+\left(\frac{\omega_6}{12}-\frac{2\omega_2}{3}\right)\frac{\gamma\pi}{2}\,\mathrm{Kn}_0^2\,\mathrm{M}_0^2\left(\frac{p_0}{p}\right)^2(U_y)^2\right]\\
&\quad=\left(\frac{\omega_6}{6}-\frac{4\omega_2}{3}\right)\sqrt{\frac{\gamma\pi}{2}}\mathrm{M}_0\,\mathrm{Kn}_0\left(\frac{p_0}{p}\right)U_yU_{yy}\\
&\quad+\epsilon\left[-\sqrt{\frac{\gamma\pi}{2}}M_0Kn_0\left(\frac{p_0}{p}\right)\left(\frac{\omega_2}{2}\right)(UU_{yx})_y+U_{xy}+V_{yy}\right]\\
&\quad-\epsilon\left[\frac{\gamma\pi}{2}Kn_0^2M_0^2\left(\frac{p_0}{p}\right)^2\left(\left(\frac{\omega_2}{2}\right)(P_xVU_{yy}+P_xUU_{xy})\right.\right.\\
&\qquad\left.\left.+\left(\frac{\omega_1}{2}-\frac{\omega_2}{6}+\frac{\omega_6}{6}\right)P_xV_yU_y\right)\right]\\
&\quad+\epsilon\left[\frac{\gamma\pi}{2}Kn_0^2M_0^2\left(\frac{p_0}{p}\right)^2\left(\frac{\omega_2}{3}\right)P_yUU_{xy}\right].
\end{aligned}
$$

Neglecting the $\mathcal{O}(\epsilon)$ terms, the x- and y-momentum equations are reduced to (in dimensional form)

$$\frac{\partial p}{\partial x}\left[1-(\omega_2/3+\omega_6/12)\frac{\mu^2}{p^2}\left(\frac{\partial u}{\partial y}\right)^2\right]=\mu\frac{\partial^2 u}{\partial y^2}+\mathcal{O}(\epsilon),$$

$$\frac{\partial p}{\partial y}+(\omega_6/12-2\omega_2/3)\frac{\partial}{\partial y}\left[\frac{\mu^2}{p}\left(\frac{\partial u}{\partial y}\right)^2\right]=\mathcal{O}(\epsilon).$$

Furthermore, assuming a Maxwellian gas model for which the coefficients $(\omega_1, \omega_2, \omega_6) = (10/3, 2, 8)$ (from (Schamberg, 1947)) and nondimensionalizing with the reference exit conditions (p_0, u_0), we obtain

$$P_x\left[1-2/3\gamma\pi\,\mathrm{Kn}_0^2\,M_0^2\left(\frac{p_0}{p}\right)^2U_y^2\right]=U_{yy}+\mathcal{O}(\epsilon).$$

Similarly, for the y-momentum equation we obtain

$$P_y\left[1+\frac{\gamma\pi}{3}\mathrm{Kn}_0^2\,M_0^2\left(\frac{p_0}{p}\right)^2U_y^2\right]=\frac{4}{3}\sqrt{\gamma\pi/2}M_0\,\mathrm{Kn}_0\left(\frac{p_0}{p}\right)U_yU_{yy}+\mathcal{O}(\epsilon),$$

where the nondimensionalized variables are denoted by capital letters.

It is clear that the $M_0^2\,\mathrm{Kn}_0^2(p/p_0)^2$ term is relatively small for low Mach number flows in the early transition regime (i.e., $\mathrm{Kn} < 1$). In this case, for flow in a very long channel the Burnett equations reduce to

$$P_x = U_{yy}, \tag{4.17a}$$

$$P_y = \frac{4}{3}\sqrt{\frac{\gamma\pi}{2}}\mathrm{Kn}_0\,M_0\left(\frac{p_0}{p}\right)U_yU_{yy}. \tag{4.17b}$$

Therefore, the streamwise Burnett equation is reduced to the form obtained in the Navier–Stokes limit. On the other hand, the cross-flow momentum equation (4.17b) shows that the pressure gradient in that direction is balanced by the Burnett normal stresses, which in the case of continuum are identically zero for a flat surface. The cross-flow momentum equation agrees with the simplified set of equations for Couette flow in the transition regime obtained in (Schamberg, 1947). Results obtained from simulations employing the full Burnett equations do not show any significant deviations from the semianalytical results we have obtained for microchannel flows (Balakrishnan et al., 1999; Agarwal et al., 2001; Agarwal and Yun, 2002; Xu, 2003).

4.2.2 A Unified Flow Model

In this section we will develop a *unified* flow model that predicts velocity profiles, pressure drop, and mass flowrate in channels, pipes, and arbitrary aspect ratio rectangular ducts in the *entire Knudsen regime*. Our approach is divided into two main steps: First, we will analyze the nondimensional velocity profile to identify the shape of the velocity distribution. Then, we will obtain the magnitude of the average velocity, and hence obtain a prediction for the flowrate. The proposed unified model also predicts Knudsen's minimum.

Velocity Scaling

From the DSMC results and solutions of the linearized Boltzmann equation, it is evident that the velocity profiles remain approximately parabolic for a large range of Knudsen number (see Figure 4.19). This is also consistent with the analysis of the Navier–Stokes and Burnett equations in long channels, as documented in Section 4.2.1. Based on this observation, we model the velocity profile as *parabolic* in the entire Knudsen regime, with a consistent slip condition. We write the dimensional form for velocity distribution in a channel of height h,

$$U(x, y) = \mathcal{F}\left(\frac{dP}{dx}, \mu_0, h, \lambda\right)\left[-\left(\frac{y}{h}\right)^2 + \left(\frac{y}{h}\right) + U_s\right],$$

where $\mathcal{F}\left((dP/dx), \mu_0, h, \lambda\right)$ shows functional dependence of velocity on the pressure gradient, viscosity, channel height, and local mean free path. The temperature is assumed to be constant, and therefore the dynamic viscosity is also a constant. Here U_s is the velocity slip, and it satisfies equation (2.43), which after substitution in the above equation (assuming $U_w = 0$) yields

$$U(x, y) = \mathcal{F}\left(\frac{dP}{dx}, \mu_0, h, \lambda\right)\left[-(\frac{y}{h})^2 + (\frac{y}{h}) + \left(\frac{2 - \sigma_v}{\sigma_v}\right)\frac{\mathrm{Kn}}{1 - b\,\mathrm{Kn}}\right].$$

Assuming this form of velocity distribution, the average velocity in the channel ($\bar{U} = \dot{Q}/h$) can be obtained as

$$\bar{U}(x) = \mathcal{F}\left(\frac{dP}{dx}, \mu_0, h, \lambda\right)\left[\frac{1}{6} + \left(\frac{2 - \sigma_v}{\sigma_v}\right)\frac{Kn}{1 - bKn}\right].$$

By nondimensionalizing the velocity distribution with the local average velocity the dependence on local flow conditions ($\mathcal{F}\left((dP/dx), \mu_0, h, \lambda\right)$) is eliminated. Therefore, the resulting relation is a function of Kn and y only. Assuming $\sigma_v = 1$ (for simplicity), we obtain

$$U^*(y, \mathrm{Kn}) \equiv U(x, y)/\bar{U}(x) = \left[\frac{-\left(\frac{y}{h}\right)^2 + \frac{y}{h} + \frac{\mathrm{Kn}}{1 - b\,\mathrm{Kn}}}{\frac{1}{6} + \frac{\mathrm{Kn}}{1 - b\,\mathrm{Kn}}}\right]. \tag{4.18}$$

A similar analysis has been performed by (Piekos and Breuer, 1995). They used the first-order slip boundary conditions and subsequently separated

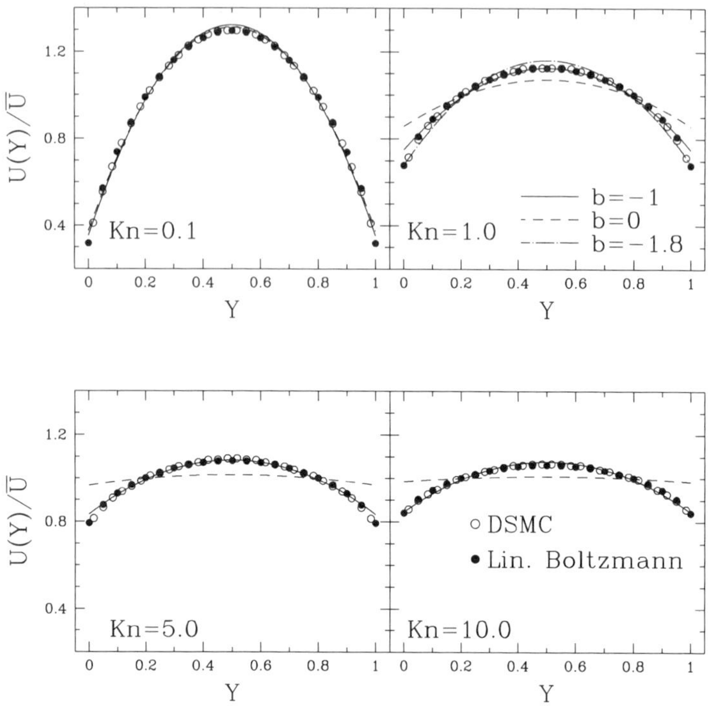

FIGURE 4.19. Velocity profile comparisons of the model (equation (4.18)) with DSMC and linearized Boltzmann solutions (Ohwada et al., 1989a). Maxwell's first-order boundary condition is shown with dashed lines ($b = 0$), and the general slip boundary condition ($b = -1$) is shown with solid lines.

this equation into an x-dependent contribution and a y-dependent contribution to investigate the breakdown of slip flow theory. However, here we will keep the form of equation (4.18) in the following analysis. Equation (4.18) solely describes the *shape* of the velocity distribution, but it does not properly model the flowrate. Flowrate modeling requires additional corrections, as shown in the subsection below.

In Figure 4.19 we plot the nondimensional velocity variation obtained in a series of DSMC simulations for Kn = 0.1, Kn = 1, Kn = 5, and Kn = 10. We also included the corresponding linearized Boltzmann solutions obtained in (Ohwada et al., 1989a). It is seen that the DSMC velocity distribution and the linearized Boltzmann solutions agree quite well. We can now use equation (4.18) and compare with the DSMC data by varying the parameter b, which for $b = 0$ corresponds to Maxwell's first-order and for $b = -1$ to the second-order boundary condition in the *slip regime* only. Here we find that for $b = -1$ equation (4.18) results in an accurate model

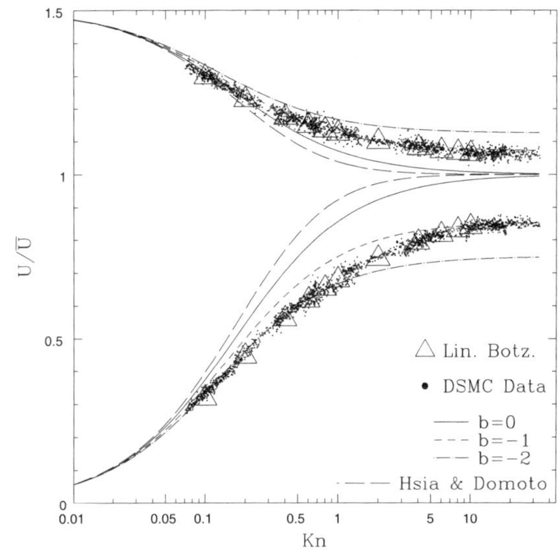

FIGURE 4.20. Velocity scaling at wall and centerline of the channels for slip and transition flows. The linearized Boltzmann solution of (Ohwada et al., 1989a) is shown by triangles, and the DSMC simulations are shown by points. Theoretical predictions of velocity scaling for different values of b, and Hsia and Domoto's second-order slip (large dashed line) boundary condition are also shown (Hsia and Domoto, 1983).

of the velocity distribution for a wide range of Knudsen number. From the figure, it is clear that the velocity slip is slightly overestimated with the proposed model for the Kn = 1 case. To obtain a better velocity slip, we varied the value of the parameter b by imposing, for example, $b = -1.8$ for the Kn = 1 case. Although a better agreement is achieved for the velocity slip, the accuracy of the model in the rest of the channel is destroyed.

In Figure 4.20 we show the nondimensionalized velocity distribution along the centerline and along the wall of the channels for the entire Knudsen number regime considered here, i.e., $0.01 \leq \text{Kn} \leq 30$. We included in the plot data for the velocity slip and centerline velocity from 20 different DSMC runs, 15 for nitrogen (diatomic molecules) and 5 for helium (monatomic molecules). The differences between the nitrogen and helium simulations are negligible, and thus this velocity scaling model is independent of the gas type. The linearized Boltzmann solution of (Ohwada et al., 1989a) for a monatomic gas is also shown in Figure 4.20 by triangles. This solution closely matches the DSMC predictions. Maxwell's first-order boundary condition ($b = 0$) (shown by a solid line) erroneously predicts a uniform nondimensional velocity profile for large Knudsen number. The breakdown of *slip flow theory* based on the first-order slip boundary condition is realized around Kn = 0.1 and Kn = 0.4 for the wall and the centerline velocity, respectively. This finding is consistent with the commonly accepted limits of the slip flow regime (Schaaf and Chambre, 1961). The prediction using $b = -1$ is shown by small dashed lines. The corresponding centerline velocity closely follows the DSMC results, while the slip velocity of the model with $b = -1$ deviates from DSMC in the intermediate range for $0.1 < \mathrm{Kn} < 5$. One possible reason for this is the effect of the Knudsen layer, a sublayer that is present between the viscous boundary layer and the wall, with a thickness of approximately one mean free path. For small Kn flows the Knudsen layer is thin and does not affect the velocity slip prediction too much. For very large Kn flows, the Knudsen layer covers the entire channel. However, for intermediate Kn values both the fully developed viscous flow (boundary layer) and the Knudsen layer exist in the channel. At this intermediate range, approximating the velocity profile to be parabolic neglects the Knudsen layers. For this reason, the model with $b = -1$ results in 10% error of the velocity slip at Kn = 1. However, the velocity distribution in the rest of the channel is described accurately for the entire flow regime.

For a comparison we also included similar predictions by the second-order slip boundary condition of (Hsia and Domoto, 1983) (large dashed line). The form of their boundary conditions is similar to Cercignani's, Deissler's, and Schamberg's, and they all become invalid at around Kn = 0.1. This boundary condition performs worse than even the first-order Maxwell's boundary condition for large Kn. Only the general slip boundary condition given by equation (2.43) predicts the scaling of the velocity profiles accurately.

Flowrate Scaling

In the previous section we analyzed the *shape* of the velocity profile. Since we have normalized the velocity profile with the *local mean velocity*, the above analysis cannot predict the volumetric flowrate. In this section we analyze the volumetric flowrate variation in the entire Knudsen regime.

The volumetric flowrate in a channel is a function of channel dimensions, fluid properties (μ_0, λ), and pressure drop, and it can be written as

$$\dot{Q} = \mathcal{G}\left(\frac{dP}{dx}, \mu_0, h, \lambda\right).$$

For a channel of height h, using the Navier–Stokes solution and the general slip boundary condition (2.43) we obtain

$$\dot{Q} = -\frac{h^3}{12\mu_0}\frac{dP}{dx}\left[1 + \frac{6\,\mathrm{Kn}}{1 - b\,\mathrm{Kn}}\right], \tag{4.19}$$

where $\mathrm{Kn} = \lambda/h$.

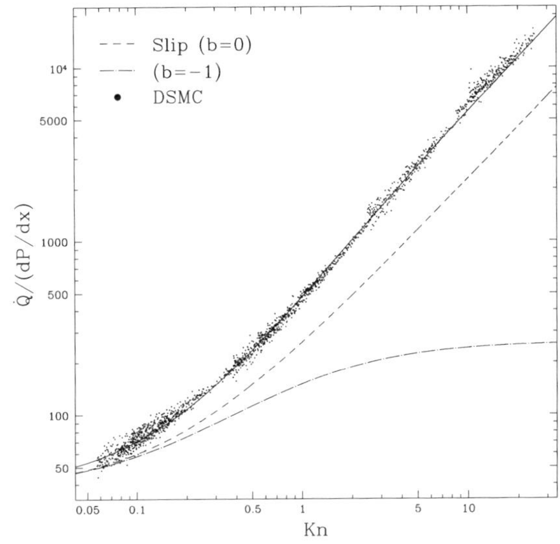

FIGURE 4.21. Volumetric flowrate (per channel width) per absolute value of the pressure gradient in $[m^3/(sPa)]$ as a function of Kn for nitrogen flow. The solid line represents the proposed model.

The flowrate for the continuum and free-molecular flows are both linearly dependent on dP/dx (Kennard, 1938), and thus we choose to normalize the flowrate with the pressure gradient. This quantity is computed based on the DSMC simulations and is shown in Figure 4.21 for nitrogen. For comparison we present the $\dot{Q}/|dP/dx|$ predictions obtained using Maxwell's first-order slip boundary condition ($b = 0$, dashed lines) and the general slip boundary condition ($b = -1$, dashed-dotted lines). In both cases the predictions are erroneous. The general slip boundary condition performs the worst for it is asymptotic to a constant value, while the DSMC data show a considerable increase with Kn. The first-order boundary condition follows the DSMC data, however with a significant error.

The slip model (2.43) gives good agreement with DSMC data and the linearized Boltzmann solutions for the *nondimensional velocity profile*, but it does not predict correctly the flowrate. This is expected, since the Navier–Stokes equations are invalid in this regime. In fact, the dynamic viscosity, which defines the diffusion of momentum due to the intermolecular collisions, must be modified to account for the increased rarefaction effects. The kinetic theory description for dynamic viscosity requires

$$\mu_0 \approx \lambda \bar{v} \rho,$$

where $\bar{v}$ is the mean thermal speed. Using mean free path λ in this relation is valid as long as intermolecular collisions are the dominant part of momentum transport in the fluid (i.e., Kn $\ll$ 1). However, for increased rarefaction, the intermolecular collisions are reduced significantly, and in the free-molecular flow regime, only the collisions of the molecules with the walls should be considered. Therefore, in free-molecular channel flow the diffusion coefficient should be based on characteristic length scale h (channel height) and thus $\mu \approx h\bar{v}\rho$ (Polard and Present, 1948). Since the diffusion coefficient is based on λ in slip or continuum flow regimes and h in the free-molecular flow regime, we propose to model the variation of diffusion coefficient with the following hybrid formula:

$$\mu \approx \rho\bar{v}\left[\frac{1}{\frac{1}{h}+\frac{1}{\lambda}}\right] = \rho\bar{v}\lambda\left[\frac{1}{1+\mathrm{Kn}}\right],$$

which can be simplified to

$$\mu(\mathrm{Kn}) = \mu_o\left[\frac{1}{1+\mathrm{Kn}}\right], \tag{4.20}$$

where μ_0 is the dynamic viscosity of the gas at a specified temperature and μ is the generalized diffusion coefficient. The **variable diffusion coefficient model** presented above is based on a simple analysis. Another point of view is to consider the ratio of intermolecular collisions of the molecules (f_g) to the total number of collisions per unit time (i.e., sum of intermolecular and wall collisions $f_g + f_w$). Following (Thompson and Owens, 1975), the frequency of wall collisions in a channel section is (with width w, height h, and length dx)

$$f_w = \frac{1}{4}n\bar{v}2(h+w)dx.$$

The intermolecular collision frequency in the flow volume is

$$f_g = \frac{\bar{v}}{\lambda}nhw\,dx,$$

where $\bar{v}$ is the mean thermal speed and n is the number density. Assuming that $w \gg h$, the ratio of intermolecular collisions to total collisions becomes

$$\frac{f_g}{f_g+f_w} = \frac{1}{1+\frac{1}{2}Kn}. \tag{4.21}$$

This analysis resulted in a correction to the continuum-based flowrate models, similar to the variable diffusion coefficient model presented earlier, with the only difference of $\frac{1}{2}$ in front of the Kn term.

In general, the increased rarefaction effects in our flowrate model can be taken into account by introducing a correction expressed as *rarefaction coefficient* $C_r(\mathrm{Kn})$, which is a function of Knudsen number. The flowrate is then obtained as

$$\dot{Q} = -\frac{h^3}{12\mu}\frac{dP}{dx}\left[1+\frac{6\,\mathrm{Kn}}{1-b\,\mathrm{Kn}}\right] = -\frac{h^3}{12\mu_0}\frac{dP}{dx}\left[1+\frac{6\,\mathrm{Kn}}{1-b\,\mathrm{Kn}}\right]C_r(Kn), \quad (4.22)$$

where $C_r(\mathrm{Kn})$ is a general function of Knudsen number. A possible model for C_r is suggested by the aforementioned analysis (equations (4.20) and (4.21)) in the form

$$C_r(\mathrm{Kn}) = 1 + \alpha\,\mathrm{Kn}, \quad (4.23)$$

where α is a parameter. If we assume that α is constant in the entire Knudsen regime, the flowrate in the slip flow regime will be erroneously enhanced, resulting in

$$\frac{\dot{M}}{\dot{M}_{\mathrm{C}}} = 1 + (6+\alpha)\,\mathrm{Kn} + \mathcal{O}(\mathrm{Kn}^2),$$

where $\dot{M}_{\mathrm{C}}$ corresponds to continuum mass flowrate. This model becomes inaccurate for a nonzero value of α in the slip flow regime. Moreover, in the free-molecular flow regime, for very long channels ($L \gg \lambda \gg h$) there are no physical values for α, since the flowrate increases logarithmically with Kn. For finite-length channels the flowrate is asymptotic to a constant value proportional to $\log_e(L/h)$ (see Figure 4.18). Therefore, for finite-length two-dimensional channel flows, the coefficient α should smoothly vary from zero in the slip flow regime to an appropriate constant value in the free-molecular flow regime. It is difficult to verify this variation using the DSMC simulations due to the statistical scatter of the DSMC method. However, progress can be made if we assume an approximate value of α, which we denote by $\bar{\alpha}$, and determine the value of it for a specific gas in a finite-length channel. Such an analysis has been performed for nitrogen flow in a channel of length to height ratio $L/h = 20$, resulting in $\bar{\alpha} = 2.2$.

Using this approximate value we compare the predictions of the model for mass flowrate versus DSMC results. By integrating equation (4.22) from the inlet to the outlet of the channel, we derive an expression for the mass flowrate per unit width:

$$\dot{M} = \frac{h^3 P_o}{24\mu_o RT_o}\frac{\Delta P}{L}\left[(\Pi+1) + 2[6\frac{2-\sigma_v}{\sigma_v} + \bar{\alpha}]\,\mathrm{Kn}_o + 12\frac{2-\sigma_v}{\sigma_v}\frac{b+\bar{\alpha}}{\Pi-1}\,\mathrm{Kn}_o^2\log_e(\frac{\Pi-b\,\mathrm{Kn}_o}{1-b\,\mathrm{Kn}_o})\right], \quad (4.24)$$

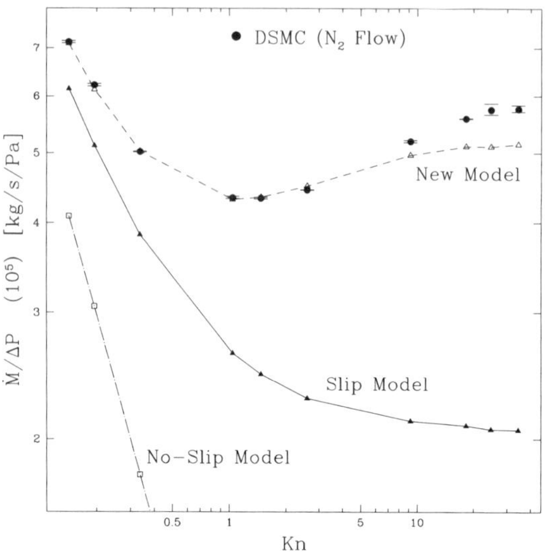

FIGURE 4.22. Variation of mass flowrate (per pressure drop ΔP) as a function of exit Knudsen number in the channel. Knudsen's minimum is captured by the model at Kn $\approx$ 1. Comparisons with the DSMC data as well as the continuum and slip models are included.

where the subscript o refers to the outlet conditions, $\Pi = P_\text{i}/P_\text{o}$ (inlet-to-outlet pressure ratio), and L is the channel length. The comparison of the corrected model with the DSMC data is given in Figure 4.22. The model predicts Knudsen's minimum obtained by the DSMC calculations quite accurately at Kn $\approx$ 1.0. Consistent with the DSMC solutions, the model predicts a flowrate independent of Kn for the free-molecular flow limit. However, this constant flowrate for larger Kn is slightly (13%) lower than the DSMC predictions. For comparisons with the model and the DSMC data, we also plot the flowrate obtained by the continuum and the first-order slip models in Figure 4.22. The continuum model behaves like 1/ Kn and gives the wrong variation, while the slip flow model yields flowrate values three times less than the DSMC calculations. Systematic investigations to test the accuracy of the mass flowrate formula for different values of the tangential momentum accommodation coefficient σ_v indicate that the error in the prediction does not exceed 15%.

The corresponding free-molecular mass flowrate of the new model can be calculated using an asymptotic expansion of equation (4.24) in 1/ Kn as Kn $\to \infty$. The result is independent of both the Knudsen number and the pressure ratio (since $\text{Kn}_\text{o} \to \infty$, $(\Pi - 1)/(b\,\text{Kn}_\text{o}) \ll 1$), i.e.,

$$\dot{M}_\text{FM} = \frac{h^2}{12}\left(\frac{\pi}{2RT}\right)^{\frac{1}{2}} \frac{\Delta P}{L}\left[\bar{\alpha}\left(1 - \frac{6}{b}\right)\right]. \tag{4.25}$$

Having obtained the mass flowrate, the corresponding pressure distribution along the channel can be obtained as

$$\tilde{P}^2 - 1 + 2(6+\bar{\alpha})\frac{2-\sigma_v}{\sigma_v}\,\mathrm{Kn}_0(\tilde{P}-1) + 2(6b+\bar{\alpha})\frac{2-\sigma_v}{\sigma_v}\,\mathrm{Kn}_0^2\log_e\left(\frac{\tilde{P}-b\,\mathrm{Kn}_o}{1-b\,\mathrm{Kn}_o}\right) = B\left(1-\frac{x}{L}\right), \quad (4.26)$$

where B is a constant such that $\tilde{P}(0) = (P_i/P_o) = \Pi$. Here we have defined

$$\tilde{P}(x) = P(x)/P_0,$$

i.e., the pressure at a station x normalized with the exit pressure. The above equation provides an implicit relation for $\tilde{P}$. The pressure distribution for a first-order boundary condition is obtained explicitly by neglecting the second-order terms $\mathcal{O}(\mathrm{Kn}^2)$ in the above pressure equation.

We also examine the differences in pressure between DSMC and μ**Flow** predictions by plotting the curvature in the pressure distribution, i.e., the deviation from the corresponding linear pressure drop $(P - P_{\mathrm{IC}})/(P_o))$ in Figure 4.23, where P_{IC} denotes the pressure of corresponding incompressible flow. The μ**Flow** simulation using the slip boundary condition (2.26) predicts larger curvature in pressure distribution than the DSMC results. The pressure distribution obtained by the first-order boundary condition is shown by dashed lines and lies between the μ**Flow** and DSMC results. Our second-order slip model without the correction of the rarefaction coefficient ($C_r = 1 + \bar{\alpha}\,\mathrm{Kn}$) gives identical results to μ**Flow** predictions. The corresponding continuum (no-slip) pressure distribution is also given in the figure. The *reduction in the curvature* of the pressure distribution with rarefaction is clearly demonstrated. Finally, the model, including the rarefaction coefficient $C_r(\mathrm{Kn})$ shown by solid lines, gives results closest to the DSMC solution. This demonstrates the ability of the new model in predicting the pressure distribution for channel flows. At higher Knudsen number, the curvature in the pressure distribution is much smaller, with linear pressure drop observed as $\mathrm{Kn} \to \infty$.

Model for Pipe and Duct Flows

The asymptotic value of flowrate for pipe and duct flows at high Knudsen number is constant, and this offers the possibility of obtaining a model for the rarefaction coefficient $C_r(\mathrm{Kn})$ and in particular the coefficient α. The objective is to construct a unified expression for $\alpha(\mathrm{Kn})$ that represents the transition of α from zero in the slip flow regime to its asymptotic constant value in the free-molecular flow regime. We will verify the velocity and new flowrate models using available experimental data for pipes (S. Tison, NIST, private communications) as well as numerical results obtained in

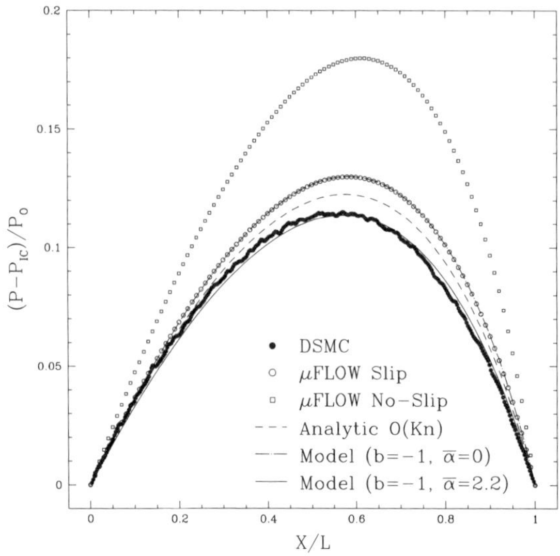

FIGURE 4.23. Deviations from linear pressure drop for nitrogen channel flow ($\Pi = 2.28$, $\mathrm{Kn_o} = 0.2$). Comparison of μ**Flow** with the DSMC predictions and the new slip model ($b = -1, \bar{\alpha} = 2.2$).

(Loyalka and Hamoodi, 1990; Sone and Hasegawa, 1987), for pipe and duct flows, respectively.

Pipe Flow

First, we derive a similar model for the pipe flow. Assuming a parabolic velocity profile with the slip amount given by (2.43), it is possible to obtain the following equation for the nondimensionalized velocity scaling in a pipe flow:

$$U^*(r, \mathrm{Kn}) = \frac{U(x,r)}{\overline{U}(x)} = \left[\frac{-\left(\frac{r}{a}\right)^2 + 1 + 2\frac{\mathrm{Kn}}{1-b\,\mathrm{Kn}}}{\frac{1}{2} + 2\frac{\mathrm{Kn}}{1-b\,\mathrm{Kn}}}\right], \tag{4.27}$$

where a is the pipe radius, and the Knudsen number is defined as $\mathrm{Kn} = \lambda/a$. We compare the nondimensional velocity scaling with the linearized Boltzmann solution of (Loyalka and Hamoodi, 1990) in Figure 4.24. The general slip coefficient is found to be $b = -1$ as before, consistent with the velocity profiles given in (Loyalka and Hamoodi, 1990), for cases $\mathrm{Kn} = 0.1$, $\mathrm{Kn} = 1.0$, and $\mathrm{Kn} = 10$. The velocity profiles predicted with the first-order slip model and the general slip model (2.43) are also shown in Figure 4.24. It is seen that the first-order model gives erroneous velocity distributions in the transitional and free-molecular flow regimes. For example, for the

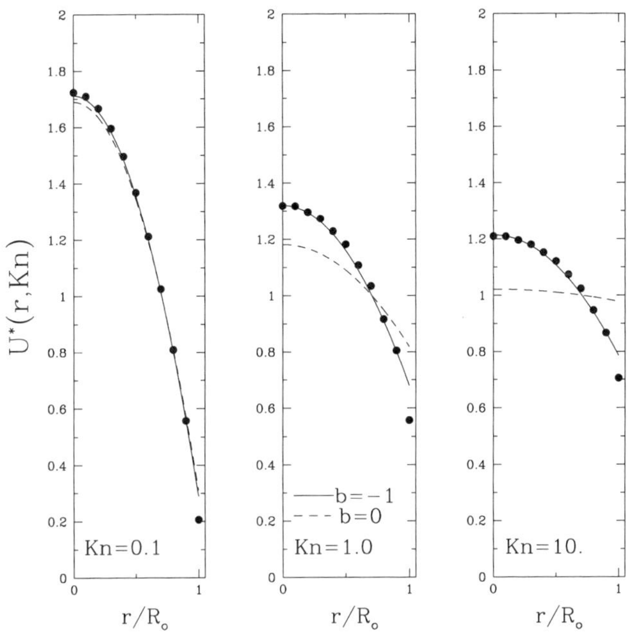

FIGURE 4.24. Velocity profile scaling in rarefied pipe flows (nondimensional). The first-order model corresponds to $b = 0$, and the general slip model corresponds to $b = -1$. The dots correspond to the linearized Boltzmann solution of (Loyalka and Hamoodi, 1990).

Kn = 10 case an almost uniform velocity distribution is predicted. However, the model of equation (2.43) predicts accurately the velocity distribution in most of the pipe with a small error in the velocity slip.

Next, we compare the flowrate model corrected as before by the rarefaction coefficient $C_r(\text{Kn})$ in a similar form obtained for channel flows with

$$C_r(\text{Kn}) = 1 + \alpha\,\text{Kn}\,.$$

The volumetric flowrate for a pipe is

$$\dot{Q} = -\frac{\pi a^4}{8\mu_o}\frac{dP}{dx}(1 + \alpha\,\text{Kn})\left[1 + \frac{4\,\text{Kn}}{1 - b\,\text{Kn}}\right], \tag{4.28}$$

and the corresponding mass flowrate is

$$\dot{M} = \frac{\pi a^4 P_o}{16\mu_o R T o}\frac{\Delta P}{L}\left[(\Pi + 1) + 2(4 + \alpha)\,\text{Kn}_o \right. \\ \left. + \frac{8(\alpha + b)}{\Pi - 1}\text{Kn}_o^2 \log_e\left(\frac{\Pi - b\,\text{Kn}_o}{1 - b\,\text{Kn}_o}\right)\right]. \tag{4.29}$$

Since $b = -1$ is already determined from the linearized Boltzmann solution, the only parameter to be determined in the model is α. However, α should vary from *zero* in the slip flow regime to a *constant* asymptotic value (α_o) in the free-molecular flow regime. It is possible to obtain the constant asymptotic value of α (as Kn $\rightarrow \infty$) by using the theoretical mass flowrate in the free-molecular flow regime given by equation (4.16) and the asymptotic value for the mass flowrate obtained by (4.30) while Kn $\rightarrow \infty$, as

$$\alpha_{\text{Kn}\rightarrow\infty} \equiv \alpha_o = \left(\frac{64}{3\pi(1-\frac{4}{b})}\right). \tag{4.30}$$

We can also compare the results with the formula derived by (Knudsen, 1909), which we normalize here with the corresponding free-molecular flow limit of equation (4.16):

$$\frac{\dot{M}_{\text{Kn}}}{\dot{M}_{\text{FM}}} = \frac{3\pi}{64\overline{\text{Kn}}} + \left(\frac{1+2.507(1/\overline{\text{Kn}})}{1+3.095(1/\overline{\text{Kn}})}\right), \tag{4.31}$$

where $\overline{\text{Kn}}$ is computed at the average pressure $\overline{P} = (P_i + P_o)/2$. The constants 2.507 and 3.095 are taken from (Loeb, 1961), where details of derivation of Knudsen's formula are presented. The same formula has also been used in (Tison, 1993), and (Loyalka and Hamoodi, 1990).

Due to the lack of detailed experimental data, we do not have exact values for the pressure ratio Π, and thus we cannot use equation (4.30) directly. Instead, we approximately integrate the volumetric flowrate equation (4.28) by multiplying it by the average density in order to obtain the mass flowrate. The flow conditions are evaluated at an average state. For example, the average pressure is defined as $\overline{P} = (P_i + P_o)/2$ so that $(dP/dx) = \Delta P/L$, and Kn is evaluated at $\overline{P}$. The corresponding mass flowrate becomes

$$\dot{M} = -\frac{\pi a^4 \overline{P}}{8\mu_o RT}\frac{\Delta P}{L}(1+\alpha\,\text{Kn})\left[1+\frac{4\,\text{Kn}}{1-b\,\text{Kn}}\right]. \tag{4.32}$$

Nondimensionalizing this relation with the theoretical free-molecular flow limit (4.16), we obtain the following relation

$$\frac{\dot{M}}{\dot{M}_{\text{FM}}} = \frac{3\pi}{64\overline{\text{Kn}}}(1+\alpha\overline{\text{Kn}})\left(1+\frac{4\overline{\text{Kn}}}{1-b\overline{\text{Kn}}}\right). \tag{4.33}$$

Comparison of equations (4.31) and (4.33) shows that both models predict the same limit in the free-molecular flow regime (Kn $\rightarrow \infty$) if the value of α_o is chosen according to equation (4.30) ($\alpha_o = 1.358$ for pipe flows, i.e. $a \ll \lambda \ll L$).

Knudsen's formula is often used to describe the flow for the entire flow regime, including the slip flow regime. Considering that

$$\frac{\dot{M}_{\text{C}}}{\dot{M}_{\text{FM}}} = \frac{3\pi}{64}\overline{\text{Kn}},$$

Knudsen's formula can be written for the slip flow regime ($\overline{\mathrm{Kn}} \leq 0.1$) as

$$\frac{\dot{M}_{\mathrm{Kn}}}{\dot{M}_{\mathrm{C}}} = 1 + \frac{64\overline{\mathrm{Kn}}}{3\pi}\left(\frac{2.507}{3.095}\right) + \mathcal{O}(\mathrm{Kn}^2),$$

where the subscript "C" stands for *continuum* predictions. This relation shows that Knudsen's formula is *not* accurate for the slip flow regime, since the first-order variation of flowrate from the corresponding continuum limit should be

$$\frac{\dot{M}_{\mathrm{slip}}}{\dot{M}_{\mathrm{C}}} = 1 + 4\overline{\mathrm{Kn}}.$$

If we used a *constant* α in the entire flow regime, the model would have resulted in an incorrect form similar to Knudsen's model in the slip flow regime. In order to obtain the variation of α as a function of Knudsen number for the transitional flow regime, we can solve for α from equation (4.33) to obtain

$$\alpha = \frac{64\frac{\dot{M}}{\dot{M}_{\mathrm{FM}}}}{3\pi(1+\frac{4\overline{\mathrm{Kn}}}{1-b\overline{\mathrm{Kn}}})} - \frac{1}{\overline{\overline{\mathrm{Kn}}}},$$

where $\dot{M}$ is the flowrate data obtained numerically or experimentally (and normalized with $\dot{M}_{\mathrm{FM}}$). The $1/\mathrm{Kn}$ behavior in this analytical expression makes it difficult to predict the value of α for small Kn. Therefore, we must rely on accurate numerical or experimental data. For this purpose we use linearized Boltzmann solutions of (Loyalka and Hamoodi, 1990) and experimental data of S. Tison (NIST, private communications). In Figure 4.25 we present the variation of α as a function of Kn (symbols). The value of α is initially small (close to zero), and it gradually increases with Kn, reaching a constant value in the free-molecular flow regime. The physical meaning of this behavior is that the dynamic viscosity remains the standard diffusion coefficient in the early slip flow regime. The value of α increases slowly with Kn in the slip flow regime, and therefore the effect of change of the diffusion coefficient is second-order in Kn. For this reason the experimental slip flow results are accurately predicted by the slip flow theory, which does not require change of the diffusion coefficient length scale from λ to channel height h. Variation of α as a function of Kn found by the numerical and experimental data can be represented accurately with the following relation:

$$\alpha = \alpha_0 \frac{2}{\pi} \tan^{-1}\left(\alpha_1 \overline{\mathrm{Kn}}^{\beta}\right), \tag{4.34}$$

where α_0 is determined to result in the desired free-molecular flowrate using (4.30) and $\alpha_1 = 4.0$, $\beta = 0.4$. This model is shown in Figure 4.25 with lines.

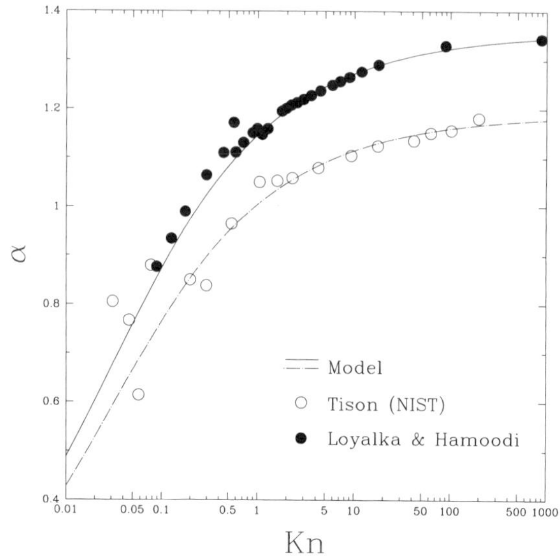

FIGURE 4.25. Variation of α as a function of Knudsen number. Analytical fit to this variation is obtained by $\alpha = \alpha_0 \frac{2}{\pi} \tan^{-1}(\alpha_1 \,\mathrm{Kn}^{\beta})$, ($\alpha_1$ and β are free parameters).

Note that the values for α_1 and β are the same for both the experimental and numerical results presented in the figure, and that these are the only two undetermined parameters of the model.

In Figure 4.26 we present the flowrate variation as a function of Kn. The data are obtained by the solution of linearized Boltzmann equations by (Loyalka and Hamoodi, 1990) for a very long pipe, so that $h \ll \lambda \ll L$ is maintained for all values of Kn. Knudsen's two-parameter model is also presented. The experimental data presented in the figure were obtained by Tison for helium flow in finite-length pipes ($L/a = 200$). There are differences between the experimental and numerical data. For example, the experimental data have not reached the corresponding free-molecular flowrate limit. At Kn $= 200$, which is the highest Kn value in the experiments, $L \propto \lambda$, and therefore end effects are important and the expected mass flowrate is less than the theoretical free-molecular flowrate (Polard and Present, 1948). Since the analytical and experimental data show some differences, in the case of the experiments we found the value of $\alpha_0 = 1.19$ by using the experimental data at Kn $= 200$. Also, for the linearized Boltzmann solution we obtained $\alpha_0 = 1.358$ using equation (4.30). The model's predictions for linearized Boltzmann solution and experimental data are also presented in the figure. The model describes the variation of the data very accurately, and it is successful in predicting the Knudsen's minimum.

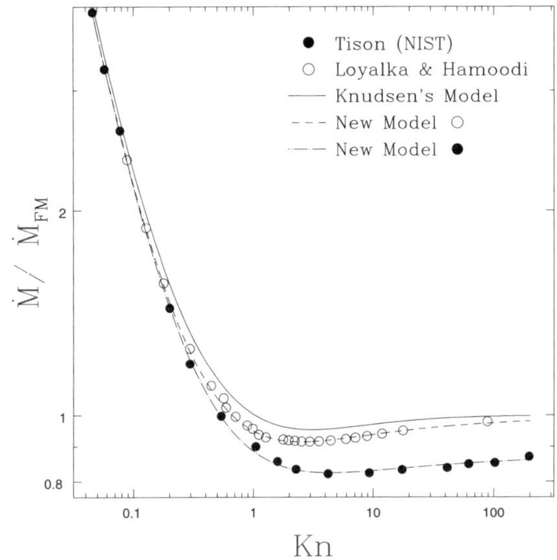

FIGURE 4.26. Free-molecular scaling of Loyalka and Hamoodi's linearized Boltzmann solutions (Loyalka and Hamoodi, 1990) and Tison's experiments (Tison, 1993). Comparisons with the proposed model for both cases and Knudsen's model are also presented.

In Figure 4.27 we present the mass flowrate variation (normalized with the corresponding no-slip value) as a function of Kn, up to Kn $= 0.5$. This covers the *slip* and the *early transitional* flow regimes. We see that Knudsen's model is not accurate in this regime. Linearized Boltzmann solutions and experimental data both start with a slope of 4. Hence,

$$\frac{\dot{M}}{\dot{M}_C} = 1 + 4\,\mathrm{Kn} + \mathcal{O}(\mathrm{Kn}^2);$$

then the slope increases gradually with Kn. The model predicts this transition very accurately for the numerical and the experimental data. The increase in slope was observed by (Sreekanth, 1969) and explained as a change in the slip coefficient in Maxwell's slip boundary conditions from 1.0 to 1.1466. If the change in the slope of the data is to be explained by an increase in the slip coefficient, the velocity scaling results shown in Figures 4.24 and 4.19 should be affected. However, it is clearly seen that such an effect is not present; a more appropriate explanation of the slope change is the change in the diffusion coefficient with Kn as presented in Section 4.2.2.

Duct Flow

We present the extensions of the new model for duct flows in the entire

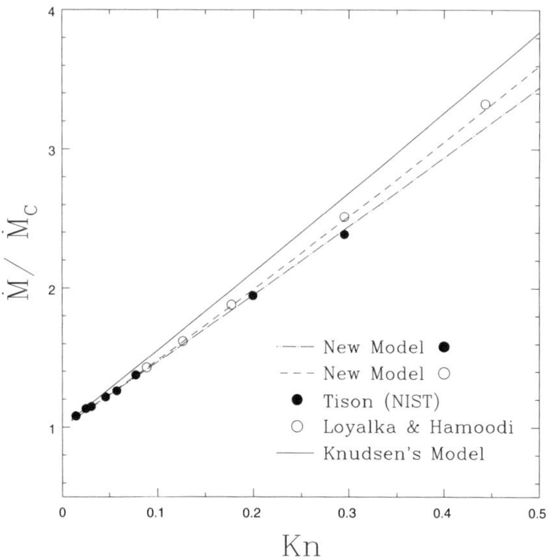

FIGURE 4.27. Normalized flowrate variation in the slip and early transitional flow regimes for pipe flows. Symbols correspond to the linearized Boltzmann solutions of (Loyalka and Hamoodi, 1990) and experimental results of (Tison, 1993). Comparisons with the proposed model for both cases and Knudsen's model are also presented.

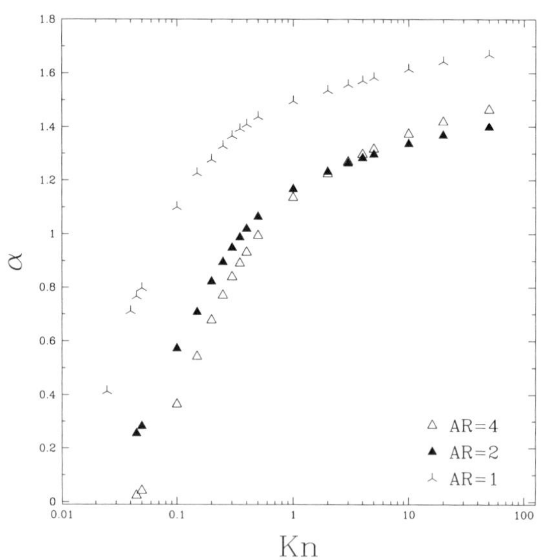

FIGURE 4.28. Variation of α as a function of Kn for various aspect ratio ducts.

Knudsen regime. We consider flows in ducts with aspect ratio ($\mathrm{AR} = w/h \equiv$ width/height) of 1, 2, and 4. The data are obtained by linearized Boltzmann solution in ducts with the corresponding aspect ratios (Sone and Hasegawa, 1987). Our previous analysis was valid for the two-dimensional channels, where we reported flowrate per channel width. For duct flows, three-dimensionality of the flow field (due to the side walls of the duct) must be considered. In no-slip duct flows the flowrate formula developed for two-dimensional channel flows is corrected in order to include the blockage effects of the side walls. According to this, the volumetric flowrate in a duct with aspect ratio AR for no-slip flows is (see (White, 1991), p. 120)

$$\dot{Q} = C(\mathrm{AR})\frac{wh^3}{12\mu}\left(-\frac{dP}{dx}\right),$$

where $C(\mathrm{AR})$ is the correction factor given as

$$C(\mathrm{AR}) = \left[1 - \frac{192(\mathrm{AR})}{\pi^5}\sum_{i=1,3,5,\ldots}^{\infty}\frac{\tanh(i\pi/2(\mathrm{AR}))}{i^5}\right]. \tag{4.35}$$

With this correction, aspect ratios of 1, 2, and 4 ducts correspond to 42.17%, 68.60%, and 84.24% of the theoretical two-dimensional channel volumetric flowrate for no-slip flows, respectively. According to the new model, the volumetric flowrate for rarefied gas flows in ducts is

$$\dot{Q} = C(\mathrm{AR})\frac{wh^3}{12\mu_0}\left(-\frac{dP}{dx}\right)(1+\alpha\,\mathrm{Kn})\left(1+\frac{6\,\mathrm{Kn}}{1-b\,\mathrm{Kn}}\right),$$

where the correction factor $C(\mathrm{AR})$ is independent of Knudsen number. The variation of α as a function of Kn is calculated by using the correction factors ($C(\mathrm{AR})$), the linearized Boltzmann solutions in (Sone and Hasegawa, 1987), and our model. This variation is given in Figure 4.28. The rarefaction coefficient ($C_r(\mathrm{Kn}) = 1 + \alpha\,\mathrm{Kn}$) was introduced in order to model the reductions in the intermolecular collisions of the molecules as Kn is increased. In duct flows, both the height and the width of the duct are important length scales, and comparison of these length scales to the local mean free path is an important factor in the variation of α. It is seen in Figure 4.28 that the transition in α occurs later for high aspect ratio ducts, as expected.

Similar to the pipe flow case, an approximate analytical formula can be derived to describe the mass flowrate in ducts of various aspect ratios as

$$\frac{\dot{M}}{\dot{M}_{\mathrm{C}}} = C(\mathrm{AR})\left(1+\alpha\overline{\mathrm{Kn}}\right)\left[1+\frac{6\overline{\mathrm{Kn}}}{1-b\overline{\mathrm{Kn}}}\right],$$

where $\overline{\mathrm{Kn}}$ is evaluated at average pressure as before. In Figure 4.29 we present the variation of flowrate nondimensionalized with the corresponding no-slip value as a function of Kn in the slip and early transitional flow

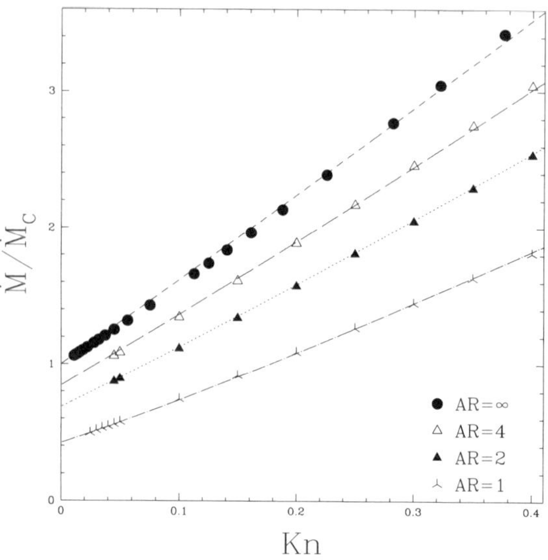

FIGURE 4.29. Normalized flowrate variation in the slip and early transitional flow regimes for various aspect ratio (AR) duct flows. Symbols are the linearized Boltzmann solution of (Sone and Hasegawa, 1987). Comparisons with the proposed model are also presented by lines.

TABLE 4.3. Parameters of the model for various aspect ratio duct flows. The only free parameters are α_1 and β, as α_0 is determined from the asymptotic constant limit of flowrate as $\text{Kn} \rightarrow \infty$.

$(\text{AR}) = w/h$	$C(\text{AR})$	α_0	α_1	β
1	0.42173	1.7042	8.0	0.5
2	0.68605	1.4400	3.5	0.5
4	0.84244	1.5272	2.5	0.5

regimes. The linear increase of the flowrate with Kn and complete description of rarefied duct flows with the introduction of the correction factor $C(\text{AR})$ are observed. The slope of the nondimensionalized mass flowrate increases gradually with Kn. This is attributed to the gradual change in the rarefaction coefficient as presented in Figure 4.28.

For the free-molecular scaling of the data we nondimensionalized the flowrate with

$$\dot{M}_{\text{FM}} = \frac{h^2 w}{\sqrt{2RT_o}} \frac{\Delta P}{L},$$

which gives the correct order of magnitude for the flowrate. The exact value of the free-molecular flowrate in rectangular ducts is given by (Thompson

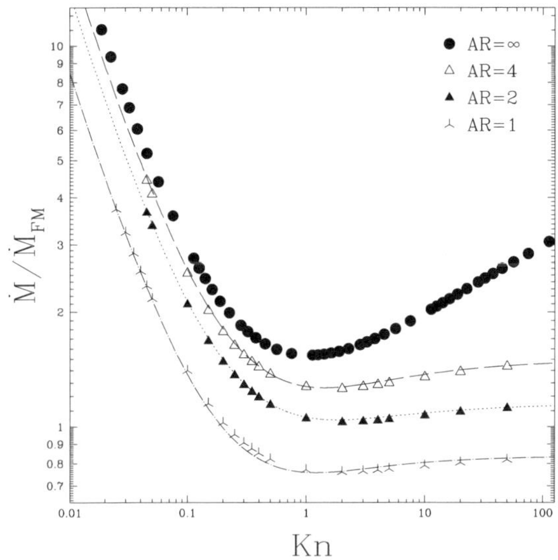

FIGURE 4.30. Free-molecular scaling of linearized Boltzmann solutions of (Sone and Hasegawa, 1987) for duct flows of various aspect ratio. Comparisons with the proposed model are also presented by lines corresponding to different aspect ratios.

and Owens, 1975)

$$\dot{M}_{\mathrm{FM}}(h, w) = \Gamma \dot{M}_{FM},$$

where

$$\Gamma = h^2 w \log_e \left(\frac{h}{w} + \sqrt{1 + \left(\frac{h}{w}\right)^2} \right) + w^2 h \log_e \left(\frac{w}{h} + \sqrt{1 + \left(\frac{w}{h}\right)^2} \right) - \frac{(h^2 + w^2)^{3/2}}{3} + \frac{h^3 + w^3}{3}.$$

Here, h and w denote the height and the width of the rectangular duct. For the aspect ratios (AR) of 1, 2, and 4 the above relation results in 0.8387, 1.1525, and 1.5008 times the free-molecular mass flowrate $\dot{M}_{\mathrm{FM}}$, respectively.

Nondimensionalizing thc model with the free-molecular mass flowrate ($\dot{M}_{\mathrm{FM}}$), we obtain

$$\frac{\dot{M}}{\dot{M}_{\mathrm{FM}}} = \frac{C(\mathrm{AR})}{6\overline{\mathrm{Kn}}} \left(1 + \alpha\overline{\mathrm{Kn}}\right) \left[1 + \frac{6\overline{\mathrm{Kn}}}{1 - b\overline{\mathrm{Kn}}}\right].$$

In Figure 4.30 we present the variation of the nondimensionalized flowrate as a function of Kn. The duct flow data are due to (Sone and Hasegawa,

1987), and the two-dimensional channel data (shown by AR $= \infty$) are due to Sone (for $Kn \leq 0.17$) and Cercignani (Kn > 0.17) (Fukui and Kaneko, 1990). Comparisons are made against the linearized Boltzmann solutions. For duct flows, good agreement of the model with the numerical data in the entire flow regime is obtained. The model is also able to capture Knudsen's minimum accurately. The parameters used in the model are given in Table 4.3. Note that α_o is determined from the asymptotic constant limit of flowrate (4.36) as Kn $\rightarrow \infty$.

4.2.3 Summary

In Section 4.2.2, we developed a *unified* flow model that can accurately predict the volumetric flowrate, velocity profile, and pressure distribution in the *entire* Knudsen regime for pipes and rectangular ducts. The new model is based on the hypothesis that the velocity distribution remains parabolic in the transition flow regime, which is supported by the asymptotic analysis of the Burnett equations in section 4.2.1. The general velocity slip boundary condition (equation (2.43)) and the rarefaction correction factor (equation (4.23)) are the two primary components of this unified model.

- The general slip boundary condition (equation (2.43)) gives the correct nondimensional velocity profile, where the normalization is obtained using the *local channel averaged velocity*. This eliminates the flowrate dependence in modeling the velocity profile. For channel flows, using equation (2.39), we obtain $b = -1$ in the slip flow regime. Evidence based on comparisons of the model with the DSMC and Boltzmann solutions shows that $b = -1$ in the *entire* Knudsen regime.

- In order to model the flowrate variations with respect to the Knudsen number Kn, we introduced the rarefaction correction factor as $C_r = 1 + \alpha\,\mathrm{Kn}$. This form of the correction factor was justified using two independent arguments: first, the apparent diffusion coefficient; and second, the ratio of intermolecular collisions to the total molecular collisions. We must note that α cannot be a constant. Physical considerations to match the slip flowrate require $\alpha \rightarrow 0$ for Kn ≤ 0.1, while $\alpha \rightarrow \alpha_0$ in the free molecular flow regime. The variation of α between zero and a known α_0 value is approximated using equation (4.34), which introduced two empirical parameters α_1 and β to the new model.

Therefore, the unified model employs two empirical parameters (α_1 and β) and two known parameters $b = -1$ and α_0. Although this empiricism is not desired, the α value in C_r varies from zero in the slip flow regime to an *order-one* value of α_0 as Kn $\rightarrow \infty$. Finally, the model is adapted to the finite aspect ratio rectangular ducts using a standard aspect ratio correction given in equation (4.35).

5

Thermal Effects in Microscales

In this chapter we consider heat transfer in gas microflows. In the first section we concentrate on the thermal creep (transpiration) effects that may be important in channels with tangential temperature gradients on their surfaces. For example, a microchannel surface with a prescribed heat flux is subject to temperature variations along its surface, and this results in thermal creep flow. We analyze thermal creep with numerical simulations to demonstrate the main concept, and subsequently we describe a prototype experiment. In the second and third sections we study other temperature-induced flows and investigate the validity of the heat conduction equation in the limit $\text{Kn} \to 0$. In the fourth and fifth sections we investigate the combined effects of thermal creep, heat conduction, and convection in pressure-, force-, and shear-driven channel flows.

5.1 Thermal Creep (Transpiration)

It is possible to start rarefied gas flows due to tangential temperature gradients along the channel walls, where the fluid starts creeping in the direction *from cold toward hot.* This is the so-called thermal creep or transpiration phenomenon. We explain this counterintuitive effect with the following example: Consider two containers filled with the same gas that are kept at the same pressure

$$P_1 = P_2$$

but at different temperatures

$$T_1 > T_2.$$

If these two containers are connected with a relatively thick channel ($\lambda \ll h$), the equilibrium condition requires no-flow in the channel. If the channel thickness (h) becomes comparable to the mean free path (λ), rarefied gas effects have to be taken into account. In such a case the local equilibrium mechanism is very complex, and interaction of the gas molecules with the walls must also be considered. Here, we consider free-molecular flow conditions (i.e., $\lambda \gg h$) to simplify the discussion. In this flow regime, the intermolecular collisions are negligible compared to the interaction of molecules with the surfaces. If we assume that molecule–wall interactions are *specular* (i.e., $\sigma_v = 0$), then the following analysis is valid. We assume that the density of the fluid is proportional to the number density (number of molecules per unit volume),

$$\rho \propto n,$$

and the temperature of the fluid is proportional to the square of average molecular speed,

$$T \propto \bar{c}^2.$$

The mass fluxes at the hot and the cold ends of the channel are

$$mn_1\bar{c}_1 \quad \text{and} \quad mn_2\bar{c}_2,$$

respectively; here m is the mass of the gas molecules. Then

$$\frac{mn_1\bar{c}_1}{mn_2\bar{c}_2} \approx \frac{\rho_1}{\rho_2}\left(\frac{T_1}{T_2}\right)^{0.5} = \frac{P_1}{P_2}\left(\frac{T_2}{T_1}\right)^{0.5} = \left(\frac{T_2}{T_1}\right)^{0.5} \leq 1,$$

where we have used the equations of state

$$P = \rho RT \quad \text{and} \quad \frac{P_1}{P_2} = 1.$$

The above analysis indicates a **flow creeping from cold to hot**. Osborne Reynolds was the first to realize this phenomenon, and he coined the term *thermal transpiration* (Reynolds, 1879). At about the same time, Maxwell developed independently a theory for thermal creep. In the early 1900s, Knudsen built a molecular compressor based on the thermal transpiration idea by connecting a series of tubes with constrictions arranged between each tube (Knudsen, 1910a; Knudsen, 1910b). The constrictions were very small, so that the rarefaction effects became important in the constrictions. By heating the same side of these constrictions to very high temperatures (773 K), Knudsen was able to maintain considerable pressure gradients

(Loeb, 1961). According to (Ebert and Albrand, 1963), Gaede also described a *thermal pump* in his unpublished notes. Other developments in thermal creep driven vacuum pumps can be found in (Vargo et al., 1998).

A detailed derivation of thermal creep boundary condition for rarefied flows with $\lambda < h$ is given in (Kennard, 1938; Loeb, 1961). It can also be derived directly from the Boltzmann equation (see Section 15.4.2). In order to accommodate the thermal creep effects, the wall velocity is enhanced with the following term:

$$U_c = \frac{3}{4} \frac{\mu R}{P} \frac{\partial T}{\partial s}, \tag{5.1}$$

where U_c is the creep velocity, and $\frac{\partial T}{\partial s}$ is the tangential temperature gradient along the surface. Therefore, the high-order velocity slip boundary condition is modified as

$$U_s = \frac{1}{2}\left[(2-\sigma)U_\lambda + \sigma U_w\right] + U_c.$$

The velocity profile for a pressure-driven channel flow of thickness h, including the thermal creep effects, is then given by equation (4.5) with (5.1) added on to the right-hand side. Integrating this profile, we obtain the mass flowrate:

$$\dot{M} = -\frac{h^3 P}{12\mu RT}\frac{dP}{dx}\left[1 + 6\frac{2-\sigma_v}{\sigma_v}(\mathrm{Kn} - \mathrm{Kn}^2)\right] + \frac{3}{4}\frac{\mu h}{T}\frac{dT}{dx}. \tag{5.2}$$

We conclude that thermal creep can change the mass flowrate in a channel. If the pressure gradient and the temperature gradient along the channel walls act along the same direction, the flowrate is decreased; otherwise, the flowrate is increased.

- *Therefore, it is possible to have nonzero flowrate in a microchannel even in the case of zero pressure gradient.*

5.1.1 Simulation Results

An interesting aspect of thermal creep is that it causes zero net mass flowrate in channels where thermal creep and pressure gradient balance each other. To demonstrate this we simulated air flow in microchannels of various dimensions connecting two tanks kept at different conditions with

- Temperatures at 300 K and 400 K, and
- Knudsen number at Kn = 0.365, 0.122, 0.052.

The pressure in both tanks is initially atmospheric. Thermal creep effects cause pumping of the fluid toward the hot tank, increasing the pressure in the hot tank and lowering the pressure in the colder one. This pressure

TABLE 5.1. Pressure differences due to the thermal creep effects obtained by numerical simulation and from the analytical formula.

Gas	$\overline{\text{Kn}}$	$\Delta P_{\text{analy}}(Pa)$	$\Delta P_{\text{numer}}(Pa)$
Air	0.052	342.0	336. 0
Air	0.122	1482.0	1409.0
Air	0.365	9151.0	8832.0

difference eventually starts flow in the middle of the channel in the direction from hot to cold (high pressure to low pressure), resulting in zero average mass flowrate in the channel as the steady state is reached. For zero net mass flowrate, equation (5.2) can be written as

$$\frac{dP}{dx} = \left(\frac{9\mu^2 R}{h^2 P}\right) \frac{\frac{dT}{dx}}{1 + 6[\frac{2-\sigma}{\sigma}(\text{Kn} - \text{Kn}^2)]}.$$

It is possible to integrate this equation approximately using *average* values of viscosity ($\bar{\mu}$), pressure ($\bar{P} = \frac{P_1+P_2}{2}$), and Knudsen number ($\overline{\text{Kn}}$), resulting in

$$P_1 - P_2 \approx \left(\frac{9\bar{\mu}^2 R}{h^2 \bar{P}}\right) \frac{(T_1 - T_2)}{1 + 6\left(\frac{2-\sigma}{\sigma}\left(\overline{\text{Kn}} - \overline{\text{Kn}}^2\right)\right)}, \tag{5.3}$$

where viscosity and Knudsen number are evaluated at average temperature ($\bar{T} = \frac{T_1+T_2}{2}$) and average pressure ($\bar{P}$). Equation (5.3) shows that the pressure drop between two tanks can be increased by either decreasing the channel thickness (h) or the average pressure ($\bar{P}$). In other words:

- *Thermal creep effects can be significant in rarefied flows where the pressure is low or in microflows in atmospheric pressures where the typical dimensions are on the order of a micron or lower.*

The steady-state pressure distribution along the channel center and in the reservoirs, normalized with atmospheric pressure (initial pressure at both tanks), is given in Figure 5.1 for three different channel sizes. It is seen that the pressure change due to thermal creep for high Kn flows is nonnegligible. In Table 5.1, we compare the pressure differences predicted by equation (5.3) with the numerical values obtained by simulations; the agreement is very good.

The above numerical experiment describes an unsteady problem for which the relevant time scales are governed by two different transient processes. The *first transient process* is due to the fluid starting to creep along the channel surface. As time goes on, the creeping fluid layer starts interacting with the stagnant fluid layers above it, creating a boundary layer similar to the impulsively started wall problem, the so-called Rayleigh problem (Batchelor, 1998). Of course, formation of the boundary layer creates

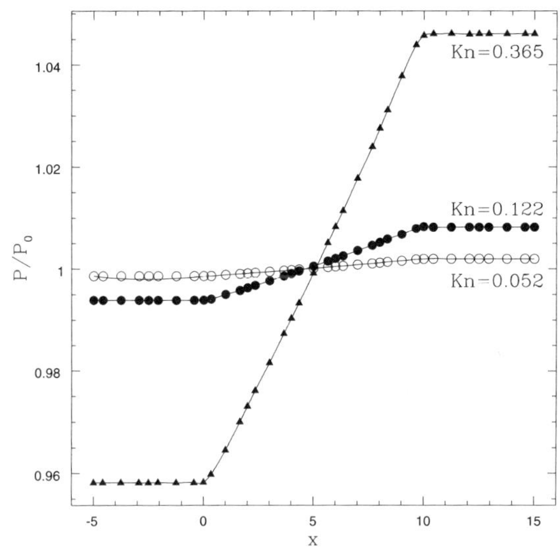

FIGURE 5.1. Normalized pressure variation along the channel center and reservoirs for different rarefaction parameters ($P_0 = 1\,atm$). Triangles and circles show the location of calculation points.

shear stresses, which, in turn, activate the velocity slip mechanism (Kennard, 1938). This is the initial transient process with time scale

$$t \propto \frac{h^2}{\nu}.$$

Considering microchannels with typical height of about one micron, this transient is very fast.

The *second time scale* of the problem corresponds to the time it takes to get from initial transients to a steady-state solution where the net mass flowrate is zero. This time scale is based on the creep velocity and tank dimensions. In particular, this time scale increases as the tank size is increased. In the limit where the tanks are reservoirs of infinite dimensions, the fluid steadily creeps from the cold to the hot tank, and the pressure at the two reservoirs remains practically the same.

Figure 5.2 shows the flow field for early times as well as for a steady-state behavior. A uniform flow is obtained in the channels initially (past the end of the aforementioned first process). As more fluid is introduced into the hot tank, the pressure there increases while the pressure at the cold tank decreases. Initially, the pressure built up by this process is not sufficient to reverse the flow in the middle of the channel, and more fluid is being transported into the hot tank (see Figure 5.2 (a)). When the pressure in the hot tank is sufficiently high to overcome the flow due to thermal creep, the

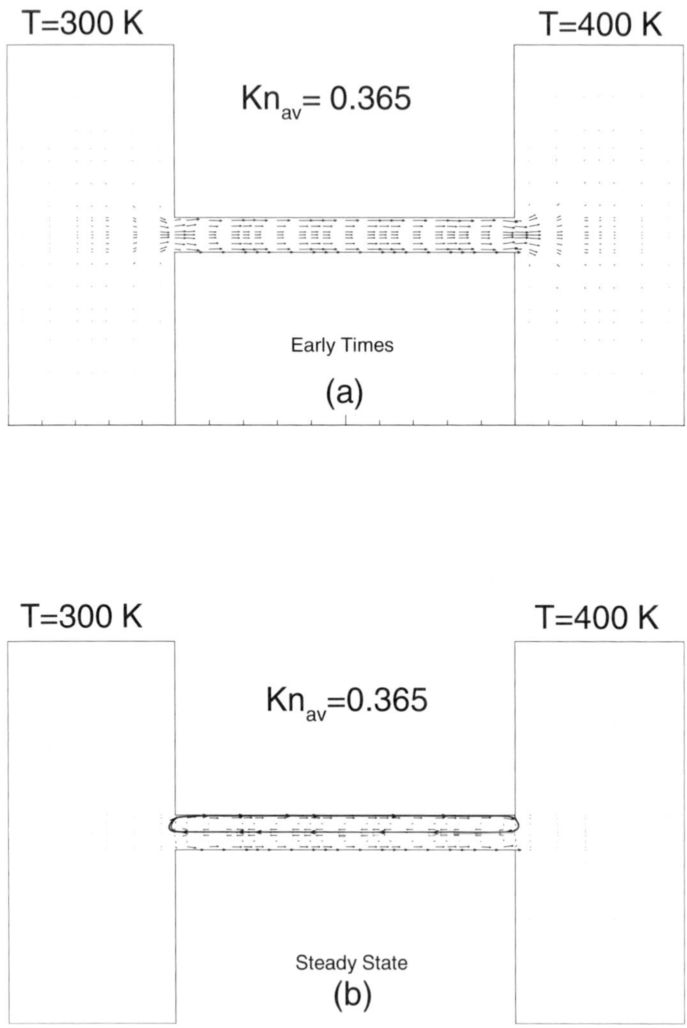

FIGURE 5.2. Demonstration of the thermal creep effects: Flow field for early times (a), and steady-state solution (b) ($\overline{\mathrm{Kn}} = \mathrm{Kn}_{\mathrm{av}} = 0.365$).

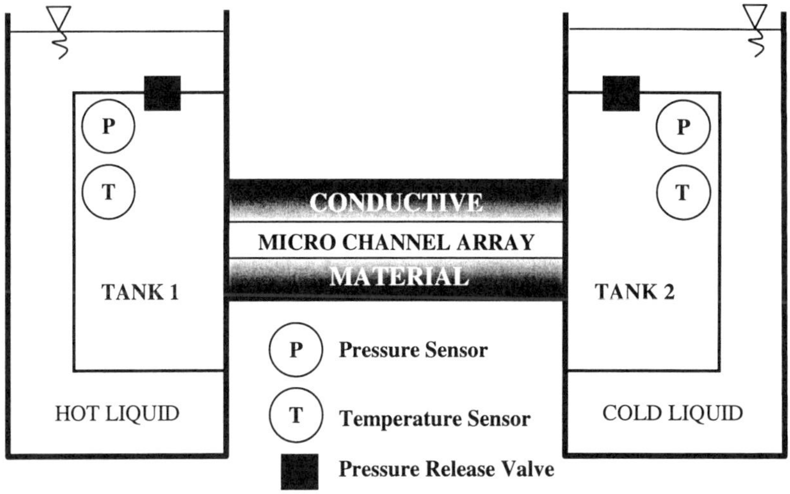

FIGURE 5.3. Proposed experimental apparatus for studying thermal creep effects.

net mass flowrate across the channel becomes zero, and the steady solution is obtained; this is shown in Figure 5.2 (b). A single streamline passing through the channel center is also shown in the plot. It is seen that the fluid recirculates in the channel and there is no net flow to either tank.

This pumping effect of thermal creep has been exploited by (Sone et al., 1996b), who designed a channel with periodic ditches, similar to the grooved channel of Figure 3.25 but with the groove present on both walls. They imposed a periodic temperature variation, whereby the temperature *decays* linearly on the groove side, but it *increases* linearly on the channel wall. This device produces a unidirectional flow with maximum mass flowrate at Kn ≈ 0.3, and it works both for channels as well as pipes. While the original studies in (Sone et al., 1996b), were based on DSMC, subsequently an experiment was conducted to verify the pumping effect (Sone et al., 2001).

5.1.2 A Thermal Creep Experiment

The purpose of this experiment is to examine the importance of thermal creep effects in microchannel flows. It is based on the numerical example presented above, also in (Beskok et al., 1995), and ideas similar to Knudsen's experiments (Loeb, 1961), but it differs from the early experiment of (Vargo and Muntz, 1996), where porous diaphragms were used rather than microchannels.

The objective is to test the thermal creep effects in an apparatus presented schematically in Figure 5.3. It consists of two reservoirs connected through a series of microchannels made out of silicon. The typical dimensions of the microchannels should be $1\,\mu$m thick, $100\,\mu$m wide, and few millimeters long. Using a series of microchannels increases the net flow area without effectively increasing the channel area to perimeter ratio.

A possible experimental procedure is as follows: *First*, both of the pressure release valves are open, and therefore, the system is in equilibrium with the ambient conditions. Then, the pressure release valves are closed. The temperature and the pressure of the system are recorded to ensure that the two reservoirs are at identical thermodynamic state. *Second*, the reservoirs are dipped into constant-temperature fluid baths at different temperatures T_1 and T_2. The pressure and the temperature in the reservoirs should be recorded in time. If the continuum hypothesis is valid, the pressure in the reservoirs should be unchanged. If thermal creep effects are present, the pressure in the cold reservoir should decrease, and the pressure in the hot reservoir should increase. The experiments should run until a stationary state is observed. The time scale of the experiment is directly related to the size of the reservoirs. Therefore, the reservoirs should be designed as small as possible. However, they should be large enough to maintain continuum description for the gas in them (i.e., $\mathrm{Kn} < 0.001$). It is possible to increase the rarefaction effects in the experiments by performing the experiment at lower pressures than atmospheric conditions. Therefore, a systematic study of thermal creep as a function of Kn can be performed. Also, the temperature of the fluid baths can be changed from one experiment to another in order to verify the sensitivity of thermal creep to temperature gradients for a given Kn.

5.1.3 Knudsen Compressors

Micromolecular compressors are useful for various microscale gas pumping applications. For example, compressors pumping gas samples through micromass spectrometers can be used to detect pollutants and various chemical or biological agents. MEMS-based thermal transpiration Knudsen compressors were proposed in (Pham-van-Diep et al., 1995), and in (Beskok et al., 1995). The idea in (Pham-van-Diep et al., 1995), is based on utilization of a cascade of multiple stages to obtain large pressure variations. Each stage consists of an array of capillaries and a connector section. The temperature increase imposed along the capillary pumps the gas from cold toward the hot direction, resulting in pressure increase in the capillary section. The gas is cooled in the connector section, and thus the temperature drops to the value corresponding to the inlet of the capillary section. This creates periodically repeatable temperature variations in each stage of the compressor. Since the pressure in the connector section drops only slightly, it is possible to have a net pressure built up with multistage units.

Large-scale Knudsen compressors have low volumetric flowrate and inefficient energy usage; however, their microscale counterparts eliminate these disadvantages and result in low-power gas-pumping systems with nonmoving components. (Vargo et al., 1998) built a microvacuum pump delivering 5×10^{14} molecules per second with a pump volume of 0.16 mL at inlet pressure of 10 mTorr and power input of 28.5 mW. Another micro pump with the same flowrate was built to operate at 1 mTorr. It required a volume of 13.9 mL and power input of 2.4 W. A tenfold decrease in the inlet pressure resulted in almost a hundred times increase in the power input and the volume of the system. The flow in the capillary section of the Knudsen compressor is in transition or free-molecular flow regime, and thus the capillary radius is small, and thermal transpiration effects are dominant in this section. However, the dimensions of the connector section are considerably larger than the local mean free path, so that slip or continuum flow is present in this section. In the compressor built by Vargo et al., the capillary and the connector radii were 0.225 μm and 10.795 mm, respectively. The bottleneck in the design of a Knudsen compressor is usually in determining the dimensions of the connector section, which becomes a limitation for low-pressure applications where the mean free path is already quite large.

In the experiments of Vargo et al., the maximum temperature variation in one stage of the compressor was about 20 to 30 K, with average temperature of 282 K. The inlet pressure and the working gas were varied. Good device performance for pressures as low as 1 mTorr was reported. Finally, Vargo et al. have developed a numerical model for predicting flowrate and pressure buildup in Knudsen compressors operating in the transition flow regime. The numerical predictions were validated using the experimental data, and the new algorithm was employed to study compressor performance under various flow conditions.

5.2 Other Temperature-Induced Flows

In addition to the flow induced by thermal creep, which is an $\mathcal{O}(Kn)$ effect, there are other possibilities for setting up a flow in rarefied flows in the absence of any external forces, e.g., gravity. Here we list a few representative cases and provide a short explanation:

- thermal stress slip flow,
- nonlinear thermal stress flow,
- flow induced around the edge of a heated plate, and
- flow induced by a temperature discontinuity.

Most of these cases were discovered by Sone, Aoki, and their collaborators by analyzing the Boltzmann equation (see Section 15.4.2). They are most probably present in microsystem flows, but they are difficult to isolate and be detected as individual effects. Here we give a brief summary for each following the work in (Sone, 2002; Sone, 2000).

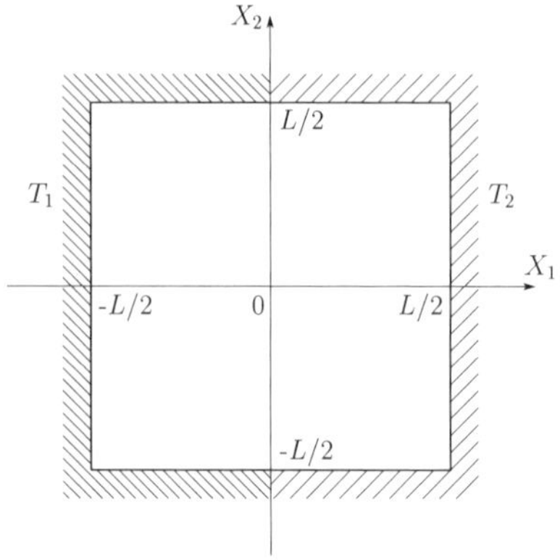

FIGURE 5.4. Flow domain showing the temperature discontinuities at the upper and lower walls. (Courtesy of K. Aoki.)

The **thermal stress slip flow** is an $\mathcal{O}(\mathrm{Kn}^2)$ effect, and it is derived following a consistent second-order expansion in Kn of the Boltzmann equation. In that case, in addition to the thermal creep that is present in the first-order expansion of the boundary terms, there is a correction term for the slip velocity proportional to the *normal gradient* of the temperature. A nonuniform normal gradient can cause a slip velocity even if the boundary is isothermal. To explain this in more detail, we follow the argument of Sone (see Section 15.4.2 and (Sone, 2002)). Let us consider a gas between two eccentric circular cylinders with different uniform temperatures T_1 and T_2. No thermal creep flow is induced; however, there is, in general, a nonuniform normal temperature gradient on each cylinder, that is,

$$\frac{\partial^2 T}{\partial x_i \partial x_j} n_i t_j \neq 0,$$

where n_i and t_j are components of the unit vectors in the normal and tangential directions, respectively. A global flow is then set up circulating between the cylinders along the clockwise direction if the temperature of the outside cylinder T_2 is higher. This phenomenon was discovered by (Sone, 1972), who termed it "thermal stress slip flow."

The **nonlinear thermal stress flow** is an $\mathcal{O}(\mathrm{Kn})$ effect and was discovered by (Kogan et al., 1976) as a new type of convection. When the temperature gradient in a gas is so large that its nonlinear effect is not negligible, the thermal stress does not balance by itself and remains in the momentum equation (see momentum equations in Section 15.4.2). This causes a flow when the distance between isothermal lines or surfaces varies along them. In this flow, in contrast to the thermal creep flow and thermal stress slip flow, a solid boundary plays only an indirect role in setting up the isothermal contours.

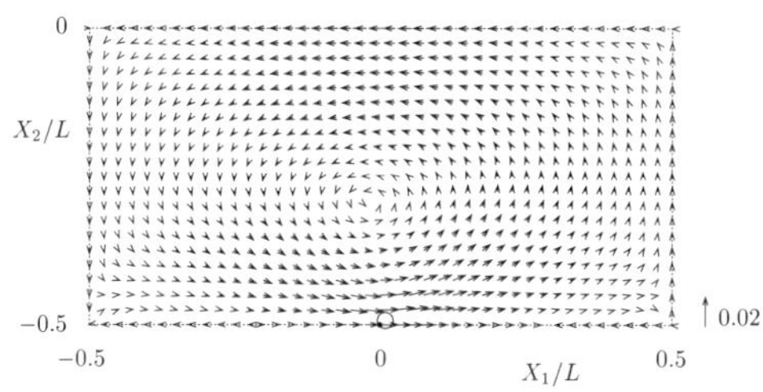

(a) Kn = 0.2

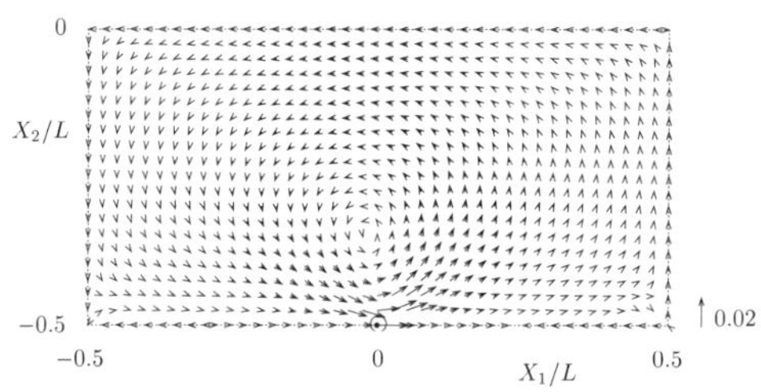

(b) Kn = 0.02

FIGURE 5.5. Flow induced in the lower half of the domain for $T_2/T_1 = 2$. The scale $\sqrt{u^2 + v^2}/\sqrt{2RT_1} = 0.02$ is indicated next to the plots. The upper plot corresponds to Kn = 0.2 and the lower plot to Kn = 0.02. (Courtesy of K. Aoki.)

The **flow induced around the sharp edge of a heated plate** cannot be predicted by the asymptotic analysis of the Boltzmann equation as in the previous two cases. This phenomenon was discovered more recently in numerical and experimental work by (Aoki et al., 1995) and (Sone and Yoshimoto, 1997). If a uniformly heated plate is placed in the middle of gas contained in a tank of uniform temperature, flow is induced around the edge of the plate for a wide range of Knudsen number. There is no flow induced by thermal creep, since the temperature is uniform everywhere. The induced velocity scales approximately as $\mathcal{O}(\sqrt{\mathrm{Kn}})$.

Lastly, **flow induced by a temperature discontinuity** has been studied by (Aoki, 2001), who set up a flow in a square enclosure, half of which is at temperature T_1 and the other half at temperature T_2 (see Figure 5.4). A flow is induced from the colder to the hotter part along the one wall with the discontinuity at the middle, and this sets up a global circulating flow (see Figure 5.5). The maximum flow speed tends to a constant value at low Kn values and decays for Kn ≥ 0.1.

5.3 Heat Conduction and the Ghost Effect

We examine here the possible breakdown of the heat conduction equation in the limit that Kn $\to 0$. This phenomenon was studied by (Sone et al.,

1996a) and shows a fundamental inconsistency in the momentum and energy (Navier–Stokes) equations. We consider a gas at rest contained in a tank. According to the continuum description, the gas temperature field is described by the heat conduction equation, i.e., the energy equation with all the convective terms absent. Below, we follow the argument of (Sone, 2002; Sone, 2000), that demonstrates that the heat equation is not always appropriate, e.g., in microscales.

Let us consider the energy equation in the continuum limit and examine only the relevant terms (for a monoatomic gas) as follows:

$$\frac{5}{2}\rho u_i \frac{\partial(RT)}{\partial x_i} = \cdots + \frac{\partial}{\partial x_i}\left(k\frac{\partial T}{\partial k_i}\right), \tag{5.4}$$

where k is the thermal conductivity. The corresponding heat conduction equation for a gas at rest is

$$\frac{\partial}{\partial x_i}\left(k\frac{\partial T}{\partial x_i}\right) = 0.$$

The thermal conductivity (k) of a gas is a function of its mass as well as its temperature. Specifically, k (divided by the density of the gas) is proportional to the mean free path with the proportionality coefficient a general function of temperature, i.e.,

$$\frac{k}{\rho} = \mathcal{F}\left(\frac{T}{T_0}\right)\sqrt{2RT}R\lambda.$$

Therefore, in the energy equation the conduction term divided by the density is $\mathcal{O}(\lambda)$. For a gas flow of the order of the mean free path (λ), which is neglected in classical gas dynamics, its convection term is of the same order as the conduction term and cannot be neglected. More specifically, the order of magnitude for the *convection* term is

$$\text{Convection} \propto \mathcal{O}(\rho V RT/L),$$

and the order of magnitude for the *conduction* term is

$$\text{Conduction} \propto \mathcal{O}(\rho(RT)^{3/2}\lambda/L^2),$$

where V and L are, respectively, the characteristic flow speed of the gas and the length scale of variation of the temperature of the gas. Thus, the two terms are comparable when

$$V \propto \mathcal{O}(\sqrt{RT}\lambda/L) \propto \mathcal{O}(\mathrm{Kn}).$$

The question then is whether there are flows of such magnitude; however, we have already seen in this section that both the thermal creep flow and the nonlinear thermal stress flow are possible candidates. These flows are always present, even at extremely small Knudsen numbers, unless all surfaces in the device have uniform temperature *and* all temperature isocontours are parallel. These two conditions are difficult to satisfy simultaneously except in the trivial case of two infinite plates at exactly uniform temperatures with no end effects! Therefore, according to (Sone, 2002) the heat-conduction equation is inappropriate for describing the temperature field of a stationary gas in the continuum limit.

Sone termed this the "ghost effect" in the sense that something that does not exist in the world of a gas in the continuum limit has a finite effect in the molecular world. In (Sone, 2000) an example of the ghost effect is given for a channel with small periodic temperature variation at the two walls. No experiments in microscales have reported the ghost phenomenon, and it is difficult to assess its practical significance; nevertheless, it is a significant result from the fundamental principles point of view.

5.4 Heat Transfer in Poiseuille Microflows

It is possible to create Poiseuille flow in a microchannel using either a pressure drop or a body force. Although the continuum descriptions for both cases are similar, kinetic theory predictions and statistical simulations for force-driven flows show that the Navier–Stokes solutions fail to predict the important temperature minimum effects, even for $\mathrm{Kn} \approx 0.01$ flows with $M < 0.1$ (Zheng et al., 2002). Therefore, we divide this section into two parts, addressing the pressure- and force-driven flows separately.

5.4.1 Pressure-Driven Flows

Heat transfer in pressure-driven gas microflows has important engineering applications. There have been several studies of microflows under constant wall temperature and heat flux conditions. For example, thermally developing slip flows under constant heat flux have been presented for circular and rectangular microchannels in (Ameel et al., 1997; Yu and Ameel, 2002), respectively. A similar problem is studied at the entrance of a conduit by including the heat transfer within the Knudsen layers (Chen and Xu, 2002). Thermally developing flows in circular tubes under constant wall temperature have been studied in the slip flow regime (Larrodé et al., 2000). The constant-temperature convective heat transfer under hydrodynamically and thermally fully developed conditions is considered for various geometries and flow regimes, such as slip flow in rectangular channels (Tunc and Bayazitoglu, 2002) and flow in two-dimensional channels for $0 < \mathrm{Kn} \leq 10$ (Simek and Hadjiconstantinou, 2002).

In this section, we neglect gas compressibility effects and present an analytical treatment of heat transfer for pressure-driven flows. We consider the slip flow regime where the continuum description is valid. Such results were first obtained by (Inman, 1964), who showed that *reduced* heat transfer rates are obtained in the presence of slip flow. Specifically, Inman obtained the following equation for the Nusselt number Nu for constant heat flux q at the wall

$$\frac{1}{\mathrm{Nu}(\mathrm{Kn})} = \frac{17}{140} - \frac{9\,\mathrm{Kn} + 48\,\mathrm{Kn}^2}{35(1+6\,\mathrm{Kn})^2} + \frac{\gamma}{\gamma+1}\frac{\mathrm{Kn}}{\mathrm{Pr}}, \tag{5.5}$$

where the Nusselt number is defined as

$$\mathrm{Nu} \equiv \frac{qD_H}{k\Delta T}.$$

Here D_H is the hydraulic diameter ($D_H = 2 \times (2h) = 4h$ for a channel of total width $2h$), and ΔT is the temperature difference between the wall and the bulk of the fluid. Also, $\mathrm{Nu}(\mathrm{Kn}=0) = 8.235$ is the value corresponding to no-slip conditions. The above equation is based on Maxwell's first-order slip condition and neglects the effect of thermal creep.

Next, we analyze the combined effects of convection and thermal creep. The momentum equation subject to slip boundary conditions with a specified tangential temperature variation (see equation (2.19)) can be solved analytically. The rarefaction effects on momentum transfer can be investigated either by analyzing the volumetric flowrate increase in a pressure-driven channel or by analyzing the change in the skin friction coefficient for a fixed volumetric flowrate, under an appropriately specified pressure gradient. The nondimensional velocity distribution in a channel extending from $y = -h$ to $y = h$ is obtained as

$$\begin{aligned} U(y/h) = {} & \frac{\mathrm{Re}}{2}\left|\frac{\partial p}{\partial x}\right|\left[1 - \left(\frac{y}{h}\right)^2 + 2\left(\frac{2-\sigma_v}{\sigma_v}\right)\frac{\mathrm{Kn}}{1+\frac{1}{2}\mathrm{Kn}}\right] \\ & + \frac{3}{2\pi}\frac{(\gamma-1)}{\gamma}\frac{\mathrm{Kn}^2\,\mathrm{Re}}{\mathrm{Ec}}\frac{\partial T_s}{\partial x}, \end{aligned} \tag{5.6}$$

where $\frac{\partial T_s}{\partial x}$ denotes the tangential temperature variation along the channel surface, and we defined $\mathrm{Kn} = \lambda/h$, with h the half-channel width. Given this parabolic velocity profile we obtain the coefficient of high-order boundary condition from equation (2.39) to be $b = -\frac{1}{2}$. Correspondingly, the volumetric flowrate $\dot{Q}$ through the channel, in nondimensional form, becomes

$$\dot{Q} = \frac{4}{3} + 4\left(\frac{2-\sigma_v}{\sigma_v}\right)\frac{\mathrm{Kn}}{1+\frac{1}{2}\mathrm{Kn}} + \frac{3}{\pi}\frac{(\gamma-1)}{\gamma}\frac{\mathrm{Kn}^2\,\mathrm{Re}}{\mathrm{Ec}}\frac{\partial T_s}{\partial x}, \tag{5.7}$$

since $|\frac{\partial P}{\partial x}| = \frac{2}{\mathrm{Re}}$. The leading-order variation in the volumetric flowrate under fixed $\frac{\partial p}{\partial x}$ is *linear* in Kn due to velocity slip, and *quadratic* in Kn due to thermal creep effects (for fixed Eckert number). However, since $\mathrm{Kn} \propto M/\mathrm{Re}$ and $\mathrm{Ec} \propto M^2$, then we see that the thermal creep term is linear in Kn, i.e., proportional to Kn/M.

In order to maintain zero average flowrate in a channel under a prescribed pressure gradient for an incompressible flow, the following condition should be maintained:

$$\frac{\partial p}{\partial x} = \frac{\frac{9}{2\pi}\frac{(\gamma-1)}{\gamma}\mathrm{Kn}^2 \frac{\partial T_s}{\partial x}}{\mathrm{Ec}(1 + 3(\frac{2-\sigma_v}{\sigma_v})\frac{\mathrm{Kn}}{1+\frac{1}{2}\mathrm{Kn}})}. \tag{5.8}$$

In this case, if

$$\frac{\partial T_s}{\partial x} > 0,$$

the flow creeps from cold to hot along the channel surface, where a positive pressure gradient creates back-flow in the middle of the channel (Kennard, 1938; Loeb, 1961). With regard to the effects of thermal creep on Fanning friction coefficient of the flow for a fixed volumetric flowrate, the ratio of the friction coefficient of a slip surface C_f to the friction coefficient C_{f_0} of a no-slip surface is given by

$$\frac{C_f}{C_{f_0}} = \frac{1 - \frac{3}{\pi\dot{Q}}\mathrm{Kn}^2 \frac{\mathrm{Re}}{\mathrm{Ec}}\frac{\partial T_s}{\partial x}}{1 + 3\frac{2-\sigma_v}{\sigma_v}(\frac{\mathrm{Kn}}{1+\frac{1}{2}\mathrm{Kn}})}. \tag{5.9}$$

It is seen that for fixed flowrate $\dot{Q}$, Eckert number Ec, and Reynolds number Re, the ratio of Fanning friction coefficients of slip flow to the no-slip flow changes significantly by varying the Knudsen number Kn. For flows without thermal creep effects (i.e., $\frac{\partial T_s}{\partial x} = 0.0$), the extra terms in the numerator of equation (5.9) are absent, and the formula is further simplified; see more details in (Beskok and Karniadakis, 1992; Beskok and Karniadakis, 1994).

The above analytical results can be used to validate computer programs for microfluidic applications. Here we present computations obtained with the spectral element program μ**Flow**. Comparisons are performed up to $\mathrm{Kn} = 0.15$, and the results are presented in Figure 5.6. The dashed line and the solid line show the *drag reduction* predicted by the first- and high-order slip flow theory without thermal creep effects, respectively. The triangles correspond to numerical predictions with high-order slip flow theory, and the circles correspond to numerical predictions with high-order slip flow theory including in this case the thermal creep effects (here $\mathrm{Ec} = 1.0$, $\mathrm{Re} = 1.0$, and $\frac{\partial T_s}{\partial x} = 1.0$). The differences between the analytical and numerical results are negligible.

The aforementioned simplified analysis can also be used to explain the drag reduction observed in the experiments reported in (Pfahler et al., 1991). For comparison, the experimental results are also plotted in Figure 5.6. The ratio predicted from equation (5.9) for $\mathrm{Kn} = 0.088$ corresponding

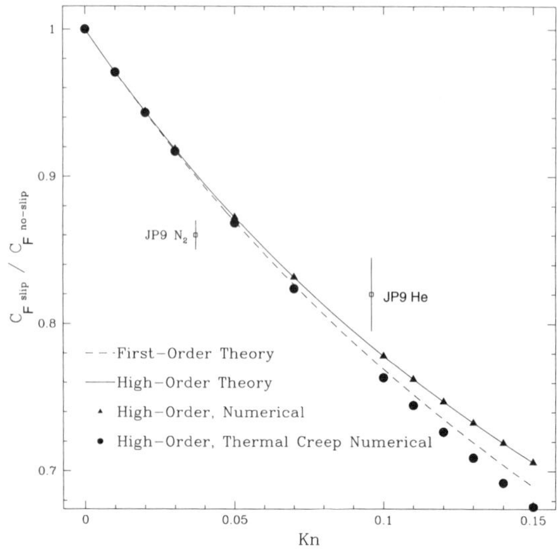

FIGURE 5.6. Ratio of Fanning friction coefficients of slip flow to no-slip flow in a pressure-driven channel. (Parameters for thermal creep contribution are Ec = 1.0, Re = 1.0, and $\frac{\partial T_s}{\partial x} = 1$.)

to the helium flow (case JP9 in (Pfahler et al., 1991)) is 0.79 in reasonable agreement with the measured value 0.8 to 0.85. The nitrogen flow gives a slightly greater drag reduction of about 0.86 compared to the theoretical predictions of 0.9 for Kn = 0.04. Our predictions assume accommodation coefficient $\sigma_v = 1$ and that compressibility effects in the channels are neglected. Thus, the Knudsen number variation in the channels due to compressibility effects is not taken into account, and the Knudsen number is calculated by taking the arithmetic average of the inlet and outlet Knudsen numbers of the microchannel. Furthermore, isothermal flow conditions are assumed, and thermal creep effects are neglected. For channel thicknesses significantly smaller, corresponding to Kn > 0.1, the experimental results show a strong dependence of the ratio of drag coefficients on the Reynolds number, which is not predicted by the above analysis.

Regarding heat exchange in microdomains, it is interesting to note that the viscous heating terms are quite significant; see also (Hadjiconstantinou, 2003b), for a discussion on shear work on solid boundaries for Kn $\leq$ 3. For example, if the reference temperature T_0 is taken to be the room temperature and the specified temperature difference of the domain ΔT is small, the viscous heating effects become nonnegligible for $M \geq 0.05$ (see equation (2.22)). An analytical solution of general heat convection problem for steady and thermally fully developed planar microchannel flows under specified uniform heat flux ($\dot{q}$) on the boundaries can be obtained by

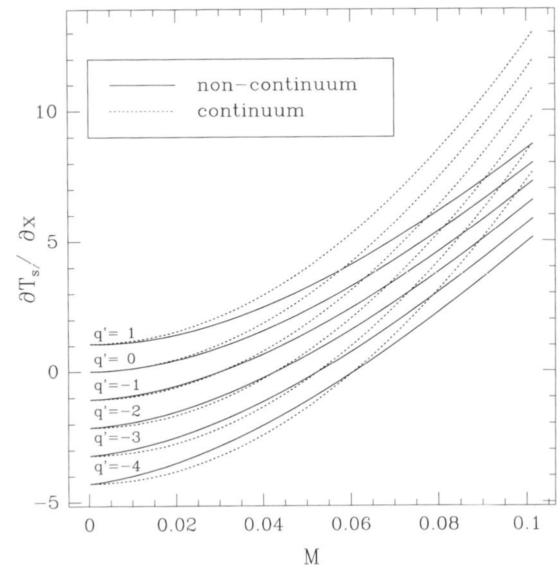

FIGURE 5.7. Variation of tangential temperature gradient ($\frac{\partial T_s}{\partial x}$) along the surface of a pressure-driven channel as a function of Mach number for different levels of heat fluxes ($\dot{q}$) ($\mathrm{Re} = 1.0, \Delta T = 1\,\mathrm{K}$, and $T_0 = 300\,\mathrm{K}$).

decomposing the temperature profile into two parts,

$$T(x, y) = \frac{\partial T_s}{\partial x} x + \theta(y), \tag{5.10}$$

where $\frac{\partial T_s}{\partial x} x$ and $\theta(y)$ show the axial and cross-flow temperature variations, respectively. The coordinates x and y are also nondimensionalized here with the reference length scale. A global energy balance in the domain with an *insulated top surface* and a specified *constant heat flux* on the bottom surface gives the following relation for the tangential temperature gradient along the channel:

$$\frac{\partial T_s}{\partial x} = \frac{1}{\mathrm{RePr}\dot{Q}} \left(\dot{q} + \frac{8}{3}\mathrm{EcPr} \right). \tag{5.11}$$

The corresponding cross-flow temperature distribution in the channel is

$$\theta(y) = \mathrm{RePr}\frac{\partial T_s}{\partial x} \left(B\frac{y^2}{2} - \frac{y^4}{12} \right) - \mathrm{EcPr}\frac{y^4}{3} + Cy + D, \tag{5.12}$$

where

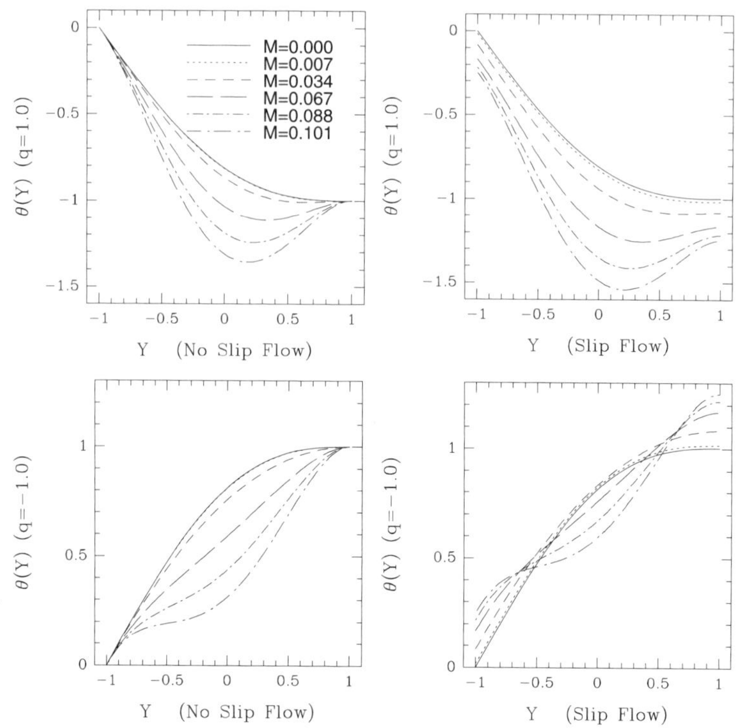

FIGURE 5.8. Variation of temperature profiles in a pressure-driven channel flow for continuum and rarefied flows, with specified heat flux on the bottom surface ($Y = -1$), as a function of Mach number ($\mathrm{Re} = 1.0$ and $\mathrm{Pr} = 0.7$).

$$B = 1 + 2\left(\frac{2-\sigma_v}{\sigma_v}\right)\frac{\mathrm{Kn}}{1+\frac{1}{2}\mathrm{Kn}} + \frac{3}{2\pi}\frac{(\gamma-1)}{\gamma}\frac{\mathrm{Kn}^2\,\mathrm{Re}}{\mathrm{Ec}}\frac{\partial T_s}{\partial x},$$

$$C = \mathrm{RePr}\frac{\partial T_s}{\partial x}\left(\frac{1}{3} - B\right) + \frac{4}{3}\mathrm{EcPr},$$

$$D = \theta_0 - \frac{2\gamma}{\gamma+1}\frac{\mathrm{Kn}}{\mathrm{Pr}}\dot{q} + \frac{5}{3}\mathrm{EcPr} - \mathrm{RePr}\frac{\partial T_s}{\partial x}\left(\frac{3}{2}B - \frac{5}{2}\right),$$

with θ_0 the reference temperature. The modifications to the coefficients B and D due to Kn shows the thermal creep, velocity slip, and temperature jump effects. The continuum solution is recovered as the rarefaction effects diminish (i.e., $\mathrm{Kn} \to 0$).

A quadratic equation for $\frac{\partial T_s}{\partial x}$ can be obtained by combining equations (5.7) and (5.11). The solution for $\frac{\partial T_s}{\partial x}$ for specified heat fluxes is shown in Figure 5.7 as a function of Mach number. Equation (2.22) is used to specify the Eckert number variation for both the continuum and the rarefied flow cases. The Knudsen number variations are specified by equation (2.21). It is seen that the heat flux required to maintain $\frac{\partial T_s}{\partial x} = 0$ is the same for both continuum and rarefied flow curves. The physical significance of this result is that for a specified Mach number there is only a single value of the heat flux required to compensate the viscous heating effects (see equation (5.11)). Another significant result is the reduction in the magnitude of $\frac{\partial T_s}{\partial x}$ in rarefied flows, which implies that microchannels sustain smaller tangential temperature gradients compared to the large-scale channels. Examining equation (5.7) and Figure 5.7, we see that the volumetric flowrate of a heated microchannel increases due to thermal creep effects. However, cooled microchannels allow less volumetric flowrate compared to the continuum case. If the rarefaction effects are increased further, the viscous heating effects will dominate. Under this condition $\frac{\partial T_s}{\partial x}$ may become positive, which will result in increase of the volumetric flowrate beyond the predictions of continuum theory even for cooled channels.

Temperature profiles under *different* heat flux conditions are shown in Figures 5.8 and 5.9. The temperature jump diminishes if both surfaces of the channel are insulated (see Figure 5.9, top). In this case, the maximum temperature occurs near the walls, where shear stresses are more dominant and the tangential temperature variation becomes positive (see Figure 5.7). This suggests significant changes in the volumetric flowrate of the microchannel, which is the main reason for the differences in the temperature profiles of two cases. Thermal creep in a microchannel can be avoided if the channel is carefully cooled at a cooling rate of $\dot{q} = -\frac{8}{3}\text{PrEc}$ (see equation (5.11)); this results in constant temperature along the channel wall. The temperature distribution for a microchannel, without thermal creep effects, is also given in Figure 5.9 (bottom). This result suggests that the temperature of the insulated surface will be greater than its counterpart modeled by the continuum theory.

The analytical results presented for pressure-driven and shear-driven channels (Section 5.5) are also verified with numerical simulations. Since the temperature variation is in the form of a fourth-order polynomial, convergence of the numerical scheme to the exact solution can be obtained with very coarse discretizations. Typically, fourth-order polynomial expansion in μ**Flow** simulation is sufficient to resolve the spatial variations; see Section 14.1.

Remark: The analytical results presented in this section are based on the incompressibility approximation. Therefore, they are valid for *small* temperature and pressure variations in the microchannels. Small temperature variations can be a good approximation, but it has been shown experimentally that the pressure drop in microchannels could be large. In the experiments of (Pfahler et al., 1991), a total pressure drop of about one order of magnitude is reported. Even though the inlet Mach numbers of

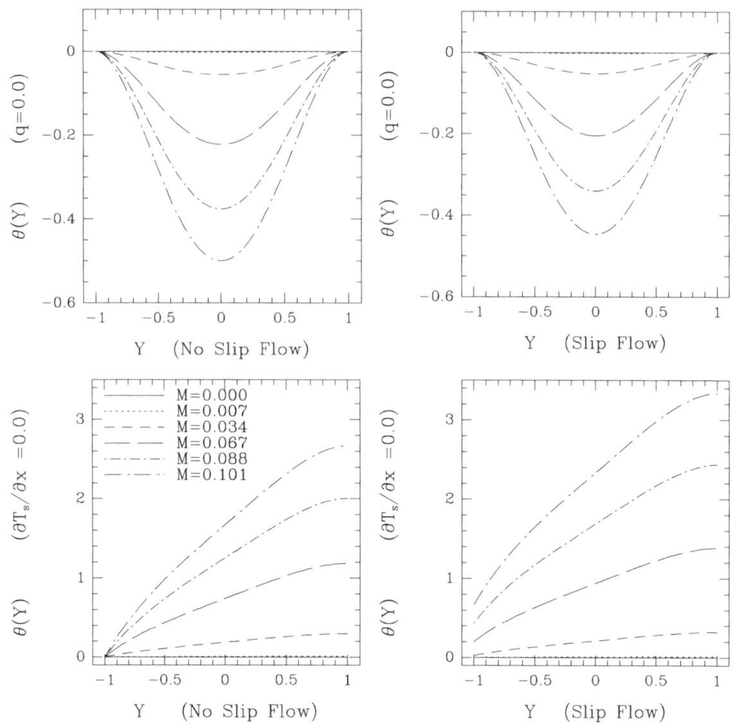

FIGURE 5.9. Temperature profiles in a pressure-driven channel flow for continuum and rarefied flows as a function of Mach number. The top row shows the insulated channel, and the bottom row shows specified heat flux that counterbalances the viscous heating effects, so that overall, $\partial T_s/\partial x = 0$, and therefore there are no thermal creep effects; Re = 1.0 and Pr = 0.7.

the flows are very small, exit Mach numbers up to $M = 0.70$ have been observed. For such situations, the flow in the microchannels cannot be assumed incompressible, and thus the above analysis will not be strictly valid. In general, it is theoretically inconsistent (Aoki, 2001) to use the incompressible flow model with slip boundary conditions caused by rarefaction but the analysis here is meant to highlight approximately the heat transfer effects.

5.4.2 Force-Driven Flows

It is possible to drive channel flows using a body force. The force- and pressure-driven flows are hydrodynamically similar. In either case, the pressure gradient or the force field will be balanced by the viscous shear on the channel walls, and for compressible flows part of the force will be used to accelerate the fluid in the streamwise direction. However, the two driving forces are very different at the microscopic level. The external force accelerates individual particles, while the pressure gradient induces a collective flow (Zheng et al., 2002). The energy equation for these two cases also shows some differences. For example, the pressure creates a cooling effect by flow expansion (the first term on the right-hand side in equation (2.7)), while the body force affects the kinetic energy of the sytstem (Panton, 1984). For pressure-driven compressible flows the expansion cooling negates the viscous heating (last term in equation (2.7)), while viscous heating may play a crucial role in force-driven flows.

(Zheng et al., 2002) performed extensive DSMC and Navier–Stokes studies of pressure- and force-driven Poiseuille flows in the slip flow regime, and compared these with each other and with the kinetic theory predictions. They have shown that the compressible Navier–Stokes equations do not predict the correct flow physics for the force-driven flow case even for Kn $= 0.01$ flows. Specifically, the Navier–Stokes solutions failed to predict the central minimum in the temperature profile and the nonconstant pressure distribution. In Figure 5.10 we present the temperature distribution predicted across the channel using the DSMC and Navier–Stokes solutions. The temperature distribution predicted by the Navier-Stokes solution gives a temperature maximum in the center of the channel, while the DSMC and the kinetic theory predicts two off-center maxima, with a local temperature minimum at the center of the channel. The inability of the Navier–Stokes equations to predict this behavior indicates a global failure, which cannot be corrected by modifications of the transport coefficients, equation of state, or the slip/jump boundary conditions. (Zheng et al., 2002) have shown that such discrepancies between the DSMC and Navier-Stokes predictions do not exist for the pressure-driven Poiseuille flow.

The temperature minima can also be predicted using the *entropic* lattice Boltzmann method (LBM), specifically the so-called minimal kinetic model; see Section 15.5. In this nonisorthermal case, the entropic LBM does a good job in capturing the correct temperature variation. Here we present results of (Ansumali et al., 2003), who performed an extensive study of this flow using the 16-velocity minimal kinetic model. The Knudsen number was varied from Kn $= 0.001$ to Kn $= 0.5$, while the Mach number was varied from $Ma = 0.01$ to $Ma = 0.3$. Some typical simulation results are presented in Figure 5.4.2. Similar to the DSMC results of (Zheng et al., 2002), the minimal kinetic model predicts the temperature minimum. Some important conclusions, which we can draw from the results of (Ansumali et al., 2003) in Figure 5.4.2 are:

- Even at very low Mach numbers, where the flow is often assumed to be isothermal, the temperature variations can be nonnegligible. For example, the simulation results predict that at $Ma = 0.12, \mathrm{Kn} = 0.2$, the temperature variation can be as large as 7 K within a distance of

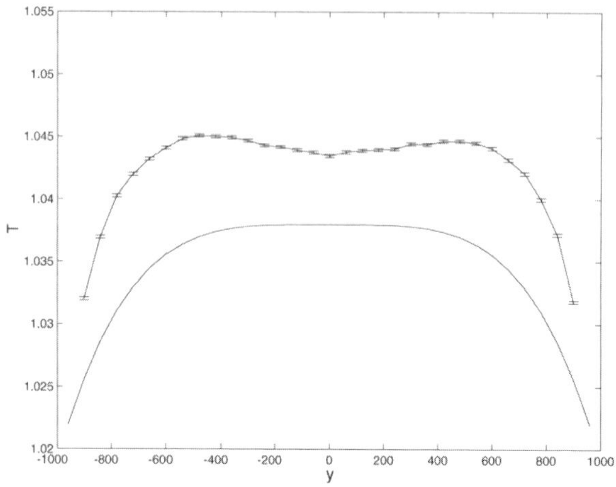

FIGURE 5.10. Temperature profile across a microchannel for the force-driven Poiseuille flow. The curve with the error bars is the DSMC, while the solid line shows the Navier–Stokes results. (Courtesy of A.L. Garcia.)

a few microns.

- The amplitude of the deviation from the isothermal condition is a function of the Knudsen and the Mach number. The deviation increases as the Knudsen or the Mach number increases.
- At a fixed Knudsen number, changing the Mach number does not lead to any qualitative change in the temperature profile.
- At a fixed Mach number the influence of the increase of the Knudsen number is to increase the depth of temperature minimum.
- At a fixed Mach number, as the Knudsen number is increased, the (two symmetric) maxima in the temperature profile move near to the wall.

5.5 Heat Transfer in Couette Microflows

In this section a two-dimensional channel extending from $y = 0$ to $y = h$ is considered. The flow is driven by moving the top wall of the channel with a specified velocity U_0. Thus, this case can be a prototype for a microrotor. Assuming a two-dimensional fully developed flow, the Navier–Stokes equations can be simplified to give linear velocity distribution in the channel. This can be written in nondimensional form

$$\frac{u}{U_0} = U(y/h) = \frac{y/h + \frac{2-\sigma_v}{\sigma_v}\,\mathrm{Kn}}{1 + 2\frac{2-\sigma_v}{\sigma_v}\,\mathrm{Kn}} + \frac{3}{2\pi}\frac{(\gamma-1)}{\gamma}\frac{\mathrm{Kn}^2\,\mathrm{Re}}{\mathrm{Ec}}\frac{\partial T_s}{\partial x}. \tag{5.13}$$

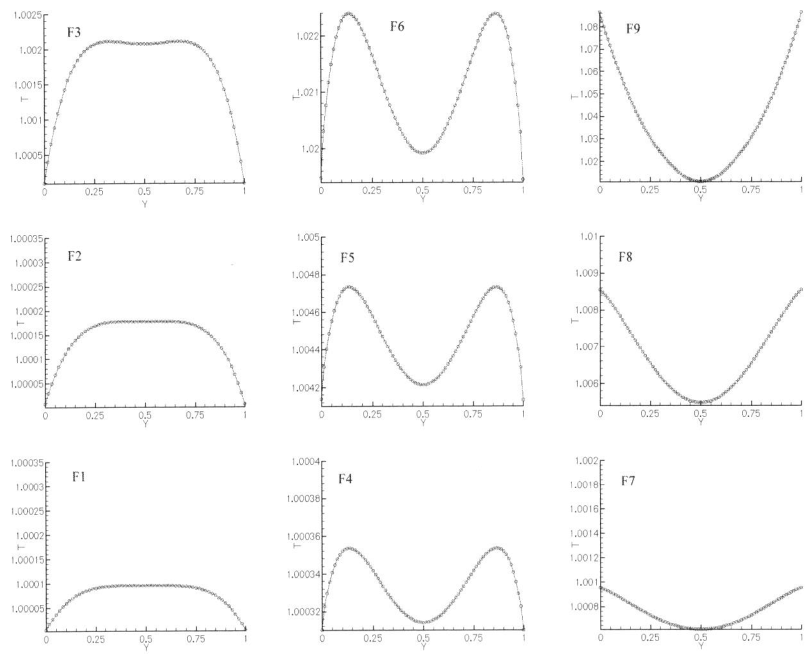

FIGURE 5.11. Temperature profile across the channel as a function of the Knudsen and the Mach numbers. In the figure, the Knudsen number increases going from left to the right, while the Mach number increases going from the bottom to the top. The dimension of the simulated channel was $L/W = 20$. The Knudsen and Mach numbers (Kn, M) used in the simulations were: Plot F1: $(0.001, 0.01)$, Plot F2: $(0.001, 0.02)$, Plot F3: $(0.001, 0.05)$, Plot F4: $(0.20, 0.01)$, Plot F5: $(0.20, 0.05)$, Plot F6: $(0.20, 0.12)$, Plot F7: $(0.45, 0.01)$, Plot F8: $(0.45, 0.04)$, Plot F9: $(0.45, 0.17)$, respectively. (Courtesy of I. Karlin and S. Ansumali.)

The corresponding volumetric flowrate per channel width is

$$\dot{Q} = \frac{1}{2} + \frac{3}{2\pi} \frac{(\gamma - 1)}{\gamma} \frac{\mathrm{Kn}^2\, \mathrm{Re}}{\mathrm{Ec}} \frac{\partial T_s}{\partial x}. \tag{5.14}$$

It is seen that thermal creep effects result in change of the flowrate of the channel. The ratio of friction coefficients of the shear-driven slip flow to a continuum flow is given by

$$\frac{C_f}{C_{f_0}} = \frac{1}{1 + 2\frac{2-\sigma_v}{\sigma_v}\,\mathrm{Kn}}. \tag{5.15}$$

The above equation is obtained for constant mass flowrate in the channel. If thermal creep effects are considered, the driving velocity U_o of the channel must be modified to keep the volumetric flowrate constant. Therefore, the

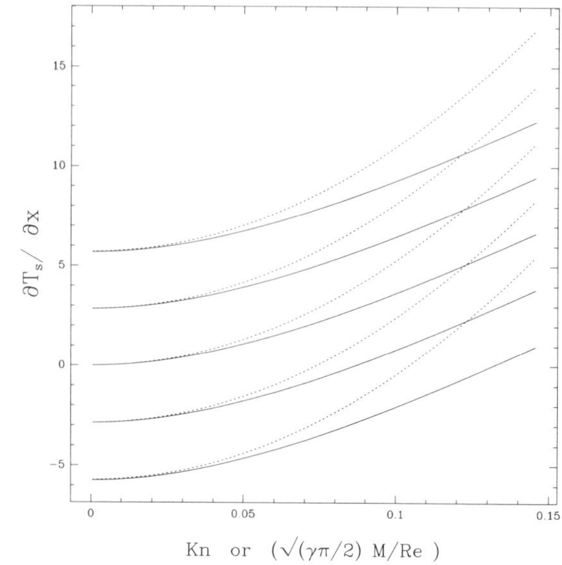

FIGURE 5.12. Variation of tangential temperature gradient ($\frac{\partial T_s}{\partial x}$) along the surface of a shear-driven channel as a function of Mach number for different levels of heat fluxes ($\dot{q}$). Dashed line: continuum; solid line: rarefied. ($\mathrm{Re} = 1.0, \Delta T = 1\,\mathrm{K}$, and $T_0 = 300\,\mathrm{K}$).

thermal creep effects are not included in the derivation of equation (5.15).

Heat convection analysis for a steady and thermally fully developed shear-driven microchannel is obtained by decomposing the temperature profile into two parts, as given in equation (5.10). The channel is assumed to have an insulated top surface and a bottom surface with a specified heat flux ($\dot{q}$). With this decomposition, the temperature variation across the channel becomes a third-order polynomial given by

$$\theta(y) = \frac{A}{6}y^3 + \frac{B}{2}y^2 - \left(\frac{A}{2} + B\right)y + C, \tag{5.16}$$

where

$$A = \frac{\mathrm{RePr}\frac{\partial T_s}{\partial x}}{1 + 2\,\mathrm{Kn}},$$

$$B = \frac{\mathrm{RePr\,Kn}\,\frac{\partial T_s}{\partial x}}{1 + 2\,\mathrm{Kn}} + \frac{3}{2\frac{\partial T_s}{\partial x}}\,\frac{(\gamma - 1)}{\gamma}\,\frac{\mathrm{Kn}^2\,\mathrm{Re}^2\mathrm{Pr}}{\mathrm{Ec}}\left(\frac{\partial T_s}{\partial x}\right)^2 - \frac{\mathrm{EcPr}}{(1 + 2\,\mathrm{Kn})^2},$$

$$C = \theta_0 - \frac{2\gamma}{\gamma + 1}\,\frac{\mathrm{Kn}}{\mathrm{Pr}}\dot{q},$$

and

$$\frac{\partial T_s}{\partial x} = \frac{2}{\mathrm{RePr}\dot{Q}}\left[\dot{q} + \frac{\mathrm{EcPr}}{(1 + 2\,\mathrm{Kn})^2}\right]. \tag{5.17}$$

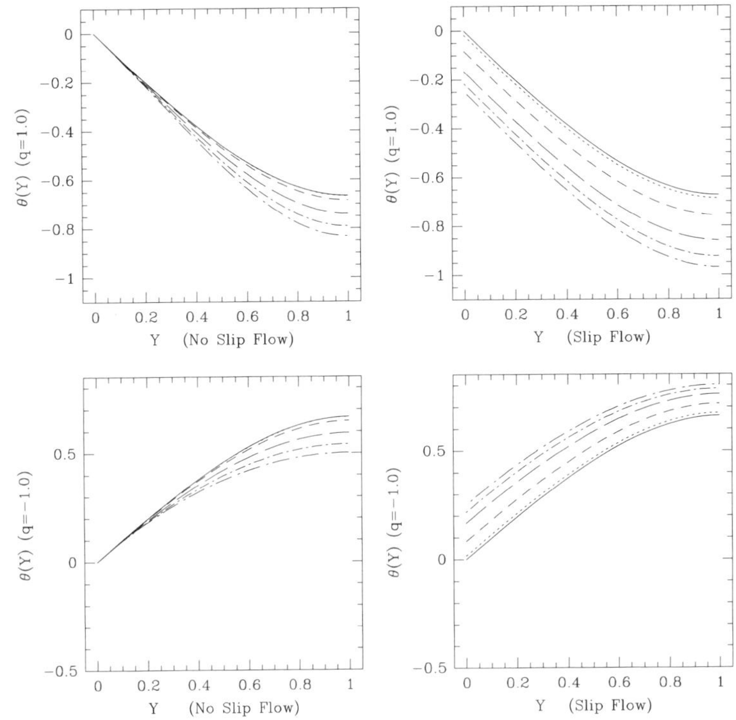

FIGURE 5.13. Variation of temperature profiles in a shear-driven channel flow for continuum and rarefied flows, with specified heat flux at the bottom surface, as a function of Mach number. $Y = 0$ corresponds to a stationary wall, and $Y = 1$ corresponds to a moving wall (Re = 1.0 and Pr = 0.7).

A quadratic equation for $\frac{\partial T_s}{\partial x}$ can be obtained by combining equations (5.14) and (5.17). The solution for $\frac{\partial T_s}{\partial x}$ for specified heat fluxes is shown in Figure 5.12 as a function of Knudsen number. Equation (2.22) is used to specify the Eckert number variation for both the continuum and the rarefied cases. The Knudsen number variations are also specified by equation (2.21). It is seen that the heat flux required to maintain $\frac{\partial T_s}{\partial x} = 0$ for a specified Mach number is smaller in microchannels compared to the continuum case. The viscous heating effects are more dominant in the continuum case compared to microscales; see also (Wendl and Agarwal, 2002). This leads to different results for cooled and heated channels with respect to thermal creep effects.

Variation of temperature profiles across the channel is given in Figure 5.13 as a function of Mach number. Both the continuum and the rarefied flow cases are presented. Here, the nondimensional heat flux of $\dot{q} = \pm 1$ is specified at the bottom surface of the channel. The temperature jump

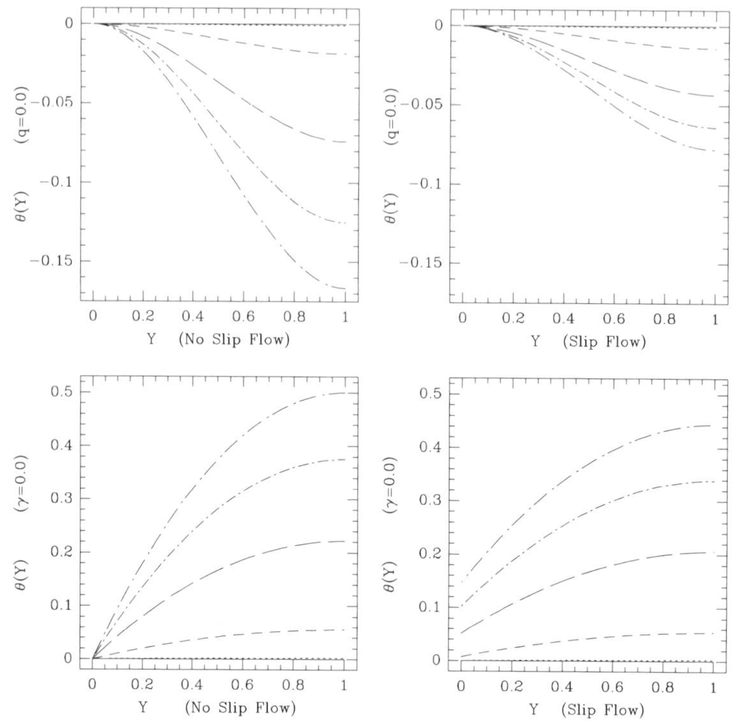

FIGURE 5.14. Variation of temperature profiles in a shear-driven channel flow for continuum and rarefied flows. The top row shows insulated channels, and the bottom row shows cooled channels with equation (5.18) such that there are no thermal creep effects (i.e., $\frac{\partial T_s}{\partial x} = 0$). $Re = 1.0$ and $Pr = 0.7$.

effects are clearly seen in the rarefied flow case. In a microchannel the temperature of the insulated surface is less than that of the continuum predictions for heated channels, while the opposite is true for cooled channels. The effects of thermal creep on volumetric flowrate of the channel depends on the direction that the flow is sheared by the top surface, and whether the channel is cooled or heated. As long as the driving velocity U_0 is in the same direction with increasing $\frac{\partial T_s}{\partial x}$, the volumetric flowrate of the channel will increase due to the thermal creep effects.

The temperature jump diminishes if both surfaces of the channel are insulated (see Figure 5.14, top). For this case, the viscous heating effect in a microchannel is less than the continuum prediction. Therefore, the temperature differences of top and bottom surfaces are relatively small in microchannels compared to the continuum case. It is also possible to specify heat flux on the boundaries that will cancel the viscous heating effects. This case gives zero tangential temperature gradient on the microchannel walls.

The heat flux necessary to maintain this condition is

$$\dot{q} = -\frac{\mathrm{EcPr}}{(1 + 2\,\mathrm{Kn})^2}. \tag{5.18}$$

The temperature variation for a microchannel without thermal creep effects is also given in Figure 5.14 (bottom). This result suggests that the temperature of the insulated surface will be smaller than its counterpart modeled by the continuum theory.

Remark: The results presented in this section assume *small* temperature and pressure fluctuations compared to the reference pressure and temperature. However, planar shear-driven channels have zero pressure gradient. Thus, the only limitation on the incompressibility assumption is small temperature fluctuations, which is satisfied, since $\Delta T \approx \mathcal{O}(1)$ in this study.

6

Prototype Applications of Gas Flows

In this chapter we consider rarefied gas flows encountered in applications other than simple microchannels. In the first section, we consider the lubrication theory, focusing mainly on the slider bearing and squeezed film problems. The slider bearing problem is relevant for magnetic storage units, where the head levitates about 50 nm above the media, corresponding to transition flow regime. The squeezed film flows are important for design of micromechanical accelerometers, where the gas is used as the damping medium, and the desired frequency response is obtained by controlling carefully the gas pressure.

In the second and third sections, we consider separated flows in internal and external geometries in the *slip flow* regime. To this end, we employ continuum modeling in conjunction with slip effects. The objective of this study is to investigate the validity of continuum-based slip models under flow separation. Such conditions are encountered in typical components of microsystems such as microchannels and microprobes, as well as in low-pressure applications, e.g., flow inside instruments operating at high altitude.

In the fourth section, we present theoretical and numerical results for Stokes flow past a sphere. We review the classical Stokes drag for external flows, and include rarefaction effects in the slip flow regime. We present drag formulas for the pressure-driven flow past a stationary sphere confined in a pipe. We verify these with numerical simulations in the slip flow regime, which shows drastic variations in the drag coefficient as a function of the Knudsen number, and the cylinder/sphere blocking ratio. The continuum flow ($\mathrm{Kn} \rightarrow 0$) results of this section are equally applicable to liquid flow past solid, electrically neutral spheres.

In the fifth part we summarize recent findings on gas flows through microfilters. Such devices are utilized for capturing and detecting airborne biological and chemical particles as well as for environmental monitoring applications.

In the last section, we consider high-speed rarefied flows in micronozzles, which are used for controlling the motion of microsatellites and nanosatellites. Nozzles with a throat as small as 20 μm have been fabricated with outlet Mach number exceeding four!

6.1 Gas Damping and Dynamic Response of Microsystems

The design of some MEMS devices requires good understanding of dynamic response characteristics of the system under various actuation conditions. This response can be analyzed by different methods, ranging from the most simplistic mass–spring–damper *lumped analysis* to the coupled interaction models, which resolve both the geometric and physical complexity of the problem (see Section 18.2). Since most of the MEMS devices work under ambient conditions, the gas behavior can have significant effects in the dynamic response of these systems. Modeling of gas damping effects is difficult for the following reasons:

- Geometric complexity of MEMS devices,
- Compressibility effects resulting in complicated frequency response, and
- Onset of rarefaction effects in micron- and submicron-scale clearances.

An example demonstrating the geometric complexity of MEMS is the Digital Micromirror Device of Texas Instruments (DMD) shown in Figure 6.1. The DMD is a MEMS device used in Digital Light Processing applications, and it consists of half a million (848×600) to over one million (1280×1024) micromirror pixels. The top left figure shows an SEM picture of nine mirrors, while the underlying hinge structure hidden below the mirrors is shown in the top right plot. The SEM images of the mirror substructure for several pixels and a close-up view of a single mirror substructure are presented in the bottom left and right figures, respectively. The DMD has $16\,\mu\text{m} \times 16\,\mu\text{m}$ pixel geometry with $1\,\mu$m pixel separation. Each mirror oscillates ± 10 degrees with a response time of $15\,\mu$s.

The DMD is packaged in a mostly nitrogen environment under ambient conditions. The presence of gas affects the device performance significantly. A typical response of a row of 840 19 μm size mirror array subject to a

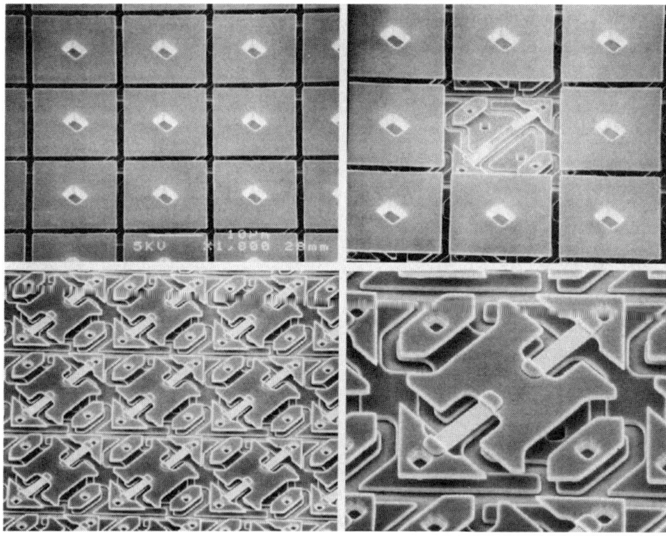

FIGURE 6.1. Texas Instruments Digital Micromirror Device. The upper left view shows nine mirrors, while the upper right view shows the central mirror removed to expose the underlying structure. The lower right shows a close-up view of the mirror substructure, and the lower left view shows several pixels with the mirror removed. (Courtesy of Texas Instruments.)

step pulse under various air pressures is shown in Figure 6.2. Operation in atmospheric pressure results in the desired fast optical response, while reductions in pressure correspond to oscillatory response with slower time decay. The decay rate is an indication of the gas damping effects, which are reduced with decreased pressure. Since the DMD dimensions are fixed and the reference pressure is reduced, the lower pressure data correspond to increased Knudsen regimes, indicating a strong rarefaction effect on the dynamic response of the DMD mirror.

Similarly, the micromotors rotating on a substrate or microcomb drives (discussed in Section 1.1) oscillating with very high frequencies also experience gas damping effects. Although the primary motion in such devices is lateral, depending on the allowable degrees of freedom, up and down motion and tilting in the pitch and yaw angles are also possible (Freeman et al., 1998). Also, MEMS torsion mirrors with high frequency pitch oscillations experience significant squeezed film effects (Pan et al., 1998). In this section, we will address the fluid-mechanical aspects of various types of motion experienced by moving MEMS components, which can be summarized as one or a combination of the following:

1. Lateral motion between two surfaces: oscillatory Couette flow (see Section 3.3).

2. Normal oscillations between two surfaces: squeezed film.

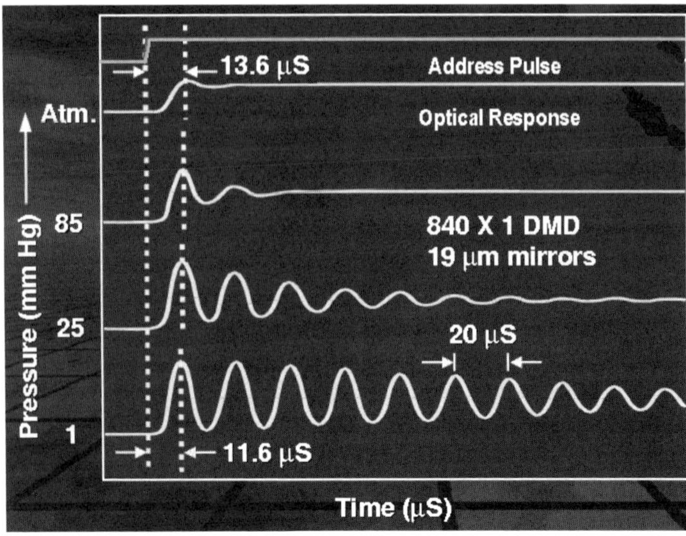

FIGURE 6.2. Dynamic response of the DMD mirrors subject to a step pulse under various air pressures. (The data were obtained by Dr. Larry Hornbeck (1988); Courtesy of Texas Instruments.)

3. Lateral motion with slight inclination: slider bearing.

4. Pitch-angle oscillatory motion between two plates: squeezed film.

All of these situations, with the exception of the first case, can be analyzed using the **Reynolds equation** for thin-film lubrication presented below.

The first case corresponds to fluid motion between oscillating parallel plates. Detailed analysis of this motion and the associated damping characteristics in the entire Knudsen regime were presented in Section 3.3. In oscillatory Couette flows, the lateral plate motion does not generate streamwise pressure variations. Hence, this case cannot be described by the Reynolds equation. In contrast to the squeezed film and the slider bearing problems, there will be no lift. We identify this case as the oscillatory Couette flow, which is a variation of the classical Stokes second problem (Batchelor, 1998) (flow over a laterally oscillating surface) with inclusion of a wall over a thin fluid gap of h_o. The lateral vibrations of comb-drives (Section 3.3) and the corresponding viscous drag (damping) can be approximately analyzed using the oscillatory Couette flow model. In its simplest form, the governing equation is reduced to

$$\frac{\partial u}{\partial t} = \nu \frac{\partial^2 u}{\partial y^2}, \tag{6.1}$$

where ν is the kinematic viscosity, and u is the streamwise velocity. For oscillatory flows with a specified frequency ω_o and corresponding boundary conditions $u = U_o \exp(i\omega_0 t)$ and $u = 0$ on the oscillating and stationary surfaces, respectively, an analytic solution of the above equation can be found in (Panton, 1984; Sherman, 1990). The results can be summarized as a function of the normalized penetration distance given by

$$Y = \frac{y}{\sqrt{\nu/\omega}}.$$

If the normalized penetration distance between the two walls is $Y > 7$, then the plates are sufficiently away from each other, and thus they do not interact. Equation (6.1) can also be solved using various slip conditions. A study of damping models for laterally moving microstructures with gas rarefaction effects can be found in Section 3.3 and in (Veijola, 2000; Park et al., 2004).

6.1.1 Reynolds Equation

Background material for thin-film lubrication theory can be found in (Szeri, 1998). For squeezed film or slider-type applications, the Navier–Stokes equations are reduced to the Reynolds equation (Sherman, 1990):

$$\nabla \cdot \left[\left(\frac{\rho h^3}{\mu} \right) \nabla p \right] = 12 \frac{\partial (\rho h)}{\partial t} + 6 \nabla \cdot (\rho h \mathbf{U}_o), \tag{6.2}$$

where ρ and p are the local gas density and pressure, respectively. The local film thickness is h, μ is the dynamic viscosity, t is time, and $\mathbf{U}_o$ is the lateral velocity of the moving plate. The first and second terms on the right-hand side correspond to the normal and lateral plate motions, respectively.

Derivation of the Reynolds Equation

The Reynolds equation is derived starting from the Navier–Stokes equations presented in Chapter 2. For simplicity, we analyze two plates with a small gap between them. The upper plate is placed at a slight angle with respect to the lower plate, while the latter is moving from left to right with velocity $\mathbf{U}_o$, as shown in Figure 6.3. For simplicity, we assume that the plate length L is much larger than the plate separation distance h_o and that the angle of inclination α is small. We also assume that the plate width W in the spanwise direction is larger than h_o, and that the plate motion is solely in the x direction, as shown in Figure 6.3. Therefore, the flow is two-dimensional in the (x, y) plane. Details of the derivation of the Reynolds equation can be found in (Panton, 1984). Here, we briefly outline this derivation and discuss the underlying assumptions and limitations.

The ratio of the inertial forces to the viscous forces in the Navier–Stokes equations is given by

$$\frac{\rho u \partial u/\partial x}{\mu \partial^2 u/\partial y^2} \approx \frac{\rho \mathbf{U}_o^2/L}{\mu \mathbf{U}_o/h_o^2} = \frac{\rho \mathbf{U}_o L}{\mu} \cdot \left(\frac{h_o}{L} \right)^2 = R^*. \tag{6.3}$$

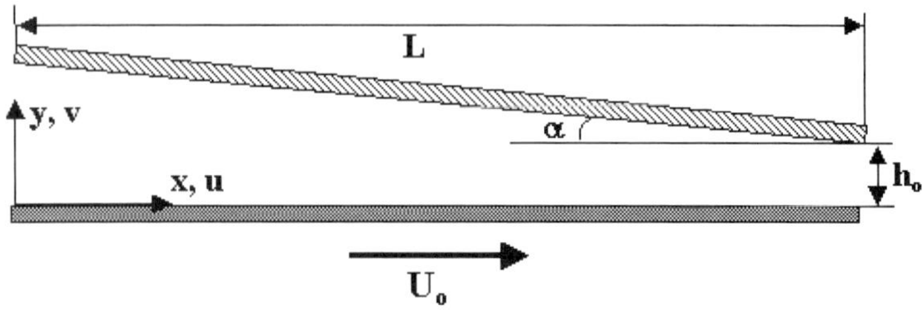

FIGURE 6.3. A schematic view for the slider bearing problem.

Therefore, the inertial forces can be neglected with respect to the viscous forces if $R^* \equiv \text{Re} \cdot \left(\frac{h_o}{L}\right)^2 \ll 1$. In most MEMS applications, due to the very small Reynolds number ($\text{Re} = \frac{\rho U_o L}{\mu}$) of the flow and the small h_o/L ratio, this condition is satisfied. A comparison of the streamwise and the cross-flow momentum equations reveals that pressure variation across the channel is very small compared to the pressure variation along the streamwise direction. Similarly, the cross-flow velocity (v) is negligible compared to the streamwise velocity component (u). Hence, the leading-order steady-state solution is determined from the streamwise momentum equation simplified in the following form:

$$\frac{dp}{dx} = \mu \frac{\partial^2 u}{\partial y^2}. \tag{6.4}$$

This equation is identical to the fully developed channel flow equation. However, in the slider problem, equation (6.4) is an *approximation* valid for $R^* \ll 1$. Here, the flow is not fully developed but varies slowly in the streamwise direction due to gradual changes in the channel area. Since

$$\frac{\partial u}{\partial x} \bigg/ \frac{\partial u}{\partial y} \approx \frac{h_o}{L} \ll 1,$$

the streamwise velocity is a function of y only to the leading-order approximation. The velocity distribution $u(y)$ is obtained by integrating equation (6.4) using either the no-slip or slip boundary conditions. If temperature variations in the system are neglected, the density becomes a function of pressure, which varies mainly in the streamwise direction. Using the velocity distribution and the local density (ρ), the following equation for the mass flowrate (per channel width) is obtained for no-slip flows:

$$\dot{M} = \frac{1}{2} U_o \rho h - \frac{h^3 \rho}{12\mu} \frac{dp}{dx}. \tag{6.5}$$

Since the flow is assumed isothermal, the density can be written as a function of the pressure using the equation of state $\rho = p/RT$. In equation (6.5), the left-hand side is constant, but the right-hand side is a function of x. Furthermore, the mass flowrate is a function of the pressure gradient, which is unknown. Taking the gradient of (6.5), we obtain the Reynolds equation for one-dimensional steady flow, i.e.,

$$\frac{\partial}{\partial x}\left(h^3 p \frac{dp}{dx}\right) = 6\mu \frac{\partial}{\partial x}\left(phU_o\right). \tag{6.6}$$

Since the spatial variation of the channel height (h) is known, the pressure is the only unknown in this equation. The boundary conditions correspond to ambient pressure at the two ends of the slider bearing geometry, shown in Figure 6.3. The terms in parentheses on the left and right of equation (6.6) are proportional to the mass flowrate per channel width (divided by RT) in plane Poiseuille and Couette flows, respectively. Furthermore, in this form the Reynolds equation neglects the spatial variations across the slider bearing. Hence, a two-dimensional problem is reduced to a one-dimensional equation for pressure.

The Reynolds equation can be nondimensionalized by normalizing the pressure using the ambient pressure p_o, and the length scales in the x and y directions with channel lengths L and h_o, respectively. This results in (Alexander et al., 1994)

$$\frac{\partial}{\partial X}\left(H^3 P \frac{dP}{dX}\right) = \Lambda \frac{\partial}{\partial X}\left(PH\right), \tag{6.7}$$

where $X = x/L$, $H = h/h_o$, $P = p/p_o$ and

$$\Lambda = \frac{6\mu U_o L}{p_o h_o^2}$$

is the *bearing number.*

The Reynolds equation gives good results for low-speed flows. For high-speed flows, viscous heating may affect the isothermal flow approximation, and thus some deviations from the predictions of the Reynolds equation are expected. Due to the small length scales of MEMS devices, typical surface speeds correspond to low subsonic flow conditions, and therefore viscous dissipation is negligible. However, very high angular speeds present in certain MEMS devices can lead to near-sonic conditions. We conclude then that the analysis presented here will be invalid for cases with significant heat transfer and viscous dissipation effects.

Reynolds Equation in the Slip Flow Regime

For gas microflows the rarefaction effects can be incorporated into the Reynolds equation using the first-order velocity slip condition. The following equation has been obtained (Burgdorfer, 1959):

$$\frac{\partial}{\partial X}\left(\left[1 + 6\frac{2-\sigma_v}{\sigma_v}\mathrm{Kn}\right] H^3 P \frac{dP}{dX}\right) = \Lambda \frac{\partial}{\partial X}\left(PH\right), \tag{6.8}$$

where $\mathrm{Kn} = \lambda/h$ is the local Knudsen number. This relation is valid strictly in the slip flow regime ($\mathrm{Kn} < 0.1$). Since for air, $\lambda_{\mathrm{air}} \simeq 65$ nm under ambient conditions, a gap height of $h_o = 0.6\,\mu$m is the smallest clearance for which equation (6.8) can be used. Although this is a reasonably small clearance for many MEMS applications, it does not cover all applications. For example, the distance between the read/write head and the media in current computer hard disk drives is of order 50 nm. The design goal of the next generation hard disk drives is to reach a read/write density of 100 Gbit/in^2, which requires 5 to 10 nm separation distance between the head and the media. Hence, the current hard drives already operate in the transition flow regime, while the next-generation drives will push this limit toward the free-molecular flow regime. Therefore, successful design of such microfluidic systems requires development of lubrication equations valid in a wide range of Knudsen number.

Reynolds Equation in the Transition Flow Regime

Gas lubrication effects had been a primary interest for scientists long before the advent of the MEMS technology. For example, the work of (Burgdorfer, 1959) was motivated by lubrication characteristics in the slip flow regime. Similarly, (Hsia and Domoto, 1983) performed a series of experiments for $0.04 \leq \mathrm{Kn} \leq 2.51$ using different gases in order to change the mean free path, with bearing gaps of $0.075\,\mu\mathrm{m} \leq h_o \leq 1.6\,\mu$m. These authors also derived a second-order slip boundary condition given in Table 2.2, and compared experimental results with the predictions of the Reynolds equation employing their second-order slip model for a Winchester type slider mechanism. In the analysis of Hsia and Domoto, the second-order slip model based on the continuum approximation was applied at the limit of the transition flow regime. Although a reasonable match between the numerical and experimental results for integral quantities like the bearing load capacity was obtained, deviations in the squeezed film damping characteristics between the slip and the transition flow regimes exist. For example, the streamwise momentum equation (6.4) is based on the continuum-based constitutive models of viscous stress tensor. This is the starting point of the (Navier–Stokes-based) Reynolds equation, and it should be *modified* in the transition flow regime, using the viscous stress tensor of the Burnett equations given in equation (2.25). An alternative approach is to utilize Grad's 13-moment equations (Grad, 1949), as demonstrated in (Chan and Sun, 2003).

The squeezed film problem can be thought of as a combination of pressure-driven and shear-driven flows. Therefore, one can foresee the differences between the Navier–Stokes and Burnett level analyses by separately analyzing the pressure- and shear-driven flow conditions. In Section 4.2.1 we presented the asymptotic analysis of pressure-driven flows in large aspect ratio channels under isothermal conditions as a function of $\epsilon = h_o/L \rightarrow 0$.

This resulted in a modified streamwise momentum equation, which could be reduced to the Navier–Stokes level only for low subsonic ($M \ll 1$) isothermal flows in large aspect ratio channels. One important deviation here is in the cross-flow momentum equation, where we have shown that the pressure variation in this direction is balanced by the viscous normal stresses of the Burnett equations (see Section 4.2.1). Since the (Navier–Stokes-based) Reynolds equation neglects the pressure variations in the cross-flow direction, we may expect some deviations between the Burnett- and Navier–Stokes-based models. However, these deviations are proportional to the Mach number in equation (4.17b), and thus they can be neglected for most low Mach number applications.

The linear Couette flow in the transition flow regime has been investigated in (Schamberg, 1947), using the Burnett equations. Schamberg showed that the streamwise momentum equation can be reduced to the Navier–Stokes level, resulting in linear velocity variation across the channel under negligible heat transfer and compressibility effects (Schamberg, 1947). However, the velocity distribution must be subject to a second-order slip condition (see also Section 3.2). Recalling that

- *the flowrate of linear Couette flow is independent of the Knudsen number,*

we can conclude that for low-speed isothermal flows in long channels, the Reynolds equation with second-order slip corrections is similar to the Reynolds equation derived using the Burnett equations. Hence, the agreement of experimental and numerical results presented by Hsia and Domoto in integral quantities can be explained. Fukui and Kaneko (1988) have shown that the second-order (slip based) Reynolds equation is valid for $\mathrm{Kn} \leq 1$. This is expected, since the Burnett equations and the corresponding second-order slip models are derived via a second-order Chapman–Enskog expansion, which is a perturbation expansion in Kn, and it is valid for $\mathrm{Kn} < 1$.

Generalized Reynolds Equation

A generalized gas film lubrication analysis can be obtained using the Boltzmann equation, which is valid in the entire Knudsen regime. Based on this idea, Fukui and Kaneko analyzed gas film lubrication using the linearized Boltzmann equations (Fukui and Kaneko, 1988; Fukui and Kaneko, 1990), and presented their results as a function of the *local inverse Knudsen number*

$$D = \frac{\sqrt{\pi}}{2\,\mathrm{Kn}} = D_o P H, \tag{6.9}$$

where $D_o = p_o h_o / (\mu \sqrt{2RT_o})$ is the characteristic inverse Knudsen number defined using the minimum film thickness (h_o) and the ambient pressure

(p_o). The primary assumptions in linearizing the Boltzmann equations are (see also Section 15.4):

1. Small flow velocities and thermal fluctuations; hence, local isotropic equilibrium conditions are assumed.

2. Small lubrication film thickness and small cross-flow velocity components.

Based on these two assumptions, the following nondimensional Reynolds equation is obtained (Fukui and Kaneko, 1988):

$$\frac{d}{dX}\left[\left(\overline{Q_P}(D_oPH)\frac{dP}{dX} - \overline{Q_T}(D_oPH)P\frac{dT_w}{dX}\right)PH^3 - \Lambda PH\right] = 0, \quad (6.10)$$

where $\overline{Q_P}(D_oPH)$ and $\overline{Q_T}(D_oPH)$ show the relative volumetric flowrate of pressure-driven and thermal creep channel flows, normalized by the no-slip Poiseuille flowrate. The last term in equation (6.10) corresponds to the linear Couette flow volumetric flowrate, which is *independent of the Knudsen number*. The thermal creep flow $\overline{Q_T}(D_oPH)$ is present under tangential temperature gradients, as shown in Section 5.1. The $\frac{dT_w}{dX}$ term in equation (6.10) is the nondimensional *temperature* variation along the surface. Since the thermal creep effects are present only under surface temperature gradients, the term $\overline{Q_T}(D_oPH)$ is zero for isothermal surfaces.

DSMC studies of slider bearing flows were presented in (Alexander et al., 1994). For low subsonic slider motion, good agreement between the DSMC and the predictions of equation (6.10) is shown. One peculiar difference between the results of the Reynolds equation and DSMC is in the load-carrying capacity of the slider bearing for high-speed flows. For near-sonic conditions, calculation of pressure on the slider surface using the ideal gas law agreed well with the Reynolds-equation-based calculations. However, the load capacity calculated from the time-averaged change in the momentum of particles striking the wall predicted 20% lower load capacity. This difference was attributed to the increased nonequilibrium effects for high-speed rarefied gas flows (Alexander et al., 1994).

The Reynolds equation derived from the linearized Boltzmann equation has a striking resemblance to the Navier–Stokes-based Reynolds model. Both models utilize the mass flowrate of linear Poiseuille and Couette flows (divided by RT). This enables a *generalized Reynolds equation*, with smooth transition between the continuum (Kn $\rightarrow 0$) and the free-molecular flow (Kn $\rightarrow \infty$) limits. Fukui and Kaneko obtained a *uniformly valid* Reynolds equation by employing the Boltzmann solutions for $\overline{Q_P}$ and $\overline{Q_T}$ for different Knudsen number regimes (Fukui and Kaneko, 1988; Fukui and Kaneko, 1990). A uniform approximation of $\overline{Q_P}$ is given in (Alexander et al., 1994):

$$\overline{Q_P}(\text{Kn}) \approx 1 + \frac{3\sqrt{\pi}A}{D} + \frac{6}{\sqrt{\pi}D}\log\left(1 + \frac{B\sqrt{\pi}}{2D}\right), \quad (6.11)$$

where $A = 1.318889$ and $B = 0.387361$. The above equation is asymptotically correct in the no-slip and free-molecular flow limits. However, the model looses its accuracy in the transition flow regime with nominal error of $\pm 5\%$. Also, equation (6.11) does not include the accommodation coefficient dependence on the flowrate. Fukui and Kaneko developed a database using analytical and numerical approximations for $\overline{Q_P}$ and $\overline{Q_T}$ valid for various Knudsen number regimes and accommodation coefficients (Fukui and Kaneko, 1990). They employed this database in the generalized Reynolds equation. Also, Veijola et al. (1998) developed a uniform approximation for Fukui and Kaneko's database, including the effects of the accommodation coefficients in the following form:

$$Q_P(D, \sigma_v) = \frac{D}{6} + \frac{1}{\sigma_v^{1.34}\sqrt{\pi}} \ln\left(\frac{1}{D} + 4.1\right) + \frac{\sigma_v}{6.4} + \frac{1.3(1-\sigma_v)}{1+0.08D^{1.83}} + \frac{0.64\sigma_v D^{0.17}}{1+1.2D^{0.72}},$$

where $Q_P(D, \sigma_v)$ is the Poiseuille flowrate, not normalized with the continuum flowrate ($D \to \infty$) limit, and σ_v is the accommodation coefficient. The above formula is valid for $D \geq 0.01$ and $0.7 \leq \sigma_v \leq 1$ with maximum error of $\pm 1\%$. Another important contribution of this work is its inclusion of possible differences in the accommodation coefficients on the top and bottom walls of the slider geometry. This feature allows the study of sliders with metal and silicon top and bottom surfaces, respectively, corresponding to different accommodation coefficients (Veijola et al., 1998).

Previous models for the generalized Reynolds equation predict the pressure distribution and load capacity on the slider. However, it is impossible to predict the skin friction distribution and viscous drag on the runner (stationary) and slider (moving) surfaces, which are important in predicting the pitch and roll moments. In addition, accurate prediction of viscous forces enables robust head suspension design and calculation of actuator power consumption. To address this need, Bahukudumbi and Beskok (2003) developed a modified Reynolds equation to accurately predict the *velocity and shear stress distribution, pressure profile, and load capacity* in slider bearings for a wide range of Knudsen numbers ($\mathrm{Kn} \leq 12$). Their approach is a superposition of the unified Poiseuille flow model of Section 4.2.2 with the linear Couette flow model of Section 3.2. Neglecting the thermal creep effects $\overline{Q_T}$, the one-dimensional generalized Reynolds equation becomes

$$\frac{\partial}{\partial X}\left[\overline{Q_p} H^3 P \frac{dP}{dX}\right] = \Lambda \frac{\partial}{\partial X}(PH), \tag{6.12}$$

where $\overline{Q_p}$ is due to the pressure-driven flow, and it is dependent on the local Kn. Following the model presented in Section 4.2.2, we have

$$\overline{Q_P} = \frac{\dot{Q}_P}{Q_{\mathrm{con}}} = \left[1 + \frac{6\,\mathrm{Kn}}{1+\mathrm{Kn}}\right](1+\alpha\,\mathrm{Kn}), \tag{6.13}$$

where α is the rarefaction parameter. In Section 4.2.2 we determined α using the volumetric flowrate for duct and pipe flows as a function of Kn. In this section, we choose α by matching the volumetric flowrate data for *two-dimensional pressure-driven flows in the entire Knudsen regime*, obtained from the solutions of the linearized Boltzmann equation. Therefore, the current modified Reynolds equation (6.12) is *identical* to that of Fukui and Kaneko (1988). However, there is an important difference between the two approaches. The model presented in (Bahukudumbi and Beskok, 2003), is able to predict the local velocity profile and shear stress. For example, the local velocity distribution for Poiseuille flow is given by

$$u_P(y) = -\frac{h^2}{2\mu_o}\frac{dp}{dx}(1+\alpha Kn)\left[\frac{\mathrm{Kn}}{1+\mathrm{Kn}} + \left(\frac{y}{h}\right) - \left(\frac{y}{h}\right)^2\right]. \qquad (6.14)$$

This results in the desired volumetric flowrate in equation (6.13). Figure 6.4 shows the normalized velocity profiles (U^*) for $k = \sqrt{\pi}/2\ \ Kn = 0.1, 1.0,$ and 10.0 flows, where the velocity profiles are normalized by $\left(\sqrt{\frac{RT}{2}}\frac{h}{P_a}\frac{dp}{dx}\right)$. This way of normalization preserves both the magnitude and shape of the velocity profile, as shown in the figure. Consistent with the approach of Section 4.2.2, the velocity profiles uniformly match the linearized Boltzmann solutions obtained in (Ohwada et al., 1989a). This empirical unified model also predicts the shear stress, as demonstrated in (Bahukudumbi and Beskok, 2003).

Figure 6.5 compares the Poiseuille flow-rate coefficient ($\overline{Q_p}$) calculated using the linearized Boltzmann equation, continuum, first- and second-order slip flow models. The model developed in (Bahukudumbi and Beskok, 2003) is shown by Q_{curr}. We must note that the free-molecular flow solution between the two parallel plates results in a logarithmic increase in the flowrate. Although this behavior is questionable, all the aforementioned models utilized the same flowrate database, which is the current industry standard for air-bearing design. In the following, we present the lubrication results for several different slider configurations.

In Figure 6.6 (a), we present the solution of the generalized Reynolds equation for a slider bearing with minimum magnetic spacing of 50 nm at ambient conditions, and compare the results with the DSMC simulations in (Alexander et al., 1994), which was obtained for a slider bearing of length $L = 5\ \mu$m and a slider speed $U_0 = 25$ m/s, resulting in bearing number $\Lambda = 61.6$ and $M = 0.08$. As evident from Figure 6.6, very good agreement is found between the generalized Reynolds equation and the DSMC results. The pressure profile predicted by the first-order slip model exhibits significant deviations from the DSMC data. Predictions using a second-order slip model (not shown on the figure) also result in considerable errors (Liu and Ng, 2001). In Figure 6.6 (b), we consider a slider bearing with an identical geometric configuration as in case (a). However, the bearing number is increased to $\Lambda = 758$ by increasing the plate speed to sonic conditions

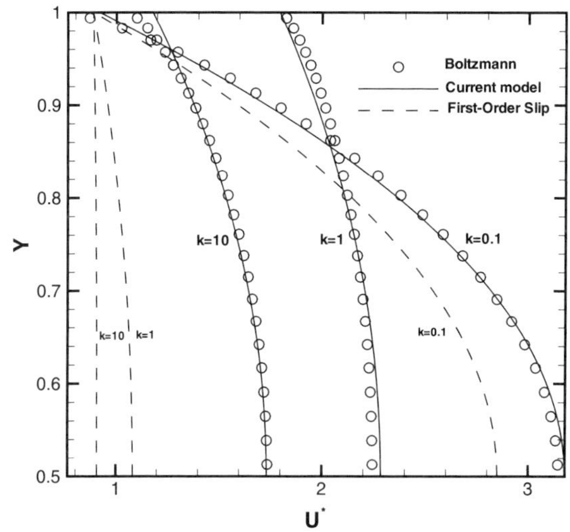

FIGURE 6.4. Nondimensional Poiseuille flow velocity profiles in the upper half of a channel for $k = \sqrt{\pi}/2$ Kn $= 0.1, 1.0$, and 10.0 flow. The "current model" in the figure is from (Bahukudumbi and Beskok, 2003).

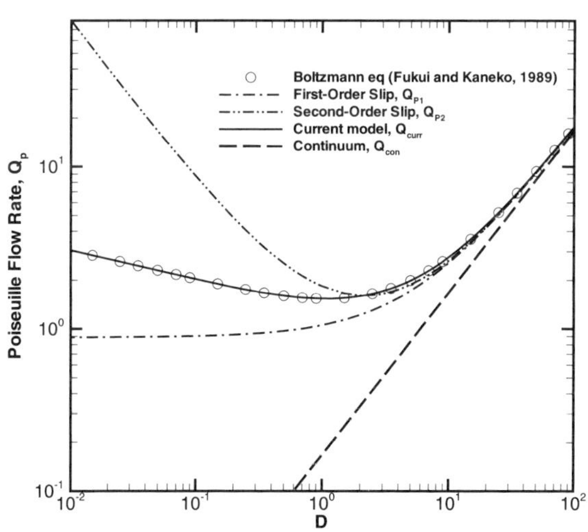

FIGURE 6.5. Variation of the Poiseuille flowrate coefficient Q_P as a function of the inverse Knudsen number at the exit of the channel. The "current model" in the figure is from (Bahukudumbi and Beskok, 2003).

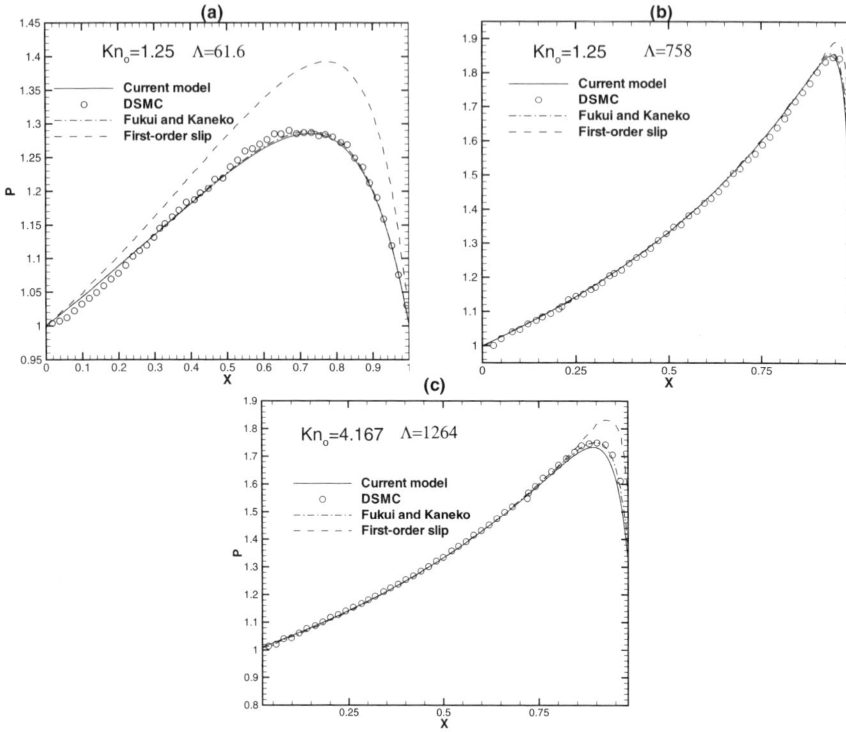

FIGURE 6.6. Slider bearing pressure profiles for different Knudsen number and bearing number combinations. The DSMC results are from (Alexander et al., 1994), and the Fukui and Kaneko data are from (Fukui and Kaneko, 1988). The "Current model" in the figure is from (Bahukudumbi and Beskok, 2003).

(M = 1.0). Surprisingly, the pressure distributions predicted by different models are very similar. This is because of the high bearing number. Since the bearing number is the ratio of the Couette and Poiseuille flow rates, the Couette flow is dominant at high Λ values. In Figure 6.6 (c), we consider highly rarefied gas flow with $\text{Kn}_\text{o} = 4.167$ and a high $\Lambda = 1264$ value. Under these conditions, the minimum magnetic spacing is $h_\text{o} = 15$ nm and the Mach number corresponding to the platter speeds is $M = 0.5$. The next-generation hard drives with ultrahigh storage densities would require slider bearings with a similar configuration as in case (c). Again, the solution of the generalized Reynolds equation is in good agreement with the DSMC data.

In summary, the generalized Reynolds equation for *multidimensional* problems is given by

$$\nabla \cdot \left[\left(\overline{Q_P}(D_o PH) \nabla P - \overline{Q_T}(D_o PH) P \nabla T_w \right) PH^3 \right] = \mathbf{\Lambda} \cdot \nabla(PH) + S \frac{\partial(PH)}{\partial t^*}, \tag{6.15}$$

where $\mathbf{\Lambda}$ is the bearing number vector defined by the lateral and streamwise components of the slider plate velocity, and t^* is the nondimensional time normalized with a characteristic frequency ω_o in case of time-harmonic excitation of the slider plate. The *squeeze number* S is given by

$$S = \frac{12\mu\omega_o L^2}{p h_o^2}.$$

The generalized Reynolds equation is analogous to the Navier–Stokes based model given in equation (6.2), and we presented several different ways to model the $\overline{Q_P}$ term in equation (6.15).

The Reynolds equation in the free-molecular flow regime employs the free-molecular flow solution between two parallel plates, which increases logarithmically as the inverse Knudsen number D apporaches 0 (Fukui and Kaneko, 1988):

$$Q_P(D) \rightarrow (-1/\sqrt{\pi}) \log D.$$

We have shown in Section 4.2 that the volumetric flowrate obtained by the linearized Boltzmann equations in microducts reaches finite asymptotic values in the free-molecular flow regime; see also (Sone and Hasegawa, 1987). The distinction between the two-dimensional channel and rectangular duct flows can be seen in Figure 4.30. This striking difference brings up the question of the *finite dimension effects* on the slider problem:

- Does the finite width (W) of the device make the flowrate behave more like a duct flow than a channel flow?
- What are the effects of the finite slider length (L) for high Kn transport?
- What are the finite length and width effects on Knudsen's minimum?

We can identify multiple operation conditions in the transition and free-molecular regimes using the relative ratios of λ to h, W, and L. However, such studies require extensive analysis of the finite length slider problem using the Boltzmann equation or DSMC. Data for internal rarefied gas flows through finite-length channels and orifices can provide guidance in determining these finite-length scale effects (see Section 15.4 and also (Sharipov and Sleznev, 1998)).

6.1.2 Squeezed Film Effects in Accelerometers

Squeezed gas film effects in microaccelerometers have been studied by various researchers (van Kampen, 1995; da Silva et al., 1999; Veijola, 1999; Chen and Kuo, 2003). In a series of papers, Veijola et al. employed the generalized Reynolds equation to study rarefaction effects on the dynamic response of microaccelerometers (Veijola et al., 1995a; Veijola et al., 1995b; Veijola et al., 1998). A lumped parameterization of the micromechanical accelerometer shown in Figure 6.7 leads to the mass–spring–damper system represented in the following form (see also Chapter 17 and Section 18.2):

$$M\frac{d^2x}{dt^2} + \gamma\frac{dx}{dt} + \kappa x = F_{\text{ext}},$$

where M is the mass, κ and γ are the spring and damping coefficients, x is the displacement of the mass, and t is time. The driving force F_{ext} includes the applied acceleration, electrostatic, and gas damping force effects. The electrostatic force depends on the proximity between the seismic mass and the substrate (h), and it is determined by

$$F_e = \frac{\epsilon A(\Delta V)^2}{2h^2},$$

where ϵ is the dielectric constant of the gas, A is the plate area, and ΔV is the electric potential difference between the moving mass and the fixed electrode. (A similar expression was used for the force in comb-drives, see Section 1.1.) The gas damping forces are obtained using the linearized Reynolds equation assuming small normal motion, small pressure variations, and also isothermal conditions under various reference pressures (Veijola et al., 1995a). The linearized Reynolds equation for oscillatory normal motion of the micromechanical accelerometer shown in Figure 6.7 is (Veijola et al., 1995a)

$$\frac{p_o h^2}{12\mu_e}\nabla^2 P - \frac{\partial P}{\partial t} = \frac{\partial h}{\partial t}, \tag{6.16}$$

where μ_e is the *effective viscosity* coefficient that models the local (linearized) rarefaction effects on the volumetric flowrate Q_P.

For rectangular surfaces and normal-only motion, equation (6.16) is solved analytically, and the result is formulated as a *mechanical admittance.* This admittance is approximately implemented as an electric equivalent circuit. The accelerometer model is a combination of this damping circuit and the equivalent circuit of the mass–spring system. Efficient time- and frequency-domain accelerometer simulations can be performed with the circuit simulation program APLAC, similar to SPICE, discussed in Chapter 18. More details of the lumped parameterization and the equivalent electric circuit model of the mechanical structure, including the squeezed gas

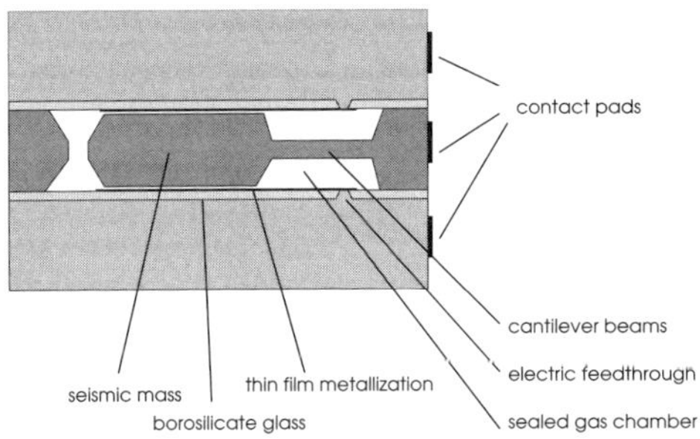

FIGURE 6.7. Cross-section of a micromechanical accelerometer. The cantilever beams support the moving mass, which forms two varying capacitances with the thin film electrodes. The cross-section is not drawn to scale. (Courtesy of T. Veijola.)

film effects are given in Chapter 17. The simulated steady-state frequency-domain response of the system was compared with experimental measurements, and a transient time-domain response to a step acceleration was also obtained. The results point to the following behavior in the frequency domain as a function of the actuation frequency:

- *Below a certain frequency identified as the cut-off frequency, the gas has enough time to flow away from the closing gap, inducing dissipation to the system. However, above this frequency the gas is trapped in the gap, and it is squeezed between the moving plates, introducing a springlike behavior with low dissipation.*

This behavior is shown in Figure 6.8 for three different reference pressure conditions. The maximum amplitude of the system is a strong function of the base pressure, especially for high frequency excitations. The maximum amplitude of the frequency response of the system increases with the increased base pressure, and the shift in the phase of the system shows transition from mostly dissipative gas film behavior toward the springlike gas response. The cut-off frequency is defined as the frequency at which the dissipative and spring effects of gas are equal to each other. Figure 6.9 shows the transient time-domain response of the system to a step acceleration of 0.5 g applied for 0.5 ms under various pressures. The response of the system at 30 Pa shows slowly damped oscillations at the resonance frequency. However, higher pressure of 300 Pa eliminates these oscillations, and the mass reaches its final position within 3 ms. The distinction between these two cases is a good indication of gas damping effects as a function of the base pressure (for fixed geometry). Of course, these results can also be represented as a function of the reference Knudsen number. Using the reference gap size between the seismic mass and the substrate as $3.95\,\mu$m, we find the reference Knudsen numbers in these simulations to be Kn $= 60, 6,$

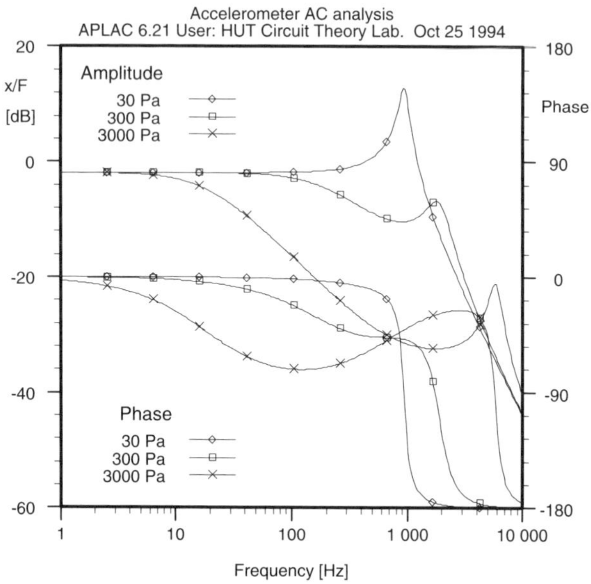

FIGURE 6.8. Simulated accelerometer amplitude and phase responses at different pressures. (Courtesy of T. Veijola.)

and 0.6 for $P = 30, 300$, and 3000 Pa, respectively. Therefore, Figures 6.8 and 6.9 both show the rarefaction effects on gas film damping. These results qualitatively agree with the time response of the DMD given in Figure 6.2.

The rarefaction effects in (Veijola et al., 1995a), were incorporated into the linearized Reynolds equation (6.16) using an *effective viscosity model* defined as

$$\mu_e = \frac{\mu_o}{1 + f(\text{Kn})}, \tag{6.17}$$

where μ_o is the tabulated viscosity coefficient of the gas, and $f(\text{Kn})$ is a function that models the flowrate changes with respect to the no-slip Poiseuille flow. The function $f(\text{Kn})$ is obtained either by asymptotic analytical solutions at a certain Knudsen regime or via a curve fit to the volumetric flowrate of linearized Boltzmann solutions. The flowrate relations used by Veijola have similarities to our *unified flowrate model* presented in Section 4.2. The effective viscosity model adopted by Veijola employing data in (Fukui and Kaneko, 1988), is also similar to the model introduced in Section 4.2.2 using the concept of rarefaction coefficient.

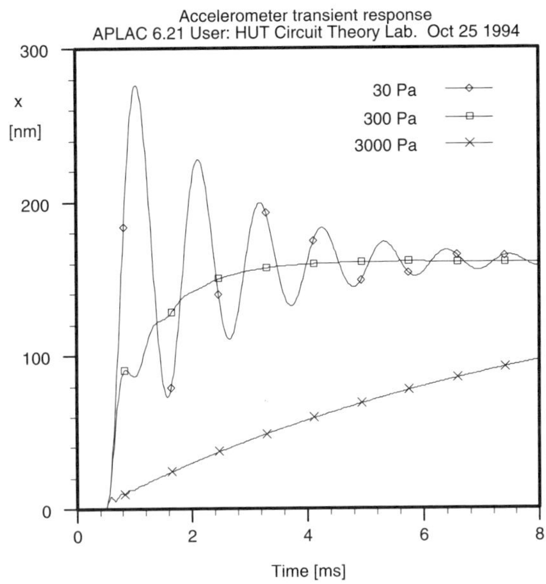

FIGURE 6.9. Simulated accelerometer displacement step response at pressures 30, 300, and 3000 Pa. The parameters are the same as in the frequency domain simulation in Figure 6.8. (Courtesy of T. Veijola.)

Squeezed Film Damping in Complex Geometries

Complex geometries are a challenge for accurate numerical modeling of squeezed film damping effects in various microsystems. A simplified rectangular capacitive accelerometer with holes in the middle portion was simulated by Veijola et al., who solved the two-dimensional Reynolds equation with a finite difference method (Veijola et al., 1995b). The ambient pressure conditions were specified in the hole region, where air is free to escape. This approach could be used for engineering modeling purposes, and it gives very reasonable results for thin structures. In the case of thick seismic mass, the hole(s) will act as finite-length flow suction/ejection channels (depending on the motion) with considerable pressure variations through the thick hole region. Figure 6.10 shows the simulated pressure distribution in the air gap of a tilting rectangular accelerometer with a hole in its middle. The tilting of the plate creates asymmetric pressure distribution in the system. The results are obtained by finite differences on a mesh of 22×22 grid points. The effect of the number of holes on the steady-state frequency and transient time responses of the system is shown in Figure 6.11. In the simulations, the base pressure is 1900 Pa, and the reference gap width is $2\,\mu$m. The transient time response to a step acceleration applied for $10\,\mu$s

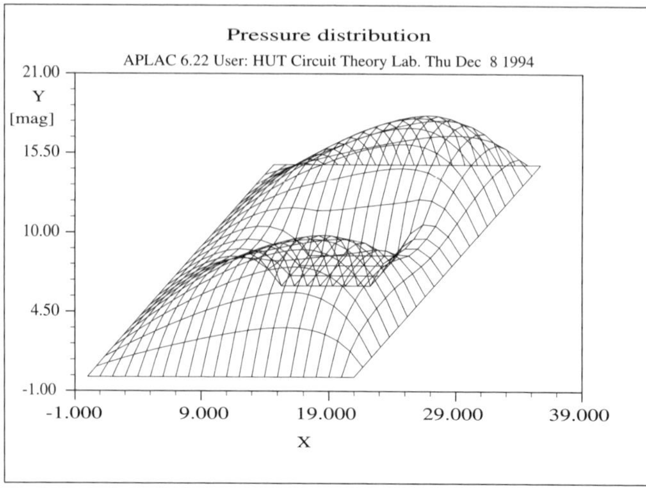

FIGURE 6.10. Simulated pressure distribution in the air gap of an accelerometer. In this example there is a hole in the middle of the rectangular plate, and the plate is tilting, which creates an asymmetric pressure distribution. (Courtesy of T. Veijola.)

is shown in the top figure. The case without holes experiences oscillations corresponding to the springlike response, and the 3-hole case gives an overshoot due to the lack of sufficient damping. However, the 2-hole case gives the optimum response, reaching steady conditions with in $300\,\mu$s (Veijola et al., 1995b). The steady frequency response of the system presented in the bottom figure also shows drastic variations in the amplitude and phase of the system. The maximum amplitude is shifting toward the low frequencies with increased number of holes. The cut-off frequency indicating the phase shift of the system from a mostly dissipative gas film behavior toward the springlike gas response also decreases with the increased number of holes (Veijola et al., 1995b).

6.2 Separated Internal Flows

In this section we investigate rarefied internal flows subject to separation. Such conditions are encountered in complex networks of microchannels where the flow is forced to turn or expand (Lee et al., 2000). As a prototype geometry we consider the backward-facing step studied extensively in continuum fluid dynamics (see (Kaiktsis et al., 1996), and references therein). The objective here is to investigate the validity of slip models considered in Chapter 2. The sudden changes in the flow conditions in the backward-facing step result in significant variations of the mean free path of the gas molecules ($\lambda \propto 1/P$), with corresponding variations in the wall

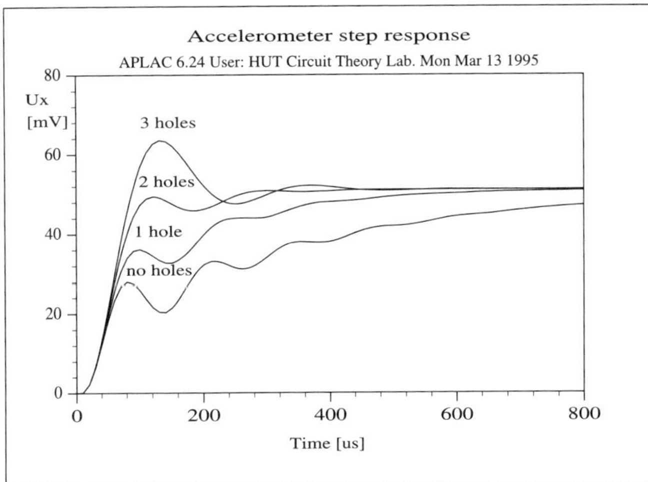

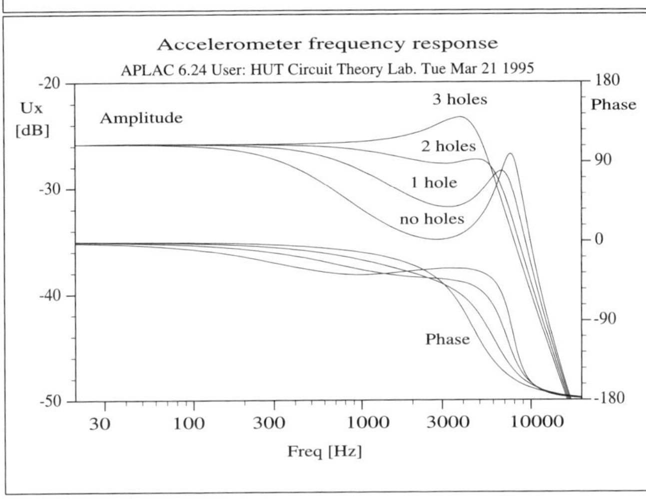

FIGURE 6.11. Simulated step (top) and frequency (bottom) responses of an accelerometer. The number of holes in the mass greatly affects the damping properties and the settling time. (Courtesy of T. Veijola.)

shear stress affecting the velocity slip directly ($U_s \sim \partial U/\partial n$). In contrast to the straight channel flow, *cross-flow variations are significant* here, so that the rarefaction effect is truly two-dimensional. In addition, the change of characteristic length scale from the inlet channel height to the larger length downstream presents extra difficulties in choosing the proper scaling.

We first investigate compressibility effects in backward-facing step flow and compare with corresponding incompressible flow simulations. The geometry used in this study is given in Figure 6.12 along with a typical spectral element discretization (see Section 14.1). It corresponds to $S/h = 0.467$, where S is the step height and h is the height of the channel.

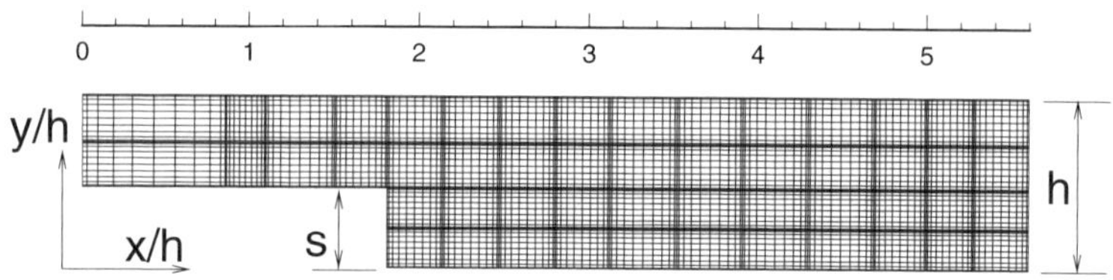

FIGURE 6.12. Backward-facing step geometry. Spectral element discretization with 52 elements and tenth-order polynomial expansions in each direction (see Chapter 14).

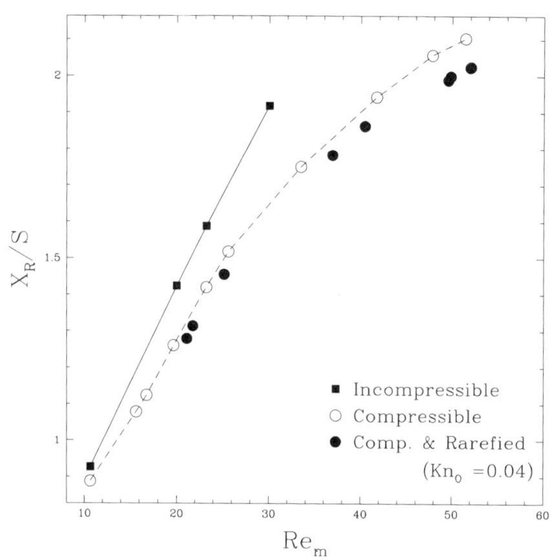

FIGURE 6.13. Variation of nondimensionalized separation distance X_R/S (S is the step height) as a function of $\mathrm{Re}_{\dot{M}}$ for incompressible, compressible, and rarefied flows. For the rarefied flow simulations $\mathrm{Kn}_o = 0.04$ based on the channel exit conditions.

The high-order program μ**Flow** that solves the compressible Navier–Stokes equations with and without slip at the wall is employed (see chapter 2). The computational domain is discretized with 52 spectral elements, where each element is further discretized with Nth-order polynomial expansions in each direction; here we use $N = 10$.

In Figure 6.13 we present the variation of nodimensionalized separation distance (X_R/S) as a function of Reynolds number $\mathrm{Re}_{\dot{M}}$ for *no-slip* flows, where $\mathrm{Re}_{\dot{M}}$ is based on the mass flowrate ($\dot{M}$) per unit width (W) of the channel

$$\mathrm{Re}_{\dot{M}} = \frac{\dot{M}}{W\mu}.$$

For comparison purposes the predictions of incompressible version of μ**Flow** are also included in Figure 6.13. For large pressure ratios, locally transonic conditions are achieved. It is seen in Figure 6.13 that the separation distance increases nonlinearly with increased $\mathrm{Re}_{\dot{M}}$, unlike the case of incompressible flow. The differences between the incompressible and compressible simulations become more dominant as the mass flowrate is increased further, corresponding to cases in which locally transonic conditions are achieved.

In Figure 6.13 we also include predictions of μ**Flow** for rarefied flows with $\mathrm{Kn_o} = 0.04$. Here the value of the Knudsen number Kn is based on exit conditions. In the simulations we used the slip model given by equation (2.26), where the slip information is obtained at a distance λ away from the wall surface. The exit pressure of the channels and the channel thickness h were fixed in the simulations. Therefore, the value of Knudsen number $\mathrm{Kn_o}$ is constant at the exit of the channel for all cases. However, the distribution of Kn varies from simulation to simulation as the inlet to exit pressure ratio is varied. One limitation in the compressible flow simulations is the possibility of *choking* the flow in the channels. Here, we limit the simulations to subsonic cases, i.e.,

$$\bar{M} = \frac{\bar{U}}{\sqrt{\gamma R\bar{T}}} \leq 1,$$

where the overbar conditions indicate quantities averaged across the channel. Therefore, the maximum possible Kn in the simulations can be estimated by

$$\overline{\mathrm{Kn}} = \frac{\bar{\lambda}}{h} = \frac{\bar{\mu}}{\bar{\rho}h(R\bar{T}2/\pi)^{\frac{1}{2}}} = \left(\frac{\gamma\pi}{2}\right)^{\frac{1}{2}} \frac{\bar{\mu}}{\bar{U}\bar{\rho}h} \frac{\bar{U}}{(\gamma R\bar{T})^{\frac{1}{2}}} = \left(\frac{\gamma\pi}{2}\right)^{\frac{1}{2}} \frac{\bar{M}}{\mathrm{Re}_{\dot{M}}}. \quad (6.18)$$

For a nonnegligible separation region we need $\mathrm{Re}_{\dot{M}} \geq 10$, and thus for $\bar{M} \leq 1$, the maximum possible value is $\mathrm{Kn} \leq 0.1$.

Next, we simulate *nitrogen flow* in a backward-facing step using the DSMC method (see Section 15.1). The DSMC simulations employed a total of 28,000 equally spaced cells (700 cells in the streamwise and 40 cells in the cross flow direction). The simulations employed 420,000 *simulated* molecules, and the results are sampled for 1×10^5 time steps. Numerical convergence is verified by monitoring the mass balance, with the maximum errors being approximately 1%. Inlet section to the channel is also included in the simulation, with the inlet located at $x/h = 0.86$. A uniform flow stream corresponding to $M = 0.45$ approaches the inlet of the channel

TABLE 6.1. Abbreviation, location, and corresponding symbols for the streamwise cuts taken in the step geometry. The TW and BW locations correspond to the DSMC cell centers adjacent to the walls, and can be treated as the wall quantities.

Abbreviation	Description	Location	Symbol
BW	Bottom Wall	$y/h = 0.01675$	Solid Triangle
BC	Bottom Center	$y/h = 0.25$	Hollow Square
C	Center	$y/h = 0.48325$	Hollow Circles
CE	Center of Entrance	$y/h = 0.75$	Solid Circles
TW	Top Wall	$y/h = 0.9875$	Solid Circles

with a free-stream static temperature of $T = 330$ K. All wall surfaces are maintained at 300 K. The Reynolds number of the simulations based on the mass flowrate per unit width is $\mathrm{Re}_{\dot{M}} = 80$, the Knudsen number at the channel outlet is $\mathrm{Kn_o} = 0.04$, and the Prandtl number is $\mathrm{Pr} = 0.7$. The temperature fields obtained by the DSMC program and μ**Flow** are shown in Figure 6.14. The temperature of the gas molecules in the DSMC are determined by combination of the translational, rotational, and vibrational energy of the diatomic nitrogen molecules. Since the overall gas temperature is not far from ambient, the vibrational energy mode is negligible here. The temperature contours calculated by both methods are quantitatively the same. Elevated temperatures occur near the recirculation zone ($2.8 \leq x \leq 4.5$). This phenomenon can be attributed to the viscous dissipation effects, i.e., the term

$$\frac{\partial}{\partial x_j}(\sigma_{ji}u_i)$$

in equation (2.16). A detailed analysis of viscous dissipation terms in the Navier–Stokes equations may be necessary for quantification of the viscous heating effects. The flow separation and recirculation zones predicted by the continuum and atomistic simulations agree well. The flow is locally transonic at the step expansion. This is due to the acceleration of the fluid and reduction of the static temperature of the fluid near the step expansion.

In Figures 6.15 and 6.16 we plot the streamwise variations of the pressure and streamwise velocity, obtained at five different y/h locations. The values of pressure and velocity are nondimensionalized with the corresponding freestream dynamic head and the local sound speed, respectively. The specific y/h locations are selected to coincide with the DSMC cell centers to avoid interpolation or extrapolation of the DSMC data, and are given in Table 6.1. In Figures 6.15 and 6.16 we first observe an increase in static pressure at C and TW locations near the entrance at $x/h = 0.86$. Such an increase in pressure is accompanied with deceleration of the fluid near the walls, while a sudden decrease of fluid temperature near the walls is

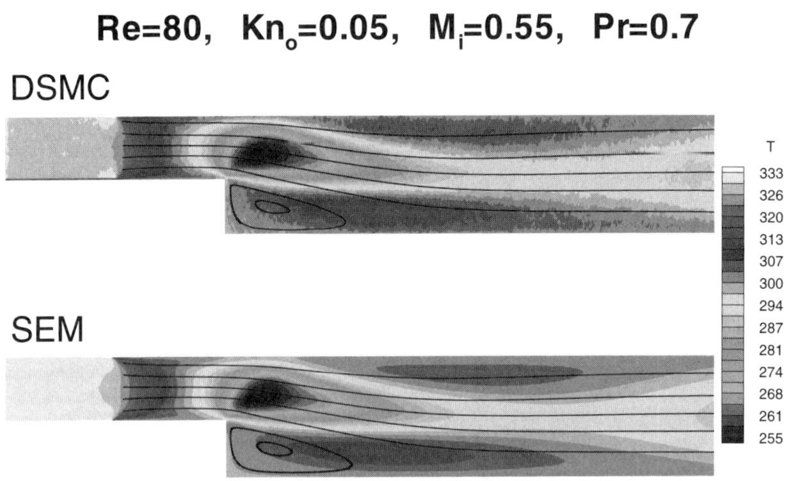

FIGURE 6.14. Temperature contours and streamlines obtained by the DSMC and continuum slip flow simulations.

obtained (see Figure 6.14). Such temperature decrease indicates large heat transfer to the channel surfaces, which are kept at isothermal conditions (at 300 K). Relatively large temperature jump and velocity slip occur in this region.

After the entrance section, the pressure drops very rapidly with corresponding fluid acceleration. The pressure drop in this section ($1.2 < x/h < 1.5$) is almost uniform across the channel, as seen in Figure 6.15. We also identify that the velocity increases along the channel, similar to a compressible straight channel flow. In this section the fluid temperature at the center of the entrance region (CE) decreases substantially, reaching a minimum at around $x/h = 2.0$. This is accompanied with acceleration of fluid in the streamwise direction, representing transformation of thermal energy into kinetic energy. Around the step expansion we also observe an increase in the cross-flow component of velocity. Therefore, the thermal energy transformation affects both the streamwise and the cross-flow velocities.

The sudden expansion in the geometry creates adverse pressure gradients across the entire cross section (see Figure 6.15) until $x/h = 3.25$ at BW, and $x/h = 3.65$ at TW. Due to the adverse pressure gradients, the flow at the bottom wall (BW) separates and reattaches at $x/h = 2.8$, before

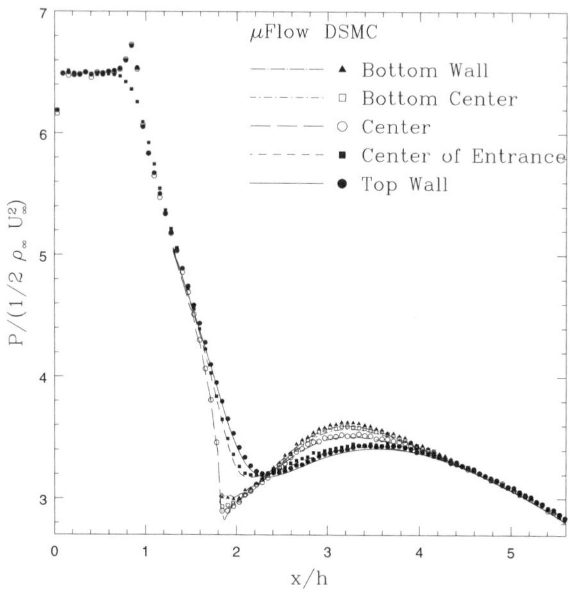

FIGURE 6.15. Pressure distribution along backward-facing channel at five selected locations. Predictions of both DSMC (symbol) and μ**Flow** (lines) are presented (see Table 6.1 for description of the symbols).

the pressure gradient at the bottom wall becomes zero (at $x/h = 3.25$). The early reattachment under slightly adverse pressure gradients is due to the compressibility effects, where conversion of thermal energy into kinetic energy enhanced the tangential momentum of the fluid.

Beyond $x/h = 3.65$, favorable pressure gradients are established. The flow goes through a developing region, followed by a typical compressible channel flow behavior far downstream in the channel, as shown in Figure 6.15 for $x/h > 4.2$. In contrast to the low Mach number flows, a decrease in the temperature of the fluid near the center of the channel is observed. This shows that the thermal energy of the fluid is converted into kinetic energy, and there is considerable heat transfer from the walls to the fluid.

Detailed observations on separated internal flows, and comparisons of the continuum-based slip flow simulations with the DSMC, can be found in (Beskok, 2001). DSMC predictions for a backward-facing step in the slip and transition flow regimes can be found in (Xue and Chen, 2003). Both studies indicate substantial separation zones for $\mathrm{Kn} \leq 0.1$ flows. In their studies, Xue and Chen (2003) did not observe flow separation for $\mathrm{Kn} > 0.1$. They attributed this to the rarefaction effects. However, this behavior may be due to the reduction in the Reynolds number of the flow, described by equation (6.18). Increasing the Knudsen number to the transition flow regime ($\mathrm{Kn} > 0.1$) requires $\mathrm{Re}_{\bar{M}} < 10$, since $\bar{M} \leq 1$. However, the separa-

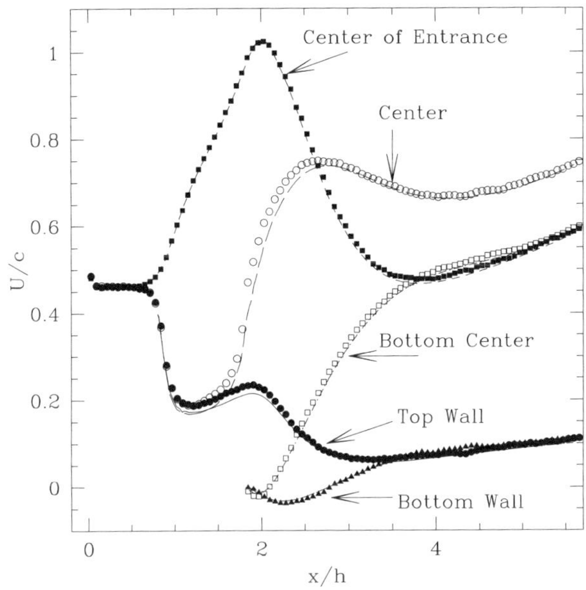

FIGURE 6.16. Streamwise velocity distribution along the backward-facing channel at five selected locations (normalized with the speed of sound). Predictions of both DSMC (symbol) and μ**Flow** (lines) are presented (see Table 6.1 for a description of the symbols).

tion zone becomes quite small for $\mathrm{Re}_{\dot{M}} < 10$ flows, even in the continuum flow regime.

6.3 Separated External Flows

In this section, we present simulations of slip flow past a circular cylinder, as a prototype of an external flow around a microprobe. Uniform flow past a cylinder with a slip surface has also been studied in (Gampert, 1978), for attached flows using an approximate boundary layer analysis. Following (Beskok and Karniadakis, 1994), we simulate both attached and separated flows; the simulations are performed at two values of Knudsen number: $\mathrm{Kn} \to 0$ (corresponding to the no-slip) and $\mathrm{Kn} = 0.015$, at $\mathrm{Re} = 10$. Separation of flow is observed with a small circulation bubble; the slip flow direction is reversed inside the separation zone. In Figure 6.17 we plot the magnitude of velocity slip distribution along the cylinder periphery for Reynolds number $\mathrm{Re} = 1$ (attached flow; triangles) and $\mathrm{Re} = 10$ (separated flow; circles). The velocity slip increases with the Reynolds number, but it decreases substantially in the separated (almost *stagnant*) region. This velocity slip is proportional to the shear stress τ_s, which is plotted in Figure

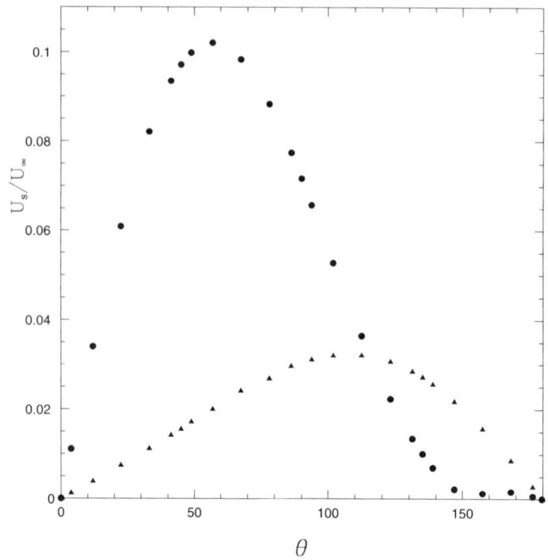

FIGURE 6.17. The magnitude of velocity slip distribution along the periphery of a circular cylinder at Kn = 0.015. Triangles, Re = 1; 0 degrees corresponds to the *rear* stagnation point. Circles, Re = 10; 0 degrees corresponds to the *front* stagnation point.

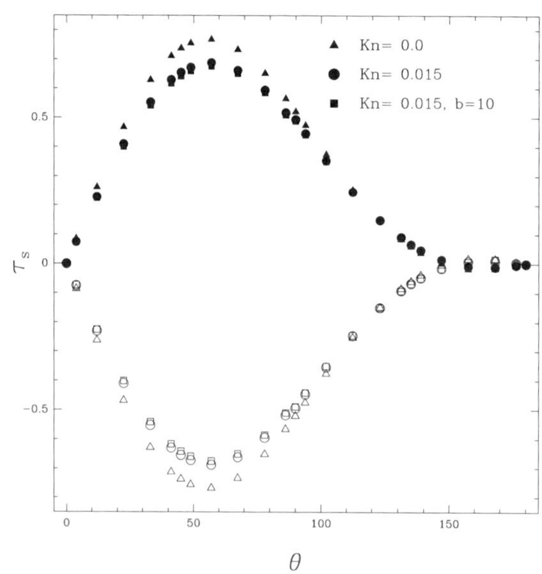

FIGURE 6.18. Distribution of tangential stresses along the upper and lower surfaces of circular cylinder at Re = 10 and Kn = 0 (triangles, no-slip case), Kn = 0.015 (circles), Kn = 0.015 and slip coefficient $b = 10$ (squares). Solid and blank symbols show upper and lower surfaces, respectively. Shear stresses are zero at about $\theta = 147°$, where the flow separates.

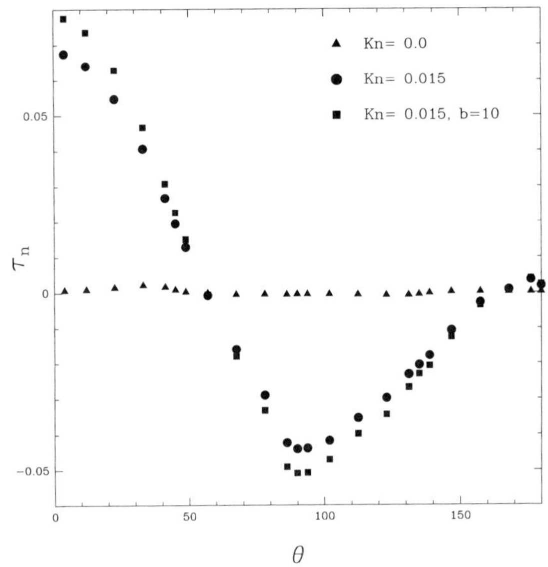

FIGURE 6.19. Distribution of viscous normal stresses along the cylinder periphery at Re = 10 and Kn = 0 (triangles, no-slip case), Kn = 0.015 (circles), and Kn = 0.015 and slip coefficient $b = 10$ (squares).

6.18 for the case of Re = 10. From this plot it is evident that separation occurs at an angle approximately 147° from the front stagnation point.

A simulation corresponding to the high-order slip boundary condition is also included (squares). The high-order expansion coefficient b is taken to be a constant ($b = 10.0$) for convenience. For the range of Knudsen number ($0 < \text{Kn} \leq 0.015$) investigated, no difference in the separation angle is observed between the slip and no-slip flow cases. For comparison of shear stress variations, we also plot in Figure 6.18 the shear stresses corresponding to the no-slip case. As expected, a reduction in skin friction is obtained especially in the front part of the cylinder where the flow accelerates.

For incompressible flows over *flat* no-slip surfaces the viscous normal stress components τ_n are identically zero. However,

- in slip surfaces the viscous normal stresses achieve finite values and increase substantially proportional to the Knudsen number.

This effect is demonstrated in Figure 6.19, where we plot the viscous normal stress distribution around the cylinder periphery. In this case, the viscous normal stresses for no-slip flow are nonzero due to the curvilinear boundary. However, they are considerably smaller compared to the slip flow even for this relatively small value of Knudsen number (Kn = 0.015). Levels of the pressure distribution are also reduced compared to the no-slip case in agreement with the results of (Gampert, 1978). The reduction in the pressure levels is balanced by the increase in the viscous normal stresses. Therefore:

- The total normal stresses (i.e., combined pressure and viscous normal stresses) do not vanish as the rarefaction effects increase (Gampert, 1978).

6.4 Flow Past a Sphere: Stokes Flow Regime

In this section, we present analytical and numerical results for flow past a sphere in the Stokes flow regime (Re $\to 0$). We will examine both the unbounded external flow and the sphere-in-a-pipe flow problems. The results in the continuum flow regime are equally applicable to the gas and liquid flows. However, the slip flow results are relevant to the gas flows. The external flow problem may be useful in estimating the drag force on airborne microparticles; see Section 14.3.2. The sphere-in-a-pipe problem is a prototype rarefied gas dynamics problem. Large-scale, low-pressure applications of this flow configuration are commonly used in spinning rotor gauge (SRG) devices (Fremerey, 1985), where damping of an electromagnetically supported rotating sphere is used for pressure, viscosity, molecular weight, and rate of molecular impingement measurements (Fremerey, 1982). A spherical particle moving in a pipe is also observed in Coulter counter devices (Ito et al., 2003). However, particle surface charge and electric double layer play an important role in these cases (see Section 7.5 for details).

6.4.1 External Flow

For external flow past a sphere in the Stokes flow regime (Re $\to 0$), the drag on the sphere is given by the Stokes drag:

$$F_D = 6\pi\mu U R, \tag{6.19}$$

where μ is the absolute viscosity of the fluid, U is the external flow velocity, and R is the sphere radius. Nondimensionalizing the drag by the dynamic head and the cross-sectional area of the sphere gives the drag coefficient C_D:

$$C_D = \frac{F_D}{\frac{1}{2}\rho U^2 \pi R^2} = \frac{12\mu}{\rho U R} = \frac{12}{\mathrm{Re}}, \tag{6.20}$$

where ρ is the fluid density, and $\mathrm{Re} = \frac{\rho U R}{\mu}$. This classical result shows that the drag coefficient for the sphere is inversely proportional to the Reynolds number in the Stokes flow regime (Re $\to 0$). For Re ≈ 1, corrections to the Stokes formula due to increased inertial effects are necessary. A well-known approximation for this case is given by Oseen (Batchelor, 1998).

In the slip flow regime ($\mathrm{Kn} \leq 0.1$), the total drag on the sphere is given by (Barber and Emerson, 2002)

$$F_D = 6\pi\mu UR \left(\frac{1 + 4\frac{2-\sigma}{\sigma}\,\mathrm{Kn}}{1 + 6\frac{2-\sigma}{\sigma}\,\mathrm{Kn}} \right). \tag{6.21}$$

The total drag can be decomposed into three components due to the skin-friction, viscous normal stresses and the pressure drag.

Skin-friction drag:

$$F_{\mathrm{SF}} = 4\pi\mu UR \left(\frac{1}{1 + 6\frac{2-\sigma}{\sigma}\,\mathrm{Kn}} \right).$$

Drag force due to the viscous normal stresses

$$F_{\mathrm{SN}} = 4\pi\mu UR \left(\frac{4\frac{2-\sigma}{\sigma}\,\mathrm{Kn}}{1 + 6\frac{2-\sigma}{\sigma}\,\mathrm{Kn}} \right).$$

Pressure or form drag:

$$F_{\mathrm{PD}} = 2\pi\mu UR \left(\frac{1 + 4\frac{2-\sigma}{\sigma}\,\mathrm{Kn}}{1 + 6\frac{2-\sigma}{\sigma}\,\mathrm{Kn}} \right).$$

We note that despite the geometric differences between a sphere and a cylinder, a substantial drag force due to the viscous normal stress is observed for both geometries for increased Kn values.

6.4.2 Sphere-in-a-Pipe

For a sphere located along the axis of the pipe, we consider the cases of a moving sphere in a stationary pipe, and Poiseuille flow past a stationary sphere in a pipe. The flow is axisymmetric for both cases. Considering the sphere diameter D and pipe diameter H, the *moving sphere case* results in the following total drag force (Haberman and Sayre, 1958):

$$F_D = \frac{6\pi\mu UR\left(1 - 0.75857(\frac{D}{H})^5\right)}{\left(1 - 2.1050\frac{D}{H} + 2.0865(\frac{D}{H})^3 - 1.7068(\frac{D}{H})^5 + 0.72603(\frac{D}{H})^6\right)}, \tag{6.22}$$

while the pressure-driven flow past a *stationary sphere-in-a-pipe* results in

$$F_D = \frac{6\pi\mu UR\left(1 - \frac{2}{3}(\frac{D}{H})^2 - 0.20217(\frac{D}{H})^5\right)}{\left(1 - 2.1050\frac{D}{H} + 2.0865(\frac{D}{H})^3 - 1.7068(\frac{D}{H})^5 + 0.72603(\frac{D}{H})^6\right)}, \tag{6.23}$$

where U is the sphere velocity in equation (6.22), and it is the pipe maximum velocity in equation (6.23).

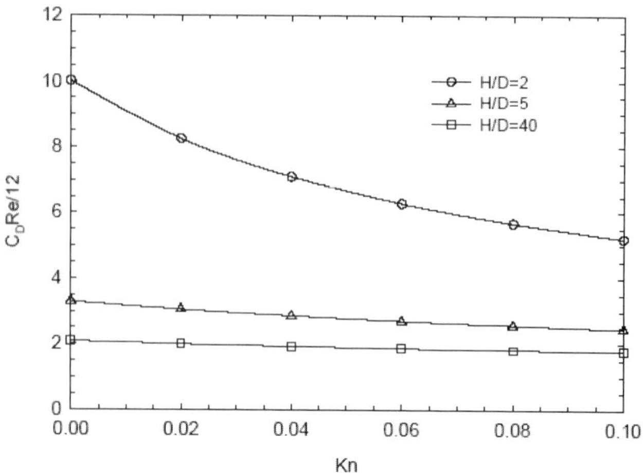

FIGURE 6.20. Variation of the normalized total drag coefficient for a confined sphere-in-a-pipe as a function of the Knudsen number in the slip flow regime. Results for various blockage ratios (H/D) are presented. (Courtesy of R.W. Barber and D.R. Emerson.)

Flow past a stationary sphere in a pipe has been investigated by (Liu et al., 1998) using μ**Flow** and DSMC simulations. In a later study, Barber and Emerson (2001) presented numerical results for the same problem using continuum-based slip models. Figure 6.20 presents the drag coefficient variation as a function of the Knudsen number for various H/D ratios. For convenience, the results are normalized by the drag coefficient of the external Stokes flow in equation (6.20). The drag force increases with decreasing H/D, due to the increased blockage effects. In the continuum flow limit, there is a tenfold increase in the drag force for $H/D = 2$. However, the drag force increases by a factor of two for $H/D = 40$. Numerical results in (Barber and Emerson, 2002) for Kn $= 0.1$ flow at $H/D = 2$ show about 50% drag reduction due to the slip flow effects. However, for $H/D = 40$, drag reduction due to velocity slip is about 10%.

In Figure 6.21 we present the normalized total drag coefficient variation as a function of the blockage ratio H/D for a sphere confined in a pipe in the continuum (Kn $= 0$) and slip flow Kn $= 0.1$ regimes (Barber and Emerson, 2004). The analytical solution of (Haberman and Sayre, 1958) is also shown in the figure. The results clearly indicate that velocity slip reduces the blockage effects, and drag reduction due to the rarefaction is more dominant for large blockage ratios (small H/D).

Wen and Lai (2003) presented analytical solutions for the cases of a sphere moving along the centerline of a micropipe, and a sphere moving parallel to the centerline of a micropipe. Analytical solutions of the Stokes equations with first-order slip boundary condition are obtained using the

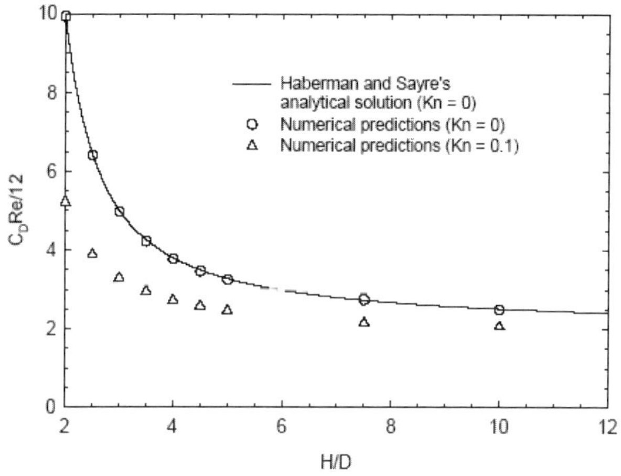

FIGURE 6.21. Variation of the normalized total drag coefficient for a confined sphere in a pipe as a function of the blockage ratio (H/D). Results for no-slip and Kn = 0.1 flows are shown. (Courtesy of R.W. Barber and D.R. Emerson.)

streamfunction method for the former case, while the method of reflections is utilized for the latter case. This enabled analytical expressions for the resultant force on the sphere, pressure drop over the sphere, and terminal velocity of the sphere (Wen and Lai, 2003).

6.5 Microfilters

Gas microfilter systems can be used for filtering and detection of airborne biological and chemical entities and for environmental monitoring applications. Recent advancements in microfabrication technologies enabled development of sufficiently thin filters that are strong enough to provide useful flowrates under large pressure drops (Chu et al., 1999). Motivated by these developments, several research groups have investigated gas flows through microfilter systems (Yang et al., 1999b; Yang et al., 2001; Mott et al., 2001). Analysis of gas flows through microfilters requires consideration of rarefaction, compressibility, and geometric complexity effects.

The geometric complexity of microfilters is very important. In the simplest form, the microfilters are very short channels or sudden constrictions in the flow field. Therefore, simple analysis based on the fully developed flow assumption cannot be used. Furthermore, the filter holes have complicated shapes, such as rectangular, hexagonal, circular, elliptic, or square cross-sections. The filters may also have geometric variations along the filter thickness, sharp or smooth inlets and exits. For example, the side-wall geometry is shown to affect the overall pressure drop across the microfilter devices (Yang et al., 2001).

Empirical formulas for pressure drop in conventional filters were obtained in earlier studies (Weighardt, 1953; Derbunovich et al., 1998). However, these scaling laws were valid for high Reynolds number flows ($\mathrm{Re} \gg 100$), and they cannot be applied to microfilters (Yang et al., 2001). The initial work on microfilters was done by (Kittilsland et al., 1990), who fabricated a filter that consisted of two silicon membranes with holes. By changing the membrane separation distance, they were able to build filters for separation of particles as small as 50 nm. Later, (Yang et al., 1999b) developed a MEMS-based microfilter using a micron-thick silicon-nitride membrane coated with Parylene, which was used both to control the opening area of the filter and to provide strength. Experimental and numerical studies have shown that the flow in the microfilters depends strongly on the opening factor β (the ratio of the hole area to the total filter area). The power requirements and the pressure drop through the microfilters have also been studied in (Yang et al., 1999b). It has been shown that

- *the power dissipation is a function of the opening factor, the ratio of the filter thickness to hole diameter, and the Reynolds number.*

(Yang et al., 1999b) were able to fit their experimental and numerical simulation data to an empirically determined scaling law, given in the following (Yang et al., 1999b):

$$\kappa = \beta^{-2} \left(\frac{t}{h} \right)^{0.28} \left(\frac{73.5}{\mathrm{Re}} + 1.7 \right), \tag{6.24}$$

where κ is the pressure drop nondimensionalized with the dynamic head, i.e.,

$$\kappa = \frac{\Delta P}{\frac{1}{2}\rho_\infty U_\infty^2},$$

and h and t are the hole diameter and thickness of the filter, respectively. For three-dimensional cases the Reynolds number is defined as

$$\mathrm{Re} = \frac{\rho_\infty U_\infty h}{\mu \beta},$$

where the subscript ∞ indicates upstream conditions. For a two-dimensional geometry the opening factor becomes the ratio of the hole opening length to the center-to-center filter separation distance L ($\beta = h/L$), resulting in

$$\mathrm{Re} = \frac{\rho_\infty U_\infty L}{\mu}.$$

Using the scaling law presented by equation (6.24) leads to higher pressure drops than the experimentally determined values. The assumptions and the parameters utilized in development of this scaling law were:

1. 2-D geometry,
2. Opening factor: $0.1 < \beta < 0.45$,
3. t/h ratio: $0.08 < t/h < 0.5$,
4. Reynolds number: $1 < \text{Re} < 100$,
5. Knudsen number: $0.005 < \text{Kn} < 0.015$.

In a follow-up study, (Yang et al., 2001) proposed the following modified scaling law based on detailed studies of the filter geometry, experimental data, and three-dimensional numerical simulations:

$$\kappa = \beta^{-2}\left(3.5\frac{t}{h} + 3\right)\left(\frac{10}{\text{Re}} + 0.22\right). \tag{6.25}$$

Although most of the experimental data fit this relation, there were some deviations for low Reynolds number flows. This scaling law did not explicitly incorporate the rarefaction effects as a function of the Knudsen number.

(Mott et al., 2001) employed two-dimensional DSMC to study gas transport through microfilters in the slip and early transition flow regimes. They have shown that Yang's model becomes inaccurate in the transition flow regime and also for high Reynolds number flows. Based on the experimental data of (Yang et al., 2001) and their DSMC results, (Mott et al., 2001) proposed the following modified scaling law, which includes the Knudsen number as an additional parameter:

$$\kappa = \beta^{-2}\left(3.5\frac{t}{h} + 3\right)\left(\frac{10}{\text{Re}} + 0.22\right)\left[\frac{0.0577}{0.0577 + \text{Kn}}\right]. \tag{6.26}$$

The assumptions and parameters in their model were:

1. 2-D geometry,
2. Opening factor; $\beta = 0.25$,
3. t/h ratio: 3,
4. Reynolds number; $0.2 < \text{Re} < 1$,
5. Knudsen number; $0.0309 < \text{Kn} < 0.233$.

It is noteworthy to mention that the t/h ratio used in Mott's studies was substantially larger than the values presented in (Yang et al., 1999b; Yang et al., 2001).

The scaling laws presented in equations (6.24)–(6.25) are developed starting from the fully developed Poiseuille flow assumption with added pressure

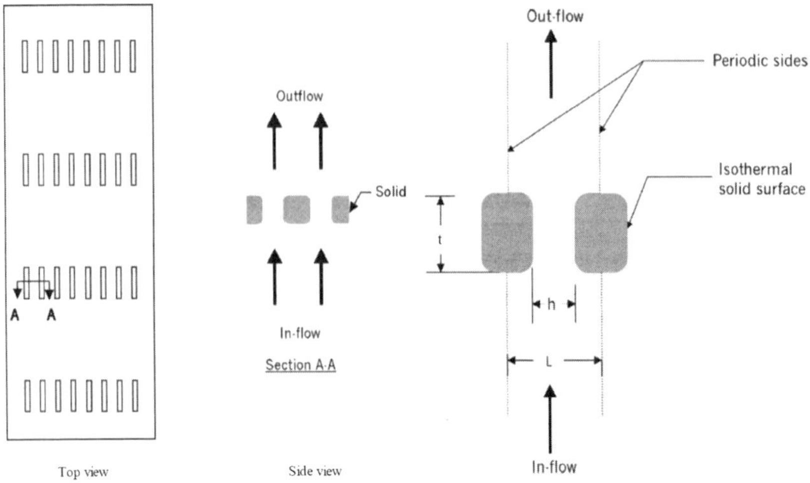

FIGURE 6.22. Left: Schematic view of the microfilter (filter front and cross-sectional views). Right: Schematic view of a simplified two-dimensional filter model. Various length scales are identified in the figure.

drop corrections due to the inlet and exit effects. (Mott et al., 2001) describe the evolution of these scaling laws, including the slip flow corrections given as a function of the Knudsen number in equation (6.26). The coefficients used in equations (6.24)–(6.26) are modified from the predictions given by simplified analytical models in order to match the experimental and numerical data (Yang et al., 1999b; Yang et al., 2001; Mott et al., 2001).

Ahmed and Beskok (2002) studied gas flows through a microfilter array using μ**Flow**. Isothermal conditions for the filter surface were assumed. A schematic view of a rectangular microfilter array is shown in Figure 6.22. In their study the filter width (w) was significantly larger than the filter height (h), so that the flow was approximated as two-dimensional. Considering that the filter holes repeat in a periodic fashion, they simulated gas flow through only one hole by imposing periodicity conditions in the spanwise direction. (see Figure 6.22). In addition, $h/t = 1.5$ and $\beta = h/L = 0.6$ were used; these dimensions are labeled in Figure 6.22 (right). Using this fixed aspect ratio geometry, they varied the reference length scale (L), creating a series of geometrically similar filters ranging from $L = 6$ μm to $L = 1$ μm (hence, h varied from 3.6 μm to 0.6 μm).

Figure 6.23 shows a comparison of numerical results with the scaling laws of (Yang et al., 2001) and (Mott et al., 2001). The results are presented as a function of a new parameter κ', which is defined in the figure. Significant scatter in the data is observed using Mott's model, while a constant offset is observed using Yang's model. Ahmed and Beskok's numerical solutions have consistently shown smaller pressure drops than the values

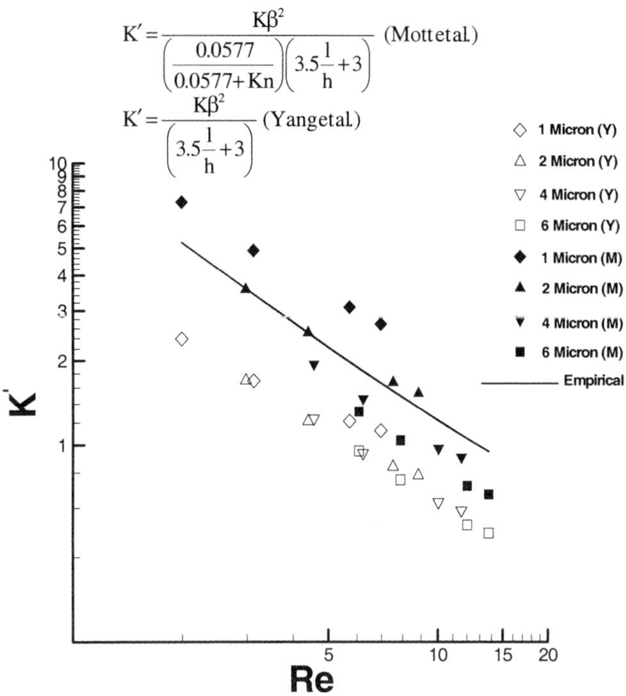

FIGURE 6.23. Comparisons of simulations of Ahmed and Beskok (2002) with the empirical scaling laws developed by (Mott et al., 2001) indicated by M, and (Yang et al., 2001) indicated by Y.

reported in (Yang et al., 2001). The reason for this was attributed to the smoothened filter geometry shown in Figure 6.22. This filter model had finite radius of curvature at the inlet and exit sections. The smooth entrance and exit shape with finite radius of curvature ($r/h = 0.1$) reduced the pressure drop at these locations, resulting in a reduced pressure drop compared to the experiments and numerical calculations of (Yang et al., 2001). Strong dependence of the data on the side-wall shape has been demonstrated in (Yang et al., 2001), by comparisons of numerical simulations with the experimental data. In addition to the differences in the side-wall shapes, three-dimensional flow effects can also be a possible explanation for the differences between Yang's model and results by (Ahmed and Beskok, 2002). Here we must mention that the inability to match Mott's scaling law can be due to the extremely small opening factor ($\beta = 0.25$) and large (t/h) ratio used in Motts's studies. The simulation parameters in (Ahmed and Beskok, 2002), were closer to those of Yang's experiments than the values used by Mott et al.

Using the simulation data, a modified relation for the scaling law was developed, as shown in Figure 6.24. Since the (t/h) ratio is fixed and a finite radius of curvature at the channel inlet and exit ($r/h = 0.1$) are

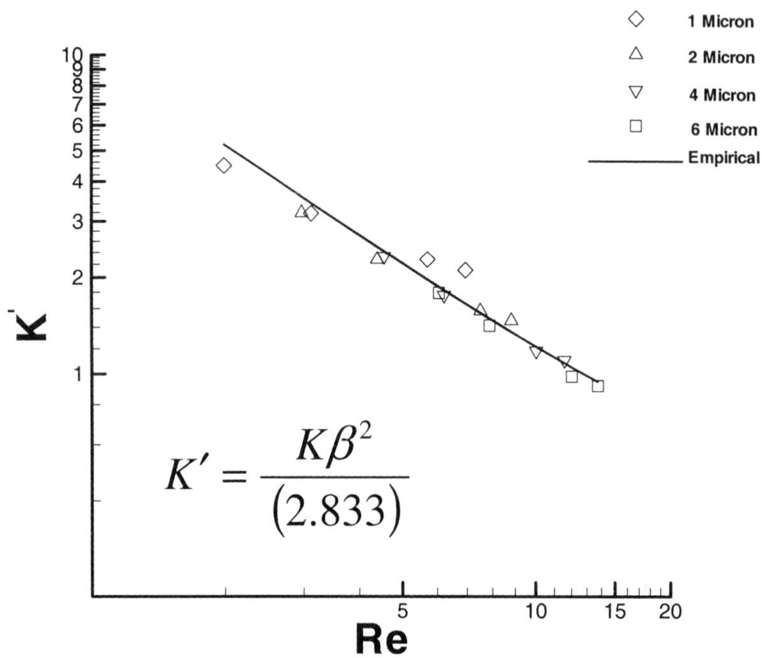

FIGURE 6.24. Modified scaling law that fits the data from numerical simulations in (Ahmed and Beskok, 2002).

utilized, the following scaling law was obtained:

$$\kappa = 2.833\beta^{-2}\left(\frac{10.0}{\mathrm{Re}} + 0.22\right). \tag{6.27}$$

For a general filter geometry, the constant 2.833 should be a function of (t/h) and (r/h). One also expects an explicit Knudsen number dependence in the model, which requires further studies beyond the slip flow regime ($\mathrm{Kn} > 0.1$). Filter performance in the transition flow regime can be investigated using the DSMC method (Mott et al., 2001; Aktas et al., 2001; Aktas and Aluru, 2002), as we discuss in Section 6.5.3.

6.5.1 Drag Force Characteristics

The drag force has two components: *viscous*- and *form-drag*. The viscous-drag is due to skin friction distribution on the body in the streamwise direction. The form-drag is due to the differences between the fore and aft pressure distributions on the body, in the streamwise direction. Due to the symmetry of the filter geometry and steady flow, there are no lift forces. In the top and bottom plots of Figure 6.25, we plot the slip and no-slip form-drag and viscous-drag data as a function of the Reynolds number.

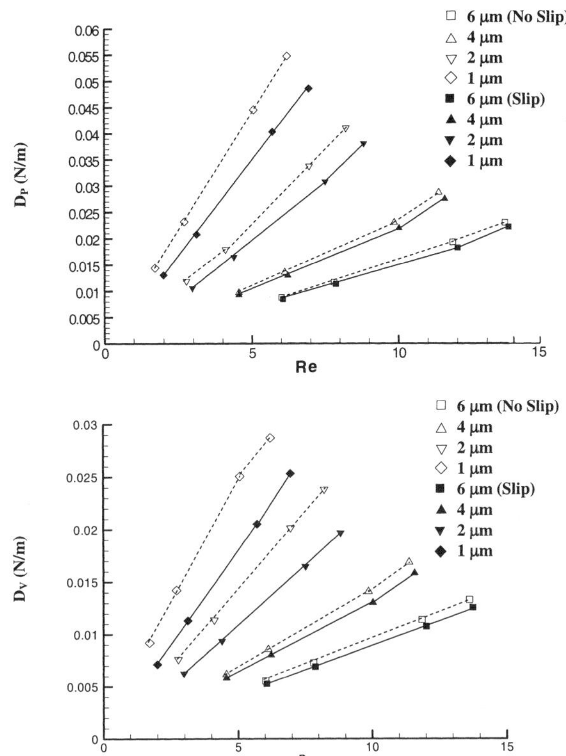

FIGURE 6.25. Form-drag (upper) and viscous-drag (lower) variation in microfilters as a function of the Reynolds number.

A comparison between the no-slip and slip data for the 1-micron versus 6-micron cases shows enhanced drag reduction in the smaller geometry, where the Knudsen number is higher. Hence, drag reduction increases by rarefaction. This is an important result, which indicates that the microfilters have the potential to allow more mass flowrate than their macroscale counterparts under fixed inlet and exit pressure conditions, and hence they require less power to operate.

Further examination of Figure 6.25 shows that for a fixed geometry, both the viscous and form-drag increase with the Reynolds number. The increase is mostly linear for low Reynolds numbers and starts to increase faster than linearly for Re > 10. We also observed that the viscous-drag is consistently about 50% of the form-drag. This is important in designing microfilter devices, because large drag forces may lead to bursting of thin filter-membranes.

6.5.2 Viscous Heating Characteristics

Work done by the viscous stresses usually becomes important for high-speed flows. For example, in the case of the micronozzles viscous heating effects cannot be neglected. In fact, work done by the viscous stresses causes heat generation, which acts as a volumetric source term in the energy equation. The viscous heating is characterized by

$$\int_{\mathrm{CS}} (n_j \tau_{ji}) \cdot u_i \, ds,$$

where τ_{ji} represents (ij)th component of the viscous stress tensor, n_j corresponds to the jth component of the outward surface normal, u_i shows the ith component of the velocity on the control surface (CS) with differential area ds. Figure 6.26 shows the viscous heating as a function of the Reynolds number for various filter sizes. Smaller filters have shown substantial viscous heating effects. For example, for the 1-micron filter with $\mathrm{Re} = 6.95$ ($M = 0.51$), the viscous heating can be as large as 1.2 W/m. In order to keep the Reynolds number within a certain range, the inlet flow speeds are increased substantially, which in return increase the Mach number for small filter dimensions. A comparison of the viscous heating effects shown in Figure 6.26 with the reference exit Mach numbers reveals that the viscous heating effects are important for high-speed flows. Furthermore, increase in the viscous-heating as a function of the Reynolds number (or Mach number, due to the increase in the reference speed) seems to be faster than linear.

6.5.3 Short Channels and Filters

In this section we summarize rarefied flow through microfilters and short channels following the DSMC analysis in (Aktas et al., 2001; Aktas and Aluru, 2002). In Figure 6.27 we present the geometry, where the physical filter surfaces are indicated with the bold lines (G and J). The length and height of the filter channel are denoted by l_c and h_c, respectively. The geometry includes an input region of length l_{in} and an output region of length l_{out}. Periodic boundary conditions are applied on surfaces C, D, E, and F to simulate a periodic array of filter elements. Dimensions of various filter elements used in this work are listed in Table 6.2.

Constant-pressure boundary conditions are applied along the surfaces A and B. The temperature of the input and output streams are kept at 300 K. In the simulations of (Aktas et al., 2001), the outlet pressure (at B) is held at atmospheric conditions, while the input pressure is varied to investigate the flowrate dependence on the inlet to exit pressure difference. In the following results, we present only the cases in which a pressure difference of 0.3 atm is applied across the filter elements. The specifics of the DSMC are presented in (Aktas et al., 2001; Aktas and Aluru, 2002). In summary,

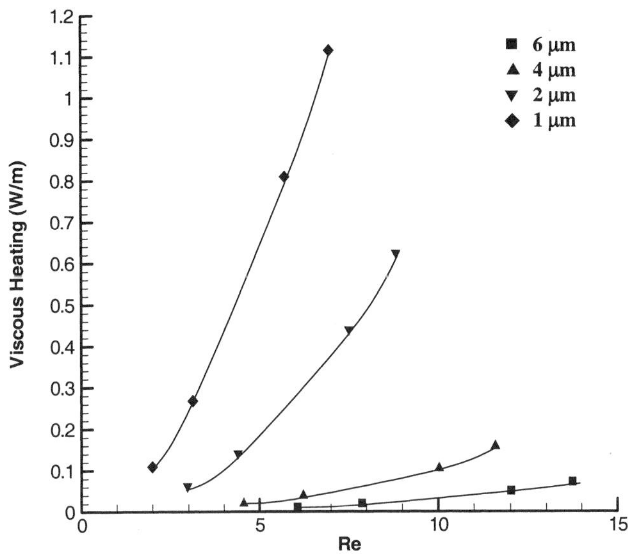

FIGURE 6.26. Viscous heating per unit filter width.

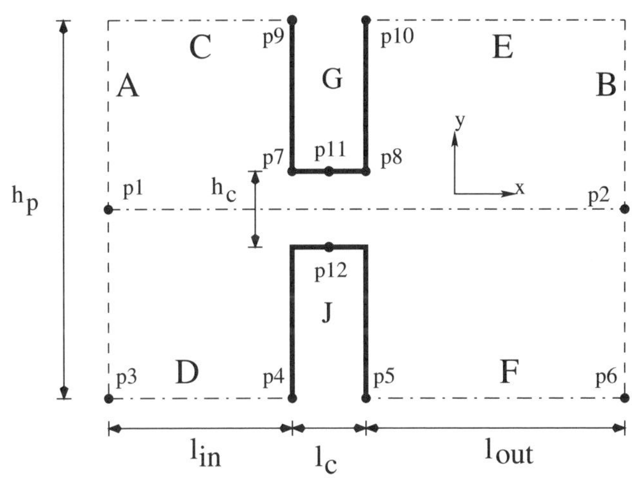

FIGURE 6.27. Description of the filter geometry studied in (Aktas et al., 2001).

nitrogen flow is simulated using DSMC for 10 ps time steps and diffuse reflection boundary conditions. The number of particles are consistent with the requirements given in Section 15.1. The flow is simulated for 1.8 μs before the time averaging started. The flow is sampled and time-averaged for 0.6 to 1.7 μs depending on the flow speed, where longer time averaging is employed for low subsonic flows.

TABLE 6.2. Dimensions of the filters and the corresponding flow rates at $\Delta P = P_{\text{in}} - P_{\text{out}} = 0.3$ atm.

	Filter 1	Filter 2	Filter 3	Filter 4	Filter 5
$h_c(\mu m)$	1	1	0.2	1	0.05
$l_c(\mu m)$	1	5	1	10	1
$h_p(\mu m)$	5	5	1	5	1
$l_{in}(\mu m)$	4	6	4	4	4
$l_{out}(\mu m)$	7	7	5	7	7
Kn_{out}	0.54	0.54	2.7	0.54	11
DSMC Flowrate $\left(\frac{m^3}{sm^2}\right)$	16.2	6.10	2.59	3.26	0.682
Flowrate using equation (4.6) $\left(\frac{m^3}{(sm^2)}\right)$	32.9	6.58	2.47	3.29	0.432
Estimated Re	5.7	2.1	0.18	1.1	0.048

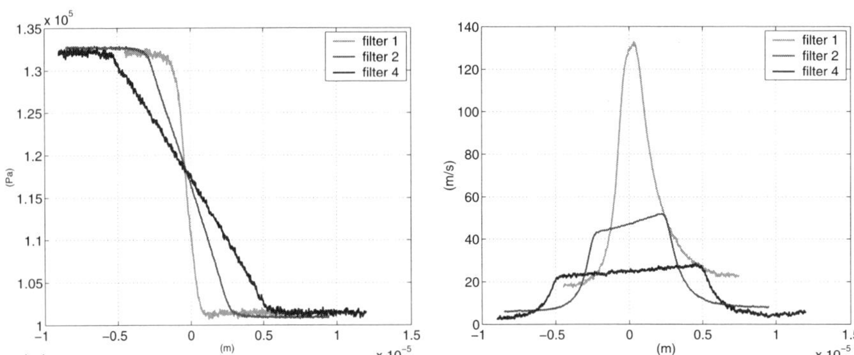

(a) (b)

FIGURE 6.28. Pressure (a), streamwise velocity (b) variations along the midline of the channel (line p1–p2 of Figure 6.27) for filters 1, 2, and 4.

Effect of Channel Length

Dependence of the flow characteristics on channel length was investigated using filters 1, 2, and 4, which have identical channel height (1 μm), but different channel lengths (1, 5 and, 10 μm) (see Table 6.2). A pressure difference of 0.3 atm was applied through these filter elements in all simulations. In Figure 6.28, we present the pressure (a) and channel center velocity (b) variation along the midline of the channel (line p1–p2 of Figure 6.27). To facilitate comparisons between the different cases, the x and y axes are shifted so that the center point for each channel is at $x = y = 0$.

The pressure drop in filters 1, 2, and 4 is shown in Figure 6.28 (a). Filter 1 has the smallest l_c/h_c ratio of the studied cases. Thus, the presence of undeveloped flow can be easily observed in the left figure. For this filter, the pressure drop is nonlinear and extends beyond the membrane channel. The flows in filters 2 and 4 exhibit typical developing compressible flow for large sections of the channel, which results in almost linear pressure drop, and a smoothly increasing velocity profile. The nonlinear pressure distribution, which is observed in compressible channel flows with large pressure drop, is not observed due to the small input-to-output pressure ratio. As a result, it can be concluded that for these filters the pressure distribution is very close to linear, unless the filter length-to-height ratio introduces undeveloped flows.

Flowrate Comparisons

From a system designer's point of view, the flowrate obtained from a filter is one of its most important characteristics. It is necessary to know when and to what extent simple analytical formulas are adequate, when the Navier–Stokes solvers are valid, or when the DSMC approach is required. Here, DSMC results are compared with the analytical results for rarefied long-channel flows in order to gain insight into validity of simple analytical formulas. Volumetric flowrate at the inlet of the channels is found using the mass flowrate equation (4.6) by considering only the first-order slip effects (see Section 4.1.1). Predictions from this analytical relation as well as the volumetric flowrates obtained from the DSMC simulations are presented in Table 6.2.

For filters 1, 2, and 4, the simulated flowrates are smaller than the analytical estimates by 50.8%, 7.3%, and 0.9%, respectively. The analytical formula will tend to overestimate the flowrate because it ignores the entrance effects, and to underestimate the flowrate because it assumes a fully developed flow. For filters 1, 2, and 4, the analytical estimates of flowrate are larger than that of the DSMC predictions. This indicates that the entrance effects are dominant, and they result in a decreased flowrate. Inspection of the Knudsen number at their outlets show that filters 3 and 5 are in the transition and free-molecular flow regimes, respectively. For filters 3 and 5, the flowrates obtained by the DSMC are larger than those calculated using the analytical formula by 4.8% and 57.9%, respectively. This difference can be attributed to the rarefaction effects. The slip velocity and the rarefaction effects are underestimated by the two-dimensional analytical model. For $\text{Kn} > 0.10$, disagreements between the continuum and DSMC predictions are expected. These simulations support the result that for $\text{Kn} < 0.10$, the analytical approach is adequate for membranes with long channels, and that in other cases, Navier–Stokes solvers with slip boundary conditions will be adequate. For $\text{Kn} > 0.10$, DSMC simulations are necessary to capture the rarefaction effects in short channels. In the case of long channels, the unified model outlined in Section 4.2 can be used.

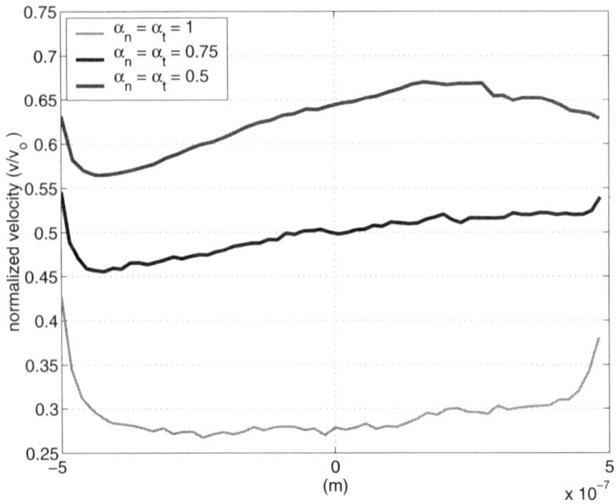

FIGURE 6.29. Slip velocity on channel walls normalized with the local mean velocity v_o (line p7–p8 of Figure 6.27).

Effects of the Accommodation Coefficients

Incomplete accommodation is observed on very clean and smooth surfaces, or for cases in which the surface atoms are much heavier than the gas molecules (see Section 2.2.2). Microfabrication techniques can provide methods to engineer the surface material for achieving lower accommodation coefficients. To predict the effects of incomplete accommodation, flow characteristics in microfilters are simulated using the Cercignani–Lampis model for surface scattering (see Section 15.4). For the simulations, tangential and normal energy accommodation coefficients are taken to be equal ($\sigma_t = \sigma_n = \sigma$) and uniform on all physical surfaces (G and J in Figure 6.27). Simulations for filter 1 are performed using $\sigma = 0.75$ and $\sigma = 0.5$. The reduction in the accommodation coefficients results in various changes in the flow. We observe in Figure 6.29 that incomplete accommodation increases the slip velocities on channel surfaces; this is also the case on the filter membrane surface, i.e., lines p4–p9 and p5–p10 in Figure 6.27. The ratios of the slip velocity for $\sigma = 0.75$ and $\sigma = 0.5$ to the slip velocity for fully accommodating surfaces ($\sigma = 1$) are 2.5 and 3.5, respectively.

The averge velocity obtained for filter 1 at $\sigma = 1.0$, $\sigma = 0.75$, and $\sigma = 0.5$ are $16.2, 18.0$, and 19.5 m^2/s), respectively. Therefore, the volumetric flowrates for $\sigma = 0.75$ and $\sigma = 0.5$ are 11% and 20% larger than that for the $\sigma = 1.0$ case, while the average velocity obtained for filter 3 at $\sigma = 1.0$, $\sigma = 0.75$, and $\sigma = 0.5$, are 2.59, 4.35, and 5.74 m^2/s), respectively. Therefore, the volumetric flowrates for $\sigma = 0.75$ and $\sigma = 0.5$ are 65% and 120% larger than that for the $\sigma = 1.0$ case. These results indicate that for a given value of σ, the flowrate of filter 3 increases by a much larger percentage compared to the increase in flowrate of filter 1. The weak dependence of flowrate on accommodation coefficients for filter 1 can be explained by its small length-to-height ratio, which results in losses due to the flow entrance effects. However, for filter 3, with larger length-to-height ratio, the flowrate is limited by the interaction of the gas with the walls, and an increase in the slip velocity increases the flowrate. The large increase in the flowrate for low values of incomplete accommodation can be utilized to design more efficient filters.

6.5.4 Summary

Gas flows through microfilters and short channels show significant differences compared to the flows through long microchannels, (see Section 4.1.1). Microfilters often experience rarefaction, compressibility, and viscous heating effects simultaneously. Since typically $\mathrm{Re} < 100$, the inertial and viscous forces are equally important. For example, simulations performed in the $1 < \mathrm{Re} < 15$ range have shown that the values of viscous-drag are about 50% of the form-drag. In addition, the filters have various thicknesses, cross-sectional areas, side-wall shapes, and surface roughnesses. DSMC simulations have shown strong dependence of the volumetric flowrate on the accommodation coefficients and the length-to-height ratio of the filters. Despite the diverse geometric characteristics and complicated physics of microfilter flows, the general form of the scaling laws presented in equations (6.24)–(6.26) is useful. Similar scaling models, with different coefficients determined by numerical simulations and validated by careful experiments, can be employed for further developments.

6.6 Micropropulsion and Micronozzle Flows

The development of microspacecraft by NASA was motivated by the desire to reduce launch masses, to increase launch rates, and reduce mission risk (Mueller, 1997). A microspacecraft may contain only a single instrument. If a fleet of several microspacecraft are launched instead of a single larger spacecraft, mission flexibility is greatly increased, and the loss of one or more microvehicles would not render the mission unsuccessful. The microspacecraft could also be very useful for the next generation of global positioning system and other low Earth orbit communicating systems.

Three different classes of microspacecraft have been proposed by (Mueller, 1997) depending on their size, mass, and power ranges. *Class I* spacecraft have mass in the range 4–20 kg and use mostly conventional hardware components with some possible MEMS components. The power density for Class I is of order 1 W/kg, and characteristic dimensions are in the range of 0.3–0.4 m. *Class II* microspacecraft are based on MEMS components, and

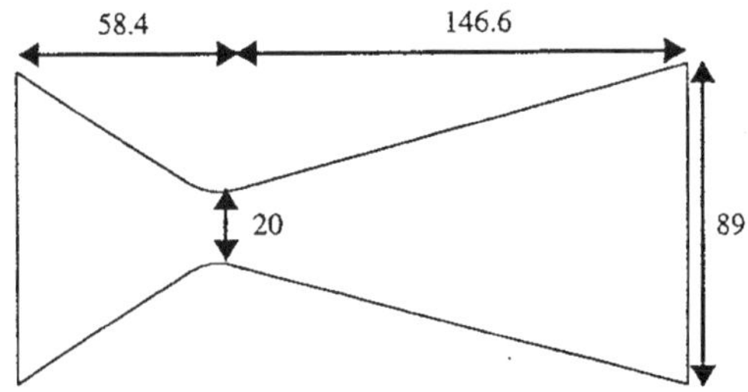

FIGURE 6.30. Sketch and typical dimensions in microns of a micronozzle.

they have mass of 1 kg and characteristic dimensions of order 10 cm. They have extremely miniaturized propulsion systems, both for delta-v maneuvers and for altitude control. Finally, *Class III* refers to microspacecraft with mass less then 1 kg and characteristic dimensions of order 3 mm or less.

The most important component for such microspacecraft is the propulsion system, which in the new mission scenarios plays an even more important role than previously. For example, propulsion capability is required on board the microspacecraft to adjust its trajectory. This points to the need for low thrust and quick response. The latter is expressed quantitatively by the *impulse bit* (I_{bit}). This is the minimum impulse obtained once the thruster is given the command to fire, and it is the integrated thrust over the faster valve cycle time.

There are a number of micropropulsion designs built or proposed that are reviewed by (Mueller, 1997) and (London, 1996). They range from ion engines to hall thrusters to pulse plasma thrusters. These technologies have been used mostly in the macro domain, but cold gas thrusters have been recently miniaturized successfully, and similar plans exist for microplasma thrusters (microPPTs) (Gulczinski et al., 2000). In particular, the cold gas thruster offers a low I_{bit} and a thrust in the range from 1 μN to 0.2 N (Bayt, 1999).

6.6.1 Micropropulsion Analysis

Micropropulsion subsystems consist primarily of thrusters, but they also contain other MEMS components such as valves, tanks, and pressure regulators. In this section, we consider the performance of a micronozzle that primarily determines the performance of the micropropulsion system. A micronozzle has a converging section, a throat, and an expanding section similar to the De Laval nozzles used in macro domain applications (see Figure 6.30). The value of Reynolds number at the nozzle throat is important,

where n_0 is the stagnation number density, T_0 is the stagnation temperature, and A_t is the nozzle area at the throat. The corresponding free-molecular mass flow rate through the nozzle is

$$(\dot{M})_{\rm FM} = m\frac{n_0 \bar{v}}{4} A_t = m\frac{n_0 A_t}{4}\sqrt{\frac{8k_{\rm B}T_0}{\pi M}},$$

where $\bar{v}$ is the average thermal speed of the propellant molecules, and m is the mass of the propellant molecule. We then obtain the specific impulse in the limit of free-molecular flow as

$$(I_{\rm sp})_{\rm FM} = \frac{(F_t)_{\rm FM}}{(\dot{M})_{\rm FM}\, g_0} = \frac{\sqrt{\frac{\pi}{2}\frac{k_{\rm B}}{M}T_0}}{g_0}.$$

We can compare this value to the maximum possible value of $I_{\rm sp}$ by considering the limit in *equilibrium* expansion, i.e.,

$$(I_{\rm sp})_L = \frac{\left[\frac{2\gamma}{\gamma-1}\left(\frac{k_{\rm B}}{M}\right)T_0\right]^{1/2}}{g_0},$$

and form the ratio

$$\frac{(I_{\rm sp})_{,\rm FM}}{(I_{\rm sp})_L} = \left[\frac{\pi(\gamma-1)}{4\gamma}\right]^{1/2}.$$

This ratio is equal to 0.474 for $\gamma = 1.4$ and 0.562 for $\gamma = 1.67$.

This analysis has motivated Ketsdever et al. (1998) to design and optimize a free-molecule microresistojet (FMMR); see Figure 6.32. The FMMR is based on the transfer of energy into the propellant gas through molecular collisions with thin-film heated elements. The inlet conditions into the expansion slot correspond to a Knudsen number of approximately one. It was found in (Ketsdever et al., 1998), that the intrinsic $I_{\rm sp}$ of the FMMR is about 60% of the intrinsic efficiency of a typical micronozzle. However, the proposed FMMR operates at very low stagnation pressures (50 to 500 Pa), and from the systems-efficiency point of view this may be advantageous, since, for example, the common problem of microvalve leakage is avoided. An FMMR has been fabricated, and heat transfer and total power input data have been obtained and presented in (Ketsdever et al., 2000a). For highly rarefied flows, small expansion ratios are preferable to avoid large shear losses. Different propellants can be used in FMMR, and the above equation shows that $I_{\rm sp}$ is highest for helium, followed by ammonia and water vapor and then by argon. Specifically, DSMC results for argon reported in (Ketsdever et al., 1998), show that a thrust of $\frac{1}{4}$ mN is obtained at $I_{\rm sp} \approx 45$ sec. In the experiments presented in (Ketsdever et al., 1998), argon was used as propellant, and the plenum pressure was 50 Pa. The thrust chip was fabricated using DRIE with 40 expansion slots of $w = 100$ μm.

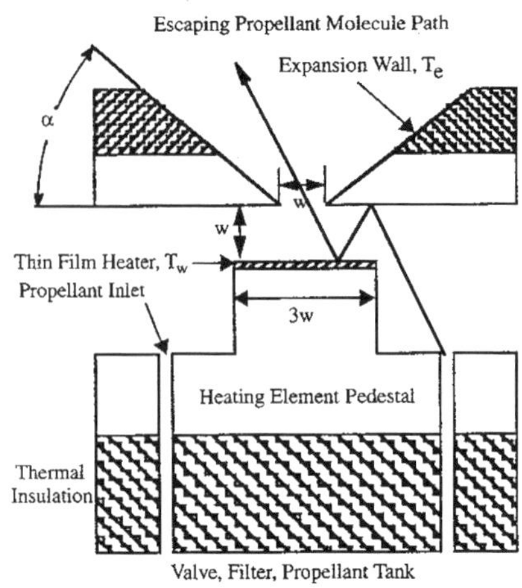

FIGURE 6.32. Diagram of the free molecule microresistojet. The width of the expansion slot w is much less than the mean free path in the stagnation chamber; $w = 100\ \mu$m. (Courtesy of A. Ketsdever.)

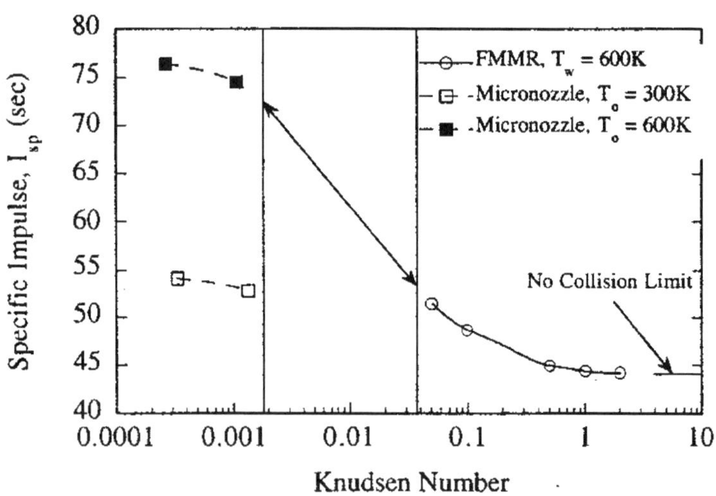

FIGURE 6.33. Free-molecule microresistojet: Specific impulse as a function of Knudsen number for argon. (Courtesy of A. Ketsdever.)

TABLE 6.3. Typical operating conditions and propulsion characteristics for different micronozzles.

Fabrication	p_0 (psi)	Throat (μm)	Re	F_t (mN)	I_{sp} (secs)
DRIE (Bayt, 1999)	85	19	3492	9.91	66.63
Conventional (Grisnik et al., 1987)	3.32	640	3000	9.65	61.96
KOH (Janson and Helvajian, 1997)	25	250	3258	9	< 56

A typical result for the specific impulse obtained using DSMC (Ketsdever et al., 1998) is shown in Figure 6.33 at two different temperatures as a function of Knudsen number. A comparison is made with results from a micronozzle with a 20 μm throat, and inlet and outlet have ratios 16.8 and 19.8, respectively. The micronozzle results were obtained with the continuum-based program GASP (Aerosoft Inc., Blacksburg Virginia, 1996), which solves the compressible Navier–Stokes equations. GASP overpredicts values of specific impulse I_{sp}, since rarefaction effects typically cause lower values of I_{sp}.

6.6.2 Rarefaction and Other Effects

There are several phenomena that affect the performance of micronozzle, including rarefaction, viscosity, three-dimensionality, roughness, nonequilibrium, and plume contamination. The relative effect of a specific phenomenon depends primarily on the thrust level targeted. Therefore, to make these contributions clearer we will refer to specific designs available in the literature.

We start with the DRIE-fabricated planar micronozzles of (Bayt, 1999) and (Bayt and Breuer, 2000). A baseline configuration has a nozzle throat of 20 μm, expansion ratio 25:1, etch depth 300 μm, and a nozzle divergence angle of 20°. In the ideal case of no losses, such a thruster operating at stagnation pressure of $p_0 = 100$ psi will produce thrust of $F_t = 7$ mN at specific impulse $I_{sp} = 75$ secs, while at $p_0 = 14.7$ psi the thrust reduces to $F_t = 1$ mN. However, these ideal operating points will be affected by losses. For example, viscous losses could be substantial at $p_0 = 14.7$ psi, where the corresponding Reynolds number is 300, and less so at $p = 300$ psi, where the corresponding Reynolds number is 2040. In fact, significant performance degradation was reported in (Bayt, 1999; Bayt and Breuer, 2000), for Re ≤ 1500 for most of the designs that they tested with nitrogen as the propellant. Table 6.3 (taken from (Bayt, 1999)) summarizes results of different micronozzles at similar thrust levels and in the same Reynolds number range.

The nozzles fabricated by (Bayt and Breuer, 2000) were planar (see Fig-

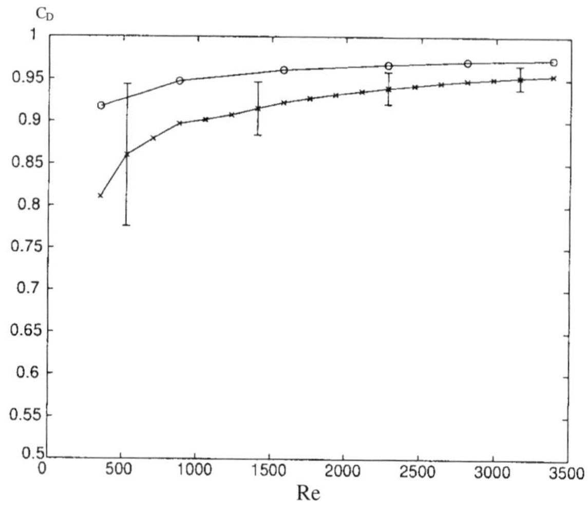

FIGURE 6.34. Mass discharge coefficient: comparisons between DSMC results (circles) and experimental results in (Bayt and Breuer, 2000). (Courtesy of M. Ivanov.)

ure 6.31), while the ones presented by (Grisnik et al., 1987) were axisymmetric, and the one by (Janson and Helvajian, 1997) had a square throat. In all cases a high Reynolds number is achieved, and the chamber pressure is relatively high. This, in turn, implies that the rarefaction effects are not dominant, with the possible exception in the results of (Grisnik et al., 1987) corresponding to low stagnation pressure. However, in these experiments the characteristic length scale is large, which implies reduced values of Knudsen number as well.

Despite the continuum operating regime, both DSMC and Navier–Stokes simulations overpredict significantly the thrust efficiency as well as the mass discharge coefficient. This is shown in Figures 6.34 and 6.35 for a nozzle with 34 μm throat and 7:1 expansion ratio fabricated by (Bayt, 1999) and (Bayt and Breuer, 2000). The computational results in (Markelov and Ivanov, 2001), utilized both two-dimensional and three-dimensional simulations with the program SMILE, which is described below. Independent two-dimensional simulations based on compressible Navier–Stokes equations performed in (Bayt, 1999) gave similar results.

More specifically, a comparison between two- and three-dimensional results by (Markelov and Ivanov, 2001) shows a decrease of the discharge coefficient from 99.6% to 89.3%, and of thrust efficiency from 87.1% to 80.6%. However, in addition to these adverse three-dimensionality effects there are other physical factors and experimental uncertainties that contribute to this large overprediciton by simulation. The following is a list of possible effects investigated systematically in (Bayt, 1999):

- Roughness,

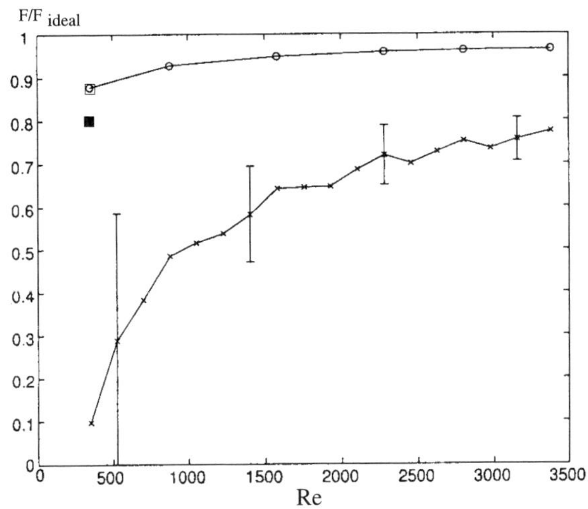

FIGURE 6.35. Thrust coefficient defined as the ratio of actual to ideal thrust force: comparisons between DSMC results (circles/squares) and experimental results in (Bayt and Breuer, 2000). The solid square corresponds to a three-dimensional simulation. (Courtesy of M. Ivanov.)

- Transition to turbulence,
- Nonequilibrium, and
- Outflow conditions.

The main question regarding *roughness* is whether it can induce shocks. This was examined by (Bayt, 1999) by considering the Mach number based on the roughness height h, where

$$M_h \propto \mathrm{Re}_h \cdot \mathrm{Kn}_h \,.$$

Therefore, if the product $\mathrm{Re}_h \cdot \mathrm{Kn}_h$ is greater than one, then shocks may form. In the experiments of Bayt, Re_h ranges from 0.1 to 10, and Kn varies between 0.1 and 5, and thus it is possible for this product to be greater than one. However, for a shock to appear, the roughness height h has to be larger than the height of the subsonic region, which does not seem to be the case for Bayt's nozzles, as corresponding simulations show. In addition, transition to turbulence does not seem to occur, since the critical Re_{δ^*} for supersonic boundary layers is 250 (based on momentum thickness δ^*), which is much larger than the Re_{δ^*} of the experiments. In Section 15.4, we will show the effects of roughness corresponding to different models on micronozzle performance.

The effects of *nonequilibrium* and associated "frozen losses" may come about because of the short length of the nozzles, which results in a short residence time for particles. There are two time scales associated with this process. The first relates to the *relaxation time*, which is the time for gas particles to reach equilibrium. It is defined as

$$t_r = nA_c\bar{c},$$

where n is the number density of the gas, A_c is the collisional cross-sectional area, and $\bar{c} = \sqrt{3RT}$ is the mean molecular speed. At standard conditions $t_r \approx 1$ nanosecond, and the residence time from the throat to exit is about 2 microseconds for a typical micronozzle tested by (Bayt, 1999; Bayt and Breuer, 2000), so in this case the flow is in equilibrium. However, in the supersonic portion of the nozzle the relaxation time could be comparable to residence time for some conditions, and this may lead to strong nonequilibrium effects. This, in turn, translates into losses of thermal energy, which is converted into kinetic energy as the gas expands. Similar effects can be induced by rotational nonequilibrium associated with a second time scale, the *rotational relaxation rate*. This time scale may also be comparable to the residence time and is larger inside the boundary layer. A rotational frozen flow could have 15% less kinetic energy.

Finally, the effect of *exit losses* and of *outflow boundary conditions* may be quite pronounced. Ivanov et al. (1999) have considered the effect of extrapolated boundary conditions used at the outflow and noted a 5–10% loss in efficiency if the plume is not modeled explicitly; similar results were obtained for large nozzles where back-flow may occur (plume contamination) (Gatsonis et al., 2000). Clearly, standard extrapolation boundary conditions are questionable in micronozzles, with the core of inviscid flow relatively small compared to external aerodynamics applications, where such outflow boundary conditions are employed routinely.

We now turn to micronozzles with lower thrust level (< 1 mN). These nozzles correspond to lower pressures, and therefore *rarefaction effects* have to be accounted for in simulations. In general, in this regime, continuum approaches overpredict the specific impulse even for high Reynolds numbers of order 1000. We start by considering the *Rothe nozzle*, which is similar to the nozzle shown in the sketch of Figure 6.30, but it has a throat of 2.5–5 mm and a chamber pressure less than 1 atm, with the Reynolds number in the range 55 to 550. The subsonic and supersonic portions of the Rothe nozzle are cones with half-angles 30° and 20°, respectively, and with longitudinal radius of curvature at the throat equal to half of the throat radius. The gas in Rothe's experiments was nitrogen at $T = 300$ K. In this axisymmetric nozzle, measurements of temperature and density were obtained along the centerline and along the radial directions at various stations (Rothe, 1971). At Re < 150, the exit temperature was increasing due to viscous shear force, causing the thermalization of flow energy. Although this is not strictly a micronozzle, the availability of detailed experimental results (Rothe, 1971) has made this a benchmark case for validating DSMC programs before other micronozzles are tested.

In the following we present simulations performed by (Ivanov et al., 1999)

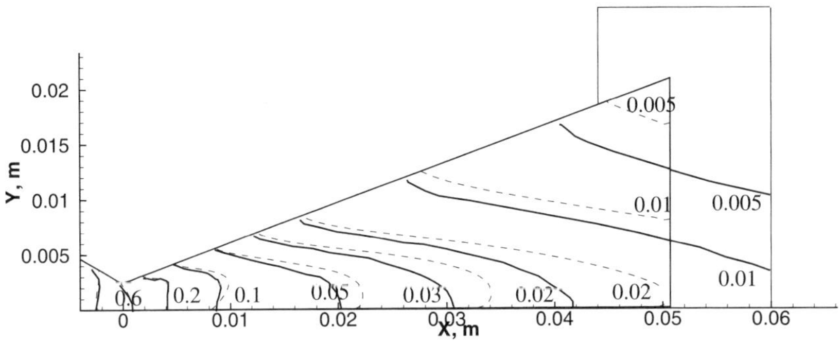

FIGURE 6.36. Rothe nozzle ($Re_t = 270$): density isocontours based on atomistic (solid lines: SMILE) and continuum (dashed lines: GASP) simulations. (Courtesy of M. Ivanov.)

for throat Reynolds numbers in the range 120–270. The DSMC program SMILE developed by Ivanov and collaborators combines majorant and free cell schemes (see also Section 15.1). Each cell has its own interaction parameter that governs local collisions via an adaptive scheme. It employs two different grids, one for collisions and another one for sampling the gas dynamics parameters, with the latter having fewer cells than the former. In addition, other features such as spatial weights for axisymmetric domains that distribute molecules closer to the axis, adaptive and asynchronous domain decomposition, and particle doubling make SMILE a particularly effective program for simulating micronozzle flows. The last feature introduces successively an increased number of particles, so it accelerates convergence to stationary states.

The density contours obtained with SMILE for $\mathrm{Re}_t = 270$ are plotted in Figure 6.36 (Ivanov et al., 1999). The dashed lines show corresponding results from the program GASP (Aerosoft Inc., Blacksburg Virginia, 1996) based on the compressible Navier–Stokes equations. Corresponding Mach line contours and differences between continuum and rarefied flows are shown in Figure 6.37. In Figure 6.38 a comparison is included with Rothe's experimental data that shows good agreement with the simulations but deviations from the ideal case.

Ivanov et al. (1999) have also examined the performance of a micronozzle with an about 27-micron throat and 15° diverging half-angle. It was designed so that at $p_0 = 10$ atm gave 1 mN thrust at stagnation temperature $T_0 = 297$ K. The stagnation pressure was varied from 1 to 10 atm, and correspondingly, the Reynolds number varied from 130 to 1300. This tenfold increase in Re_t results in specific impulse efficiency change from 90% to 95% compared with the ideal case. A typical result obtained by Ivanov et al. is shown in Figure 6.39 in terms of the specific impulse along the micronozzle axis. Again, comparisons between continuum and atomistic predictions are made, similar to those for the results for the Rothe nozzle. Simulations pre-

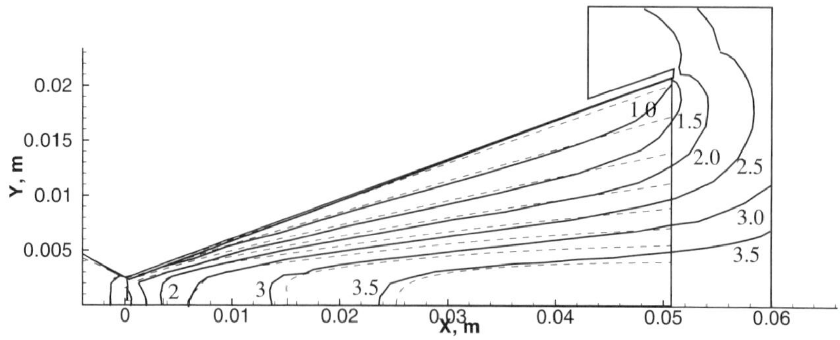

FIGURE 6.37. Rothe nozzle ($Re_t = 270$): Mach isocontours based on atomistic (solid line: SMILE) and continuum (dashed line: GASP) simulations. (Courtesy of M. Ivanov).

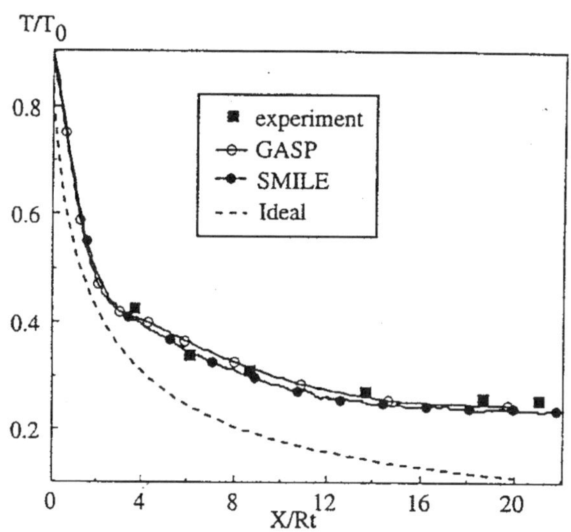

FIGURE 6.38. Temperature distribution along the Rothe nozzle axis based on atomistic (SMILE) and continuum (GASP) simulations ($\text{Re}_t = 270$). (Courtesy of M. Ivanov.)

dict that I_{sp} is greater than the ideal for low Reynolds number ($\text{Re}_t = 130$), which is a nonphysical result. Even at large pressures there seems to be a large effect of rarefaction. For low pressure, the maximum I_{sp} occurs upstream of the nozzle exit. As pointed out by Ivanov *et al.*(1999), this is due to relatively long diverging section which leads to excessive viscous losses. A shorter expansion nozzle would be advantageous, and would also reduce back-flow contamination due to plume at the nozzle exit.

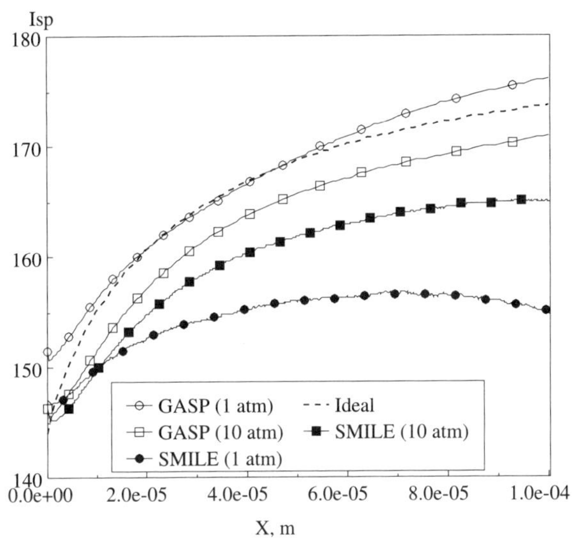

FIGURE 6.39. Atomistic (SMILE) and continuum (GASP) predictions for the specific impulse along the micronozzle axis at different stagnation pressures. (Courtesy of M. Ivanov.)

Part II:
Liquid Flows

7
Electrokinetic Flows

Rapid developments in microfabrication technologies have enabled a variety of microfluidic systems consisting of valves, pumps, and mixers to be utilized effectively for medical, pharmaceutical, defense, and environmental monitoring applications. Examples of such applications are drug delivery, DNA analysis/sequencing systems, and biological/chemical agent detection sensors on microchips. These microfluidic systems require seamless integration of sample collection, separation, biological and chemical detection units with fluid pumping, flow control elements, and the necessary electronics on a single microchip. The reliability and compliance of these components are important for successful design and operation of the entire microfluidic system. In particular, subsystems like microvalves and micropumps with moving components are complicated to design and fabricate, and they are prone to mechanical failure due to fatigue and fabrication defects.

In this chapter, we review and explore ideas of *microflow control elements* using electrokinetic flow control schemes, which do not require any moving components. We cover electroosmotic and electrophoretic transport in detail. We also present dielectrophoresis, which enables separation and detection of similar size particles based on their polarizability. Theoretical treatments of these electrokinetic transport mechanisms are kept at an introductory level for brevity. For further information, the reader is referred to classical textbooks (Probstein, 1994; Hunter, 1981; Melvin, 1987; Righetti, 1983; Shaw, 1980; Westermeier, 1990; Pohl, 1978). Other reviews of electrokinetically driven liquid microflows can be found in (Stone et al., 2004; Gad-El-Hak, 2001; Nguyen and Wereley, 2003; Morgan and Green, 2003). Also, in Chapter 12 we focus on electroosmotic flows in nanochannels.

7.1 Electrokinetic Effects

The electrokinetic phenomena can be divided into the following four categories (Probstein, 1994):

- *Electroosmosis* is the motion of ionized liquid relative to the stationary charged surface by an applied electric field.
- *Electrophoresis* is the motion of the charged surfaces and macromolecules relative to the stationary liquid by an applied electric field.
- *Streaming potential* is the electric field created by the motion of ionized fluid along stationary charged surfaces (the opposite of electroosmosis).
- *Sedimentation potential* is the electric field created by the motion of charged particles relative to a stationary liquid (the opposite of electrophoresis).

Electrokinetic effects are important for micro- and nanoscale transport applications. For example, electroosmosis enables fluid pumping and flow control using electric fields, eliminating the need for mechanical pumps or valves with moving components. Furthermore, comparisons between the electroosmotically and pressure-driven flows reveal several important differences. First, the volumetric flowrate (per channel width) in electroosmotic flows varies linearly with the channel height (h), enabling useful flowrates in microchannels and nanochannels with low-driving electric fields. However, the flowrate (per unit width) in pressure-driven flows varies as the cube of the channel height (h^3). This requires unrealistically large pressure drops, making pressure-driven flows impractical for nanochannel applications. Second, the velocity profiles for electroosmotic flows are (mostly) uniform (plug-like), while the pressure-driven flow velocity profiles are parabolic. This difference has significant effects in species transport and dispersion in microfluidic applications (see Section 7.5.3).

Typical physical and electrochemical conditions employed in electrokinetic flows are presented in Table 7.1. Since we are mostly interested in bulk fluid and particle/species transport, we cover electroosmotic and electrophoretic transport in detail. We also present dielectrophoresis, which enables separation and detection of similar-size particles based on their polarizability.

In the rest of this section we present a brief review of historical developments and recent research on electrokinetic flows. Liquid flows in capillary porous systems under the influence of external electric fields have attracted the attention of many scientists since the discovery of electrokinetic transport (Reuss, 1809). In 1870, Helmholtz developed the electric double layer theory, which relates the electrical and flow parameters for electrokinetic transport. Electroosmosis has been used for chemistry applications

Parameter	Parameter range
Typical channel thickness, h (μm)	$0.01 \sim 300$
Electrolyte concentration, n_o (mM)	$10 \sim 0.001$
Debye length, λ_D (nm)	$1 \sim 100$
Zeta potential, ζ (mV)	$\pm 1 \sim \pm 100$
Electric field, $\mathbf{E}$ (V/mm)	$1 \sim 100$
Electroosmotic Velocity, U (mm/s)	< 2
Reynolds number, Re	$10^{-4} \sim 1$

TABLE 7.1. Typical physical and electrochemical parameters utilized in electrokinetically driven phenomena.

since the late 1930s. Modern theoretical developments include solution of mixed electroosmotic/pressure-driven flows in very thin two-dimensional slits (Burgreen and Nakache, 1964; Ohshima and Kondo, 1990), as well as in thin cylindrical capillaries (Rice and Whitehead, 1965; Lo and Chan, 1994; Keh and Liu, 1995). In 1952, Overbeek proposed irrotationality of internal electroosmotic flows for arbitrarily shaped geometries (Overbeek, 1952). This was followed by the ideal electroosmosis concept, i.e., electroosmotic flow in the absence of externally imposed pressure gradients, which results in the similarity between the electric and velocity fields under *specific* outer field boundary conditions (Cummings et al., 2000). Also, Santiago has shown that ideal electroosmosis is observed for low Reynolds number steady flows. However, unsteady or high Reynolds number flows violate this condition (Santiago, 2001). The analytical solution of unsteady electroosmotic flows obtained in Section 7.4.2 confirms these predictions.

Next we review some of the experimental work; (Molho et al., 1998) presented measurements of electroosmotically driven microcapillary flows, and they have shown that Joule heating (see Section 7.4.6) and the corresponding changes in fluid viscosity are secondary effects compared to the streamwise pressure gradients. (Paul et al., 1998) have used a caged dye fluorescence technique to capture the flow patterns in mixed electroosmotic/pressure driven microchannel flows. (Cummings et al., 1999) have used microparticle image velocimetry (μPIV) techniques to obtain velocity distribution in straight channels and cross-flow junctions. (Kim et al., 2002) have also developed a μPIV technique to measure mixed electroosmotic/pressure-driven flows in cross-flow and T-junctions. (Singh et al., 2001) developed unilamellar liposome particles to trace microflows. Unlike the traditional latex flow marker, these contain fluorescent material both in the core and at the surface of the marker; hence this technique provides higher fluorescence intensity. (Herr et al., 2000) reported velocity and dispersion rate measurements for electroosmotic flows through cylindrical capillaries with nonuniform surface-charge distribution. They have used various surface materials as well as polymeric coatings to obtain different surface-charge distributions by modifying the local zeta potential. Experiments performed using the caged-dye fluorescence technique indicated strong dependence of fluid velocity and dispersion rate on the surface charge distribution (Herr et al., 2000). In electrokinetic flows, fluid dispersion may also be caused by a mismatch in the electroosmotic flow rate and electric field. It can also be induced by Joule heating (see Section 7.4.6).

There have been several studies on numerical simulation of electroosmotic transport. (Yang and Li, 1998) developed a numerical algorithm based on the Debye–Hückel linearization and studied electrokinetic effects in pressure-driven liquid flows. (Patankar and Hu, 1998) developed a finite-volume technique and studied electroosmotic injection at the intersection of two channels. Their numerical results for Re $>$ 1 flows showed significant inertial effects, which is in agreement with the theoretical work of (Santiago, 2001). (Bianchi et al., 2000) used a finite-element method to model electroosmotic flow in a T-channel junction. Beskok and colleagues have developed a spectral element algorithm for solution of mixed electroosmotic/pressure-driven flows in complex geometries (Dutta et al., 2002a; Dutta et al., 2002b). Aluru and coworkers developed meshless methods as well as compact methods to study steady electroosmotic flows in complex microchannel geometries (Mitchell et al., 2000; Qiao and Aluru, 2002). They also investigated the validity of the Poisson–Boltzmann equation in nanochannels using molecular dynamics (MD) simulations and coupled these MD results with a modified hydrodynamic continuum model; see Chapter 12 and also (Qiao and Aluru, 2003b; Qiao and Aluru, 2003d), for details.

7.2 The Electric Double Layer (EDL)

Electrokinetic phenomena are present due to the *electric double layer* (EDL), which forms as a result of the interaction of ionized solution with static charges on dielectric surfaces (Hunter, 1981). For example, when silica is in contact with an aqueous solution, its surface hydrolyzes to form silanol surface groups. These groups may be positively charged as Si-OH_2^+, neutral as Si-OH, or negatively charged as Si-O^-, depending on the pH value of the electrolyte solution. If the channel surface is negatively charged (such as in the case of deionized water), the positive ions are attracted toward the surface, and the negative ions are repelled from the surface, keeping the bulk of the liquid, far away from the wall, *electrically neutral.* A schematic of ion distribution in the buffer solution is shown in Figure 7.1. The ions of opposite charge cluster immediately near the wall, forming the *Stern layer*, a layer of typical thickness of one ionic diameter. The ions within the Stern layer are attracted to the wall with very strong electrostatic forces; hence they are immobilized near the charged surface, as demonstrated also by molecular dynamics studies (Lyklema et al., 1998). Immediately after the

Stern layer there forms the *diffuse layer*, where the ion density variation obeys the Boltzmann distribution, consistent with the derivation based on statistical-mechanical considerations (Feynman et al., 1977). Hence, the EDL consists of two distinct zones: Stern and diffuse layers. The extent of the EDL can be approximately predicted by the Debye length (λ_D), which is defined as the distance from the wall, where the electrokinetic potential energy is equal to the thermal energy. The Debye length depends on the molar concentration of the ionized fluid, and its thickness λ_D can be estimated using the Debye–Hückel parameter (ω):

$$\omega = \frac{1}{\lambda_D} = \sqrt{\frac{e^2 \Sigma_i n_i z_i^2}{\epsilon \epsilon_o k_B T}}, \tag{7.1}$$

where n is the concentration, k_B is the Boltzmann constant, e is the electron charge, z is the valence, T is the temperature, ϵ_o is the dielectric permittivity of vacuum, and ϵ is the dielectric constant of the solvent. The subscript i indicates the ith species. Using equation (7.1) and assuming symmetric electrolyte at $n_o = 1.0 \times 10^{-2}, 1.0 \times 10^{-5}$, and 1.0×10^{-6}M($\equiv$ Mole/liter), we obtain Debye lengths of $\lambda_D = 3$ nm, 100 nm and 300 nm, respectively.

Ion distribution due to the EDL can be characterized using an electrokinetic (electric) potential ψ. Since the oppositely charged ions in the Stern layer shield some of the electric charges on the surface, the electrokinetic potential drops rapidly across the Stern layer. The value of ψ at the edge of the Stern layer is known as the zeta potential (ζ). For most practical cases, we can employ the zeta potential to describe electrokinetic flows rather than the wall potential ψ_0 (See Figure 7.1). Ion distribution in the diffuse layer results in a net electric charge, which can be related to the electrokinetic potential as follows:

$$\nabla^2 \psi = \frac{-\rho_e}{\epsilon \epsilon_o}. \tag{7.2}$$

The electric charge density ρ_e is given by

$$\rho_e = F \sum z_i n_i, \tag{7.3}$$

where F is Faraday's constant. Here we emphasize that:

- *The net electric charge contained in the diffuse layer is the primary reason for electrokinetic effects, where charged ions or particles can be mobilized by externally applied electric fields.*

If we assume a symmetric electrolyte of equal valence that is in *equilibrium* with the charged surface, equation (7.3) leads to a Boltzmann distribution, resulting in the **Poisson–Boltzmann equation**

$$\nabla^2(\psi^*) = \beta \sinh(\alpha \psi^*), \tag{7.4}$$

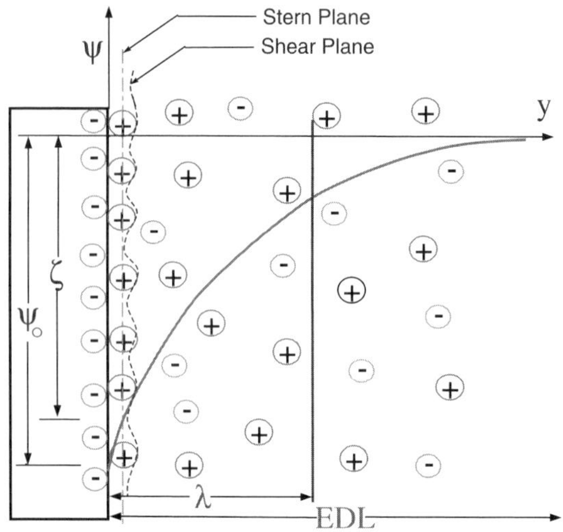

FIGURE 7.1. Schematic diagram of the electric double layer (EDL) next to a negatively charged solid surface. Here ψ is the electrokinetic potential, ψ_o is the surface electric potential, ζ is the zeta potential, and y is the distance measured from the wall. The Debye length and the EDL thickness are shown by λ_D and EDL, respectively.

where $\psi^*(= \psi/\zeta)$ is the electrokinetic potential *normalized* with the zeta potential ζ, and α is the ionic energy parameter given by

$$\alpha = ez\zeta/k_B T. \tag{7.5}$$

At 20°C, $\alpha = 1$ corresponds to 25.4 mV. The variable β relates the ionic energy parameter α and the characteristic channel length (flow dimension) h to the Debye–Hückel parameter ω in equation (7.1) as follows:

$$\beta = \frac{(\omega h)^2}{\alpha}.$$

We must note that for small zeta potential ($\zeta \leq 25$ mV), it is possible to linearize the right-hand side of the Poisson–Boltzmann equation (7.4) via a Taylor series expansion. This results in the Debye–Hückel approximation

$$\nabla^2(\psi^*) = \beta\alpha\psi^*. \tag{7.6}$$

In equation (7.4), we presented the Poisson–Boltzmann equation in non dimensional form. Let us consider a two-dimensional channel and assume that the zeta potential ζ is known, and that it remains constant along the channel. Under these conditions equation (7.4) can be simplified in the following form:

$$\frac{d^2\psi^*}{d\eta^2} = \beta \sinh(\alpha\psi^*), \tag{7.7}$$

where $\eta = y/h$ and h is the half-channel height. Multiplying both sides of this equation by $(2\frac{d\psi^*}{d\eta})$, and integrating with respect to η, the following relation is obtained:

$$\frac{d\psi^*(\eta)}{d\eta} = \sqrt{\frac{\beta}{\alpha}}\left[2\cosh(\alpha\psi^*) - 2\cosh(\alpha\psi_c^*)\right]^{\frac{1}{2}}, \tag{7.8}$$

where both the electric potential and its spatial gradient at point η are represented as a function of the electric potential at the channel center (i.e., $\psi_c^* = \psi_{\eta=0}^*$).

(Burgreen and Nakache, 1964) obtained an analytical solution of (7.7) in terms of an elliptic integral of the first kind. Their work presents the potential distribution as a function of the Debye length λ_D and the ionic energy parameter α. It was shown by (Dutta and Beskok, 2001a) that for $\alpha \geq 1$ and $\lambda_D \ll h$ the electric potential in the middle of the channel is practically zero. Hence, as $\psi_c^* \to 0$ the last term in equation (7.8) is simplified, and using the identity $\cosh(p) = 2\sinh^2(p/2) + 1$, equation (7.8) can be integrated once more, resulting in the solution

$$\psi^*(\eta^*) = \frac{4}{\alpha}\tanh^{-1}\left[\tanh\left(\frac{\alpha}{4}\right)\exp\left(-\sqrt{\alpha\beta}\ \eta^*\right)\right], \tag{7.9}$$

where η^* is the distance from the wall (i.e., $\eta^* = 1 - |\eta|$).

In Figure 7.2, a numerical solution of the electroosmotic potential distribution as a function of various α and β values is presented. The left and right figures show the potential distributions for $\alpha = 1$ and $\alpha = 10$, respectively, for various values of β. For $\alpha = 1$ and $\beta < 100$ the EDL is quite thick, and it covers the entire channel. As the value of β is increased, the electric double layer is confined to a zone near the channel walls, resulting in sharp variations in the electric potential. Comparisons of $\alpha = 1$ and $\alpha = 10$ cases at the same value of β show faster decay of the electroosmotic potential for increased values of α.

7.2.1 Near-Wall Potential Distribution

The potential distribution in equation (7.9) can also be represented as a function of the near-wall parameter $\chi = y'\omega$, where $y' = h - y$ is the distance from the wall, and ω is the Debye–Hückel parameter given by equation (7.1). Since $\omega h = \sqrt{\alpha\beta}$, the near-wall scaling parameter (χ) and the nondimensional distance from the wall ($\eta^* = 1 - \eta$) can be represented in terms of each other (i.e., $\chi = \sqrt{\alpha\beta}\eta^*$). Based on this, equation (7.9) can be simplified to

$$\psi^* = \frac{4}{\alpha}\tanh^{-1}\left[\tanh\left(\frac{\alpha}{4}\right)\exp\left(-\chi\right)\right]. \tag{7.10}$$

It is clear that the *inner layer scaling* of the potential distribution is independent of β for $\lambda_{\mathrm{D}} \ll h$. In Figure 7.3 the near-wall potential distribution

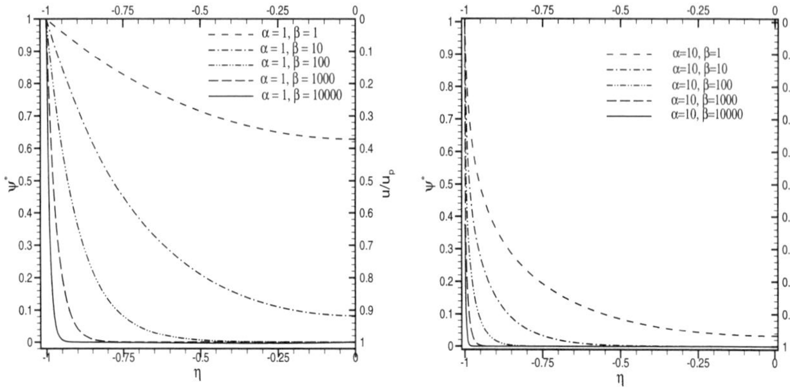

FIGURE 7.2. Variation of normalized electroosmotic potential ψ^* across half of a channel for various values of α and β.

is presented as a function of χ for several α values. We observe that the electroosmotic potential decays to zero with increased χ for all these cases. The decay rate can be quantified by presenting a logarithmic plot of the electroosmotic potential in the near-wall region as a function of χ, as shown in Figure 7.3 (right). A careful examination of Figure 7.3 (right) shows *exponential decay* of the electroosmotic potential with slope -1 for $\chi > 2$. This result can be easily verified by expanding equation (7.10) for $\chi > 2$, where $\tanh(\alpha/4) \leq 1$, and $\exp(-\chi) \ll 1$. Under these conditions

$$\psi^*(\chi) \approx \frac{4}{\alpha} \tanh\left(\frac{\alpha}{4}\right) \exp(-\chi). \tag{7.11}$$

In analogy to the 99% boundary layer thickness in traditional fluid mechanics, an effective EDL thickness (δ_{99}) can be defined as the distance from the wall (in terms of λ_D) at which the electroosmotic potential decays to 1% of its original value (Dutta and Beskok, 2001a). The *effective EDL thickness* as a function of the ionic energy parameter α is presented in Table 7.2. We can calculate the value of δ_{99} in terms of the η^* coordinates by dividing the value of δ_{99} given in Table 7.2 by $\sqrt{\alpha\beta}$.

TABLE 7.2. Variation of the effective EDL thickness δ_{99} and the EDL displacement thickness δ^* as a function of the ionic energy parameter α. The values of δ_{99} and δ^* are given in terms of the Debye length λ_D.

α	1	3	5	7	10
δ_{99}	4.5846	4.439	4.2175	3.9852	3.6756
δ^*	0.98635	0.891567	0.75670	0.627027	0.47731

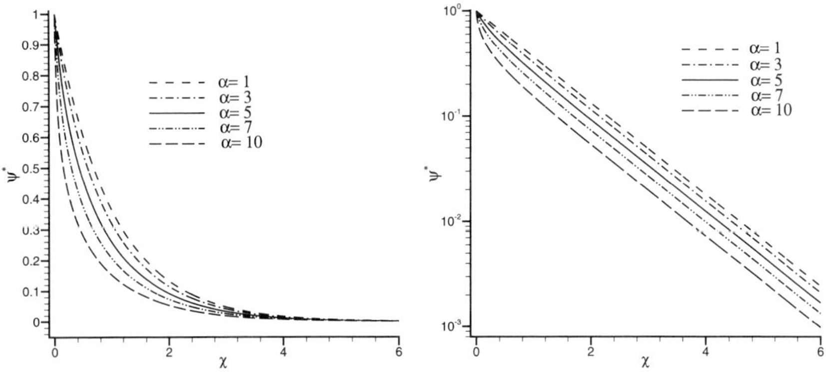

FIGURE 7.3. Electrosmotic potential distribution within the electric double layer (left) and its logarithmic scaling (right) as a function of the inner-layer scale $\chi = \omega y'$.

7.3 Governing Equations

Assuming incompressible Newtonian fluid with constant viscosity, the bulk fluid motion is governed by the incompressible Navier–Stokes equations

$$\rho_{\mathrm{f}} \left(\frac{\partial \mathbf{u}}{\partial t} + (\mathbf{u} \cdot \nabla)\mathbf{u} \right) = -\nabla p + \mu \nabla^2 \mathbf{u} + \mathbf{f}_{\mathrm{EK}}, \tag{7.12}$$

where p is the pressure, $\mathbf{u}$ is a divergence-free velocity field ($\nabla \cdot \mathbf{u} = 0$) subject to the no-slip boundary conditions on the walls, ρ_{f} is the fluid density, and $\mathbf{f}_{EK}$ is the electrokinetic body force. A general equation for the electrokinetic force per unit volume is given by (Stratton, 1941)

$$\mathbf{f}_{\mathrm{EK}} = \rho_e \mathbf{E} - \frac{1}{2}\mathbf{E} \cdot \mathbf{E} \epsilon_{\mathrm{o}} \nabla \epsilon + \frac{\epsilon_{\mathrm{o}}}{2} \nabla \left(\rho_f \frac{\partial \epsilon}{\partial \rho_{\mathrm{f}}} \mathbf{E} \cdot \mathbf{E} \right), \tag{7.13}$$

where $\mathbf{E}$ is the externally applied electric field. The last term shows permittivity variations with density, and it is especially important at liquid/gas interfaces as well as in ionized gas flows (Stratton, 1941). For our purposes, we will assume an incompressible medium with constant electric permittivity. Hence we consider only the contribution of the first term ($\rho_e \mathbf{E}$).

The species conservation equation for a multicomponent fluid, in the absence of chemical reactions, can be expressed as

$$\frac{\partial n_i}{\partial t} + \nabla \cdot \mathbf{j}_i = 0, \tag{7.14}$$

where n_i is the concentration of the ith species flux, given by

$$\mathbf{j}_i = -D_i \nabla n_i + n_i [\mathbf{u} + \mu_{\mathrm{EK},i} \mathbf{E}], \tag{7.15}$$

where D_i is the diffusion coefficient and μ_{EK} is the electrokinetic mobility. The first term on the right-hand side corresponds to molecular diffusion flux due to the concentration gradient, while the second term corresponds to convection due to bulk fluid motion with velocity $\mathbf{u}$. The following term represents transport due to the electrokinetic effects. In general, the electrokinetic mobility (μ_{EK}) includes both the electroosmotic and electrophoretic effects (Cummings, 2001; Ermakov et al., 1998). Mobility is related to the electrokinetic migration velocity $\mathbf{u}_{\text{EK}}$ by

$$\mathbf{u}_{\text{EK}} = \mu_{\text{EK}}\mathbf{E}, \tag{7.16}$$

and it depends on the physical and chemical properties of the particle or surface and the suspending fluid.

The electroosmotic mobility for infinitesimally thin EDL is given by the Helmholtz–Smoluchowski relation (Hunter, 1981)

$$\mu_{\text{EO}} = \frac{-\zeta\epsilon}{\mu}, \tag{7.17}$$

where ζ is the zeta potential. Using the electroosmotic mobility, we obtain the Helmholtz–Smoluchowski electroosmotic velocity (also indicated as $\mathbf{u}_{\text{HS}}$ in equation (7.24))

$$\mathbf{u}_{\text{EO}} = \mu_{\text{EO}}\mathbf{E} = \frac{-\zeta\epsilon}{\mu}\mathbf{E}. \tag{7.18}$$

The negative sign is due to the use of surface zeta potential. For example, for a negatively charged surface ($\zeta < 0$), the EDL will be positively charged, and the resulting electroosmotic motion will be toward the cathode, as shown schematically in Figure 7.4. To give an idea of the typical mobility magnitudes, we present in Figure 7.5 the electroosmotic mobility of sodium tetraborate buffer as a function of the buffer concentration (Sadr et al., 2004). The experiments were performed in fused silica and quartz microchannels, with heights ranging from $5\,\mu$m to $25\,\mu$m, under electric fields $E \leq 4.8$ kV/m and buffer concentration (C) of $0.19 \leq C \leq 36$ mM. The experimental results indicate significant variations in the electroosmotic mobility as a function of the buffer concentration, consistent with the theoretical predictions in (Conlisk et al., 2002).

The electrophoretic mobility for spherical colloidal particles with uniform zeta potential is (Hunter, 1981)

$$\mu_{\text{EP}} = \frac{2\zeta\epsilon}{3\mu}, \tag{7.19}$$

where the 2/3 coefficient is appropriate for infinitesimally thin EDL conditions. The reader is referred to (O'Brien and White, 1978), for further

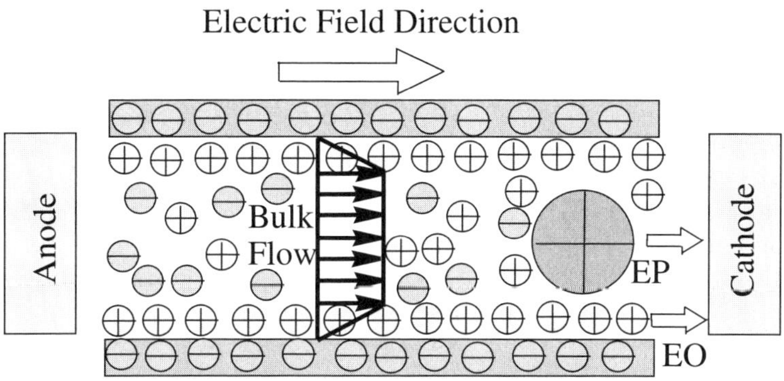

FIGURE 7.4. A schematic view of electroosmotic and electrophoretic transport processes.

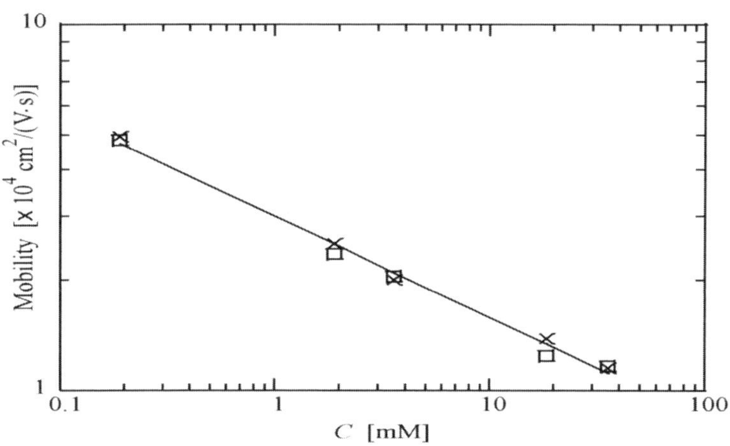

FIGURE 7.5. Electroosmotic mobility variation as a function of molar concentration (C) for sodium tetraborate buffer (Sadr et al., 2004). Experimental results are shown by squares, while the theoretical predictions are indicated by $\times$. (Courtesy of A.T. Conlisk.)

details about the electrophoretic mobility and its relation to the zeta potential and the particle charge. For most practical applications, the electrophoretic mobility is determined experimentally, and for a given electric field, it results in an electrophoretic migration velocity of

$$\mathbf{u}_{\mathrm{EP}} = \mu_{\mathrm{EP}} \mathbf{E}. \tag{7.20}$$

Based on this, a positively charged particle ($\zeta > 0$) that is free to move will migrate towards the cathode, as schematically shown in Figure 7.4.

7.4 Electroosmotic Flows

Electroosmotic flow is generated due to the interactions of ions in the EDL with an externally applied electric field (**E**). Nonzero ion density within the EDL results in net ion migration toward the oppositely charged electrode, dragging the viscous fluid with it. This effect is characterized by the electroosmotic body force term in the Navier–Stokes equation (7.12). The externally applied electric field can be represented as

$$\mathbf{E} = -\nabla\phi,$$

where ϕ is the electrostatic potential. Assuming the laws of electrostatics, the potential obeys

$$\nabla \cdot (\sigma \nabla \phi) = 0, \tag{7.21}$$

where σ is the electric conductivity. The right-hand side of the equation is zero, since the electric charge density (ρ_e) contained in the EDL is already included in equation (7.2). The external electric field is subject to the insulating boundary conditions ($\nabla\phi \cdot \mathbf{n} = 0$) on the walls. The zeta potential is assumed to be uniform for all surfaces.

The main simplifying assumptions and approximations are as follows:

1. The fluid viscosity is *independent of the shear rate*; i.e., Newtonian fluid is assumed.

2. The fluid viscosity is *independent of the local electric field strength*. This condition is an *approximation*. Since the ion concentration and the electric field strength within the EDL are increased, the viscosity of the fluid may be affected. However, such effects are neglected in the current analysis, which considers only dilute solutions; in Chapter 12 we consider such effects for nanochannels.

3. The Poisson–Boltzmann equation (7.4) is valid. Hence the *ion convection effects* are negligible.

4. The solvent is continuous, and its permittivity is not affected by the overall and local electric field strength.

5. The ions are point charges.

7.4.1 Channel Flows

In this section mixed electroosmotic/pressure-driven flows in straight microchannels are analyzed for channel heights (h) much smaller than the channel width (W). Therefore, the flow can be treated as two-dimensional,

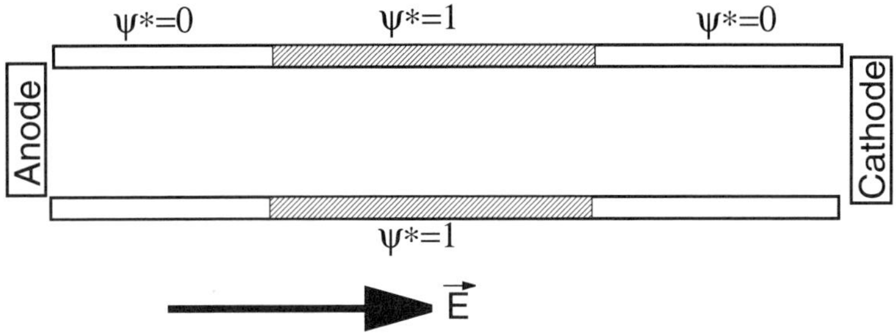

FIGURE 7.6. Schematic view of a mixed electroosmotic/pressure-driven flow channel. The inlet and exit portions of the channel have negligible electroosmotic effects.

as shown in Figure 7.6. For simplicity, fully developed steady flow with no-slip boundary conditions is assumed. The streamwise momentum equation is

$$\frac{\partial p}{\partial x} = \mu \frac{\partial^2 u}{\partial y^2} + \rho_e E_x, \tag{7.22}$$

where u is the streamwise velocity and $E_x = -d\phi/dx$. Using equation (7.7) and (7.4) for ρ_e we obtain

$$\frac{\partial p}{\partial x} = \mu \frac{\partial^2 u}{\partial y^2} - \epsilon E_x \frac{d^2 \psi}{dy^2}. \tag{7.23}$$

This equation is linear, and thus we can decompose the velocity field into two parts:

$$u = u_{\text{Pois}} + u_{\text{EO}},$$

where u_{Pois} corresponds to the pressure-driven channel flow velocity (i.e., plane Poiseuille flow), and u_{EO} is the electroosmotic flow velocity. In the absence of externally imposed pressure gradients, $u_{\text{Pois}} = 0$. Hence, the viscous diffusion terms are balanced by the electroosmotic forces. This leads to the Helmholtz–Smoluchowski electroosmotic velocity u_{HS} (Probstein, 1994):

$$u_{\text{HS}} = -\frac{\zeta \epsilon E_x}{\mu}. \tag{7.24}$$

From nondimensionalizing equation (7.23), the streamwise momentum equation becomes

$$\frac{\partial P^*}{\partial \xi} = \frac{\partial^2 U}{\partial \eta^2} + \frac{d^2 \psi^*}{d\eta^2}, \tag{7.25}$$

where, $U = \frac{u}{u_{\text{HS}}}$, $P^* = \frac{p}{\mu u_{\text{HS}}/h}$, and $\xi = x/h$. Here the pressure is normalized by viscous forces rather than the dynamic head, consistent with the Stokes flow formulation (see Section 2.1).

In the case of zero net pressure gradient we integrate equation (7.25) to obtain (Burgreen and Nakache, 1964)

$$U(\eta) = 1 - \psi^*(\eta). \tag{7.26}$$

Figure 7.2 (right vertical axis) shows the velocity variation in pure electroosmotic flows for various values of α and β. As shown in the plot, in the limit of small Debye lengths the electroosmotic potential ψ^* decays very fast within the thin electric double layer, and a uniform pluglike velocity profile is obtained in most of the channel. The plug flow behavior has been observed in various experiments (Paul et al., 1998; Herr et al., 2000).

For the mixed electroosmotic/pressure-driven flows, the superposition principle for linear equations is used to obtain the following nondimensional velocity profile (Dutta and Beskok, 2001a):

$$U(\eta) = -\frac{1}{2}\frac{dP^*}{d\xi}\left(1-\eta^2\right) + 1 - \psi^*(\eta), \tag{7.27}$$

where $\frac{dP^*}{d\xi}$ corresponds to the pressure gradient in the mixed electroosmotic/pressure-driven flow regime. Substituting the solution for ψ^* from equation (7.9), we obtain an analytical formula for the velocity distribution. In Figure 7.7 velocity profiles under various pressure gradients are shown. The case for $\frac{dP^*}{d\xi} = 0$ corresponds to a pure pluglike flow, and the cases $\frac{dP^*}{d\xi} < 0$ and $\frac{dP^*}{d\xi} > 0$ correspond to flow with favorable and adverse pressure gradients, respectively.

In order to obtain the mass flowrate, we integrate the velocity and the electroosmotic potential distribution across the channel (see equation (7.27)). This can be cumbersome in the η-coordinate system, where ψ^* is a function of both α and β, but in the χ-coordinate system, ψ^* is a function only of α. In (Dutta and Beskok, 2001a), the *electric double layer displacement thickness* defined δ^* was in analogy with the boundary layer displacement thickness in fluid mechanics in the following form:

$$\delta^* = \int_0^{\hat{\chi}} \psi^* d\chi, \tag{7.28}$$

where $\hat{\chi}$ is a large enough distance to include variations in ψ^* as observed from Figure 7.3. For example, $\hat{\chi} \approx 10$ is sufficient to accurately define δ^*. Typical values of δ^* as a function of α are presented in Table 7.2.

- *The physical meaning of δ^* is that it expresses the volumetric flowrate defect due to the velocity distribution within the EDL.*

Integration of the ψ^* term in equation (7.27) is performed using equation (7.28), where

$$\int_{-1}^{1} \psi^* d\eta = 2\int_0^1 \psi^* d\eta = \frac{2}{\sqrt{\alpha\beta}}\int_0^{\hat{\chi}} \psi^* d\chi = \frac{2\delta^*}{\sqrt{\alpha\beta}}.$$

The resulting *volumetric flowrate per channel width*, normalized by $hu_{\rm HS}$, becomes

$$\dot{Q} = -\frac{2}{3}\frac{dP^*}{d\xi} + 2\left(1 - \frac{\delta^*}{\sqrt{\alpha\beta}}\right). \quad (7.29)$$

Since most of the microfluidic experiments are performed by imposing a certain amount of pressure drop along the microchannel, we use equation (7.29) to correlate the volumetric flowrate with the imposed pressure drop. Also, for applications with specified volumetric flowrate, one can always obtain the resulting pressure variation along the channel using equation (7.29).

The shear stress on the wall for the mixed electroosmotic/pressure-driven flow region is found by differentiating (7.27) with respect to η, and utilizing equation (7.8), which results in

$$\tau_w^* = \sqrt{\frac{\beta}{\alpha}}\sqrt{2\cosh(\alpha) - 2\cosh(\alpha\psi_c)} - \frac{dP^*}{d\xi}. \quad (7.30)$$

This is an implicit *exact* relation under the assumptions of our analysis, which requires ψ_c^*. Assuming that $\psi_c^* = 0$ (valid for $\alpha \geq 1$ and $\lambda_D \ll h$), the following approximate relation can be found:

$$\tau_w^* = \sqrt{\frac{\beta}{\alpha}}\sqrt{2\cosh(\alpha) - 2} - \frac{dP^*}{d\xi}. \quad (7.31)$$

The first term on the right-hand-side is due to the variation of velocity within the EDL, while the second term is due to the parabolic velocity profile. The shear stress in the mixed electroosmotic/pressure-driven flow region is enhanced due to the presence of the EDL.

The aforementioned analytical results can be used to validate the numerical computations. In the following we demonstrate the mixed electroosmotic/pressure-driven flows in a channel that is made out of two different materials. The first material exhibits negligible electroosmotic effects, and it is used at the entry and exit portions of the channel. The second material, on the other hand, exhibits strong electroosmotic effects, and it is used in the middle section of the channel. This configuration is shown in Figure 7.6. It may be possible to fabricate such a microchannel by using different materials on various portions of the channel surface. In practice, it is also possible to obtain variations in the wall potential due to contamination in the capillary walls, variations in the wall coating, or gradients in the buffer pH, as discussed in Section 7.4.7. Therefore, the proposed configuration has some practical relevance, and it is a suitable testbed to study mixed electroosmotic/pressure-driven flows.

In the simulations, volumetric flowrate at the channel entrance is specified, and the corresponding velocity and pressure distributions in the rest of the channel are calculated (Dutta et al., 2002b). In order to eliminate the channel entrance effects, a parabolic velocity profile at the inlet with a maximum inlet velocity of $U_{\text{in}} = u_{\text{in}}/u_{\text{HS}}$ is specified. It is possible to generate the desired pressure gradients in the mixed electroosmotic/pressure-driven zone using specific values of U_{in}. The numerical simulations are performed for Re $= 0.005$, where Re is based on the average channel velocity and the channel half-height. In the results that follow, the streamwise electric field strength and the EDL properties are constant at $\alpha = 1$ (corresponding to $\zeta = 25.4$ mV) and $\beta = 10{,}000$. Therefore, the Debye length in the simulations is *one-hundredth* of the channel half-height. The entire flow domain, including the EDL, is resolved in the simulations.

Figure 7.7 presents the nondimensional pressure distribution along the channels for various values of U_{in}. This numerical modeling employs zero gauge pressure at the channel outlet. Therefore, all numerical results show zero gauge pressure at the exit. The entrance and exit portions of the channels are purely pressure-driven, and the electroosmotic forces are present only for $3.1 \le \xi \le 6.2$. The effective electric field is in the positive streamwise direction. Using equation (7.29) we estimate the theoretical value of U_{in}, which results in a desired pressure gradient in the mixed region. For example, the theoretical value of $U_{\text{in}} = 1.485$ for $\alpha = 1$ and $\beta = 10{,}000$ gives zero pressure gradient in the mixed electroosmotic/pressure-driven flow region.

The corresponding velocity profiles across the channel at $\xi = 4.5$ are presented in Figure 7.7 (right plot). A pluglike velocity profile is observed for $U_{\text{in}} = 1.485$, as predicted by the theory. Setting $U_{\text{in}} = 2.5$ corresponds to a favorable pressure gradient case, which is a combination of a pluglike flow with a parabolic profile in the bulk of the channel. The corresponding pressure variation shown on the left indicates significant pressure drop at the entrance and exit portions of the channel. However, in the mixed zone, the pressure drop is relatively low due to the electroosmotic pumping.

The adverse pressure gradient case of $U_{\text{in}} = 0.5$ is an *electrokinetically driven micropump*. For this case, the inlet and the exit pressures are the same, corresponding to a *laboratory-on-a-chip* device that is exposed to atmosp pressure at both ends. The entire flow is driven by the electrokinetic forces, which overcome the drag force within the entire channel system. The pressure drop at the inlet and exit portions of the channel ($\xi \le 3.1$ and $\xi \ge 6.2$) is due to the shear stress. A micropump must be able to raise the system pressure in order to drive the flow. The electroosmotic pump is doing precisely this. The net pressure gradient is positive within the pump, as shown in Figure 7.7. Here we note that:

- *In a purely electroosmotic system, pluglike velocity profiles with zero-pressure gradient will be obtained. In the case of* $U_{\text{in}} = 0.5$, *the adverse pressure gradient is present to overcome the pressure drop at the inlet and exit sections.*

Therefore, any mixed flow system should exhibit a behavior similar to the

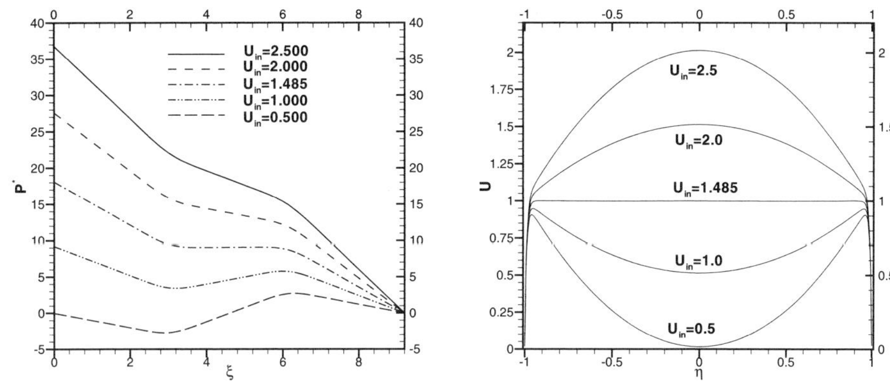

FIGURE 7.7. Normalized pressure and velocity distribution in a mixed electroosmotic/pressure-driven channel for various values of U_{in}. The case $U_{\text{in}} = 1.485$ corresponds to the pluglike flow. The electroosmotic forces are present only for $3.1 \leq \xi \leq 6.2$. Simulation results are for $\alpha = 1$, $\beta = 10{,}000$ and $\text{Re} = 0.005$.

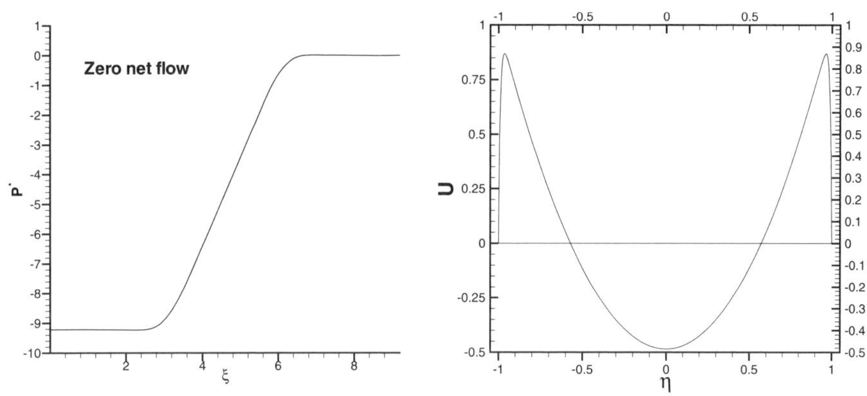

FIGURE 7.8. Pressure build-up along the microchannel for zero net flow (left), and corresponding velocity distribution across the channel (right).

simulation results presented in Figure 7.7. The velocity profile for this case indicates a combination of plug and adverse pressure gradient channel flow behavior, and the net volumetric flowrate is positive, as shown in Figure 7.7.

For a closed system it is possible to create large pressure gradients using electroosmotic forces. This can be used for actuation of micropistons or microbellows mechanisms. Such a configuration is simulated by closing the exit of the channel. Due to the presence of electroosmotic forces, the pressure rises linearly within the electroosmotic region, as shown in Figure 7.8.

This pressure rise is accompanied by the electroosmotic flow near channel walls and a reverse flow in the middle of the microchannel, as shown in Figure 7.8.

7.4.2 Time-Periodic and AC Flows

In this section we present an overview of the AC electroosmotic flows and time-periodic electroosmosis, where the flow is driven by an alternating electric field. The primary difference between these two flows is that in AC electroosmosis the electric field is nonuniform, and it creates a nonzero time-averaged flow (Morgan and Green, 2003), while the time-periodic electroosmosis utilizes a uniform electric field, and results in zero time-averaged flow (Dutta and Beskok, 2001b).

AC Electroosmosis

AC electroosmosis is an important electrokinetic effect that can be utilized for particle manipulation and separation, as well as for flow pumping, and mixing enhancement applications. An advantage of AC electroosmosis is observed in designing electroosmotic pumps. Although DC electroosmotic pumps utilize large electric fields (see Table 7.1), AC electroosmotic pumps utilize low voltages (< 10 V) due to the close proximity of the electrodes, and they can maintain flowrates on the order of 1 to 10 mm^3/s (see Chapter 8 in Morgan and Green, 2003). For particulate flows, AC electroosmosis is often accompanied by dielectrophoresis (see Section 7.6).

AC electroosmosis is observed when an AC electric field is applied on two electrodes that are placed next to each other, as shown in Figure 7.9. Electric field lines near the electrode surfaces display tangential components, especially in the vicinity of the two electrodes. Interactions of the tangential electric field with the induced charges on each electrode (schematically shown on the figure with + and − signs) result in AC electroosmotic force and fluid flow. Since the tangential electric field is larger near the two neighboring edges of the electrodes, the electroosmotic velocity is also large at these locations, and the velocity decays away from these neighboring edges. Here we must note that changing the electrode polarity in Figure 7.9 does not change the direction of the AC electroosmotic force. Therefore, AC electroosmosis results in unidirectional flow. In a series of papers, Green and coworkers presented experimental measurements, theoretical analysis, and numerical simulation of AC electroosmotic flows (Green et al., 2000; Gonzalez et al., 2000; Green et al., 2002). The experimental results have shown that AC electroosmosis is effective in the 10 Hz to 100 kHz frequency range, after which electrothermal/Joule heating effects (Section 7.4.6) become dominant. The electric field frequency greatly affects the electroosmotic velocity. For example, using a field of 1.25 V at electrolyte conductivity of 8.6 mS/m, velocity magnitudes of 75, 300, and 50 μm/s

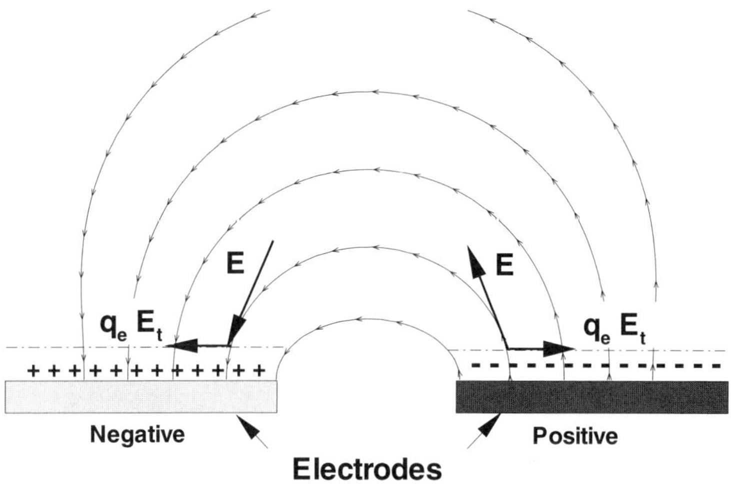

FIGURE 7.9. Schematic diagram for AC electroosmosis. The tangential electric field (E_t) near the neighboring electrodes interacts with induced charges (q_e) on the electrodes and creates electroosmotic force and velocity in the horizontal direction, shown by the $q_e E_t$ vector. Changing the electrode polarity does not alter the flow direction. The figure is adopted from (Morgan and Green, 2003).

were measured 5 μm away from the electrode edge at 100 Hz, 2 kHz, and 20 kHz, respectively (Green et al., 2000). Numerical modeling of these cases resulted in quantitatively consistent predictions only after the numerical results were corrected by a scaling factor of 0.24 (see Figure 8.9 in Morgan and Green, 2003). Green and coworkers also reported that for increased conductivity of the fluid, the AC electroosmosis can be overwhelmed by the increased dielectrophoretic effects.

Time-Periodic Electroosmosis

Unlike the steady electroosmotic flows, time-periodic or unsteady electroosmosis has been addressed in relatively few publications. For example, numerical results for impulsively started electroosmotic flow have been reported in (Dose and Güiochon, 1993). Analytical solutions of starting electroosmotic flows for a number of geometries, including the flow over a flat plate and two-dimensional microchannel and microtube flows, have been presented in (Soderman and Jonsson, 1996). Effects of a sinusoidally alternating electric field superimposed onto a steady electroosmotic flow have been reported in (Barragan and Bauza, 2000). Time-periodic electroosmotic flows can be combined with steady electroosmotic or pressure-driven flows to induce temporal and spatial flow modifications. This has the potential to produce continuous-flow chaotic mixers (see Chapter 9).

In the rest of this section, we study channel flows driven by time-periodic axial electric fields. Although we are essentially dealing with an AC electric field, we assume that the external electric field obeys the laws of electrostatics. Furthermore, the ion distribution in the EDL region is determined by the zeta potential and the ion density of the buffer solution, which are both constants. Hence, the electrokinetic potential ψ is not affected by the oscillatory external electric field. Neglecting the channel entry and exit effects, the flow is *fully developed*. Hence, we do not expect any streamwise velocity gradients. This enables us to neglect the inertial terms in the Navier–Stokes equations (7.12), resulting in the unsteady Stokes equation. In the absence of externally imposed pressure gradients, the flow is driven purely due to the electrokinetic effects. Following (Dutta and Beskok, 2001b), the momentum equation can be simplified as

$$\rho_f \frac{\partial u}{\partial t} = \mu \frac{\partial^2 u}{\partial y^2} + \rho_e E_x \sin(\Omega t), \tag{7.32}$$

where E_x is the magnitude, and Ω is the frequency of the unsteady external electric field $\mathbf{E}$. We note here that due to the straight channel geometry, the cross-stream velocity and electric field components are zero, and equation (7.32) is used to determine the streamwise velocity component due to the time-periodic electric field. The electric charge density ρ_e can be expressed as

$$\rho_e = -2n_o ez \sinh\left(\frac{ez\psi}{k_B T}\right) = -2n_o ez \sinh(\alpha\psi^*), \tag{7.33}$$

Assuming that the EDL thickness is much smaller than the channel-half gap, the electroosmotic potential variation can be written as a function of the distance from the wall ($\chi = y/\lambda_D$), using equation (7.10). Utilizing the Deby–Hückel parameter (ω), we can rewrite the electroosmotic body force on the fluid in the following form:

$$-2n_o ezE_x \sin \Omega t = \frac{\omega^2 \mu u_{HS}}{\alpha} \sin \Omega t, \tag{7.34}$$

where $u_{HS} = -\epsilon\zeta E_x/\mu$, is the Helmholtz–Smoluchowski velocity. Therefore, equation (7.32) can be written as

$$\rho_f \frac{\partial u}{\partial t} = \mu \frac{\partial^2 u}{\partial y^2} + \frac{\omega^2 \mu u_{HS}}{\alpha} \sinh(\alpha\psi^*) \sin(\Omega t). \tag{7.35}$$

We are interested in the solution of the above equation under no-slip and symmetry boundary conditions on the wall and the channel center, respectively. Nondimensionalization of equation (7.35) using characteristic time $(1/\Omega)$ and length (λ_D) results in the following equation:

$$\rho_f u_{HS} \Omega \frac{\partial U}{\partial \theta} = \mu u_{HS} \omega^2 \left[\frac{\partial^2 U}{\partial \chi^2} + \frac{\sin(\theta)}{\alpha} \sinh(\alpha\psi^*)\right], \tag{7.36}$$

where $\theta = \Omega t$ is the nondimensional time, $\chi = y/\lambda_D$ is the nondimensional distance, and $U = u/u_{HS}$ is the nondimensional velocity. Rearranging the dimensional components in equation (7.36) results in the following equation:

$$\frac{\partial U}{\partial \theta} = \frac{1}{\kappa^2}\left[\frac{\partial^2 U}{\partial \chi^2} + \frac{\sin(\theta)}{\alpha}\sinh(\alpha\psi^*)\right], \tag{7.37}$$

where $\kappa = \sqrt{\Omega\lambda_D^2/\nu}$ is a normalized parameter that is a function of the Debye length (λ_D), the kinematic viscosity, and the electric field excitation frequency.

- *The parameter κ can be interpreted as the ratio of the Debye length λ_D to a diffusion length scale (l_D), based on the kinematic viscosity and the excitation frequency.*

This diffusion length scale can be estimated from the unsteady Stokes equations using dimensional analysis as $l_D \approx \sqrt{\nu/\Omega}$. Hence, the diffusion length scale is related to the Debye length and κ in the following form:

$$l_D \approx \lambda_D/\kappa.$$

Time-periodic solution of equation (7.37) is obtained using the separation of variables technique (Dutta and Beskok, 2001b). Here we summarize the basic results. Time-periodic velocity distributions across the channel for three different κ values are shown in Figure 7.10 at time $\theta = \pi/2$. These results are obtained for a channel half-height $h = 100\lambda_D$. Since the electroosmotic forces are confined within the EDL, the dynamics of fluid motion in the bulk of the channel are mostly determined by an unsteady diffusion process. Therefore, the magnitude of the channel half-height (h) relative to the diffusion length scale (l_D) plays an important role in determining the dynamics of the bulk flow region. For the $\kappa = 0.001$ case, the diffusion length scale ($l_D = 1000\lambda_D$) is an order of magnitude larger than the channel half-height ($100\lambda_D$). This results in a quasi-steady velocity distribution that resembles the "plug velocity" obtained in steady electroosmotic flows, as shown in Figure 7.10. On the other hand, for $\kappa = 0.01$ the channel half-height and the diffusion length scales are of the same order of magnitude, and the bulk flow velocity deviates from the plug profile. For $\kappa = 0.1$ the velocity becomes practically zero in the channel center at any time, as shown in Figure 7.11.

For the $\kappa = 0.01$ case, there is *vorticity* in the *entire* bulk flow region (see Figure 7.10). From the $\kappa = 0.01$ and $\kappa = 0.1$ cases, we see that the time-periodic electroosmotic flows are *rotational* when the diffusion length scale is comparable to or less than the channel half height ($l_D \leq h$). The vorticity is put into the problem on the walls, and its magnitude alternates in time due to the time fluctuations of the external electric field. Unlike the steady electroosmotic flows, vorticity diffuses deeper into the channel, while its

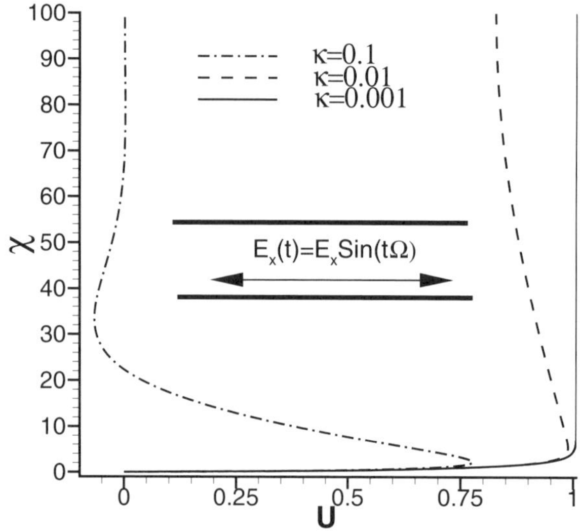

FIGURE 7.10. The velocity distribution of time-periodic electroosmotic flow for various values of κ at time $\theta = \pi/2$. Here, the ionic energy parameter α equals 5.

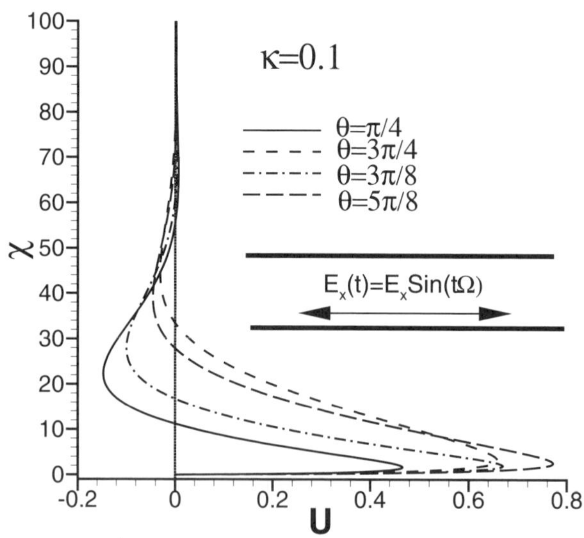

FIGURE 7.11. The velocity distribution for time-periodic electroosmotic flows at different times. Here, $\kappa = 0.1$ and the ionic energy parameter α equals 5.0.

value on the wall is cyclically changing. The high-frequency excitation case ($\kappa = 0.1$) presented in Figure 7.11 shows exponentially damped vorticity waves traveling into the bulk flow domain, penetrating the channels as much as 80 Debye lengths. A discussion on vorticity creation mechanisms for unsteady electroosmotic flows is given in (Santiago, 2001).

The velocity profiles presented in Figure 7.11 resemble the classical solution of a flat plate oscillating in a semi-infinite flow domain, also known as Stokes's second problem. In Figure 7.12, time-periodic electroosmotic velocity profiles are compared with the solution of Stokes's second problem. Unlike the time-periodic electroosmotic flow, the fluid is driven here by an oscillating plate with velocity $u_w = u_{\rm HS} \sin(t\Omega)$, where Ω is the frequency of the plate oscillations and $u_{\rm HS}$ is the amplitude of the plate velocity. The governing equation for Stokes's second problem becomes

$$\rho_f \frac{\partial u}{\partial t} = \mu \frac{\partial^2 u}{\partial y^2}, \tag{7.38}$$

and the system is subjected to the following boundary conditions:

$$u(y = 0, t) = u_{HS} \sin(t\Omega),$$
$$u(y \to \infty, t) = 0.$$

These equations yield the following similarity solution (Panton, 1984):

$$U(\chi, \theta) = \exp\left[-\frac{\kappa\chi}{\sqrt{2}}\right] \sin\left[\theta - \frac{\kappa\chi}{\sqrt{2}}\right], \tag{7.39}$$

where U is the velocity normalized with $u_{\rm HS}$, similar to our solution of time-periodic electroosmotic flow. The solution for both equations is practically the same for $\chi \geq \delta_{99}$ (See Figure 7.12). Therefore, the bulk flow dynamics are adequately described by the Stokes solution for a flat plate oscillating with frequency Ω and amplitude $u_{\rm HS}$. However, the velocity distribution within the effective electric double layer ($\chi \leq \delta_{99}$) differs from the Stokes solution significantly, since the velocity on the wall needs to obey the no-slip condition at all times. A zoomed view of the velocity distribution of Figure 7.12 is presented in Figure 7.13.

We note that the Stokes solution in the bulk flow and the analytical solution outside the effective EDL thickness match without any phase lag. This enables us to conclude that the Stokes solution with a prescribed wall velocity of $u_{\rm HS} \sin(t\Omega)$ can be used to describe the bulk flow region for sufficiently large κ values, showing that

- *the instantaneous Helmholtz–Smoluckowski velocity is the appropriate wall slip condition for time-periodic electroosmotic flows.*

We must note that this claim is also valid for low-frequency (small κ) flows, as can be deduced from Figure 7.10 by extending the velocity profiles from the bulk flow region on to the wall.

In summary:

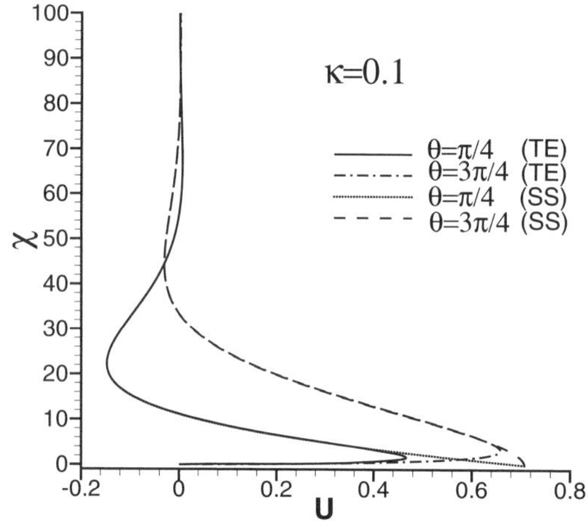

FIGURE 7.12. Comparisons between the time-periodic electroosmotic velocity (TE) and Stokes's second problem (SS) for $\kappa = 0.1$ at various instances in time. The results are obtained for ionic energy parameter $\alpha = 5.0$ and times $\pi/4$ and $3\pi/4$.

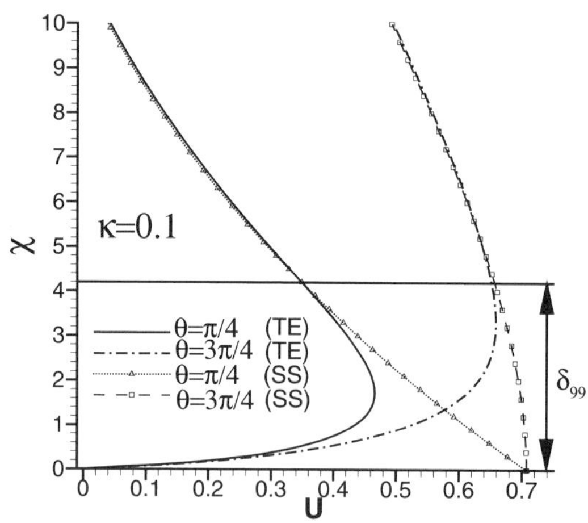

FIGURE 7.13. Near-wall velocity distribution for time-periodic electroosmotic flow and Stokes's second problem for $\kappa = 0.1$ at various times. See the legend of Figure 7.12 for details.

- The flow dynamics are determined by a normalized parameter κ, which can be interpreted as the ratio of the EDL thickness λ_{D} to a characteristic diffusion length scale l_{D}. Based on the values of κ and the half channel height, various flow conditions, ranging from the oscillatory plug flow to flows resembling the flat plate oscillating in a semi-infinite domain with velocity $u_{\mathrm{HS}} \sin(t\Omega)$ (Stokes's second problem) are observed.

- For large values of κ there are similarities between Stokes's second solution and the analytical solution in the bulk flow region. This leads to identification of the instantaneous Helmholtz–Smoluchowski velocity as the appropriate electroosmotic slip condition for time-periodic flows.

7.4.3 EDL/Bulk Flow Interface Velocity Matching Condition

In this section, we present the velocity matching condition between the EDL and the bulk flow regions. This is important in assessing the interaction of high-vorticity fluid in the EDL with the vorticity of the bulk flow region. It seems that the often-used Helmholtz–Smoluchowski velocity (7.24) as the "matching condition" at one Debye length (λ_{D}) away from the wall is inadequate for the following two reasons:

- First, such a matching condition should be implemented at the *effective EDL thickness* ($\delta_{99}\lambda_{\mathrm{D}}$), which is considerably larger than the Debye length.

- The second limitation arises due to the variation of the bulk velocity across the EDL.

If we examine the velocity distribution at the edge of the EDL in Figures 7.13 and 7.14, it is clear that the matching velocity changes with the velocity gradient of the bulk flow region. Hence, the appropriate velocity matching condition (u_{match}) at the edge of the EDL ($y = \delta_{99}\lambda_{\mathrm{D}}$) should become

$$u_{\mathrm{match}} = \lambda\delta_{99}\frac{\partial u}{\partial y}|_w + u_{\mathrm{HS}}, \tag{7.40}$$

where $\frac{\partial u}{\partial y}|_w$ corresponds to the *bulk flow gradient obtained on the wall.* The appropriate matching distance is taken to be the effective EDL thickness ($\delta_{99}\lambda_{\mathrm{D}}$). Here, the first term in (7.40) corresponds to a *Taylor series expansion* of bulk flow velocity at the edge of the EDL from the wall. Equation (7.40) is analogous to slip velocity in rarefied gas flows given in Section 2.3. It is noteworthy to mention that for finite Debye layers with large bulk flow gradients, the velocity matching using equation (7.40) will give *considerable deviations* from the Helmholtz–Smoluchowski prediction.

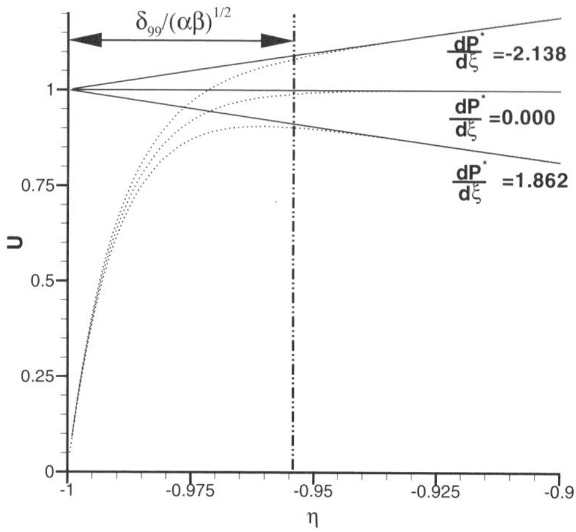

FIGURE 7.14. A magnified view of the velocity distribution in a mixed electroosmotic/pressure-driven flow near a wall for $\alpha = 1$, $\beta = 10{,}000$. Extrapolation of the velocity using a parabolic velocity profile with constant slip value U_{HS} on the wall are shown by the solid lines. The analytical solution is shown by the dashed lines.

7.4.4 Slip Condition

The electroosmotic forces are concentrated within the EDL, which has an effective thickness on the order of 1 nm to 100 nm. On the other hand, the microchannels utilized for many laboratory-on-a-chip applications have a typical height of 100 μm to 1 μm This two to five orders of magnitude difference in the EDL and the channel length scales is a great challenge in numerical simulation of electroosmotically driven microflows. Therefore, it is desired to develop a *unified slip condition*, which incorporates the EDL effects by specifying an appropriate velocity slip condition on the wall. Examining Figures 7.13 and 7.14, and equation (7.40), it is seen that the bulk velocity field extended onto the wall has a constant slip value equivalent to u_{HS}. Hence,

- *the appropriate slip condition at the wall is the Helmholtz–Smoluchowski velocity u_{HS}, even for finite EDL thickness conditions.*

For a general numerical algorithm, implementation of slip velocity u_{HS} at the walls *overpredicts* the volumetric flowrate, since the velocity distribution within the EDL is neglected. This flowrate error can be corrected by subtracting $2\delta^*/\sqrt{\alpha\beta}$ (in nondimensional form) using the EDL displacement thickness δ^* given in Table 7.2. For engineering applications with $\alpha = 1$ and $\beta = 10{,}000$, corresponding to a 0.1 mM buffer solution in a 6 μm glass channel, the error in the conservation of mass equation due to this slip condition is about 4.5%.

With regard to the errors in the momentum equation, neglecting shear stresses due to the velocity distribution within the EDL, given by equation (7.31), will be in gross error. In the next section we present a method that can be used to predict the extra drag force due to the EDL, in the postprocessing stage of computations.

7.4.5 A Model for Wall Drag Force

For steady incompressible flows, conservation of linear momentum in integral form is given by

$$\int_{\Gamma} \mathbf{v}\rho\mathbf{v} \cdot \mathbf{n}\, d\Gamma = \mathbf{F}_{\text{sur}} + \mathbf{F}_{\text{body}}, \tag{7.41}$$

where $\mathbf{F}_{\text{sur}}$ and $\mathbf{F}_{\text{body}}$ are the net surface and body forces, respectively. The control surface is shown by Γ, and the outward unit normal is denoted by $\mathbf{n}$. Depending on the device geometry and the control volume, one must properly apply equation (7.41), which also includes the electrokinetic body force terms. In this section, we propose a method to calculate the additional drag force due to the electrokinetic effects at the *postprocessing stage*. Hence, we first solve the flow system using the Helmholtz–Smoluchowski slip velocity u_{HS}, and then calculate the total drag force by superposition of the calculated hydrodynamic drag with the cumulative effects of the electroosmotic *body forces* concentrated on *domain boundaries.*

The additional drag force acting on a control volume due to the electrokinetic effects can be expressed as

$$\mathbf{F}_{\text{EK}} = \int_{\text{CV}} \rho_e \mathbf{E}\, d\Omega.$$

Substituting ρ_e from the Poisson–Boltzmann equation and $\mathbf{E} = -\nabla\phi$, we obtain

$$\mathbf{F}_{\text{EK}} = \int_{\text{CV}} \epsilon\nabla^2\psi \left[\frac{\partial\phi}{\partial n}\mathbf{e}_n + \frac{\partial\phi}{\partial l}\mathbf{e}_l + \frac{\partial\phi}{\partial s}\mathbf{e}_s\right] d\Omega,$$

where n, l, and s are the normal, streamwise and spanwise coordinates, respectively, and $d\Omega = dn\, ds\, dl$. This volume integral is complicated to evaluate in general. However, some simplifications can be made when $\lambda_D/h \ll 1$. Also, for a general complex geometry, we further assume that the radius of curvature R is much larger than the Debye length λ_{D}. The latter condition is required to exclude application of the forthcoming procedure in the vicinity of sharp corners. Based on these assumptions, $\nabla^2\psi$ can be approximated to be $\frac{d^2\psi}{dn^2}$. Also, $\frac{\partial\phi}{\partial n} \approx 0$ across the entire EDL, which is approximately valid due to the small EDL thickness and the no-penetration boundary condition of the externally applied electric field on the surfaces. This enables us to separate the volume integral into the following two components:

$$\mathbf{F}_{\mathrm{EK}} = \epsilon \left[\int_0^{2h} \frac{d^2\psi}{dn^2} dn \right] \int_0^W \int_0^L \left[\frac{\partial \phi}{\partial l} \mathbf{e}_l + \frac{\partial \phi}{\partial s} \mathbf{e}_s \right] dl\, ds,$$

where L and W are the streamwise and spanwise lengths of the domain, respectively. Also, for a general geometry we assumed the separation distance between the two surfaces to be $2h$. The second integral in the above equation can be obtained in the postprocessing stage from the solution of the electrostatic problem. Numerical solution for the first integral requires resolution of the EDL region, which requires *enhanced near-wall resolution* and results in the *numerical stiffness*. However, this integral can be evaluated analytically in the following form:

$$\int_0^{2h} \frac{d^2\psi}{dn^2} dn = \int_0^{2h} d\frac{d\psi}{dn} = \frac{\zeta}{h} \left[\frac{d\psi^*}{d\eta} \right]_{-1}^{1} = 2\frac{\zeta}{h} \sqrt{\frac{\beta}{\alpha}} \sqrt{2\cosh(\alpha) - 2}.$$

Hence, the contribution of $\mathbf{F}_{\mathrm{EK}}$ can be evaluated as

$$2\epsilon \frac{\zeta}{h} \sqrt{\frac{\beta}{\alpha}} \sqrt{2\cosh(\alpha) - 2} \int_0^W \int_0^L \left[\frac{\partial \phi}{\partial l} \mathbf{e}_l + \frac{\partial \phi}{\partial s} \mathbf{e}_s \right] dl\, ds.$$

Therefore, the additional drag force due to the electrokinetic effects can be calculated in the *postprocessing stage*, under the approximation of decoupling between the directions of the electroosmotic and external electric fields. This approach is valid for $\lambda_{\mathrm{D}}/h \ll 1$ and $\lambda_{\mathrm{D}}/R \ll 1$. The primary advantage is that we do not need to resolve the flow and the corresponding electroosmotic body forces in the EDL region. Hence, there is no need to solve for the Poisson–Boltzmann equation, and the numerical stiffness in the momentum equation is reduced.

7.4.6 Joule Heating

Large electric fields utilized in electrokinetic flows often result in Joule heating, due to the electrical current and the resistivity of the electrolyte. Using Ohm's law, Joule heating can be characterized as a volumetric heat source

$$\dot{q} = \frac{I^2}{\sigma},$$

where σ is the electric conductivity, and I is the electric current density. In absence of fluid flow, $I = \sigma\|\mathbf{E}\|$, and hence $\dot{q} = \sigma\|\mathbf{E^2}\|$. However, in the presence of fluid flow, current density should also include the charge convection effects (Tang et al., 2004a). Therefore, for electroosmotic flows,

$$I = \rho_e u_{\mathrm{EO}} + \sigma\|\mathbf{E}\|,$$

where ρ_e is the charge density, and u_{EO} is the axial electroosmotic flow velocity. We must note that the $(\rho_e u_{\mathrm{EO}})$ term can be substantial for thick/overlapping EDL situations. However, for thin EDL cases, electric current density variations are confined to a very thin region, since the net charge density (ρ_e) vanishes outside the EDL region. Therefore, we can assume $\dot{q} \approx \sigma \|\mathbf{E^2}\|$ in the entire channel domain. Analysis of Joule heating in electroosmotic flows and dielectrophoresis can be found in (Tang et al., 2004a; Tang et al., 2004b; Sinton and Li, 2003), and (Morgan et al., 1999).

Joule heating has adverse effects on microfluidic device performance. For example, local temperature increases result in local reductions in the absolute viscosity of the fluid. If we neglect temperature-dependent variations of any other properties, then the local viscosity reductions lead to increased electroosmotic slip velocity. For constant volumetric flowrate such local variations are compensated by onset of local pressure-driven flow. Therefore, under substantial Joule heating, it will not be possible to maintain pluglike velocity distribution in electroosmotic flows. This adversely affects the species transport, and it results in enhanced dispersion (see Section 7.5.3). However, we must note that miniaturization of device components reduces Joule heating and its adverse effects for the following reasons:

- First, large electric field gradients can be achieved with relatively smaller potential differences between the electrodes; hence $\|\mathbf{E}\|$ is reduced.

- Second, reduction in volume reduces the total heat generation.

- Third, increased surface area to volume ratio enhances heat loss to the environment through heat conduction.

7.4.7 Applications

In this section, we present microfluidic applications of electroosmotic flows. We first present suppression of electroosmosis, which becomes important in certain applications. Then we present mixing enhancement with electroosmotic flows, followed by electroosmotic flow control examples.

Suppression of Electroosmosis with Zeta Potential Modifications

Although electroosmosis is an attractive technique for microfluidic pumping, it may need to be suppressed or modified for certain applications, such as the capillary isoelectric focusing (IEF) and on-chip IEF, by altering the zeta potential. This can be obtained by various techniques, including polymer coatings and embedded surface electrodes. There are two basic polymer coating techniques: static and dynamic coating. The static coating is based on covalent bonding between the coating material and capillary surface, but dynamic coating relies on ionic interactions (Horvath and Dolnik,

2001; Righetti et al., 2001). Dynamic coating is an active field of research in chemistry. (Liu et al., 2000) utilized different polymer bilayers, such as a cationic layer of polybrene and, an anionic layer of dextran sulphate, to change the direction and magnitude of the flow. Both of these coatings are shown to reproduce electroosmotic flow for a wide range of pH. In a hydrodynamically driven capillary zone electrophoresis, electroosmotic flow needs to be suppressed. Kaniansky and coworkers worked on eight different electroosmotic flow suppressors and have shown influences of these on electrophoretic separation efficiencies at different pH values (Kaniansky et al., 1997). Ramsey and coworkers have developed a microfluidic device to detect *Escherichia coli* (*E. coli*) using pure electrophoretic transport. They were able to suppress electroosmosis, using poldimethylacrylamide (McClain et al., 2001). (Barker et al., 2000) have used polyelectrolyte multilayers to alter the electroosmotic flow direction in polystyrene and acrylic microfluidic devices. They were able to achieve complex flow patterning, and flow in the opposite directions of the same channel.

Embedded surface electrodes can also be used to locally alter the zeta potential. Schasfoort et al. have built microchannels using conducting material, and have covered these with a thin layer of insulator. They have shown that the electroosmotic mobility can be altered by applying electrostatic potential on the walls (Schasfoort et al., 2001). A similar approach for direct zeta potential control has also been implemented in (Buch et al., 2001).

Active Mixing Using Electroosmotic Flows

Mixing in microfluidic systems is difficult due to the minute inertial effects ($\mathrm{Re} \ll 1$) and small molecular diffusion coefficients ($\mathrm{Sc} \gg 1$). Convective/diffusive mixing in microscales requires large length and/or time scales; see Chapter 9. Given these limitations, electroosmotically induced mixing has attracted the attention of several research groups. Jacobson et al. have developed parallel and serial mixing mechanisms using microcapillary networks (Jacobson et al., 1999).

In a series of experiments, Santiago and coworkers have observed an electrokinetic instability for flows that are practically in the Stokes flow regime (Oddy et al., 2001; Chen and Santiago, 2002; Chen et al., 2003). Their extensive studies have shown that the electrokinetic instability is due to the conductivity gradients in the fluid, imposed either by the concentration or temperature gradients. In Figure 7.15, we present the time evolution of the electrokinetic flow instability in a microchannel of lenth, 40 mm, width, 1 mm and depth, 100 μm. The image area shown in the figure is 1 mm in the vertical direction and 3.6 mm in the streamwise direction. The channel is filled with two streams of 10 mM HEPES-buffered aqueous solution. The top stream (shown by gray) also included potassium chloride, which increased its conductivity to 50 μS/cm, while the bottom stream (shown by black) had conductivity of 5 μS/cm. This created a conductivity gradient in the spanwise direction. It has been found that the electrokinetic instability initiates after applying a *critical* electric field in the streamwise direction (perpendicular to the conductivity gradients). Figure 7.15 (a) shows time evolution of the electrokinetic instability for a streamwise (horizontal) electric field of 50 V/mm. These experimental results show rapid growth of small-amplitude waves, resulting in fast stirring of the initially distinct buffer streams. Reproduction of dynamics from simplified two-dimensional nonlinear numerical computations are also shown in Figure 7.15. The numerical model reproduces the main features of the instability observed in experiments, including the wave number and time scales. However, the two-dimensional approximation shows this rapid mixing at a lower field value of 17.5 V/mm. Details of this electrokinetic instability physics are given in (Lin et al., 2004). Considering the extensive use of electrokinetically driven flows in recent microflow applications, electrokinetic instability can be utilized to design efficient micromixers. For example, the work described in (Oddy et al., 2001), has demonstrated that time-periodic electroosmotic flows obtained by oscillatory electric fields can potentially result in fast mixing.

The possibility of local and temporal zeta potential variations has opened a new direction for extensive flow control and mixing enhancement applications. Erickson and Li have simulated microfluidic mixing induced by electroosmotic flow with local zeta potential variations, resulting in enhanced mixing efficiency (Erickson and Li, 2002). Qian and Bau (2002) have developed a theoretical model that induces chaotic mixing by electroosmotic stirring. For details on chaotic advection and mixing, see Chapter 9, where we also present some details about the chaotic electroosmotic stirrer (Qian and Bau, 2002).

Electroosmotic Flow Control

Electroosmotic forces can be selectively applied for flow control in complex microgeometries either by utilization of local electric fields or by modification of surface zeta potential (ζ). In this section, we will primarily study flow control in flow junctions using multiple electric fields by keeping the zeta potential unaltered.

The simulations are performed for a dielectric material of $\zeta = -25.4$ mV and half-channel-height of $h = 3\,\mu$m, corresponding to $\lambda_D/h = 0.01$. The magnitude of the externally imposed electric field $\|\mathbf{E}_o\|$ corresponds to 950 V/cm, resulting in a Helmholtz–Smoluchowski velocity of $u_{HS} = 1.6$ mm/s. We assumed that the buffer solution is water and the ion concentration density is $n_o = 0.1$ mM. The Reynolds number based on the average channel velocity and half-channel height is $Re = 0.005$. These simulation parameters are selected according to the data given in (Hunter, 1981). Here we must note that for $Re = 0.005$, we practically have Stokes flow with

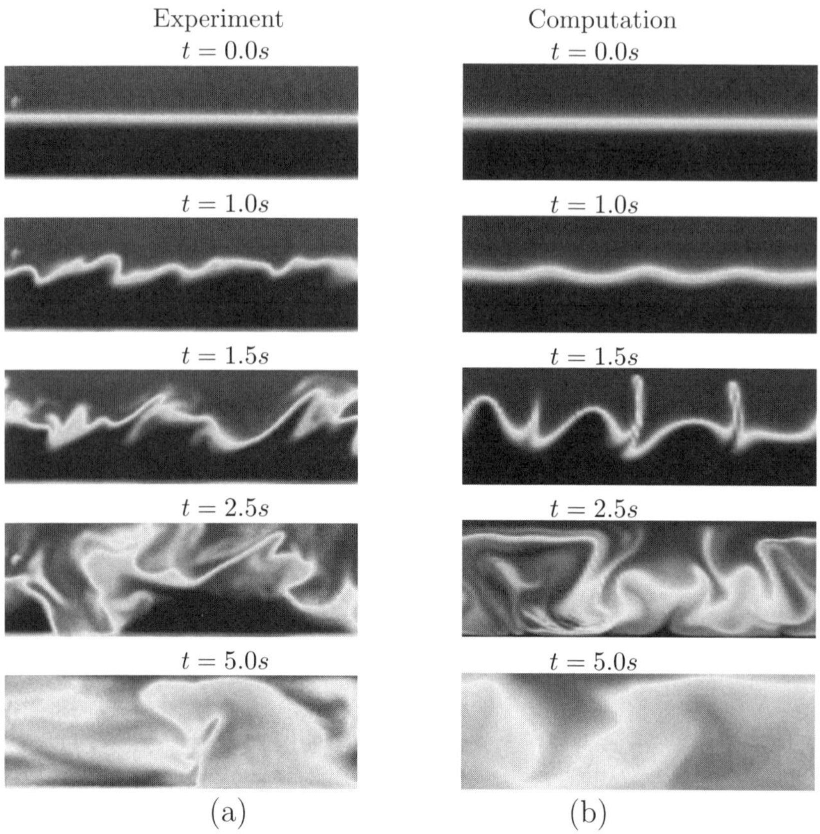

FIGURE 7.15. Time evolution of electrokinetic flow instability. (a) Experimental results show instability mixing of HEPES-buffered 50 μS/cm (gray) and 5μS/cm (black) conductivity streams under a horizontally applied static electric field of 50 V/mm. (b) Computational results reproduce the main features of the instability observed in the experiments. (Courtesy of H. Lin, M.H. Oddy, and J.G. Santiago.)

electroosmotic body forces, where the inertial forces are negligible and the velocity field is insensitive to the Reynolds number. Therefore, parametric studies as a function of the Reynolds number are not necessary in the Stokes flow limit.

Cross-Flow Junctions: The cross-flow junction of two microchannels has many important applications in electrophoretic separation (Polson and Hayes, 2000; Culbertson et al., 2000), serial and parallel mixing (Jacobson et al., 1999), and species-transport control (Cummings et al., 1999). In this section, we apply two different electric fields in a cross-flow junction to demonstrate the flow behavior as a function of the applied electric field strength. The externally imposed electric potential is determined by equation (7.21) subject to the specified electric potential at the entrance/exit of the branches, and zero flux conditions on the walls. The electric field is constant and one-dimensional at the entry and exit regions of the channel. However, near the cross-junction the electric field is two-dimensional. In Figure 7.16 we show the electric field lines obtained in the cross-junction under various horizontal and vertical electric field strengths. The figure shows equipotential contours with dashed lines and the electric field lines with solid lines. The electric field lines are orthogonal to the equipotential lines. For the equal horizontal and vertical electric field strength case, shown in the top figure, the electric field lines are symmetric with respect to the F-G plane. The electric field lines and equipotential contours for the case of horizontal electric field being twice the vertical electric field ($E_{\text{hor}} = 2E_{\text{ver}}$) is shown in the bottom figure. Under bias electric field, the electric potential distribution is no longer symmetric. Interaction of two electric fields in the cross-junction creates dips in the electric field lines near points F and G. The curvature in the electric field lines near these corners are physical, and similar qualitative trends exist in numerical results obtained for a T-junction under bias electric field (Mitchell et al., 2000), and cross-flow junctions (Patankar and Hu, 1998).

In Figure 7.17 (top) we show the velocity vector field and the streamlines obtained for equal horizontal and vertical electric fields. The inlet channels A and B are subjected to equal flow rates. Uniform, i.e., pluglike, velocity profiles are observed in all branches, with localized two-dimensional flow in the cross-junction. The velocity distribution within the EDL is also visible in the figure. The streamlines clearly show that the flow entering from inlet A turns toward outlet channel D, and the flow entering from B leaves from outlet C. Comparison of the streamlines with electric field lines given in Figure 7.16 (top) shows similarities between the velocity and the electric fields in the bulk flow region. Since the flow is driven by the electroosmotic forces, the pressure is uniform in the entire flow system. These results agree qualitatively with the experimental μ–PIV results (Cummings et al., 1999) for equal horizontal and vertical electric field strengths, presented in Figure 7.18.

In Figure 7.17 (bottom) we show the velocity vectors and the streamlines obtained using a bias electric field with horizontal field being twice the vertical field ($E_{\text{hor}} = 2E_{\text{ver}}$). Inlet conditions to sections A and B satisfy pure electroosmotic flow in each branch. Comparison between the streamlines and the electric field lines shown in Figure 7.16 (bottom) reveals similarities between the velocity and the electric fields in the bulk flow region. In Figure 7.17 (bottom) the velocity vectors at exit branch C show uniform plug profile of magnitude $2u_{\text{HS}}$, similar to that of inlet branch A, while the exit branch D reaches a uniform plug profile of magnitude u_{HS}, similar to that of inlet branch B. Under the bias electric field, the cross-flow junction creates unique opportunities for flow control. If we examine the stream-

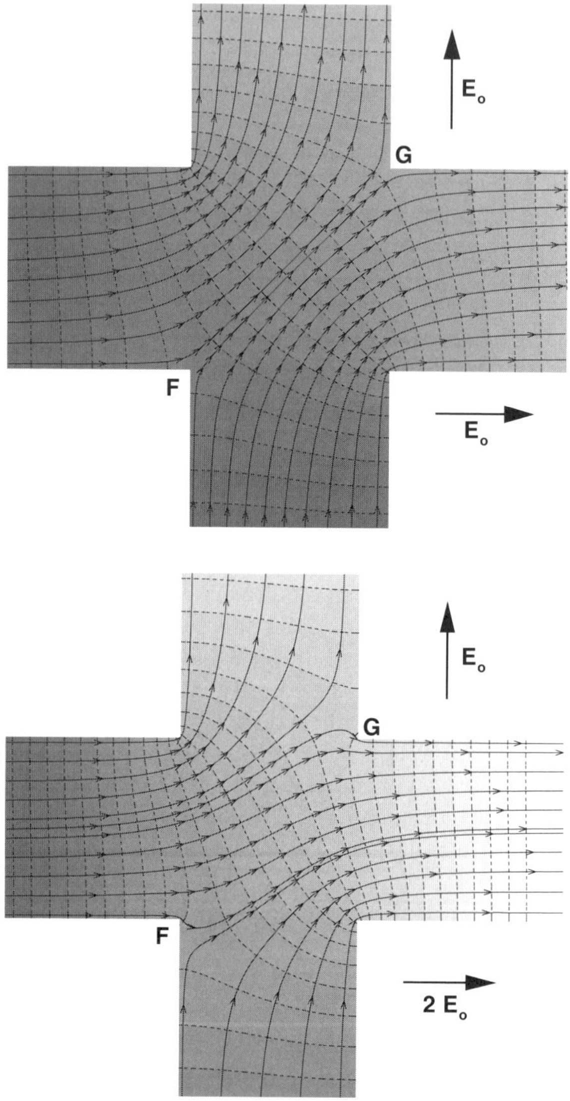

FIGURE 7.16. Electric field lines (solid) and equipotential contours (dashed) in a cross-junction geometry, under various electric fields. Top: $E_{\text{hor}} = E_{\text{ver}} = E_{\text{o}}$; Bottom: $E_{\text{hor}} = 2E_{\text{ver}} = 2E_{\text{o}}$.

lines in the figure, it is clear that 50% of the fluid leaving from channel C is coming from inlet channel A. This is required by the bias electric field strength and conservation of mass in the microfluidic system. Using this, we conclude that it is possible to control the amount of fluid in exit channel C that is coming from inlets A and B by controlling the ratio of the

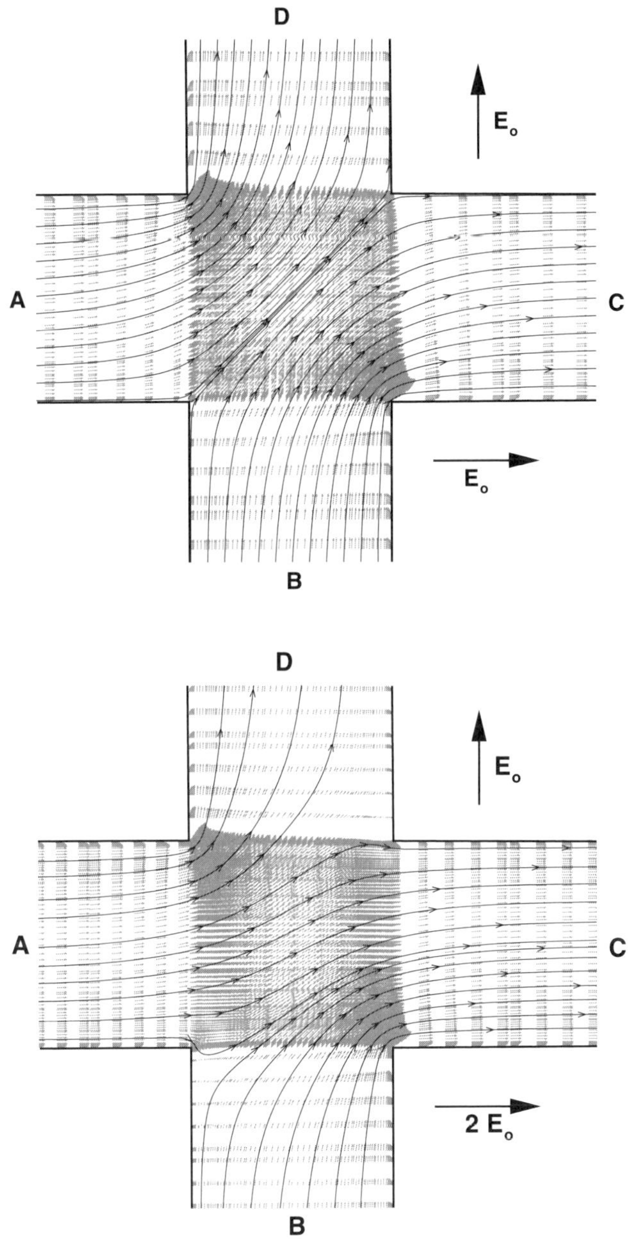

FIGURE 7.17. The streamlines and velocity vectors for pure electroosmotic flow in a cross-flow junction (only 25% of the vectors are shown for clarity of the figure). Top: $E_{\text{hor}} = E_{\text{ver}} = E_{\text{o}}$; Bottom: $E_{\text{hor}} = 2E_{\text{ver}} = 2E_{\text{o}}$.

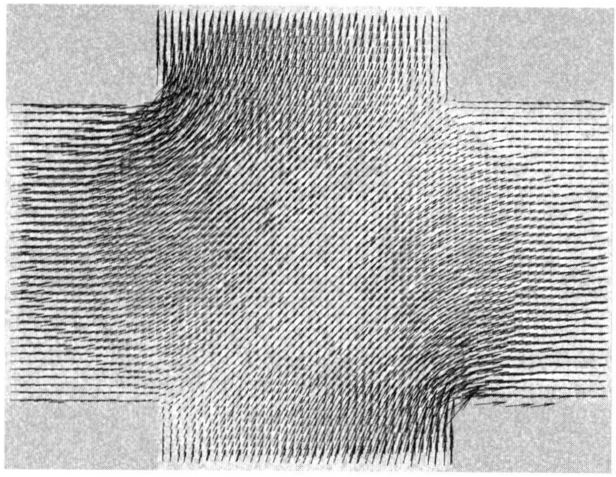

FIGURE 7.18. The μ–PIV velocity measurements of electroosmotic flow in a cross-flow junction, where $E_{\text{hor}} = E_{\text{ver}}$. (Courtesy of E. Cummings.)

horizontal to vertical electric fields ($E_{\text{hor}}/E_{\text{ver}}$). The ratio of the flowrates from inlets A and B at the exit channel C can be written in the following form:

$$\frac{\dot{Q}_{\text{AC}}}{\dot{Q}_{\text{BC}}} = \frac{E_{\text{hor}}}{E_{\text{ver}}} - 1, \tag{7.42}$$

where $\dot{Q}$ shows the flowrate, and the subscripts AC and BC show the contributions of flow from inlets A and B to the total flowrate in channel C, respectively. The above formula is subject to the restriction

$$\dot{Q}_{\text{A}} = \dot{Q}_{\text{C}} = \dot{Q}_{\text{AC}} + \dot{Q}_{\text{BC}} \tag{7.43}$$

due to the mass conservation requirements.

In summary:

- *Locally imposed electroosmotic forces* in the Stokes flow regime enable *linear flowrate control* in the branches of a microchannel network system. This linear response can be utilized in the design of various electroosmotically actuated micropump/valve systems and flow switches. For example, the cross-flow junction geometry presented above can be used for dispensing a *precise amount of fluid* from one channel to another in the *absence* of valves or pumps with *moving components.*

Array of Circular and Square Posts: Measurements of electroosmotic flow in arrays of circular and square posts have been obtained by Cummings in the absence of externally imposed pressure gradients, by maintaining zero elevation difference between the upstream and downstream reservoirs (Cummings, 2001). The microfluidic system consisted of uniformly distributed

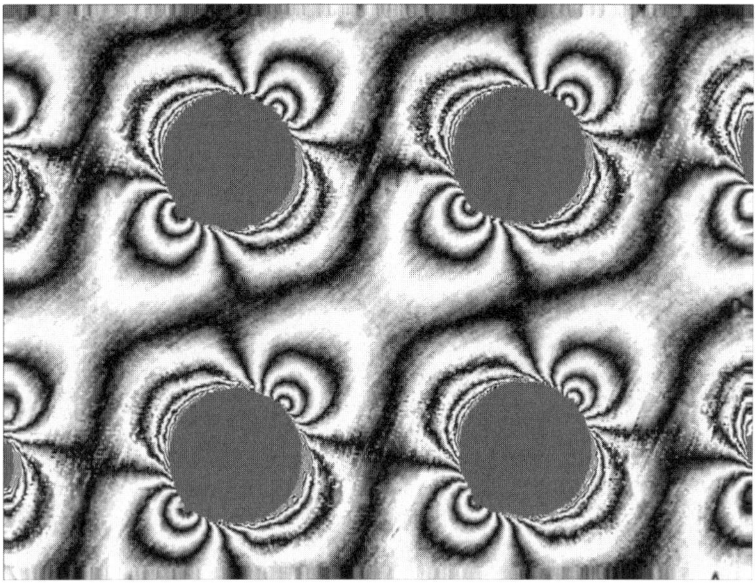

FIGURE 7.19. Electrokinetic speed contours in an array of circular posts at 45° with respect to the applied electric field of 2 V/mm. (Courtesy of E. Cummings.)

post arrays that are isotropically etched in glass with a thermally bounded glass cover slip. The circular posts have diameter 93 μm and the square post dimensions are 104 μm, with center-to-center separation of 200 μm. Phosphate-buffered saline solution of 1 mM, resulting in buffer pH = 7.7 is used. An external electric field was applied in various angles to the post arrays, and the electric field value was kept low to avoid particle dielectrophoresis. MicroPIV measurements of the velocity field were performed, and the results were presented in the form of a simulated interferogram (Cummings, 2001).

Figure 7.19 shows an electrokinetic speed field in an array of circular posts at a 45° angle with respect to the applied electric field of 2 V/mm. The flow is from lower left toward upper right. Lines of constant gray scale are contours of constant speed. The magnitude of speed at any point can be estimated by counting and interpolating the fringes starting at a stagnation point. The interferogram fringe spacing in the figure corresponds to 24.5 μm/s, and the stagnation points on each post is at 45° and 225° from the horizontal axis, aligning with the applied electric field.

Figure 7.20 shows an electrokinetic streamwise velocity field in an array of square posts. The electric field is applied from left to right at a value 1 V/mm, creating flow in the electric field direction. The interferogram fringe spacing in the figure corresponds to 9.8 μm/s. The uniform flow between the top and bottom posts is pure electroosmotic flow. Two-dimensional flow is observed in the region between the two posts, where the flow expands

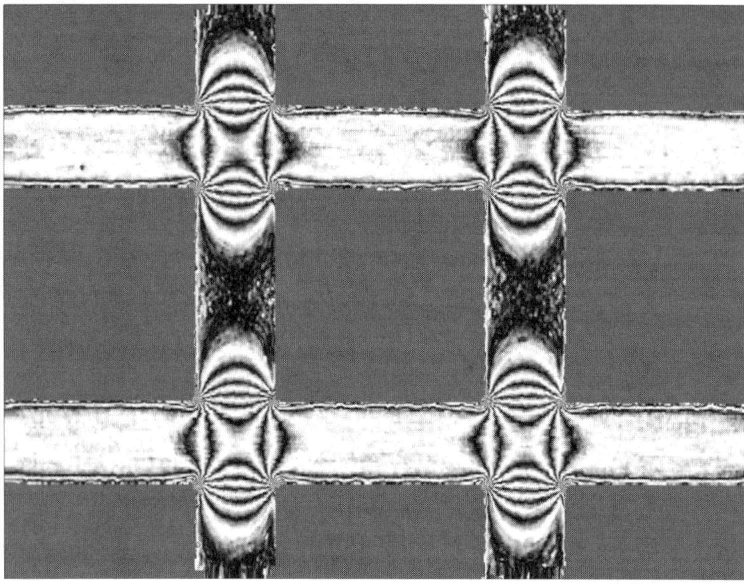

FIGURE 7.20. Electrokinetic streamwise velocity contours in an array of square posts. The electric field is from left to right (1 V/mm). (Courtesy of E. Cummings.)

and contracts; the velocity contours are symmetric in this zone.

In the flow examples given above, the EDL is infinitesimally small compared to the flow dimensions, and the total pressure is constant at the entry and exit ports. Hence, the flow conditions obey the ideal electroosmosis requirements (Cummings et al., 2000), with the following result:

- The flow field outside the EDL is proportional to the external electric field. Therefore, the bulk flow is *potential flow.*

The velocity and speed contours presented in Figures 7.19 and 7.20 closely follow the numerical solution of potential flow past circular and square array posts, as demonstrated in (Cummings, 2001).

7.5 Electrophoresis

Electrophoresis is the process of inducing motion of charged particles relative to a stationary liquid using an applied electric field, where the liquid acts as a conducting medium. The velocity at which charged particles/molecules move toward the anode or cathode is known as the *electrophoretic migration velocity.* This motion is determined by a balance between the net particle charge under the EDL shielding and the opposing viscous drag. Therefore, electrophoretic migration velocity is proportional to the applied electric field and the net charge of the particle, and it is inversely proportional to the viscosity of the liquid.

A particle in viscous fluid experiences strong hydrodynamic interactions. Assuming Stokes flow around a spherical particle of radius a, the velocity field around the particle decays as (a/r), r being the radial distance away from the particle. This gives very long interaction distances for particle/particle and particle/fluid interactions; see Section 14.3.1. However, in electrophoretic motion, this interaction distance scales as $(a/r)^3$, provided that the material properties, sizes, and zeta potentials of particles are the same, and the EDL thickness relative to the particle radius is very small ($\lambda/a \ll 1$). (Reed and Morrison, 1976) studied hydrodynamic interactions in electrophoresis, and obtained analytical solutions for two spherical particle interactions. Here we summarize *approximate results* for two particles moving along and normal to their line of centers. In the first case, the electrophoretic mobility (μ_{EP}) is

$$\mu_{\mathrm{EP}} = \frac{\mathbf{u}_{\mathrm{EP}}\mu}{\epsilon\zeta_{\mathrm{o}}\mathbf{E}} = \frac{1 + \frac{\zeta_1}{\zeta_{\mathrm{o}}}\left(\frac{a}{H}\right)^3}{1 + \left(\frac{a}{H}\right)^3},$$

where H is the center-to-center particle separation, and subscripts o and 1 identify the two particles. The electrophoretic mobility of two particles moving normal to their line of centers is

$$\mu_{\mathrm{EP}} = \frac{\mathbf{u}_{\mathrm{EP}}\mu}{\epsilon\zeta_{\mathrm{o}}\mathbf{E}} = \frac{1 - \frac{1}{2}\frac{\zeta_1}{\zeta_{\mathrm{o}}}\left(\frac{a}{H}\right)^3}{1 - \frac{1}{2}\left(\frac{a}{H}\right)^3}.$$

These approximate relations are valid for large separation distances, and $\mathbf{u}_{\mathrm{EP}}$ and $\mathbf{E}$ vectors are in the same direction. The terms in the denominator are corrections due to the electric field in the presence of particles, and the numerator terms are due to the particle/particle interactions. A striking outcome of these approximate results (equally valid for the analytical/exact results) is that there are no particle/particle interactions if the *particle size and zeta potential are the same.* For particle size or zeta potential mismatch, the interactions decay like $(a/d)^3$. These conditions enable a *similarity* between the electric field and the velocity field, so that the flow outside the EDL region obeys potential flow conditions, which also satisfy the Stokes equations. Implications of this result are quite important:

- Electrophoretic particle motion *does not* disturb the surrounding fluid to a great extent (Reed and Morrison, 1976).

Therefore, one can model the *steady* particle motion using a simple mobility concept, where the particle velocity can be found by

$$\mathbf{u}_{\mathrm{EP}} = \mu_{\mathrm{EP}}\mathbf{E}.$$

The reader is referred to (Swaminathan and Hu, 2004), for results on particle interactions in electrophoresis due to inertial effects.

The mobility concept can also be applied to a large number of particles and dilute species. Reed and Morrison (1976) compared their findings with several experimental results. They reported that for the thin EDL cases, mobilities of groups of particles remain constant for a wide range of particle concentrations. In the following section, we will review the governing equations for charged species transport and present experimental results and numerical modeling efforts for electrophoresis using the aforementioned mobility concept.

7.5.1 Governing Equations

A detailed formulation of electrophoretic transport can be found in (Ermakov et al., 1998). In this section, we will present a simpler model that is based on the electric neutrality (i.e., $\rho_e = 0$) and uniform electric conductivity (σ) assumptions. Using these simplifications, the species transport equation is reduced to (Palusinski et al., 1986; Ermakov et al., 1998)

$$\frac{\partial n_i}{\partial t} + \nabla(n_i \mathbf{u} + n_i \mathbf{u}_{\mathrm{EP},i}) = D_i \nabla^2 n_i, \tag{7.44}$$

where $\mathbf{u}_{\mathrm{EP},i} = \mu_{\mathrm{EP},i}\mathbf{E}$ is the electrophoretic migration velocity. The electric field ($\mathbf{E}$) is determined using equation (7.21), and the Poisson–Boltzmann equation (7.4) is solved for the electrokinetic potential. The fluid velocity $\mathbf{u}$ is found by solution of the incompressible Navier–Stokes equations (7.12), subject to the electrokinetic body force terms ($\rho_e\mathbf{E}$) and the no-slip boundary condition on channel surfaces. This formulation *ignores* interaction of the charged species with ions in the EDL region (where the net electric charge is nonzero), and possible zeta potential changes.

A simplification of this model neglects the electrokinetic forces in the EDL region, and replaces them with the Helmholtz–Smoluchowski slip condition (7.24) to drive the flow. This approach greatly simplifies the problem of mesh generation and the numerical stiffness associated with the resolution of the thin EDL region. In addition, solution of the Poisson–Boltzmann equation is also not required. Despite these simplifications, the electroosmotic slip velocity (7.24) should be calculated, and imposed as the boundary condition on dielectric surfaces. In flows with complex-geometry, this approach requires imposing spatially varying slip velocity on every grid point on the channel surface, which may not be trivial, depending on the numerical solution methodology.

Other assumptions and simplifications used in this simplified formulation are (Ermakov et al., 1999):

- Sample concentration is small compared to the buffer concentration, and the buffer concentration and buffer pH value are uniform.

- Conductivity of the solution (σ) is uniform throughout the liquid volume.
- Concentration of the buffer solution is large enough so that the EDL thickness on channel walls is negligible, compared to the channel dimensions.
- The temperature of the solution is uniform, and Joule heating is insignificant (for details on Joule heating see Section 7.4.6).
- Thermophysical parameters, such as the diffusion coefficients, fluid viscosity, electrokinetic mobilities, and dielectric properties are constant.

Despite its limitations, the simplified model is employed by most research groups for modeling the electrophoretic transport (Palusinski et al., 1986; Ermakov et al., 1992; Grateful and Lightfoot, 1992; Ermakov et al., 1994; Ermakov et al., 1998; Ermakov et al., 2000; Giridharan and Krishnan, 1998; Krishnamoorthy and Giridharan, 2000). However, this model requires revisions for finite EDL effects, observed in low ionic concentrations in nanoscale channels. Such revisions should incorporate local charge distribution effects on electrophoretic transport; see Chapter 12. In addition, isoelectric focusing (IEF), which utilizes pH gradients, cannot be modeled using the current formulation. Numerical modeling of isoelectric focusing can be found in (Baygents et al., 1997; Mosher and Thormann, 2002).

7.5.2 *Classification*

Electrophoresis is one of the most extensively utilized techniques for separation and/or characterization of charged particles, as well as biological molecules (Tseng and Chang, 2001; Kleparnik et al., 2001; Saur et al., 2001). For example, proteins, amino acids, peptides, nucleotides, and polynucleotides can be separated using electrophoretic techniques. Electrophoresis can be divided into three major categories. These are the moving-boundary, steady-state, and zone electrophoresis (Melvin, 1987).

Moving-boundary electrophoresis is a widely used technique in commercial and laboratory setups, where the solution containing positively and negatively charged particles is subjected to electric fields, and the particles move toward the oppositely charged electrode. This results in motion of the solution boundary, which is commonly detected using Schleiren optical techniques.

Steady-state electrophoresis is obtained when the positions of separated components do not change in time. Steady-state electrophoresis is commonly observed in isoelectric focusing (IEF) applications. In IEF, charged particles migrate under electrophoretic forcing and pH gradients to a location in the buffer, where they experience zero net charge. This location

FIGURE 7.21. Isoelectric focusing of proteins PE, APC, and GFP in a 2 cm ×300 μm ×5 μm PDMS microchannel. Three proteins are completely resolved with band thicknesses below 100 μm, and the GFP is subfractionated into two bands. (Courtesy of P. Dutta and C.F. Ivory.)

is known as the isoelectric point (Righetti, 1983; Macounova et al., 2000; Macounova et al., 2001; Cabrera et al., 2001). In Figure 7.21 we present isoelectric focusing of three different proteins in a polydimethyl siloxane (PDMS) microchannel (Horiuchi et al., 2003). The results, obtained in 5 minutes in a 25 V/cm electric field, show separation of phycoerythrin (PE), allophycocyanin (APC), and green fluorescent protein (GPF) mixture with band thickness below 100 μm. The GFP has three constituents with very close focusing points, and the figure shows that the method enables subfractionation of GFP into two bands (on the right side of the figure). Overall, the IEF can be used to separate as well as concentrate charged species under pH gradients. The design shown in (Horiuchi et al., 2003), has the advantage of having approximately 4,500 theoretical bands of 300 μm in a 2 cm channel, enabling detection and concentration of a large number of species.

Zone electrophoresis utilizes a supporting medium to hold the sample, while an external electric potential is applied at the end of the supporting media. Typically, filter paper, cellulose, cellulose-acetate, and gel are used as the supporting media (Westermeier, 1990).

Capillary electrophoresis is applied in capillaries and microchannels (Janos, 1999). It has utmost potential for development of automated analytical equipment with fast analysis time and on-line detection possibilities. Today, many separation techniques rely on combined capillary electrophoresis and electroosmotic flow to pump solutes toward the detector. Through a set of experiments, (Polson and Hayes, 2000) demonstrated flow control using external electric fields in capillary electrophoresis. (McClain et al., 2001) have developed a microfluidic array for *E. coli* detection, which utilized pure electrophoretic transport.

In the following, we demonstrate numerical simulations of electrophoretic motion in a microchannel. The electrode configuration and channel surface conditions are consistent with the schematic shown in Figure 7.4. The electric field is from left to right, the zeta potential on the channel surface is negative ($\zeta < 0$), and the sample is positively charged. Simulation parameters are chosen such that the electroosmotic mobility of the buffer is twice the electrophoretic mobility of the sample ($\mu_{\text{eo}} = 2\mu_{\text{ep}}$), and the Debye length is (1/100) of the channel height. The electric field generates pure electroosmotic flow with Helmholtz–Smoluchowski velocity (u_{HS}) given by equation (7.18), resulting in Re = 0.03 and Pe = 500 flow. Figure 7.22 shows sample motion with an initial Gaussian distribution under electrophoresis and electroosmosis. Since we assume a positively charged sample, the electrophoretic motion is toward the cathode (right), which is augmented by the electroosmotic flow in the same direction. Figures (a–c) show snapshots of sample contours at various times. The sample maintains its initial distribution in the bulk flow region. In Figure (d), we present the sample distribution along the channel center at various times. Minute diffusion effects are visible for this Pe = 500 flow. Figure (e) shows the velocity distribution across the channel, which corresponds to the pluglike velocity distribution of "pure electroosmotic flows." It is this pluglike velocity that maintains the initial sample distribution in the bulk flow region, as shown in Figures (a–c). Velocity distribution within the EDL region is visible in Figure (e), where the velocity rapidly decays from the bulk flow value in the edge of the EDL region to zero on the wall. Hence, we observe slower sample motion within the EDL region. This generates a retardation and smearing of sample distribution near the walls due to the mixed convective/diffusive transport. The effect of convective/diffusive transport on the cross-section-averaged sample shape is known as the *Taylor dispersion*; see the next section. In most electrokinetic flow applications, the EDL thickness is three to five orders of magnitude smaller than the channel height. Hence, the EDL/electrophoresis interactions become negligible with increased disparities between the EDL and channel length scales. For such cases, the retarded sample in the vicinity of walls quickly diffuses due to the small diffusion length scales that are on the order of the Debye length. Hence, sample transport in electroosmotic flows in straight channels experiences "minimal" Taylor dispersion effects.

7.5.3 Taylor Dispersion

Taylor dispersion has adverse effects in identifying the species type using capillary electrophoresis measurements. The electrophoretic motion of the sample experiences both convection and diffusion effects. For the cases in which the radial diffusion (or diffusion across the microchannel) is more dominant than the axial diffusion of the species Pe $\gg 1$, we can analyze the species transport equation to obtain a cross-sectionally averaged species transport equation in the following form (see (Probstein, 1994), Section 4.6,

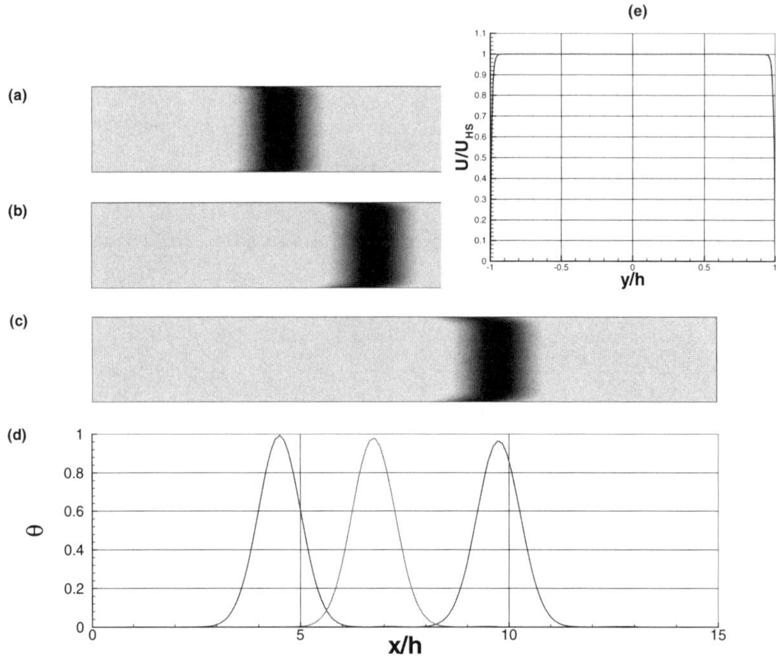

FIGURE 7.22. Numerical simulation results of electrophoretic transport in a microchannel under electroosmotic flow. Figures (a–c) show concentration contours at various times, while (d) shows the sample distribution along the channel center, and (e) shows the streamwise velocity distribution for "pure electroosmotic flow." The velocity is normalized with the Helmholtz–Smoluchowski velocity (7.18), and simulations are performed for $\mathrm{Re} = 0.03, \mathrm{Pe} = 500$, and $\lambda/h = 0.01$ conditions.

for details):

$$\frac{\partial \bar{n}}{\partial t} + U\frac{\partial \bar{n}}{\partial x} = D_e \frac{\partial^2 \bar{n}}{\partial x^2}, \tag{7.45}$$

where $\bar{n}$ is the cross-section-averaged species concentration, U is the channel-averaged velocity, and D_e is the effective diffusion coefficient. For pressure-driven cylindrical capillary flows, the effective diffusion coefficient becomes

$$D_e = D\frac{\mathrm{Pe}^2}{48}, \tag{7.46}$$

where $\mathrm{Pe} = Ua/D$ is the Peclet number based on the capillary radius (a), and D is the species-molecular diffusion coefficient. This relation is valid specifically for $\mathrm{Pe} \gg 7$. For moderate Pe values the streamwise diffusion cannot be neglected to obtain the "Taylor–Aris dispersion coefficient"

$$D_e = D\left(1 + \frac{\mathrm{Pe}^2}{48}\right). \tag{7.47}$$

We must emphasize that the aforementioned results are valid for pressure-driven capillary flows. For the pure electroosmotic flow shown in Figure 7.22, the effective diffusion coefficient can be written as

$$D_e = D(1 + \alpha_o \mathrm{Pe}^2),$$

where α_o depends on the channel geometry, ratio of the EDL thickness to the channel hydraulic diameter, and the channel surface zeta potential. For pure electroosmotic flow in a tube or a two-dimensional channel, the value of α_o is approximately given by (Griffiths and Nilson, 1999)

$$\frac{4}{\alpha_o} \approx 192 + \frac{4}{(\lambda_D^*)^{3/2}} + \frac{8}{(\lambda_D^*)^2}, \tag{7.48}$$

where λ_D^* is the Debye length (see equation (7.1) normalized with the tube radius or the channel half-height. This equation is valid for *small surface potential*, since it is derived from the solution of the Debye–Hückel equation (7.6). In the limit of large Debye length ($\lambda_D^* \gg 1$), the EDL on both sides of the channel (or tube) overlaps, resulting in a parabolic velocity profile for the *pure* electroosmotic flow. For such cases, equation (7.48) gives $\alpha_o \to 48$, as expected. For large zeta potential, the effective diffusion coefficients in a tube or a two-dimensional channel have been presented in (Griffiths and Nilson, 2000). Dispersion coefficients for pure electroosmotic flows in various cross-section channels were presented in (Zholkovskij et al., 2003). In a subsequent work, hydrodynamic dispersion for mixed electroosmotic/pressure-driven flows in arbitrary cross-section channels was presented for electric double layers that are much smaller than the channel dimensions (Zholkovskij and Masliyah, 2004). This study was valid for a relatively small contribution of the pressure-driven flow.

In Figure 7.23, we demonstrate electrophoretic transport under mixed electroosmotic and pressure-driven flow conditions. The electrochemical conditions for this case are identical to those presented in Figure 7.22, with the exception of the favorable pressure gradient imposed on the bulk flow by regulating the channel flowrate. We have deliberately increased the mass flowrate in the channel by a factor of two. Analysis of the electroosmotic flow for mixed electroomostic/pressure-driven flows was presented in Section 7.4.2. In Figure 7.23, we show snapshots of sample contours at various times. Although the initial sample distribution was uniform across the channel, due to the mixed electroosmotic/pressure-driven flow, the sample distribution across the channel quickly becomes parabolic. This parabolic profile is constantly stretched as the material points move with different streamwise velocities. Hence, Taylor dispersion effects are dominant for this flow. Here we must note that in addition to the convective transport due to the local velocity field, the sample also experiences electrophoretic transport, as shown by equation (7.44). In Figure (d), we present the sample distribution along the channel center at various times at which diffusion ef-

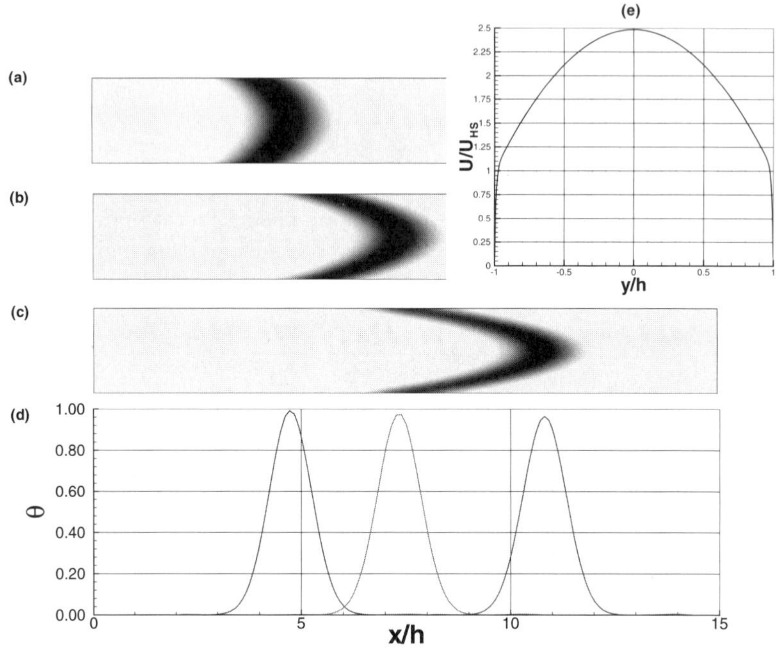

FIGURE 7.23. Numerical simulation results of electrophoretic transport in a microchannel under mixed electroosmotic/pressure-driven flow. Figures (a–c) show concentration contours at various times, while (d) shows the sample distribution along the channel center, and (e) shows the streamwise velocity distribution. The velocity is normalized with the Helmholtz–Smoluchowski velocity of equation (7.18), and simulations are performed for $\mathrm{Re} = 0.03, \mathrm{Pe} = 500$, and $\lambda/h = 0.01$ conditions.

fects are visible. Figure (e) shows the velocity distribution across the channel for mixed electroosmotic/pressure-driven flows. The velocity profile is merely a superposition of pure electroosmotic flow with pluglike velocity distribution and a parabolic velocity profile of pressure-driven flow. Sharp velocity variation across the EDL is also visible in the figure.

The initial sample shape plays a key role in determining the species type using capillary electrophoresis. If we can prescribe the sample shape at the entry of the capillary, it is possible to determine the sample shape at any time (and place) in the channel. For example, an injection with a Gaussian distribution (with σ_o initial standard deviation) can be written as

$$\bar{n}(x, 0) = \frac{\bar{n}_o}{\sigma_o \sqrt{2\pi}} \exp\left(-\frac{x^2}{2\sigma_o^2}\right).$$

Solution of equation (7.45) using this initial distribution results in (Bharadwaj et al., 2002)

$$\bar{n}(x,t) = \frac{\bar{n}_{\text{o}}}{\bar{\sigma}\sqrt{2\pi}} \exp\left(-\frac{(x-Ut)^2}{2\bar{\sigma}^2}\right),$$

where, $\bar{\sigma}^2 = \sigma\text{o}^2 + 2tD_e$. Therefore, knowing the initial distribution of the injection, we can easily deduce the electrophoretic mobility of the species using $\mu_{\text{EP}} = U/E$, where $U = x/t$ is found by measuring the location (x) of the sample at a given time (t). Noise associated with the initial sample shape or satellite sample droplets creates difficulties in identification of the sample type using this technique. To this end, there have been numerous experimental and numerical studies on initial sample focusing. For example, (Ermakov et al., 1999) utilized electroosmotic flow in a cross-channel to focus and pinch the sample for electrophoretic detection in a straight channel. Due to the convective–diffusive transport nature of electrophoretic detection, channel length plays an important role. One way to fit a very long channel in a microchip is to use spiral or serpentine channels, which result in severe dispersion effects due to the channel curvature. Several studies were conducted to understand and eliminate this effect by modification of the channel geometry (Molho et al., 2001; Dutta and Leighton, 2002), as well as local modifications of the channel zeta potential (Qiao and Aluru, 2003a). In addition, there have been efforts to modify the hydrodynamic dispersion in pressure-driven or mixed electroosmotic/pressure-driven flows by modification of the channel geometry (Dutta and Leighton, 2001; Dutta and Leighton, 2003).

The results presented in Figures 7.22 and 7.23 characterize electrophoretic transport in simple channels. Similar techniques can be implemented in complex microchannel networks to enable flow and species transport control applications. (Ermakov et al., 1998) have developed a two-dimensional numerical model for electroosmotic/electrophoretic transport and species diffusion, which enabled them to analyze electrokinetic transport in two basic chip elements: cross-channel geometry for sample focusing and injection, and T-channel for sample mixing. Numerical results of electrokinetic sample focusing, injection, and separation steps are compared with the experimental data (Ermakov et al., 1999). We must note that the results shown in Figures 7.22 and 7.23 differ from the numerical approach in (Ermakov et al., 1998), since the EDL region is fully resolved in the results presented here using spectral element discretization; see Section 14.1.

Finally, most practical applications of electrokinetically driven flows may experience variations in the wall charge (and hence the zeta potential) or variations in the channel cross-section. For example, proteins and peptides in capillary zone electrophoresis adsorb on the capillary walls, changing the zeta potential. This mismatch on the zeta potential locally changes the electroosmotic flow, and pressure-driven flow is initiated to conserve mass (see Figure 7.7). Since the flow conditions vary axially in the channel and there may be secondary flows, Taylor dispersion increases and becomes nonuniform. Such conditions have been investigated by Ghosal in a series of papers. Electroosmotic flow in a straight channel of arbitrary cross-sectional geometry and wall charge distribution is solved using the lubrication approximation (Ghosal, 2002c), while stepwise zeta potential variations are presented in (Ghosal, 2002a). Analyte adsorption and the corresponding changes in the flow pattern and the analyte dispersion are presented in (Ghosal, 2002b; Ghosal, 2003).

7.5.4 Charged Particle in a Pipe

Electrophoretic motion of a spherical colloidal particle moving along the axis of a pipe has been studied by (Keh and Anderson, 1985) using the method of reflections. Assuming thin EDL on the particle and the pipe surfaces, the electrophoretic velocity of the particle is given up to the $\mathcal{O}\left(\frac{d}{D}\right)^8$ terms as

$$\mathbf{u}_{\mathrm{EP}} = \frac{\epsilon \mathbf{E}(\zeta_p - \zeta_w)}{\mu} \left(1 - 1.28987\left(\frac{d}{D}\right)^3 + 1.89632\left(\frac{d}{D}\right)^5 - 1.02780\left(\frac{d}{D}\right)^6\right),$$

where d and D are the sphere and pipe diameters, respectively. The zeta potentials on the particle and wall surfaces are shown by ζ_p and ζ_w, respectively. The minus sign in front of ζ_w characterizes the electroosmotic flow. For example, for a negatively charged surface, the EDL will be positively charged, and the electroosmotic motion will be toward the cathode. At the same time, a positively charged particle will also move toward the cathode (in the electric field direction), and the electroosmotic flow will enhance the particle migration speed. The equation above can be utilized to characterize blockage effects of the pipe on particle velocity. To a leading order, $\frac{d}{D} = 0.5$ gives 16% variation in the electrophoretic velocity. This becomes important in nanoflow applications. For example, the nanotube-Coulter-counter (Ito et al., 2003), utilized to determine the size and surface charge of the nanoparticles, should require such corrections when the particle size becomes comparable to the nanotube diameter. Another case that may need further attention is the blockage effects under finite EDL thickness, where the EDL from the pipe surface and that from the sphere overlap.

7.6 Dielectrophoresis

Dielectrophoresis is the motion of polarizable particles that are suspended in an electrolyte and subjected to a spatially nonuniform electric field (Pohl, 1978). The particle motion is produced by the dipole moments induced on the particle and the suspending fluid due to the nonuniform electric field. When the induced dipole moment on the particles is larger than that of the fluid, the particles move toward regions of high electric field density. This is known as *positive dielectrophoresis*. In the case of the fluid being more polarizable than the particles, the particles move away from the high electric field density, which is known as *negative dielectrophoresis* (Cummings and Singh, 2000).

The time-averaged dielectrophoretic force is given by

$$\mathbf{F}_{\text{DEP}} = 2\pi r^3 \epsilon_m \Re[K(\omega)] \nabla \|E_{\text{rms}}\|^2, \tag{7.49}$$

where E_{rms} is the rms electric field, r is the particle radius, ϵ_m is the dielectric permittivity of the medium, ω is the electric field frequency, and $\Re[K(\omega)]$ indicates the real part of the Clausius–Mossotti factor ($K(\omega)$), which is a measure of the effective polarizability of the particle, given by (Morgan et al., 1999)

$$K(\omega) = \frac{(\epsilon_p^* - \epsilon_m^*)}{(\epsilon_p^* + 2\epsilon_m^*)}, \tag{7.50}$$

where ϵ_p^* and ϵ_m^* are the complex permittivities of the particle and the medium, respectively. The complex permittivity is defined by

$$\epsilon^* = \epsilon - \sqrt{-1}\frac{\sigma}{\omega}, \tag{7.51}$$

where ϵ is the permittivity, and σ is the conductivity of the dielectric medium.

Ignoring the Brownian motion, the buoyancy force, and the motion of the buffer solution, the equation of motion for a suspended particle can be written as

$$m_p \frac{d\mathbf{v}}{dt} = \mathbf{F}_{\text{DEP}} - \mathbf{F}_d,$$

where $\mathbf{F}_d$ is the instantaneous drag force acting on the particle. For particle sizes smaller than $10\,\mu$m in buffer solutions with viscosity close to that of water, the Reynolds number based on the particle size is smaller than unity. Hence, the inertial effects on particle motion can be neglected. If we assume a dilute solution, so that particles do not interact and spherical particles with radius r, we can use Stokes's formula for the drag force:

$$\mathbf{F}_d = 6\pi\mu r \mathbf{v},$$

where μ is the dynamic viscosity and $\mathbf{v}$ is the velocity of the particle. Since the inertial effects are negligible, one can assume that the instantaneous velocity of a particle $\mathbf{v}$ is proportional to the instantaneous dielectrophoretic force. This results in the dielectrophoretic velocity of a particle given by (Morgan et al., 1999)

$$\mathbf{v} = \frac{r^2 \epsilon_m \Re[K(\omega)] \nabla \|E_{\text{rms}}\|^2}{3\mu}. \tag{7.52}$$

Since the surface area of the particle is proportional to r^2,

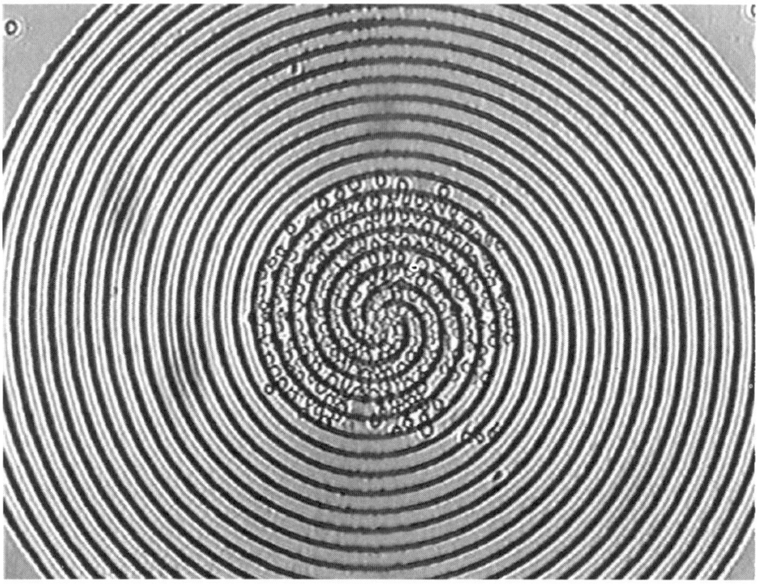

FIGURE 7.24. A 50 kHz AC electric signal induces electric polarization on human leukemia cells and moves them to the center of four spiral electrodes, while the normal cells are trapped on the electrode surfaces. (Courtesy of P. Gascoyne and X. Wang.)

- the *dielectrophoretic particle velocity is proportional to the surface area of the particle.*

Further examination of equation (7.52) also reveals that the particle velocity is determined by the *square of the rms electric field.* Therefore,

- *dielectrophoresis can be maintained by either DC or AC electric fields.*

Positive or negative dielectrophoresis (i.e., motion of particles toward or away from the large electric field gradients) is obtained according to whether $Re[K(\omega)] > 0$ or $Re[K(\omega)] < 0$. These properties of dielectrophoresis enable highly controlled selective microfluidic particle/cell separation methodologies.

7.6.1 Applications

In the rest of this chapter we will present various biomedical and microfluidic applications of dielectrophoresis. Green and Morgan were the first to show that it is possible to separate a population of nanoparticles (latex beads of 93 nm) into two subpopulations due to the differences in their dielectrophoretic properties, by using microfabricated electrode arrays (Green and Morgan, 1997). This initiated many applications of separation of small, yet similar-size, particles with different biological properties, including chromosomes, viruses, DNA, and other macromolecules. For example, Gascoyne et al. were able to separate human breast cancer cells from blood using (AC electric field) dielectrophoretic separation (Gascoyne and Wang, 1997). Their technique utilized the *frequency dependence* of dielectrophoretic (DEP) properties of blood and cancer cells and worked in the following sequence: First, trapping and accumulation of both blood and cancer cells on microfabricated dielectric affinity chambers (electrodes) using DEP collection at 500 kHz. Second, reducing the DEP collection to 50 kHz, where the blood cells are released and only the cancer cells are trapped on the electrodes. This is followed by washing the released blood cells with pressure-driven flow, where the blood cells are convected downstream, while the cancer cells remained on the electrode tips (Gascoyne and Wang, 1997).

In a somewhat different subsequent design, Gascoyne and Wang used four spiral microfabricated electrodes for dielectrophoretic separation of human leukemia cells from the normal cells. Figure 7.24 shows the leukemia cells concentrated toward the center of four spiral electrodes. In this design the normal cells are trapped on the electrode surfaces, and human leukemia cells are washed toward the center. Several other applications of dielectrophoresis can be found in (Gascoyne et al., 1992; Markx and Pethig, 1995; Markx et al., 1996; Fiedler et al., 1998; Cheng et al., 1998; Morgan et al., 1999).

In a series of papers Gascoyne and coworkers have also utilized combined dielectrophoretic/gravitational field flow fractionation for cell separation on microfabricated electrodes (DeGasperis et al., 1999; Yang et al., 1999a; Yang et al., 2000). The gravitational field flow fractionation utilizes balance between the vertically applied dielectrophoretic forces and the gravitational forces for levitation of different particles to different heights in a miniaturized channel flow system. The bulk flow is pressure-driven in the axial direction, and it splits into two different outlet ports at a desired channel height, separating the heavier particles in the bottom exit port from the lighter ones in the top exit port. This particle separation system is free from any moving mechanical components.

Most of the dielectrophoretic applications utilize AC electric fields. However, as we have stated earlier in the chapter, it is possible to utilize a DC electric field. In this case the Clausius–Mossotti factor given by equation (7.50) is real, and there is no frequency dependence in the electrophoretic force. Cummings and Singh built arrays of insulated circular and square posts without embedded electrodes (as shown earlier in Figures 7.19 and 7.20). The flow is driven by electrodes outside the post arrays with a DC electric field at a desired angle to the post row orientation (Cummings and Singh, 2000). Under a *weak electric field*, dielectrophoretic effects are overwhelmed by the electrokinetic effects and diffusion, since dielectrophoresis is a second-order effect in the applied electric field. When the electric field is increased, two additional distinct flow phenomena are observed. The first occurrs when dielectrophoresis starts to dominate diffusion over a certain magnitude of electric field, where "filaments" of low and high particle concentration appear in the flow. Depending on the angle between the electrodes and the post arrays as well as the shape of the posts (circular or square), flow with various concentration gradients is observed, which is identified as *filamentary dielectrophoresis* (Cummings and Singh, 2000; Cummings, 2001). The second phenomenon occurs at higher electric fields, where dilectrophoresis is comparable and greater than the advection and electrokinetic effects. These experiments have shown "trapping" of reversibly immobilized particles near the insulator surfaces, and this flow regime is identified as *trapping dielectrophoresis* (Cummings and Singh, 2000; Cummings, 2001). Figure 7.25 shows particle fluorescence image of filamentary (upper plot) and trapping (lower plot) dielectrophoresis in arrays of circular posts. Other experiments of Cummings, obtained under DC electric fields, were used to systematically analyze the electrokinetic, filamentary dielectrophoretic, and trapping dielectrophoretic transport in complex microgeometries (Cummings, 2001).

The experimental results in (Cummings and Singh, 2000; Cummings, 2001), can also be used for verification of numerical models for electrokinetic and dielectrophoretic transport. The simulation domain is periodic; hence, it is free of external boundary conditions, with the exception of the imposed electric field, which is determined by Laplace's equation.

In a combined experimental and numerical study, Ho and coworkers utilized AC filamentary dielectrophoresis to induce chaotic mixing in a microchannel (Deval et al., 2002). Their mixer consisted of a straight channel with two grooves on the top and bottom surfaces, see Figure 9.9 in Section 9.3. They have applied $\pm$ 10 V AC electric field on the groove surfaces to induce weak dielectrophoretic forces on the particles flowing through the channel. Although the flow was laminar and two-dimensional, particle trajectories exhibited chaotic motion that enhanced mixing in this continuous-flow mixer.

An interesting nanotechnology application of DEP has been demonstrated by Velev and coworkers in self-assembly of microscale wires using metallic nanoparticles (Hermanson et al., 2001; Bhatt and Velev, 2004). Figure 7.26 (a) shows an optical micrograph obtained during the wire growth process. Due to the DEP, the gold nanoparticles are highly concentrated at the end of the wire tip, which enable extension of the wire in the electric-field gradient direction. The authors reported wire growth exceeding 50 micrometers per second with lengths on the oder of 5 millimeters. They have used planar electrodes and AC electric fields ranging from 50 to 200 V at frequencies 50 to 200 Hz. The wires are automatically assembled using 15 to 30 nm diameter gold particles, and the wires exhibited good ohmic conductance. Wire thickness can be controlled, resulting in high surface-to-volume ratio structures. The assembly process is simple and self-repairing, and the wires automatically form electrical connections to conductive islands or particles, as shown in Figure 7.26. These properties make the microwires promising for *wet electronic* and bioelectronic circuits (Hermanson et al., 2001; Bhatt and Velev, 2004).

(Jones et al., 2001) utilized dielectrophoresis to create water droplet volumes ranging from microliters down to nanoliters. They have shown that dielectrophoresis provides a means for manipulation of small water volumes, against overwhelmingly strong surface tension and capillary effects. Their method enables rapid motion of water in capillaries with smooth substrates, and creation of multiple nanoliter droplets in less than 0.1 s. A great potential of this method is that droplet motion across a substrate for diagnostics, mixing, separation, and dispensing purposes is possible. However, the foremost advantage of this new method is "division of an initial liquid inventory into discrete droplets before processing, which avoids sample cross contamination" (Jones et al., 2001). In addition to the DEP and the opposing capillary and surface tension effects (see Chapter 8), wetting, transient fluid motion, Joule heating (see Section 7.4.6), and RF discharge are also important. More research is required for further understanding and accurate modeling of this phenomenon (Jones, 2001). In other work, Velev and coworkers described dielectrophoretic manipulation of freely suspended droplets in a liquid–liquid microfluidic system (Velev et al., 2003). Unlike the previous work, this new approach eliminates droplet contact with the electrodes by suspending the droplets in an immiscible fluid. They have reported manipulation and mixing of micro to nanoliter volume droplets. These microfluidic devices can be utilized in a range of applications, including the synthesis of new materials and development of biological microassays (Velev et al., 2003).

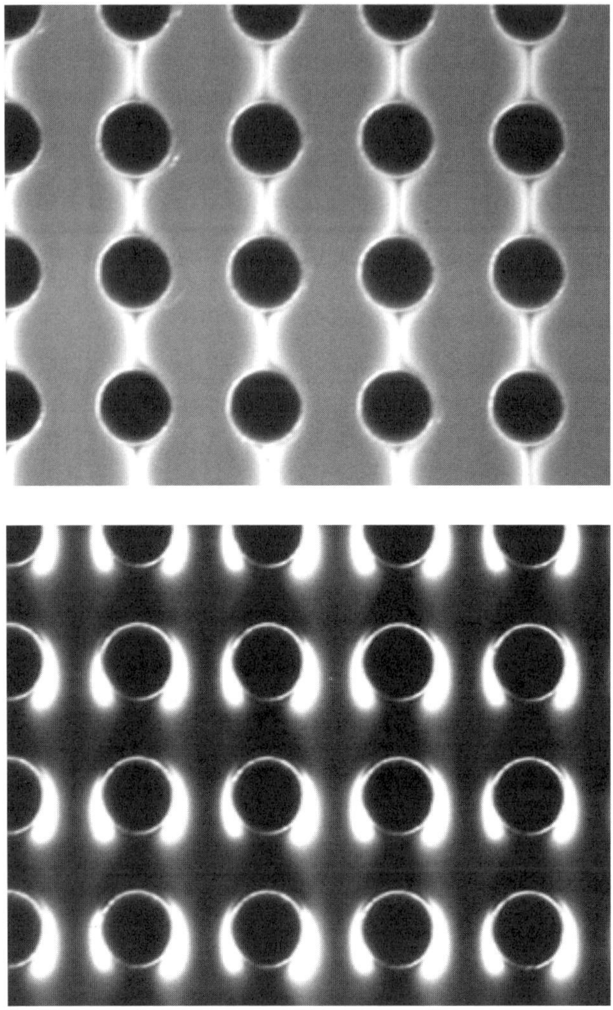

FIGURE 7.25. Particle fluorescence image of filamentary (upper) and trapping (lower) dielectrophoresis. Regions of high particle concentration emit intense fluorescence. The flow is from top to bottom produced by an applied field of 25 V/mm and 100 V/mm, for the upper and lower figures, respectively. The circular posts have diameter 33 μm with center separation of 63 μm. (Courtesy of E. Cummings.)

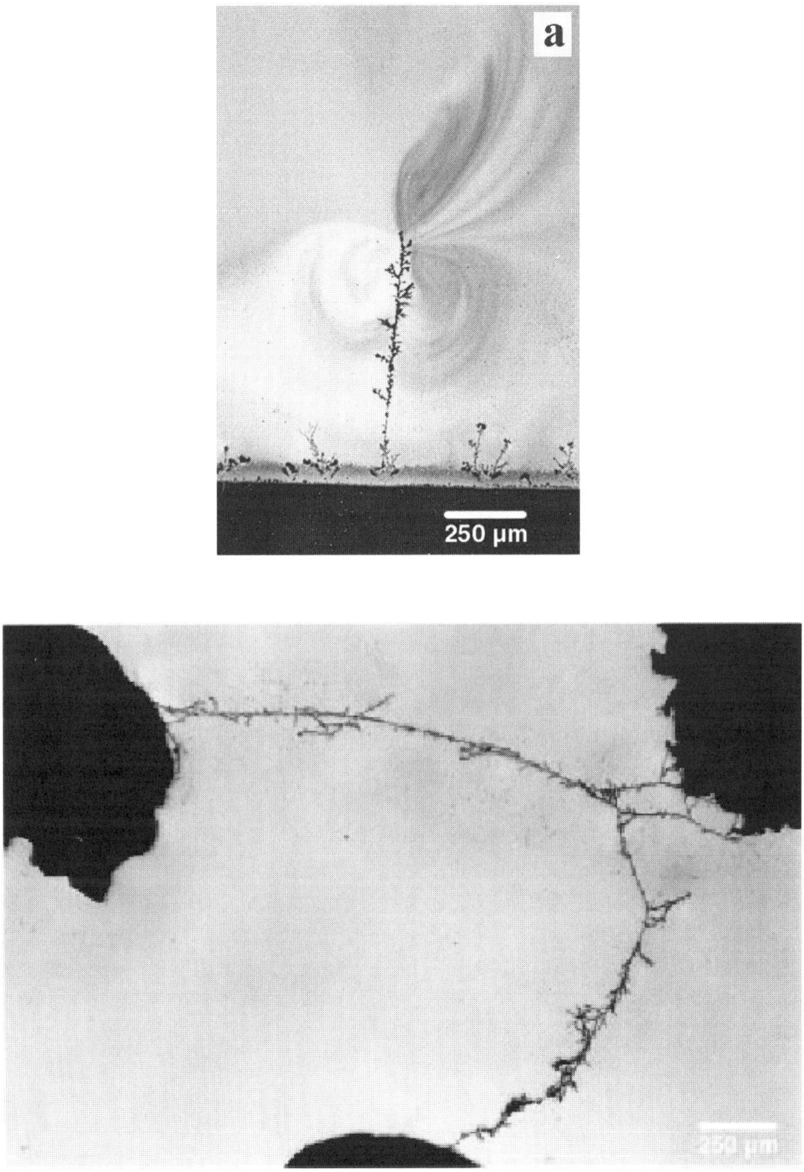

FIGURE 7.26. Optical micrograph of a growing microwire. Contours show nanoparticle concentration around the wire (a). The microwires generate complex connections with conductive islands (b), which is achieved by electric field disturbance imposed by the conductive surfaces. (Courtesy of O. Velev.).

8
Surface Tension-Driven Flows

Capillary phenomena involving wetting and spreading of liquid thin films and droplets have been studied for a long time for modeling of classical engineering applications such as coating and lubrication. This same body of theoretical and experimental knowledge can be of great use in microfluidic research and design. For microfluidic delivery on open surfaces, electrowetting and thermocapillary along with dielectrophoresis have been employed to move continuous or discrete streams of fluid, for example droplets along specified paths on glass surfaces (Sammarco and Burns, 1999; Kataoka and Troian, 1999; Lee et al., 2002). A new method of actuation exploits optical beams and photoconductor materials in conjunction with electrowetting (Ichimura et al., 2000). Such *electrically* or *lithographically* defined paths can be reconfigured dynamically using electronically addressable arrays that respond to electric potential, temperature, or laser beams and control the direction, timing, and speed of fluid droplets. Mixing two fluids in T-junctions creates interesting structures that can be manipulated in a controlled fashion. An example is shown in Figure 8.1, which shows droplet breakup by extensional flow, i.e., a flow in which fluid elements are only stretched in a network of T-junctions. Specifically, water slugs are broken into droplets in oil, a scaled-up version of the experiment shown in Figure 1.24. Here the microchannels have a rectangular cross-section $50\,\mu\text{m} \times 25\,\mu\text{m}$. Exploiting this concept, one can produce more than a few hundred droplets per second. The size of the droplets can be controlled by the lengths of the arms of the corresponding T-junction (Link et al., 2004).

In this section, we study microfluidic transport mechanisms based on capillary phenomena, taking advantage of the relative importance and sen-

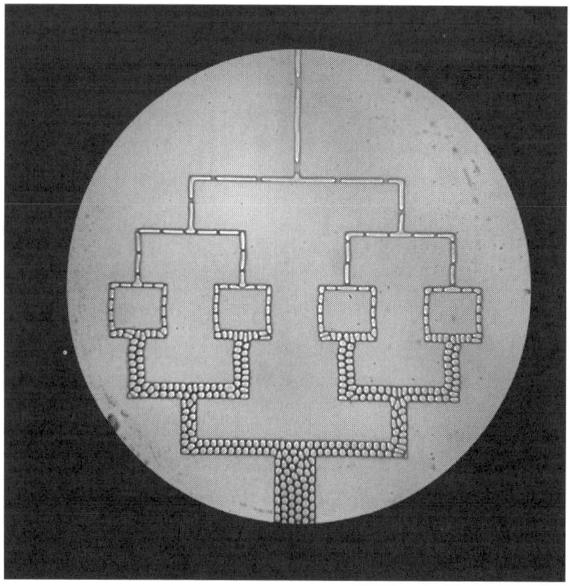

FIGURE 8.1. Droplet breakup by extensional (i.e., stretched) flow in T-junctions. (Courtesy of D.A. Weitz.)

sitivity of surface tension in microscales. In particular, we will study how temperature, electric potential, and light can affect the value and possibly the sign of surface tension, which is the dominant driving force in the cases we consider in this section. Its strong dependence on the type of surface material can be exploited to design scalable and controllable *wet circuits* for diverse functions in microdomains. A related topic, important both for engineering and biomedical applications, is bubble transport in microducts; we study this topic in some detail in the last section of this chapter. For background reading on the fundamental concepts presented in this chapter and for a more mathematical treatment we recommend to the interested reader the book (deGennes et al., 2004), and also the classical book on physicochemical hydrodynamics of (Levich, 1962) (Chapters 7 to 12).

8.1 Basic Concepts

We review some basic concepts from fluid mechanics that we will use in the following sections. We need to know how droplets are created, how they are transported, and the governing equations for droplets and thin films in general, under various conditions.

First, we discuss *droplet formation*, since in most applications we need to create and transport droplets. A long cylinder of liquid in air (e.g., water jet emanating from a tap) is unstable and breaks up into small droplets. This instability was first studied by Lord Rayleigh, who found that infinitesimal axisymmetric disturbances grow exponentially if their wavelength λ exceeds a critical value, which is equal to the circumference of the undisturbed jet (radius R_0), i.e., if

$$\lambda/(2\pi R_0) > 1.$$

Similarly, if the same column of liquid resides on a solid surface with a contact angle θ between the liquid and the surface, a similar hydrodynamic instability appears if

$$\lambda/(2\pi R_0) > \sqrt{1 - \cos\theta}.$$

Another situation, often encountered in microfluidics is the case of a liquid column trapped between two parallel plates. Unlike the two aforementioned cases, this case is hydrodynamically stable irrespective of the value of the wavelength λ. This, in turn, implies that it is very difficult to create droplets in a confined environment unlike the open surfaces where droplets are created spontaneously by hydrodynamic instabilities. Details of the theoretical derivation for these cases can be found in (Isenberg, 1992). In applications, if we transport microdroplets over an *open area*, evaporation may be very high, so often a cover glass is used to prevent this. However, due to the aforementioned enhanced hydrodynamic stability, creating droplets from a liquid column sandwiched between two plates requires large amounts of energy.

We now discuss some useful concepts and the governing laws related to *surface tension*. Surface tension, denoted by γ, is *force per unit length*. For a system at equilibrium in the presence of an interface of area A, it contributes γA to the system's energy. Correspondingly, the work done on the system to change the interface (for relatively small changes in the system dynamics) is $\gamma \delta A$. Surface tension can have either positive or negative values, reflecting the fact that it is associated with the free energy of the system, which can increase or decrease. When a liquid is in contact with a gas, the surface tension is positive. For example, the surface tension between water and air at 20° is 72.8 dyn/cm, or 72.8 mN/m in SI units; it depends strongly on impurities and on the temperature. The surface tension of a liquid decreases linearly with temperature in the range from the melting point to the boiling point. For example, for the liquid polydimethylsiloxane (PDMS), a silicone oil, an empirical fit to available data was obtained in (Darhuber et al., 2003), as follows:

$$\gamma(T) = 21.5 - 0.06T \quad \text{mN/m},$$

where T is in degrees Celsius. This indicates that the variation of surface tension with temperature is linear; also, the thermal coefficient $d\gamma/dT$ is constant in the range of 0° to 100°, unlike the dynamic viscosity of liquids, which also decreases with temperature, but nonlinearly.

Surface tension is created by intermolecular cohesive forces. Liquid molecules that are close to a liquid–gas interface and interact with gas molecules experience a cohesive force directed toward the bulk of the liquid and away from the interface. This imbalance of forces creates an apparent contraction and a corrugation of the interface. For interactions with other media such as solids or other liquids, the surface tension sign depends on many other factors and can be either positive or negative. For immiscible liquids we have $\gamma > 0$; for example, for an oil–water interface, $\gamma = 20$ mN/m or (dyn/cm) at 20°. Typical values of surface tension for different substances interacting with air or water are listed in Table 8.1.

The presence of an interface creates a jump in the pressure, which can be easily computed for systems in equilibrium; it depends on the surface tension and the deformation of the interface. Following (Batchelor, 1998), we compute a tension force at a small segment $\mathbf{dx}$ on the interface lying in a plane (x-y) as follows:

$$-\gamma \int \mathbf{n} \times \mathbf{dx},$$

where $\mathbf{n}$ is the normal vector to the surface at the point described by $z = \zeta(x, y)$. Upon substituting $\mathbf{n} = \nabla\zeta$, we obtain

$$\gamma \int \left(\frac{\partial \zeta}{\partial x} dy - \frac{\partial \zeta}{\partial y} dx \right) = \gamma \nabla_2^2 \zeta \delta A, \tag{8.1}$$

where ∇_2 denotes the two-dimensional Laplacian operator. We see that the curvature of the interface determines the value of the force, which is equal to a pressure of magnitude $\gamma\nabla_2^2\zeta$ acting on the infinitesimal surface δA. We can rewrite the above expression in terms of reference-independent variables by introducing the principal radii of curvature, i.e.,

$$\nabla_2^2 \zeta = \frac{1}{R_1} + \frac{1}{R_2},$$

and hence we can express the pressure jump at the interface induced by the surface tension as

$$\Delta p = \gamma \left(\frac{1}{R_1} + \frac{1}{R_2} \right). \tag{8.2}$$

This equation is usually referred to as the *Laplace–Young equation.*

We note here that in the absence of equilibrium, i.e., in moving interfaces, the full stress balance equation should be employed instead, which has the form

$$\sigma_{ij}^{(1)} \eta_j - \sigma_{ij}^{(2)} \eta_j = -\gamma \int \left(\frac{1}{R_1} + \frac{1}{R_2} \right) \eta_i, \tag{8.3}$$

where $\sigma_{ij}^{(n)}$ denotes the stress tensor in medium (n), and η_i is the normal unit vector. Also, the radius R_i is positive if its center of curvature is on the side toward the direction defined by η_i. Some known examples in which the equilibrium equation is realizable are bubbles and droplets, which are spherical. In this case $R_1 = R_2 = R$, and thus $\Delta p = 2\gamma/R$.

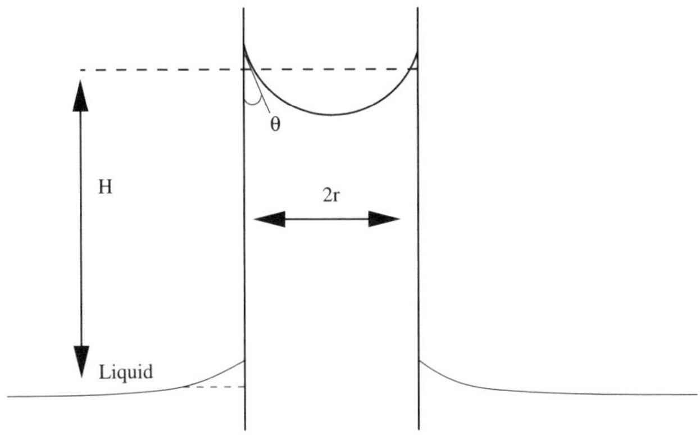

FIGURE 8.2. Capillary flow in a micropipe.

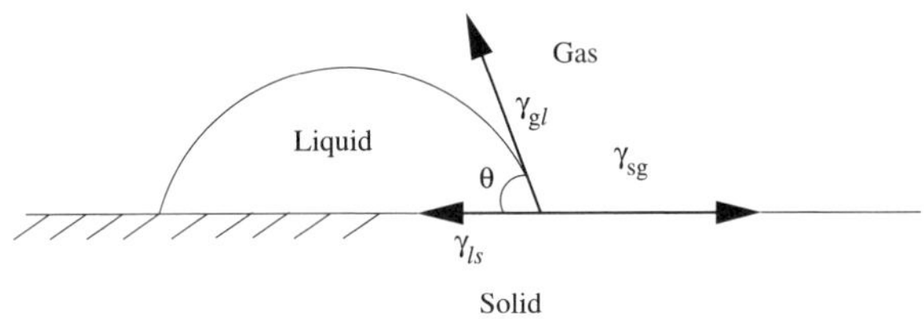

FIGURE 8.3. Surface tension balance at equilibrium.

We have assumed here that the effect of gravity is negligible; this is valid if the radius of the microbubble is less than the characteristic length scale $l_s = \sqrt{\frac{\gamma}{\rho g}}$. For pure water bubbles (at room temperature) we get $l_s \approx 2.5$ mm; thus all microbubbles are spherical. Similarly, for a silicone oil droplet we have that $l_s \approx 1.5$ mm, and thus micron-size oil droplets are spherical as well. The *Bond number* defined as

$$\text{Bo} \equiv \frac{L^2}{l_s^2}, \tag{8.4}$$

where L is a characteristic length (here the bubble diameter), determines the relative importance of gravity. For $\text{Bo} \leq 10^{-3}$ the effects of gravity can be neglected.

When gravity is important, the condition for equilibrium at points belonging to the interface is

$$\rho g z - \gamma \left(\frac{1}{R_1} + \frac{1}{R_2} \right) = C, \tag{8.5}$$

which expresses hydrostatic equilibrium with C a constant. The radii of curvature are taken positive if the centers of curvature are located on the gas side of the interface. This equation is useful in studying the rise of a liquid in a micropipe partially immersed in a liquid. In particular, depending on

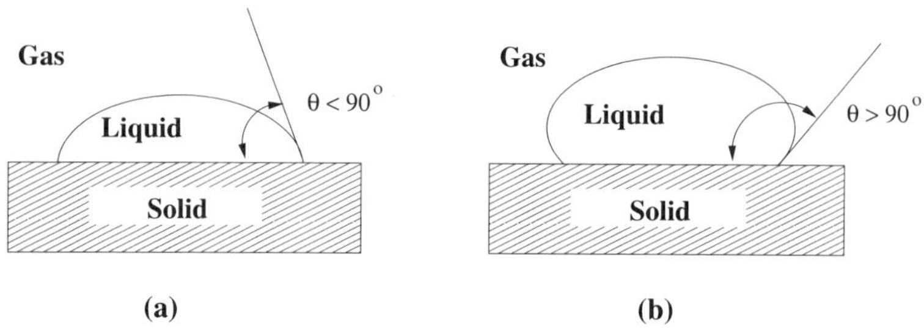

FIGURE 8.4. A sketch illustrating the definition of the contact angle. Panel (a) corresponds to a wetting surface ($\theta < 90°$), and panel (b) corresponds to a nonwetting surface ($\theta > 90°$).

the liquid–pipe interaction, the free surface of the liquid may rise or fall into the pipe. This phenomenon is called *capillarity.* We can compute the height of liquid rise H by a simple hydrostatic balance following the sketch of Figure 8.2. The hydrostatic pressure is $\Delta p = \rho g H$, and for a micropipe of radius r we assume that the radius of curvature is $R = r/\cos\theta$, where θ is the contact angle; see below. Therefore, equilibrium implies that

$$\rho g H = \frac{2\gamma\cos\theta}{r},$$

and thus the rise of liquid is inversely proportional to the micropipe radius. For example, for a 10-micron glass pipe, there may be a rise of water of up to about 100 microns.

As we have seen in the example of Figure 8.2, the contact angle θ is a significant factor in capillarity. When a liquid droplet is in contact with a solid surface, the value of the contact angle reflects the type of interaction between the liquid and the solid. The shape of a small liquid droplet on a homogeneous substrate is usually a spherical cap. The contact angle is approximately constant if there are no impurities, temperature gradients, or surface defects, and it is determined by the Laplace–Young equation (8.2). The point (or line) of contact, as shown in the sketch of Figure 8.3, is subject to three different surface tensions. Because of equilibrium at the contact point (or line), we have

$$\gamma_{\mathrm{sg}} = \gamma_{\mathrm{ls}} + \gamma_{\mathrm{gl}}\cos\theta, \tag{8.6}$$

where γ_{sg} denotes the surface tension between solid and gas, γ_{ls} between liquid and solid, and γ_{gl} between gas and liquid. This equation is often referred to as *Young's law.*

- When $0 < \theta < 90°$, we say that the liquid is partially wetting or that the solid is *hydrophilic.*

Substance	Air		Water
	°C	dyn/cm γ or mN/m	dyn/cm γ or mN/m
Water	10	74.2	
	20	72.8	
	30	71.2	
	40	69.6	
Benzene	20	28.9	35
Ethyl ether	20	17	10.7
Glycerol	20	63.4	< 0
Mercury	20	487	375
n-Hexane	20	18.4	51.1
n-Octane	20	21.8	50.8
Olive Oil	20		20
Silicone Oil (PDMS)	25		20

TABLE 8.1. Surface tension values for some liquids (1 mN/m = 1 dyn/cm).

- When $\theta > 90°$, the liquid is nonwetting, and thus the solid is called *hydrophobic*.
- When $\theta = 0$, we have *complete wetting*.

A schematic is shown in Figure 8.4. In the following we will present a more general version of this equation.

8.2 General Form of Young's Equation

The triple-line force balance expressed in equation (8.6) is valid for simple cases, but it is inadequate in describing situations in which internal bulk forces in the droplet are present, e.g., due to gravity or electric fields. In such cases these bulk forces have to be balanced against the interfacial forces. An effective approach to obtaining the governing equations is to consider the total energy of the system and minimize it in order to obtain equilibrium solutions. Such procedures have been developed and applied in (deGennes, 1985; Dussan, 1979; Digilov, 2000), and also in (Shapiro et al., 2003a); here we follow the analysis of (Shapiro et al., 2003a), which is directly applicable to microfluidic problems.

The starting point is to assume that the droplet is spherical, characterized fully by its radius R and the contact angle θ. Other system parameters will depend on the specific application, e.g., gravity, temperature, voltage. Having the energy $E = E(R, \theta)$ as a function of two independent variables, at equilibrium we have that $dE(R,\theta) = 0$, and thus

$$\frac{\partial E}{\partial R} dR + \frac{\partial E}{\partial \theta} d\theta = 0.$$

Neglecting any evaporation, the above equation states that at equilibrium, if the contact angle increases, then the radius must also change. Specifically, assuming a constant droplet volume, this implies that the radius should decrease. This volume v also depends on R, θ, and thus $dv = \frac{\partial v}{\partial R} dR + \frac{\partial v}{\partial \theta} d\theta$. From geometric considerations we have that, $v = \pi R^3(2/3 - (3/4)\cos\theta + (1/12)\cos 3\theta)$, and since volume is constant

$$dv = 0 \Rightarrow dR = R\left(-\frac{2\cos^2(\theta/2)\cot(\theta/2)}{2+\cot\theta}\right) d\theta = Rq(\theta)\, d\theta, \qquad (8.7)$$

where the last equation defines the function $q(\theta)$. By substituting equation (8.7) into the equilibrium equation ($dE = 0$), we obtain the *generalized Young's equation*:

$$\left(-\frac{2+\cos\theta}{2\pi R^2 \sin\theta}\right)\left(\frac{\partial E}{\partial R} Rq(\theta) + \frac{\partial E}{\partial \theta}\right) = 0. \qquad (8.8)$$

This equation can accommodate modeling of any energy function with multiple contributions due to interfacial effects, temperature or voltage effects, gravity, etc. In the case in which E contains contributions only due to interfaces between gas–liquid–solid, we recover the standard Young equation (8.6). To illustrate this we compute this interfacial energy, assuming constant surface tensions, from

$$E_i = \gamma_{\mathrm{gl}} A_{\mathrm{gl}} + (\gamma_{\mathrm{ls}} - \gamma_{\mathrm{sg}}) A_{\mathrm{ls}},$$

where $A_{\mathrm{gl}} = 2\pi R^2(1 - \cos\theta)$ is the surface area of the gas–liquid interface and $A_{\mathrm{ls}} = \pi R^2 \sin^2\theta$ is the surface area of the liquid–solid interface. The interfacial potential energy is then

$$E_i = R^2[2\pi\gamma_{\mathrm{gl}}(1-\cos\theta) + \pi\sin^2\theta(\gamma_{\mathrm{ls}} - \gamma_{\mathrm{sg}})]. \qquad (8.9)$$

By substituting equation (8.9) into the generalized Young's equation (8.8), we obtain the classical Young's equation (8.6).

If *gravity* is important, then the total energy of the system consists of the interfacial as well as the gravitational potential energy. The latter is given by $E_g = R^4 a_g(\theta)$, where $a_g(\theta)$ is a shape function. This expression was derived in (Shapiro et al., 2003a), using scaling arguments, where the shape function was also calculated. The total energy in this case is

$$E = E_i + E_g = R^2[2\pi\gamma_{\mathrm{gl}}(1-\cos\theta) + \pi\sin^2\theta(\gamma_{\mathrm{ls}} - \gamma_{\mathrm{sg}})] \qquad (8.10)$$
$$+ R^4 \frac{2\pi}{3}\rho g(3+\cos\theta)\sin^6(\theta/2).$$

By minimizing this energy, as before, we obtain the corresponding Young equation with gravity

$$\cos\theta - \frac{\gamma_{\mathrm{sg}} - \gamma_{\mathrm{ls}}}{\gamma_{\mathrm{gl}}} + \frac{R^2 \rho g}{\gamma_{\mathrm{gl}}} \left(\frac{\cos\theta/3 - \cos(2\theta)}{12} - \frac{1}{4} \right) = 0.$$

We see that in this case the contact angle depends on the radius R of the droplet, unlike the case without gravity.

8.3 Governing Equations for Thin Films

We have already presented some basic concepts for droplet formation and equilibrium equations involving surface tension as the dominant mechanism. We will present more modeling details in the following sections, but in the current section we deal with thin films. The evolution equation of a liquid film on a solid surface is derived from the incompressible Navier–Stokes equations. The fundamental assumption is that the mean thickness is much smaller than the characteristic length of the interfacial disturbance, the so-called long-wave approximation. Let us denote by z the direction along the film thickness. Then the equation for the film thickness $z = h(x,t)$ in two dimensions is

$$\mu h_t + \frac{1}{2}\left[h^2(f^x + \gamma_x)\right]_x + \frac{1}{3}\left\{h^3[(f^z - \Phi_h)_x + \gamma h_{xxx}]\right\}_x = 0, \qquad (8.11)$$

where μ is the viscosity. Here, f^x and f^z denote external interfacial forces, and Φ_h is a potential evaluated at the liquid–air interface, i.e., at $z = h$. Physically, the terms in the above equation represent the action: external forces, surface tension and its nonuniformity, gravity, and viscous damping, respectivbely starting from the second term. Specifically, the potential Φ_h may include any conservative force, e.g., gravity, centrifugal forces, or van der Waals forces. In the case of an isothermal film with constant surface tension, the above equation reduces to (Oron, 2001)

$$\mu h_t - \frac{1}{3}\rho g(h^3 h_x)_x + \frac{1}{3}\gamma(h^3 h_{xxx})_x = 0, \qquad (8.12)$$

which expresses a balance of viscous damping, gravity, and capillary forces. This is a nonlinear equation for diffusion enhanced by a dissipation term due to surface tension. It is stable, which implies that any disturbance imposed at the interface will decay very fast and the film will return ito its original shape. On the other hand, a change in the sign of the gravity term leads to a very unstable system, the well-known Rayleigh–Taylor instability, which will eventually lead to rupture of the thin film.

The *van der Waals forces* between an apolar liquid and a solid can be modeled by a standard Lennard–Jones potential of the form

$$\Phi = \alpha_3 h^{-3} - \alpha_9 h^{-9},$$

where α_j are the positive Hamaker coefficients. A different potential appropriate also for rough solid substrates, derived in (Oron and Bankoff, 1999), is

$$\Phi = \alpha_3 h^{-3} - \alpha_4 h^{-4}.$$

A simpler potential for *apolar liquids* is

$$\Phi = \frac{\alpha}{6\pi} h^{-3},$$

where positive α corresponds to an attractive force, while negative α corresponds to a repulsive force driving the interface toward a flat profile. In the case of constant surface tension and neglecting gravity, the above potential leads to the following film evolution equation:

$$\mu h_t + \frac{1}{6\pi}\alpha(h^{-1}h_x)_x + \frac{1}{3}\gamma(h^3 h_{xxx})_x = 0. \tag{8.13}$$

This equation generates solutions with very small wavelength at least initially, due to van der Waals forces. However, asymptotically, surface tension acts to cut off the smaller scales and stabilize the interface (Williams and Davis, 1982). The attractive van der Waals force included in the equation establishes naturally a precursor film, which removes the stress singularity encountered in problems with moving contact lines. Using different contact line models to relieve the stress singularity leads to some differences in the initial evolution of the thin film, but its asymptotic stability does not seem to depend on the particular model. Thefore, using a flat precursor film or a slip boundary condition or employing van der Waals forces will not change the basic dynamics obtained in long-time integration (Diez et al., 2000; Davis and Troian, 2003).

A more general equation regarding the upward thermocapillary spreading of a Newtonian liquid film in an inclined plane with angle ϕ was derived in (Oron et al., 1997), and also studied in (Davis and Troian, 2003), and it is given by

$$h_t + \left(\frac{\tau h^2}{2\mu}\right)_x - \left[\frac{\rho g \sin\phi h^3}{3\mu}\right]_x + \nabla\cdot\left[\frac{h^3}{3\mu}\nabla(-\rho g h\cos\phi + \gamma\nabla^2 h + \alpha h^{-3})\right] = 0, \tag{8.14}$$

where τ is the shear stress and x indicates the flow direction. The dynamic viscosity is assumed variable, since in some applications, such as thermocapillary pumping, the viscosity may be changing due to temperature variation. We will discuss this topic in some detail in Section 8.5.

The process of *dewetting* of a solid surface is of great interest in microfluidic applications and has been the subject of many fundamental computational and experimental studies. Theoretical work supports the idea that the dominant mechanism driving film evolution is the so-called *spinodal dewetting*, and molecular dynamics simulations support this thesis (Koplik and Banavar, 2000); however, some experiments point to nucleation of holes from defects of the surface. The governing equations in this case should include both gravity and capillary forces but also intermolecular repulsive and attractive forces. The general equation (8.11) reduces in this case to

$$\mu h_t - \frac{1}{3}\left[h^3(\rho g h - \gamma h_{xx} + \Phi)_x\right]_x = 0. \tag{8.15}$$

Small disturbances usually decay fast, but when the van der Waals forces overcome the stabilizing effect of gravity, rupture may take place.

The above modeling assumes isothermal conditions. However, surface tension is a strong function of *temperature*. Assuming a constant ambient temperature T_∞, then if the temperature at the bottom of the film T_0 is greater than T_∞, then the surface tension (difference) $\Delta\gamma$ is positive. On the other hand, if $T_\infty > T_0$, then $\Delta\gamma > 0$. The equation for the evolution of the film thickness should now include the thermocapillary stress, so equation (8.11) becomes

$$\mu h_t - \frac{1}{3}\rho g(h^3 h_x)_x + \frac{1}{3}\gamma(h^3 h_{xxx})_x + \frac{\alpha_T \Delta\gamma}{2k}(h^2 h_x)_x = 0, \tag{8.16}$$

where k, α_T are the thermal conductivity and diffusivity, respectively. In general, the effect of the thermocappilary stress is to produce an unstable interface. However, thermocapillary stresses due to internal heat generation have a stabilizing effect (Oron and Peles, 1998).

8.4 Dynamics of Capillary Spreading

In this section we consider *isothermal conditions* and examine in more detail the spreading of liquids on homogeneous substrates with roughness as well as heterogeneous substrates consisting of hydrophobic and hydrophilic stripes. Of interest is the evolution of the spreading front, i.e., its location and speed as a function of time. A classical result due to (Tanner, 1979) concerns the spreading of a Newtonian liquid droplet on a homogeneous smooth substrate. The radial advance $r(t)$ of the liquid of volume V is

$$r(t) \propto \left(\frac{\gamma}{\mu}V^3 t\right)^{1/10},$$

which shows a weak dependence on time. However, careful analysis shows that a thin precursor film of molecular dimensions is advancing at a rate proportional to $\sqrt{t}$, i.e., it follows standard diffusion dynamics.

In *nonsmooth surfaces* the spreading of liquids follows different dynamics, since *capillary wicking* of small amounts of liquids into microgrooves occurs.

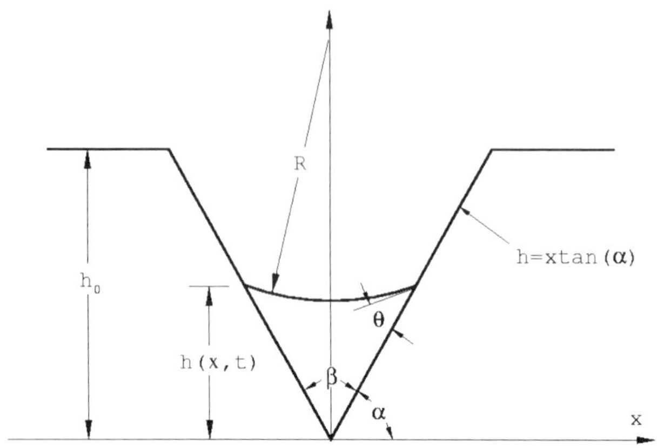

FIGURE 8.5. Sketch for liquid spreading in a V microgroove.

Capillary wicking is a well-known phenomenon that has been studied extensively, first in the pioneering work of (Washburn, 1921). Also, (Romero and Yost, 1996) performed a systematic analytical study of capillary flow into a V-shaped microgroove. A typical configuration is shown in Figure 8.5 with the flow out of the page; the groove has height h_0, the height of the liquid is denoted by $h(x,t)$, and the equilibrium contact angle is θ_0.

The pressure drop along the groove is $\Delta p = p(x) - p_0 = \gamma\kappa(x)$, where $\kappa(x)$ is the curvature and p_0 is the constant pressure above the liquid. This expression is valid if the *capillary number* $\mathrm{Ca} = U\mu/\gamma \ll 1$, which implies that surface tension forces dominate over viscous forces. For a long microgroove the curvature parallel to the flow direction is neglected, and thus

$$\kappa(x) = -\frac{1}{R(x)} = -\frac{\sin(\alpha-\theta)\tan(\alpha)}{h(x,t)},$$

computed from the sketch of Figure 8.5. Following a quasi-one-dimensional flow analysis, Romero and Yost (1996) found that the flowrate is

$$Q = -\frac{h^4(x,t)}{\mu}\Gamma(\theta,\alpha)\frac{\partial p}{\partial x},$$

which is an expression similar to that for Poiseuille flow. Also, $\Gamma(\theta,\alpha)$ is a positive function that can be approximated numerically by

$$\Gamma(\theta,\alpha) \approx \Gamma(\alpha,\alpha)\left[\frac{h_c(\theta,\alpha)}{h_0}\right]^3\left[\frac{\hat{A}(\theta,\alpha)}{\cot(\alpha)}\right]^{1/2},$$

where h_c is the height of the fluid at the middle of the groove given by

$$h_c(\theta, \alpha) = h_0 \left[1 + \cot \alpha \frac{\cos(\alpha - \theta) - 1}{\sin(\alpha - \theta)} \right].$$

The constant $\Gamma(\alpha, \alpha)$ was approximated analytically in (Romero and Yost, 1996), for the case $(\theta < \alpha)$ as follows:

$$\Gamma(\alpha, \alpha) \approx \frac{1}{6} \frac{\cot^3 \alpha + 3.4 \cot^4 \alpha + \cot^5 \alpha}{1 + 3.4 \cot \alpha + 4 \cot^2 \alpha + 3.4 \cot^3 \alpha + \cot^4 \alpha}.$$

Finally, the function $\hat{A}(\theta, \alpha)$ is related to the cross-sectional area $A(x, t)$ of the liquid inside the groove, i.e.,

$$A(x, t) = h^2(x, t) \hat{A}(\theta(x, t), \alpha),$$

where

$$\hat{A}(\theta, \alpha) = \frac{\sin^2(\alpha - \theta) \tan \alpha - (\alpha - \theta) + \sin(\alpha - \theta) \cos(\alpha - \theta)}{\tan^2 \alpha \sin^2(\alpha - \theta)}.$$

The following equation for the height of the liquid in the groove is derived from the quasi one-dimensional continuity and momentum equations:

$$\frac{\partial h^2(x, t)}{\partial t} = \frac{D}{h_0} \frac{\partial}{\partial x} \left[h^2(x, t) \frac{\partial h(x, t)}{\partial x} \right], \qquad (8.17)$$

where the diffusion coefficient is $D = \gamma h_0 / \mu \kappa(\theta_0, \alpha)$ with

$$\kappa(\theta_0, \alpha) = \frac{\Gamma(\theta_0, \alpha) \sin(\alpha - \theta_0) \tan \alpha}{\hat{A}(\theta_0, \alpha)}.$$

This is a nonlinear diffusion equation for $h^2(x, t)$. The diffusion coefficient is positive if $\alpha > \theta_0$, which is equivalent to having a concave free surface; for $\alpha < \theta_0$ no capillary wicking takes place.

Equation (8.17) was solved in (Romero and Yost, 1996), using similarity variables for various conditions. For the simple case $h(0, t) = h_0$ we have

$$h(x, t) = h_0 \Phi(\eta) \quad \text{and} \quad \eta = \frac{x}{\sqrt{Dt}}.$$

An approximate solution was obtained for this case in the form

$$\Phi(\eta) = -\frac{1}{2} \eta_0 (\eta - \eta_0) - \frac{1}{6} (\eta - \eta_0)^2 - \frac{1}{108 \eta_0} (\eta - \eta_0)^3 + \cdots,$$

where η_0 is the location where the solution goes to zero and stays zero. It was found by (Romero and Yost, 1996) that for small positive values of $(\alpha - \theta_0)$, we have $\eta_0 \approx 2.272$, while for large values of $(\alpha - \theta_0)$ the value of

η_0 can be increased up to 40%. The main result of the analysis of Romero and Yost for V-grooves as well as grooves of more complex cross-sectional areas is that diffusion dynamics dominate, and thus the leading edge (i.e., spreading front) of the fluid propagates as

$$x_0(t) = \eta_0(\theta_0, \alpha)\sqrt{Dt}.$$

This expression shows that spreading increases proportionally with the depth of the groove h_0, since $D \propto h_0$. Also, the self-similar solution implies that the free surface of the liquid spreading remains self-similar downstream.

Similar work was presented in (Darhuber et al., 2001), where the dynamics of capillary spreading along hydrophilic microstripes were studied numerically and experimentally. The surface was smooth but was processed chemically in order to create narrow hydrophilic stripes on a hydrophobic background. Following an analysis similar to the work of (Romero and Yost, 1996), it was found that self-similar solutions also exist for the microstripes, and these can be obtained from the equation

$$\frac{d}{d\eta}\left(\Phi^3 \frac{d\Phi}{d\eta}\right) + \frac{\eta}{2}\frac{d\Phi}{d\eta} = 0,$$

where the similarity variables are defined in the same way as before. The approximate self-similar solution in this case is

$$\Phi(\eta) \approx (1 - \eta/\eta_0)^{1/3}.$$

where $\eta_0 \approx 0.87$. Clearly, the spreading front advances as $x(t) \propto \sqrt{Dt}$, where $D = (64\gamma h_0^3)/(35\mu w^2)$ with w the width of the microstripe. The average streamwise velocity, however, is proportional to

$$U \propto \frac{\gamma w^4}{\mu}\frac{1}{x}.$$

Therefore, the spreading speed is proportional to w^4 and decays downstream; experimental results with microstripes of widths varying from 200 to 800 μm verified the theoretical results; see (Darhuber et al., 2001), for more details.

8.5 Thermocapillary Pumping

We now discuss capilary spreading in the presence of thermal gradients. The development of thermocapillary stress at the air–liquid interface of a thin film gives rise to fluid motion. Since $d\gamma/dT$ is constant and negative

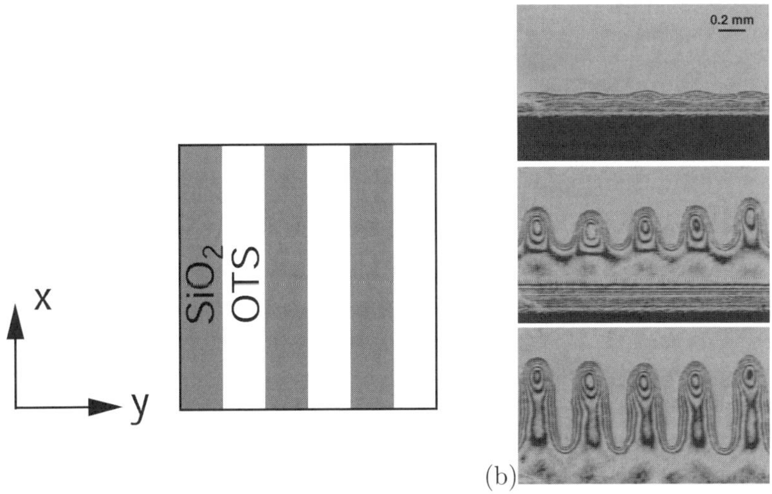

FIGURE 8.6. Schematic of the alternating SiO_2/OTS microstripes in (a), and visualizations of a silicone-oil film spreading on the patterned silicon wafer in (b). The three interferograms were taken five minutes apart. (Courtesy of S. Troian.)

for most liquids, by applying a constant thermal gradient a *constant* shear stress is produced given by

$$\tau = \frac{d\gamma}{dx} = \frac{d\gamma}{dT}\frac{dT}{dx}$$

along the x-direction from cold to hot regions. Such a spreading of the fluid on a homogeneous substrate is subject to *fingering*-type instabilities similar to the phenomena observed in flows driven by gravity or centrifugal forces. In particular, the spreading front develops a capillary ridge, which becomes unstable in the presence of infinitesimal disturbances of a certain wavelength. The parallel small rivulets that form after the instability occurs have a characteristic wavelength

$$\lambda_c = 18\left(\frac{2\gamma h_0^2}{3\tau}\right)^{1/3} \propto \frac{h_0}{(3\mathrm{Ca})^{1/3}},$$

where h_0 is the film thickness, and the *capillary number* Ca is defined here as $\mathrm{Ca} = \tau h_0/(2\gamma)$. In practice, this instability is realizable, but because of surface defects or impurities the rivulets are not stationary, with a large variation in the value of λ_c up to 30%; see (Kataoka and Troian, 1999), and references therein.

In order to drive the flow along specified pathways it is possible to pattern the surface with microstripes that have a different contact angle from that of the rest of the surface. This was attempted in (Kataoka and Troian, 1999), where alternating stripes of bare or oxidized SiO_2 coated with a bonded monolayer of octadecyltrichlorosilane (OTS) were patterned on a wafer using a spin-coating technique; see the sketch of Figure 8.6. Of fundamental interest here is to examine whether the wavelength imposed by fabrication dominates over the wavelength of the instability λ_c. In (Kataoka and Troian, 1999), the thermocapillary spreading of a silicone oil was investigated on stripes varying from 100 μm to 500 μm. A typical result from the experiments is shown in Figure 8.6 displaying visualizations of the silicone oil on alternating 200 μm stripes subject to $\tau = 0.8$ dyn/cm^2. The film thickness observed was less than 1 μm. Unlike spreading over homogeneous substrates that requires a long incubation period, spreading over heterogeneous substrates is spontaneous, and the fluid is immediately channeled into the hydrophilic stripes. The average speed of tips and troughs for the flow visualization shown in Figure 8.6 was 2.60 μm/s, and 1.73 μm/s on the hydrophilic stripe with the troughs advancing more slowly on the OTS stripes. This is typically three orders of magnitude *less* than the induced speed in an electrocapillary; see the next section. It was found in (Kataoka and Troian, 1999), that on the patterned surface the periodicity of the spreading front follows the periodicity imposed by the pattern unless the width of the hydrophilic stripe is below about 50 μm.

In general, for *heterogeneous* substrates the wetting of a liquid on a solid is quite complex. If a liquid droplet is residing partially on a hydrophilic stripe and partially on a hydrophobic patch, there may be a variation of the contact angle along the contact line. Typically, the droplet may migrate to regions with small contact angle. This was demonstrated in the experiments of (Darhuber et al., 2000) with glycerol on patterned silicon surfaces. The chemical modulation leading to alternating hydrophilic and hydrophobic stripes was achieved using a self-assembled monolayer of octadecyltrichlorosilane (OTS) 3 nm in thickness. The contact angle of glycerol on OTS was measured in (Darhuber et al., 2000), to be about 95°, while that on SiO_2 was less than 5°. The hydrophilic SiO_2 stripes in the experiments were 7 to 15 μm in width, while the width of the microstructures formed varied between 10 and 47 μm. If the heterogeneous stripes are much larger than the droplet radius, then the contact angle will take the appropriate value defined by Young's equation for the relevant stripe.

In (Gau et al., 1999), the equilibrium shape of a liquid droplet in contact with a completely hydrophilic surface was examined as a function of the liquid volume per unit length. It was found that if it exceeds a critical value, the droplet forms a pronounced bulge along the contact line. In general, the shape and distribution of liquid droplets on a microchannel wall depend on the contact angle, the surface tension, and as the volume of liquid deposited. If the amount of liquid deposited on a relatively long microchannel is subcritical, then the contact lines will stay within the hydrophilic stripe, and it will behave as if the substrate were homogeneous; i.e., it is susceptible to the natural instability we described earlier manifested as capillary

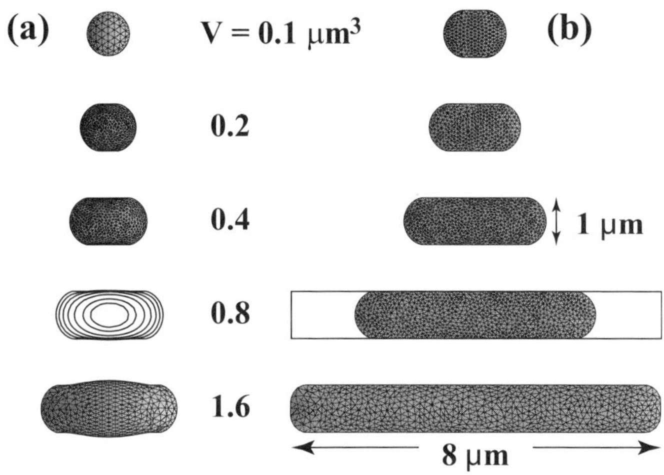

FIGURE 8.7. Top view profiles of liquid droplets as a function of the liquid volume for $\theta = 60°$ in (a) and $\theta = 30°$ in (b). (Courtesy of S. Troian.)

breakup. The critical value of the volume per unit length of the stripe is (Darhuber et al., 2000)

$$\frac{V}{l} \leq \frac{w^2}{4\sin^2\theta}\left(\theta - \frac{1}{2}\sin 2\theta\right),$$

where w is the channel width and the contact angle θ is measured in radians. Exceeding the critical value of V/l will effectively increase the value of the contact angle, and thus it may delay breakup, but it may create bulges, as in the work of (Gau et al., 1999).

Such effects were studied by (Darhuber et al., 2000), who performed numerical simulations using the program *SURFACE EVOLVER*, which is based on energy minimization techniques; this program was developed by (Brakke, 1992). The liquid surface was triangulated, and the total energy of the system was expressed as a function of the vertices of the triangular elements. Standard minimization techniques were employed (e.g., conjugate gradients), and different boundary conditions were incorporated in the minimization process. Curved interfaces produce a pressure excess inside the liquid (see equation (8.2)). Figure 8.7 shows profiles from the simulations of Darhuber et al. for a single microdroplet residing on a $1\,\mu\text{m} \times 8\,\mu\text{m}$ hydrophilic region for contact angles $\theta = 60°$ and $\theta = 30°$, while the rest of the surrounding area is completely hydrophobic, i.e., $\theta = 180°$. For the same volume of liquid, the smaller liquid with contact angle spreads faster, as expected. As the volume of the microdroplet increases the liquid fills up the stripe but remains confined within the stripe, although it forms a bulge for the larger contact angle. This implies incomplete wetting of the hydrophilic strip, which can be interpreted as surface energy imbalance between the liquid–vapor and the liquid–solid systems, with the latter acquiring higher energy levels. The morphological features of liquid microdroplets residing over heterogeneous substrates depend critically on the surface tension value. In (Darhuber et al., 2000), parametric studies for different surface tensions but constant liquid volume were also performed. By reducing the surface tension, complete spreading of the microdroplet over the hydrophilic patch is obtained. This is expected, since reduction in the surface tension leads to a reduction of the corresponding contact angle; see equation (8.6).

Using the above concepts, self-assembled monolayer chemistry was used in (Zhao et al., 2001), to pattern surface-free energies in microchannel networks. In particular, it was found that when the pressure was maintained at sufficiently low levels, the liquid followed the hydrophilic pathways but above a threshold, the liquid crossed the boundary between the hydrophilic and hydrophobic regions. Two liquid streams separated by a gas membrane were transported side by side, allowing reactive components to be diffused from one stream to the other. This type of configuration can be used as a network of microchannels with "virtual" walls. The condition for rupture of the virtual wall is that the angle θ_b at the hydrophilic–hydrophobic boundary be equal to the (advancing) contact angle of the liquid on the nonpolar surface θ_n. The maximum pressure that these virtual walls can sustain for a straight stream is

$$\Delta P_{\max} = \frac{2\gamma}{h} \sin(\theta_n - 90°),$$

where h is the film thickness. For curved pathways there is a limit on the curvature of the flow determined by the virtual wall rupture condition and the extra pressure difference due to curvature. An ultraviolet photopattering method was developed in (Zhao et al., 2001), to pattern surface-free energies inside microchannels in situ within minutes, and applications for gas–liquid reactions in microchips as well as for pressure-sensitive switches were demonstrated. Virtual walls can lead to multiple and diverse functionality on microchips that may be difficult to achieve with other methods.

8.6 Electrocapillary

Similar to thermocapillary, where temperature is the controlling mechanism, in electrocapillary, or electrowetting electric potential can be used to change the surface tension and thus cause flow motion. Compared to thermocapillary, electrocapillary is much more energy efficient (power consumption is about 10 mW), and various applications have demonstrated induced speeds over 100 mm/s, in contrast to less than about 1 mm/s in thermocapillary. Some of the applications have demonstrated addressable liquid handling, i.e., droplet routing but also droplet cutting and merging

(Cho et al., 2003). Other applications include optical switches (Beni et al., 1982), rotating micromotors (Lee and Kim, 2000), and liquid lenses (Kwon and Lee, 2001). From the fundamental standpoint, electrowetting is the low-frequency limit of the electromechanical response of an aqueous liquid to an electric field. In contrast, dielectrophoresis can be thought of as the high-frequency limit. (We note that, in general, the phenomenon of dielectrophoresis requires only a nonuniform field, and it exists also under DC conditions.) An analysis of the two mechanisms and a related experiment were reported in (Jones et al., 2003).

The basic electrowetting mechanism was discovered over a century ago by (Lippmann, 1875), who observed that electrostatic charge may modify capillary forces. This principle was introduced to microfluidics by (Matsumoto and Colgate, 1990), who attempted to design micropumps based on controlling surface tension using electric potentials. Perhaps the simplest implementation is *continuous electrowetting* (CEW), which uses liquid metal droplets (e.g., mercury) in a filler liquid, thus requiring two liquids in the system. CEW exploits changes in the surface tension of the liquid metal–liquid (electrolyte) system at the liquid–solid–gas interface. This is demonstrated in the sketch of Figure 8.8(a) adopted from the work of (Lee et al., 2002). Flow motion is induced due to the different surface tension at the two menisci caused by the electric potential.

Two other forms of electrocapillary are electrowetting (EW) and electrowetting-on-dielectric (EWOD). Unlike CEW, these methods can be used to move aqueous liquid in microchannels without the presence of a second liquid medium, so the droplet in this case is in contact with air, just as in thermocapillary. EW and EWOD exploit *changes* in the solid–liquid surface tension, which in turn changes the contact angle. Specifically, in EW an electrical double layer (EDL) is formed between the electrode and aqueous solution that is between 1 nm and 10 nm thick. Applying a voltage difference as shown in the sketch of Figure 8.8(b) may cause a hydrophobic surface to behave like a hydrophilic one. In essence, the electric energy counterbalances the free surface energy and lowers the surface tension $\gamma_{\rm sl}$. In EWOD there is no electric double layer, but the change in the energy balance takes place in the hydrophobic dielectric layer; in (Lee et al., 2002), a Teflon layer 0.8 μm thick was used. The basic principle is demonstrated in the sketch of Figure 8.8(c) adopted from the work of (Lee et al., 2002).

Thermodynamic analysis at the interface leads to the so-called *Lippmann's* equation

$$\gamma = \gamma_0 - \frac{1}{2} c V^2, \tag{8.18}$$

where V is the voltage difference, γ_0 is the surface tension at zero voltage, and c is the capacitance per unit area. This equation is applicable to all three forms of electrowetting, but c relates to the EDL for CEW and EW, and to the dielectric layer for EWOD. Also, for CEW, γ is the interfacial surface tension between the two media, whereas for EW and EWOD,

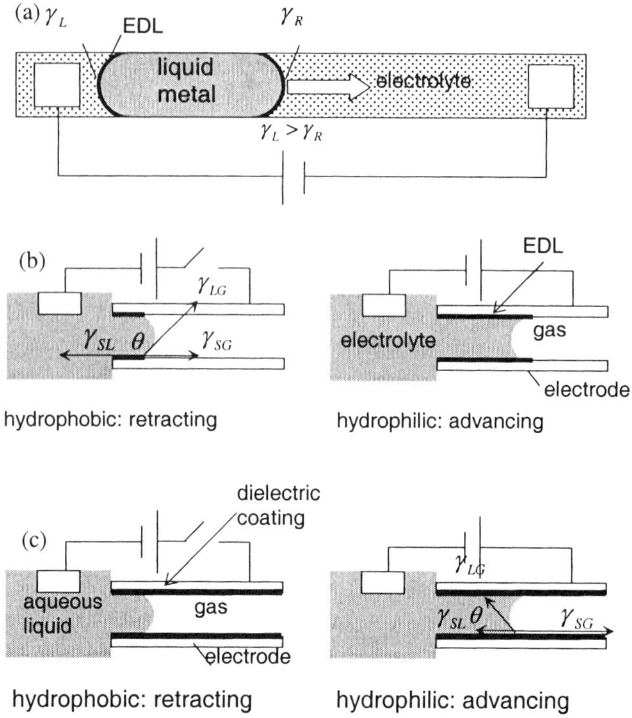

FIGURE 8.8. Electrocapillary principles for (a) continuous electrowetting; (b) electrowetting, and (c) electrowetting-on-dielectric (Lee et al., 2002). (Courtesy of C.-J. Kim.)

$\gamma = \gamma_{sl}$ is the solid–liquid surface tension. In the following, we will derive the generalized Young–Lippmann equation that governs equilibrium based on energy minimization principles. An alternative derivation for constant surface tension and neglecting gravity is obtained using Young's equation of equilibrium at the triple contact line (equation (8.6)) and substituting the Lippmann equation (8.18) to get

$$\cos\theta = \cos\theta_0 + \frac{cV^2}{2\gamma_{\text{gl}}}, \tag{8.19}$$

where θ_0 is the zero-voltage contact angle. We note the strong dependence of the contact angle on the voltage, which may lead to a switch between a hydrophobic and hydrophilic surface, and also its independence on DC or AC current. The above equation is a special case of the generalized

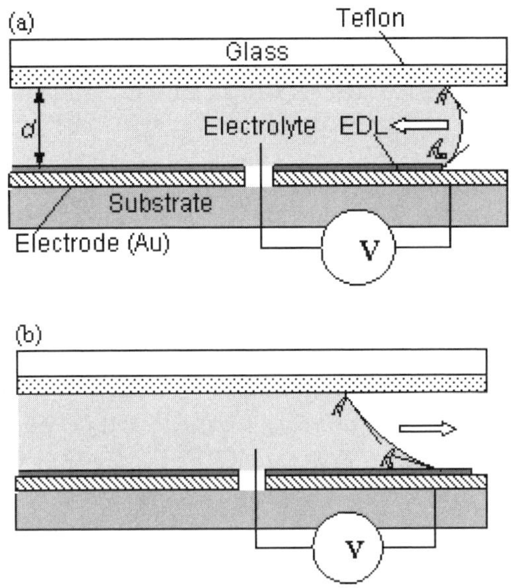

FIGURE 8.9. Electrowetting (EW) device with electric potential *off* (a) and *on* (b) corresponding to a hydrophobic and hydrophilic lower surface, respectively (Lee et al., 2002). (Courtesy of C.-J. Kim.)

Young–Lippmann equation that we will study in the next section.

A proof-of-concept experiment for EW was conducted in (Lee et al., 2002), with a liquid droplet (Na_2SO_4) squeezed between an electrode and a cover glass, as shown in the schematic of Figure 8.9, subject to about 1 volt. Without the cover glass, microdroplets evaporate very quickly. The contact angle on the cover glass (coated with a hydrophobic layer) is not changed, but the wettability of the lower surface is changed by the electric potential. The pressure difference inside the liquid droplet and the air $(p - p_a)$ can be computed from Laplace's equation, assuming a spherical geometry, to obtain

$$p - p_a = -\frac{\gamma_{\rm gl}}{d}(\cos\theta + \cos\theta_b),$$

where d is the microchannel height.

Similarly, a proof-of-concept experiment was conducted in (Lee et al., 2002), to test EWOD. A water droplet was placed on a Teflon-coated surface (hydrophobic), and it was demonstrated that upon applying a voltage of about 100 V, the surface switched to hydrophilic, causing spreading of the droplet. To compute the pressure build-up in this case we refer to the schematic of Figure 8.10 adopted from (Lee et al., 2002). On the left end of the meniscus, we have

$$p_L - p_a = -\frac{\gamma_{\rm gl}}{d}(\cos\theta + \cos\theta_{b0}) > 0,$$

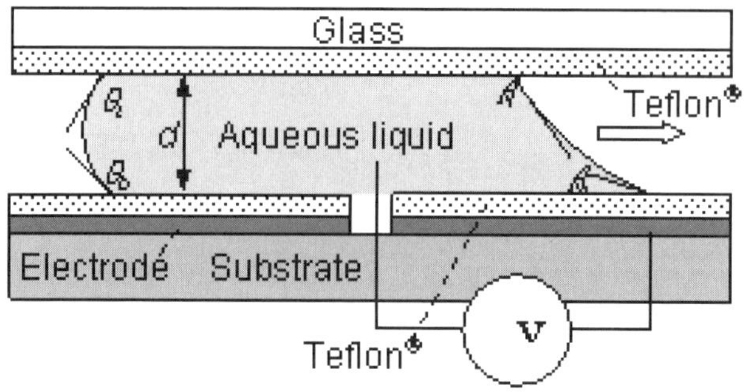

FIGURE 8.10. Electrowetting-on-dielectric (EWOD) device with electric potential *on* corresponding to a hydrophilic lower surface (Lee et al., 2002). (Courtesy of C.-J. Kim.)

while on the right end of the meniscus, we have

$$p_R - p_a = -\frac{\gamma_{\rm gl}}{d}(\cos\theta + \cos\theta_b) < 0,$$

where d is the channel height, which is equal to the droplet height. The pressure difference within the liquid droplet is then

$$p_L - p_R = \frac{\gamma_{\rm gl}}{d}(\cos\theta_b - \cos\theta_{b0}),$$

which shows that the surface angle on the cover glass does not influence the spreading process. Using the Young–Lippmann equation (8.19) we can eliminate the contact angles alltogether to obtain

$$p_L - p_R = \frac{cV^2}{2d} = \frac{\epsilon_0 \epsilon V^2}{2td},$$

where t is the thickness of Teflon (1 μm in (Lee et al., 2002)), $d = 10\,\mu$m, $\epsilon_0 = 8.854\times10^{-12}$ C/V is the vacuum permittivity, and $\epsilon = 2.0$ is the Teflon permittivity. As pointed out in (Lee et al., 2002), it is somewhat surprising to see, at least for this application, that the surface tension is not present in the above equation, although flow motion occurs due to surface tension changes between the liquid and the solid! Clearly, if hysteresis effects are present, then $\gamma_{\rm sl}$ will be involved explicitly in the above pressure equation. With regards to efficiency, a large voltage is required in EWOD (about 100 volts compared to EW of about 1 volt) due to the relatively thick layer of the Teflon dielectric layer. For a very thin dielectric layer on the order of 0.1 μm or less, the required voltage is about 20 volts (Lee et al., 2002).

In summary, the EW method is appropriate for electrolytes and is energetically quite efficient. In contrast, the EWOD method can handle any aqueous liquids, but it requires substantially higher voltage. Switching between hydrophobic and hydrophilic surfaces using electric potential is, in principle, a reversible process, but other parasitic effects may affect reversibility. It was reported in (Verheijen and Prins, 1999), that charge trapping may degrade the surface and affect electrowetting adversely. However, it was recommended that silicone oil be used to treat the surface in order to minimize the undesired contact angle hysteresis caused by the trapped charge. Another limitation of electrowetting is the saturation in the change of contact angle caused by applying an electrical potential. We discuss this in the next section, but we first present a more rigorous derivation of the Young–Lippmann equation.

8.6.1 Generalized Young–Lippmann Equation

The Young–Lippmann equation can be derived rigorously based on the energy minimization equation that leads to the generalized Young equation (8.8). To this end, following the analysis of (Shapiro et al., 2003a), we consider a conducting liquid droplet residing on a dielectric solid and obtain the total energy of the system. It consists of the *interfacial energy* given in equation (8.9) as well as the *dielectric energy* stored in the solid and the externally applied charging source. The latter is twice the energy stored in the dielectric solid but with opposite sign. Therefore, we need to compute the potential energy only in the very thin solid dielectric layer. An electric field almost perpendicular to the surface area A_{ls} is proportional to the voltage V and inversely proportional to the layer thickness h. The energy stored in the solid dielectric is then

$$E_{de} = \frac{1}{2}\epsilon_s (V/h)^2 h A_{\mathrm{ls}} = \pi R^2 \frac{\epsilon_s V^2}{2h} \sin^2\theta,$$

where ϵ_s is the dielectric constant of the solid. The total energy in this case is then

$$E(R,\theta) = R^2 \left(\pi \sin^2\theta(\gamma_{ls} - \gamma_{\mathrm{sg}} - \epsilon_s V^2/(2h)) + 2\pi\gamma_{\mathrm{gl}}(1-\cos\theta)\right),$$

which upon minimization leads to the Young–Lippmann equation (Lippmann, 1875)

$$\cos\theta - \frac{\gamma_{\mathrm{sg}} - \gamma_{\mathrm{ls}}}{\gamma_{\mathrm{gl}}} - \epsilon_s \frac{V^2}{2\gamma_{\mathrm{gl}} h} = 0. \qquad (8.20)$$

We note that this equation does not depend on the radius of the droplet R, as in the standard Young equation that expresses triple-line force balance (equation (8.6)).

The above analysis is not valid for apolar liquids, e.g., a silicone oil atop a conducting solid surface. This case is analyzed in (Shapiro et al., 2003a), assuming that the droplet is an insulator with dielectric constant ϵ_l. The electric field varies in this case as V/R, so the stored electric energy is

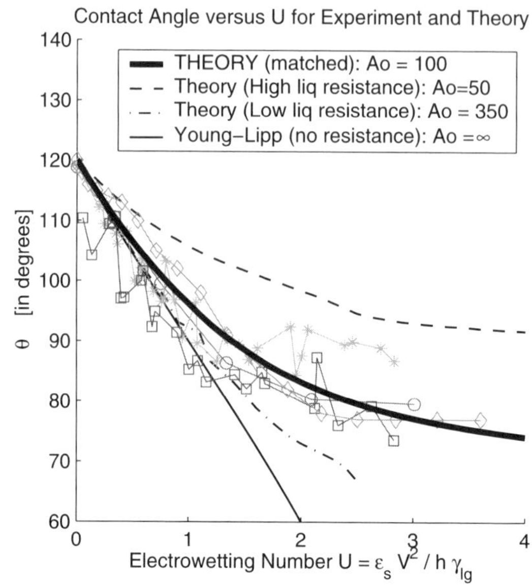

FIGURE 8.11. Contact angle saturation: The experimental data are taken from four different devices (see (Moon et al., 2002)). The theoretical curves are taken from (Shapiro et al., 2003a). They show energy-minimization-based results for three different values of the electrical resistivity in the liquid. The middle value is appropriate for water and closely matches the observed experimental results. (Courtesy of B. Shapiro.)

$1/2\epsilon_l R^3(V/R)^2 a_{\rm de}(\theta) \propto R$. Here, $a_{\rm de}(\theta)$ is the shape factor, which cannot be computed analytically as for the previous cases; however, the following empirical fit was developed in (Shapiro et al., 2003a):

$$a_{\rm de}(\theta) \approx 0.0592 + 0.0012\theta + 0.0022\tan(1.71 - \theta),$$

where θ is in radians. The following Young equation is obtained for this case by energy minimization:

$$\cos\theta - \frac{\gamma_{\rm sg} - \gamma_{\rm ls}}{\gamma_{\rm gl}} + \epsilon_l \frac{V^2}{2\gamma_{\rm gl} R} \frac{2 + \cos\theta}{2\pi \sin\theta} \left(a_{de}(\theta) q(\theta) + \frac{da_{\rm de}}{d\theta} \right) = 0. \qquad (8.21)$$

The quantity $q(\theta)$ has been defined in equation (8.7). We note that there is $1/R$ dependence in this equation, unlike all the other cases we have presented so far. In general, the contact angle variation with the voltage is very weak for dielectric droplets.

A more realistic model for the microdroplets is that of an imperfect conductor residing on an imperfect insulator. This case was studied numerically in (Shapiro et al., 2003a), and it leads to the *contact angle saturation* phenomenon observed in experiments. Many physical mechanisms can cause such saturation, including charge-trapping, liquid resistivity, and electrolysis. The model developed in (Shapiro et al., 2003a), identifies *liquid resistivity* as the leading cause of contact angle saturation. For sufficiently small θ the interfacial energy beats the electrical energy, and thus the total energy goes to infinity; i.e., a minimum is never achieved around the complete wetting state. This model agrees very well with the experimental data reported in (Moon et al., 2002), where a single and a double layer of Teflon and silicon dioxide were used for the solid dielectric layer. In Figure 8.11 a comparison of the numerical results of Shapiro et al. is presented against several sets of experimental data for a certain value of the solid-to-liquid resistance. If the value of this ratio is infinity, then we recover the perfect conducting droplet described by the standard Young–Lippmann formula; however, no contact angle saturation is predicted for that case. The nondimensional voltage used in the plot is defined as $U = \epsilon_s V^2/(h\gamma_{\rm gl})$, where h is the thickness of the dielectric layer. Also, A_0 measures the resistivity ratio of solid to liquid.

8.6.2 Optoelectrowetting

Implementation of electrowetting for multichannel liquid networks requires a large number of electrodes, often leading to a wiring bottleneck in two-dimensional arrays. An alternative but conceptually similar approach is *optoelectrowetting* (OEW), which was first studied in (Ichimura et al., 2000; Chiou et al., 2003). The basic mechanism is shown in the sketch of Figure 8.12, where a photoconductive material is placed under the dielectric layer of a standard electrowetting setup. The contact angle can still be computed by the Young–Lippmann equation, i.e.,

$$\cos\theta = \cos\theta_0 + \frac{\epsilon V^2}{2\gamma d},$$

where d is the thickness of the dielectric layer. In the *dark state*, i.e., no light source, the frequency of the AC curent is controlled so that the photoconductor dominates, and thus through a voltage divider the voltage drop will occur across the photoconducting layer. When a light source is present, the conductivity of photoconductor increases by orders of magnitude, and consequently, the voltage drop is mainly in the dielectric layer. The material used in the experiments of (Chiou et al., 2003) was *amorphous silicon* because of its low dark conductivity and visible light response. Its conductivity increased by almost two orders of magnitude with light intensity of 65 mW/cm^2.

The liquid droplet in the demonstration experiments of (Chiou et al., 2003) was deionized water, which was placed between a top hydrophobic surface (indium tin oxide coated with Teflon), while the bottom surface of amorphous silicon was also coated with Teflon. An electrode grid was placed under the photoconducting layer. By applying light on one end of the microdroplet, the contact angle decreases and motion is induced. Velocities up

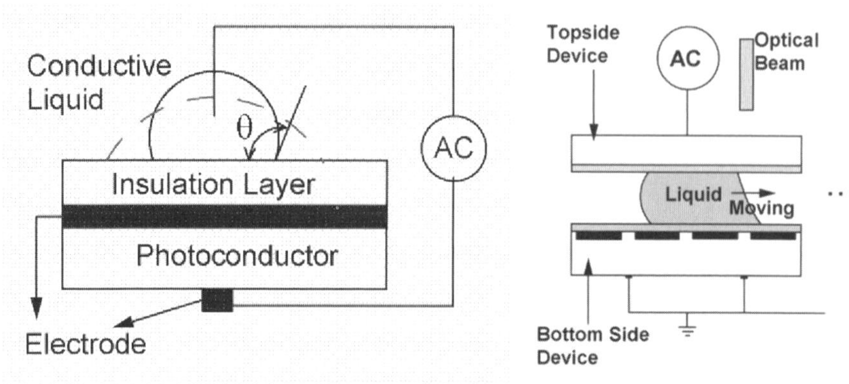

FIGURE 8.12. Sketches illustrating the basic principle of light actuation of liquid droplets (Optoelectrowetting). (Courtesy of M.C. Wu.)

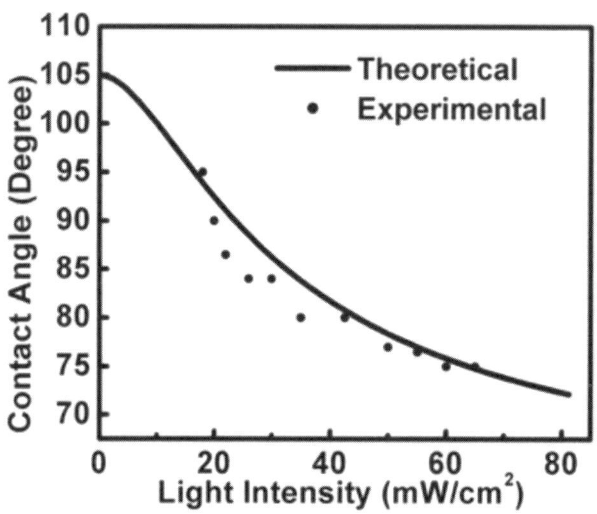

FIGURE 8.13. Optoelectrowetting: Contact angle of a water droplet versus light intensity (Chiou et al., 2003). (Courtesy of M.C. Wu.)

to 7 mm/s were recorded, which are much higher than velocities achieved by thermocapillary but lower than standard electrowetting. The liquid microdroplet follows the path of the laser beam (4 mW power). A change in contact angle up to 30° was demonstrated in (Chiou et al., 2003). However, the saturation phenomenon encountered in conventional electrowetting was present here too. In Figure 8.13 we plot the measured contact angle as a function of the light intensity obtained in the experiments of (Chiou et al., 2003). The results are in agreement with the Young–Lippmann formula if

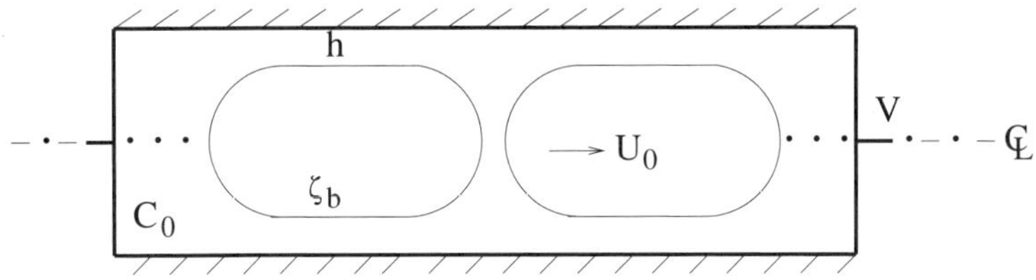

FIGURE 8.14. Sketch of bubble trains and details in the gap. Here V denotes voltage, h is the annular gap size, ζ_b is the zeta potential, U_0 is the bubble velocity, and C_0 is the KCL concentration.

we substitute $V = 70$ (rms value), $d = 1\ \mu$m, $\epsilon = 4 \times 8.854 \times 10^{-12}$ F/m, and $\gamma = 73$ N/m; the saturation angle is about 75°.

This technique, which exploits light actuation combined with electrowetting, enables a large number of electrodes to be addressed for multifunctioning lab-on-a-chip operation without wiring bottlenecks.

8.7 Bubble Transport in Capillaries

Transport of long bubbles and organic liquid droplets is required in microgenerators, microreactors, and drug delivery applications. The droplets, for example, can be drugs, while air bubbles can be used as spacers to separate samples along a pathway in a network of microchannels. The air bubbles can also be used as pumps, e.g., as pistons that drive the flow, or as valves as in ink-jet printers. The motion of droplets and bubbles in pipes has been studied in the classical work of (Bretherton, 1961) and in many subsequent papers; see, for example, (Ratulowski and Chang, 1989), and references therein. In microfluidic applications it is interesting to determine the maximum speed of transporting bubbles in microchannels and to understand the physical mechanisms that control this transport.

Let us consider a bubble in a micropipe immersed in a wetting liquid, assuming also that gravity is negligible (i.e., small Bond number limit). The bubble is then axisymmetric within the capillary, and a wetting annular film is formed around the bubble and between the capillary walls. The film thickness h is very small, so we can employ the standard lubrication limit to approximate the flow in the film. Thus, the velocity profile is

$$u(y) = -\frac{\gamma}{\mu}\frac{\partial^3 h}{\partial x^3} y(y/2 - h),$$

where y is measured from the wall of the micropipe; also, the pressure is constant across the film thickness but varies in the longitudinal direction x. In the presence of (insoluble) surfactants the interface between the bubble and the film behaves, as rigid wall, and thus in this case the velocity profile should be modified accordingly, i.e.,

$$u(y) = -\frac{\gamma}{\mu}\frac{\partial^3 h}{\partial x^3} y \left(\frac{y}{2} - \frac{h}{2}\right),$$

so it satisfies no-slip boundary conditions. Typically, the bubble is elongated so the length to diameter ratio l_b/d is grater than 1, and the bubble consists of two caps and a middle parallel section, as shown in the sketch of Figure 8.14. If the bubble is symmetric, i.e., the front and back caps are identical, then there is no capillary pressure difference, and thus the bubble is not moving. Therefore, for mobile bubbles there is a large pressure in the back cap and a smaller pressure in the front, and this pressure difference pushes the bubbles into the annular liquid wetting film.

In order to estimate the corresponding pressure drop we employ the Laplace–Young equation. The two curvatures in this problem are the curvature in the caps, which is approximately 1/R (with $R = d/2$ the capillary radius), and the film (axial) curvature h_{xx}, which is nonzero in the transition regions, i.e., between the caps and the middle parallel section of the bubble. Balancing the curvatures in the transition region, we obtain that

$$x \propto \sqrt{Rh}.$$

On the other hand, using the momentum equation and balancing viscous stresses with pressure drop, we obtain the classical *Bretherton scaling*. Specifically,

$$\frac{\mu U}{h^2} \propto \frac{dp}{dx} \propto \frac{\gamma}{Rx} \Rightarrow \frac{h^2}{Rx} \propto \frac{\mu U}{\gamma} = \mathrm{Ca},$$

where the last equation defines the *capillary number*. Combining the above two equations, we obtain

$$h \propto R\mathrm{Ca}^{2/3} \quad \text{and} \quad x \propto R\mathrm{Ca}^{1/3}.$$

Typical bubble speeds are in the range of 1 μm/s to 1 mm/s, and thus for aqueous solutions we have that the corresponding capillary number is 10^{-7} to 10^{-4}. This, in turn, implies that using the above Bretherton scalings, the film thickness is about 10,000 times smaller then the radius of the micropipe R, while the transition layer is about 100 times less than R.

Bretherton has used asymptotic analysis to obtain accurate expressions for the above scalings including the coefficients. The starting point is to use the quasi-one-dimensional mass conservation equation

$$\frac{\partial h}{\partial t} + \frac{\partial Q}{\partial x} = 0,$$

where $Q = (\gamma h^3 h_{xxx})/(3\mu)$ is the flowrate across the film obtained from the parabolic velocity profile. In the Lagrangian reference frame moving with the bubble velocity we have that $h_t = -U h_x$, and integrating the above equation from the middle point (where $h \to h_\infty$) to the transition region, we obtain

$$\mathrm{Ca}^{-1} h^3 h_{xx} = 3(h - h_\infty).$$

Defining now $\tilde{h} = h/h_\infty$ and $\xi = (x/h_\infty)(3\mathrm{Ca})^{1/3}$, we obtain the *Bretherton equation*

$$\tilde{h}^3 h_{\xi\xi\xi} = \tilde{h} - 1. \tag{8.22}$$

This nonlinear equation can be solved either analytically or numerically, and its solutions need to be matched with the static solutions of the Laplace Young equation (8.2) at the ends of the transition region, i.e., at the two caps. The solution is different at the two limits, i.e., as $\xi \to \pm\infty$, and this asymmetry at the back and front caps produces the pressure difference that drives the bubbles into the wetting annular liquid film.

The above solution gives the film thickness at the midsection of the bubble $h_\infty = 0.64R(3\mathrm{Ca})^{2/3}$, and also the pressure difference between the back and front caps

$$\Delta p = 10\frac{R}{\gamma}\mathrm{Ca}^{2/3}.$$

This equation has been modified for large values of capillary number, $10^{-1} > \mathrm{Ca} > 10^{-2}$, in (Ratulowski and Chang, 1989), as follows:

$$\Delta p = \frac{R}{\gamma}[10\mathrm{Ca}^{2/3} - 12.6\mathrm{Ca}^{0.95}],$$

but this correction is negligible for microfluidic applications, where the capillary number is very small. The pressure estimate helps us in quantifying the length scales in applications in which a *train of spacer bubbles* is used to transport slugs or drops of liquid drugs. Clearly, the pressure drop Δp along the length L of the liquid slug is equal to the difference between the pressure at the front cap and the pressure at the back of the leading bubble, which is given by the above expression. We also have that

$$\frac{\Delta p}{L} \propto \frac{\mu U}{R^2} \Rightarrow L \propto R\mathrm{Ca}^{-1/3},$$

so the above scaling shows that the size of the liquid slug is about 100 times the size of the micropipe radius. The aforementioned scalings are realizable in experiments. However, surfactant contaminants, which are particularly effective at low Ca, may affect the results. In (Ratulowski and Chang, 1990), a comparison between theory and experiments is presented, and corrections due to Marangoni traction at the liquid–air interface are proposed. Also, for noncircular cross-section, the Bretherton–Chang theoretical solutions are not valid, and numerical solutions should be obtained using the formulation outlined in (Ratulowski and Chang, 1989).

Transport of microbubbles can be achieved either by pressure-driven or electrokinetically driven flows. It may be more efficient to build up back pressure on the bubbles using the latter. However, it has been found in experiments in (Takhistov et al., 2002), that air bubbles were stationary even though the electrokinetic velocity was 1 mm/s. Therefore, it is important to understand the physical mechanisms that are in place so that we can modify them to induce bubble motion. Here, we follow the work of (Takhistov et al., 2002), who developed the theory of *asymmetric double layers.* The reason that the bubble may stay stationary despite the fast surrounding flow can be explained by simple scalings. Specifically, the electrokinetic velocity u_e is proportional to E, where E is the electric field, which scales in inverse proportion to the cross-sectional area; hence, the flowrate is independent of the cross-sectional area. This, in turn, implies that the flowrate in the annular region (between the bubble and the capillary) is the same as that behind the bubble, and thus there is no extra pressure buildup due to the lack of liquid accumulation, so the liquid simply flows around it.

There are many possibilities of circumventing this difficulty, e.g., by disturbing the gap flow and thus reducing the flowrate in the annular region. This can be achieved by changing the conductivity of the film, i.e., by adding appropriate *ionic surfactants,* which will cause the development of an electric double layer (EDL) at the liquid–air interface in addition to the EDL on the capillary wall. The relative size of these two EDLs is a major factor in determining the motion of the bubble.

In the experiments of (Takhistov et al., 2002), air bubbles were created in solutions of KCl/H_2SO_4 in a micropipe of diameter $d = 2R = 0.55$ mm and length 3 cm. The glass walls were treated properly so that a positive zeta potential ζ_c was produced, although typical glass surfaces are negatively charged. A small amount (2×10^{-5} mol/l) of an anionic surfactant SDS (sodium dodecyl sulfate) was used to create the interface EDL. Voltages in the range of 10 to 120 volts were applied, and the concetration of KCl varied from 10^{-6} to 10^{-2} l/m. The main finding of these experiments was that there exists a window of operation parameters within which bubble motion occurs, and it can reach even 3 mm/s for short bubbles. Specifically, this window is defined by the voltage range (20 to 80 volts) and the KCl concentration C_0 (between 10^{-5} to 10^{-3} mol/l).

A summary of the results is shown in Figure 8.15, taken from (Takhistov et al., 2002). The measured bubble velocity also depends on the aspect ratio of the bubble l_b/d, as shown in Figure 8.16, also taken from (Takhistov et al., 2002). Overall, a strong dependence of the bubble velocity u_0 is obtained in terms of the applied voltage V, the electrolyte concentration C_0, and the bubble length l_b. Below the critical value of C_0 and above the critical value of voltage, the electric current stops completely. In addition to the experiment with the anionic surfactant, Takhistov et al. also conducted experiments with a cationic surfactant, but no bubble motion was observed.

It is important to understand the reasons for this behavior, which produces a mobile versus a stationary bubble. Takhistov et al. argued that the most dominant mechanism is the enhancement of the film conductivity

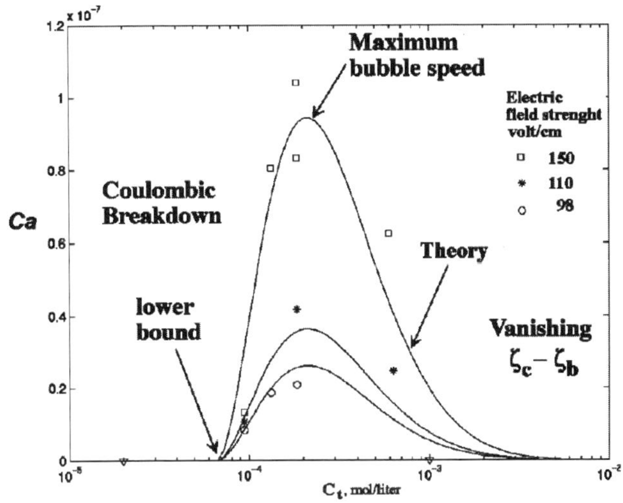

FIGURE 8.15. Nondimensional bubble velocity expressed as the capillary number Ca versus ion concentration for different values of voltage drop. (Courtesy of H.-C. Chang.)

that the anionic surfactant produces. In particular, SDS introduces negative charges in the liquid–air interface, and thus a negative zeta potential ζ_b is created, in contrast to the positive capillary potential ζ_b at the wall. The corresponding electric double layers contain charge q of opposite sign, and thus they drag the flow along opposite directions in the gap region due to the corresponding force qE. If $\zeta_c = |\zeta_b|$, then the net flow in the gap is zero and the bubble will move as in the case of the pressure-driven flow, thus at rather low speeds. However, if $|\zeta_b| > \zeta_c$, then net reverse flow occurs at the gap, which can cause liquid accumulation in the back of the bubble and eventually an ejection of the bubble. In contrast, the addition of a cationic surfactant will produce a positive ζ_b thus producing the opposite effect from before, i.e., no bubble motion.

The above arguments are in general valid, but they do not explain the observed window of bubble motion and specifically the upper and lower limits. The EDL scales inversely proportional to the square root of the ion concentration (see equation (7.1)), and thus at high ionic concentrations the EDL almost vanishes. Therefore, the aforementioned mechanism for flow reduction in the film gap is not there, and this explains the upper limit in the window of Figure 8.15. On the other hand, at very low ionic concentrations the two EDLs overlap, and while the two zeta potentials ζ_c and ζ_b increase, ζ_c becomes greater than $|\zeta_b|$. This change in the relative strength of the potentials at the liquid–air interface and at the capillary wall reverses the aforementioned flow reduction effect at the film gap, and

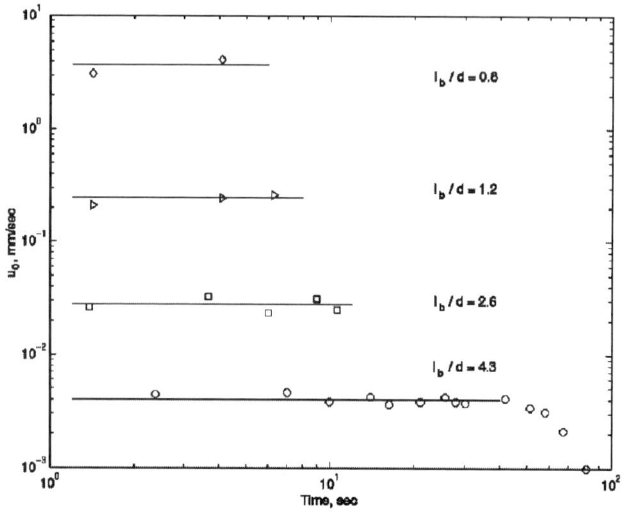

FIGURE 8.16. Time history of the bubble velocity as a function of the bubble length. $C_0 = 10^{-4}$ mol/l; $V = 42$ volts. (Courtesy of H.-C. Chang.)

the bubble moves much slower and eventually stops.

In summary, in transporting air bubbles in micropipes using electroosmosis, asymmetric electric double layers are desirable in order to produce a flow reversal in the annular region around the bubble. This, in turn, will cause a pressure buildup at the back of the bubble with an eventual forward ejection of the bubble. This occurs in a certain range of ionic concentration, which causes significant enhancement of the conductivity of the film. The resulting bubble velocities are of order 1 mm/s, but only for short bubbles; longer bubbles (length to diameter around 5) may travel much slower, at about 1 μm/s. These results, however, are not valid for noncircular microducts, since corners create complications such as very thick films, which basically are responsible for much of the voltage drop in electrokinetically driven flows. Therefore, for microchannels with general cross-sectional shapes, pressure-driven flow may be more efficient in transporting bubbles.

9

Mixers and Chaotic Advection

Mixing is an old subject in fluid mechanics, and it is typically associated with high-speed flows, since it occurs naturally due to turbulent diffusion. For example, in channels with height above 1 cm, water flow is typically turbulent for speeds of 10 cm/s or higher, and mixing occurs rapidly due to turbulent fluctuations. In microchannels, the height is more than an order of magnitude smaller, and thus the flow is laminar. Therefore, transport is controlled solely by the diffusion coefficient D of the medium. In many biological applications, e.g., mixing a stream of proteins in an aqueous buffer, the diffusion coefficient is very small, of order 10^{-10} m^2/s, and thus mixing by laminar diffusion is a very slow process. In order to enhance mixing and thus reduce the corresponding time, various subcritical excitation techniques have been developed, some of them resembling heat transfer enhancement methods used in microelectronic cooling applications. A more systematic procedure based on rigorous theory is to exploit the concept of chaotic advection or Lagrangian chaos that can be achieved for low Reynolds number flows or even for Stokes flows. In this chapter, we present the basic ideas behind chaotic advection and give analytical solutions for prototypical problems. We then discuss examples of passive and active mixers that have been used in microfluidic applications. Finally, we provide some quantitative measures of characterizing mixing based on the concept of Lyapunov exponent from chaos theory as well as some convenient ways to compute them.

9.1 The Need for Mixing at Microscales

Mixing of fluids in microchannels is important in many applications, including homogenization of solutions of reagents in chemical and biological reactions, sequencing of nucleic acids, and drug solution dilution. Mixing reduces longitudinal dispersion, which is important in determining performance in pressure-driven chromatography, which is the transfer of fractions from a separation column to a point detector, where it leads to peak broadening (Stroock et al., 2002). G.I. Taylor has studied longitudinal dispersion of a scalar in a pipe, demonstrating that the effective diffusion coefficient is inversely proportional to the transverse diffusivity (see Section 7.5.3). In microchannels the flow is laminar, with the typical Reynolds number at least two orders of magnitude lower than the critical value for laminar-to-turbulence transition. Mixing of tracers is then based on molecular diffusion, with mixing times of order h^2/D, where h is the channel height and D is the molecular diffusivity. Even though h is less than $100\,\mu$m in most applications, tracers with large molecules have very small D, leading to intolerably large mixing times. In Table 9.1 we present the molecular diffusivity of some relevant substances in water. Diffusivities of solutions containing large molecules (e.g., hemoglobin, myosin, or viruses) are typically two orders of magnitude lower than for most liquids. For example, at room temperature myosin's coefficient of diffusion in water is 10^{-11} m^2/s, which for $h = 100\,\mu$m requires a time of about 1000 s, which is unacceptable in practice (Bau et al., 2001).

Solute	D ($\times 10^{-9}$ m^2/s)	**Solute**	D($\times 10^{-9}$ m^2/s)
Nitrogen	1.88	Oxygen	2.10
Hydrogen	4.50	Propane	0.97
Acetone	1.16	Hemoglobin	0.069
Ovalbumin	0.078	Fibrinogen	0.020

TABLE 9.1. Diffusivity D of different substances in water at 25°.

The objective is therefore to achieve rapid mixing using the minimum amount of space. More specifically, the objective of mixing is to maximize the interfacial area in two coflowing fluids using the minimum amount of space, time, or energy. As we will see in the next section, this implies *stretching* of the interfacial area as much as possible. Mathematically, this is expressed simply in terms of the Peclet number Pe. Relying on clasical diffusion only, the mixing length L is proportional to $h\,\mathrm{Pe}$, where h is the microchannel height. Achieving *complete mixing* using the concepts of Lagrangian chaos or chaotic advection, as we will discuss in the next section, leads to $L \propto h \ln \mathrm{Pe}$. For solutions based on large molecules, the Peclet number achieves high values, and hence the logarithmic scaling becomes very important.

Micromixers are devices that have been proposed and tested in microfluidics research to enhance mixing rates. In the literature, both the mixing process and the mixing microdevices are characterized as either *passive* or *active*. In particular, passive mixing refers to a process whereby the interface follows the flow without any back effect on the flow (noninteracting case), whereas active mixing involves interfaces interacting with the flow. On the other hand, micromixers are also broadly classified as *passive* and *active* or static and dynamic, respectively. Most of the active mixers involve moving parts, and they expend additional energy resources, unlike the passive mixers. In terms of robustness, passive mixers are superior, but they are topologically very complex and hence difficult to fabricate. In addition, once they are incorporated in a microfluidic design they cannot be used for any other function. Active micromixers are reconfigurable and simpler to fabricate, although perhaps more complicated to operate.

The first generation of mixers sought to increase the mixing surface area between two streams of fluids by splitting them into n substreams and rejoining them again in a single stream at a certain location downstream. This mixing method is called *parallel lamination*, and it can potentially reduce the mixing time by a factor proportional to n^2. Similarly, *sequential lamination* has been used based on vertical and horizontal splittings in many stages, with the potential of reducing the mixing time by a factor of 4^{n-1}, where n is the number of splitting stages. Passive mixers involve appropriate geometric modifications, while active mixers rely on the unsteady action of a stirring force using a mechanical, acoustic, magnetic, or electroosmotic actuation (Bau et al., 2001; Oddy et al., 2001; Moroney et al., 1991). A magnetohydrodynamic (MHD) mixer was designed and tested in (Bau et al., 2001); it uses Lorentz forcing to produce cellular advection patterns (with induced speed up to 1.5 mm/s), and this can enhance mixing in microchannels. Such an MHD micromixer can be useful in biological fluids that are partially conductive. However, the mixing rate achieved in (Bau et al., 2001), for a steady magnetic field was only slightly faster than linear, suggesting that no complete mixing was possible with this excitation. A time-varying magnetic field could potentially increase the mixing rate significantly. Good mixing with exponentially fast rate is achieved if *chaotic advection* is implemented properly using either time-dependent forcing or appropriate geometric modifications. We present some details of this concept and examples in the following section, focusing in particular on techniques that have the potential of achieving *complete* mixing. Representative designs of both passive and active mixers are studied in some detail in the special volume on *mixing at microscales* edited by (Ottino and Wiggins, 2004).

9.2 Chaotic Advection

Homogenization of a tracer by a flow involves two processes: *stirring* and *mixing*. Stirring is the mechanical stretching of a material interface, while mixing is the diffusion of a substance across this interface. Mathematically, this process can be expressed by describing the passive advection of fluid particles given by

$$\frac{dx}{dt} = u(x, y, z, t), \quad \frac{dy}{dt} = v(x, y, z, t), \quad \frac{dz}{dt} = w(x, y, z, t), \tag{9.1}$$

where $\boldsymbol{v}(x, y, z, t) = (u, v, w)$ is the velocity vector. When the flow is turbulent, the particle paths described by equation (9.1) are chaotic. Research in dynamical systems in the mid 1980s has shown that chaotic fluid particle motion can also be generated with simple velocity fields either two-dimensionally with time-dependent excitation or three-dimensionally with or without time dependence. This concept was introduced by (Aref, 1984), who coined the term *chaotic advection*. Streamlines and pathlines in steady three-dimensional flow coincide. However, they are not closed curves, and they are not confined to smooth surfaces unless the Lamb vector $\omega \times \mathbf{v}$ is nonzero everywhere in the field. In fact, in chaotic advection, particle paths diverge exponentially in time, which is simply translated in practice to very large *residence time* for fluid particles. This, in turn, offers the possibility for enhanced transport in the mixing of two coflowing liquid streams, a typical setup in applications. For unsteady flows, a necessary condition for chaos is the *crossing* of streamlines at two consecutive time instants. This is expressed mathematically by the concept of *link twist maps* (Wiggins and Ottino, 2004).

The stirring and mixing of a passive scalar described by the nondimensional concentration θ is governed by the advection–diffusion equation

$$\frac{\partial \theta}{\partial t} + \boldsymbol{v} \cdot \nabla \theta = \mathrm{Pe}^{-1} \nabla^2 \theta,$$

with $\mathrm{Pe} = UL/D$ is the Peclet number defined as the ratio of the diffusion time (L^2/D) to the advection time (L/U), where D is the molecular diffusivity. With regard to numerical discretization of the above equation, algorithms that work well for low Reynolds number but also for high Peclet number are required. To this end, a semi-Lagrangian method can be used to deal with the advection–diffusion equation (Xiu and Karniadakis, 2001). It can effectively bypass the strict time-step restriction imposed in explicit time-stepping integration, and it is stable for very large values of Pe.

Advection dominates for large values of the Peclet number, and a steepening of concentration gradients occurs. However, at much later times molecular diffusion smooths out these steep gradients. Of interest is the characteristic time for which $\nabla \theta$ is maximum, which we denote by t_m. This time can be estimated by equating the diffusion length $l_d \propto \sqrt{Dt}$ to the striation thickness (transverse dimension) of the tracer l_a associated with advection. Stretching in applications achieving chaotic advection is a self-similar process at small scales following a lognormal distribution. Here, we refer to the *average* striation thickness in order to carry out the following order-of-magnitude analysis.

Let us first consider a *simple shear flow* $\mathbf{v} = (\dot{\gamma} y, 0)$, where $\dot{\gamma}$ denotes the deformation rate, and regular advection for which we have that $l_a = l_0/\sqrt{1 + (\dot{\gamma} t)^2}$. Therefore, the characteristic time t_m^s in this case is

$$t_m^s \propto \frac{\mathrm{Pe}^{1/3}}{\dot{\gamma}},$$

where $\mathrm{Pe} = \dot{\gamma} l_0^2/D$ and l_0 denotes the initial thickness of the layer. On the other hand, for a field with characteristic velocity and length scales U and h that generates *chaotic advection*, the striation thickness of the tracer is $l_a = l_0 \exp(-\lambda t)$. Here λ is the stretching rate of the tracer, equal to the asymptotic value of the Lyapunov exponent (Ottino, 1997). In this case, the characteristic time, t_m^{ca}, for long times is estimated as

$$t_m^{\mathrm{ca}} \propto \frac{L}{U} \ln \mathrm{Pe}.$$

Comparison of the above estimates for the mixing time scale implies that the diffusion time increases much faster as Pe increases in linear shear flow compared to chaotic advection. It turns out that this is a general result with the exception of the pure strain flow $\mathbf{v} = (\dot{\gamma} x, \dot{\gamma} y)$, which although integrable (i.e., nonchaotic) also has mixing times proportional to $\ln \mathrm{Pe}$. This anomaly is a consequence of an exponential separation associated with this flow and the infinite acceleration for $|x| \to \infty$; for a detailed explanation see (Jones, 1991).

The long mixing times imply long mixing lengths, which are impractical in microfluidic applications. To this end, many efforts have been made to design and operate effective mixers that exploit the chaotic advection concept. One of the first theoretical designs suggested for general viscous flows was based on *twisted pipes*, i.e., pipe segments that are not all in the same plane (Jones et al., 1989). It is well known that in a curved pipe a secondary motion is induced, giving rise to longitudinal vortices. The flow of particles in a twisted pipe with a pitch angle χ is represented by a sequence of Dean solutions augmented by a rotation of particles through an angle $-\chi$ between the successive segments. The specific value of the angle is very important, since it will make the system nonintegrable and generate chaotic advection.

Perturbation solutions for the Dean flow have been derived in (Jones et al., 1989). Referring to the sketch of Figure 9.1 and using polar coordinates (r, ϕ) in the transverse x-y plane, we follow Dean's formulation for

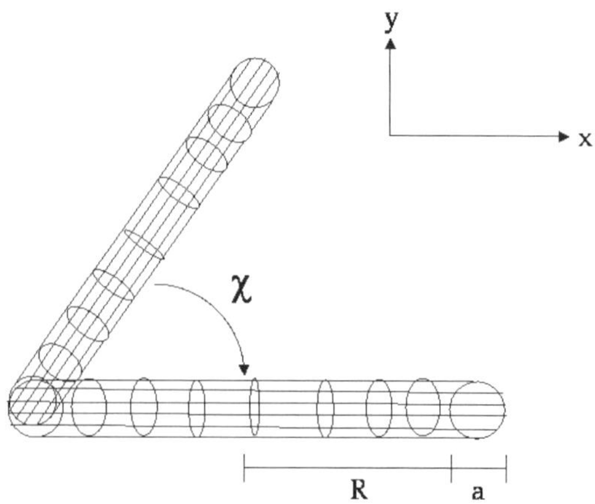

FIGURE 9.1. Sketch of a twisted pipe and notation used in the Dean flow solution. This basic unit is composed of two 180° curved pipe segments of constant curvature.

the streamfunction ψ and axial velocity w:

$$\nabla^2 w = \frac{1}{r}\left(\frac{\partial\psi}{\partial r}\frac{\partial w}{\partial\phi} - \frac{\partial\psi}{\partial\phi}\frac{\partial w}{\partial r}\right) - C,$$
$$\nabla^2 \psi = \frac{1}{r}\left(\frac{\partial\psi}{\partial r}\frac{\partial}{\partial\phi} - \frac{\partial\psi}{\partial\phi}\frac{\partial}{\partial r}\right)\nabla^2\psi + 2D_e w\left(\frac{\sin\phi}{r}\frac{\partial w}{\partial\phi} - \cos\phi\frac{\partial w}{\partial r}\right).$$

Here $D_e = W^2a^3/(R\nu^2)$ is the Dean number with W the average axial velocity, a the pipe radius, and R the radius of curvature of the bend. Therefore, the Dean number is proportional to the square of the Reynolds number. Also, C is a nondimensional pressure gradient defined by

$$C = -\frac{a^2}{RW\mu}\frac{\partial p}{\partial\theta}.$$

The perturbation solution is obtained in a power series in D_e; at the lowest order the standard Poiseuille flow for a straight pipe is recovered. The first-order equations give the following equations for the particle motion

$$\dot{x} = \frac{\alpha}{1152}\left[h(r) + y^2\frac{h'(r)}{r}\right], \quad \dot{y} = -\frac{\alpha}{1152}\frac{xy}{r}h'(r), \quad \dot{\theta} = \frac{1}{4}\beta(1-r^2), \tag{9.2}$$

where $\alpha = D_eC^2$, $\beta = D_eC/\mathrm{Re}$, and

$$h(r) = \frac{1}{4}(4-r^2)(1-r^2)^2.$$

The angle θ is used to describe the three-dimensional motion of the flow along the curved pipe together with the coordinates (x, y).

Since the above equations are steady, the independent variable can be changed from time t to angle θ to obtain

$$\frac{dx}{d\theta} = \frac{c}{1152}\left[4 - 5x^2 - 23y^2 + x^4 + 8x^2y^2 + 7y^4\right], \tag{9.3}$$

$$\frac{dy}{d\theta} = \frac{c}{192}xy\left[3 - x^2 - y^2\right], \tag{9.4}$$

where $c = \alpha/\beta$. These equations define the mapping of fluid particles from one cross-section to another (Jones et al., 1989). They produce chaotic solutions for the particle paths for certain ranges of the parameters c and χ. For example, for $c = 100$ the most chaotic pattern is obtained if $\chi = 90°$, but as $\chi \to 180°$ a regular pattern is achieved. In (Jones et al., 1989), it was shown that transverse chaotic transport occurs when

$$\tan\frac{\chi}{2} \leq \sinh\frac{\pi c}{192},$$

which provides a criterion for the design of the twisted pipe or other three-dimensional channels in order to realize, at least partially, the effects of chaotic advection on longitudinal dispersion. Finally, we note that the stochastic patterns of the particle paths are similar to the instantaneous ones seen in a corresponding turbulent flow, but with the important difference that in the chaotic advection these paths are stationary in time.

9.3 Micromixers

The first implementation of the twisted pipe concept in microfluidics was presented in (Liu et al., 2000). A three-dimensional *serpentine microchannel* was designed as shown in the sketch of Figure 9.2. The basic building block is a "C-shaped" section; the planes of two successive such sections are perpendicular to each other. The channel height is 300 μm, and its total length 7.5 mm. The microchannel was fabricated in a silicon wafer using a double-sided KOH wet-etching technique. The mixing streams were phenolphthalein and sodium hydroxide solutions, which typically mix fast compared to other large-molecule solutions. In the experiments of (Liu et al., 2000), after ten "C-shaped" sections at least 98% of the maximum intensity of reacted phenolphthalein was observed. This mixing was superior to that in a straight channel (a factor of 16 better) and about 60% better than in a similar square-wave channel that has all "C-shaped" sections on the same plane. However, full chaotic advection was established only at a relatively high Reynolds number, i.e., Re = 70; this value is higher than in typical microfluidic applications.

A passive chaotic mixer for microchannels that exploits the aforementioned "complete mixing" concept and verifies the logarithmic Peclet number dependence was developed by (Stroock et al., 2002). To generate transverse secondary flows, microriblets similar to rifling in a gun barrel were

FIGURE 9.2. Schematic of a serpentine channel used in the experiments of (Liu et al., 2000).

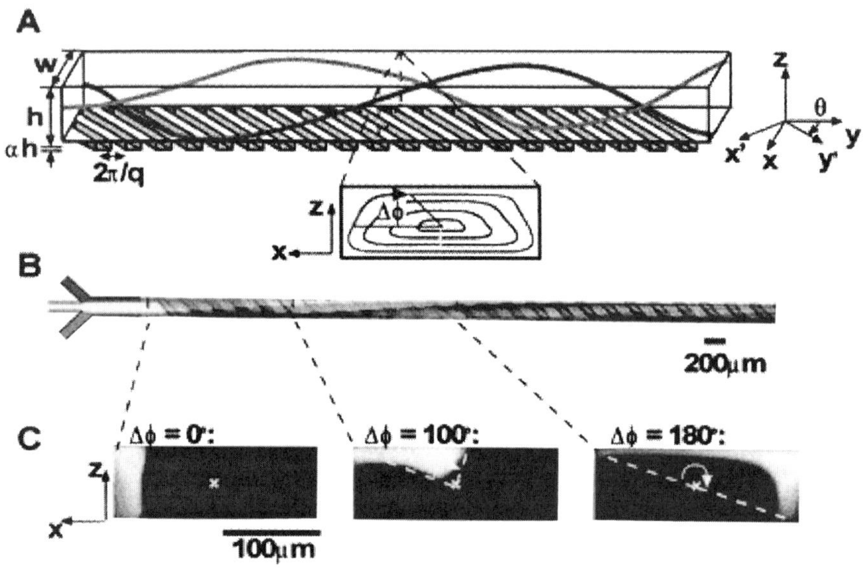

FIGURE 9.3. A: Schematic of microchannel with riblets. B: Optical micrograph showing the two streams flowing on either side of a clear stream. C: Fluorescent confocal micrographs at three different cross-sections showing the rotation and distortion of a stream of fluorescent solution that was injected upstream. $h = 70\,\mu$m; $w = 200\,\mu$m, $\alpha = 0.2, q = 2\pi/200\,\mu\text{m}^{-1}, \theta = 45°$. (Courtesy of H.A. Stone.)

placed on the floor of the channel at an oblique angle θ with respect to the long axis y of the channel, as shown in Figure 9.3. They were fabricated using two steps of photolithography. These microriblets, whose height is typically less than 30% of the channel height, present an anisotropic resistance to flow, with less resistance along the main flow direction than in the orthogonal direction. A transverse component of the flow (along the x-direction in the figure) is then produced that is initiated at the riblet surface, with the flow circulating back across the top of the channel. Optical micrographs used in the experiments of (Stroock et al., 2002) show that the flow has helical trajectories, as shown in Figure 9.3(A).

A somewhat different design was also tested in the experiments of (Stroock et al., 2002); it uses V-grooves to make a structured surface, instead of straight riblets, as shown in Figure 9.4. The objective here is to subject the fluid to a repeated sequence of rotations and extensions in order to realize stronger chaotic transport. This is effectively analogous to the *baker's transformation*, i.e., a repetitive action involving stretching, cutting, and fusion that can, in principle, achieve the best achievable mixing (Ottino, 1997). In this design the shape of the grooves is changing along the flow in each half-cycle, as shown in Figure 9.4. The efficiency of mixing is primarily controlled by two parameters: the measure of asymmetry and the amplitude of rotation of the fluid in each cycle. For symmetric microgrooves or zero amplitude of rotation the flow becomes deterministic, but optimum values for both parameters have been obtained experimentally. The degree of mixing was quantified using the standard deviation of the intensity distribution in confocal images of the cross-section of the flow. The results suggest that the V-grooves achieve mixing with zero standard deviation only 1 cm downstream, while the microchannel with the straight riblets, although effective, does not quite achieve complete mixing. These results were obtained for a Peclet number range from 2×10^3 to 9×10^5, and they confirmed the theoretical logarithmic dependence of the mixing length on the Peclet number.

For applications, it was argued in (Stroock et al., 2002), that the mixing of a stream of proteins in an aqueous buffer ($D = 10^{-10}$ m^2/s) with $U = 1$ cm/s and $h = 100\,\mu$m would require a channel of length $L = 100$ cm for regular advection compared to $L = 1$ cm for chaotic advection achieved using a V-grooved microchannel. This improvement simply reflects the difference in the *linear* versus *logarithmic* dependence on the Peclet number for regular and chaotic advection, respectively.

The first *active* micromixer designed to exploit the good mixing of chaotic advection was presented in (Evans et al., 1997). It attempted to establish chaotic advection using a sink/source system, the so-called blinking vortex, first presented in (Aref, 1984). Unmixed fluid is drawn into a mixing chamber, and subsequently two sink/source systems are alternately pulsed. Here we present a somewhat different concept and design, which was also developed by the same research group at UC Berkeley. It is a pulsatile micromixer based on a bubble micropump developed in (Deshmukh et al., 2000). The use of *pulsatile flow* creates a greatly lengthened interface that leads to faster mixing. A schematic of the bubble micromixer is shown in Figure 9.5. Two pulsatile pumps are operated out of phase to mix two streams of fluid in a mixing channel. Fluid 1 is pushed into the mixing channel while fluid 2 is drawn from the inlet for half of the cycle, and the process is reversed for the other half of the cycle. The pumps consist of a bubble chamber and two check valves. When a bubble is created, it acts as a piston and drives the fluid out. The check valves control the direction of the fluid. When the bubble collapses, fluid is drawn in only from the

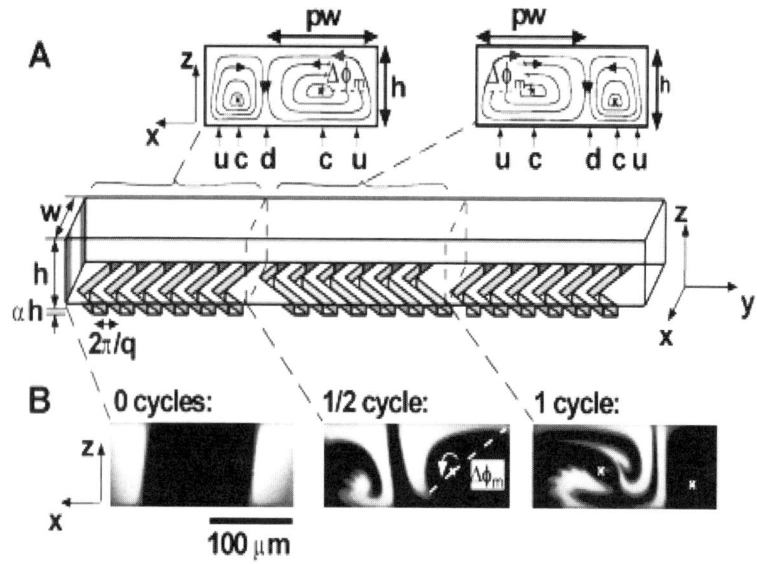

FIGURE 9.4. A: Schematic of microchannel with the V-grooves and of streamlines showing the secondary induced motion at the two halves of the periodic array of grooves. Here $h = 77\ \mu$m, $\alpha = 0.23, q = 2\pi/100\ \mu\text{m}^{-1}$. B: Confocal micrographs of vertical cross-sections at different distances from the inlet. (Courtesy of H.A. Stone.)

inlet. By cycling bubble generation and collapse, a net pumping action is achieved. Steam bubbles can be generated by heat; upon heat removal the steam condenses and the bubble collapses. Electrolysis bubbles can also be used. This micromixer was fabricated using SOI (silicon on insulator) and quartz wafers bonded together using epoxy. Bubbles were created using polysilicon resistors on quartz, which act as heaters. The designs tested in (Deshmukh et al., 2000), use 1 to 4 watts per pump, which is quite inefficient. Another practical issue is that the valves and substrate, which are made of bare silicon, sometimes stick together, since water is a poor lubricant on silicon surfaces.

A novel concept of an *active* micromixer was introduced in (Volpert et al., 1999). It is based on mixing a sequence of shear flows at different angles, and it is called the "shear superposition micromixer," or SSM. In contrast to the pulsatile micromixer we described above, the SSM is a continuous through-flow micromixer consisting of the main channel and three cross-flow side channels. The side-channels produce a time-dependent shear flow in the direction transverse to the main flow. The frequency of actuation in the three side-channels increases downstream to accommodate the decreasing scales that need to be further mixed downstream. A sketch of SSM and a micrograph of the actual device is shown in Figure 9.6; a typical range of

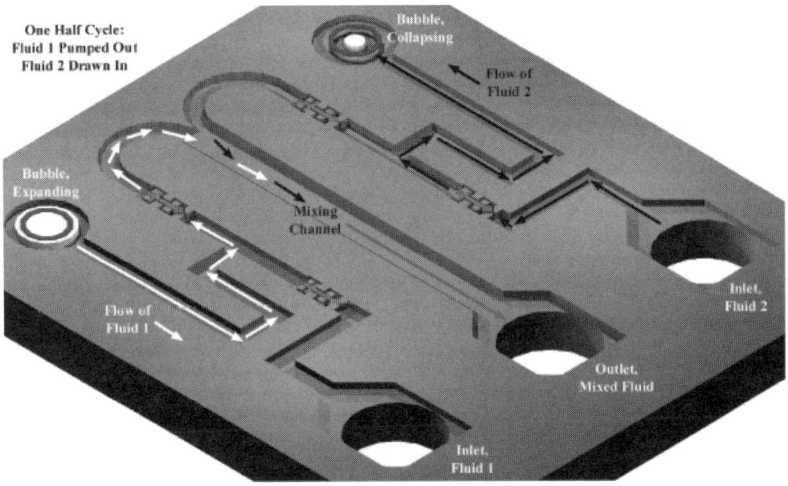

FIGURE 9.5. Schematic of the bubble micromixer. (Courtesy of D. Liepmann.)

the half-channel height is $50\,\mu$m to $150\,\mu$m.

The flow from the side-channels is set to oscillate out of phase at phases $(0, \pi, 0)$, while both amplitudes and frequencies are different for the three pairs. Both simulations and experiments were performed to evaluate the performance of this mixer (Bottausci et al., 2004). In the experiment, velocity profiles were measured using PIV, and also the mixing variance coefficient (MVC) was computed from photographs of the concentration field (side views (x-y) only). Typical results for MVC are shown in Figure 9.7(a) for the case in which only one side-channel is activated at 28 Hz. On the horizontal axis, S denotes the base-2 logarithm of the number of small boxes used to compute the mixing variance coefficient. The mixing variance coefficient, or MVC, is normalized such that a value of zero corresponds to complete mixing, whereas a value of 0.25 corresponds to no mixing. The experimental results are in agreement with the numerical predictions, but at small scales the experiments show better MVC value; see Figure 9.7(a). The quality of mixing decreases with the scale size. Figure 9.7(b) shows numerical results for the MVC, with all three pairs of side-channels activated at oscillation frequencies of 26.025 Hz for the first, 59.34 Hz for the second, and 92.925 Hz for the last side-channel. We see that overall, the MVC values decreased compared to the actuation of the single side-channel in Figure 9.7(a). This indicates significantly improved mixing and validates the theory that the oscillation frequency should be increasing downstream. Proper optimization techniques can be incorporated to select the set of frequencies automatically for a given application.

In the next example, we consider the *chaotic electroosmotic stirrer* of Qian and Bau (2002). The device consists of a spatially-periodic mixing

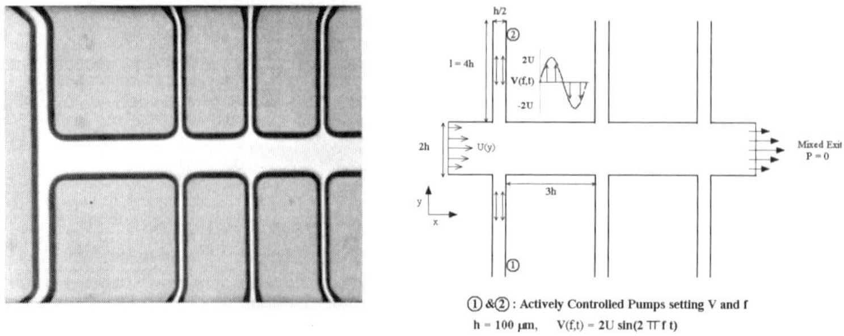

FIGURE 9.6. Schematic of the shear superposition micromixer and corresponding micrograph of the actual device. (Courtesy of I. Mezic.)

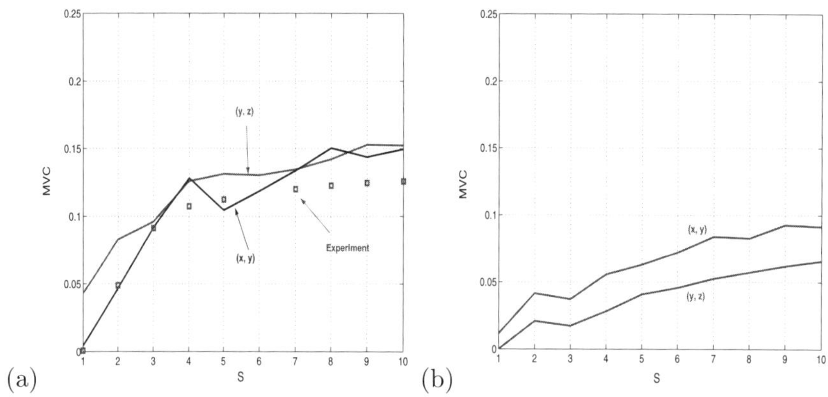

FIGURE 9.7. (a) Experimental measurements of MVC for an $(x\text{-}y)$ plane and numerical simulations for an $(x\text{-}y)$ and a $(y\text{-}z)$ plane for a single side-channel activated. (b) Numerical simulations for all three pairs of side-channels activated. (Courtesy of I. Mezic).

chamber, where the bottom and top surfaces each have two surface electrodes that are covered with a thin insulator. The zeta potential on the insulated surface can be altered by applying electrostatic potential on these electrodes (Schasfoort et al., 2001); see also Section 7.4.7. It is possible to create various flow patterns in the mixing chamber by using different zeta potentials under horizontal electric field. For simplicity, Qian and Bau utilized zeta potential combinations of $\pm\zeta_0$, and obtained analytical solutions of Stokes flow in the rectangular chamber using the thin EDL approximation (i.e., the Helmholtz–Smoluchowski slip velocity (equation (7.24)) is assumed on electrode surfaces). Figure 9.8 shows four steady Stokes flow patterns obtained under different zeta potentials (shown by patterns A, B, C, and D). The arrows on the top and bottom of the mixing chambers show the electroosmotic flow direction. The main idea in the electroosmotic stirrer is to alter the zeta potential, and hence the flow patterns vary peri-

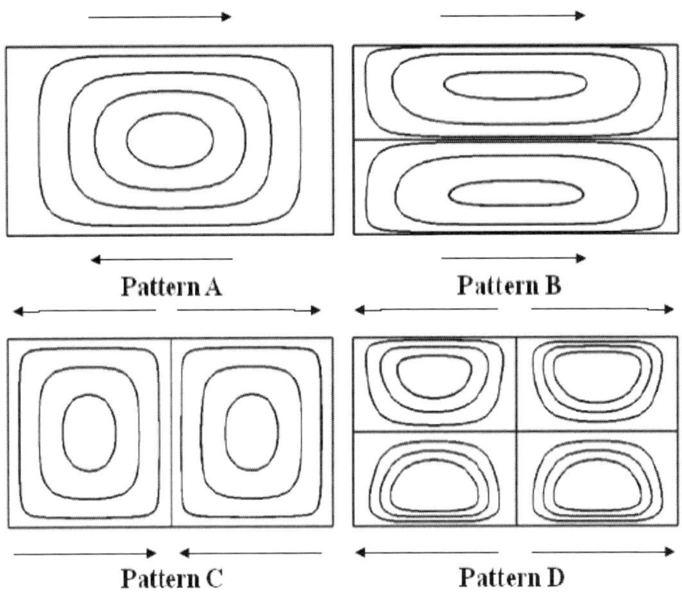

FIGURE 9.8. Four different electroosmotic flow patterns obtained by zeta potential alterations in (Qian and Bau, 2002). Arrows show the electroosmotic flow direction on electrode surfaces. The flow is maintained by a steady horizontal electric field.

odically to induce chaotic mixing. Although the flow is time-periodic here, the particle paths for two-dimensional unsteady flow may become nonintegrable, resulting in chaotic advection/mixing even in the Stokes flow regime. We must indicate that the actual device requires precise control over the zeta potential magnitude and the time scales for zeta potential alterations. In addition, the results presented in (Qian and Bau, 2002), ignore possible interactions between the insulated electrodes and the flow transients.

In Section 7.6 we have briefly described applications of dielectrophoresis to mixing. The final example of this section describes one such micromixer developed by (Deval et al., 2002). It is appropriate for microflows that contain charged or polarizable particles that can move under the influence of a nonuniform AC electric field. The spherical particles used in (Deval et al., 2002), are polystyrene spheres, but in principle, bacteria or cells may be present in the flow.

Assuming a spherical particle of radius a and conductivity σ subject to an electric field with rms value E_{rms}, the time-averaged force on the particle is given by equation (7.49). The real part of the Clausius–Mossotti factor, $K(\omega)$, is in the range $[-1/2, 1]$, and thus both attractive and repulsive forces can be induced simply by changing the frequency ω. The conductivity of polysterene spheres is on the order of 10 mS/m, and correspondingly, the

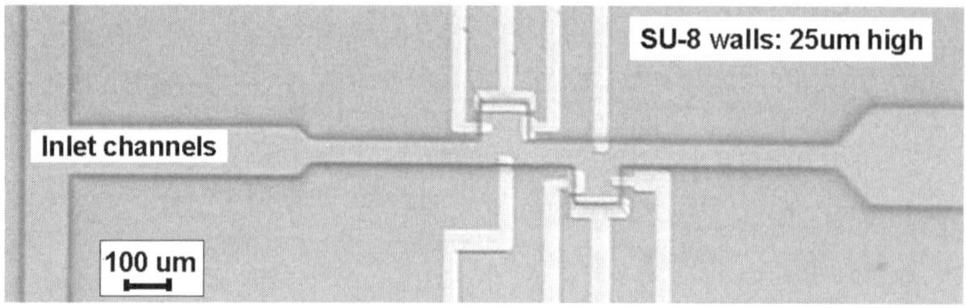

FIGURE 9.9. Top view of the DEP micromixer. (Courtesy of C.-M. Ho.)

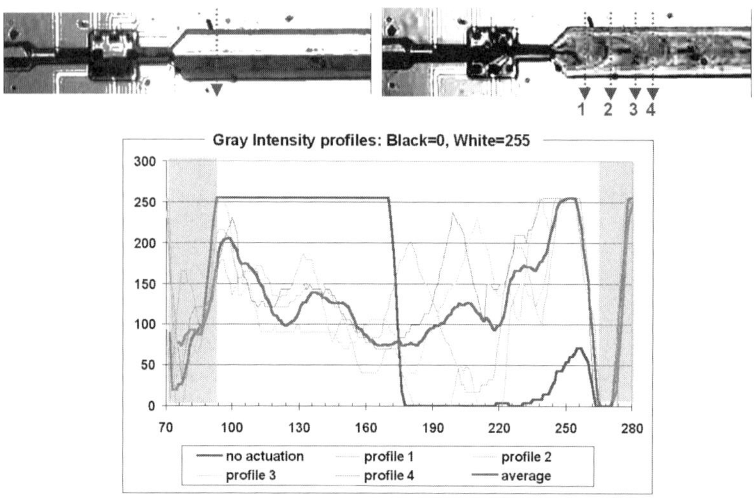

FIGURE 9.10. The top left picture shows a nonactuated mixer with flat interface and sharp intensity profile. The top right picture corresponds to a mixing regime. The lower plot shows intensity profiles of mixing at the stations indicated on the upper right plot. (Courtesy of C.-M. Ho.)

crossover frequency from attractive to repulsive induced forces is a few MHz, with high frequencies generating negative forces. Therefore, particles are attracted toward the electrodes at low frequencies and are pushed away at higher frequencies. This is the principle that is exploited in the DEP micromixer fabricated in (Deval et al., 2002). A top view of the micromixer test section is shown in Figure 9.9. The chamber dimensions are $200 \times 200 \times 25\,\mu$m. The electric field is created by a 1–15 MHz, 10 V AC voltage applied between selected pairs of micromachined electrodes located on the walls of the two cavities. Visualizations of the particle motion revealed that as the particles enter the first cavity, positive DEP forcing attracts them into the low-velocity region, while frequency switching repels them back to the main flow. The combined competing motion generates folding and stretching, thereby producing enhanced mixing. Typical results are shown in Figure 9.10, with the background velocity being approximately $420\,\mu$m/s.

9.4 Quantitative Characterization of Mixing

In many experimental and numerical studies on micromixers mixing is characterized only qualitatively by snapshots of a passive tracer or two differently colored fluids. Clearly, crossing of streamlines is an indication of chaotic mixing, but how exactly do we quantify the *degree* of mixing? The Lyapunov exponent (LE) is a possible accurate measure, since it is related to the stretching rate. It is defined by the equation

$$\lambda_\infty = \lim_{t\to\infty} \frac{1}{t} \ln \frac{d(t)}{d(0)}, \tag{9.5}$$

where $d(t)$ is the distance between two points that are initially very close to each other. An n-dimensional system has at most n LEs, and it is characterized as chaotic if at least one of the LEs is positive. One of the problems in applying the above definition in microfluidic systems is that we have only a finite length and corresponding mixing time. In addition, it is quite expensive to compute the LE from this definition, although it has been applied with success in quantifying mixing in an active micromixer, similar to the SSM presented earlier, in (Lee, 2002).

A more convenient measure for quantifying chaotic mixing is the finite-time Lyapunov exponent (FTLE). It is given by a similar definition to the one in equation (9.5) but without the limit $t \to \infty$, i.e.,

$$\lambda_F(\xi, t) = \frac{1}{T} \ln \frac{d(t+T)}{d(t)}. \tag{9.6}$$

It clearly depends on the time t and also on the Lagrangian position ξ, and it converges to LE as $T \to \infty$. In addition, as shown in (Tang and Boozer, 1996), the FTLE satisfies the following equation:

$$\lambda_F(\xi, t) = \tilde{\lambda}(\xi)/t + f(\xi, t)/\sqrt{t} + \lambda_\infty, \tag{9.7}$$

where $\tilde{\lambda}(\xi)$ is a smooth function of geometry only, and the function $f(\xi, t)$ satisfies

$$\lim_{t\to\infty} f(\xi, t)/\sqrt{t} = 0.$$

Although the distribution of FTLE is strongly space- and time-dependent, it has been shown in (Tang and Boozer, 1996), that the *mean* FTLE converges very rapidly to the actual value of LE. Its effectiveness in micromixers has been demonstrated in (Niu et al., 2003), for both active and passive mixers.

The dimensionless advection time of the system, within which complete mixing can be achieved, can be written in terms of the FTLE as

$$t_n = \frac{\ln(2\Omega) - 1}{2\lambda_F},$$

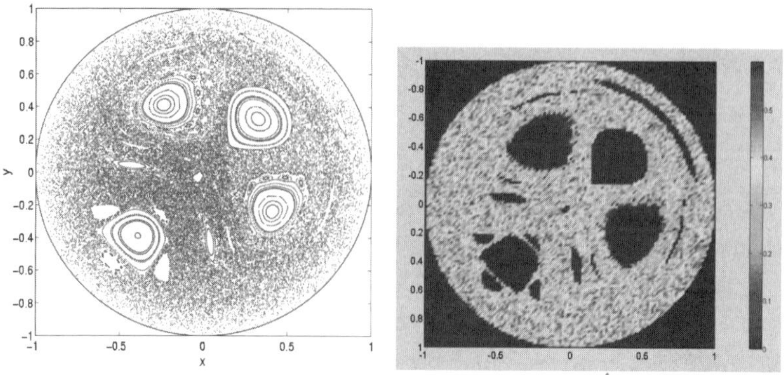

FIGURE 9.11. Poincaré section of the twisted pipe (left) and FTLE contour at $\chi = \pi/2; c = 100$. (Courtesy of Y.K. Lee.)

where $\Omega = \lambda_F L^2/D \gg 1$ for micromixing. A criterion for *complete mixing* was derived in (Niu et al., 2003), as follows:

$$\int_0^{t_n} \frac{2\lambda_F(t)}{\ln(2\lambda_F(t)L^2/D) - 1} dt = 1.$$

The above approach was used in (Niu et al., 2003), to analyze the twisted pipe of (Jones et al., 1989), that we presented earlier, see equations (9.3) and (9.4). In particular, FTLE contours in the (x,y) planes were computed and compared to Poincaré sections for different values of the parameters χ and c. Figure 9.11 (left) shows the Poincaré section for $\chi = \pi/2$ and $c = 100$. Clearly, mixing is poor in regions where islands appear. Figure 9.11 (right) shows the FTLE contour for the same parameters, which is topologically similar to the Poincaré section. The *mean* FTLE was computed for many values of (χ, c), shown in Figure 9.12. The highest value corresponds to best mixing, which is achieved at χ slightly greater than $\pi/2$. However, for realistic values of the parameter c the twisted pipe turns out not to be a very good mixer.

We now study the *electroosmotic stirrer* of Qian and Bau (2002), described in the previous section, in order to quantify its mixing effectiveness. Using the flow patterns shown in Figure 9.8 for half a period ($T/2$) each, we obtain six different pattern combinations (A-B, A-C, A-D, B-C, B-D, and C-D). We computed the FTLE for all of these patterns for nondimensional periods of $T = 4, 6$, and 8, where T is normalized by the convective time scale (based on the half-channel height and the electroosmotic slip velocity from equation (7.24)). The variation of the FTLE, as a function of time for $T = 4, 6$, and 8 is shown in Figure 9.13. For all the cases, the initial particle location was at $(x, y) = (0.5, 0.1)$, and the virtual particle was initially offset by a distance of 10^{-5} (channel half-height). The results show that the pattern B-C at $T = 6.0$ has the largest FTLE, corresponding

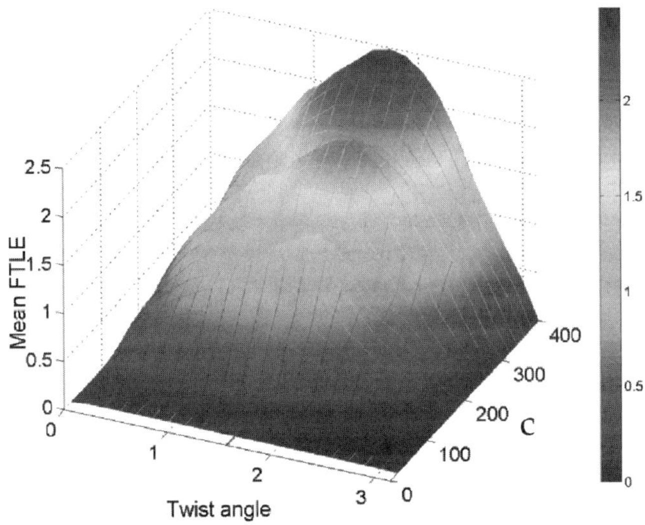

FIGURE 9.12. Mean finite time Lyapunov exponent (FTLE) for the twisted pipe: optimization study. (Courtesy of Y.K. Lee.)

to $\lambda_F = 0.31$, while for $T = 8.0$ and $T = 4.0$, $\lambda_F = 0.2$ and $\lambda_F = 0.25$, respectively.

In Figure 9.14, we present the Poincaré sections obtained for pattern B-C at periods $T = 1, 2, 4$, and 6. The Poincaré sections are obtained by tracking the motion of 121 particles for 100 time periods. Islands of bad mixing zones are observed for $T = 1, 2$ cases. The island boundaries, also known as the Kolmogorov–Arnold–Moser (KAM) boundaries, separate the chaotic and regular regions of the flow (Ottino, 1997). In the figure, we also present the KAM boundaries, obtained by tracking 20 particles that were initially located on the KAM boundaries, for 300 periods. We observe reduction in the bad mixing zone with increasing T. For example, the islands disappear for $T \geq 6$. The Poincaré section for $T = 8$ is qualitatively similar to that of the $T = 6$ case, and is not shown in the figure. Destruction of KAM boundaries is desired for enhanced mixing, but is not a sufficient condition for the best mixing case. For example, the FTLE for $T = 6$ is considerably larger than that of the $T = 8$ case, and it corresponds to the best mixing case among the flow patterns and frequency ranges studied in (Kim, 2004). We must indicate that the FTLE values presented in Figure 9.13 were obtained for particles that were *outside* the bad mixing zones.

Calculation of the FTLE is often computationally expensive, and it requires accurate knowledge of the flow field. Qian and Bau (2002) have shown that the flow solution often requires accuracy levels comparable to the computer evaluation of the analytical solution. The same is true for calculation of the Poincaré sections, especially for determination of the

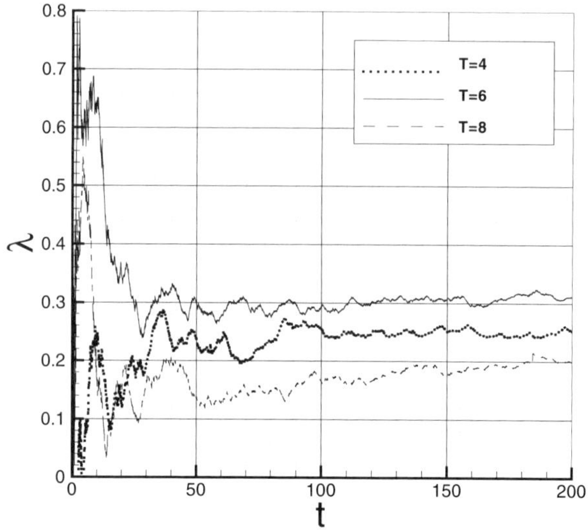

FIGURE 9.13. Time variation of FTLE for pattern B-C with $T = 4, 6$, and 8.

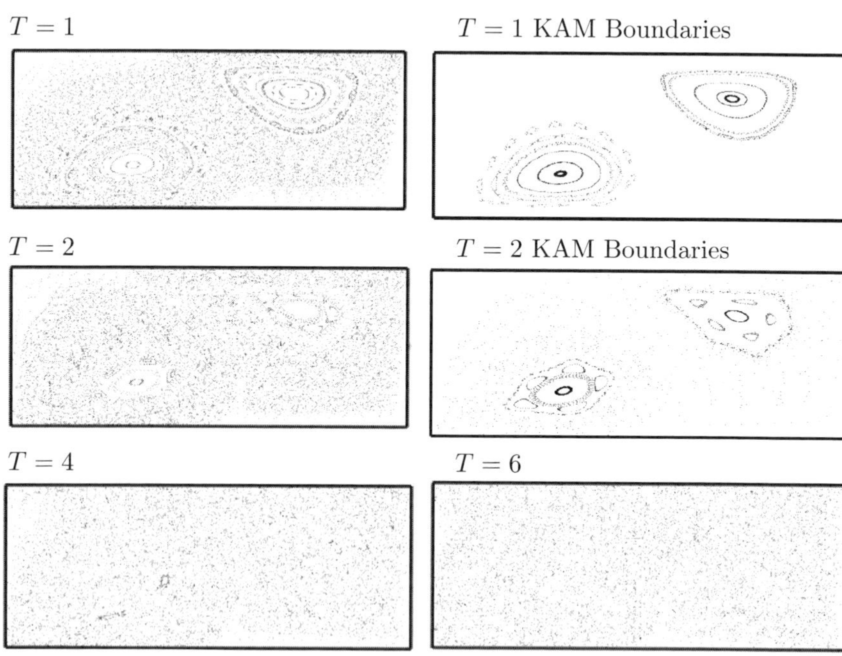

FIGURE 9.14. Poincaré sections for pattern B-C at four different periods.

KAM boundaries. However, we note that the Poincaré sections can be obtained experimentally by long-time tracking of noninertial particles. Given the computational difficulties of obtaining FTLE and Poincaré sections, it is desirable to develop other computation- and experiment-friendly methods to quantify the mixing quality. Dispersion of particles that are initially confined to a zone can be utilized to characterize mixing. In particular, uniform dispersion of particles to the entire mixing region can be considered as a homogeneously mixed state. However, such observations need to be quantified to obtain a reasonable measure of the mixing quality.

To this end, we employ the box counting method to quantify the rate at which particles are dispersed by the flow into small uniform boxes (Liu et al., 1994). In this method, selection of an appropriate box size is important, and it is related to the number of tracked particles. Jones recognized that a perfectly randomized population of particles has a Poisson bin-occupancy distribution, and used this information to determine the box size such that on average there was one particle per cell (Jones, 1991). If the box size is chosen such that for a perfectly random distribution of N particles, 98% of the boxes contain at least one particle, then the box size s for a unit-square domain (of length 1×1) is approximately given by

$$s \approx 2N^{-1/2}.$$

Once the number of particles (N) is selected, the domain is divided into boxes of size s, and the motion of each particle is calculated for a long time. Then the dispersed particles in each box are counted, and a decision is made regarding the mixing state of each box. For example, tracking 1600 particles requires 400 boxes in the entire domain. This results in a maximum of 4 particles per box in a homogeneous mixing state (i.e., $n_{\max} = 4$). For such conditions one can define a *Mixing index* for each box (M_i) as

$$M_i = \begin{cases} \dfrac{n_i}{n_{max}} & \text{if } n_i \leq n_{max} \\ 1 & \text{if } n_i \geq n_{max}, \end{cases}$$

where n_i is the number of particles in box i. There will be empty boxes in the domain for each overpopulated box ($n_i > n_{\max}$). Therefore, assuming the overpopulated boxes well mixed is reasonable, since impact of this decision is compensated by the low M values of underpopulated boxes in the domain. Another alternative is to characterize particle dispersion by assuming that boxes that contain at least one particle are well mixed (Liu et al., 1994). This is useful, since it is practically impossible to observe a homogeneous mixing state ($n_i = n_{\max}$) for all the boxes in the domain. Using either definition for M_i, the total mixing index M_T is defined as

$$M_T = \frac{1}{N} \sum_{i=1}^{N} M_i. \tag{9.8}$$

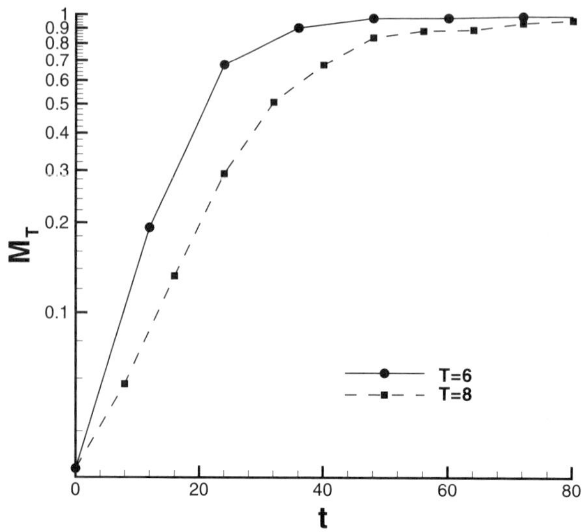

FIGURE 9.15. Time variation of the mixing index M for pattern B-C at $T = 6$ and $T = 8$, obtained using equation (9.8).

In Figure 9.15, we present the time variation of the mixing index for pattern B-C at $T = 6$ and $T = 8$, using the method of Liu et. al (1994). The mixing index for both cases increases exponentially at early times, then reaches the asymptotic limit of unity. The mixing index for $T = 6$ reaches its asymptotic limit much faster than in the $T = 8$ case, thus indicating that the case $T = 6$ corresponds to a better mixer than the $T = 8$ case. Overall, the behavior of M_T for both cases is consistent with the FTLE results, indicating that time variation of the mixing index can be utilized as a consistent measure for monitoring mixing quality.

In order to demonstrate the relation between the dispersion of particles and the mixing index we present in Figure 9.16 the time evolution of dispersion of 1600 particles using pattern B-C with $T = 6$ and $T = 8$. Snapshots of dispersed particles are shown in the figure at the same nondimensional time. Comparisons between the figures show that the $T = 6$ case is better stirred than the $T = 8$ case, as also indicated in Figure 9.15.

Finally, we address mixing-efficiency quantifications using numerical solutions of the species transport equation (see Section 14.1). Unlike the previous methods that utilize the Lagrangian motion of pseudo-particles, the species transport equation involves diffusion, and it can better match the experimental mixing results. In both numerical simulations and experiments (that utilize fluorescent dyes), it is possible to define an *alternative mixing index* by dividing the flow domain into N boxes, and recording the

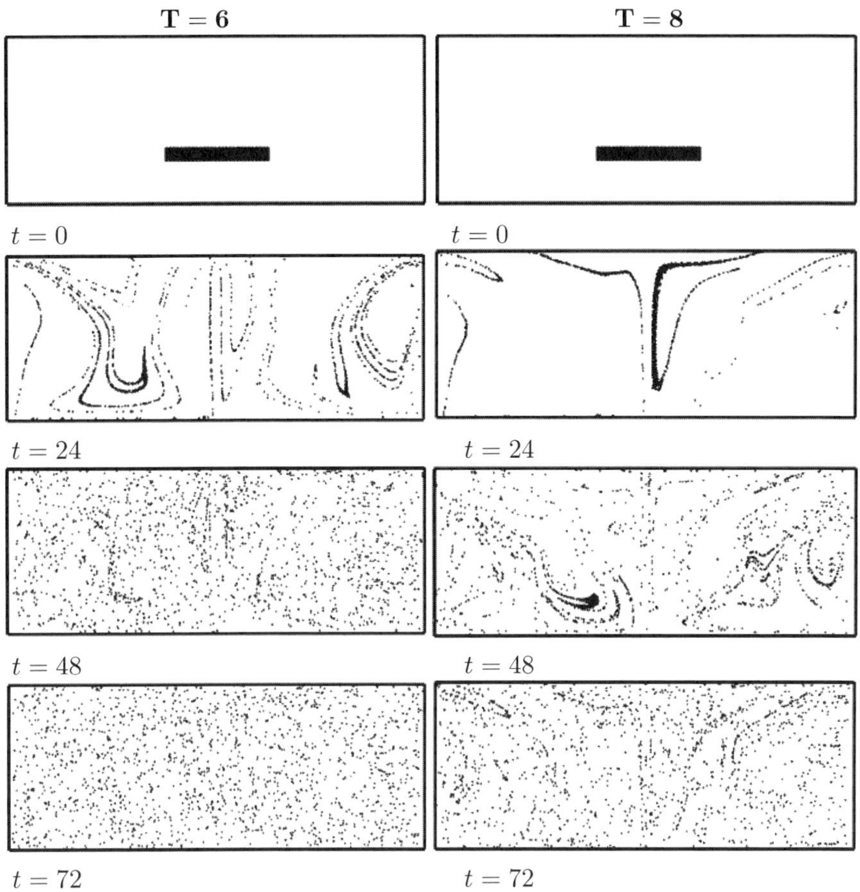

FIGURE 9.16. Dispersion of 1600 particles as a function of time, obtained using pattern B-C at $T = 6$ (left) and at $T = 8$ (right).

fluorescence intensity/concentration value in these boxes. Using this, an alternative mixing index can be defined as

$$M = \frac{1}{N}\sqrt{\sum_{i=1}^{N}\left(\frac{\theta_i}{\theta_o} - 1\right)^2}, \tag{9.9}$$

where θ_i is the concentration/fluorescence value in box i, and θ_o is the mean concentration/fluorescence value of the mixture. This definition is analogous to the Euclidean error norm, and measures deviations from a perfect mix. Therefore, smaller M values show better mixing in the domain, with $M \to 0$ for a perfect mix. It is often useful to plot the time variation of M (or M^{-1}) in logarithmic scale, which enables accurate comparisons between various mixing states.

10
Simple Fluids in Nanochannels

With the growing interest in the development of faster, smaller, and more efficient biochemical analysis devices, nanofluidic systems and hybrid micro/nano fluidic systems have attracted considerable attention in recent years. In nanoscale systems, the surface-to-volume ratio is very high, and the critical dimension can be comparable to the size of the fluid molecules. The influence of the surface and the finite-size effect of the various molecules on fluid transport needs to be understood in detail, while such effects may be largely neglected for liquid flows in macroscopic channels. In this chapter, we discuss the analysis of simple fluids such as Lennard–Jones liquids in confined nanochannels. A key difference between the simulation of the fluidic transport in confined nanochannels, where the critical channel dimension can be a few molecular diameters, and at macroscopic scales is that the well-established continuum theories based on Navier–Stokes equations may not be valid in confined nanochannels. Therefore, atomistic scale simulations, in which the fluid atoms are modeled explicitly or semiexplicitly and the motion of the fluid atoms is calculated directly, shed fundamental insights on fluid transport. The most popular technique for atomistic simulation of liquid transport is molecular dynamics (MD), which is discussed in detail in Chapter 16. After presenting some details on the atomistic simulation of simple fluids, we discuss density profiles, diffusion transport, and validity of the Navier–Stokes equations for simple fluids in confined nanochannels. In the last section we discuss in detail the slip condition at solid–liquid interfaces and present experimental and computational results as well as conceptual models of slip. We also revisit the lubrication problem, first discussed in Chapter 7, and we present the Reynolds–Vinogradova

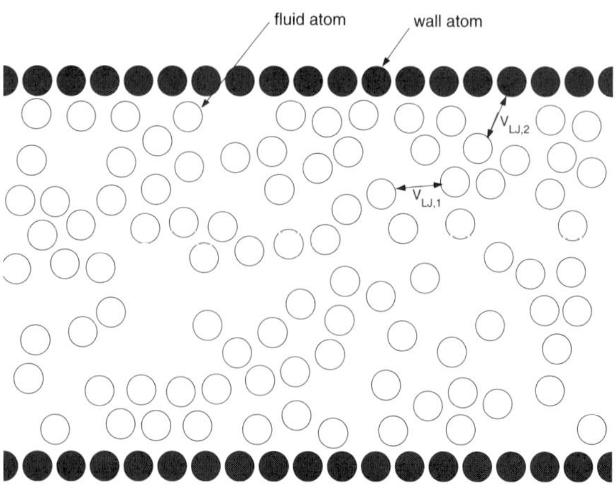

FIGURE 10.1. A sketch of a nanochannel filled with a simple fluid. The filled circles denote the channel wall atoms, and the open circles denote the fluid atoms. The fluid atoms interact with each other by a Lennard–Jones potential $V_{LJ,1}$, and the fluid atoms interact with the wall atoms by a Lennard–Jones potential $V_{LJ,2}$.

theory for hydrophobic surfaces.

10.1 Atomistic Simulation of Simple Fluids

Atomistic simulation of nanofluids has gained considerable attention over the last two decades. Much of the existing literature has been devoted to understanding "simple fluids" in nanochanels. Though there is no consensus on the precise definition of a simple fluid in the literature, in most cases, it is simply taken as a collection of atoms that interact via the Lennard–Jones potential and the dynamics of which follow the classical mechanics described by Newton's law. In practice, some noble gases (e.g., argon) can be modeled fairly accurately as a simple fluid. Figure 10.1 shows a schematic of a nanochannel filled with a simple fluid.

The investigation of simple fluids, in contrast to the study of complex fluids such as water (discussed in Chapter 11) or electrolytes (discussed in Chapter 12), has many advantages. First, the computational cost of atomistic simulation involving simple fluids is much lower compared to that of complex fluids, since it is much cheaper to evaluate the Lennard–Jones potential describing simple fluids compared to the evaluation of the electrostatic interactions that are required in the study of most complex fluids. Second, despite its simplicity, the investigation of simple fluids can provide deep insight into the physics of fluid transport in nanochannels, and such insight can guide the study of more complex fluids. For example, the study of simple fluids indicated that the classical Navier–Stokes equations breakdown in a channel as narrow as 4 fluid atomic diameters (Travis et al., 1997), and later, a similar finding was reported for electroosmotic transport in a silicon nanochannel that is also about 4 water diameters wide (Qiao and Aluru, 2003b). Third, the investigation of simple fluid transport provides data for the validation of theories describing fluid transport in nanochannels. Due to the complicated interactions involved in complex fluids, most of the nanofluid transport theories that have been developed so far are limited to simple fluids.

As mentioned above, simple fluids can be described using a Lennard–Jones (LJ) potential (LJ and other potentials are discussed in detail in Section 16.1; here we quickly recap the salient features of the LJ potential and introduce the reduced units that are used in this chapter) of the form

$$V_{LJ} = 4\epsilon \left[\left(\frac{\sigma}{r} \right)^{12} - \left(\frac{\sigma}{r} \right)^{6} \right], \tag{10.1}$$

where ϵ, σ are the Lennard–Jones parameters that depend on the atoms involved in the interaction. Note that:

1. ϵ is related to the interaction strength, and a higher ϵ corresponds to a higher interaction energy between the atoms.
2. σ corresponds to the distance at which the potential between the two atoms goes to zero, which can be approximately taken as the diameter of a fluid atom.

Since the Lennard–Jones potential describes the interactions between non polar molecules quite well (Talanquer, 1997) and the force corresponding to the Lennard–Jones potential can be evaluated efficiently numerically, it is the most popular interaction potential used in MD simulations. In the MD simulation of Lennard–Jones fluids, the physical quantities are typically computed using reduced units. Table 10.1 summarizes the units for various quantities, e.g., length, temperature, and density. In the table, ϵ and σ are as defined in equation (10.1), $k_{\rm B}$ is the Boltzmann constant, and m is the mass of a Lennard–Jones atom. Unless otherwise mentioned, all the quantities are measured in reduced units in the next two sections.

The studies on Lennard–Jones fluids have indicated that depending on the critical length scale of the channel (typically the channel width/height or the diameter), the fluidic transport behavior (e.g., convection and diffusion phenomena) can either deviate significantly from the classical continuum theory prediction or be very similar to the transport of a bulk fluid described by the classical theory. These observations follow from the fact that when the fluid atoms are confined to molecular channels, the fluid can no longer be taken to be homogeneous, and strong oscillations in fluid density occur near the solid–fluid interface. Therefore, the dynamic behavior of the fluid becomes significantly different from that of the bulk. Some

TABLE 10.1. Units for various quantities in Lennard–Jones fluids.

Length	σ	Velocity	$(\epsilon/m)^{1/2}$
Mass	m	Shear rate	$(\epsilon/m\sigma^2)^{1/2}$
Energy	ϵ	Stress	ϵ/σ^3
Time	$(m\sigma^2/\epsilon)^{1/2}$	Viscosity	$(m\epsilon)^{1/2}/\sigma^2$
Number density	σ^{-3}	Diffusivity	$\sigma(\epsilon/m)^{1/2}$
Temperature	ϵ/k_B		

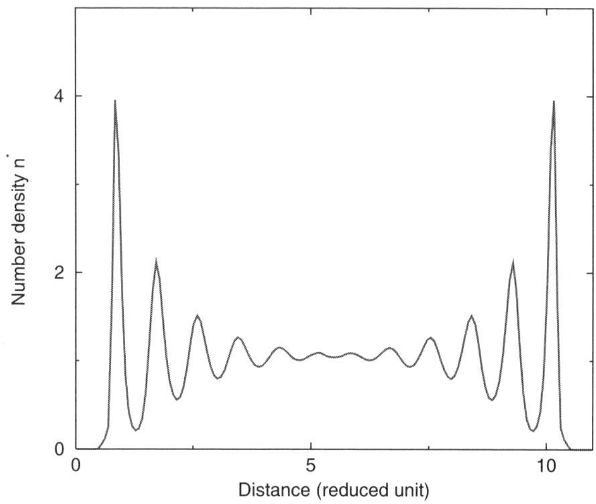

FIGURE 10.2. Density profile of a Lennard–Jones fluid. Simulations are performed in an 11-fluid-atomic-diameter channel.

significant results that have been observed when LJ liquids are confined in nanochannels are summarized below.

10.2 Density Distribution

The strong density oscillations of fluid atoms near the fluid/solid interface is a universal phenomenon, and it has been observed in almost all MD simulations of nanofluidic flows and been verified experimentally (Chan and Horn, 1985; Zhu and Granick, 2002; Zhu and Granick, 2001). Figure 10.2 shows the density profile of Lennard–Jones fluid atoms in a 11-fluid-diameter-wide channel (see also Figure 1.7, which shows density fluctuations of an LJ liquid in a larger channel). Density fluctuations near a channel wall can be explained using the concept of a radial distribution function (RDF). A radial distribution function (or the pair correlation function), typically denoted

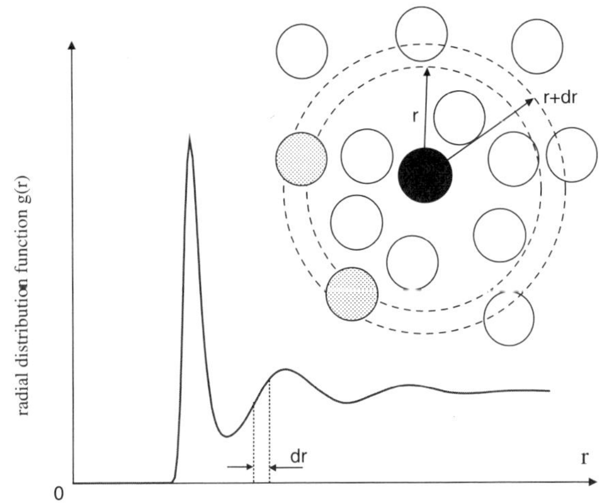

FIGURE 10.3. Sketch of a typical radial distribution function (RDF) $g(r)$. RDF measures the probability density of finding a particle at a distance r from a given particle ($r = 0$ corresponds to the position of the given particle).

by $g(r)$, is a basic measure of the structure of a liquid. RDF measures the probability density of finding a particle at a distance r from a given particle position. Figure 10.3 is a sketch of a typical radial distribution function. At a short distance from the given particle position, $g(r)$ is essentially zero because of the strong repulsion between the particles; i.e., particles cannot get too close to each other. As r increases, $g(r)$ shows a first peak, which is mainly caused by the attractive interactions between the particles. At a short distance from the first peak, a depletion of the particles is observed because of repulsive forces, and this gives rise to a minimum in $g(r)$. The combination of the attractive and the repulsive forces between the particles leads to the various peaks and valleys observed in the radial distribution function shown in Figure 10.3. At a distance farther away from the given particle position, the distribution of particles is no longer influenced by the given particle, and $g(r)$ approaches a constant. The fluid layering near the channel wall is mainly induced by the structure of the fluid radial distribution function and the structure of the solid wall. Here the position of the solid wall is similar to the position of the given particle in the radial distribution function, and the fluid density oscillations are similar to the oscillations in the radial distribution function.

Simple fluids in nanochannels are inhomogeneous because of the strong layering of fluid atoms near the channel wall. Classical fluid transport theories do not account for the inhomogeneity of the fluid, and transport parameters such as diffusivity and viscosity are strongly influenced by the fluid layering in nanochannels (Thompson and Troian, 1997). Fluid lay-

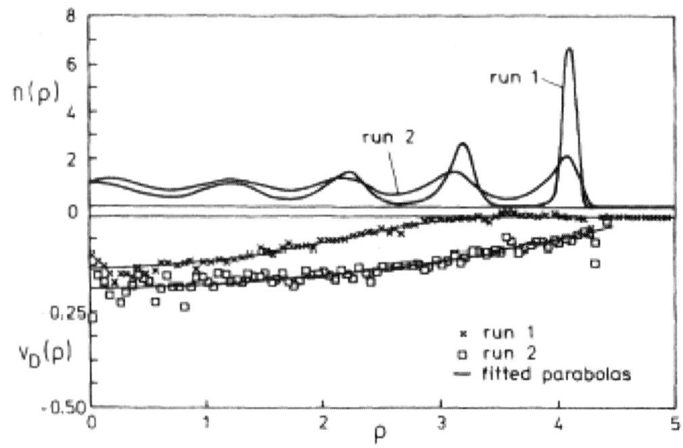

FIGURE 10.4. Density (upper panel) and velocity (lower panel) profiles in a 5.0σ-radius cylindrical pore for two separate runs with different wall–fluid interactions. In the first run, $\epsilon_{\rm wf}$ is 3.5 times larger than ϵ, and in the second run, $\epsilon_{\rm wf}$ is equal to ϵ. (Courtesy of J. Fischer.)

ering can be influenced by various parameters such as the wall structure, fluid–wall interactions, and channel width, and these issues are discussed below.

Effect of Fluid–Wall Interactions

The interaction between a fluid atom and a wall atom is usually modeled by the Lennard–Jones potential. The Lennard–Jones parameters for fluid–fluid and fluid–wall interactions are denoted by (ϵ, σ) and $(\epsilon_{\rm wf}, \sigma_{\rm wf})$, respectively. A higher $\epsilon_{\rm wf}$ corresponds to a stronger interaction between the fluid and the wall atoms. (Heinbuch and Fischer, 1989) found that the fluid layering becomes stronger when $\epsilon_{\rm wf}$ increases. Figure 10.4 shows the number density and velocity profiles for two separate runs with different fluid–wall interaction parameters in a 5.0-σ radius cylindrical pore. In the first run, $\epsilon_{\rm wf}$ is 3.5 times larger than ϵ, and in the second run, $\epsilon_{\rm wf}$ is equal to ϵ. In both runs, the average number density of the fluid in the pore is 0.8 and the temperature of the fluid is 0.835. Clearly, the layering effect is much more distinct in the first run compared to the second run. This can be explained by the fact that as $\epsilon_{\rm wf}$ increases, the attractive force exerted by the wall atoms on the fluid atoms increases, and the tendency of a fluid atom to stay near the wall increases. A similar observation has been made by (Thompson and Robbins, 1990) in their study of Couette flow in a slit channel of width 12.8σ.

(Travis and Gubbins, 2000) further investigated the variation of the density profile in a nanochannel when the attractive part of the Lennard–Jones interaction is turned *on* and *off* by shifting the Lennard–Jones potential. Specifically, the Poiseuille flow in a 4.0σ-wide slit channel was investigated for three different systems, A, B, and C. In system A, fluid–fluid and fluid–wall interactions are described by the purely repulsive part of the Lennard–Jones, or Weeks–Chandler–Andersen (WCA) potential (see Section 16.1 for details on WCA potential). In system B, the fluid–fluid and fluid–wall interactions are described by the full 12-6 Lennard–Jones (including both the attractive and the repulsive interactions) potential. In system C, the fluid–fluid interactions are described by the WCA potential, and the interactions between the fluid and the wall are described by the full 12-6 Lennard–Jones potential. Figure 10.5 (a) shows a comparison of the density profiles for the three different systems. It is observed that the presence of attractive fluid–wall forces (system B and C) leads to the formation of boundary liquid layers of higher density than in the case of repulsive wall–fluid interactions (system A). It is also observed that the density of the layers is higher in system C compared to that of system B. This can be explained by the fact that compared to system B, the fluid atoms in system C have a greater affinity for the wall atoms and less affinity for other fluid atoms. In addition, the number of density peaks (i.e., the number of fluid layers in the channel) is also different for the three systems. These results indicate that the density distribution of fluid atoms in the channel is sensitive to both the fluid–wall and fluid–fluid interactions, and care should be taken in choosing the best potential to depict a particular fluidic system. Figure 10.5 (b) shows the average number density of fluid atoms along the channel length direction. We see that the fluids are highly structured in all three systems. The density oscillates with a wavelength of order σ. Clearly, the wall structure has been imposed upon the fluid. Similar behavior has also been observed by (Zhang et al., 2001) in the simulation of n-decane confined between two Au(111) surfaces.

Effect of Structure and Thermal Motion of the Wall Atoms

Smooth walls (Toxvaerd, 1981; Somers and Davis, 1992) as well as walls with atomistic structure (Travis and Gubbins, 2000; Sokhan et al., 2001; Somers and Davis, 1992) have been widely used in the MD simulation of fluids confined in nanoscale channels. For a smooth wall, the wall–fluid potential depends only on the normal distance between the fluid atom and the channel wall, while for a wall with atomistic structure, the wall–fluid potential depends on the relative distance between the fluid atom and each atom in the wall. Typically, only the first fluid layer is significantly influenced by the wall structure, and the rest of the fluid layers are not significantly affected by the structure of the wall. In many simulations, the wall atoms are either frozen to their lattice sites (Heinbuch and Fischer, 1989; Zhang et al., 2001) or constrained to their lattice sites by a spring (Thompson and Robbins, 1990; Travis and Gubbins, 2000; Sokhan et al., 2001). The former

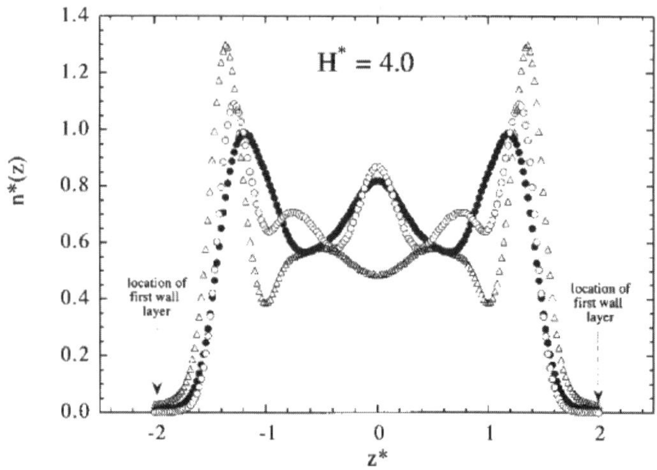

(a)

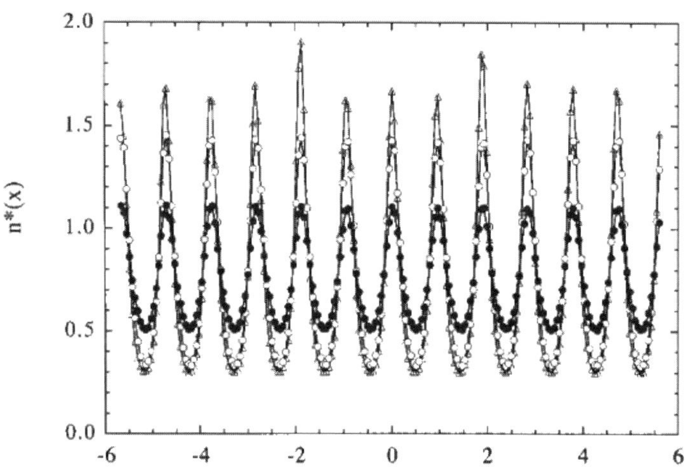

(b)

FIGURE 10.5. Number density profiles across (a) and along (b) a 4.0σ-wide slit channel for three different cases where the fluid–fluid and fluid–wall interactions are modeled differently (System A: WCA system, filled circles, system B: LJ system, open circles, and system C: WCA-LJ system, open triangles). (Courtesy of K. P. Travis.)

enables the use of a larger time step in MD, since the thermal vibration of the solid atoms is not resolved, while the latter seems to be more realistic. The thermal oscillation of wall atoms introduces further corrugations into the potential felt by the fluid atoms near the wall, and therefore leads to a reduced density oscillation near the channel wall (Thompson and Robbins, 1990; Sokhan et al., 2001).

Effect of Channel Width

The density oscillations in the channel also depend on the channel width. Somers and Davis investigated the variation of density profiles in slit channels of different widths ranging from 2.0σ to 8.0σ (Somers and Davis, 1992). Figure 10.6 shows the density profiles obtained with various channel widths. The results indicate that for channel widths that can accommodate integral fluid layers, e.g., 2σ, 2.75σ, and 4σ, distinct peaks in the density profile are observed. For channel widths, such as 2.25σ and 3.25σ, the channel is wide enough to accommodate distinct fluid layers: hence additional layers begin to develop, and the new layers appear at channel widths of 2.5σ and 3.5σ. When the channel width is 8σ, seven distinct peaks are observed, and as the channel width increases further, more peaks are observed, but these additional peaks are much weaker compared to the peaks very close to the channel wall. For channels wider than 10σ, the fluid layers near the wall are independent of the channel width, and the fluid in the central portion of the channel behaves more like a bulk fluid.

Effect of Fluid Flow

The fluid flow is found to have negligible effect on the density distribution. (Bitsanis et al., 1987) conducted simulations on Couette flow in a nanometer-slit pore using shear rates of 10^{10} to 10^{11} s^{-1} and found that the density distribution is not significantly influenced by the flow; i.e., the density profile in the channel in the presence of bulk transport of fluid is the same as that obtained from an equilibrium MD simulation. This result was also verified by other reported MD simulations. Since experimental techniques or real processes that deal with fluid flow in nanopores are likely to employ smaller shear rates than those that were employed in MD, we can conclude that in simple fluids the density profile will be the equilibrium density profile.

While all the results presented above on the calculation of density profiles in nanochannels were based on MD simulations, the density profiles can also be calculated using analytical methods. (Fischer and Methfessel, 1980) as well as (Bitsanis et al., 1988) have used the Yvon–Born–Green (YBG) theory (McQuarrie, 1973) of inhomogeneous fluids with the Fischer–Methfessel approximation for the fluid pair-correlation functions (Fischer and Methfessel, 1980) to calculate the density distribution of Lennard–Jones atoms confined in a nanoscale channel. In order to obtain an equation for number

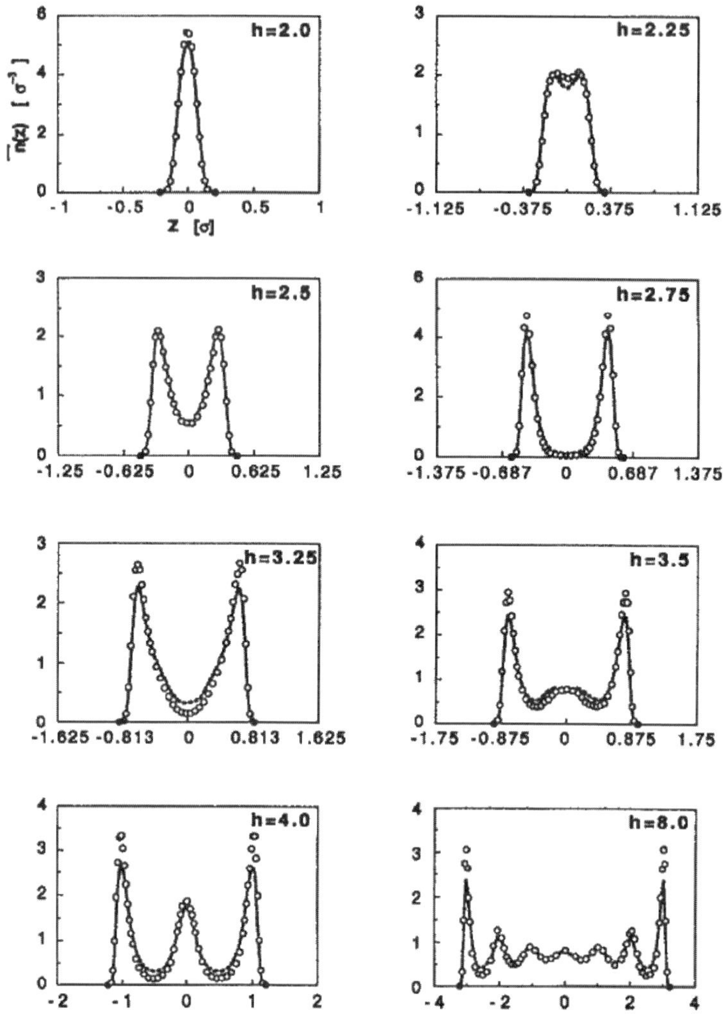

FIGURE 10.6. Fluid density distribution in smooth-slit channels of different widths. The results with reservoir are shown as a line, and the results with no reservoir are shown as circles. A reservoir is introduced to simulate the Couette flow. (Courtesy of H. T. Davis.)

density one has to approximate the pair correlation function, which is the Fischer–Methfessel approximation. This is the only approximation needed, and in this respect the YBG theory is superior to the free-energy theories. The YBG theory has been shown to predict the fluid densities near smooth walls with good accuracy.

10.3 Diffusion Transport

Diffusion transport is typically important in most nanofluidic systems. This can be understood by calculating the Peclet number, $P_e = UL/D$, which measures the ratio of the bulk transport (convection) to the diffusion transport. In most nanofluidic systems, the characteristic length L ranges from a nanometer to a micrometer, and the bulk velocity ranges from a micrometer per second to a millimeter per second. For a fluid with a diffusivity of $D = 1.0 \times 10^{-9}$ m/s^2, the Peclet number ranges from 10^{-6} to 1, indicating that diffusion either dominates the transport or is as important as the bulk transport.

The diffusion coefficient must be generalized in order to describe transport in confined nanochannels. For *homogeneous* and *equilibrium* systems, the diffusion coefficient can be calculated using either the Green–Kubo equation

$$D = \frac{1}{3}\int_0^\infty \langle \boldsymbol{v}(0)\cdot\boldsymbol{v}(t)\rangle\,dt, \tag{10.2}$$

where $\boldsymbol{v}$ is the atom velocity and $\langle\rangle$ denotes the ensemble average, or by the Einstein equation

$$D = \frac{1}{6}\lim_{t\to\infty}\frac{\langle[\boldsymbol{r}(t_0+t)-\boldsymbol{r}(t_0)]^2\rangle}{t}, \tag{10.3}$$

where $\boldsymbol{r}$ is the atom position. The Green–Kubo expression given in equation (10.2) is strictly valid only for homogeneous and equilibrium systems. However, it is generally accepted that, at least for the calculation of the average diffusivity in nanochannels, the Green–Kubo expression given in equation (10.2) or the Einstein relationship given in equation (10.3) is adequate. For example, Bitsanis and coworkers (Bitsanis et al., 1987) computed pore-averaged diffusitives and found that the diffusivities under flow and the equilibrium diffusivity agree within the limits of statistical uncertainity. Moreover, the diffusivities calculated from the Green–Kubo formula and the Einstein relationship agree quite well. It is important to note that in the calculation of the diffusivities under flow, the drift contribution to either equation (10.2) or equation (10.3) has to be excluded. In summary, both the Green–Kubo formula and the Einstein relationship are widely used in the calculation of diffusivity of fluids in nanochannels.

The diffusion of fluids confined in nanoscale channels has been studied extensively in slit and cylindrical pores. In a slit pore, diffusion is different in the direction parallel (the x- and y-directions) and normal (z-direction) to the pore wall, especially for narrow pores. This is because, unlike the diffusion parallel to the pore wall, the diffusion in the direction normal to the pore wall is inherently transient; i.e., in the long time limit, the diffusivity in the direction normal to the pore wall is zero due to the geometrical limit. To circumvent this problem, the diffusion in the z-direction is usually characterized by a mean-square displacement $\Delta z^2(t)$ that can be calculated for a short time. In this section, we will discuss the results for diffusion parallel to the pore wall (characterized by $D_{\|}$ or D_x and D_y) and the diffusion normal to the pore wall (characterized by $\Delta z^2(t)$) separately.

(Magda et al., 1985) studied diffusion in slit pores with smooth pore walls using equilibrium MD simulations. Figure 10.7 shows the variation of the pore-averaged diffusivity parallel to the pore ($D_{\|}$) with the pore width. The plot indicates that:

1. Even for the smallest pore width ($h = 2$), where the fluids are highly confined, the fluid atoms maintain considerable mobility.

2. When the channel width is small ($h < 4$), the average $D_{\|}$ in the pore fluctuates with the channel width, and when the channel width increases beyond $h = 5$, the average $D_{\|}$ increases smoothly toward the asymptotic bulk value.

3. For a channel width of $h = 11.57$, the average $D_{\|}$ is almost the same as the bulk diffusivity.

The second observation can be attributed to the average density variation with the change in channel width. As shown in Figure 10.7, when the channel width is small ($h < 4$), the average density fluctuates with the channel width, and when the channel width increases beyond $h = 5$, the average density decreases smoothly toward the asymptotic bulk value. The dependence of the diffusivity on density in the pore region is much weaker compared to the quadratic dependence observed in the bulk (Levesque and Verlet, 1970). This means that the variation of diffusivity with density follows a quadratic dependence as the pore width increases and the properties of the confined fluid approach that of the bulk. However, when the pore width is low (lower than 6σ), then the layering effect dominates and affects the variation of diffusivity. This leads to a weaker dependence of diffusivity on density for narrow pores. A possible explanation for this is the structured, almost solid-like form of the density profile in narrow pores.

To investigate the effect of the local fluid density on the diffusivity parallel to the pore ($D_{\|}$), the pore ($h = 11.57$) has been divided into five slices parallel to the solid–liquid interface, and the diffusivity $D_{\|}$ is calculated inside each slice. Figure 10.8 shows the density profile and the diffusivity in each slice. Clearly, even though there is a significant variation in the fluid density, the diffusivity in each slice is within the statistical error of those of the others. To understand this result in more detail, an empirical theory, *local average density method* (LADM), has been developed to describe the transport coefficient (e.g., diffusivity) of a fluid confined in a nanochannel

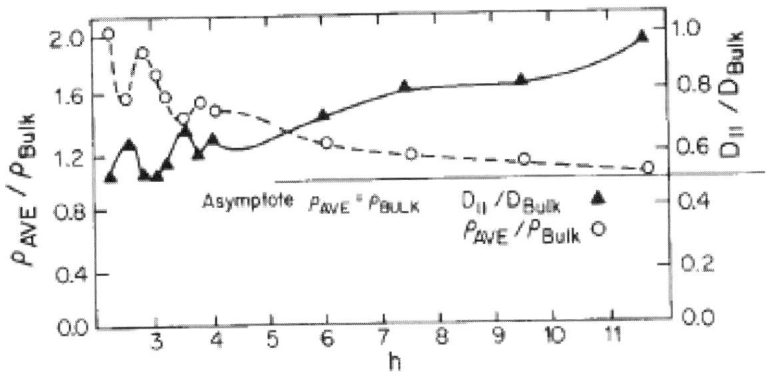

FIGURE 10.7. Correlation of the pore-averaged diffusivity parallel to the wall ($D_{\parallel}$) with the average fluid density. Here h is the pore width (reduced unit); $\rho_{\text{ave}}/\rho_{\text{bulk}}$ is the average density of the occupied pore volume divided by the density of the bulk liquid. (Courtesy of H.T. Davis.)

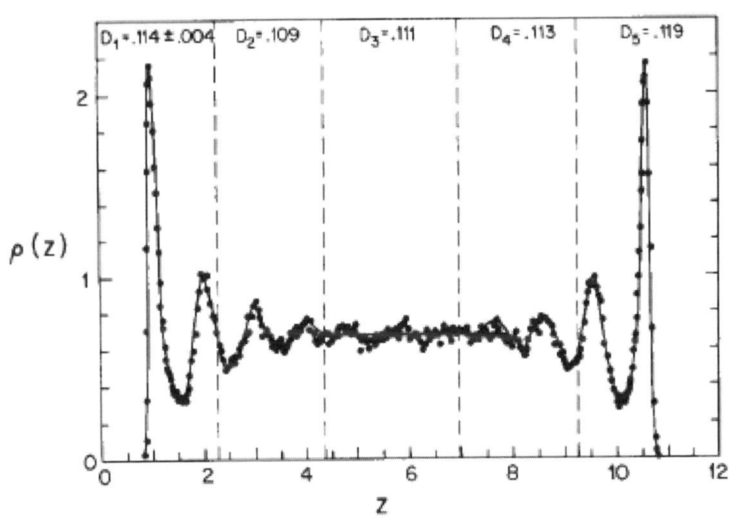

FIGURE 10.8. Diffusivity as a function of the distance from the pore walls. Here D_i is the diffusion coefficient parallel to the pore walls averaged over the ith slice parallel to the interface. The pore width is 11.57σ. (Courtesy of H.T. Davis.)

(Bitsanis et al., 1988). In the LADM theory, the diffusivity of the fluid at a position $\boldsymbol{r}$ depends on the *local average* density $\overline{\rho}(\boldsymbol{r})$ of the fluid instead of the *local* density $\rho(\boldsymbol{r})$. The local average density at $\boldsymbol{r}$ is defined as the average density inside a sphere with its center at $\mathbf{r}$ and with diameter equal

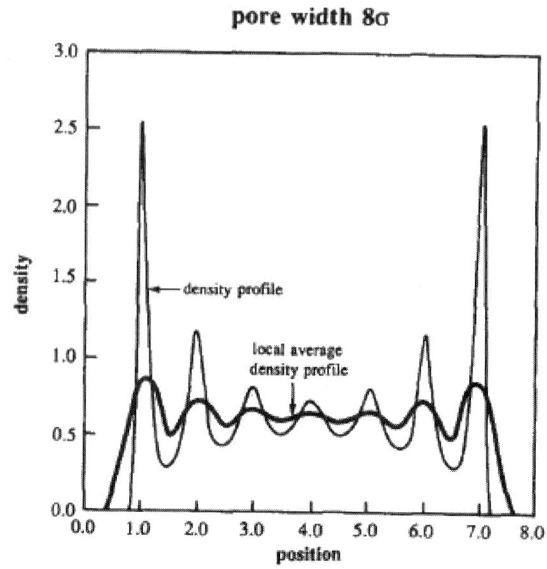

FIGURE 10.9. Density and local average density profiles in an 8σ-wide slit pore. (Courtesy of H.T. Davis.)

to the diameter of the fluid molecules σ, i.e.,

$$\overline{\rho}(\boldsymbol{r}) = \frac{1}{\pi\sigma^3/6} \int_{s<\sigma/2} \rho(\boldsymbol{r} + \boldsymbol{s})\, d^3 s.$$

Figure 10.9 shows the local average density distribution in an 8.0σ channel (Bitsanis et al., 1988). Clearly, even though the oscillations in the local fluid density are significant, the local average density shows very little oscillation. This explains the small diffusivity variation in the pore (see Figure 10.8) even though there is a significant variation in the local fluid density. From the above discussion, we can conclude that in confined nanopores the parallel diffusivity at a given position is determined primarily by the average density in the pore and not by the local density.

To investigate the effect of the wall structure on the diffusivity parallel to the pore wall, (Somers and Davis, 1992) performed diffusivity calculations by considering structured and smooth walls. Figures 10.10 (a) and (b) show the variation of the diffusivity $D_{\|}$ with the pore width for the structured and smooth walls, respectively. The results for the smooth-pore wall (panel (b)) agree qualitatively with those reported in (Magda et al., 1985), and the results for the structured wall show some interesting differences from the results for the smooth wall; i.e., as the pore width increases, the diffusivity in the structured pore approaches the bulk value more slowly compared to the smooth wall. This can be explained by the added fluid ordering in the structured pore. Figure 10.10 also shows the variation of the diffusivity with bulk transport of the fluid. For both the smooth and structured pore walls, the presence of Couette flow does not change the diffusivity noticeably for a shear rate less than $0.20\sqrt{\epsilon/m\sigma^2}$, but the diffusivity increases considerably for shear rates higher than $0.20\sqrt{\epsilon/m\sigma^2}$. It is likely that shear thinning may be responsible for the observed increase in the diffusivity. Similar results have also been observed by (Bitsanis et al., 1987). In addition, the inset in Figure 10.10 (a) shows that the wall registry (characterized by the wall registry index α) also affects the diffusivity significantly in narrow pores. This is likely to be caused by the changed fluid structure when the wall structure is varied. The wall registry is a measure of the variation in surface roughness of the wall (Somers and Davis, 1992). When the axial coordinate of the surface atoms on both the walls (i.e., the upper wall and the lower wall of the slit) are the same, then the wall registry index α is 0. However, when the upper wall atoms and the lower wall atoms are seperated by a certain distance, then the wall registry index is nonzero.

As mentioned above, the diffusivity normal to the channel wall cannot be defined using equation (10.2) or (10.3). However, it is possible to characterize the diffusion process normal to the channel wall by a mean-square displacement $\Delta z^2(t)$ that can be calculated for a short time. Figure 10.11 shows the comparison of the mean-square displacement in the direction normal (z-direction) and parallel to the channel wall (x- and y-directions) for channel widths of 3.0σ and 4.0σ. Figure 10.11 indicates that:

1. After 2.0 ps, the diffusion process in the z-direction can be characterized by a mean-square displacement.

2. The diffusion process in the z-direction is faster in the larger channel ($h = 4\sigma$) compared to the smaller channel ($h = 3\sigma$).

3. For $h = 4\sigma$, the diffusion in the x- and y-directions is much faster compared to the diffusion in the z-direction.

These results are not surprising, since the movement of the fluid atoms is more confined in the direction normal to the channel wall compared to the movement of the fluid in the direction parallel to the channel wall.

In the discussion so far, we have assumed that the fluid molecule size is smaller than half of the slit pore width and that the diffusion is characterized by the normal-mode diffusion; i.e., the mean-square displacement of the fluid molecules obeys the Einstein relationship. In the normal-mode diffusion, one molecule can pass another molecule within the channel. However, if the pore width decreases further and the pore is cylindrical, a molecule cannot pass another molecule because of its large size relative to the pore size, and the diffusion process is then characterized by a single-file diffusion. The mean-square displacement of a fluid molecule due to single-file

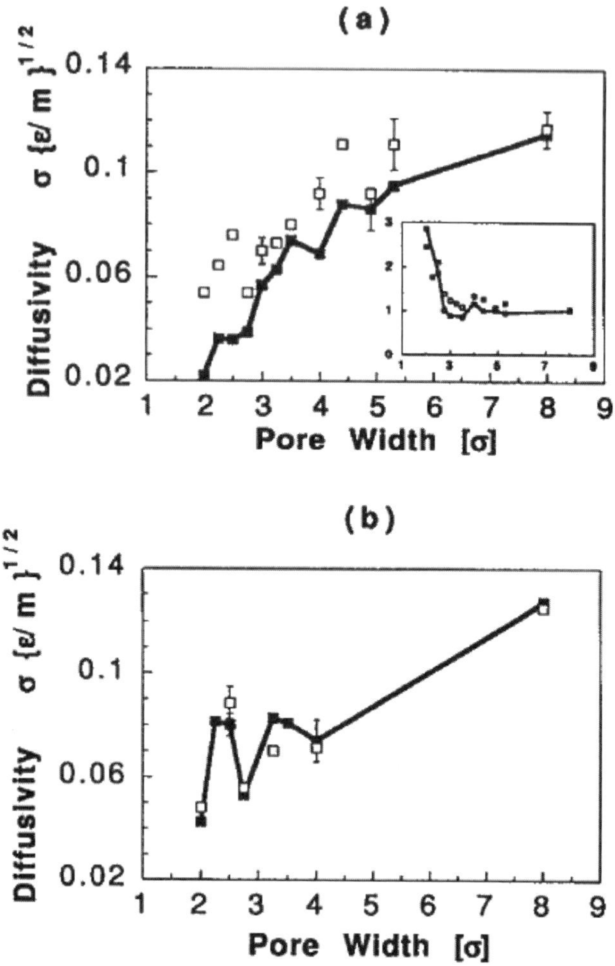

FIGURE 10.10. Diffusivity parallel to pore walls versus pore width (■ - equilibrium ($\alpha = 0$), □ - Couette flow). (a) Structured pores (inset: □ - ratio of the diffusivity in Couette flow and in equilibrium simulation ($\alpha = 0$), ■ - ratio of diffusivity in equilibrium for $\alpha = 0.71$ and $\alpha = 0$. See (Somers and Davis, 1992), for the definition of the wall registry index α). (b) Smooth pores. (Courtesy of H.T. Davis.)

diffusion can be expressed as

$$s^2 = 2Bt^{0.5},$$

where B is the diffusion mobility. For the diffusion of methane, ethane,

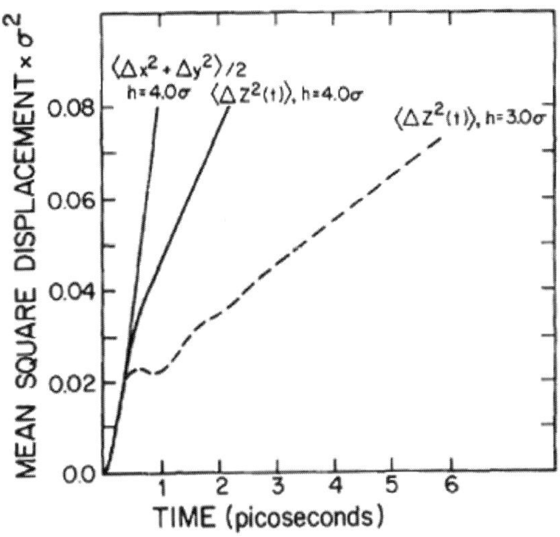

FIGURE 10.11. Comparison of the mean-square displacement parallel and normal to the channel walls. (Courtesy of H.T. Davis.)

and ethylene through carbon nanotubes (see Section 13.2 for a discussion on carbon nanotubes), (Mao and Sinnott, 2000) showed that there exists a transition-mode diffusion for which the mean-square displacement of a fluid molecule due to a single-file diffusion can be expressed as

$$s^2 = 2Ct^n.$$

Figure 10.12 shows the log-log plots of the mean-square displacement for methane, ethane, and ethylene in a $(10, 0)$ carbon nanotube (diameter: 0.8 nm). Clearly, the diffusion of methane is a normal-mode diffusion and the diffusions of ethane and ethylene are transition-mode diffusions. Mao and Sinnott further showed that the diffusion of ethane and ethylene in a $(9, 0)$ carbon nanotube (diameter: 0.72 nm) follows the single-file mode, while the diffusion of methane is still in the normal mode. Such differences are caused by the fact that the size of ethane and ethylene are larger compared to methane; therefore, the methane molecules can pass one another in a 0.72-nm diameter, while ethane and ethylene molecules cannot pass each other in such small-diameter tubes.

10.4 Validity of the Navier–Stokes Equations

In the continuum fluid transport theory governed by the Navier–Stokes equations, it is assumed that the state variables (e.g., density and tempera-

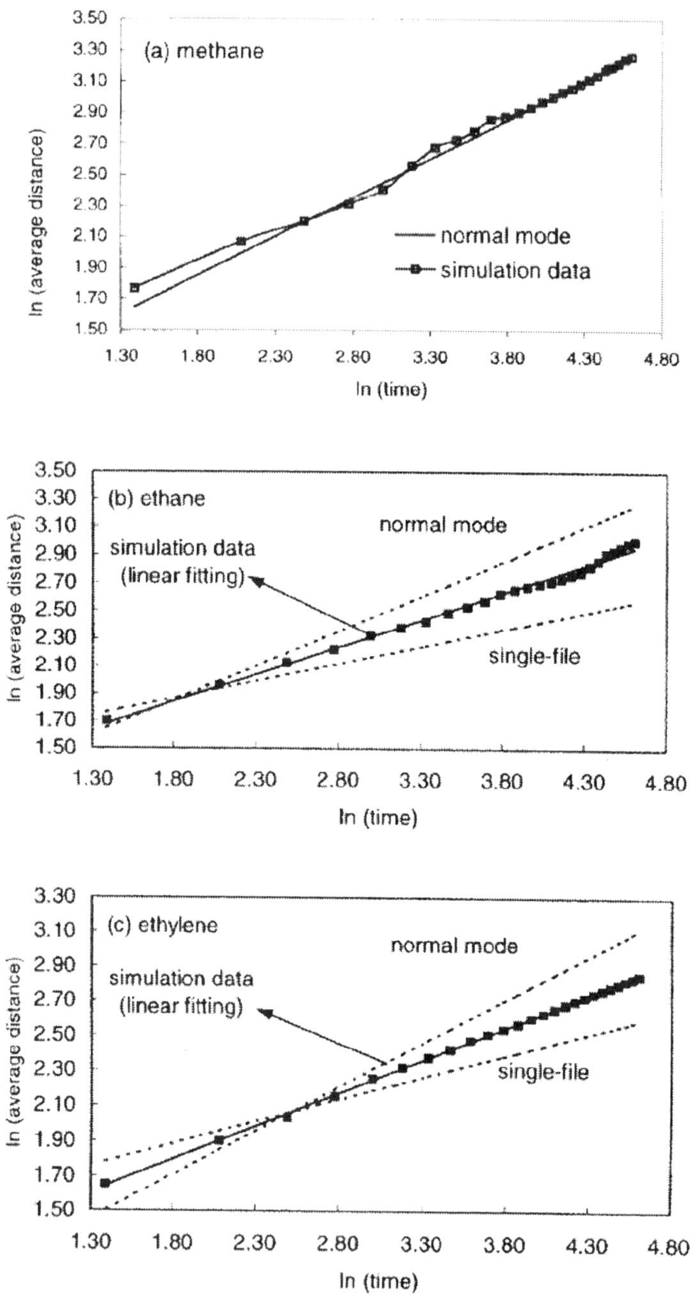

FIGURE 10.12. Log-log plot of the mean-square displacement for methane (a), ethane (b), and ethylene (c) in a (10, 0) carbon nanotube (diameter: 0.8 nm). (Courtesy of S. Sinnott.)

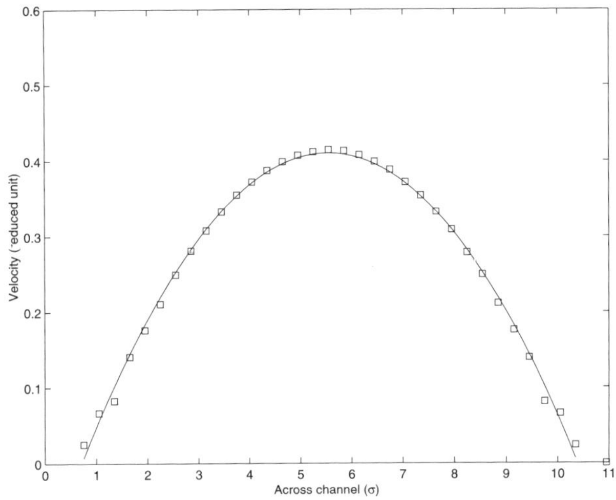

FIGURE 10.13. Velocity profile obtained from MD simulation of Poiseuille flow in an 11-fluid-diameter channel. The solid line is a quadratic fit to the velocity profile. The temperature of the fluid is set to 2.5, the average fluid density is 0.8, and a constant force of 0.1 is applied on each fluid molecule to generate the flow. All the variables are measured in reduced units.

ture) do not vary appreciably over the length and time scales comparable to the molecular free path and molecular relaxation time. However, as shown in Section 10.2, the fluid density near the solid–liquid interface can vary significantly over intermolecular distances. While these local density oscillations may not necessarily mean the breakdown of the continuum theory, it is important to understand in detail how the continuum theory works for fluids in confined nanochannels.

During the last several years, researchers have used MD simulations to test the accuracy of Navier–Stokes equations in nanochannels (Koplik et al., 1989; Koplik et al., 1987; Travis and Evans, 1996; Travis et al., 1997; Bitsanis et al., 1987; Travis and Gubbins, 2000; Pozhar, 2000). In many of these simulations, a Poiseuille flow with a constant force on each fluid molecule is used as a prototypical problem. The continuum Navier–Stokes equations predict a parabolic velocity profile across the channel for the Poiseuille flow. The velocity profiles in slit channels as narrow as 10 molecular diameters indicate that the deviation between continuum and MD predictions is very small (Travis et al., 1997). Figure 10.13 shows the velocity profile obtained from MD simulation of Poiseuille flow in an 11-fluid-diameter channel and its quadratic fit. Clearly, the deviation of the velocity profile from the Navier–Stokes equation is small. However, if the channel width is smaller than 10 fluid diameters, the deviation of the MD velocity from the continuum prediction becomes more significant (Travis and Gubbins,

2000; Travis et al., 1997). For example, Figure 10.14 shows the velocity distribution for Poiseuille flow in a 4-fluid-molecule-diameter slit channel for three systems (see Section 10.2 for a description of the potentials used in the three systems) with different interaction potentials between fluid–wall and fluid–fluid (Travis and Gubbins, 2000). For each system considered, the velocity profile obtained from MD simulations is no longer parabolic and deviates significantly from the Navier–Stokes prediction. Specifically, for system A (we will discuss only the result for system A, since the results for system B and system C are similar to that of system A), the velocity decreases in the region $0.75 < |z| < 0.97$ as we approach the channel center, and there is a local maximum for the velocity located at $|z| \approx 0.95$. The corresponding strain rate profile is shown in Figure 10.15, and the strain rate is zero at $|z| \approx 0.97$ and $|z| \approx 0.2$. The fluid viscosity inside the channel, calculated by

$$\mu(z) = \frac{\tau_{xz}(z)}{\gamma(z)}, \tag{10.4}$$

is shown in Figure 10.16. Note that in equation (10.4), $\tau_{xz}(z)$ is the shear stress at position z, and a local, linear constitutive relationship between the shear stress and the strain rate, on which the classical Navier–Stokes equation is based, is assumed. We observe that the viscosity calculated by equation (10.4) is negative in the region $0.75 < |z| < 0.97$ and $0 < |z| < 0.2$ and diverges at $|z| \approx 0.97$ and $|z| \approx 0.2$. This indicates that the viscosity in such a narrow channel cannot be described by a local, linear constitutive relation. Therefore, the classical Navier-Stokes equation is not valid for the analysis of fluid flow in a 4.0-fluid-diameter slit channel.

For fluid flow in channels larger than 10 fluid molecular diameters, the classical Navier–Stokes equation can be used to analyze the flow. However, since the fluid can be highly confined in nanochannels, the fluid viscosity may be significantly different from the bulk value. Despite the fact that the fluid viscosity is an important parameter in determining the flow characteristics, only a few papers have discussed the viscosity of fluids in nanochannels systematically. (Bitsanis et al., 1990) have calculated the effective viscosity for Couette and Poiseuille flow in slit channels ranging from 2 to 9.5 fluid molecular diameters. The effective viscosity is defined in such a way that the Navier–Stokes equation using the effective viscosity can predict the macroscopic observables correctly, e.g., the surface shear stress in a Couette flow or the flowrate in a Poiseuille flow. Figure 10.17 shows the variation of the effective viscosity for Couette and Poiseuille flow with the pore width. The plot indicates that:

1. The effective fluid viscosity is flow-dependent;

2. At small pore width, the effective viscosity increases dramatically; and

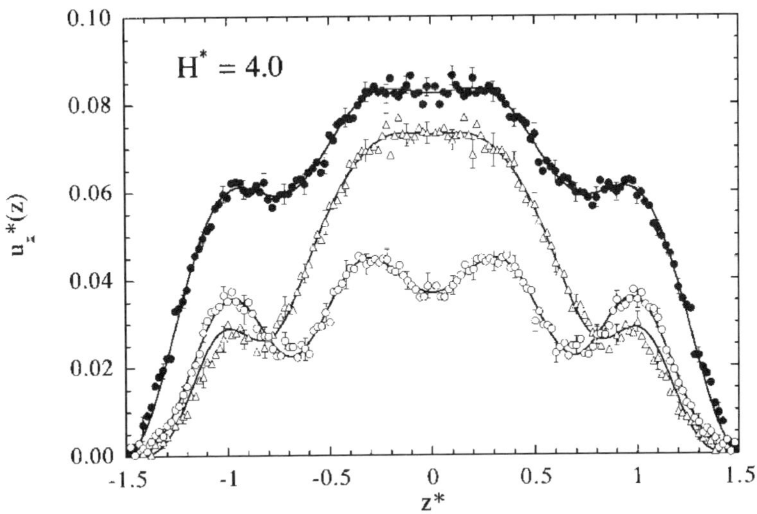

FIGURE 10.14. Velocity profiles in a 4.0-fluid-diameter slit channel for three different systems. System A: WCA system, filled circles; system B: LJ system, open circles and system; C: WCA-LJ system, open triangles. See Section 10.2 for the definition of the three systems. (Courtesy of K.P. Travis.)

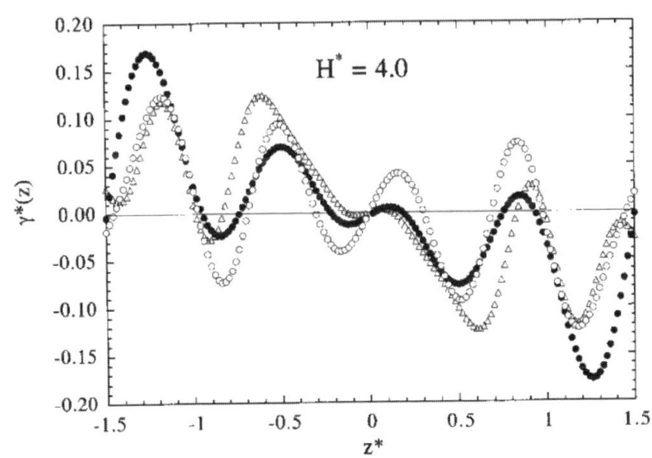

FIGURE 10.15. Strain rate profiles in a 4.0-fluid-diameter slit channel for three different systems. System A: WCA system, filled circles; system B: LJ system, open circles; and system C: WCA-LJ system, open triangles. (Courtesy of K.P. Travis.)

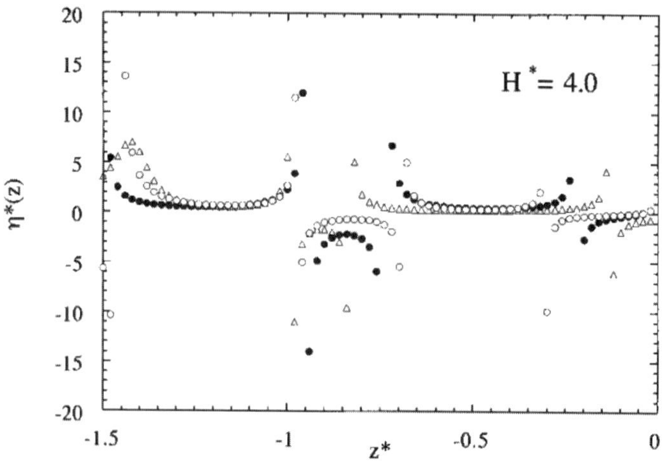

FIGURE 10.16. Shear viscosity in a 4.0-fluid-diameter slit channel for three different systems as calculated by equation (10.4). System A: WCA system, filled circles, system B: LJ system, open circles and system C: WCA-LJ system, open triangles. (Courtesy of K. P. Travis)

3. The effective viscosity converges toward the bulk viscosity as the pore width increases.

The first observation originates from the definition of the effective viscosity for the flow. The second observation is caused by the fluid layering near the pore surface. As the pore width decreases, more fluid molecules are observed within the fluid layers near the pore wall, where the movement of the fluid molecules is highly confined, thus leading to a higher effective viscosity.

Over the past decade, Pozhar has developed a rigorous statistical-mechanics approach to nonequilibrium phenomena in strongly inhomogeneous fluids (Pozhar, 1994; Pozhar, 2001; Pozhar, 2000). The approach relies on the rigorous generalization of the Mori–Zwanzig projection operator technique developed in the framework of the theory of dynamical systems. This approach, unlike the LADM method, has been realized for the most general case of strongly inhomogeneous fluids. However, due to the complicated nature of the theory, there have been very few applications of this method for the analysis of fluid flow. Therefore, the accuracy of the theory has not yet been rigorously established.

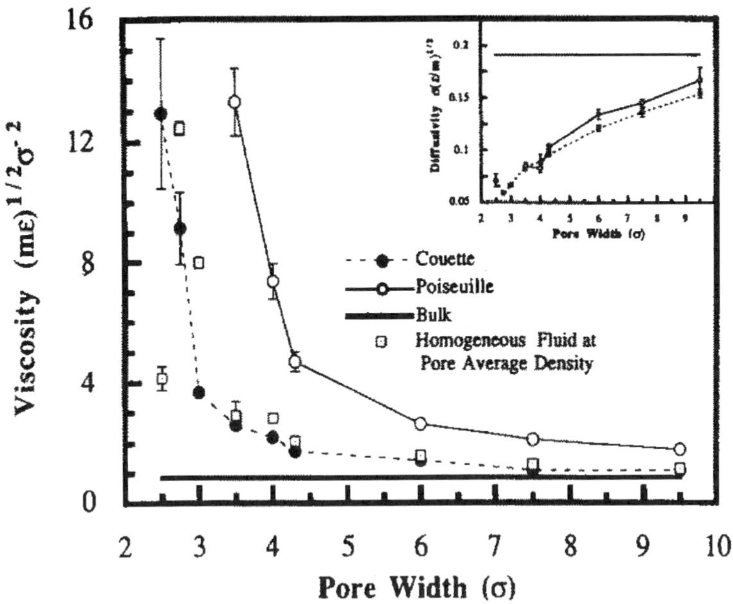

FIGURE 10.17. Effective viscosities for Couette and Poiseuille flows versus pore width. The viscosity of a homogeneous fluid at the pore average density is also shown. The inset shows the variation of the pore-averaged self-diffusion coefficients. (Courtesy of H.T. Davis.)

10.5 Boundary Conditions at Solid–Liquid Interfaces

In this section we revisit the question of slip at solid–liquid interfaces and present an in-depth review of the experimental work as well as the conceptual models derived from these findings. We then present different mathematical models that correct the no-slip condition and have been found to represent the available experimental data reasonably well.

10.5.1 Experimental and Computational Results

The first experimental work to investigate the validity of the no-slip boundary condition at a solid surface was conducted by Coulomb (1784), who concluded that it was valid even at microscopic scales. About a century later, Helmholtz and von Piotrowski (1860) found evidence of slip between a solid surface and a liquid, and later, (Brodman, 1891) verified their results. However, Couette (1890) and others used glass tubes with grease inside and concluded that the no-slip boundary condition is valid. It is

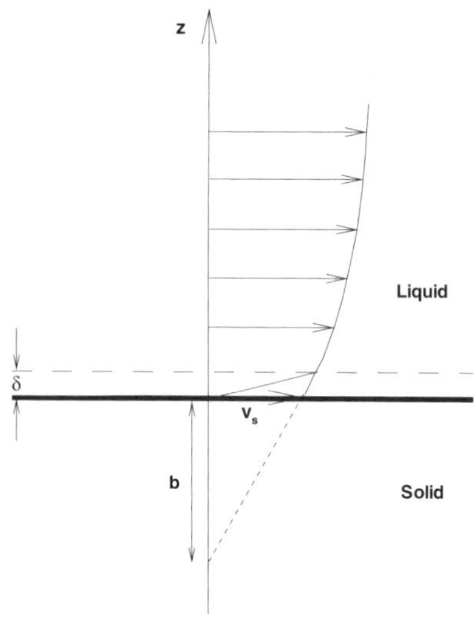

FIGURE 10.18. Definition of slip length b. The inner boundary layer δ is explained in the section on conceptual models of slip.

worth noting also that the experiments of Whetham (1890) that led to the acceptance of the no-slip condition were made on hydrophilic surfaces (Bonaccurso et al., 2003).

Navier (1823) was the first to model partial slip at the wall for liquids well before Maxwell's slip condition for gases (1879). Specifically, Navier's boundary condition at the wall is

$$\boldsymbol{v}_s = b\mathbf{n} \cdot \left(\nabla \boldsymbol{v} + \nabla^T \boldsymbol{v}\right). \tag{10.5}$$

An interpretation of the slip length b is shown in the sketch of Figure 10.18 for unidirectional flow over a flat wall. It is the distance behind the solid–liquid interface at which the velocity extrapolates to zero.

The validation of slip boundary conditions continued in the beginning of the twentieth century, focusing mostly on flow in capillaries. (Traube and Whang, 1928) reported a four- to five-fold increase in flowrate of water in a capillary treated with oleic acid. However, their results could be interpreted either as boundary slip or simply as surface-tension-induced capillary rise. In independent experiments with water flowing in capillaries treated with paraffin, (Ronceray, 1911) also studied how changes in the surface tension may affect the flowrate but concluded aganist the slip condition. The most systematic study, perhaps, of this effect was undertaken by (Schnell, 1956), who used water in glass capillaries (from 240 to of 800 μm) treated with dimethyldichlorosilane (i.e., silicone) to make them hydrophobic. He found that for a small pressure drop in the capillary the flowrate was lower in

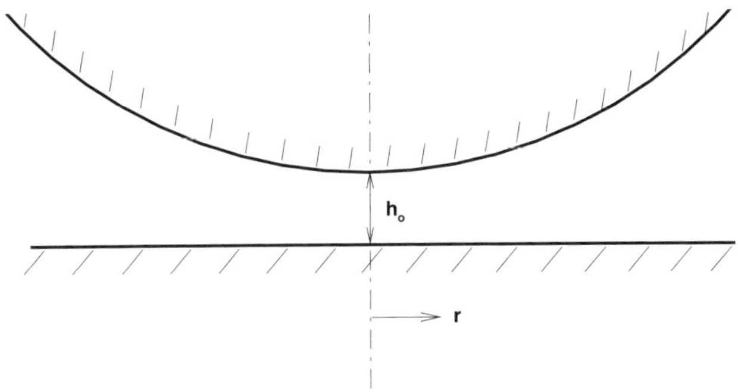

FIGURE 10.19. Schematic of the surface force apparatus (SFA).

the treated cases compared to the untreated ones, but at higher pressure drop he observed the opposite. Interestingly, at velocities past the onset of turbulence there was no discerible difference in flowrate. Overall, Schnell's experiments stood the test of time and are considered the first to prove convincingly that a boundary slip occurs for hydrophobic (i.e., water-repellent) surfaces. This result agrees with the physical intuition, i.e., that boundary slip is larger in hydrophobic surfaces, since the attractive forces between the liquid and the solid surface are less than for hydrophilic surfaces, and thus the solid–liquid interface friction is reduced. At about the same time, it was established by (Debye and Cleland, 1959) that boundary slip can also occur in liquid hydrocarbons for flow through porous Vycor glass.

In the last few decades there has been a renewed interest in determining the validity of the no-slip boundary condition for liquids due to the interest in polymers and other complex fluids but primarily due to microfluidic applications. In (Chuarev et al., 1984), both water and mercury were tested in flow through glass capillaries of diameter less than 10 μm treated with trimethylchlorosilane. It was found that for water with contact angles less than 70° the no-slip condition was valid, but for higher hydrophobicity increased flowrates were obtained corresponding to a slip length between 30 and 200 nm according to Navier's formula of equation (10.5). For mercury, a contact angle of more than 130° also led to boundary slip. These results suffer, however, from the limitation in determining the capillary diameter precisely as well as in controlling the homogeneity of the internal capillary surface.

The effective use of the surface force apparatus (SFA) in the 1990s has led to many interesting experimental results and detailed studies of boundary slip with water and other substances. SFA employs a sphere in close proximity to a plane, but other configurations are also possible, e.g., two crossed cylinders. The two surfaces are approaching at a controlled speed, causing drainage of the liquid placed within the sphere–plane gap; see sketch of figure 10.19. Typically, an oscillation of small amplitude is imposed on the sphere (or the plane), and the response force on the plane is recorded. The

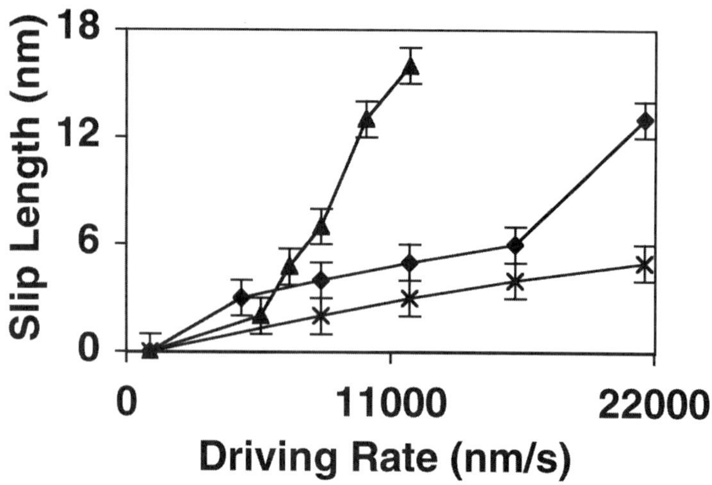

FIGURE 10.20. Slip length versus driving rate obtained in SFA experiments. Sucrose solutions of different concentrations are used to change the viscosity: 19.2 mPa s (crosses); 38.9 mPa s (diamonds); 80.3 mPa s (triangles); taken from (Craig et al., 2001). (Courtesy of V. Craig.)

ratio of the force component in-phase with the oscillation to the amplitude of the oscillation gives the normal stiffness coefficient, while the corresponding ratio for the out-of-phase component gives the damping coefficient.

In (Baudry et al., 2001), a drop of glycerol was placed between the two surfaces with roughness of about 1 nm. The surfaces of the plane and the sphere were coated with thiol and gold, respectively, in a first set of experiments, and with thin cobalt film (for both sphere and plane) in a second set of experiments; cobalt makes the surface hydrophilic, while thiol makes it hydrophobic. The measured (advancing) contact angles for thiol and cobalt were 94° and 62°, respectively. The main finding, based on the measured damping coefficient, was that at sphere–plane distances less than 300 nm the hydrophobic surface gave a large deviation from the no-slip condition with slip length $b \approx 40$ nm, while for the cobalt surface the slip was zero. The molecular size of the glycerol molecule is 0.6 nm, so the slip length is about 65 times the molecular diamater of glycerol. Interestingly, at distances smaller than 5 nm, both hydrophobic and hydroplilic surfaces gave the same response and it was assumed that at these small distances the film did not behave like a liquid.

In (Craig et al., 2001), SFA measurements were also used to infer boundary slip for water on sphere–plane surfaces with (advancing) contact angle 70°. The silica sphere and the mica flat surface were coated with a layer of gold and an additional layer of titanium to promote adhesion. Aqueous solutions of sucrose at various concentration levels were used to control the viscosity magnitude. It was found that at low viscosity and low approach rate of the surfaces, the no-slip condition was valid, whereas at higher viscosities and shear rates boundary slip was observed. The corresponding slip length was up to 18 nm, corresponding to the highest viscosity of the aqueous solutions (80.3 mPa s) and varied nonlinearly with the driving rate (up to 22 μm/s). A summary of these results is included in Figure 10.20. In a follow-up experiment in (Bonaccurso et al., 2003), the effect of roughness mounted on hydrophilic surfaces on the slip length was investigated. Roughness with rms heights of 0.7 nm, 4 nm, and 12.2 nm was tested, and the force curves showed clearly an enhanced slip at the wall as the roughness increased. The particular slip length value depended on the way that the data were fitted, but very large values of b were reported on the order of *hundreds of* nanometers.

In (Zhu and Granick, 2001), SFA measurements with molecularly smooth surfaces of mica were obtained for water (polar) and tetradecane (non polar), an oil with low viscosity close to water. In particular, three different liquid–solid systems were considered with increasing contact angle: (1) tetradecane against adsorbed surfactant; (2) tetradecane against a methyl-terminated self-assembled monolayer (SAM), and (3) water against a methyl-terminated SAM. In the last two cases a monolayer of octadecyltriethoxysiloxane (OTE) was used on which the (advancing) contact angle of water was 110° and that of tetradecane was 44°. It was found that for cases (2) and (3) with the OTE layer causing partial wetting of the surface, boundary slip was obtained for film thickness less than about 100 μm. However, case (1) gave no slip at the solid interface. The results of (Zhu and Granick, 2001) led to similar conclusions as in (Craig et al., 2001), in that the slip length depends strongly on the approach (driving) rate and is largest for water: the largest value is $b \approx 35$ nm at shear rates about 10^3 s^{-1}. Below a threshold value of the approach rate (and thus shear rate) the no-slip boundary condition is valid.

The effect of roughness on boundary slip was examined in a follow-up paper by the same researchers (Zhu and Granick, 2002). Roughnesses with rms values up to 6 nm were fabricated using self-assembled OTE monolayers and OTS (octadecyltrichlorosilane) layers. The advancing contact angle was similar for all cases, but the receding contact angle was a decreasing function of surface roughness. It was hypothesized that large roughness will *decrease* the slip length, although cases with an increase in slip length have also been reported (Bonaccurso et al., 2003; Ponomarev and Meyerovich, 2003). The results of (Zhu and Granick, 2002) are summarized in Figure 10.21, where atomic force microscopy images of roughness on a 3 μm $\times$3 μm area are also shown. The data of (Zhu and Granick, 2002) show that even the case of largest slip length ($b \approx 35$ nm) for water produces no slip at the wall if the roughness rms height exceeds 6-nm. The critical shear rate for onset of slip seems to depend linearly on the roughness *rms* height and is independent of its wavelength. For 6-nm roughness a value of shear rate

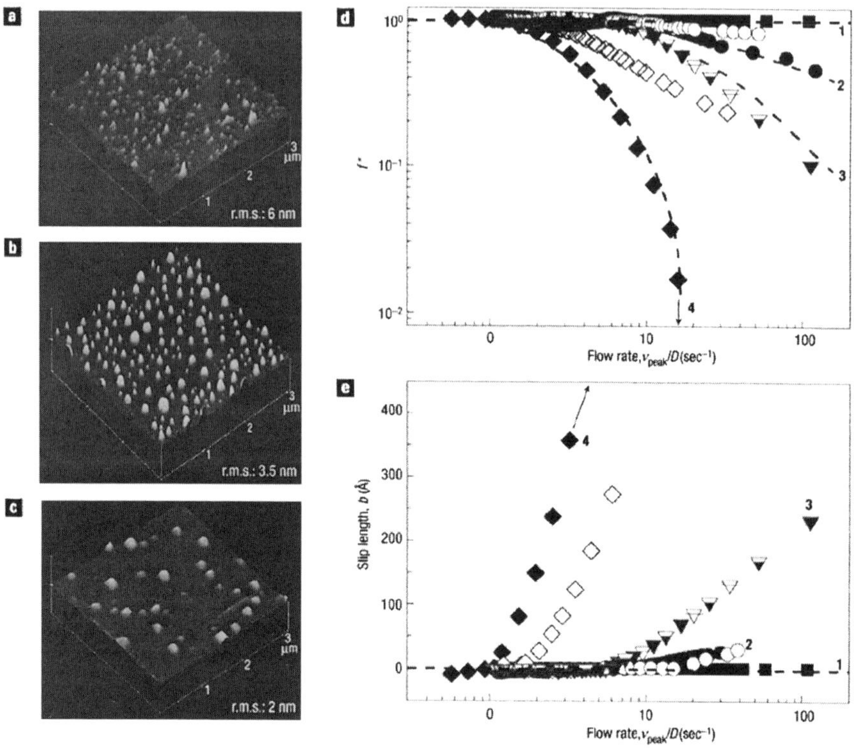

FIGURE 10.21. Effect of roughness and flow rate (or driving rate) on boundary slip. Plots (a), (b) and (c) show images of roughness at 6 nm, 3.5 nm, and 2 nm, respectively. Plot (d) shows the correction factor in the Reynolds–Vinogradova theory (equation (10.11)), and plot (e) the corresponding slip length. Filled symbols correspond to *deionized water* and open symbols to *tetradecane*. The various curves correspond to roughness of: squares - 6 nm; circles - 3.5 nm; triangles - 2 nm; and diamonds – atomically smooth. (Courtesy of S. Granick.)

of 10^5 s^{-1} is required to cause onset of boundary slip. These results are in disagreement with the results of (Bonaccurso et al., 2003) for hydrophilic surfaces.

There has been some skepticism regarding the findings of boundary slip based on the SFA measurements. However, similar conclusions were obtained in (Bonaccurso et al., 2002), using a colloidal probe technique to measure forces between hydrophilic surfaces (mica and glass) for water. In particular, spherical borosilicate glass particles of radius 10 μm were sintered to atomic force microscope (AFM) cantilevers. The hydrophilic surface was periodically moved up and down, and the cantilever deflection was measured optically. A slip length of up to 9 nm was measured for shear

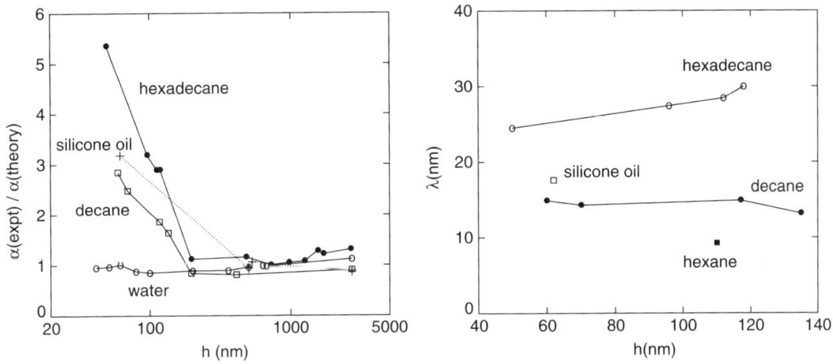

FIGURE 10.22. Normalized flowrate (left) and slip length versus channel height for several fluids. (Courtesy of N. Giordano.)

rates of 10^4 s^{-1}. Electrokinetic effects, which could render the data erroneous by causing an increase of the effective viscosity in the electric double layer (EDL), were found insignificant. In another study in (Pit et al., 2000), a novel technique was employed to test boundary slip for hexadecane flowing over a hydrocarbon/lyophobic smooth surface. This technique was adapted from an experimental setup used to investigate boundary slip in polymers. In particular, fluorescent probes of the size of the hexadecane molecules were used as traces in a capillary formed between two parallel disks, only one of which was rotating. A combination of fluorescence recovery after photobleaching and total internal reflection at the solid–liquid interface was employed to directly probe the velocity within 80 nm of a solid wall. The surface was made of modified sapphire treated with OTS ($\gamma > 21$ mJ/m^2) and FDS (perfluorodecanetrichlorosilane, $\gamma < 13$ mJ/m^2) in order to change the interfacial energy in a controlled manner. It was found that for the bare sapphire surface a slip length of 175 nm (±50 nm) was obtained, while for a dense OTS layer the slip length was 400 nm (±100 nm) independent of the shear rate in the range from 200 s^{-1} to 2000 s^{-1}. For the FDS surface no boundary slip was observed. In agreement with most of the other investigators, (Pit et al., 2000) also hypothesized that roughness decreases slip, and thus it is in competition with the strength of the fluid–surface interaction.

With regard to microfluidic applications directly involving microchannels, it is not clear how much the slip boundary condition depends on the way that the flow is driven, although the majority of the experimental work clearly points to a strong dependence on the shear rate (an exception is the work of (Pit et al., 2000)). In the work of Bau and collaborators (Urbanek et al., 1993), a pressure-driven flow was considered, and boundary slip was reported for channels with the smallest height of 20 μm and silicone oil as well as isopropol alcohol (see Figure 1.16 in Section 1.2). A comprehensive study of pressure-driven flows was undertaken in (Cheng and Giordano, 2002), with several fluids for very small microchannels fabricated lithographically down to 40 nm. The channel width was 20 μm, and the length was in the range of 100 to 900 μm; the roughness was about 0.5 nm. The flowrate was measured using a macroscopic capillary in series with the outlet side of the sample, but for smaller flowrate values a microchannel was fabricated and the flow was measured with photomicroscopy. One of the surfaces of the channel was glass, and the other one was coated with photoresist; no measurements of the contact angle were made.

In Figure 10.22 (left) the flowrate normalized by the theoretical prediction for no-slip Poiseuille flow is plotted, taken from (Cheng and Giordano, 2002). Specifically,

$$\alpha_{\text{theory}} = \frac{h^3 w}{12\mu L}$$

(where h, w, L denote the height, width, and length of the microchannel) is used in the normalization. We see that for all the liquids tested, the flowrate increases except for water. In particular, hexadecane (the fluid with the largest molecular size) exhibits the largest deviation from the no-slip theory. This is in agreement with the results in (Pit et al., 2000), for the capillary hexadecane flow, although the slip length values reported in (Cheng and Giordano, 2002), are much smaller. In general, the experimental evidence given in (Cheng and Giordano, 2002), indicates a monotonic increase of the slip length with the molecular size but for channel height $h < 300$ nm; above this value the no-slip theory seems to be valid according to the results of (Cheng and Giordano, 2002). This dependence on the molecular structure is shown in Figure 10.22 (right), indicating also that there is some weak dependence of the slip length on the channel height in the slip regime. However, this effect may be associated with the uncertainties in measuring the very small values of h.

Other experiments with larger microchannels for pressure-driven flows revealed boundary slip for water, in contrast to the aforementioned results of (Cheng and Giordano, 2002). For example, in (Tretheway and Meinhart, 2002), microPIV (300-nm diameter fluorescent polystyrene spheres) was used to measure velocity profiles of water in a $30 \times 300\,\mu$m channel. The channel surfaces were treated with a 2.3 nm OTS layer. The velocity profiles were measured in a $25 \times 100\,\mu$m plane to within 450 nm of the channel wall. A slip velocity of about 10% of the maximum velocity was measured, which corresponds to slip length of about 1 μm. This is a very large value for the slip length, of the order of magnitude that is typically encountered in polymer flows. For the untreated glass surface, which is hydrophilic, no-slip conditions were observed. Similar reults were also reported in (Choi et al., 2003), in smaller hydrophobic microchannels of 0.5 μm and 1 μm height. The channels were 500 μm wide and 9 mm long, while the surfaces were coated with OTS layers to make them hydrophobic. In both the experiments of (Tretheway and Meinhart, 2002) and (Choi et al., 2003), roughness was negligible. The slip length was found to depend linearly on the shear rate with $b = 30$ nm at a shear rate of 10^5 s^{-1} for hydrophobic surfaces, while for hydrophilic surfaces $b = 5$ nm at the same shear rate. The corresponding slip velocity was of order 3 mm/s for the hydrophobic case and 0.5 mm/s for the hydrophilic case.

In *MD simulations* the slip length predicted is typically much lower due to the substantial pressure imposed, which can modify the wetting properties of the surface. Specifically, MD simulations with hexadecane were performed in (Stevens et al., 1997), and dependence on the strength of the liquid–wall interaction was established similar to that in the experiments. However, a realistic representation of the surface, i.e., to account accurately for the glass or sapphire or other surfaces tested experimentally, is not available. In (Cieplak et al., 2001), MD simulations were performed for a simple molecule as well as a chainlike molecule. They were described by a shifted Lennard–Jones potential for two atoms for the former and for ten atoms for the latter. The consecutive atoms along the chain were tethered by the finitely extensible nonlinear elastic potential (FENE) used often in polymer modeling; it has the form

$$V_{\text{FENE}} = -\kappa/2r_0^2 \log[1 - (r/r_0)^2], \tag{10.6}$$

where $\kappa = 30\epsilon$ and $r_0 = 1.5\sigma$. The crucial wall–fluid interaction was modeled by a distinct Lennard–Jones potential of the form

$$V_w = 16\epsilon[(r/\sigma)^{-12} - c_{\text{FS}}(r/\sigma)^{-6}],$$

where c_{FS} determines the wall type, so that $c_{\text{FS}} = 1$ corresponds to a thermal (attractive) wall and $c_{\text{FS}} = 0$ corresponds to a specular (repulsive) wall. The narrowest channel simulated had dimensions of $13.6\sigma \times 5.1\sigma \times 12.75\sigma$, with the the last dimension denoting the distance between the two walls (channel height).

The results for Couette flow in (Cieplak et al., 2001), suggest that the slip length is independent of the type of flow or the channel height, but that it is a strong function of the wall type. When $c_{\text{FS}} = 0$ there is a relatively large slip (about 10σ), but for $c_{\text{FS}} = 1$ the slip length is equal to the negative of the distance between the wall and the second layer (about -1.7σ). In the case of a chain molecule, the slip length depends more strongly on the value of c_{FS}, and for $c_{\text{FS}} = 1/4$ the slip length is 20.6σ, in qualitative agreement with the MD simulation results of (Stevens et al., 1997) for hexadecane and also with the experimental results of (Pit et al., 2000).

Similar results were reported in (Priezjev and Troian, 2004), for shear polymer films in a Couette flow with a gap height of 24.57σ. In particular, N-mer polymer chains with N up to 16 were simulated for various levels of shear rate $\dot{\gamma}$. The scaling law obtained in the earlier simulations of

(Thompson and Troian, 1997) for $N = 1$, see equation (1.2), was extended for polymer chains; that is, the normalized slip length is given by

$$\frac{b}{b_s^0} = \left(1 - \frac{\dot{\gamma}}{\dot{\gamma}_c}\right)^{-1/2} \tag{10.7}$$

for shear rates $\gamma \geq 5 \times 10^{-3}\tau^{-1}$; γ_c is the critical level of shear rate above which divergence behavior is observed. It was also found that beyond $N = 10$ the molecular weight dependence of the slip length is mostly associated with the bulk viscosity.

Also, in (Barrat and Bocquet, 1999), MD simulations were performed with a Lennard–Jones potential of the form

$$V_{\mathrm{LJ}} = 4\epsilon[(r/\sigma)^{-12} - c_{ij}(r/\sigma)^{-6}],$$

where c_{ij} controls the type of interaction. The interactions between fluid atoms were defined by $c_{\mathrm{FF}} = 1.2$, while the fluid–solid interaction coefficient c_{FS} was varied between 0.5 and 1. For $c_{\mathrm{FS}} = 0.9$ a contact angle of 100° is obtained, whereas for $c_{\mathrm{FS}} = 0.5$ a contact angle of 150° is obtained, which is close to the case of mercury on a glass surface, a flow configuration studied in the experiments of (Chuarev et al., 1984) with mercury flowing through glass capillaries. For this hydrophobic case a slip length of about 15 molecules was computed, whereas for the 100° case the no-slip condition was valid all the way to the wall boundary. Taking into account that the molecule size is about 0.25 nm, the maximum computed slip length in the MD simulations of (Barrat and Bocquet, 1999) is about 4 nm, which is still much smaller than any of the experimental data, especially for hydrophobic surfaces.

10.5.2 Conceptual Models of Slip

We have seen that the MD simulations systematically underpredict the slip length deduced from the various experiments either based on force curves in the surface force apparatus or based on the flowrate measured in capillaries and microchannels. Specifically, the MD simulations predict a slip length roughly *ten times smaller* than in the experiments, and such a large discrepancy implies that there are some other physical phenomena not included in the simulation. Here we present some of the physical models that have been proposed in the literature that can justify the large values of slip length and the origin of slippage.

1. Molecular slippage: This theory first appeared in the Russian literature and is due to Tolstoi (1952); it was analyzed more thoroughly in (Blake, 1990). It provides a link between the mobility of the molecules in the first few layers closer to the wall with the interfacial energy associated with the liquid–solid interface. Therefore, it relates directly the contact angle to the slip length.

2. Gaseous film: This model was first suggested by (Ruckenstein and Rajora, 1983) and was also discussed by (Vinogradova, 1999). It assumes that there may be a film of gas at the interface between the solid surface and the liquid. The origin of this film may be due to external dissolved gases up to metastable concentrations. According to (deGennes, 2002), this gas film nucleates bubbles preferentially near the wall at contact angles greater than 90°, i.e., on hydrophobic surfaces. This mechanism can take place above a threshold in shear rate, a fact consistent with the experimental observations. Evidence of nanobubbles on a hydrophobic glass surface in water using an atomic force microscope was reported in (Tyrrell and Attard, 2002). Another possibility, consistent with MD simulations that reveal a depletion of the first layer of molecules, is that a flat vapor bubble is generated at the solid–liquid interface due to thermal fluctuations. In either case, the gaseous film is assumed to be small, e.g., less than 0.5 nm.

A simple mathematical model was proposed by (deGennes, 2002) for this case. He assumed that the gas in the gap is in the molecular regime (since the mean free path satisfies $\lambda \gg h$, where h is the film thickness), and thus the only collisions are with the wall. Correspondingly, a molecule leaving the liquid has a Gaussian velocity distribution for the tangential velocity component with the corresponding peak at the slip velocity v_s. Denoting by ρ, m the density and molecular mass of the gas, respectively, the average momentum transmitted to the solid by the gas is mv_s, and thus the shear stress σ at the wall is

$$\sigma = mv_s \frac{\rho}{m} \bar{v}_y = \rho v_s \bar{v}_z,$$

where $(\rho/m)\bar{v}_z$ is the average number of collisions with the wall, and the normal to the surface average velocity $\bar{v}_z$ is

$$\bar{v}_z = \frac{v_{th}}{\sqrt{2\pi}} \quad \text{with} \quad v_{th} = \sqrt{kT/m}.$$

On the other hand, $\sigma = \mu \partial v / \partial z = \mu v_s / b$, and thus by comparing with the above expression, we obtain the slip length

$$b = -h + \frac{\mu}{\rho \bar{v}_z} \approx \frac{\mu}{\rho \bar{v}_z},$$

where the thickness h of the gaseous film is assumed negligible. Upon substitution of typical values for water vapor in the above expression, we obtain a slip length of a few microns, which is clearly much higher than any of the available experimental data. We note that the model of de Gennes shows that the slip length increases with the viscosity and thus with the molecular weight, which is consistent with the measurements in (Cheng and Giordano, 2002).

3. Viscosity model: This model, proposed by (Vinogradova, 1999), is inspired by the slip mechanism in polymer melts. It provides a connection between the slip length and a decrease in viscosity within a very thin

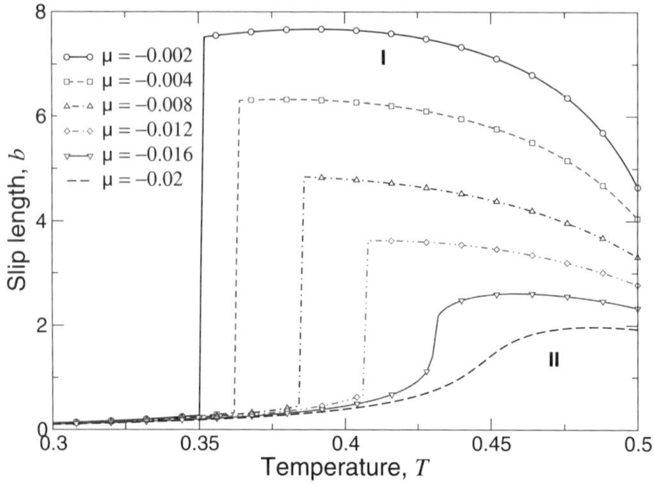

FIGURE 10.23. Slip length versus temperature in prewetting transition for different values of the chemical potential μ. (Courtesy of O.I. Vinogradova.)

boundary layer δ close to a hydrophobic surface. Assuming a *bulk viscosity* μ_b and a *near-wall viscosity* μ_s, then the slip length is

$$b = \delta \left(\frac{\mu_b}{\mu_s} - 1 \right).$$

This expression shows that there are two mechanisms for obtaining a large slip length, i.e., either by increasing δ or by increasing the viscosity ratio in the bulk and the surface. For example, for $\mu_b/\mu_s = 21$ and $\delta = 10$ nm, a slip length of $b = 200$ nm can be obtained, but a more realistic viscosity ratio is $\mu_b/\mu_s = 3$, which corresponds to $b = 20$ nm.

The above arguments suggest that there may be another mechanism in place that produces thick films (i.e., large δ), and that is why in some experiments large values of the slip length have been reported. To this end, in (Andrienko et al., 2003), a new model that accounts for prewetting transition was developed. It takes into consideration the structure of the binary mixture in the region near the solid surface and allows for a temperature dependence of the thickness in the form $\delta \propto -\ln(|T_w - T|)$, where T_w is the wetting temperature of the surface.

The governing equation for this case is obtained in terms of the *order parameter* Φ, which is defined as

$$\Phi = \frac{n_1 - n_2}{n_1 + n_2},$$

where n_i denotes the number density of each species. This order parameter changes very fast very close to the interface, but it is almost constant in the bulk. The viscosity of the binary mixture can then be expressed as a linear combination of its two components, i.e.,

$$\mu_m(z) = \mu_s \frac{1+\Phi(z)}{2} + \mu_b \frac{1-\Phi(z)}{2}.$$

The thermodynamics of the binary mixture are described via a free-energy functional plus other contributions to account for surface effects. The corresponding *semigrand potential* proposed in (Andrienko et al., 2003), is given by

$$U(\Phi) = \frac{1}{\sigma^3} \int dV \left(\frac{k}{2} \sigma^2 (\nabla\Phi)^2 + f(\Phi) - \mu\Phi \right) + \Psi_s,$$

where σ is a length scale characteristic of the molecule size, $f(\Phi)$ is the Helmholtz free-energy density, Ψ_s is the surface energy, and μ is the chemical potential. The governing equation is then obtained by minimizing the above functional to obtain

$$k\frac{\partial^2\Phi}{\partial z^2} + \frac{\Phi}{2} - \frac{1}{2} T \ln \frac{1+\Phi}{1-\Phi} + \mu = 0, \tag{10.8}$$

where T is the temperature. This is a boundary value problem, which was solved in (Andrienko et al., 2003), for a channel with identical walls located sufficiently far from each other so that the film layers do not overlap. The solution of the above equation reveals a *prewetting transition* that depends on the temperature; it is sudden, and it jumps from a thin film to a thick film (first-order transition). For thin films a small slip length is obtained, but for thick films a large slip length is obtained that also depends on the chemical potential. This is shown in Figure 10.23, which is taken from (Andrienko et al., 2003), and is in nondimensional units (the molecular size σ is employed in the nondimensionalization). When a thick film is obtained the corresponding slip length depends on temperature, while below the threshold for transition the slip length is independent of the temperature.

4. No-shear/no-slip patterning: This model was first considered in (Philip, 1972), in an attempt to explain slip in porous media. The main idea is to consider the liquid–solid interface segmented into aternating stripes of *no-slip* and *no-shear* and deduce an effective slip length from this static configuration. This model was advannced more recently by (Lauga and Stone, 2003) who extended some of the exact solutions in (Philip, 1972), and hypothesized the existence of small bubbles attached to the wall as providers of the slip and thus the corresponding *stress-free* condition. As was already mentioned, there is direct experimental evidence by AFM of the existence of such nanobubbles; see, for example, (Tyrrell and Attard, 2002). In rough surfaces or surfaces with tiny cracks, air pockets may exist that act as stress-free local boundaries. Therefore, the proposed model is that of surface heterogeneities that lead to an effective or equivalent macroscopic slip.

The two basic configurations, a *longitudinal* and a *transverse* one, considered in the works of (Philip, 1972) and (Lauga and Stone, 2003), are

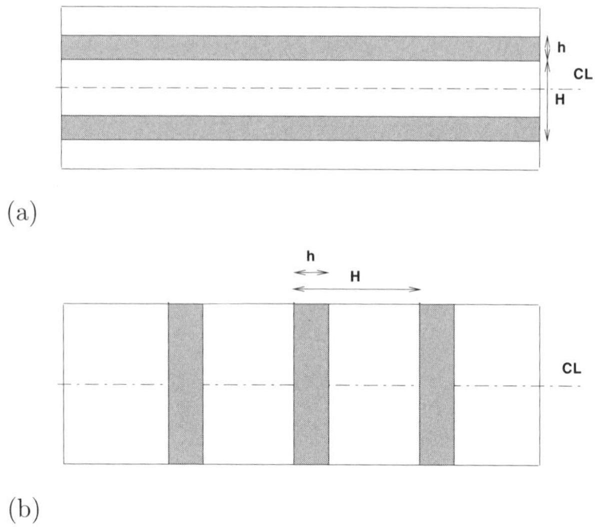

FIGURE 10.24. Longitudinal (a) and transverse (b) models of no-shear stripes.

shown in the sketch of Figure 10.24 for a capillary. Semianalytical Stokes flow solutions can be obtained for these geometries, and the effective slip length b_e can be obtained in terms of the ratios $\delta = h/H$ and $L = H/R$ and the capillary radius R. This effective slip length is defined indirectly from the flow rate as follows: Let us assume that the partial slip condition is applied everywhere, then the velocity profile is

$$u(r) = \frac{1-r^2}{4} + \frac{b}{2R},$$

and correspondingly the nondimensional flowrate is

$$Q = \frac{\pi}{8}\left(1 + \frac{4b}{R}\right).$$

Solving for the effective slip length, we obtain

$$b = \frac{R}{4}\left(\frac{8Q}{\pi} - 1\right).$$

Therefore, by obtaining the flowrate for a given configuration, we can then obtain the effective slip length from above (Lauga and Stone, 2003).

For the *longitudinal* configuration, (with m no-shear bands with half-angle α) shown in Figure 10.24, an exact solution was obtained in (Philip, 1972) for the velocity distribution

$$u(r,\theta) = -\frac{R^2}{4\mu}\frac{\partial p}{\partial r}\left[1 - (r/R)^2 + (4/m)\Im\left(\cos^{-1}\left(\frac{\cos(M)}{\cos(m\alpha/2)}\right) - M\right)\right],$$

where $M = -im/2\ln(re^{i\theta}/R)$ and $\Im$ implies imaginary part. The corresponding effective slip length, b_l, for this case is

$$\frac{b_l}{R} = \frac{2}{m}\ln\left(\sec(m\alpha/2)\right) = \frac{L}{\pi}\ln\left(\sec(\delta\pi/2)\right).$$

For the configuration with the *transverse* no-shear stripes the solution is a bit more complicated, but four asymptotic limits were obtained for the slip length b_t in (Lauga and Stone, 2003), in terms of the slip percentage δ and the separation between slip stripes L, as follows:

1. $b_t/R \propto \delta/4$ for $\delta \to 0$ and L fixed.
2. $b_t/R \propto [4(1-\delta)]^{-1}$ for $\delta \to 1$ and L fixed.
3. $b_t/R \propto (L/(2\pi))\ln\left(\sec(\delta\pi/2)\right)$ for $L \to 0$ and δ fixed.
4. $b_t/R \propto [4(1-\delta)/\delta]^{-1}$ for $L \to \infty$ and δ fixed.

For a small percentage of slip, the above limits suggest that the effective slip length decreases faster (quadratically) to zero for longitudinal stripes compared to linear decrease for transverse slip stripes. For a large percentage of slip the opposite is true. Also, for small separation between slip stripes ($L \to 0$) we have that $b_l = 2b_t$. These two configurations represent the two extreme idealized cases, since in reality we expect a random distribution of no-shear pockets mixed with no-slip pockets. In addition, the dependence of the slip length of the shear rate can also be included in this model by assuming that the inhomogeneities (e.g., nanobubbles) are elongated at large values of shear rate, hence effectively increasing the relative no-shear to no-slip regions.

10.5.3 Reynolds–Vinogradova Theory for Hydrophobic Surfaces

In this section, we derive analytical solutions for the steady-state flow between two curved hydrophobic surfaces following the work of (Vinogradova, 1995; Vinogradova, 1996). This theory is an extension of Reynolds lubrication theory appropriate for *slip* surfaces.

The theory is valid for general curved surfaces, but for simplicity here we show the main steps in the solution for two spherical rigid bodies of radii R_1 and R_2. The distance h between the two bodies is small compared to the radii, and contact is allowed only at a single point. We also assume that a hydrophilic surface is characterized by $b = 0$. A schematic of the setup is shown in Figure 10.25; a cylindrical coordinate system is employed in deriving the solution. The relative velocity is $v = |v_1 - v_2|$, where the spheres move along the line connecting their centers with velocities v_1 and v_2.

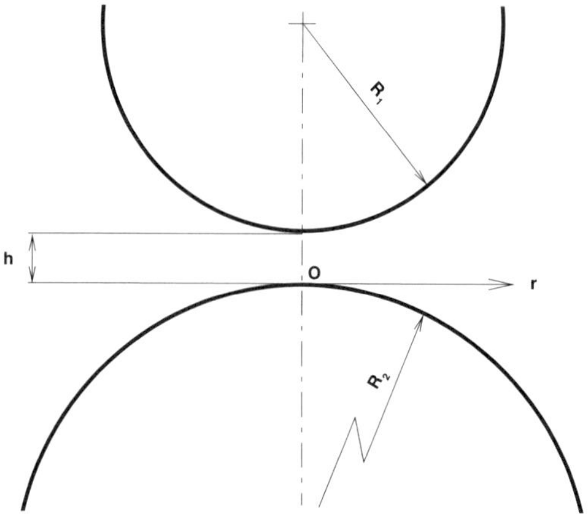

FIGURE 10.25. Geometry and notation for two spheres approaching each other.

The surfaces of the two bodies (upper and lower, respectively) can be described by paraboloid of revolution, i.e.,

$$Z = h + \frac{1}{2}\frac{r^2}{R_1} + \mathcal{O}(r^4) \quad \text{and} \quad Z = -\frac{1}{2}\frac{r^2}{R_2} + \mathcal{O}(r^4),$$

and introducing a shifted coordinate $z = Z + r^2/(2R_2)$ and $R_e = R_1R_2/(R_1 + R_2)$, we can represent the two surfaces in a new coordinate system as

$$z = h + \frac{1}{2}\frac{r^2}{R_e} + \mathcal{O}(r^4) \quad \text{and} \quad z = \mathcal{O}(r^4).$$

The governing equation is Reynolds's lubrication equation, assuming that the characteristic length is the gap between the two particles, i.e.,

$$\mu\frac{\partial^2 v_r}{\partial z^2} = \frac{\partial p}{\partial r},$$

while in the z-direction we have that $\partial p/\partial z = 0$, which implies that the pressure is a function of r only. The boundary conditions correspond to slip on the lower body, characterized by $b_2 = b$, while on the upper surface we assume that $b_1 = b(1 + k)$, where k characterizes the specific type of the interaction. Specifically,

- $k = -1$ corresponds to a hydrophilic upper surface.
- $k = 0$ corresponds to a hydrophobic upper surface.
- $k \to \infty$ corresponds to a bubble upper surface.

In addition, k can take any other value between -1 and ∞ to represent other types of interaction and surfaces. We note that contrast to the standard Reynolds problem, where the only length scale present is the gap h, here we have two additional length scales, namely b and $b(1+k)$. So the boundary conditions on the lower surface are

$$v_z = 0 \quad \text{and} \quad v_r = b\frac{\partial v_r}{\partial z}$$

and those on the upper surface are

$$v_z - \frac{r v_r}{R_e} = -v \quad \text{and} \quad v_r = -b(1+k)\frac{\partial v_r}{\partial z}.$$

The solution of the above Reynolds equation with the aforementioned boundary conditions yields

$$v_r(r,z) = \frac{1}{2\mu}\frac{\partial p}{\partial r}\left[z^2 - z\frac{H(H+2b(1+k))}{H+b(2+k)} - \frac{bH(H+2b(1+k))}{H+b(2+k)}\right],$$

where $H = h + r^2/(2R_e)$. The relative velocity v can be obtained from the continuity equation

$$\frac{\partial v_z}{\partial z} + \frac{1}{r}\frac{\partial (r v_r)}{\partial r} = 0,$$

which by integration yields

$$v = \frac{1}{r}\frac{\partial}{\partial r}\left[(r\frac{\partial p}{\partial r})\frac{1}{2\mu}\left(\frac{H^3}{3} - \frac{H^3(H+2b(1+k))}{2(H+b(2+k))} - \frac{bH^2(H+2b(1+k))}{H+b(2+k)}\right)\right].$$

This velocity is constant, and thus we can solve the above differential equation in terms of pressure by integrating twice and assuming that $p = 0$ at $r \to \infty$, and $\partial p/\partial r = 0$ at $r = 0$ due to symmetry. The equation for the pressure is then

$$p(r) = -\frac{3\mu R_e v}{H^2} p^*, \tag{10.9}$$

consisting of two factors, namely, the Reynolds part and the correction factor p^* given by

$$p^* = \frac{2AH}{BC} + \frac{2H^2}{C-B}\left(\frac{B-A}{B^2}\ln(1+B/H) - \frac{C-A}{C^2}\ln(1+C/H)\right). \tag{10.10}$$

The constants A, B, C in this expression characterize the two surfaces; they are given by

$$\begin{aligned} A &= b(2+k), \\ B &= 2b(2+k+\sqrt{1+k+k^2}), \\ C &= 2b(2+k-\sqrt{1+k+k^2}). \end{aligned}$$

The resistance forces acting on the spheres are equal in magnitude and are primarily due to the pressure, so the force can be computed exactly from $F_z = \int_0^\infty p 2\pi r dr$, to obtain

$$F_z = -\frac{6\pi\mu R_e^2 v}{h} f^*, \tag{10.11}$$

s

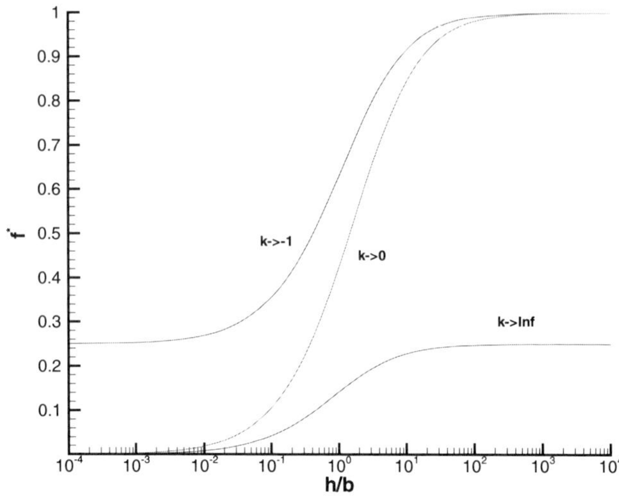

FIGURE 10.26. Correction factor f^* as a function of the gap to slip length ratio for the three asymptotic cases discussed in the text.

consisting also of two factors, namely, the Reynolds part and the correction factor f^* given by

$$\begin{aligned} f^* = & -\frac{Ah}{BC} \\ & -\frac{2h}{C-B}\left(\frac{(B+h)(B-A)}{B^2}\ln(1+B/h)\right. \\ & \left. -\frac{(C+h)(C-A)}{C^2}\ln(1+C/h)\right). \end{aligned} \tag{10.12}$$

For the aforementioned three limiting cases, the above expression reduces to

$$\begin{aligned} f^* &= \frac{1}{4}(1+\frac{3h}{2b}\left[(1+h/(4b))\ln(1+4b/h)-1\right] \quad \text{for } k=-1, \\ f^* &= \frac{h}{3b}\left[(1+h/(6b))\ln(1+6b/h)-1\right] \quad \text{for } k=0, \\ f^* &= \frac{h}{3b}\left[(1+h/(3b))\ln(1+3b/h)-1\right] \quad \text{for } k\to\infty. \end{aligned}$$

These three cases are plotted in Figure 10.26 and show, as expected, that the correction factor is always less than one; the no-slip case corresponds to $f^* = p^* = 1$. Also, both correction factors depend on the length scale ratios, namely h/b and $h/[(k+1)b]$. In the limit of very small gap, the case of

$f^* \to 0$ (corresponding to $k > -1$) represents a configuration of two bubbles approaching each other, while the case of $f^* \to 1/4$ (corresponding to $k = -1$) represents the flow resistance for a hydrophilic sphere interacting with a bubble.

In the standard Reynolds theory ($f^* = 1$) the hydrodynamic resistance is inversely proportional to the gap and diverges for $h \to 0$. However, the new physical result in the solutions of Vinogradova is for two hydrophobic surfaces, i.e., $k > -1$ and $h \to 0$, where the friction coefficient

$$\frac{f^*}{h} = -\frac{2}{C-B}\left[(1 - A/B)\ln B/h - (1 - A/C)\ln C/h\right]$$

depends logarithmically on h and is inversely proportional to the slip length b. The above result is valid for $h \ll C < B$. This dependence is more clearly seen for the case in which A and C are approximately of the same order of magnitude. In this case, (Vinogradova, 1995) has derived that

$$\frac{f^*}{h} = \frac{1}{3b}\ln\left(\frac{6b}{h}\right),$$

which again shows the logarithmic dependence on the gap h.

The above formulation for two spheres has been extended to *general curved* hydrophobic surfaces by (Vinogradova, 1996). The results are very similar to those of the case of two spheres. For example, the resistance force is given by

$$F_z = -\frac{3\pi\mu v}{hI_1\sqrt{I_2}}f^*, \tag{10.13}$$

where f^* is defined by the same expression of equation (10.12), but the geometry is now described by the curvatures of the two surfaces as follows:

$$I_1 = \frac{1}{2}\left[\frac{1}{R_1^-} + \frac{1}{R_1^+} + \frac{1}{R_2^-} + \frac{1}{R_2^+}\right],$$

$$I_2 = \frac{1}{4}\left[\frac{1}{R_1^- R_1^+} + \frac{1}{R_2^- R_2^+} + \sin^2\phi\left(\frac{1}{R_2^+ R_1^+} + \frac{1}{R_2^- R_1^-}\right)\right.$$
$$\left. + \cos^2\phi\left(\frac{1}{R_2^- R_1^+} + \frac{1}{R_2^+ R_1^-}\right)\right].$$

Here R^+ and R^- denote the maximum and minimum principal radii of the surface, and thus I_1 and I_2 are the *mean* and *Gaussian* curvatures of the effective surface, respectively. Also, ϕ defines the orientation of the two coordinate systems attached to the two surfaces. For example, we can consider the interaction of a *sphere with a plane*, a case typical in the surface force apparatus, in which case we obtain $I_1 = 1/R$ and $I_2 = 1/(4R^2)$. Similarly, we can model two *crossed cylinders* for which $R_2^+, R_1^+ \to \infty$ and $\phi = \pi/2$, so the two invariants are

$$I_1 = \frac{1}{2}\left(\frac{1}{R_2^-} + \frac{1}{R_1^-}\right) \quad \text{and} \quad I_2 = \frac{1}{4R_2^- R_1^-}.$$

The reader can try to determine the two invariants for the case of two hydrophobic cylinders with aligned axes for which $\phi = 0$.

The analytical models of Vinogradova have been used with success in fitting the force curves in several SFA measurements, e.g., (Zhu and Granick, 2001; Zhu and Granick, 2002), but the expression for f^* does not depend on the shear rate. However, in several SFA experiments it was shown convincingly that there is a strong dependence of the response on the driving speed, and thus these data do not agree with Vinogradova's theory (Spikes and Granick, 2003). To this end, a new model was proposed by (Spikes and Granick, 2003) based on the observation that the experimental results may represent *onset of slip* at a fixed shear stress $\tau_{\rm co}$ rather than slip at a constant slip length b. Because in the SFA a sphere interacts with a plane, the surface shear stress is zero at the center and also away from it with maximum values in between; see Figure 10.19. This, in turn, implies that there exists an annular region around the contact point where slip occurs. The proposed new model in (Spikes and Granick, 2003), combines both this critical shear stress and the slip at constant b, so the shear stress when boundary slip occus is

$$\tau_c = \tau_{\rm co} + \frac{\mu}{b} v_s.$$

The corresponding pressure gradient for the case of one slippery surface only is

$$\frac{dp}{dr} = -\min\left\{\frac{6\mu W r}{h^3}, \left(\frac{6\mu W r}{h^3} - \frac{6b}{(h+4b)}\left(-\frac{\tau_{\rm co}}{h} + \frac{3\mu W r}{h^3}\right)\right)\right\},$$

where min denotes the minimum of the two quantities, and W is the squeeze velocity; h is the gap height at radial distance r. The influence of $\tau_{\rm co}$ may not be realized in some applications, including cases in which it is constant, as in microchannel pressure-driven or Couette flow. However, it provides a correction for low shear stress configurations and also for the surface force apparatus and the atomic force microscope as well as in surfaces with roughness.

11

Water in Nanochannels

Water and its properties in various forms is one of the most actively investigated areas because of its importance in nature. The anomalies that exist in the bulk properties of water make it very interesting and challenging for research, and a vast deal of literature is already available. Even though water has been studied for more than 100 years now, its properties are far from understood. With the advances in fabrication of nanochannels that are only a few molecular diameters in critical dimension, there is now an opportunity for a major breakthrough in understanding the properties of water in confined nanochannels and in validating atomistic simulations. In this chapter, after introducing some definitions and atomistic models for water, we present the static and dynamic behavior of water in confined nanochannels.

11.1 Definitions and Models

Water is composed of two hydrogen atoms and one oxygen atom. Figure 11.1 shows a sketch of the water molecule and the various quantities used to characterize it. For an isolated water molecule (e.g., water in gas phase), it has a *bond length* $r_{\rm OH}$ of 0.95718 Å, and a *bond angle* $\theta_{\rm HOH}$ of 104.474°. In the liquid state, both these values are slightly modified by the water–water and/or water–ion interactions. For example, neutron diffraction experiments suggested a bond length of $r_{\rm OH} = 0.970$ Å and a bond angle of $\theta_{\rm HOH} = 106.00°$ (Ichikawa et al., 1991) for the liquid state. Because of the higher electronegativity of the oxygen atom compared to that of the

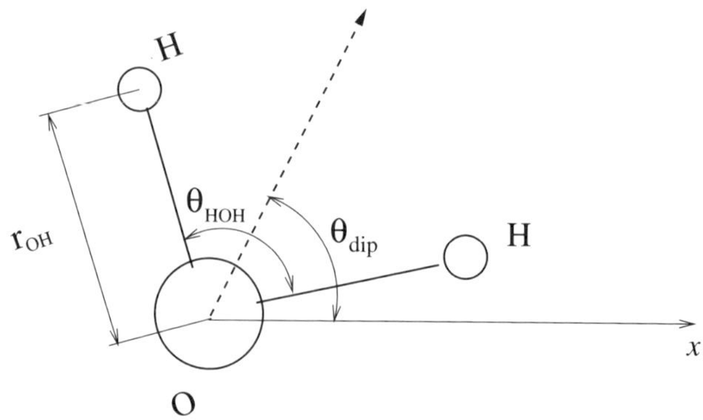

FIGURE 11.1. Sketch of a water molecule and the various quantities used to characterize the water molecule.

hydrogen atom, the oxygen site of the water molecule will appear to be "negatively charged," and the hydrogen site will appear to be "positively charged." Such a charge separation then creates an electric dipole. Because of the symmetry of the water molecule, the dipole can be represented by a line that starts from the oxygen atom and bisects the H–O–H angle (the dotted line in Figure 11.1 is the dipole). If a water molecule is represented by several point charges in an atomistic model, the dipole moment of the model water molecule is given by

$$\boldsymbol{\mu}_{\text{dip}} = \sum_{i=1}^{N} q_i \boldsymbol{r}_i,$$

where N is the number of point charges in the model, and q_i and $\boldsymbol{r}_i$ are the charge and position vector of the point charge i, respectively. The magnitude of the dipole moment, μ_{dip}, is the length of the dipole vector $\boldsymbol{\mu}_{\text{dip}}$. Experimentally, the dipole moment of a single water molecule in liquid state at 300 K is determined to be 2.95±0.2 Debye (1 Debye = 3.336×10^{-30} C m) (Gubskaya and Kusalik, 2002). Sometimes, it is advantageous to describe the water molecule simply as a dipole, e.g., when studying the response of the water molecule to an electrical field. In this case, the orientation of the water is usually characterized by the dipole orientation angle θ_{dip} with respect to a certain reference direction (see Figure 11.1). Some of the important properties of water obtained experimentally are summarized in Table 11.1.

A large number of "hypothetical" models for water have been developed in order to discover the structure of water, on the basis that if the (*known*) model can successfully predict the physical properties of water, then the (*unknown*) structure of water can be determined. They involve positioning the electrostatic sites and Lennard–Jones sites, which may or

TABLE 11.1. Important properties of water (H_2O) †

Dipole moment	2.95 ±0.2 Debye
Boiling point	100°C at 101.325 kPa
Isothermal compressibility	0.4599 GPa^{-1}
Density	997.05 kg/m^3
Dielectric constant	78.4
Diffusion coefficient	$2.27\times10^{-9} m^2/s$
Enthalpy of Vaporization	40.657 kJ/mol (100°)
Electronic polarizability (liquid)	1.48 Å^3
Specific heat capacity (C_p)	75.327 J/mol/K
Specific heat capacity (C_v)	74.539 J/mol/K
Thermal conductivity	0.610 W/m/K
Dynamic viscosity	0.8909 mPa·s

†Unless mentioned otherwise, all the properties are reported at 25° C.

may not coincide with one or more of the charged sites. Generally, each model is developed to match well a certain set of physical structures or parameters (e.g., the density anomaly, radial distribution function, or other critical parameters). A discussion of the various atomistic models for water is presented next.

11.1.1 Atomistic Models

There are a number of atomistic models for water, and a description of the majority of the models can be found in (Chaplin, 2004). In this section we summarize only the most commonly used models for water.

The SSD Model

The water molecule in the soft sticky dipole (SSD) model is treated as a Lennard–Jones sphere with an embedded point dipole plus a tetrahedral "sticky" potential, all situated at the molecular center of mass (M) located on the H-O-H bisector at 0.0654 Å from the oxygen toward the hydrogens (see Figure 11.2). The geometry of the water molecule used in this model is similar to that of any three-site model (e.g., TIP3P). The O-H length (i.e. $r_{\rm OH}$) is 0.9572 Å and the H-O-H angle (i.e. $\theta_{\rm HOH}$) is 104.52°. The H-M-H angle is 109.47°. The center of mass is the only interaction site of this model (Liu and Ichiye, 1996).

In the SSD model, the total interaction potential energy between two water molecules i and j is given by (Liu and Ichiye, 1996)

$$V_{ij} = V_{ij}^{LJ}(r_{ij}) + V_{ij}^{dp}(\mathbf{r}_{ij}, \Omega_i, \Omega_j) + V_{ij}^{sp}(\mathbf{r}_{ij}, \Omega_i, \Omega_j), \qquad (11.1)$$

where r_{ij} is the distance between the molecular centers, $\mathbf{r}_{ij}$ is the separation vector between two molecular centers, and Ω is the orientation of the water

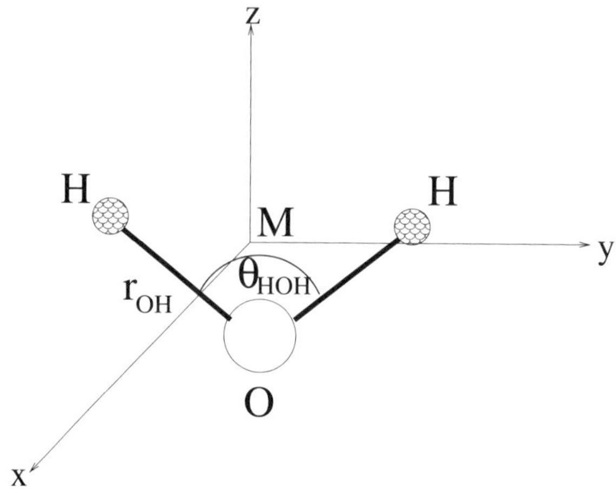

FIGURE 11.2. Schematic representation of a three-site water model. This particular model corresponds to the geometry of SSD. M is the molecular center of mass, which is taken as the origin.

molecule, which is determined by both the orientation of the dipole moment vector and the orientation of the molecular plane. The first term in equation (11.1) is the Lennard–Jones potential (see Section 16.1 for more details),

$$V_{ij}^{LJ}(r_{ij}) = 4\epsilon_w \left[\left(\frac{\sigma_w}{r_{ij}} \right)^{12} - \left(\frac{\sigma_w}{r_{ij}} \right)^{6} \right],$$

where $\sigma_w = 3.051$ Å and $\epsilon_w = 0.152$ kcal/mol (Liu and Ichiye, 1996). The second term in equation (11.1) is the point dipole–point dipole potential,

$$V_{ij}^{dp}(\mathbf{r}_{ij}, \Omega_i, \Omega_j) = \frac{\boldsymbol{\mu}_i \cdot \boldsymbol{\mu}_j}{r_{ij}^3} - \frac{3(\boldsymbol{\mu}_i \cdot \mathbf{r}_{ij})(\boldsymbol{\mu}_j \cdot \mathbf{r}_{ij})}{r_{ij}^5},$$

where $\boldsymbol{\mu}_i$ and $\boldsymbol{\mu}_j$ are the dipole moment vectors, each of whose magnitude is 2.35 D. Note that the factor $1/4\pi\epsilon_0$ has been omitted for simplicity in defining the potential due to electrostatic charges. The last term in equation (11.1) is the tetrahedral sticky potential,

$$V_{ij}^{sp}(\mathbf{r}_{ij}, \Omega_i, \Omega_j) = V^\circ \left[s(r_{ij}) w_{ij}(\mathbf{r}_{ij}, \Omega_i, \Omega_j) + s'(r_{ij}) w_{ij}^x(\theta_{ij}) \right],$$

where $V^\circ = 3.7284$ kcal/mol determines the strength of the sticky potential. The function $w_{ij}(\mathbf{r}_{ij}, \Omega_i, \Omega_j)$ is given by

$$w_{ij}(\mathbf{r}_{ij}, \Omega_i, \Omega_j) = \sin(\theta_{ij}) \sin(2\theta_{ij}) \cos(2\varphi_{ij}) + \sin(\theta_{ji}) \sin(2\theta_{ji}) \cos(2\varphi_{ji}),$$

where $(\theta_{ij}, \varphi_{ij})$ is the set of spherical polar angles of the position of molecule j in the frame fixed on molecule i and with an orientation such that the

Z-axis is parallel to the dipole moment of i and the X-axis is perpendicular to the molecular plane of i. In addition, $w_{ij}^x(\theta_{ij})$ is an empirical correction form added to the sticky potential; $s(r_{ij})$ and $s'(r_{ij})$ are the modulating functions, which interpolate between 0 and 1. The properties of water modeled by SSD are given in Table 11.3.

Evaluation of the interaction between two SSD water molecules requires computing only one distance between the two centers of mass, four spherical angles, and the angle between the dipole vectors. On the other hand, three-site models like TIP3P and SPC/E require computing nine intermolecular distances, while five-site models like ST2 and TIP5P require computing 17 intermolecular distances. Simulations with the SSD potential are about 4 times and 7 times faster compared with the three-site models in molecular dynamics (Tan et al., 2003) and Monte Carlo (Tan et al., 2003), respectively. The original parameters of the SSD water model give low water density (0.977 g/cm^3) and low heat of vaporization energy at room temperature and 1 atmosphere. However, these issues have been addressed by (Tan et al., 2003) by optimizing the parameters used in the SSD potential.

The SPC Model

The simple point charge (SPC) model is also a three-site model. It consists of a tetrahedral water model with an OH distance of 0.1 nm, H-O-H angle of 109.47°, point charges on the oxygen and hydrogen positions of $-0.82e$ and $+0.41e$ (electronic charge units), respectively, and a Lennard–Jones interaction on the oxygen positions given by

$$V_{\mathrm{LJ}} = -(A/r)^6 + (B/r)^{12}, \tag{11.2}$$

where $A = 0.37122\ (\mathrm{kJ/mol})^{1/6}\cdot$ nm and $B = 0.3428\ (\mathrm{kJ/mol})^{1/12}\cdot$ nm. The total interaction energy between two SPC water molecules consists of the Lennard–Jones potential and the Coulombic potential based on classical electrostatics,

$$V_{\mathrm{total}} = V_{\mathrm{LJ}} + V_C, \tag{11.3}$$

where V_C between two molecules i and j is represented as the sum of Coulomb interactions acting among the charged points (i.e., the O and H atoms' positions) in the following way:

$$V_C = \sum_{\mathrm{O,H}} \frac{q_I q_J}{r_{ij}^{IJ}}, \tag{11.4}$$

where r_{ij}^{IJ} is the distance between site I of the molecule i and site J of the molecule j.

The geometry is similar to the water geometry shown in Figure 11.2. However, the center of mass coincides with the oxygen atom for the SPC model. The dipole moment of the SPC model is 2.27 D, compared to 1.85 D for the isolated molecule. The diffusion coefficient of the model is 3.6×10^{-5}

$cm^2 \cdot s^{-1}$ at 300 K. The potential energy for liquid SPC water (Berendsen et al., 1987) at 300 K is -41.7 kJ/mol, and the density at 300 K and atmospheric pressure turns out to be $0.98\ g \cdot cm^{-3}$. Although the model behaves quite satisfactorily for most purposes, there is room for improvement with respect to density, radial distribution function, and diffusion constant. The properties of the SPC model are summarized in Table 11.3.

The SPC/E Model

The extended simple point charge (SPC/E) model is characterized by three point masses with O-H distance of 0.1 nm, H-O-H angle equal to 109.47°, charges on the oxygen and hydrogen equal to $-0.8476\ e$ and $+0.4238\ e$, respectively, and with Lennard–Jones parameters of oxygen–oxygen interaction according to equation (11.2) ($A = 0.37122\ (\text{kJ/mol})^{1/6} \cdot$ nm and $B = 0.3428\ (\text{kJ/mol})^{1/12} \cdot$ nm) (Berendsen et al., 1987). The interaction between two SPC/E water molecules is represented by equations (11.2), (11.3) and (11.4). The SPC/E model has a dipole moment of 2.35 D. The diffusion constant is improved considerably compared to the SPC model. The agreement of the radial density distribution with experiment is somewhat better for the SPC/E model than for the SPC model (Berendsen et al., 1987).

The ST2 model

The ST2 potential developed by (Stillinger and Rahman, 1974) is based on a rigid four-point-charge model for each water molecule. The ST2 water model is shown in Figure 11.3. The positive charges $+q$ are identified as partially shielded protons, and they have been located precisely 1 Å from the oxygen nucleus, O. The distance l from O to each of the negative charges $-q$ is 0.8 Å in the ST2 model. Pairs of vectors connecting O to the point charges are all disposed at an angle θ_t,

$$\theta_t = 2\cos^{-1}(3^{-1/2}) \approx 109°28'.$$

The ST2 molecules interact with each other through a potential function consisting of a Lennard–Jones central potential acting between oxygens, plus a Coulombic potential for the 16 (4×4) pairs of point charges. The ST2 model has a dipole moment of 2.353 D (Stillinger and Rahman, 1974).

The TIP*n*P Model

The geometry of the TIP5P model is shown in Figure 11.4, and its parameters are presented in Table 11.2, along with those for the TIP3P and TIP4P models. For all TIP*n*P models, the O-H bond length r_{OH} and H-O-H bond angle θ_{HOH} have been set to the experimental gas-phase values, i.e., 0.9572 Å and 104.52°. For TIP5P, the negatively charged interaction sites are located symmetrically along the lone-pair directions with an intervening

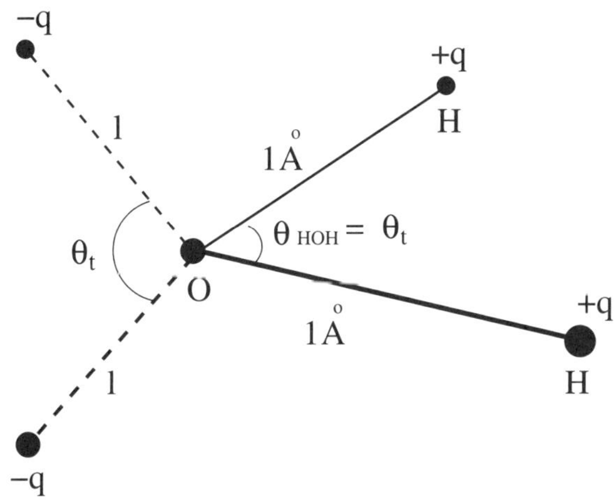

FIGURE 11.3. A sketch of an ST2 water model.

angle, θ_{LOL} , of 109.47°. A charge of +0.241 e is placed on each hydrogen site, and charges of equal magnitude and opposite sign are placed on the lone-pair interaction sites. The dipole moments are 2.35 D, 2.18 D, and 2.29 D for TIP3P, TIP4P, and TIP5P, respectively. There is no charge on oxygen for TIP5P. The Lennard–Jones potential acts on only the oxygens with a σ_0 of 3.12 Å and an ϵ_0 of 0.16 kcal/mol. The potential energy between two water molecules a and b is then given by the following equation (Mahoney and Jorgensen, 2000):

$$V_{ab} = \sum_{ij} \frac{q_i q_j e^2}{r_{ij}} + 4\epsilon_0 \left[\left(\frac{\sigma_0}{r_{OO}} \right)^{12} - \left(\frac{\sigma_0}{r_{OO}} \right)^{6} \right], \tag{11.5}$$

where i and j are the charged sites on a and b, respectively, and r_{OO} is the oxygen–oxygen distance.

The geometry of the TIP4P water model is shown in Figure 11.5. The TIP4P model is based on four interaction sites located in a planar configuration. Two of those are labeled M and O, which are associated with the oxygen nucleus, and the other two (which are labeled H) are the protons. The two distances and one angle required to fully specify the site coordinates are $r_{OH} = 0.957$ Å, $r_{OM} = 0.15$ Å (M lies on the axis of symmetry between O and the line joining the H sites), and ∠HOH = 104.5°.

The potential energy between two TIP4P water molecules consists of a double sum over the interaction sites of both molecules; the terms in the sum allow for Coulombic interactions between the electric charges assigned to the sites as well as an LJ-type contribution (equation (11.5)). The charge on the H site is equal to 0.52e, on the oxygen site is zero, and on the M site is equal to −1.04e (Rapaport, 1997). The physical properties of water based

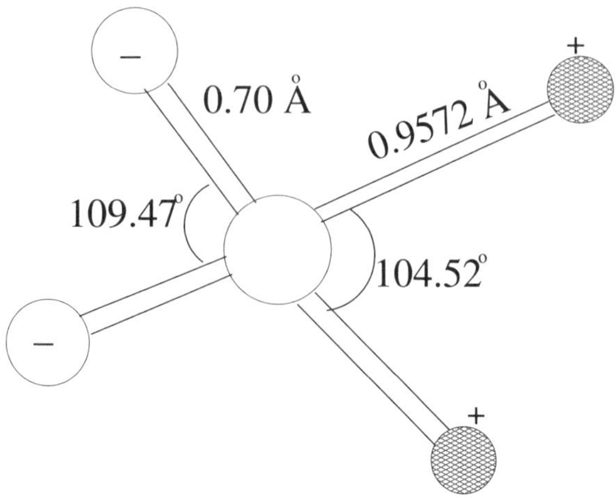

FIGURE 11.4. A sketch of the TIP5P monomer geometry.

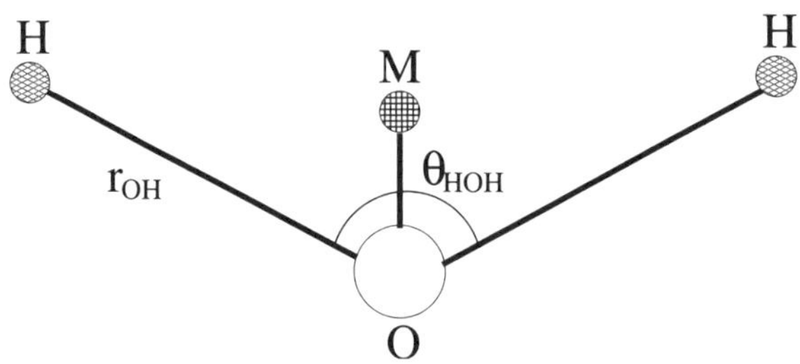

FIGURE 11.5. A sketch of the planar TIP4P water molecule.

on the TIP4P model are summarized in Table 11.3, and the geometrical and interaction parameters of the TIP4P model are summarized in Table 11.2.

The PPC Model

The polarizable point charge (PPC) model is an efficient polarizable model that retains most of the simplicity of the popular classical three-site potentials while incorporating the polarization response of the water molecule to a local electric field as determined from ab initio calculations (Svishchev et al., 1996). Studies (Kusalik et al., 1995) have shown that the PPC potential accurately describes both the virial equation of state for steam and the liquid properties, including the local structure, dielectric constant, the self-diffusion coefficient, and the Debye relaxation time (Kusalik et al., 1995), over a wide range of temperatures. The distance between O and H (i.e., r_{OH}) in the PPC model is equal to 0.943 Å and $\measuredangle$HOH $= 106°$. The ge-

TABLE 11.2. Monomer geometry and parameters for the TIPnP potential functions (Mahoney and Jorgensen, 2000).

Parameters	TIP3P	TIP4P	TIP5P
q_{H} (e)	0.417	0.520	0.241
σ_0 (Å)	3.15061	3.15365	3.12
ϵ_0 (kcal/mol)	0.1521	0.1550	0.16
r_{OH} (Å)	0.9572	0.9572	0.9572
θ_{HOH} (deg)	104.52	104.52	104.52
r_{OL} (Å)	–	0.15	0.70
θ_{LOL} (deg)	–	–	109.47

ometry of the PPC water model is similar to the geometry of the three-site water model shown in Figure 11.2. The values of the hydrogen charges are given by (Kusalik et al., 1995)

$$q_{+} = 0.486 \pm 0.03E_x + 0.02E_z,$$

where for the second term a "+" sign is used for one of the hydrogens and a "−" sign is used for the other hydrogen, and the electric field is given in V/Å. The charge on the negative site follows from charge neutrality. The short-range interaction of the PPC model is taken to be an LJ potential centered on the oxygen. The LJ parameters, $\epsilon_{\mathrm{LJ}} = 0.6$ kJ/mol and $\sigma = 3.234$ Å, are optimized to give the correct energy, self-diffusion coefficient, and structure for water at 25°C. It is observed by (Kusalik et al., 1995) that unlike other polarizable models, which typically execute at least three times slower than effective potential models, a comparable calculation with the PPC model runs only 1.5 times slower than a simulation with SPC/E water. The physical properties of the water simulated using the PPC model are summarized in Table 11.3.

The Six-Site Model

(Nada and van der Eerden, 2003) have proposed an intermolecular potential model of a rigid H_2O molecule that has six interaction sites. It has been observed that this six-site model is much more suitable for the simulation of ice and water near T_{M} (i.e., melting point of ice) than the TIP4P and TIP5P models. Figure 11.6 shows a sketch of the six-site model. A positive point charge is placed on each hydrogen (H) site, and a negative charge on each lone-pair (L) site, as in the TIP5P model. A negative charge is also placed on a site M, which is located on the bisector of $\measuredangle$HOH, as in the TIP4P model. In contrast to the TIP4P and TIP5P models, the Lennard–Jones interaction acts not only on the oxygen (O) site but also on the H sites in the present model. For the O-H length (r_{OH}) and H-O-H angle in this model, intermediate values between those of the real molecule

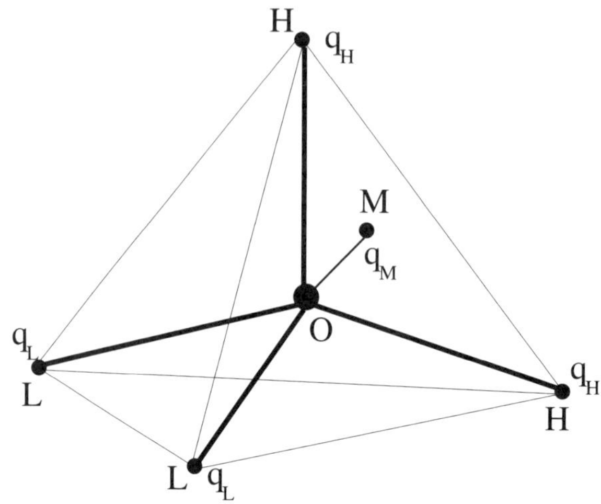

FIGURE 11.6. Schematic illustration of the water molecule in the six site model.

in the ice and vapor phases are taken (Nada and van der Eerden, 2003). The average of ∡HOH and ∡LOL in the present model is the tetrahedral angle of 109.5°. The intermolecular interaction between molecules i and j, V_{ij}, is represented as the sum of the Coulomb interactions acting among the charged points and the Lennard–Jones interactions acting on O and H atoms, in the following way:

$$V_{ij} = \sum_{\mathrm{H,L,M}} \frac{q_I q_J}{r_{ij}^{IJ}} + 4 \sum_{\mathrm{O,H}} \epsilon_{IJ} \left[\left(\frac{\sigma_{IJ}}{r_{ij}^{IJ}} \right)^{12} - \left(\frac{\sigma_{IJ}}{r_{ij}^{IJ}} \right)^{6} \right],$$

where r_{ij}^{IJ} is the distance between site I of the molecule i and site J of molecule j. Here q_I (see Figure 11.6) gives the charge amount of site I.

Summary of Physical Properties

Table 11.3 summarizes some of the important properties of water obtained from various water models.

11.2 Static Behavior

The static behavior of water in various confined states—near a surface, in a slit pore, or in a cylindrical pore—is discussed in this section. The static behavior includes density distribution, dipole orientation, hydrogen bonding and clustering, and the contact angle of water to the surface in a nano-confinement environment. These properties are of fundamental interest and of critical importance. For example, an understanding of the

TABLE 11.3. Some of the calculated physical properties of the water models †

Model	μ_{dip} (Debye)	ϵ (relative)	D_{self} (10^{-5} cm^2/s)	E_{config} (kJ/mol)	T_{max} (°C)
SSD	2.35	72	2.13	-40.2	-13
SPC	2.27	65	3.85	-41.0	–
SPC/E	2.35	71	2.49	-41.5	-38
PPC	2.52	77	2.60	-43.2	+4
TIP3P	2.35	82	5.19	-41.1	-13
TIP4P	2.18	53*	3.29	-41.8	-25
TIP5P	2.29	81.5	2.62	-41.3	+4
six-site	1.89	33	–	–	+14
Expt.	2.65, 3.0	78.4	2.30	-41.5	+3.98

†Note that μ_{dip}, ϵ, D_{self}, E_{config}, and T_{max} denote the dipole moment, dielectric constant, self diffusion coefficient, average configurational energy, and maximum density temperature. All the data are at 25° and 1 atm, except * at 20°.

density distribution and dipole orientation of the water molecules near a surface is crucial for understanding the electrochemical reaction at a surface (Henderson et al., 2001). In recent years, experimental techniques have improved dramatically, and it is now possible to probe the structure of a liquid at atomistic detail (Toney et al., 1994; Cheng et al., 2001). For example, based on x-ray scattering measurements, (Toney et al., 1994) have proposed that water is ordered in layers extended to about three molecular diameters from an electrode surface and that water density near a charged electrode is very high. Though these experiments can provide good insight into water density distribution and dipole orientation, they cannot provide detailed and direct information of these functions; rather, they provide information about the integrals involving these functions. In addition, these experiments can probe only relatively simple geometry and cannot be used easily to study how the various parameters (e.g., surface characteristics) influence the static behavior of water in confined states. To this end, atomistic simulations have been more sucessfully used, and we will review some of the interesting results here.

11.2.1 *Density Distribution and Dipole Orientation*

There is an extensive literature on the density distribution and dipole orientation of water near a one-dimensional surface (Henderson et al., 2001; Yeh and Berkowitz, 2000; Spohr et al., 1998; Galle and Vortler, 1999; Gordillo and Martí, 2003; Puibasset and Pellenq, 2003; Muller and Bubbins, 1998) (1-D confinement), inside a cylindrical pore (Henderson et al., 2001; Allen et al., 1999; Rovere and Gallo, 2003; Walther et al., 2001a; Allen et al., 2002;

Gallo et al., 2000; Lynden-Bell and Rasaiah, 1996; Leo and Maranon, 2003; Gallo et al., 2002a; Gallo et al., 2002b; Green and Lu, 1997) (2-D confinement) and inside a cavity (Levinger, 2002; Brovchenko et al., 2001; Egorov and Brodskaya, 2003) (3-D confinement). Here we focus on how the density and the dipole orientation of the water molecules are influenced by the degree of confinement (e.g., size of nanopore), by the properties of surface (e.g., hydrophobic vs. hydrophilic surface), and by the presence of surface charge.

(Allen et al., 1999) studied the water density and the dipole orientation in a cylindrical pore systematically by varying the pore size and the surface properties. Three types of pores were considered. The first is a hydrophobic wall consisting of a regular array of Lennard–Jones (LJ) 12-6 centers on a cylindrical shell; the second is a structureless one-dimensional potential function that approximates the atomic hydrophobic wall; and the third is a hydrophilic surface containing an array of bound water molecules. Figure 11.7 shows the water oxygen atom density profiles against distance from the effective wall radius for each surface type and for an effective radius $R =$ 2.1 Å to 5.6 Å. The pore is solvated by a band of water molecules centered at 1.2 to 1.4 Å from the effective channel radius. The hydrophobic and hydrophilic channels result in very different water density profiles. Inside hydrophobic channels, for $R \leq 3.6$ Å, there is only one layer of water, while there are two layers for larger cross-sections. In a hydrophilic channel, the water molecules are able to approach closer to the channel wall, resulting in a more sharply defined density peak compared to that in a hydrophobic channel. The location of the first water density peak near the channel wall is also shifted toward the channel wall for the hydrophilic surface case. In addition, in large pores, a third peak of water density appears when the channel surface is hydrophilic.

Figure 11.8 shows 20-Å segments of the center of mass corrected water oxygen atom density profiles along the channel axis (z-direction) for atomic hydrophobic (A), LJ hydrophobic (B), and hydrophilic channels (C) with $R = 2.6$ Å, 4.1 Å, and 5.6 Å. Note that for both hydrophobic channels, a transition occurs at $R = 4.1$ Å, where the water becomes highly structured, and well-defined layers emerge. The similar results for the two hydrophobic channels suggest that an explicit treatment of the atomic structure of the channel does not have an appreciable effect on water structure. Hence, the atomic hydrophobic channels are well represented by a one-dimensional potential function. The LJ 5-3 function can be used to significantly reduce the simulation times, since this gives a good reproduction of structural properties of water. In the hydrophilic channels, the water structure exhibits a markedly different behavior, with only the narrow 2.6-Å channel exhibiting some order along the channel axis.

Figure 11.9 shows the dipole orientation distribution for channels of effective radius 2.6 Å (panel A), 4.1 Å (panel B) and 5.6 Å (panel C). The left panels show the average projection of dipole orientation onto the z-

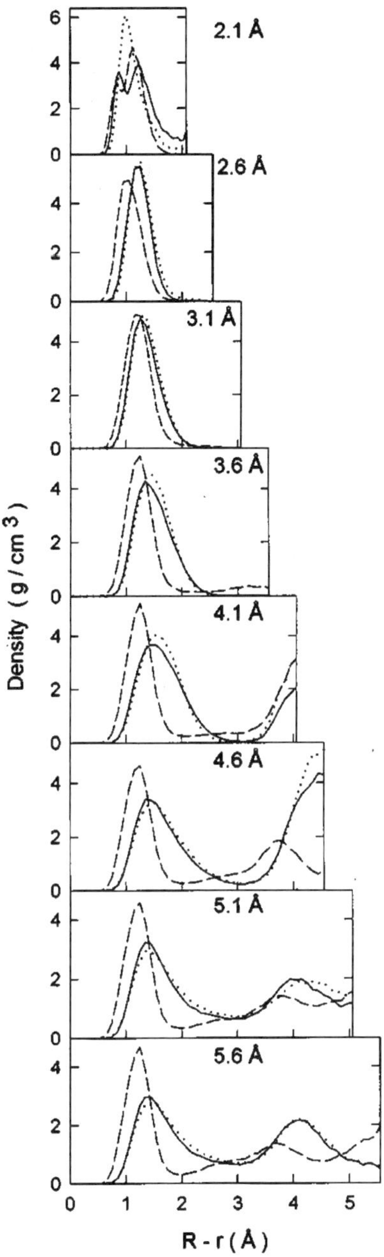

FIGURE 11.7. Water–oxygen atom radial density profiles: Density profiles for the atomic hydrophobic (solid curve), LJ hydrophobic (dotted curve), and hydrophilic (dashed curve) channels are compared for effective channel radii 2.1–5.6 Å. All profiles are plotted against the distance from the effective wall, of the channel $R-r$. The left-hand vertical axis represents the effective wall, while the right-hand axis represents the channel axis z. Each graph is labeled with the effective channel radius R. (Courtesy of S.-H. Chung.)

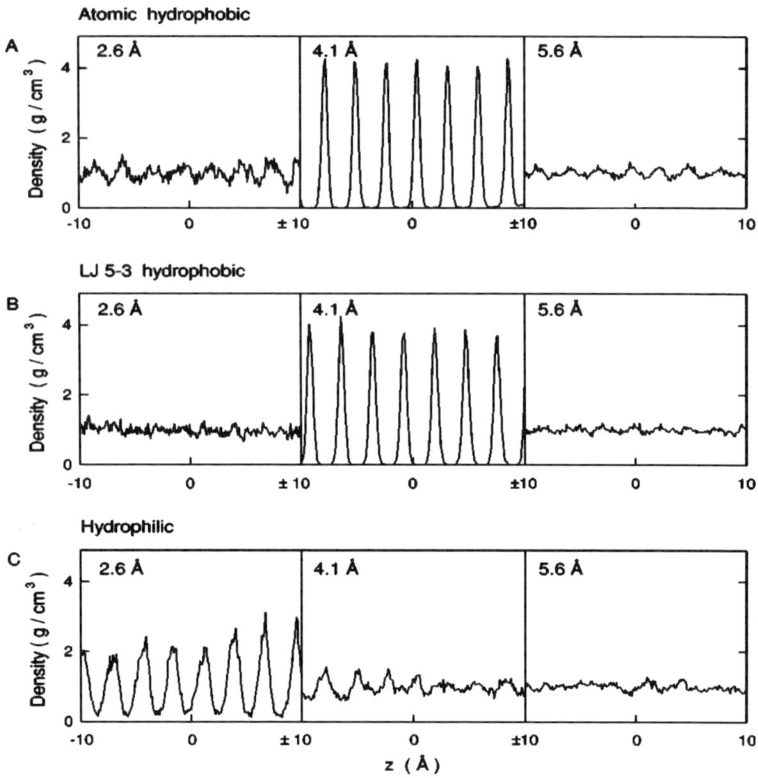

FIGURE 11.8. Water–oxygen atom density profiles along the channel axis direction: (A) Density profile for atomic hydrophobic channels of effective radii 2.6 Å (left panel), 4.1 Å (central panel), and 5.6 Å (right panel). (B) Density profiles for LJ hydrophobic surface described by the LJ 5-3 potential. (C) Density profiles for hydrophilic channels. (Courtesy of S.-H. Chung.)

direction (channel axis direction) for atomic hydrophobic (solid curve), LJ hydrophobic (dotted curve), and hydrophilic (dashed curve) channels. The right panels show the projection onto the radial vector **r**. In the hydrophobic channels, water molecules have a definite preference to point either up or down the channel axes, and such a trend is especially clear in narrow channels. Most distributions possess symmetry about the origin and result in zero net dipole moment. However, in a very narrow channel (e.g., $r =$ 2.6 Å), the symmetry is broken, and this leads to a rather large net dipole moment along the channel. In contrast, in the hydrophilic channels, the water dipoles prefer to lie in the xy-plane rather than align along the channel axis, and align themselves more toward the channel wall. The net dipole moments are zero in all the hydrophilic channels. Clearly, the geometrical confinement alters the behavior of water from that in the bulk, and the effects of different types of surfaces make its behavior further complicated.

This can have a very profound influence on various macroscopic measurables. As discussed by (Allen et al., 1999), an ion attempting to pass along a line of dipoles aligned head to tail (as in the narrow hydrophobic channel case) will be impeded less than if it were to pass alongside a line of dipoles oriented perpendicular to the line (as in a hydrophilic channel), and this can lead to very different transport behavior of the ions.

Density distribution and dipole orientation studies have also been performed on charged nanopores. (Hartnig et al., 1998) studied the density distribution and dipole orientation of water molecules in two cylindrical nanopores with a radius of 10.9 Å. In the first pore, there are 72 alternating positive and negative elementary charges on the pore surface, and the second pore has no surface charge. Figure 11.10 shows the oxygen and hydrogen density distribution for the two pores. From the graph, we observe that the water oxygen density has a slightly larger peak when the pore is charged. In addition, the density peak of the oxygen atoms is closer to the pore surface when the nanopore is charged. This can be explained by the attraction of the water molecules toward the nanopore surface. The hydrogen atom density profile shows a peak at $R = 9.2$ Å, which was not observed when the pore was not charged. This is caused by the attraction of the hydrogen atoms by the negative surface charges, which is not balanced by the short-range Lennard–Jones repulsion as in the case of oxygen. Figure 11.11 shows the average dipole moment of a water molecule in the radial direction as a function of its radial position. For both charged and noncharged pores, we observe that the water dipole tends to point toward the pore center in the region 6.7 Å $< R <$ 7.2 Å (where R is the pore radius). However, in the charged pore, there is an additional region $R >$ 8.2 Å, where the dipoles point toward the pore surface, consistent with the hump in the hydrogen density profile at $R = 9.2$ Å. From these results, we can conclude that the presence of a charge on the pore surface can influence the density distribution, dipole orientation, and the average dipole moment.

Water confined in silica pores has also been investigated. Silica and other porous materials are of great interest for a number of nanotechnology applications including catalytic and separation processes. (Gallo et al., 2000) have examined the density profiles of water confined in a cylindrical Vycor glass pore at various hydration levels (the smaller the number of water molecules in the pore, the smaller the hydration level). Figure 11.12 shows the snapshot of equilibrium configuration of confined water at various hydration levels (the pore diameter is 4.0 nm). The hydrophilic nature of the Vycor glass is evident, since all the water molecules are attracted toward the pore surface. Low amounts of water (e.g., the number of water molecules is 500) are almost completely adsorbed on the pore surface. With increasing water concentration, water molecules occupy an increasing fraction of the free volume of the pore. For about 2600 water molecules, the hydration level is found to be close to 100%.

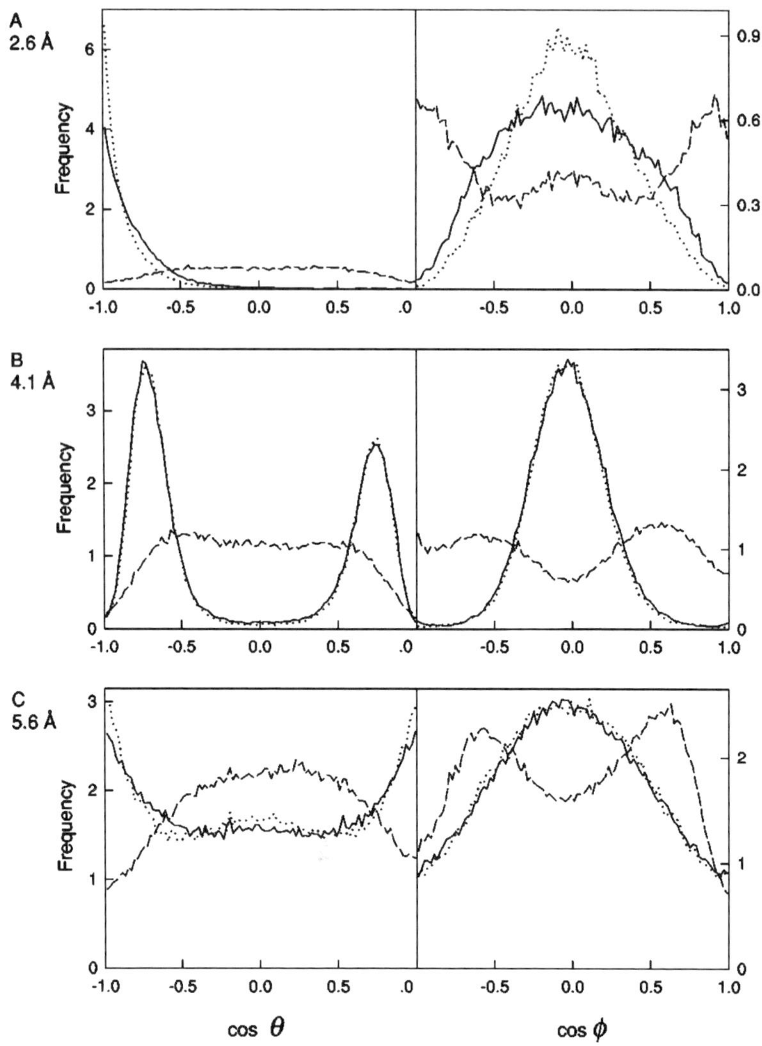

FIGURE 11.9. Water dipole orientation distributions for channels of effective radius 2.6 Å (A), 4.1 Å (B), and 5.6 Å (C). The left panels show the average projection of dipole orientation onto the z-axis ($\cos(\theta)$) for atomic hydrophobic (solid curve), LJ hydrophobic (dotted curve), and hydrophilic (dashed curve) channels. The right panels show the projections of the dipole onto the radial vector $\mathbf{r}$ ($\cos(\phi)$). (Courtesy of S.-H. Chung.)

11.2.2 Hydrogen Bonding

A hydrogen atom covalently bound to an electronegative atom (e.g., nitrogen, oxygen) has a significant positive charge and can form a weak bond

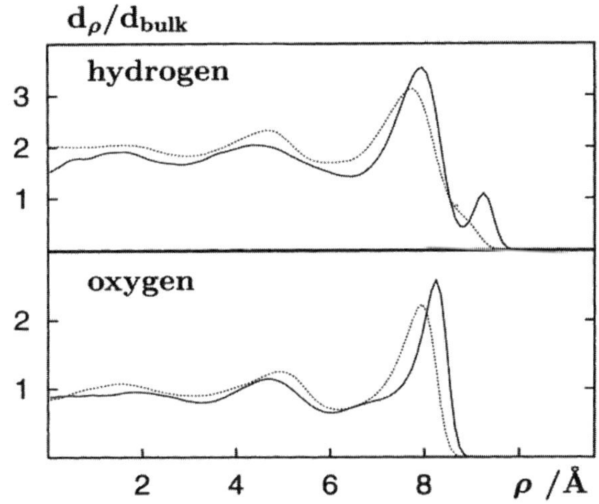

FIGURE 11.10. Oxygen and hydrogen density distribution in a cylindrical pore of radius 10.9 Å (ρ denotes the radius of the pore in Å) when the pore wall has alternating positive and negative elementary charges (solid line) and when the pore wall has no charges (dotted line). (Courtesy of E. Sphor.)

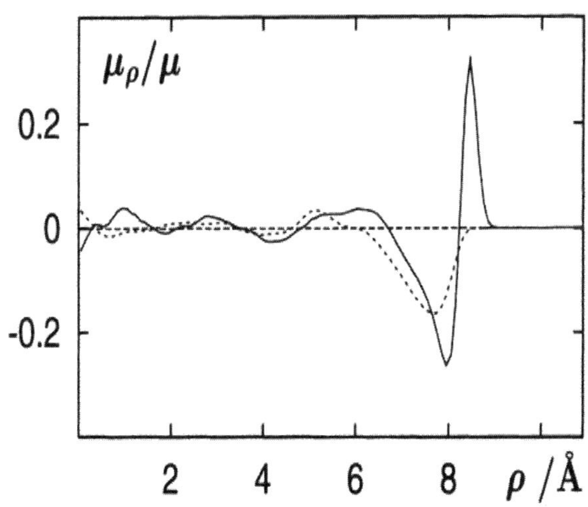

FIGURE 11.11. Average dipole moment (divided by the molecular dipole of TIP4P water model) as a function of the radial position (ρ) in the pore. Solid line: charged pore wall. Dotted line: uncharged pore wall. (Courtesy of E. Spohr.)

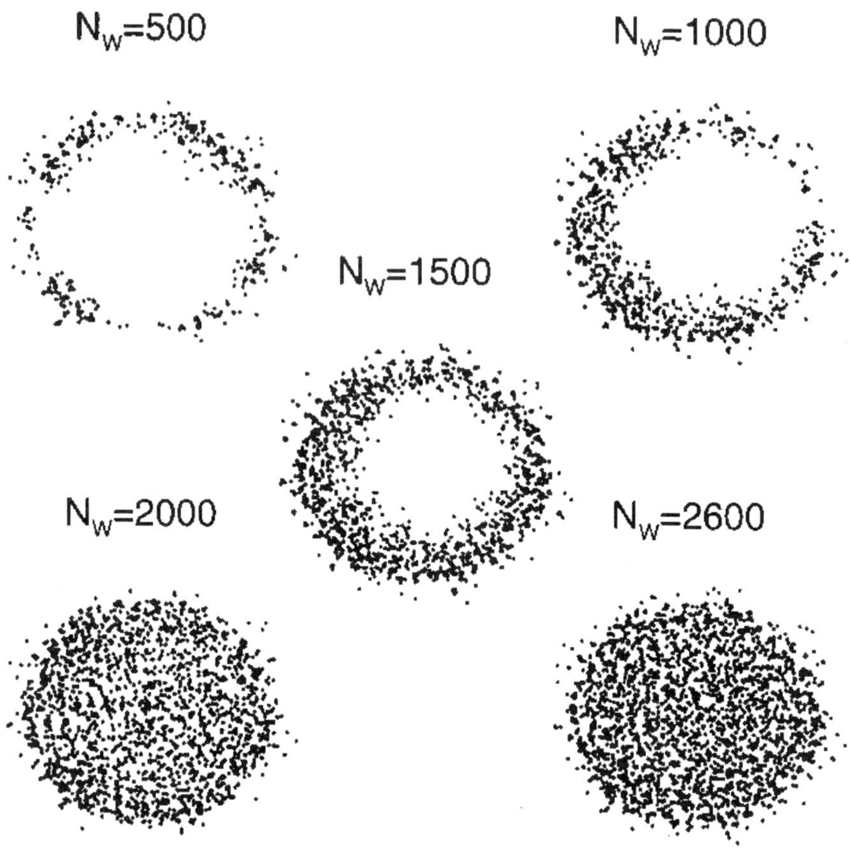

FIGURE 11.12. Snapshots of equilibrium configuration of confined water at different hydration levels. Only the oxygen atoms of water are shown projected on the xy-plane perpendicular to the axis of the confining cylinder. N_w is the number of water molecules in the pore. N_w = 500, 1000, 1500, 2000, and 2600 correspond to approximately 20%, 40%, 60%, 75%, and 100% hydration levels. (Courtesy of M. Rovere.)

to another electronegative atom, and such a bond is usually termed as a hydrogen bond. Water molecules in liquid state can form an extensive hydrogen bonding network. Figure 11.13 shows the scenario of hydrogen bonding between two water molecules. The strength of hydrogen bonds ($\approx$ 21 kJ/mol) is normally intermediate between that of the weak van der Waals interactions ($\approx$ 1.25 kJ/mol) and that of the covalent chemical bonds ($\approx$ 420 kJ/mol).

In atomistic simulations, there are two types of criteria for determining whether two water molecules form a hydrogen bond, namely, an energetic criterion and a geometric criterion. In the energetic criterion, two water

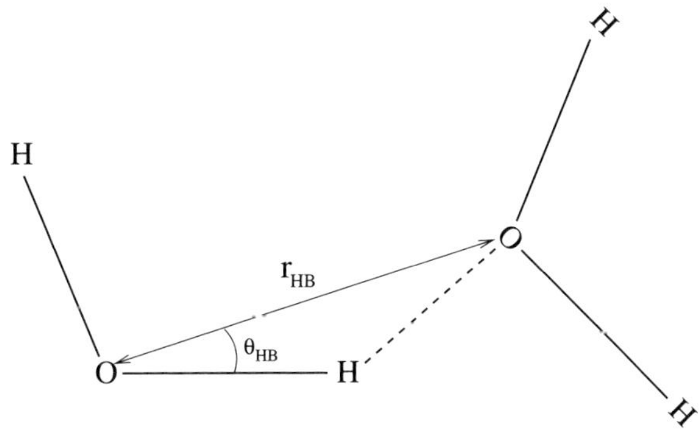

FIGURE 11.13. Schematic of the hydrogen bonding between two water molecules.

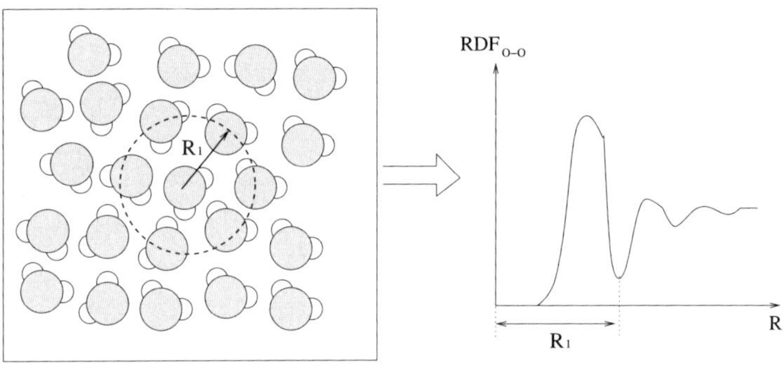

FIGURE 11.14. Schematic of bulk water molecules and the definition of the first coordination shell.

molecules are said to form a hydrogen bond if their interaction energy is less than a certain threshold energy E^c_{HB}. In the geometric criterion, two water molecules are said to form a hydrogen bond if (a) the O–O distance is less than certain distance $\mathrm{r}^c_{\mathrm{HB}}$ and (b) the O–H$\cdots$O angle is less than a certain threshold angle θ^c_{HB}. The O–O distance and O–H$\cdots$O angle are usually referred to as hydrogen bond length (r_{HB}) and hydrogen bond angle (θ_{HB}). Typically, E^c_{HB} is chosen to be -10 kJ/mol, $\mathrm{r}^c_{\mathrm{HB}}$ is chosen to be the radius of the *first coordination shell* of a water molecule, and θ^c_{HB} is chosen to be 30°. The first coordination shell of a water molecule is defined as the first water shell around that water molecule. The radius of the shell, R_1, is usually chosen to be the position of the first minimum of the oxygen–oxygen radial distribution function ($\mathrm{RDF}_{\mathrm{O-O}}$) (see Figure 11.14).

At the molecular level, the water structure is determined by the hydrogen bonding (HB) network. Since HB plays an important role in determining the transport properties of water molecules and ions, it is important to understand how the HB network is influenced when the water is confined in nanochannels. The HB of water inside slit pores (Galle and Vortler, 1999; Galle and Vortler, 2001) and cylindrical pores (Allen et al., 1999; Rovere and Gallo, 2003; Walther et al., 2001a; Rovere et al., 1998; Werder et al., 2001; Spohr et al., 1999; Mashl et al., 2003) has been studied extensively in the past decades.

Figure 11.15 shows the number of hydrogen bonds per water molecule in a cylindrical pore as a function of the pore radius and surface properties (Allen et al., 1999). For a hydrophobic pore, the extent of HB is maximized at moderate radii ($r \approx 3.6$ Å), and drops notably as the pore size decreases. Similar observations have been reported for other hydrophobic pores, e.g., inside a carbon nanotube (Mashl et al., 2003). For a hydrophilic pore, the number of hydrogen bonds per water molecule is not sensitive to the pore size. However, if we exclude the hydrogen bonds between wall atoms and water molecule (dash-dot-dot curve in Figure 11.15), i.e., by counting only the hydrogen bonds between water molecules inside the pore, we find that the average number of hydrogen bonds per water molecule is significantly lower than that for the corresponding hydrophobic pores. This is because the water molecules inside the pore sacrifice water–water hydrogen bonds for the water–wall hydrogen bonds. Since the wall atoms are fixed, the strong interactions between water molecules and the wall atoms are likely to result in a reduced transport (e.g., diffusion transport) of water molecules in a hydrophilic pore compared to that in a hydrophobic pore.

(Rovere et al., 1998) also reported on the HB of water molecules inside a 4.0 nm diameter cylindrical SiO_2 pore. Figure 11.16 shows the number of hydrogen bonds as a function of distance from the center of the pore computed as the wall–water interactions are turned "on" (solid line) and "off" (dash line). Note that the number of hydrogen bonds decreases almost monotonically as we approach the pore surface when the wall–water interaction is turned on. However, when the surface atom–water interaction is turned off, the number of hydrogen bonds is essentially constant up to 15 Å from the pore center. Comparison of these two results indicates that the pure geometrical confinement can alone be responsible for the reduction in the number of hydrogen bonds at the interface.

Finally, the dynamic properties of HB have also been investigated. (Hummer et al., 2001) investigated HB inside a carbon nanotube of 8.1 Å diameter. They found that the HB inside the carbon nanotube is much more stable and highly oriented compared to that in the bulk. For example, the average lifetime of a hydrogen bond inside a carbon nanotube is 5.6 ps, compared to 1.0 ps in the bulk. In addition, less than 15% of the H–O$\cdots$H angles inside the carbon nanotube exceeds 30°, compared to 30% in the bulk. It was also reported that the OH bonds involved in the hydrogen bonds are nearly aligned with the nanotube axis, collectively flipping direction every 2 to 3 ns on average. In summary, the HB can be significantly

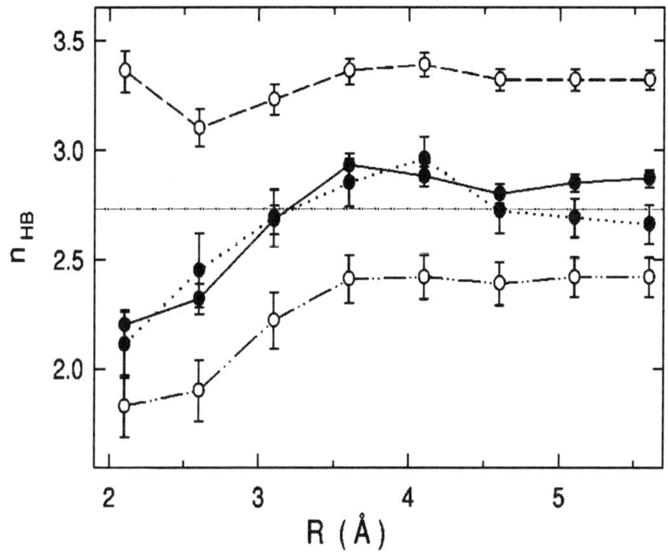

FIGURE 11.15. The effect of pore radius on the number of hydrogen bonds per water molecule for atomic hydrophobic (solid curve, filled circles), LJ hydrophobic (dotted curves, filled circles), and hydrophilic channels (dashed curve, open circles). The number of hydrogen bonds in hydrophilic pores when channel wall molecules are excluded are drawn as dash-dot-dot curves (open circles). (Courtesy of S.-H. Chung.)

influenced by the strong confinement inside a carbon nanotube, and the strongly stable hydrogen bonding inside the nanotube can change the wetting characteristic of the nanotube significantly (Hummer et al., 2001).

11.2.3 Contact Angle

The contact angle θ is a quantitative measure of the wetting of a solid by a liquid, as we have seen in Section 8.1; see Figure 8.4. At nanoscales, where the surface to volume ratio of the system can be orders of magnitude larger compared to a macroscopic system, the surface–water interactions tend to dominate the behavior of water in the system. The contact angle of water on the surface is a very important indicator of the strength of the surface–water interactions. Several authors have investigated the static wetting behavior of water on a graphite sheet (Werder et al., 2003) and on a carbon nanotube surface (Werder et al., 2001) and computed the contact angle of water on these surfaces.

(Werder et al., 2003) presented a study on the contact angle of water on a graphite surface by using molecular dynamics simulations. Figure 11.17

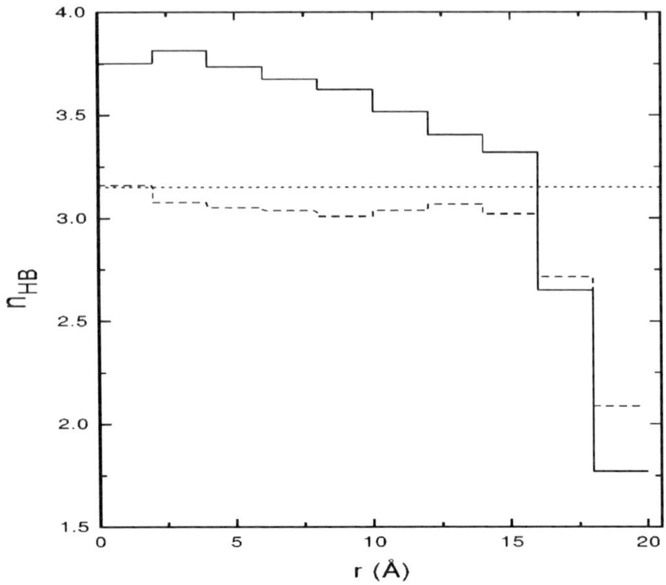

FIGURE 11.16. Number of hydrogen bonds as a function of the distance from the center of the pore, calculated with the wall–water interaction turned on (solid line) and off (dashed line). The horizontal dotted line represents the density-weighted average value of number of hydrogen bonds for confined water with the wall–water interaction turned on. (Courtesy of M. Rovere.)

shows snapshots of the system. It shows that a cubic water block will spontaneously evolve to a hemisphere as it reaches equilibrium. The contact angle of the water droplet on the graphite surface is a function of the Lennard–Jones potential between the carbon atom and the water molecule. The authors found that using a $\sigma_{CO} = 3.19$ Å and $\epsilon_{CO} = 0.392$ kJ/mol, one can reproduce the contact angle (86°) of water on a graphite surface as measured at the macroscopic scale. Such interaction parameters can then be very useful in the study of water behavior in carbon nanotubes.

Recently, (Gonnet et al., 2004) have suggested that the contact angle decreases weakly with increasing concentration of impurities like ions in water, and that chemisorption[1] of water onto the graphite surface, altering the partial charge distribution on the graphite surface, can influence the contact angle significantly.

[1] *Chemisorption*, or *chemical adsorption*, is adsorption in which the forces involved are valence forces of the same kind as those operating in the formation of chemical compounds, and the elementary step in chemisorption often involves an activation energy.

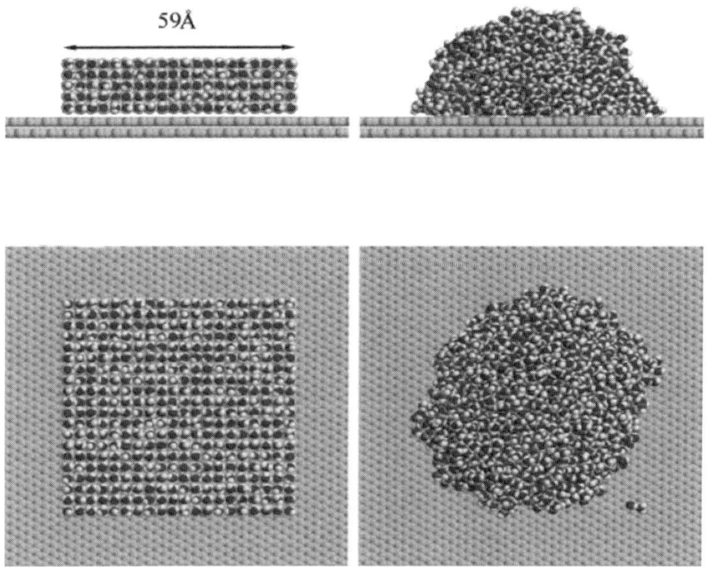

FIGURE 11.17. Side (top row) and top view (bottom row) of the initial ($t = 0$) and equilibrated ($t = 0.2$ ns) water droplet. The lateral graphite dimensions in the simulation are 119 Å ×118 Å. (Courtesy of P. Koumoutsakos.)

11.2.4 Dielectric Constant

From the results presented in this section, we can conclude that confinement can change the dipole orientation of water molecules significantly. This could lead to a change in the dielectric constant of water, which, in turn, can influence the dynamic properties and the electrostatic interactions between water molecules and between ion and water molecules. (Senapati and Chandra, 2001) studied the dielectric constant of water inside spherical cavities of various sizes using two different water models, namely, soft sticky dipole (SSD) (Liu and Ichiye, 1996) and SPC/E (Berendsen et al., 1987) models (see Section 11.1 for details). Figure 11.18 shows the dielectric constant computed using the two models as a function of the cavity size (Senapati and Chandra, 2001). Clearly, the dielectric constant in a nanocavity is significantly different from that in the bulk, and it decreases as the cavity radius decreases. Specifically, a nearly 50% decrease of the dielectric constant is observed in a cavity of about 12 Å diameter. Since the cavity surface is not charged, the reduction of the dielectric constant is purely an effect of confinement.

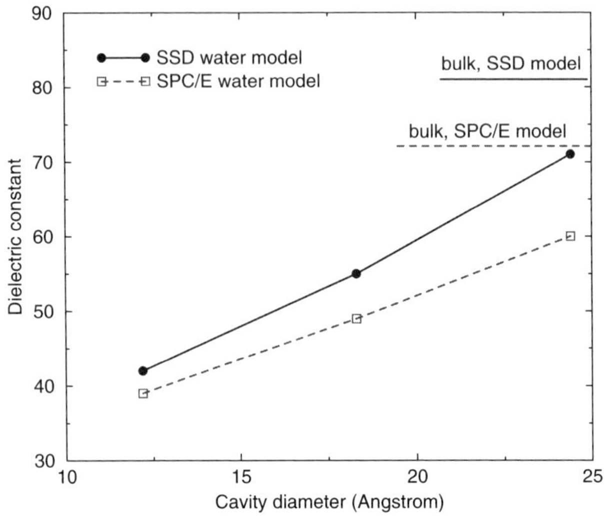

FIGURE 11.18. Dielectric constant of water confined in nanocavities of different cross sections. (Courtesy of A. Chandra.)

11.3 Dynamic Behavior

Understanding the dynamic behavior of water is critical to many biological and engineering applications. For example, the study of the diffusion of water molecules through nanochannels can help explain the operating mechanisms of the water channels, which are responsible for many important biological processes in the cell (Sui et al., 2001; Hummer et al., 2001; Beckstein and Sansom, 2003). In this section, we first review the research on the basic concepts of dynamic behavior of water and then summarize some of the simulation results on the diffusion transport of water through nanochannels. Finally, the filling and emptying kinetics of water in nanopores is discussed.

11.3.1 Basic Concepts

In this section, we focus our attention on the properties of the motion of a single water molecule in confined states, such as the reorientation dynamics, residence time, dipole correlation, and the velocity distribution. Understanding these properties can provide insight into the dynamic properties of water in more complicated scenarios such as the diffusion transport.

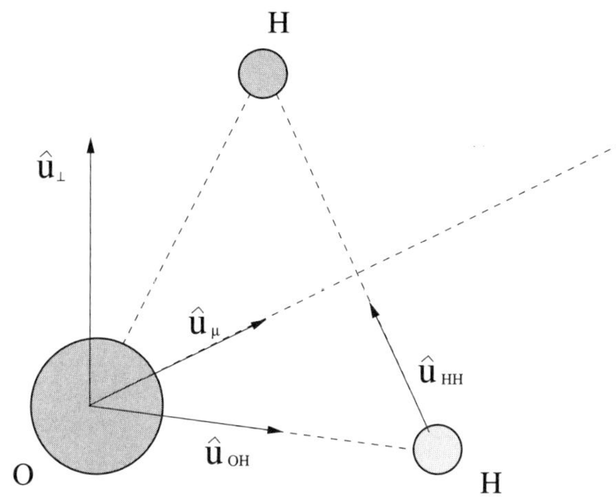

FIGURE 11.19. Definition of molecular unit vectors in a water molecule.

Reorientation Dynamics

Molecular reorientational motions in liquids are usually analyzed through the time correlation functions

$$C_{l,\alpha}(t) = \langle P_l(\hat{\boldsymbol{u}}_\alpha(t) \cdot \hat{\boldsymbol{u}}_\alpha(0)) \rangle ,$$

where P_l refers to lth Legendre polynomial and $\hat{\boldsymbol{u}}_\alpha$ is a unit vector along a given direction ($\alpha = \mu$, HH, OH, $\perp$). As shown in Figure 11.19, four different unit vectors are considered: a unit vector $\hat{\boldsymbol{u}}_\mu(t) \equiv \boldsymbol{\mu}(t)/\mu(t)$ along the molecular dipole moment direction, a unit vector $\hat{\boldsymbol{u}}_{\mathrm{HH}}(t) \equiv \boldsymbol{r}_{\mathrm{HH}}(t)/r_{\mathrm{HH}}(t)$ along the H-H direction, a unit vector $\hat{\boldsymbol{u}}_{\mathrm{OH}}(t) \equiv \boldsymbol{r}_{\mathrm{OH}}(t)/r_{\mathrm{OH}}(t)$ along the O-H direction, and a unit vector $\hat{\boldsymbol{u}}_\perp(t) = \hat{\boldsymbol{u}}_\mu(t) \times \hat{\boldsymbol{u}}_{\mathrm{HH}}(t)$ perpendicular to the molecular plane. The correlation functions associated with the Legendre polynomials are

$$C_{1,\alpha}(t) = \langle \cos\theta_\alpha(t) \rangle ,$$

$$C_{2,\alpha}(t) = \frac{1}{2} \left\langle \cos^2\theta_\alpha(t) - 1 \right\rangle ,$$

where $\cos\theta_\alpha \equiv \hat{\boldsymbol{u}}_\alpha(t) \cdot \hat{\boldsymbol{u}}_\alpha(0)$. The characteristic reorientational times $(\tau_1^\alpha, \tau_2^\alpha)$ are computed by

$$\tau_i^\alpha = \int_0^\infty C_{i,\alpha}(t)\, dt, \qquad (i = 1, 2)$$

The reorientational motion of water molecules is characterized by the dipole moment reorientation time $\tau_{1,2}^\mu$ (Martí et al., 2002). A smaller τ corresponds to a faster reorientation motion of the water molecule. Martí

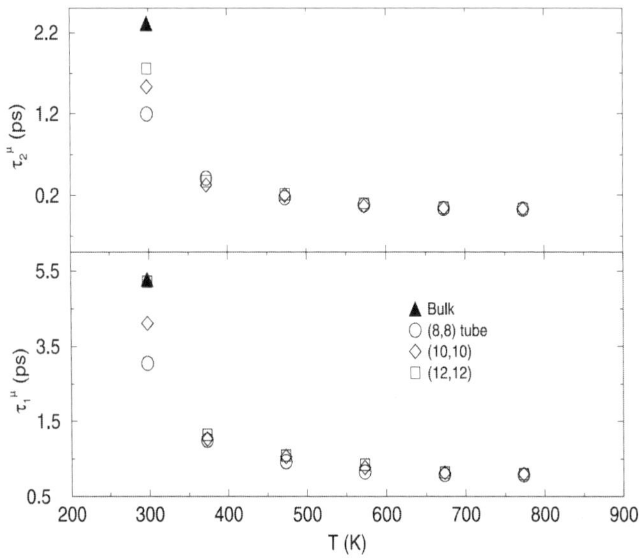

FIGURE 11.20. Molecular dipole moment reorientation time $\tau_{1,2}^{\mu}$ as a function of temperature at a density of 0.83 g/cm^3. Computation with the first (bottom) and second (top) Legendre polynomials. (Courtesy of J. Martí.)

and coworkers (Martí et al., 2002) studied the reorientation time of water molecules confined in carbon nanotubes of length 7.45 nm and internal radii of 2.65 Å ((8, 8) nanotube), 4 Å ((10, 10) nanotube) and 5.33 Å ((12, 12) nanotube); see Section 13.2.1 for details on carbon nanotubes. Figure 11.20 shows the reorientation time of the water dipole moment at different temperatures and in different-sized nanotubes. At room temperature, the smaller the nanotube diameter, the faster the reorientational motion, and in the largest nanotube, the reorientation time approaches that of the bulk water. From this result we can conclude that confinement tends to speed up molecular reorientations. The faster reorientational motion in small diameter nanotubes can be attributed to the partial breakdown of the tetrahedral hydrogen-bond network, which is typical when water is confined.

Residence Time

The mean time spent by a water molecule in its first coordination shell (see Section 11.2.2 for the definition) can be characterized by a residence time function $C_{\text{res}}(t)$. The spatial distribution of various molecules and ions closely surrounding a water molecule forms the coordinate shells of the water molecule. The first coordination shell gives rise to a noticeable

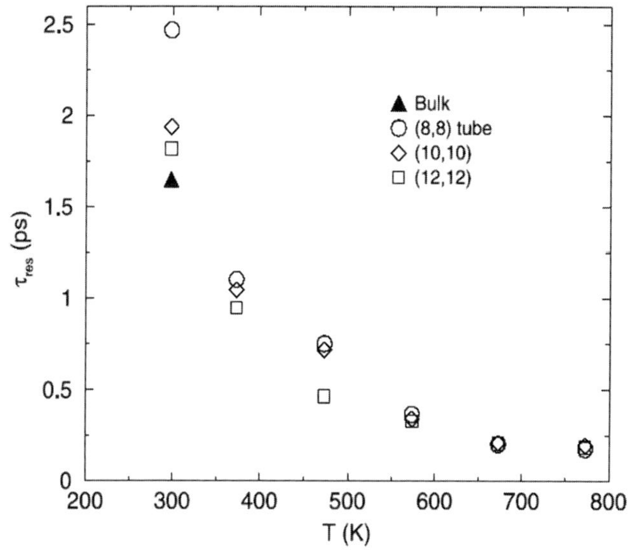

FIGURE 11.21. Residence time τ_{res} of a water molecule in its first coordination shell as a function of temperature in different-sized nanotubes at a density of 0.83 g/cm^3. (Courtesy of J. Martí.)

cluster, whereas the second coordination shell is more diffuse, and it becomes difficult to discern a pattern for subsequent coordination shells. The residence time for a given water molecule is defined as the number of water molecules in the first coordination shell of the tagged molecule during the time interval of length t. Usually, an exponential-like behavior of $C_{res}(t)$ is observed, and a residence time τ_{res} can be obtained from the fitting of $C_{res}(t)$ to a single exponential:

$$C_{res}(t) \approx c\, e^{-t/\tau_{res}}.$$

A longer residence time suggests that water molecules form a more stable structure. Figure 11.21 (Martí et al., 2002) shows the residence time τ_{res} of a water molecule in its first coordination shell as a function of the temperature in different-sized nanotubes. We observe that at room temperature, the residence time τ_{res} increases as the nanotube diameter decreases, indicating that a water molecule inside a narrow pore will spend longer time in its neighborhood than in larger pores. As the temperature increases, the residence time decreases sharply, and the deviation of τ_{res} in different-sized tubes also decreases.

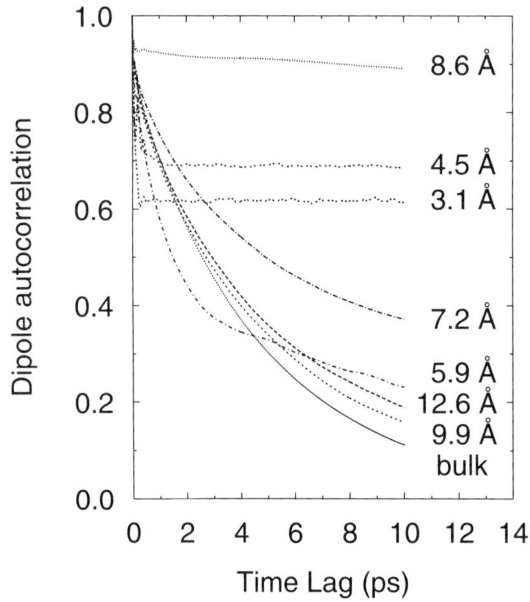

FIGURE 11.22. Water dipole correlation function inside nanotubes at 300 K.

Dipole Correlation

The reorientability of the water molecules can also be characterized by an autocorrelation function (ACF) of the water dipoles. The ACF is defined as

$$C(t) = \langle \boldsymbol{\mu}(0) \cdot \boldsymbol{\mu}(t) \rangle / \langle \mu^2 \rangle,$$

where $\boldsymbol{\mu}$ is the water dipole and the angle bracket denotes average over time and molecules. The reorientability of the water molecules plays an important role in determining the dynamic properties of water (e.g., viscosity, diffusivity).

(Mashl et al., 2003) studied the reorientability of the water molecules inside carbon nanotubes; Figure 11.22 shows the ACF obtained for water inside carbon nanotubes of various diameters. It was observed that for water in wide nanotubes, the ACFs generally take a bulk-like character. However, for smaller nanotubes, the appearance of a plateau suggests that the hydrogen bonds tend to form very rapidly and can be maintained for a much longer time compared to those in the bulk. An interesting observation is that the water molecules in the 8.6-Å nanotube show the largest degree of rotational immobilization, while the ACF of water molecules inside nanotubes that are slightly narrower or wider shows almost bulk-like behavior.

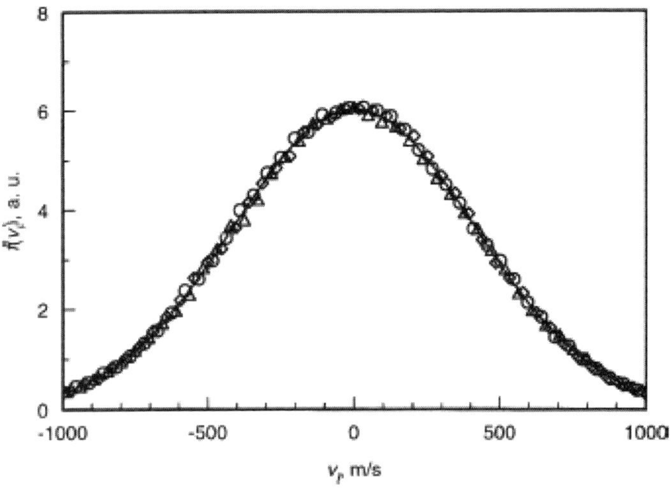

FIGURE 11.23. The distribution of the x- (circles), y- (triangles) and z-components (rhombi) of the translational velocities of water molecules at steady state. The solid line corresponds to the Boltzmann–Maxwell distribution. (Courtesy of J. Fischer.)

Velocity Distribution Function

The velocity distribution of water at a given temperature is a fundamental property, and in bulk it follows the Boltzmann–Maxwell distribution. However, whether water confined in a nanochannel would still obey the same velocity distribution is not that obvious. (Lishchuk and Fischer, 2001) studied the velocity distribution function of water confined in a 2.06 nm wide slit pore under an external microwave electric field. Figure 11.23 shows the distribution of the x-, y- and z-components of the velocity of water inside a 2.06 nm slit pore. It is observed that the velocity distribution obtained from the simulation agrees very well with the classical Boltzmann–Maxwell distribution.

11.3.2 Diffusion Transport

One of the motivations to study water behavior at nanoscale is to understand how its transport is influenced by confinement. Since diffusion is usually the dominant transport mechanism at small scales, the diffusion of water in nanochannels has been explored in detail in the past. Several researchers have investigated the diffusion of water inside artificial cylindrical nanopores (Allen et al., 1999; Lynden-Bell and Rasaiah, 1996; Zhou et al., 2003), inside silica nanopores (Rovere and Gallo, 2003; Spohr et al.,

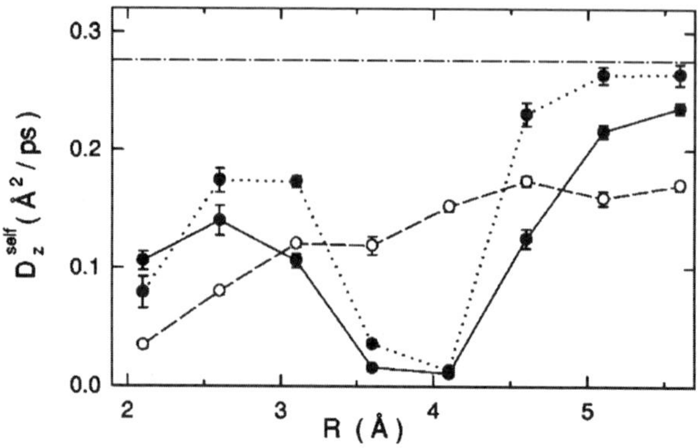

FIGURE 11.24. Self-diffusion coefficient along the axial direction (z-direction) of water molecules inside atomic hydrophobic (solid curve, filled circles), LJ hydrophobic (dotted curve, filled circles), and hydrophilic (dashed curve, open circles) pores. The horizontal dash-dot line is the bulk reference value. (Courtesy of S.-H. Chung.)

1999; Gallo et al., 1999), inside carbon nanotubes (Leo and Maranon, 2003; Mashl et al., 2003; Paul and Chandra, 2003), and inside slit nanopores (Brovchenko et al., 2001; Zhang et al., 2002).

(Allen et al., 1999) investigated the diffusion coefficient of water molecules inside artificial cylindrical pores of various radii and surface properties. Figure 11.24 shows the variation of the axial diffusion coefficient of water with the radii of the pore for atomic hydrophobic, LJ hydrophobic, and hydrophilic channels (see Section 11.2.1 for details on these channels). The result indicates that the axial (z-direction) self-diffusion of water molecules within hydrophobic pores experiences a dramatic fall for channel sizes with a radius of 3.6 to 4.1 Å. For a pore with a radius of 4.1 Å, the axial diffusion decreases to about 4% of the bulk value. This is mainly caused by the stable hydrogen-bonding network. As the pore size increases, the diffusion coefficient of water approaches its bulk value. It was also observed that the axial diffusion coefficient of water molecules inside a hydrophilic pore increases monotonically with the increase in the pore radius. The results also show that inside a small ($R < 3.0$ Å) or a large ($R > 5.0$ Å) pore, the water diffusion in a hydrophilic pore is slower than that in a hydrophobic pore. This can be explained by the preferred dipole orientation of the water molecules with the stationary wall molecules.

(Mashl et al., 2003) studied the diffusion of water inside carbon nanotubes with different diameters. Figure 11.25 shows snapshots of the system studied, and Figure 11.26 shows the mean-square displacement and axial diffusion coefficient of water molecules obtained inside different-sized nanotubes. Similar to the studies in (Allen et al., 1999), the water molecules

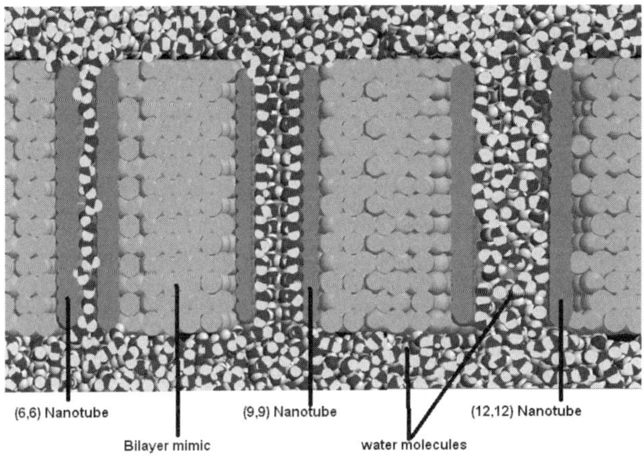

FIGURE 11.25. Snapshots from molecular dynamics simulations at 300 K with a composite image of three typical systems with nanotube size increasing from left to right. In narrow nanotubes (left), the water adopts a single-file arrangement with a 1-D hydrogen bonding but becomes more disordered in a fashion similar to that of bulk water in wider nanotubes (right). Confined within a nanotube of a "critical" diameter (center), the water spontaneously orders itself into a regular array with a 2-D hydrogen bonding.

are "immobilized" in a nanotube of 8.6 Å diameter. Panel (b) of Figure 11.25 shows the cross-sectional view of water structure inside the critically sized 8.6-Å-diameter nanotube. A stable hydrogen-bonding network is observed as expected. This is consistent with the diffusion coefficient results. Since the simulations were performed at 300 K, these results indicate that when the water is confined in a critical-diameter pore, it can behave in an ice-like manner even at room temperature. Such an effect can be very useful for a number of applications, such as proton conduction in synthetic ion channels.

11.3.3 Filling and Emptying Kinetics

In nanopores, confinement of the water can induce the so-called *drying transition* as a result of the strong hydrogen bonding between water molecules. This state can cause the water molecules to recede from the nonpolar surface to form a vapor layer separating the bulk phase from the surface (Stillinger, 1973; Lum et al., 1999; Lum and Luzar, 1997). This water behavior is important in understanding many biological processes, e.g., ion and water transport in biological channels where the channel dimension is of order a few angstroms and the channel surface is usually hydrophobic (Hummer et al., 2001). Motivated by the design of robust artificial ion and water chan-

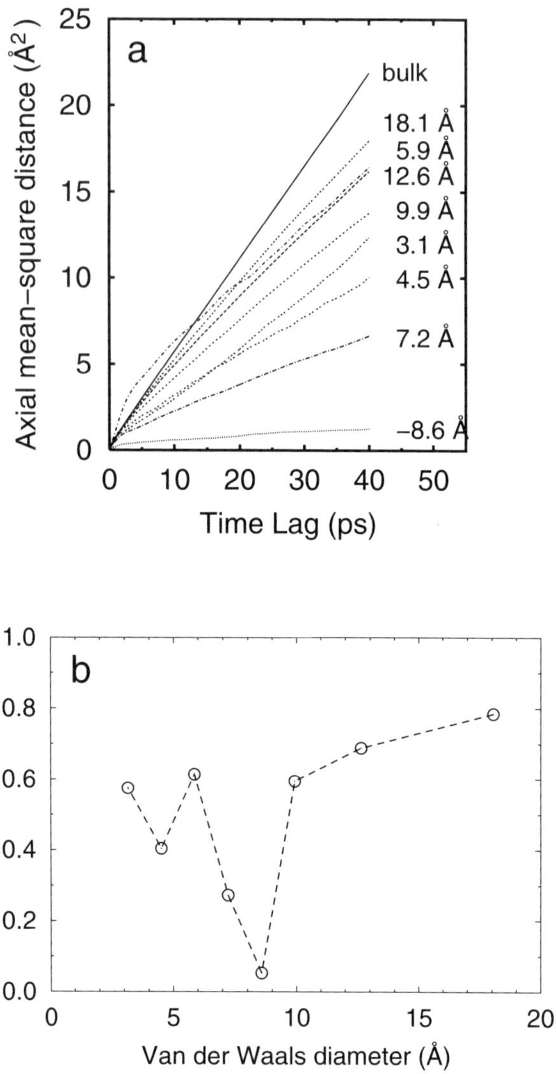

FIGURE 11.26. Water mobility inside nanotubes at 300 K. (a) Mean-square displacement (MSD) of water traveling along the nanotube axis for various nanotubes. (b) Axial diffusion coefficient of water, normalized to bulk (2.69×10^{-5} cm^2/s is the bulk value with SPC/E water), derived from panel (a).

nels (see Section 13.2 for more details on artificial ion channels), a number of researchers have investigated the filling and emptying of water in carbon nanotubes (Hummer et al., 2001; Waghe et al., 2002; Berezhkovskii and Hummer, 2002; Mann and Halls, 2003) and artificial nanopores (Allen et al., 2002; Beckstein and Sansom, 2003; Allen et al., 2003; Beckstein et al.,

2001; Kalko et al., 1995).

Water Entry in Hydrophobic Nanopores

Simulations of water conduction through carbon nanotubes (Hummer et al., 2001) indicate that even though nanotubes are hydrophobic, the channel has a steady occupancy of water molecules during the course of the simulation (Figure 11.27). Even for a 60-ns simulation (Hummer et al., 2001), the occupancies are not different. As mentioned in Section 11.2.2, the high occupancy of water inside the hydrophobic nanotube is mainly caused by the stable hydrogen bond inside the nanotube. (Hummer et al., 2001) explained this quantitatively by computing the local excess chemical potential, $\mu_{\rm NT}^{\rm ex}$, defined as the negative free energy of removing a water molecule from the channel. Such a free energy is not dominated by how strongly bound a water molecule is on average, but by how populated weakly bound states are. Even though the binding energy of water molecules inside the nanotube is unfavorable compared to bulk water, the binding energy inside the nanotube is more sharply distributed (see Figure 11.28), and high-energy states dominating the free energy are less frequently occupied. As a result, though the water molecules inside the nanotube lose about 2 kcal/mol in energy, they have a lower excess chemical potential of $\mu_{\rm NT}^{\rm ex} \approx -6.87\pm\, 0.07$ kcal/mol, compared to bulk water of $\mu_{\rm NT}^{\rm ex} \approx -6.05\pm\, 0.02$ kcal/mol. Water molecules not only penetrate the nanotube, but they also transport across the nanotube, and on average, about 17 water molecules pass through the nanotube per nanosecond (see Figure 11.29). The conduction occurs in a burst-like manner because of the tight hydrogen-bonding network inside the nanotube; rupturing the water chain is energetically costly, and so rare. However, once the rupture of the hydrogen bond occurs, the water chain moves with little resistance through the nanotube, resulting in a burst of the water flow.

Filling Mechanism

When the diameter of the nanopore is small enough for a single-file chain (in a single-file chain the water molecules move as a single chain), the mechanism of filling the nanopore can be understood as summarized below (Waghe et al., 2002):

1. Filling of water molecules in a nanopore can occur from either side of the nanopore, with hydrogens entering first, dipoles oriented outward.

2. Filling of water molecules progresses as a chain from the end; it is initiated with the orientation remaining the same.

3. Simultaneous filling from both sides is not favorable, because the dipole orientations at the ends repel each other.

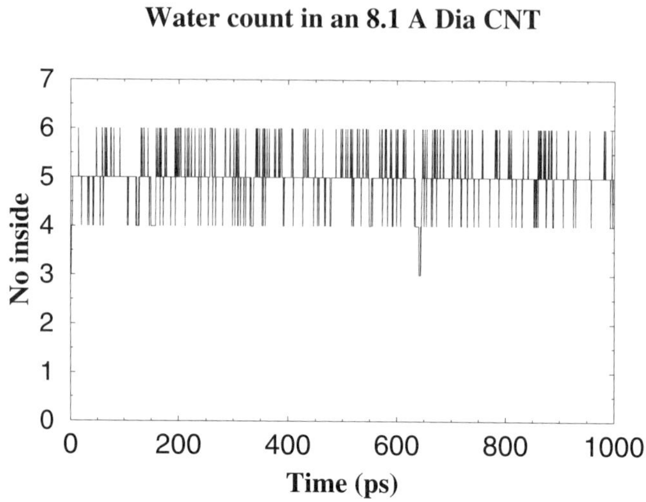

FIGURE 11.27. Water occupancy: number of water molecules inside the nanotube as a function of time.

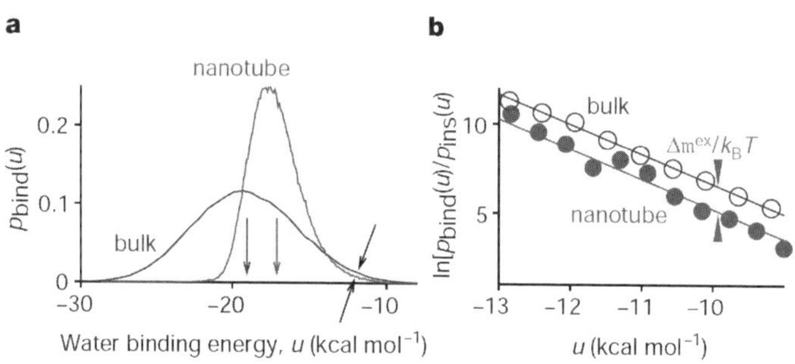

FIGURE 11.28. Water binding energies. ***a.*** Probability density $p_{\text{bind}}(u)$ of binding energies u for bulk water and water inside the nanotube. Vertical arrows indicate average binding energies. Tilted arrows indicate crossover region, in which weakly bound states are more populated in bulk water. ***b.*** $\ln[p_{\text{int}}(u)/p_{\text{ins}}(u)]$ for water inside the nanotube and in bulk TIP3P water (open circles), fitted to $\beta(\mu_w^{\text{ex}} - u)$ (lines). The vertical distance between the two parallel lines of slope $-\beta$ gives the difference in the excess chemical potentials, $\beta(\mu_w^{\text{ex}} - \mu_{\text{NT}}^{\text{ex}})$. (Courtesy of G. Hummer.)

4. Depending on the material (wall)–oxygen van der Waal's interaction forces, the chain can rupture at the ends, causing conduction in bursts as in carbon nanotubes or stable as in silicon dioxide pores.

5. Filling and emptying occurs mainly by the sequential addition of wa-

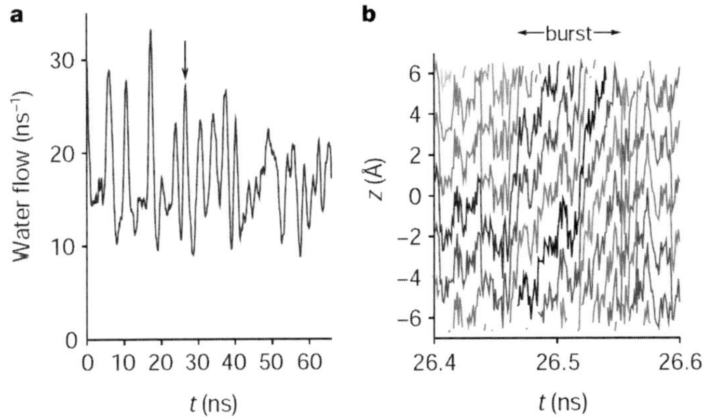

FIGURE 11.29. Water flow through the nanotube. ***a.*** Numbering of water molecules leaving the nanotube at time t that entered the nanotube from the other side. Individual water conduction events are smoothed with a 1-ns-wide triangular filter. ***b.*** motion (along the tube length) of individual water molecules during the conduction "burst" at 26.5 ns. (Courtesy of G. Hummer.)

ter to or removal from the single-file water chain inside the nanotube.

Figure 11.30 shows the trajectories of individual water molecules entering a silicon dioxide nanopore about 8 Å in diameter and 6 nm long as a function of time. With longer tubes, the dipoles at the ends do not repel each other as much as in a shorter tube, so chains form at both ends. But one of them becomes dominant, and then filling progresses sequentially through that chain till the chains meet. Figure 11.31 shows the filling as distributions of normalized dipole orientations of water molecules with respect to the axis of the tube. Here the behavior of water with respect to time does not come into the picture, but rather the fact that orientation remains the same in a chain. The orientations are predominantly either -1 or $+1$, showing that the direction of entry is along the axis of the tube from either end.

Sensitivity to Water–Nanotube Interaction Parameters

The simulations performed on different types of nanotubes show that the water entry is highly sensitive to the interaction of water with the nanotube. A small reduction of the attraction between the carbon nanotube and the water changes the hydration of the nanotube dramatically, and a sharp two-state transition between empty and filled states on a nanosecond time scale occurs in the nanopores (Waghe et al., 2002; Hummer et al., 2001; Beckstein et al., 2001). Table 11.4 summarizes the LJ parameters from various papers, all agreeing the fact that above a particular value of $\sigma_{\text{oxygen-wall}}$, the tube

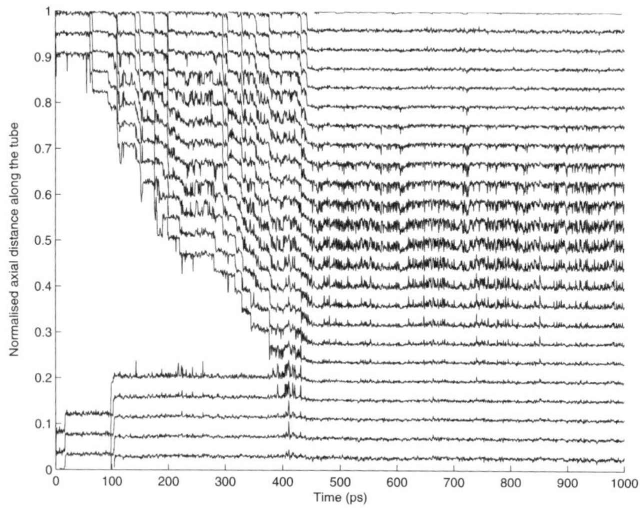

FIGURE 11.30. Trajectories of individual water molecules in a nanopore in a silicon dioxide membrane 8 Å in diameter and 6 nm long.

is empty.

Sensitivity to Nanotube Radius and Partial Charges on the Wall

(Allen et al., 2003) studied the filling and emptying in an artificial ion channel with varying diameter and different permittivities of the membrane surrounding the channel. They found that the permeation of a pore of fixed length is very sensitive to the pore radius. For very narrow pores ($R < 0.55$ nm in their simulations), water molecules are excluded from the pore. As the pore radius increases to a threshold value ($R = 0.60$ nm when the permittivities of the membrane is 1.0), intermittent permeation occurs, and the pore fluctuates between the "filled" and "empty" states stochastically. Further increase of the pore radius (e.g., $R > 0.65$ nm) then leads to the complete filling of the pore. The threshold radius is sensitive to the permittivity of the channel membrane, and using a polarizable membrane results in a decreased threshold radius for the intermittent permeation. In contrast to what was reported for a small carbon nanotube, where the water filled in the nanotube forms a one-dimensional chain, the authors found that the filled channel contains a much larger number of water molecules, and these water molecules exhibit a bulk-like behavior. Such a difference is mainly caused by the different properties of the pore surface. The simulations also reveal that the filling process is preceded by formation of a percolating chain of water molecules through the nanopore (see Figure 11.32), and the channel filling seems to "nucleate" around a percolating cluster. The proposed filling mechanism can be understood by

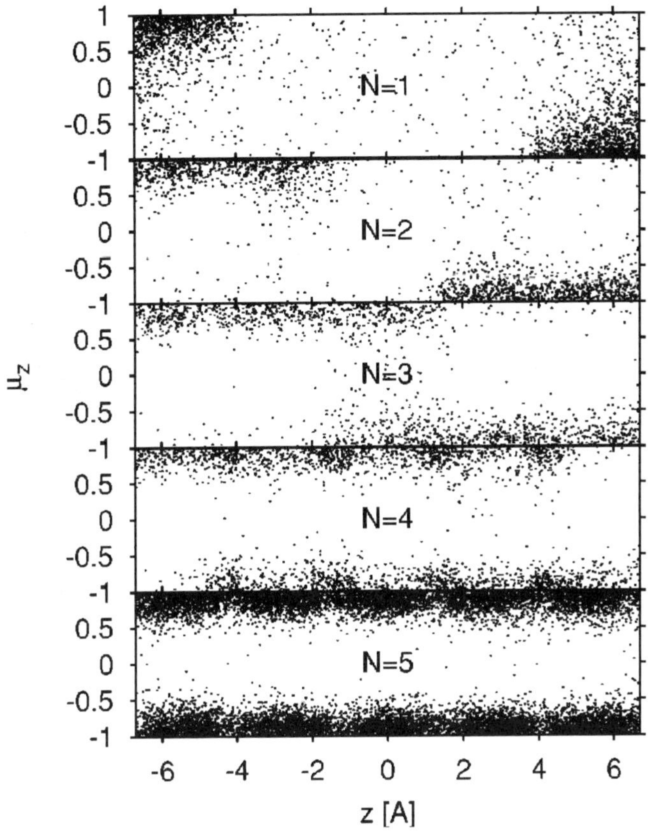

FIGURE 11.31. Normalized dipole of water molecules inside the tube projected onto the tube axis as a function of the position z of the water–oxygen atom along the tube axis. Each point corresponds to one water molecule in a saved configuration. The results are separated for $N = 1$ to 5 water molecules inside the tube (top to bottom). It was observed that for a single molecule in a tube, the water dipole moments point inward preferentially. With subsequent molecules entering into the tube, this orientational preference is maintained, such that the chain grows with all dipoles pointing inward. As a consequence, the dipolar orientations of water chains entering from the two ends simultaneously are not compatible, thus disfavoring simultaneous filling from both sides. (Courtesy of G. Hummer.)

analyzing Figure 11.33, where the average number of percolating chains in the channel is plotted as a function of the number of water molecules n inside the channel. Clearly, when n is small, there is never a percolating cluster, and when n is large enough, there is almost always one present. (Beckstein et al., 2001) show that the filling can be enhanced by placing dipoles along the walls of the nanotube at diameters at which water cannot enter the tube.

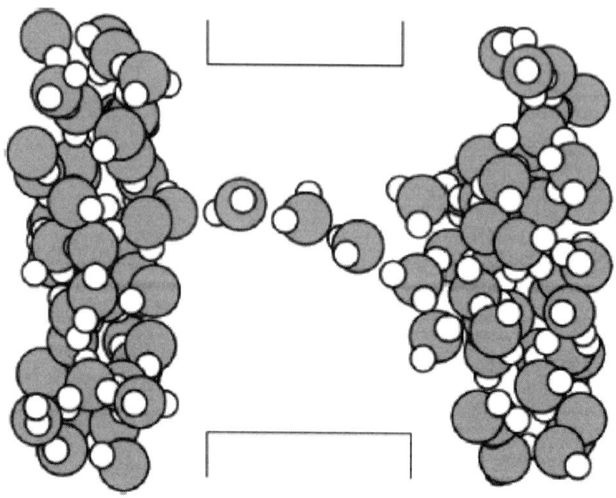

FIGURE 11.32. Snapshot of a "percolating cluster" from a biased simulation run with a pore radius of 0.6 nm and length of 0.8 nm. Only the water molecules near the pore are shown. The positions of the confining walls are indicated. (Courtesy of S. Melchionna.)

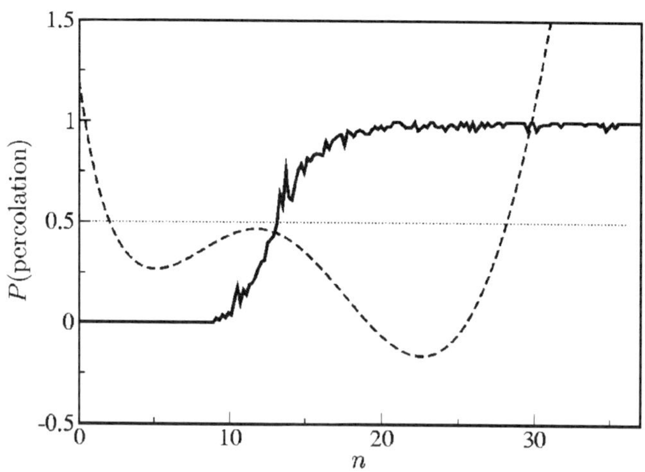

FIGURE 11.33. Probability of finding an unbroken chain of water molecules through the pore as a function of the water molecules inside the pore shown in Figure 11.32. The dashed line shows a scaled grand potential profile. (Courtesy of S. Melchionna.)

TABLE 11.4. Dependence of the full and empty states of water on the Lennard–Jones interaction parameters.

Nanotube material	Tube Diameter (Å)	Tube Length (Å)	$\sigma_{\text{oxygen-wall}}$ (Å)	Fill/Empty States
Artificial Membrane (Beckstein et al., 2001)	9.0	8.0	3.42	Both States
Carbon nanotube (Waghe et al., 2002)	8.1	27.0	3.27	Full
Carbon nanotube (Waghe et al., 2002)	8.1	27.0	3.41	Both States
Carbon nanotube (Waghe et al., 2002)	8.1	27.0	3.43	Empty
Carbon nanotube (Hummer et al., 2001)	8.1	13.4	3.23	Full
Artificial Slab	9.1	15.1	3.49	Both States
Silicon Dioxide	10.0	60.0	3.27(Si), 3.16 (O)	Full

Summary

In summary, from the various results presented in this chapter, we note that water in confined nanochannels can exhibit very interesting and different physical characteristics compared to that of bulk water. The properties of water in confined nanochannels can depend strongly on the type of surface (hydrophilic versus hydrophobic channel wall structure) and on whether the nanopore surface is charged. It is important to properly understand the merits and limitations of the various water models before they can be used for nanotechnology design.

12
Electroosmotic Flow in Nanochannels

In this chapter we discuss fundamental concepts and simulation of electroosmotic flow in nanochannels. The basic continuum theory was presented in Chapter 7, so here the limitations of the continuum theory for electroosmotic flow in nanochannels are identified by presenting a detailed comparison between continuum and MD simulations. Specifically, the significance of the finite size of the ions and the discrete nature of the solvent molecules is highlighted. A slip boundary condition that can be used in the hydrodynamic theory for nanochannel electroosmotic flows is presented. Finally, the physical mechanisms that lead to charge inversion and corresponding flow reversal phenomena in nanochannel electroosmotic flows are discussed.

12.1 The Need for Atomistic Simulation

A fundamental issue that needs to be addressed is whether continuum theories based on the Poisson–Boltzmann and the Navier–Stokes equations—which have been popularly used to understand electroosmotic flow in micron-scale channels—can be used to describe electroosmotic flow in nanometer-wide channels. For example, in the derivation of the classical Poisson–Boltzmann equation, the ions are assumed to be infinitesimal; the interactions between the ions, ion–water, and ion–wall are all considered in a mean-field fashion, and the molecular aspects of these interactions are neglected. Similarly, the continuum flow theory based on Stokes or Navier–Stokes equations assumes that the state variables (e.g., density) do not

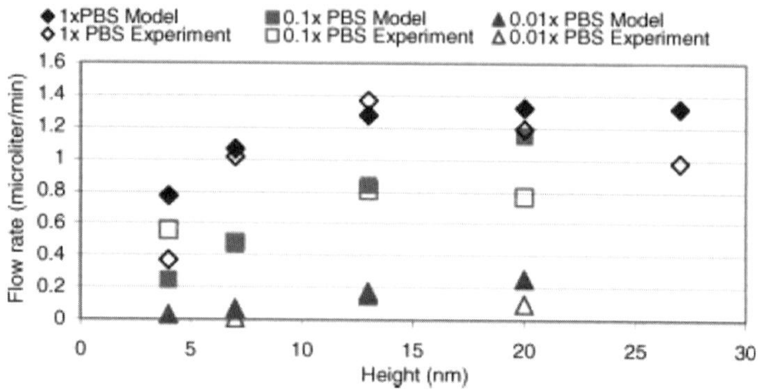

FIGURE 12.1. Comparison of the theoretical flowrates with experimental results. 1 x PBS, 0.1 x PBS and 0.01 x PBS refer to original PBS, 10-fold diluted PBS, and 100-fold diluted PBS, respectively. (Courtesy of A.T. Conlisk.)

vary significantly over intermolecular distances and that the shear stress can be related to the local strain rate by a linear constitutive relationship. However, as discussed in Chapter 10, significant fluctuations in fluid density have been observed close to the surface in molecular dynamics simulations (Travis and Gubbins, 2000) as well as in experiments (Cheng et al., 2001). Hence, it is important to understand in detail the validity of continuum theory for electroosmotic flow in nanochannels. Conlisk and coworkers (Zheng et al., 2003) presented a comparison between experimental and theoretical flow rates for electroosmotic flow in nanochannels with critical dimension (typically the width or the height) varying from 4 nm to 27 nm. This comparison is shown in Figure 12.1. The ionic solution considered is phosphate-buffered saline (PBS), which consists of Na^+, Cl^-, K^+, $H_2PO_4^-$, and HPO_4^{2-}. The pH value of PBS is 7.4. The theoretical results are obtained using the classical continuum theory for electroosomotic flow. The experimental and the theoretical flow rates agree very well except for the 4-nm case. While it is difficult to exactly pin-point the reason for the discrepancy, noncontinuum effects, such as the finite size of the ions, can be a big part of the discrepancy, since they play an important role when the critical channel dimension is just a few molecular diameters.

To understand the limitations and the various assumptions built into the continuum theory, we present detailed comparisons between continuum and MD simulations. Continuum modeling of electroosmotic flow is discussed in detail in Chapter 7. In this section, however, a simplified form of the equations presented in Chapter 7 is used to explain electroosmotic flow in nanochannels. Specifically, we focus on electroosmotic transport in straight flat channels; the channel width is in the z-direction, and the flow is along the x-direction; see Figure 12.2. The assumptions are:

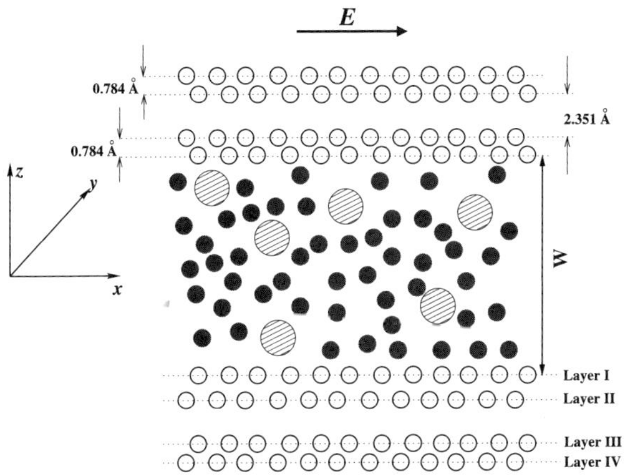

FIGURE 12.2. A schematic of the channel system under investigation. The two channel walls are symmetrical with respect to the channel center line. Each wall is made up of four layers of silicon atoms. The channel width W is defined as the distance between the two innermost wall layers. The dark dots denote water molecules, and the shaded circles denote either Cl^- or Na^+ ions. For the coordinate system chosen, $z = 0$ corresponds to the central plane of the channel system.

1. Only counterions are present in the channel, and
2. The flow is fully developed, and there is no externally applied pressure gradient.

Under these assumptions, the mathematical model presented in Chapter 7 can be simplified as

$$\frac{\partial^2 \psi(z)}{\partial z^2} = -\frac{q}{\epsilon}\tilde{z}c_0 e^{-q\tilde{z}\psi(z)/k_B T} \tag{12.1a}$$

$$\frac{d}{dz}\left(\mu \frac{du(z)}{dz}\right) + q\tilde{z}c(z)E_{\text{ext}} = 0, \tag{12.1b}$$

where $\psi(z)$ is the potential induced by the charges on the channel wall and the ions, q is the electron charge (i.e., 1.6×10^{-19} C), $\tilde{z}$ is the valency of the counterion, c_0 is the counterion concentration at the channel center, where the potential is assumed to be zero (note that is true when the double layers do not overlap), ϵ is the permittivity of the fluid in the channel, k_B is the Boltzmann constant, T is the temperature, $u(z)$ is the velocity of the fluid, μ is the dynamic viscosity of the fluid, $c(z)$ is the counterion concentration across the channel, and E_{ext} is the external electric field applied along the channel.

The boundary conditions for equation (12.1b) and equation (12.1b) are

$$\left.\frac{d\psi(z)}{dz}\right|_{z=\pm h/2} = \pm\frac{\sigma_s}{\epsilon}, \tag{12.2a}$$

$$u(z)|_{z=\pm h/2} = 0, \tag{12.2b}$$

where $z = \pm h/2$ corresponds to the location of the lower and the upper channel walls and σ_s is the charge density on the channel walls. Analytical solutions for equation (12.1b) and equation (12.1b) are available for the boundary conditions given in equations (12.2b) and (12.2b) (Israelachvili, 1992a; Eikerling and Kornyshev, 2001). However, to use the analytical solution, one needs to first solve a transcendental equation numerically, so equations (12.1b) and (12.1b) are typically solved numerically. The relative permittivity of water is typically taken as 81, since this is the reported value for SPC/E water at 300 K (van der Spoel et al., 1998). The dynamic viscosity of water is taken as 0.743 mPa·s in the continuum simulations, since this gives the best match to the velocity profile in the central portion of the channel.

Nonequilibrium molecular dynamics (NEMD) simulations, as described in Chapter 16, were performed for systems consisting of a slab of water molecules and ions sandwiched by two channel walls. Figure 12.2 shows a schematic diagram of the system under investigation. The two channel walls are symmetrical with respect to the channel center line. Each wall is made up of four layers of silicon atoms oriented in the $\langle 111 \rangle$ direction. Typical lateral dimensions of the channel wall are 4.66 nm $\times$ 4.43 nm, which corresponds to 161 silicon atoms for each layer of the channel wall. The channel width is varied from 0.95 nm to 10.0 nm in the simulations. For the simulation of electroosmotic flow, the outermost wall layers (i.e., layer I of the lower channel wall and its counterpart in the upper channel wall) are partially charged. We assume that the charges are uniformly distributed among the wall atoms; i.e., wall atoms are partially charged. The wall atoms are fixed to their original positions during the simulation. The water is modeled by using the SPC/E model (Berendsen et al., 1987) (see Chapter 16 for details). We consider two types of interaction potentials, i.e., Lennard–Jones and Coulomb potentials. The Lennard–Jones potential is considered for every atom pair (except the atom pairs that have a hydrogen atom and the Si–Si pair). The parameters for the Lennard–Jones potential are taken from the Gromacs forcefield (Spoel et al., 2001) and are summarized in Table 12.1. The Coulomb potential is considered for every charged atom pair.

The temperature of fluid is regulated to 300 K by using a Berendsen thermostat with a time constant of 0.1 ps (see Section 16.1.3 on thermostats). When setting up the simulation, the molecules were randomly positioned. An energy minimization was performed to remove the local contacts. To start the simulation, an initial velocity sampled from a Maxwellian distribu-

TABLE 12.1. Parameters for the Lennard–Jones potential $U(r) = \frac{C_{12}}{r^{12}} - \frac{C_6}{r^6}$.

Interaction	C_6 (kJ · nm^6/mol))	C_{12} (kJ · nm^{12}/mol)	σ (nm)†
O-O	0.2617×10^{-2}	0.2633×10^{-5}	0.317
O-Si	0.6211×10^{-2}	0.7644×10^{-5}	0.327
O-Cl	0.6011×10^{-2}	0.1678×10^{-4}	0.375
O-Na	0.4343×10^{-3}	0.2352×10^{-6}	0.286
Cl-Cl	0.1380×10^{-1}	0.1069×10^{-3}	0.445
Cl-Si	0.1426×10^{-1}	0.4871×10^{-4}	0.388
Cl-Na	0.9974×10^{-3}	0.1499×10^{-5}	0.339
Na-Na	0.7206×10^{-4}	0.2101×10^{-7}	0.257
Na-Si	0.1031×10^{-2}	0.6829×10^{-6}	0.295

† σ is the separation distance between atoms where the potential energy is zero.

tion at 300 K was assigned to each molecule in the system. The system was simulated for a time period of 1 ns to 2 ns, so that the system has reached steady state. A production run of 1 ns to 7 ns (depending on the system to be simulated) was performed to gather the statistics of various quantities, e.g., streaming velocity. The density and velocity profile across the channel were computed using the binning method as described in Chapter 16. The flow is driven by an external electric field E_{ext}, applied along the channel in the x-direction. Because of the extremely high thermal noise, a strong electric field is required so that the fluid velocity can be retrieved with reasonable accuracy. Note that the strong electric field results in a much larger velocity than what is practically achievable in nanochannels. The external electric field strength used ranges from 0.38 V/nm to 0.55 V/nm. Table 12.2 summarizes the various simulations investigated in this section.

TABLE 12.2. Summary of the simulations performed.

Case #	Channel width (nm)	σ_s (C/m^2)	# Water molecules	# Ions	E$_{\text{ext}}$ (V/nm)	Simulation time (ns)
1	3.49	+0.120	2246	32 (Cl^-)	−0.55	5.1
2	3.49	+0.320	2075	85 (Cl^-)	−0.55	2.5
3	3.49	−0.120	2246	32 (Na^+)	+0.55	2.8
4	2.22	+0.120	1288	32 (Cl^-)	−0.46	6.3
5	0.95	+0.120	405	32 (Cl^-)	−0.55	9.4
6	10.00	+0.124	6606	32 (Cl^-)	−0.38	4.8

12.2 Ion Concentrations

Figure 12.3 (a) shows the concentration profile of Cl^- ions and water across the channel for case 1, where the channel width is 3.49 nm and the wall charge density is +0.120 C/m^2. The ion distribution obtained from the Poisson–Boltzmann equation is also shown for comparison. To compute the ion distribution using the Poisson–Boltzmann equation, we have assumed that the position of the wall coincides with the position of the first peak of Cl^- ion concentration obtained from MD simulation. The MD simulation result deviates from the Poisson-Boltzmann equation prediction in several aspects:

1. There is no Cl^- ion within 0.24 nm from the channel wall,
2. The peak concentration of Cl^- ions occurs at a position about 0.35 nm away from the channel wall, and the peak value is about 88% higher compared to the Poisson–Boltzmann equation prediction, and
3. The ion concentration from MD does not decrease monotonically to its value in the channel center. In particular, there is a plateau located at about 0.49 nm to 0.60 nm away from the channel wall.

These deviations can be understood by looking at the molecular aspects of the ions, wall atoms, and the water molecules. First, since a bare Cl^- ion has an effective radius of about 0.18 nm (Israelachvili, 1992a), a Cl^- ion cannot approach too close to the channel wall. Second, the ion and the wall interact with each other via the Lennard-Jones potential in the MD simulation. Such an interaction can contribute to the attraction of ions toward the wall. Figure 12.3 (b) shows the potential energy of a Cl^- ion due to the Lennard–Jones potential between the Cl^- ion and the channel wall. In this calculation, we have assumed that the ion can access any position in the xy-plane with equal probability. Figure 12.3 (b) indicates that the potential energy due to the ion–wall Lennard–Jones interaction is about $-1.8\ \mathrm{k_B}\ T$ ($T = 300$ K) at a position about 0.39 nm away from the channel wall. Since a location with lower potential energy is more favorable for the ions, the potential energy valley can attract more ions toward it. The second deviation is primarily caused by this effect. Third, the molecular interaction between the ion and the water molecules also plays an important role in determining the ion concentration. Figure 12.3 (a) shows that the water concentration is not uniform across the channel, and a significant layering of water is observed near the channel wall. Such a layering effect is well known and has been already discussed in Chapter 10. Since the water molecules are less closely packed near the density "valley" than in the bulk, the energy required to insert a finite-sized ion into the density "valley" is lower compared to inserting an ion in the bulk. Hence, more ions are attracted toward the density "valley" of water. In fact, a very weak peak of Cl^- ions is observed near the second density valley of water. From these

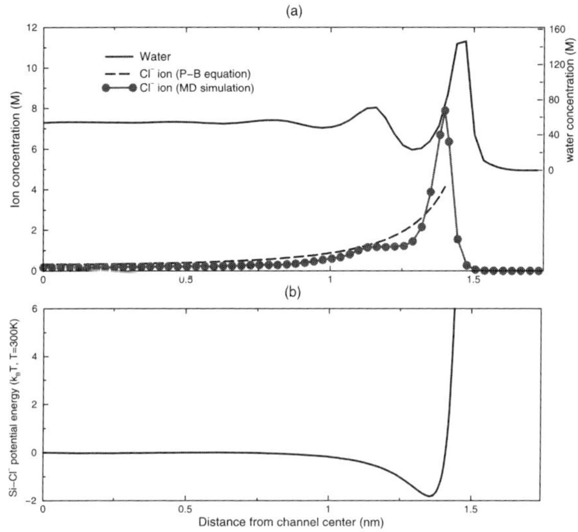

FIGURE 12.3. (a) Cl^- ion and water concentration across the channel for case 1 (W = 3.49 nm, σ_s = +0.120 C/m^2). The channel center is located at z = 0 nm, and the position of water molecules is computed as the center-of-mass position. (b) Potential energy of Cl^- ions over the channel wall computed using Lennard–Jones potential. It is assumed that the ion can access any position in the xy-plane with equal probability.

results we can infer that the molecular interactions between ion–wall and ion–water play an important role in determining the distribution of ions near the channel wall.

Figure 12.4 shows the concentration profile of Cl^- ions across the channel for case 2, where the channel width is 3.49 nm and the charge density on the channel wall is +0.320 C/m^2. Compared to the previous case, the charge density on the channel wall is very high (such a high charge density is realistic in practical systems (Poppe et al., 1996)). The Poisson–Boltzmann equation again underestimates the ion concentration near the channel wall. A clear second peak of Cl^- concentration is observed at a position about 0.45 nm away from the channel wall, whereas such a peak was very weak in case 1. The second peak is primarily caused by the fact that the ions are very densely packed (as indicated by the high concentration) near the channel wall, and the strong electrostatic repulsion between the ions makes it difficult to accommodate more ions in the region within 0.41 nm from the channel wall. As a result, a second peak is observed. Since the ions are more densely packed in the near wall region in this case compared to case 1 (in fact, the average shortest distance between two Cl^- ions within 0.41 nm from the channel wall is found to be 0.54 nm for this case, and 0.69 nm for case 1), the electrostatic repulsion between ions is much stronger

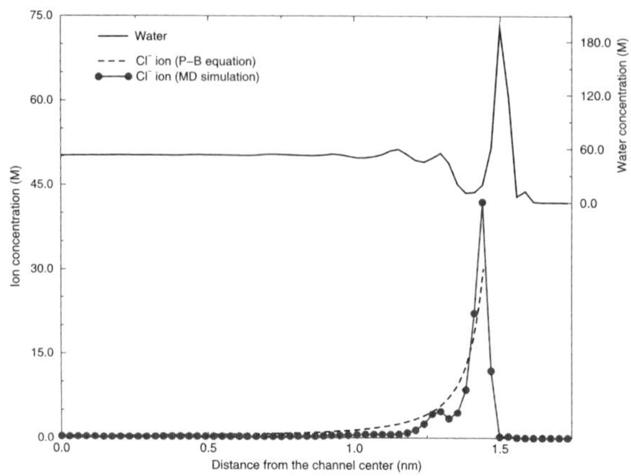

FIGURE 12.4. Cl^- ion and water concentration profile across the channel for case 2 ($W = 3.49$ nm, $\sigma_s = +0.320$ C/m^2).

compared to case 1; hence, the second concentration peak is more distinct in this case.

The water concentration profile in Figure 12.4 shows two interesting features. First, compared to the previous case (see Figure 12.3 (a)), the first concentration peak of water is about 33% higher. Second, there is an additional water concentration peak located at about 0.44 nm away from the channel wall, and the second concentration peak is very close to the second peak of the Cl^- concentration. This result indicates that under high surface charge density, the high concentration of ions near the channel wall can change the concentration of water significantly. Such a change can be partly explained by the hydration of the ions. Typically there will be several water molecules bound to each ion due to the strong charge–dipole interaction between the ion and water (Israelachvili, 1992a). Therefore, a region with very high ion concentration tends to have a high water concentration region near it.

Figure 12.5 shows the Na^+ ion concentration profile across the channel for case 3, where the channel width is 3.49 nm and the charge density on the channel wall is -0.120 C/m^2. The only difference from the first case is that the wall is oppositely charged and the Cl^- ions are replaced by Na^+ ions. We observe that (1) the first Na^+ concentration peak near the wall is about 37% lower compared to the Cl^- concentration peak in the first case, (2) the position of the first peak is located about 0.42 nm away from the channel wall, i.e., about 0.07 nm farther away from the channel wall compared to the first case, where the first Cl^- ion peak is located at 0.35 nm away from the channel wall, and (3) the second Na^+ ion concentration peak is much more distinct compared to case 1. The first ion concentration peak is lower in this case because the attractive interaction between the

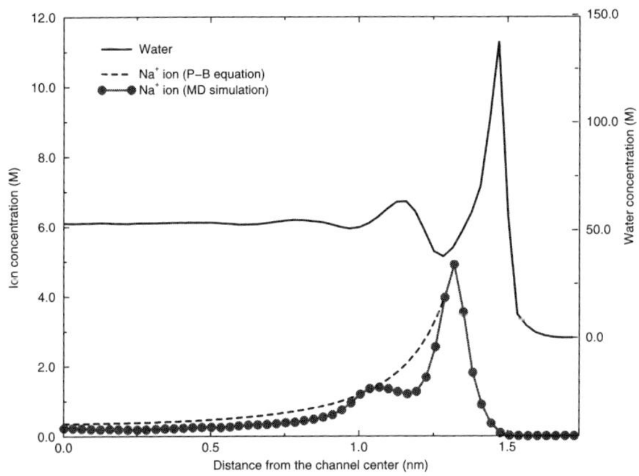

FIGURE 12.5. Na^+ ion and water concentration across the channel for case 3 (W = 3.49 nm, $\sigma_s = -0.120$ C/m^2).

Na^+ ion and the Si wall atoms—which contributes significantly to the high ion concentration near the channel wall—is much weaker compared to the Cl^- ion concentration with the Si wall atoms (see Table 12.1). The second observation can be explained by the fact that though a Na^+ ion (bare radius about 0.095 nm) is smaller compared to the Cl^- ion, its hydrated radius (0.36 nm) is bigger compared to the Cl^- ion (0.33 nm), since a Na^+ ion can attract more water molecules around it (the hydration number of a Na^+ ion is about 4 to 5, while the hydration number of Cl^- ion is about 1).

The second Na^+ concentration peak is primarily caused by the ion–water interactions. Figure 12.5 indicates that the second ion concentration peak is close to the second concentration valley of water. As explained in case 1, such a water concentration valley is energetically more favorable than the bulk for ions. Since the bare ion radius of Na^+ is much smaller than that of the Cl^- ion, the Na^+ ion can fit into the water concentration valley more easily than the Cl^- ion. Therefore, the second concentration peak of Na^+ ions is more distinct compared to the Cl^- ion case. One consequence of the higher second peak of Na^+ ions is that the water concentration valley near the second Na^+ peak is shallower compared to that observed in the Cl^- ion case (case 1). This is because the Na^+ ion has a higher hydration number. Thus, it can bring more water molecules toward the water concentration valley than the Cl^- ion.

12.2.1 Modified Poisson–Boltzmann Equation

From the results presented above, we can conclude that the wall–ion, water–ion, and ion–ion interactions are important factors influencing the ion distribution in the channel and that the ion distribution can significantly influence the water concentration in the channel. The classical Poisson–Boltzmann equation considers these interactions only in a mean-field fashion and fails to account for the molecular nature of the ion, water, and the wall (e.g., water is modeled as a continuum with a constant permittivity). In this section, we present a modified Poisson–Boltzmann equation that takes into account the wall–ion, water–ion, and ion–ion interactions in a lumped manner. The Poisson–Boltzmann equation can be modified to incorporate various effects that were neglected in the classical Poisson-Boltzmann equation, e.g., the finite ion size effects (Borukhov et al., 2000), nonelectrostatic interactions (Lue et al., 1999), dependence of the permittivity of the solution on the field strength (Woelki and Kohler, 2000), wall effects (Adamczyk and Warszynski, 1996) and discrete solvent effects (Burak and Andelman, 2001). Many of these modifications were based on statistical mechanics principles, and by incorporating all these effects into the Poisson–Boltzmann equation, it is possible to predict the ion distribution in the channel with good accuracy. However, due to the extremely complicated nature of the ion–wall, ion–water, and ion–ion interactions, a number of simplifications need to be made in the derivation of the modified Poisson–Boltzmann equation, and the calculation of the new terms can still be very difficult in many cases. Thus, though these existing modifications to the Poisson–Boltzmann equation can aid in the interpretation of various experimental observations, the development of a comprehensive, easy-to-calibrate, and accurate model is still an active area of research.

Here we introduce the concept of *electrochemical potential correction* to account for the interactions neglected in the classical Poisson–Boltzmann equation. At thermodynamic equilibrium, the electrochemical potential of an ion should be constant in the entire system, i.e.,

$$\mu_i^c = \tilde{z}_i q \psi + k_B T \ \log c_i + \phi_{\mathrm{ex},i} = k_B T \ \log c_{0,i}, \tag{12.3}$$

where $\tilde{z}_i$ is the valency of ion i, ψ is the electric potential in the system, c_i is the ion concentration, $\phi_{\mathrm{ex},i}$ is the electrochemical potential correction of ion i, and $c_{0,i}$ is the concentration of ion i when the electric potential and the electrochemical potential correction terms are zero. Based on equation (12.3), the ion concentration can be expressed as

$$c_i = c_{0,i} \, e^{-\frac{\tilde{z}_i q \psi}{k_{\mathrm{B}} T}} e^{-\frac{\phi_{\mathrm{ex},i}}{k_{\mathrm{B}} T}}. \tag{12.4}$$

Substituting equation (12.4) into the Poisson equation (12.5b), we have

$$\nabla^2 \psi = -\frac{q}{\epsilon} \sum_{i=1}^{N} \tilde{z}_i c_i, \tag{12.5a}$$

$$\nabla^2 \psi = -\frac{q}{\epsilon} \sum_{i=1}^{N} \tilde{z}_i c_{0,i} \, e^{-\frac{\tilde{z}_i q \psi}{k_{\mathrm{B}} T}} e^{-\frac{\phi_{\mathrm{ex},i}}{k_{\mathrm{B}} T}}, \tag{12.5b}$$

where N is the total number of ionic species, which is equal to 1 in all our simulations presented in this section.

The electrochemical potential correction term accounts for the deviation of the ion–water and ion–wall molecular interactions from their values at the channel center. Since the wall–ion interaction via the Lennard–Jones potential is short-ranged, and the water-ion interaction would not deviate significantly in the entire system except at positions very close to the channel wall where the water concentration is not constant, the electrochemical potential correction term is nonzero only at positions close to the channel wall. The precise contribution of ion–water and ion–wall molecular interactions to the electrochemical potential correction term depends on the specific case under study. For example, for case 1, where the ion–wall (Cl^- –Si) interaction is strong, the contribution of the ion–wall molecular interactions to the electrochemical potential correction term is significant, whereas for case 3 the contribution of ion-wall (Na^+ –Si) molecular interactions will be minor.

In principle, one can calculate ϕ_{ex} provided the wall–ion, water–ion, and ion–ion interactions can be computed explicitly. However, such a calculation, if possible, is very difficult. For example, to account for the molecular nature of water, the charge–dipole interaction between water and the ion as well as other molecular interactions (e.g., the van der Waals interaction as included in the Lennard–Jones potential) will need to be considered explicitly. In addition, since the water concentration profile in the channel is related to the ion concentration profile (as demonstrated by the correlation between the second peak of the ion concentration with the water concentration valley in Figures 12.3 (a), 12.4, and 12.5), the concentration profiles of water and the ion must be computed self-consistently. In Section 16.3, we describe a multiscale approach to computing the electrochemical potential correction and to computing the ion concentrations using the modified Poisson–Boltzmann equation.

12.3 Velocity Profiles

The comparison of velocity profiles obtained from continuum and MD simulations is presented in this section. Figure 12.6 shows the velocity profile across the channel for case 1, where the channel width is 3.49 nm. The velocity profile based on the continuum flow theory, calculated by substituting the ion concentration obtained from the MD simulation into equation (12.1b) and using the boundary condition specified by equation (12.2b), is also shown for comparison. Figure 12.6 shows that the continuum flow theory prediction using a constant viscosity of 0.743 mPa·s overestimates the velocity in the entire channel. This is because the continuum calculation fails to take into account the fact that the viscosity near the channel wall is much higher than its bulk value. It is reasonable to assume (Freund,

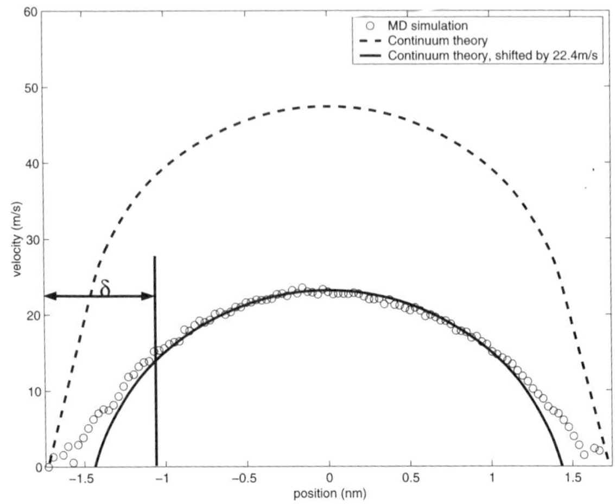

FIGURE 12.6. Comparison of water velocity profile across the channel for case 1 ($W = 3.49$ nm, $\sigma_s = +0.120$ C/m^2) as predicted by the MD simulation and by the continuum flow theory.

2002) that the viscosity of water increases dramatically in the near wall region. Such a dramatic increase in viscosity seems to be related to the high electric field strength (Hunter, 1981), layering of the fluid molecules (Lyklema et al., 1998), and the high concentration of ions near the channel wall. However, a comprehensive theory accounting for all the effects is not yet available.

The question of whether the continuum flow theory based on a constant viscosity can predict the flow behavior in the central part of the channel is an interesting one. We observe that if the predicted velocity is shifted down by about 22.4 m/s, the continuum prediction matches the MD velocity at a distance δ away from the channel wall; i.e., the continuum prediction matches the MD simulation result very well in the central portion of the channel. This is equivalent to saying that if the velocity at a position δ away from the channel wall is given as the boundary condition to the Stokes equation (12.1b), then the continuum flow theory based on a constant viscosity can still be used to predict the velocity in the central part of the channel. Figure 12.6 also indicates that the no-slip boundary condition is applicable to the case studied. However, the no-slip plane is not located at the center of mass of the innermost layer of the channel walls (i.e., layer I in Figure 12.2), but is located at approximately 0.14 nm from the channel wall, where the water concentration is almost zero. In Section 16.3, we describe a multiscale approach that can be used to calculate the velocity profile in the entire channel.

Comparison of velocity obtained from MD simulation and continuum

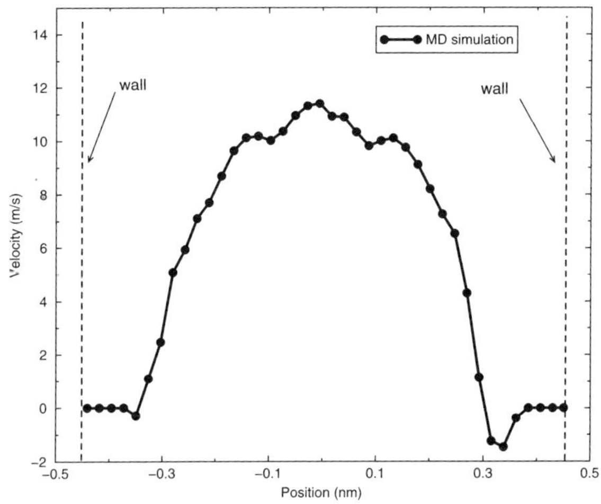

FIGURE 12.7. Velocity profile across the channel for case 5 ($W = 0.95$ nm, $\sigma_s = +0.124$ C/m^2). The negative velocity close to the channel wall is statistical noise.

flow theory for case 4, where the channel width is 2.22 nm, also indicates that though the continuum flow theory based on a constant viscosity overestimates the velocity in the entire channel, it can be used to study the flow behavior in the central part of the channel, provided that the velocity at a position δ away from the channel wall is given. It is also observed that the no-slip plane is located at approximately 0.14 nm from the channel wall. This is similar to what (Travis and Gubbins, 2000) had observed for Poiseuille flow of Lennard–Jones atoms in various channel widths where the no-slip plane is located at the position closest to the channel wall that a fluid atom can approach and is independent of the channel width.

Figure 12.7 shows the velocity profile of water across the channel for case 5, where the channel width is 0.95 nm and the surface charge density is 0.124 C/m^2. The characteristics of the velocity profile are significantly different from those of case 1 (channel width: 3.49 nm) and case 4 (channel width: 2.22 nm). Specifically:

1. The strain rate du/dy goes to zero at $|z| \approx 0.09$ nm and $|z| \approx 0.14$ nm, and

2. The velocity at $|z| \approx 0.14$ nm is higher than the velocity at $|z| \approx 0.09$ nm, and the velocity decreases from $|z| \approx 0.14$ nm to $|z| \approx 0.09$ nm. Similar behavior was also observed by (Travis and Gubbins, 2000) and (Travis et al., 1997) for Poiseuille flow of Lennard–Jones atoms when the channel width approaches 4 to 5.1 times the diameter of the fluid molecules.

This behavior suggests that the continuum flow theory is *not valid* for fluid

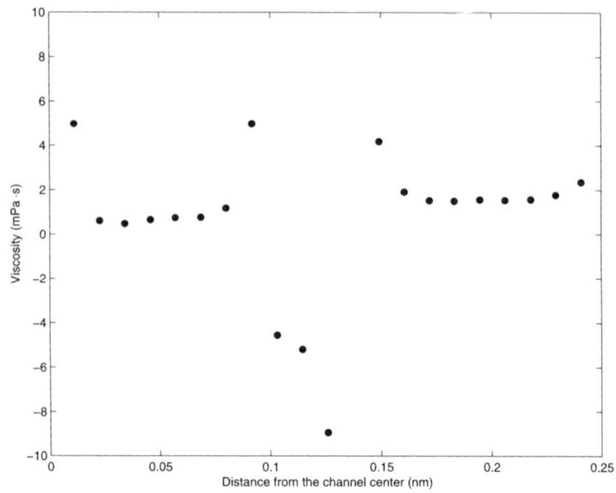

FIGURE 12.8. Shear viscosity across the channel for case 5 ($W = 0.95$ nm, $\sigma_s = +0.124$ C/m^2). Note that the shear viscosity, computed by using a linear, local constitutive relationship, diverges at $|z| \approx 0.14$ nm and becomes negative in the region 0.09 nm $< |z| <$ 0.14 nm.

flow in such narrow channels. Specifically, the velocity profile shown in Figure 12.7 indicates that the shear stress cannot be related to the strain rate by a local, linear constitutive relationship. The shear stress across the channel can be computed by

$$\tau_{zx}(z) = \int_0^z c(z)\tilde{z}qE_{\text{ext}}\, dz, \tag{12.6}$$

where $z = 0$ is the middle plane of the channel, $c(z)$ is the ion concentration, and E_{ext} is the external electric field along the channel length (x-direction). Figure 12.8 is a plot of the shear viscosity calculated by

$$\mu(z) = \frac{\tau_{zx}(z)}{du/dz|_z}. \tag{12.7}$$

Figure 12.8 indicates that the shear viscosity, computed by assuming a local, linear constitutive relationship between shear stress and strain rate, diverges at $z \approx \pm 0.14$ nm and $z \approx \pm 0.09$ nm, and becomes negative in the region 0.09 nm $< |z| <$ 0.14 nm. These *unphysical* results indicate that the continuum flow theory, which assumes that the shear stress can be related to the strain rate by a local constitutive relationship, is *not valid* for electroosmotic flow in a 0.95 nm wide channel.

12.4 Slip Condition

Marry and collaborators have presented an approach to modeling nanometer-scale electroosmotic flows with slip conditions on the channel walls (Marry et al., 2003). Starting from the Navier–Stokes equation given in Section 12.1, the velocity in the x-direction is given by

$$\frac{u(z)}{E_{ext}} = -\frac{\iint F_e(z)\,dz\,dz}{\mu E_{\text{ext}}} + a + bz,$$

where F_e is the force acting on the solution due to the external electrical field (E_{ext}), given by $F_e = c(z)qE_{\text{ext}}$. If the origin of the z-axis is chosen to be the middle of the channel (along the width direction), the symmetry of the system results in the relation

$$\frac{u(z)}{E_{\text{ext}}} = -\frac{\int_{-W/2}^{z}\int_0^z F_e(z)\,dz\,dz}{\mu E_{\text{ext}}} + a \tag{12.8}$$

In the above equation, W is the channel width. The value of a depends on the boundary conditions. In the classical Smoluchowski treatment of electroosmosis (see Chapter 7), the velocity is zero at the surface and $a = 0$. Furthermore, if we consider the Poisson–Boltzmann equation to compute the concentration profiles, then we get

$$\frac{u(z)}{E_{\text{ext}}} = \frac{q}{2\pi\mu L_{\text{B}}} \ln \frac{\cos(\alpha z)}{\cos(\alpha W/2)},$$

where $L_{\text{B}} = q^2/4\pi\epsilon_0\epsilon_r k_{\text{B}}T$, α is given by $\alpha \tan(\alpha W/2) = 2\pi L_B\sigma_s/q$, and σ_s is the opposite of the surface charge density. As discussed in the previous section, the above expression can be in large error from MD simulations.

A slip boundary condition was presented by (Marry et al., 2003) starting from the work of (Bocquet and Barrat, 1993) and (Bocquet and Barrat, 1994). The component of the hydrodynamic velocity parallel to the interface $u(z)$ is assumed to be proportional to the perpendicular derivative, namely,

$$\frac{\partial u(z_0)}{\partial z} = \frac{u(z_0)}{\delta}. \tag{12.9}$$

The no-slip boundary condition is recovered if the slip length δ satisfies $\delta = 0$. Here z_0 is the hydrodynamic position of the interface, δ and z_0 can be obtained from the microscopic Kubo relations (Kubo et al., 1991):

$$\frac{\mu}{\delta} = \frac{1}{Sk_{\text{B}}T} \int_0^{+\infty} \langle \mathcal{F}_x(t)\mathcal{F}_x(0)\rangle\,dt, \tag{12.10}$$

where $\mathcal{F}_x$ is the microscopic friction exerted by the wall on the fluid, and S is the interface area. The position of the fluid–wall interface is given by

$$z_0 = \frac{\delta}{S k_B T \mu} \int_0^{+\infty} \langle \mathcal{F}_x(t) \sigma_{xz}(0) \rangle \, dt, \tag{12.11}$$

with σ_{xz} the xz component of the microscopic stress tensor.

By introducing the above slip boundary condition, we get the slip solution of the Navier–Stokes equations. The electroosmotic profile is given by equation (12.8) with

$$a = \frac{1}{\mu E_{\text{ext}}} \left(\int_{-W/2}^{z_0} \int_0^{z} F_e(z) \, dz \, dz - \delta \int_0^{z_0} F_e(z) \, dz \right). \tag{12.12}$$

Note that the slip boundary condition shifts the electroosmotic profiles by adding a constant a that depends on δ and z_0. If z_0 is close to the channel walls, it does not influence a significantly. Assuming that $z_0 \approx -W/2$, which is a reasonable approximation, we get

$$a = \frac{\delta}{\mu E_{\text{ext}}} \int_0^{W/2} F_e(z) \, dz. \tag{12.13}$$

The expression in (12.13) implies that the slip boundary condition increases the electroosmotic flow. Using the expression for the force F_e (and an analytical solution for the concentration obtained from the Poisson–Boltzmann equation, $c(z) = \frac{1}{2\pi L_{\text{B}}} \frac{\alpha^2}{\cos^2(\alpha z)}$) in equation (12.13), we get

$$a = \frac{\delta}{\mu} \sigma_s.$$

The calculation of the slip parameters from the Kubo formula (12.10) and (12.11) cannot be performed accurately from MD simulations. For the sake of simplicity, (Marry et al., 2003) have computed the slip parameters from the electroosmotic profile itself. Since the hydrodynamic limit holds only for large channel widths, in this limit, we can assume that δ and z_0 do not vary a lot and can be taken as constant. Furthermore, we can also assume that z_0 is equal to $-W/2$. For the largest channel width shown in Figure 12.9, Marry and colleagues have estimated the slip length to be $\delta = 6$ Å. Using this slip length in equation (12.13) and using the Poisson–Boltzmann equation for the concentration profile results in the electroosmotic profile shown by the solid curve in Figure 12.9. The agreement with the macroscopic model is excellent for the three simulated channel widths shown in Figure 12.9. The influence of the hydrodynamic boundary condition can be significant even for large channel widths. As shown in Figure 12.10, even for a channel width of 100 Å, the slip boundary condition correction is not negligible.

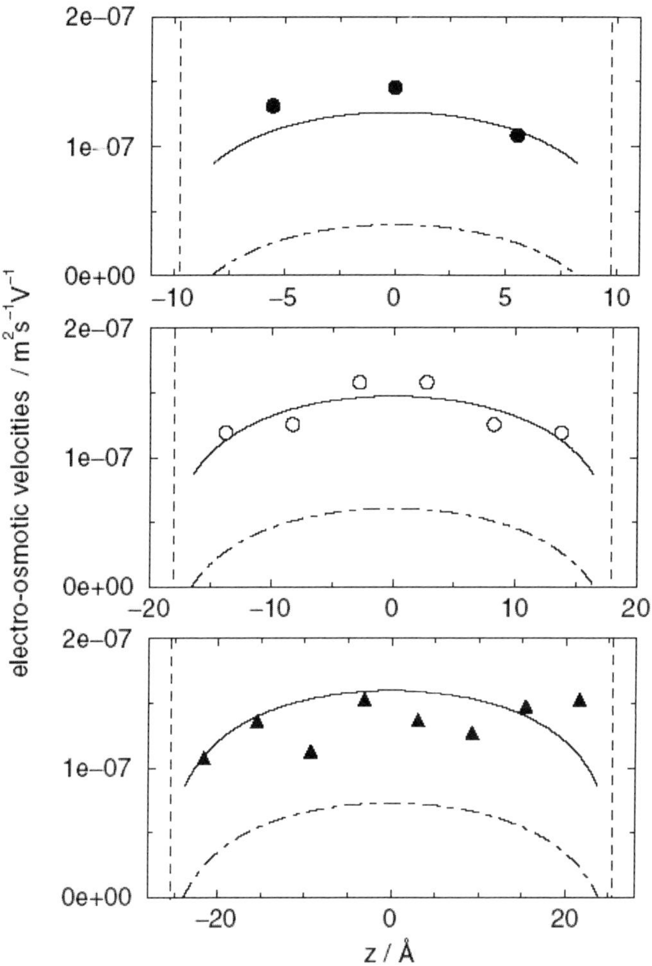

FIGURE 12.9. Comparison of electroosmotic velocity profiles $((u(z)/E_{\text{ext}}))$ obtained for different channel widths; MD results are shown by solid circles (number of solvent molecules is 200), open circles (number of solvent molecules is 400), and filled triangles (number of solvent molecules is 600); PB/NS results with no-slip boundary conditions are shown by the dotted lines; PB/NS results with slip boundary conditions ($\delta = 6$ Å, $z_0 = -W/2$) are shown by the solid lines; the vertical dashed lines denote the position of the channel walls. (PB/NS refers to Poisson–Boltzmann/Navier–Stokes equations.) (Courtesy of P. Turq.)

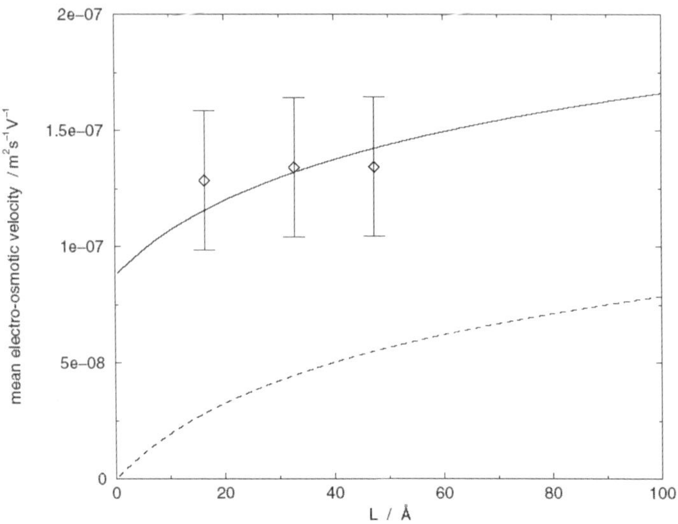

FIGURE 12.10. Mean electroosmotic velocity (u/E_{ext}) as a function of the channel width. MD data are shown as diamonds; PB/NS equation with no-slip boundary conditions is shown as a dotted line; PB/NS with slip boundary conditions are shown by the solid line; L in this plot is the channel width, and the error bars are for the MD data. (PB/NS refers to Poisson–Boltzmann/Navier–Stokes equations.) (Courtesy of P. Turq.)

12.5 Charge Inversion and Flow Reversal

Charge inversion refers to the phenomenon that the coion charge density exceeds the counterion charge density in a certain region of the electric double layer (EDL) (Qiao and Aluru, 2004). Consider again the channel system shown in Figure 12.2 with NaCl solution sandwiched between the two channel walls. A total charge of $-70e$ is evenly distributed among the atoms of the innermost wall layers, giving an average surface charge density (σ_s) of -0.285 C/m^2. Such a charge density can be considered high, but it is not impractical, since the typical charge density of a fully ionized surface can exceed 0.3 C/m^2 in magnitude. The system contains 108 Na^+ ions, 38 Cl^- ions, and 2144 water molecules. MD simulations were performed using the parameters and models discussed in Section 12.1. Starting from a random configuration, the system was simulated for 2.0 ns to reach steady state, followed by a 15-ns production run. The flow was driven by an electric field, E_{ext}, applied in the x-direction. Because of the extremely high thermal noise, a strong electric field ($E_{\text{ext}} = 0.55$ V/nm) is necessary so that the fluid velocity can be retrieved with reasonable accuracy. A strong external electric field can induce noticeable water alignment along the field direction, which can influence the ion distribution. To understand how this influences the charge inversion results presented here, simulations were also performed with zero external field. The ion distribution was found to be only slightly different from what is reported here, and the charge inversion is still observed.

To analyze the ion concentrations across the channel, the z-direction mean force acting on the ions is computed in the lower portion of the channel. The mean force $f_i(z)$ acting on an ion i, located at position z, is computed as the total force on ion i from all other particles in the system averaged over all configurations. The force $f_i(z)$ is considered negative if it drives the ion toward the lower channel wall, and positive otherwise. The potential of mean force (PMF), denoted by $w_i(z)$, for an ion i at a position z is computed by

$$w_i(z) = \int_z^{r_f} f_i(z')\, dz',$$

where r_f is the reference plane (taken as the channel center plane here) at which the PMF is taken as zero. Within the limit of classical statistical mechanics, the concentration of an ion i at a position z, denoted by $c_i(z)$, is related to the PMF by the Boltzmann distribution

$$c_i(z) = c_i^{r_f} \exp(-w_i(z)/k_{\rm B}T),$$

where $c_i^{r_f}$, $k_{\rm B}$, and T are the concentration of ion i at the reference plane, the Boltzmann constant, and the temperature, respectively. To facilitate discussion, we decompose the total mean force into an electrostatic mean force and a nonelectrostatic mean force. In certain cases, we further decompose the nonelectrostatic mean force into several components arising from the interactions of the molecule (e.g., wall atoms, ion, or water) with the ion.

To measure the screening of the surface charge by the ions, we define a screening factor

$$S_f(z) = \int_0^z F\ [c_{\rm Na^+}(z) - c_{\rm Cl^-}(z)]/\ |\sigma_s|\, dz, \qquad (12.14)$$

where F is the Faraday constant. The screeing factor $S_f(z) > 1$ corresponds to an overscreening of the surface charge.

Figure 12.11 shows the variation of water and ion concentration as well as the screening factor across the channel. Apart from the well-known layering of water near the channel wall, we also observe that:

1. a significant number of Na^+ ions are accumulated within only 0.5 nm of the channel wall,

2. the concentration of Cl^- ion is very low in the region $z < 0.5$ nm even though its access to the region 0.35 nm $< z <$ 0.5 nm is not limited by its finite size, and

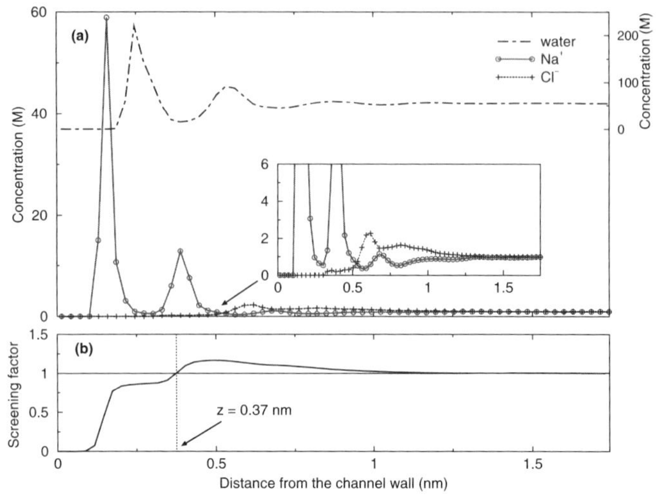

FIGURE 12.11. Water/ion concentration profiles and the screening factor $S_f(z)$ across the channel (only the lower portion of the channel is shown because of the symmetry). Note that $z = 0$ denotes the position of the lower channel wall.

3. the surface charge is overscreened at a position 0.37 nm away from the channel wall (i.e., $S_f(z) \geqslant 1$ for $z > 0.37$ nm), and charge inversion occurs in the region $z > 0.53$ nm.

The overscreening and charge inversion cannot be predicted by the classical EDL theories, e.g., the Poisson–Boltzmann equation. Observation (1) can be understood by analyzing the mean force acting on a Na^+ ion as shown in Figure 12.12. From the PMF definition and the Boltzmann distribution, it follows that the accumulation of the Na^+ ions in the region $z <$ 0.5 nm is mainly caused by the negative total mean force in regions II and IV of Figure 12.12 (a). Figure 12.12 (a) also shows that the nonelectrostatic Na^+–water interactions contribute significantly to the negative total mean force in these regions (the mean force due to the nonelectrostatic interactions between Na^+–Na^+, Na^+–Cl^-, and Na^+–wall is not shown, since its contribution to the total mean force is less significant). The magnitude and sign of the nonelectrostatic Na^+–water mean force depends mainly on the repulsive nonelectrostatic interactions between the Na^+ ion and its nearby water molecules and on the asymmetrical distribution of the water molecules around the ion (a symmetrical distribution will result in a zero mean force). For an ion located near the channel wall, the asymmetrical distribution of water molecules around the ion can be significantly influenced by the fluctuation of water concentration near the channel wall (see Figure 12.11), which then leads to a fluctuating mean force in the region $0 < z < 0.75$ nm of Figure 12.12(a).

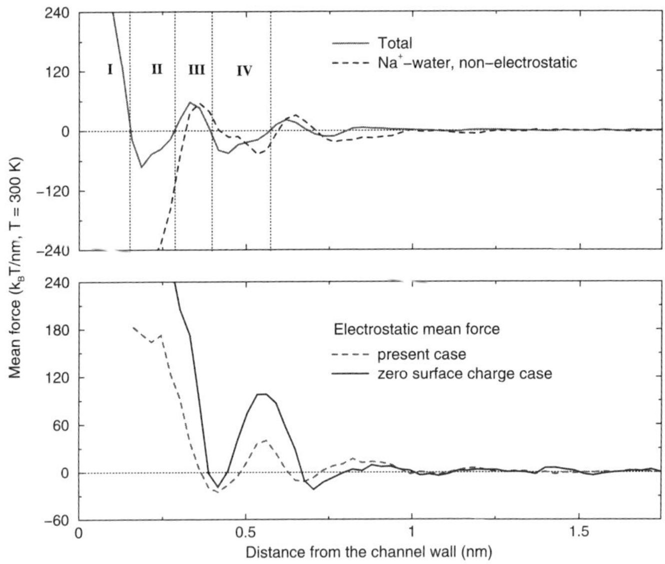

FIGURE 12.12. (a) Total mean force acting on a Na^+ ion and its component due to the nonelectrostatic Na^+–water interactions, (b) the electrostatic mean force acting on a Na^+ ion with and without the surface charge density. The zero surface charge case has a bulk ion concentration of 1.0 M.

The contribution of the electrostatic interactions is evaluated by studying the electrostatic mean force acting on the Na^+ ion, which is computed by subtracting the nonelectrostatic component from the total mean force. As shown in Figure 12.12 (b), the mean force acting on a Na^+ ion due to its electrostatic interaction with all other molecules in the system is positive for $z < 0.35$ nm. Such a positive mean force arises mainly from the fact that when the Na^+ ion is very close to the lower channel wall, the electrostatic interactions between the ion and its hydration water molecules, which are distributed mainly above the ion, tend to "pull" the ion from the surface. To understand how the surface charge contributes to the accumulation of Na^+ ions near the charged channel wall, we computed the electrostatic mean force experienced by a Na^+ ion in a reference case, where the surface charge density is zero and the bulk ion concentration is the same as in the present case. Figure 12.12(b) shows that when the surface is charged, though the electrostatic mean force acting on the Na^+ ion near the wall is positive, it is significantly lower compared to the case in which the surface is not charged. From this we see that when the molecular nature of water is considered, the electrostatic interactions contribute to the accumulation of Na^+ ions near the charged channel wall in a different way from what one would expect in a classical EDL theory.

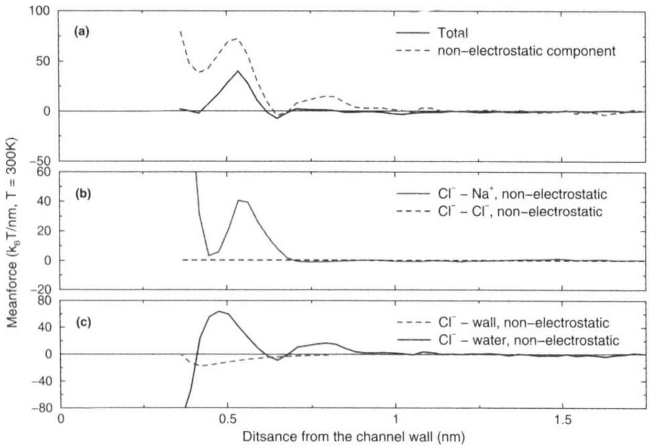

FIGURE 12.13. (a) Total mean force acting on a Cl^- ion and its nonelectrostatic component, (b) and (c) are the mean forces experienced by a Cl^- ion due to its nonelectrostatic interactions with Na^+ and Cl^- ions, wall atoms and water molecules.

Observation (2), which deals mainly with the low Cl^- ion concentration in the region $z < 0.5$ nm, can be explained by performing an analogous mean force analysis for the Cl^- ion. Figure 12.13 shows that the total mean force acting on a Cl^- ion is strongly positive in the region 0.4 nm $< z <$ 0.6 nm (see panel (a)), and it is dominated by the nonelectrostatic interactions between the Cl^-–Na^+ ions (see panel (b)) and Cl^-–water molecules (see panel (c)). This indicates that the low concentration of Cl^- ion in the region 0.35 nm $< z <$ 0.5 nm is mainly caused by the nonelectrostatic interactions between Cl^-–Na^+ and Cl^-–water. The depletion of Cl^- ion in the region $z < 0.35$ nm is caused by the finite size of the Cl^- ion.

Observation (3), which deals with overscreening and charge inversion, can be explained by the following two mechanisms. First, as discussed above, in addition to the electrostatic interactions, the nonelectrostatic Na^+–water interactions also contribute significantly to the accumulation of Na^+ ions near the channel wall. Together, these interactions bring many more Na^+ ions toward the channel wall than is predicted by the classical EDL theory, where the molecular nature of water is not considered. Second, because of the nonelectrostatic interactions between the Cl^- ions and the wall atoms and the water molecules, and because of the accumulation of Na^+ ions near the channel wall, the Cl^- ions tend to stay away from the channel wall and accumulate in the region $z > 0.5$ nm. The combination of these two mechanisms then makes it possible for the surface charge to be overscreened at a short distance from the channel wall ($z = 0.37$ nm), and for the concentration of Cl^- ions to exceed that of the Na^+ ions at a distance of 0.53 nm away from the channel wall.

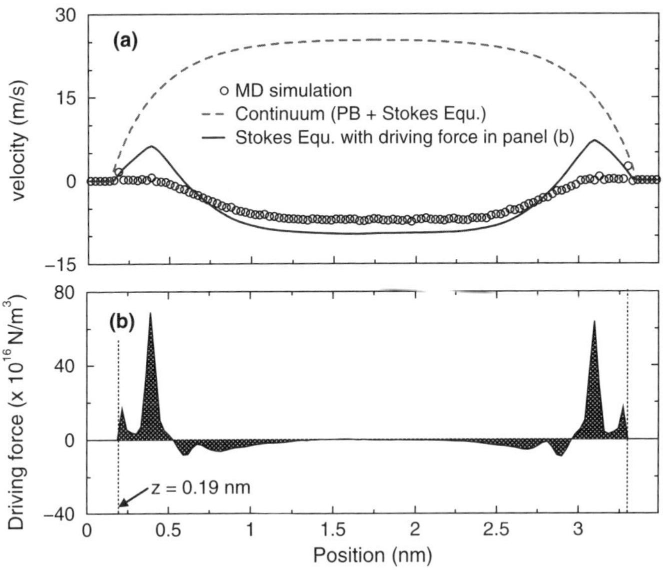

FIGURE 12.14. (a) Water velocity profile across the channel obtained from MD and continuum simulations. (b) driving force for the flow across the channel using the ion concentrations obtained from MD simulation.

Flow reversal refers to the phenomenon that the electroosmotic flow is in the opposite direction to the one typically expected (Qiao and Aluru, 2004). For example, when the surface is negatively charged and the external electrical field is in the positive x-direction, one typically expects the electroosmotic flow to be in the positive x-direction. However, an electroosmotic flow in the negative x-direction will be observed when flow reversal occurs.

For the channel system discussed above, Figure 12.14 (a) shows the water velocity profile across the channel obtained by using MD and continuum calculations. In the continuum calculation, the Poisson–Boltzmann (PB) equation is solved to obtain the ion concentration, which is then used to calculate the driving force for the flow given by the expression

$$F_e(z) = q[c_{\mathrm{Na}^+}(z) - c_{\mathrm{Cl}^-}(z)]E_{\mathrm{ext}},$$

where q is the electron charge. The flow $F_e(z)$ is used as the driving force in the Stokes equation to compute the water velocity. In solving the PB equation, the bulk concentration of the NaCl solution is taken to be 1.0 M, which is consistent with the result shown in Figure 12.11. For the Stokes equation, a no-slip boundary condition is applied at positions $z = 0.13$ and 3.36 nm, which is consistent with the MD observation. The dielectric constant and the viscosity of the water are taken as 81 (this is the relative dielectric constant) and 0.743 mPa·s, respectively. From Figure 12.14 (a), we observe that while the continuum theory based on the PB and the Stokes equations predicts a positive velocity in the entire channel, the MD simulation shows a velocity that is slightly positive near the channel wall, but is negative in the region 0.42 nm $< z <$ 3.01 nm. Clearly, a flow reversal is observed.

To understand the flow reversal, we calculate the driving force $F_e(z)$ for the flow using the ion concentration obtained from the MD simulation. Specifically, in the region where the ions are immobilized, $F_e(z)$ is taken as zero, and in the rest of the channel, $F_e(z)$ is computed by the expression given above, using the ion concentrations shown in Figure 12.11. Figure 12.14 (b) shows the calculated driving force. The flow $F_e(z)$ is zero within 0.19 nm from the channel wall, since the Na^+ ions adsorbed on the wall are immobilized. Notice that because of the charge inversion, $F_e(z)$ is negative in the region 0.53 nm $< z <$ 2.96 nm. Figure 12.14 (a) also shows the velocity computed by substituting the driving force obtained from the MD ion concentrations into the Stokes equation. Clearly, the new driving force can predict the flow reversal in the region 0.58 nm $< z <$ 2.91 nm, which indicates that the major mechanism for the flow reversal is the immobilization of the adsorbed Na^+ ion on the channel wall and the charge inversion. The velocity profile with the new driving force, however, still deviates from the MD velocity profile, especially in the regions close to the channel wall. The reason for this deviation is not clearly understood at present, but is probably caused by the high local viscosity in the near wall region.

In summary, the charge inversion and flow reversal are some of the new physical phenomena that have been observed in nanochannel electroosmotic flows. While these results indicate that the molecular nature of water and ions can lead to interesting new phenomena, the inclusion of the molecular nature of water and ions into continuum electrostatic and hydrodynamic theories remains an active area of research.

13

Functional Fluids and Functionalized Nanotubes

The possibility of targeting and precisely controlling the electrooptical as well as the mechanical properties of microstructures in a dynamic way using external fields has opened new horizons in microfluidics research, including new concepts and protocols for microfabrication. New functionalized ferrofluids and ferromagnetic particles have led to a range of new biomedical and diagnostic applications. Self-assembled magnetic matrices can find a large range of applications for the separation of DNA and other intermediate-size objects such as cells, proteins, organelles, and micro- or nanoparticles. Self-assembly of colloids can be used in a bottom-up approach to the fabrication of nanosystems; in particular, self-assembly offers a possible route to fabricating three-dimensional microsystems. Such self-assembly techniques are biomimetic; i.e., they are inspired by processes in biological systems that enable proteins and cells to arrange themselves organically in functionally beneficial ways. However, they need external forces to be imposed to facilitate the process and provide quality control by steering and tailoring target properties. To this end, magnetic and electric fields can be used for paramagnetic and charged particles, respectively. How exactly this is accomplished is the subject of the first part of this chapter. On a more fundamental level, systematic studies of paramagnetic particles or charged particles and their dynamics offer insight into the role of Brownian noise in microsystems as well as conceptual differences between deterministic and stochastic modeling.

Because of advances in nanotechnology, there is now a possibility of understanding cell behavior at a fundamental level. In particular, there is now an exciting opportunity to understand how ionic channels in cellular membranes function by taking advantage of the progress in nanotechnology. Carbon nanotubes with diameters as small as 5 Å to 10 Å are comparable to the diameters encountered in ion channels. However, the surface characteristics of a carbon nanotube may not match closely the surface properties of an ion channel. By *functionalizing* carbon nanotubes, it is possible to tune the surface properties of carbon nanotubes to investigate the function of a variety of ion channels. To enable such advances, it is important to understand how water, ions, and various electrolytes interact with carbon nanotubes and functionalized nanotubes. This is the subject of the second part of this chapter.

In particular, in the first section we present details on the physical mechanisms involved in self-assembly, and subsequently we give examples of patterns formed using magnetic fields for magnetorheological fluids and electrophoretic deposition for electrorheological fluids. In the second section, after a brief introduction to carbon nanotubes and ion channels in biological membranes, we present results on electrolyte transport through carbon nanotubes. Specifically, we present concepts and results showing that the transport of electrolytes can be augmented by using functionalized nanotubes and electric fields.

13.1 Colloidal Particles and Self-Assembly

Construction of functional materials with feature sizes ranging from 1 nm to 1000 nm is a challenging problem and an area of active research. Due to the inherent dimensional limitations of currently available mechanical forming and lithographic techniques, it is difficult to produce structures with length scales on the order of nanometers. In addition, the following factors make it difficult for colloidal particles to assemble into ordered structures:

1. The particle motion is dominated by the Brownian motion and interfacial interactions. This becomes significant in submicron scales.

2. Brownian motion is random, and therefore it is very difficult to control.

3. The assembly times of the colloidal particles are too slow for practical technologies. Due to their larger size, the motion of colloidal particles is orders of magnitude slower than that of the atom. This is one of the reasons why molecular self-assembly in nature is common and easier than colloidal self-assembly.

Despite these difficulties, major strides have been made in the construction of self-assembled monolayers (Wasserman et al., 1989). As a result, designing surfaces and structures at the molecular scale has now become a possibility. The current molecular self-assembly techniques are inspired by processes in biological systems that enable proteins and cells to arrange themselves organically in functionally beneficial ways. This involves designing molecules and supramolecular entities that will aggregate into desired structures due to their shape complementarity (Wasserman et al., 1989).

Simulation of particulate microflows could play an important role in the development and active control of dynamically reconfigurable self-assembled structures. Colloidal micropumps and colloidal microvalves are already in existence and have been used for active microfluidic control; see an example in Figure 1.6, and for details (Terray et al., 2002). Self-assembly of colloidal particles onto the interface of emulsion droplets has been exploited by (Dinsmore et al., 2002) to fabricate solid capsules with precise control of size, permeability, strength, and compatibility. The resultant structures, which are called colloidosomes, are hollow elastic shells; they can be used for cellular immunoisolation.

Electrorheological (ER) and magnetorheological (MR) fluids, that is, suspensions that exhibit sensitivity to external electric and magnetic fields, are *fluids with more than just viscosity* as their main property. ER and MR fluids can dynamically change their optical properties, anisotropy, mechanical rigidity, and electronic properties, often in a reversible way. The applications are many, but perhaps the most exciting ones are based on self-assembly of magnetic colloidal particles into chains or columnar structures (Doyle et al., 2002; Furst et al., 1998; Liu et al., 1995). In (Doyle et al., 2002), self-assembled magnetic matrices were developed for DNA separation chips. Compared to previous separation media, suspensions of paramagnetic particles have several advantages: they have a low viscosity in the absence of magnetic field, their pore size can be tuned (from 1 to 100 microns), and they do not require sophisticated microlithography.

There has been a lot of work in understanding ER fluids, and at least their field-induced structures are reasonably well understood (Gast and Zukoski, 1989; Halsey and Toor, 1990; Tao and Sun, 1991; Chen et al., 1992; Martin et al., 1992). There are, however, several complicating factors (e.g., surface charge, electrode polarization; see (Promislow et al., 1995)), which have limited the range of their application in microfluidics research today. These can be avoided with MR fluids, which exhibit an analogous field-induced aggregation and can also be controlled by a single external magnetic field. However, the behavior of particles interacting through induced magnetic fields is more complex than that of ER fluids, and their dynamics are not well understood. For example, the long-range nature of the particle interaction can persist even when they form chains or columns; the range of the interaction depends sensitively on the chain length.

There are several experimental studies focused on understanding the dynamics of magnetic chains and columns (Liu et al., 1995; Furst et al., 1998; Promislow et al., 1995), as well as a few simulation studies (Climent et al., 2004). Here we consider two main classes of problems. The first category involves flow geometries and devices with *tens* of suspended paramagnetic microparticles. The second category includes problems with *hundreds* or even *thousands* of paramagnetic microparticles. It has been observed experimentally that in the presence of an orienting external field, at low concentrations one-dimensional lattices are formed, while at high concentrations two-dimensional lattices of staggered rows of particles are obtained. Both classes are of great interest from the theoretical as well as the engineering point of view. More specifically, in the first class of problems, paramagnetic microparticles have been used as the active flow-control element to design and then optimize basic elements of a microfluidic network, e.g., pumps, valves, mixers (Hayes et al., 2001). This approach has the potential of avoiding some of the difficulties experienced by other techniques, while the materials can be synthesized in large scales. For example, colloidal silica is easily modified for dispersion in both aqueous and nonaqueous solvents and is biocompatible; see, for example, (Terray et al., 2002). Typically, fewer than one dozen microspheres are involved in these designs. In the second class of problems, we are interested in fundamental understanding of the scaling laws that govern the interaction of many chains or magnetic columns formed by hundreds or thousands of microspheres. Self-assembled structures can be used in fabricating magnetically controlled microdevices with complex functionality, for example, actively addressable arrays of microreactors that can react dynamically with fluids for sorting and mixing applications; or self-assembled magnetic matrices for DNA separation chips. A fundamental unresolved question is reversibility in static or dynamic self-assembly processes and its validity as a function of the magnetic field strength and geometry.

Modeling of ER or MR fluids is quite complicated, and in simulating such flows, it is more effective to adopt a *hierarchical simulation methodology* that performs best in a certain range of parameters in terms of both accuracy and computational complexity. It should include stochastic techniques to represent Brownian noise, geometric roughness or other uncertainties associated with the boundary conditions, particle size, and interaction forces. That is, both continuum and atomistic techniques, as we discuss in Chapters 14 and 16, respectively, should be used.

Another method to build structure and function in the nanometer scale involves the use of electric fields to assemble colloidal particles at electrode surfaces. Patterned crystalline arrays of colloidal particles can be formed by appropriately altering the surface topography of the electrode surface. Besides the disadvantage of added time and expense for altering the electrode surface (lithography and etching steps), this method cannot be used for forming patterns comprising different colloidal particles (Hayward et al., 2000). (Kim et al., 1997) exploited the presence of capillary forces in a microcontact printing mold to draw the colloidal suspension into small channels above the substrate. Evaporation of the solvent results in the colloidal particles to self-assemble into crystals. While this method eliminates the use of photolithography to form patterned surfaces, it is restricted to patterns with interconnected areas alone, since it is based on capillary flows. (Hayward et al., 2000) described a method for assembling patterned colloidal crystals using selective illumination of an optically sensitive electrode with electromagnetic radiation. This method overcomes limitations of the other methods described above.

In the following two sections we focus on the physical and modeling issues that need to be considered for understanding and simulating colloidal structures using paramagnetic beads and charged particles. We first discuss issues in magnetorheological fluids and subsequently address electrophoretic deposition.

13.1.1 Magnetorheological (MR) Fluids

In order to determine the mechanics of paramagnetic micropsheres in different flow configurations, we need first to characterize the particles and the forces that act on them. Superparamagnetic beads consisting of iron oxide crystals dispersed in a styrene polymer are readily available in a range of sizes from 0.3 μm to 5 μm in diameter (e.g., Bang Labs). Other functionalized paramagnetic beads are available in sizes down to 20 nm that are generally spherical in shape. The particles are magnetically soft and readily acquire a dipole moment when placed in a magnetic field. They also show minimal hysteresis and quickly lose their dipole moment when the magnetic field is removed. The particles are colloidal in character and can remain in suspension for a long time, with negligible sedimentation, under the action of Brownian motion and short-range electrostatic repulsion. In MR flows for the formation of self-assembled structures the volume fraction of the particles is typically low, and the particles are larger than 20 nm. These features distinguish the systems from *ferrofluids* (Rosenweig, 1987), in which individual magnetite crystals, on the order of 20 nm in size or smaller, are in random suspension at much higher volume fractions. In the experiments of (Hayes et al., 2001), (Furst et al., 1998), and (Doyle et al., 2002) on the formation of supraparticle structures, the particles were all about 1 μm to 3 μm in diameter, while in (Promislow et al., 1995), they were about 0.6 μm in diameter.

The forces acting on the particles arise from the imposed magnetic field, Brownian motion, short-range electrostatic repulsion, van der Waals forces, gravity, and fluid forces. In particular, van der Waals forces would cause particles to bind together and aggregate due to collisions from Brownian motion or in a shear flow, but these forces act only over very short distances, while electrostatic repulsion between particles acts over a relatively larger distance, comparable to the Debye layer thickness, and will prevent aggregation; see (Russel et al., 1989; Tabor, 1991). The details of the electrostatic properties depend on the preparation of the beads, how they are stabilized, and the electrolyte content of the liquid phase. The effect of gravity diminishes rapidly with particle size, and the dominant forces to

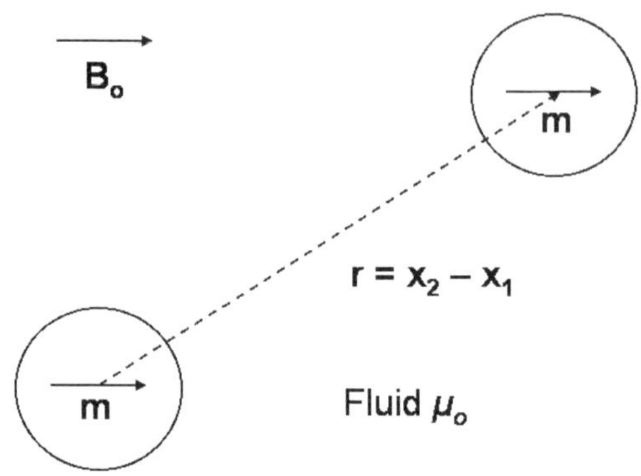

(a)

FIGURE 13.1. Sketch and definition of geometric parameters.

consider are due to the magnetic field, Brownian motion, and viscous fluid forces. The diffusivity of an isolated sphere is given by the Stokes–Einstein formula

$$D = \frac{k_B T}{6\pi a \mu}$$

for a particle of radius a, where μ is the fluid viscosity, k_B is Boltzmann's constant, and T is the absolute temperature. Some characteristic features of particles of radius 1 μm and 0.1 μm are listed in Table 13.1.

A criterion for determining whether two particles that have a magnetic dipole will form a chain can be obtained by comparing the energy of thermal fluctuations $k_B T$ with the interaction energy U of the magnetic dipoles. To this end, we define the *dipole strength ratio*

$$\lambda = \frac{|U|_{\max}}{k_B T},$$

so for $\lambda > 1$ attraction forces overcome the Brownian motion, and aggregation is expected. If r is the center–center separation and θ is the angle between the line of centers and the orientation of the dipoles (see Figure 13.1), the interaction energy (SI units) is

$$U(r, \theta) = \frac{\mu_0}{4\pi} |\mathbf{m}|^2 (1 - 3\cos^2\theta)/r^3,$$

where $\mathbf{m}$ is the dipole strength. For a magnetic field strength $\mathbf{H}$, the induced dipole is

$$\mathbf{m} = \frac{4}{3}\pi a^3 \chi \mathbf{H},$$

in terms of the effective magnetic susceptibility χ of the particle. A chain will form if $U_{\max}$ is significantly larger than $k_B T$. As discussed in (Hayes et al., 2001), the formation of supraparticle structures depends on particle

TABLE 13.1. Estimates for a paramagnetic bead, density 1.5×10^3 kg m^{-3}, suspended in aqueous solution at temperature 300 K.

		$d = 2\mu$m ($a = 1\mu$m)	$d = 0.2\mu$m ($a = 0.1\mu$m)
Stokes settling velocity	W_s	1.1 μm/sec	$1.1 \times 10^{-2} \mu$m/sec
Brownian diffusivity	$D = kT/6\pi\mu a$	0.22×10^{-12} m^2/sec	0.22×10^{-11} m^2/sec
Schmidt number	ν/D	4.5×10^6	4.5×10^5
Diffusion time scale	a^2/D	4.5 sec	4.5×10^{-3} sec
Scale for Brownian force	kT/a	4.14×10^{-15} N	4.14×10^{-14} N
Buoyancy force		2.06×10^{-14}N	2.06×10^{-17} N
Peclet number	$Pe = W_s a/D$	5	0.5×10^{-3}

size, magnetic susceptibility, field strength, temperature, channel geometry and surface properties, and volume fraction of the particles. Structures rapidly disassemble once the magnetic field is removed due to the electrostatic repulsion at short-range and Brownian motion.

We can make an order of magnitude estimate for the attractive force between two paramagnetic beads by assigning representative values of $H = 10^4$ Am^{-1} for the magnetic field strength and $\chi = 1$ for the volumetric magnetic susceptibility. The particles respond to each other's gradients of magnetic flux density, which decreases as r^{-4}. At contact this force can be relatively large, 6×10^{-11} N, for a particle of radius 1 μm or 6×10^{-13} N for a particle of radius 0.1 μm. These values may be compared to the parameters in Table 13.1. At a separation distance of $r = 10a$, characteristic of the interparticle distance in a random suspension at 0.5% volume fraction, the forces drop to 10^{-13} N and 10^{-15} N respectively for separations aligned with the dipoles. Actual field strengths may be lower than these levels, and the force estimates scale with H^2. The general indications are that Brownian motion is significant for the larger particles only when they are dispersed, while for the submicron scales Brownian motion plays an increasing role. The characteristics of the magnetic field will change once chains of beads start to form.

Hydrodynamic Forces

Fluid forces have a central part in the dynamics of particles in MR flows. In the development of microdevices for pumping, mixing, or flow control in a microchannel, we need to determine flow characteristics around the individual particles and in the channel as a system. While the flow will be low Reynolds number in character, inertial effects will influence the large scale dynamics and will be appreciable where the geometry of the channel changes.

Various methods are available for flow simulations. Specifically, we discuss the continuum-based ALE, DLM, and FCM schemes in Chapter 14. These are flexible methods suitable for particulate flows both in low Reynolds number and zero Reynolds number flows. In addition lattice–Boltzmann simulations (LBM), see Section 15.5, have been developed as a general simulation approach for dispersed two-phase flows as well as dissipative particle dynamics methods; see Section 16.4. For Stokes flows there are several methods based on multipole expansions; see (Kim and Karrila, 1991). The Stokesian dynamics method of (Brady and Bossis, 1988) uses a low-order multipole representation to construct a mobility matrix for the particles. This is supplemented with localized resistance forces and lubrication forces when particles are close to each other or near a wall. These simulations have been used to investigate dense suspensions and the rheology of particles in shear flows. New developments of the method (Ichiki and Brady, 2001) have extended the range of application to larger systems of particles and to possibly higher multipole representations. These methods provide a range of techniques to determine flow characteristics and the force and torques acting on the particles in response to the flow or their motion through the fluid.

At low Reynolds numbers, special attention is needed to determine the motion of particles that are close to each other or near a wall. As the separation distance is reduced, very large forces can be generated as fluid is displaced from the narrow gap between particles or from shearing motion across the gap. These are characterized by lubrication forces and have been calculated accurately for a number of standard configurations; see section 14.3.2 and (Jeffrey, 1982; Kim and Karrila, 1991). Even if one uses high-order multipole methods or other accurate flow simulation methods, the numerical resolution of the gap becomes insufficient to calculate these forces accurately, and in general, some approximation of these effects must be included with the simulations. The lubrication forces are usually included through a summation of discrete pairwise interaction between particles or a particle and a wall, with the forces determined by the particle positions and relative velocity. This has been done for the force coupling method (FCM) in Stokes flows; see (Dance and Maxey, 2003). There are limitations to the accuracy of summing pairwise interactions, and this is evident, for example, for a particle in a narrow channel, where both walls simultaneously can influence the particle motion. A simple model that can be used to prevent overlapping of particles is a velocity repulsion barrier. It can be activated only when particles are closer than a cutoff distance R_{ref} on the order of 20% of the radius a. This repulsive velocity is given by

$$\boldsymbol{v}_{ij} = -\frac{\boldsymbol{v}_{\text{ref}}}{2a}\left[\frac{R_{\text{ref}}^2 - r^2}{R_{\text{ref}}^2 - 4a^2}\right]^2 \mathbf{r}_{ij},$$

where $\mathbf{r}_{ij}$ is the distance connecting the particles i and j (see sketch in Figure 13.1). Calibration of the relative motion of two or three particles can help in determining the velocity scale $\boldsymbol{v}_{\text{ref}}$.

As the separation distance between particles is reduced further, particle–particle forces will act. In the context of colloids the most relevant is the short-range electrostatic repulsion between particles due to surface charges.

The effects of these charges that are naturally occurring in most systems is generally screened by ions in the suspending liquid phase, and their influence is confined to a thin Debye layer. Their action on particle motion is effectively approximated by a Derjaguin potential; see (Russel et al., 1989).

Magnetic Forces

In a permeable magnetic medium the magnetic flux density $\mathbf{B}$ (units of T) is due to the external field and the induced magnetic moments within the material. The governing equations are

$$\begin{aligned}\mathbf{B} &= \mu_0(\mathbf{H}+\mathbf{M}),\\ \nabla\cdot\mathbf{B} &= 0,\\ \nabla\times\mathbf{H} &= \mathbf{j};\end{aligned}$$

Here $\mathbf{j}$ is the macroscopic current density that is generated externally and $\mathbf{M}$ is the magnetic dipole density (Am^{-1}) induced in the material. The latter is written in terms of the magnetic field strength $\mathbf{H}$ (units of Am^{-1}) as

$$\mathbf{M} = \chi\mathbf{H},$$

where χ is the magnetic susceptibility of the material. For a particle, the overall dipole moment $\mathbf{m}$ is the volume integral of the dipole density over the particle volume,

$$\mathbf{m} = \int \mathbf{M}\, d^3\mathbf{x},$$

so for a uniform sphere of radius a subject to an external uniform magnetic field with flux density $\mathbf{B}_0$,

$$\mathbf{m} = \frac{4}{3}\pi a^3\chi\mathbf{H} = \frac{4}{3}\pi a^3\chi\frac{\mathbf{B}_0}{\mu_0}.$$

In general, the susceptibility will be a nonlinear function of the field strength, and the induced dipole moment density will saturate at some level. Saturation often occurs when the magnetic flux density in the material reaches about 1.0 to 1.5 tesla. The Langevin theory provides an estimate for the susceptibility including the effect of thermal energy. In a medium in which the dipole elements interact through their induced fields, a modified estimate for the overall dipole moment must be made that leads to a higher effective susceptibility. For the beads used in the experiments of (Promislow et al., 1995), the effective volumetric susceptibility was $\chi = 0.9$.

In vacuum, $\mathbf{B} = \mu_0\mathbf{H}$, where $\mu_0 = 4\pi\times 10^{-7}\ \mathrm{Hm}^{-1}$ is the permeability of free space, so a magnetic field strength $H = 1000\ \mathrm{Am}^{-1}$ would correspond to a magnetic flux density $B = 1.257\times 10^{-3}$ T. In (Doyle et al., 2002), an external magnetic flux density of about 10 mT was used, corresponding to a magnetic field strength $H = 10{,}000\ \mathrm{Am}^{-1}$. In (Hayes et al., 2001), the magnetic flux density imposed in the liquid phase was about 50 mT, or 500 gauss in terms of cgs units. Also, the flux densities used in (Furst et al., 1998), were around 3 mT, corresponding to 30 gauss or $H = 2300\ \mathrm{Am}^{-1}$.

The magnetic flux density due to an isolated dipole moment is

$$\mathbf{B} = -\frac{\mu_0}{4\pi}\nabla(\frac{\mathbf{m}\cdot\mathbf{x}}{r^3}).$$

In the presence of an external uniform magnetic field with flux density $\mathbf{B}_0$, we obtain

$$\mathbf{B} = -\frac{\mu_0}{4\pi}\left(\frac{3(\mathbf{m}\cdot\mathbf{r})\mathbf{r}}{r^5} - \frac{\mathbf{m}}{r^3}\right) + \mathbf{B}_0.$$

The corresponding force acting on the dipole in the presence of an ambient magnetic field is

$$\mathbf{F} = \nabla\times(\mathbf{B}\times\mathbf{m}) = \mathbf{m}\cdot\nabla\mathbf{B}.$$

This force may be written in terms of a scalar potential $\mathbf{F} = -\nabla\mathbf{U}$ with $U = -\mathbf{m}\cdot\mathbf{B}$. The torque acting on a dipole that has a fixed moment is

$$\mathbf{G} = \mathbf{m}\times\mathbf{B},$$

and the dipole will tend to turn to align with the magnetic field.

Paramagnetic beads placed in a uniform magnetic field will acquire a dipole moment aligned with the field. The dipoles will then attract each other to form a chain of aggregated particles with the dipoles joined end to end. The attraction is quite long-range but does fall off quickly with distance as r^{-4}. The uniform magnetic field will produce no force on the particles, only the dipoles will create a gradient of the magnetic flux density $\mathbf{B}$ needed to generate a net force. At large separations the force will be below the level of Brownian thermal fluctuations or other background forces. As the particles disperse due to random motion or as longer chains form, the particles will come close enough for the magnetic forces to become dominant.

The description of the particles in terms of simple magnetic dipoles captures the primary dynamics, even though a more detailed multipole representation may be warranted for local variations in the magnetic field. In (Paranjpe and Elrod, 1986), such a description has been used in terms of dipoles to effectively study the equilibrium configurations of chains of magnetic beads and to determine their stability from the overall potential U. While energy minimization principles yield information about the final states, dynamic simulations are needed to obtain the development in time and information about the time scales for chain formation and their eventual fate. This requires the solution of the coupled system of equations for the dipole strengths, which are determined by the external field and the fields generated by the other particles or other chains of particles.

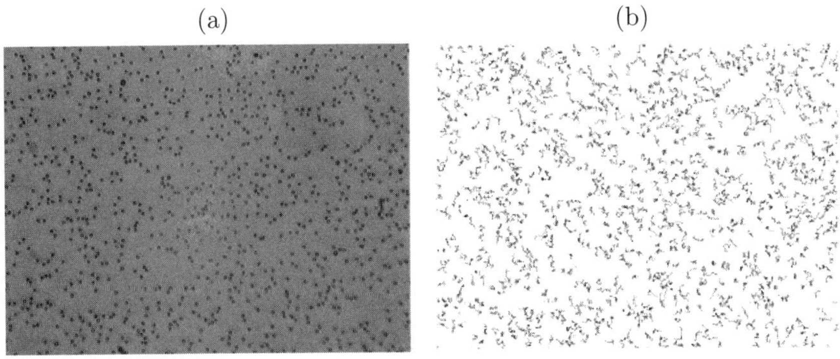

FIGURE 13.2. Brownian motion of 2.34-μm silica particles in DI water. (a) A snapshot of monolayer of particles. (b) Particle trajectories of silica microspheres for 10 seconds.

Brownian Motion

Brownian motion and the effects of thermal fluctuations become an increasingly important feature for submicron-sized particles. Figure 13.2 (a) shows a snapshot of a monolayer of 2.34-mum-sized silica particles in DI water, taken using a conventional inverted light microscope. A sequence of digital images was taken using a monochrome CCD camera to track the Brownian diffusion of the silica particles. Image processing algorithms given in (Crocker and Grier, 1996), were employed to track the particle centers and draw their corresponding trajectories. The trajectories of the silica microspheres for 10 seconds of Brownian diffusion are shown in Figure 13.2 (b). Experiments were conducted at room temperature.

The standard description of Brownian motion is based on the Langevin model, where the particles are subject to a random white-noise forcing from the thermal fluctuations. This yields the aforementioned classic Stokes–Einstein result for the diffusivity of a single spherical particle that is independent of the mass of the particle. In a dilute system the particles can be considered in isolation, as is the case of many Brownian dynamics simulations of polymer chains. However, there are long-range hydrodynamic interactions between particles in a suspension under low Reynolds number conditions. These modify the mobility of a system of particles and hence the diffusivities and particle motion in response to thermal fluctuations.

The starting point for most simulation studies of Brownian motion is the Langevin equation written in the form of an Ito stochastic differential equation for the evolving displacements of the particle. This is essentially a Monte Carlo simulation of the suspension. An alternative is to formulate the problem as a Fokker–Planck equation for the configurational distribution function. A key element of the stochastic simulations is to characterize the vector of stochastic Brownian forces $\mathbf{F}^B$ using the fluctuation–dissipation theorem in terms of the collective resistance tensor R for the system of particles. The fluctuating forces have zero mean and are uncorrelated in time, but the forces on individual particles are correlated on average, through the hydrodynamic interactions, with

$$\langle \mathbf{F}^{\mathrm{B}}(t) \cdot \mathbf{F}^{\mathrm{B}}(t')^{T} \rangle = 2k_{\mathrm{B}}T \; R \, \delta(t - t').$$

The resistance tensor R is determined by the instantaneous configuration of the particles and must be recalculated as the particles move. The fluctuating forces give rise to corresponding random displacements in the stochastic equation, which can be characterized through the corresponding mobility tensor R^{-1}. For a system of particles the diffusion tensor is $D = k_{\mathrm{B}}TR^{-1}$. The repeated computations of the resistance or mobility tensors can become lengthy as the system of particles becomes larger, and steps are needed to accelerate the computational steps. A simple representation of the Brownian motion using a fluctuating force $\mathbf{F}_{\mathrm{B}}$ with zero mean and the correlation specified above is

$$\mathbf{F}_{\mathrm{B}} = \xi \sqrt{\frac{12\pi\mu a k_{\mathrm{B}}T}{dt}},$$

where dt is the time step in the numerical integration of the Lagrangian tracking of the particles and ξ is a random vector following a Gaussian distribution.

In addition to the factors already noted, beads and chains of particles in a microchannel will involve the hydrodynamic effects of the channel walls on the resistance or mobility tensors together with the hydrodynamic interaction of spherical beads with nonspherical chains. Magnetic beads that have formed a bound pair will be subject to both random displacements and rotations.

Dynamics of Particle Chaining

We present here a simulation of many particles placed in a periodic box of width $L/a = 48$, where a is the particle radius. The simulation is based on the force coupling method (FCM); see Section 14.3.2. The resolution is 128^3 Fourier grid nodes for the solution of the fluid flow equations. Under the influence of hydrodynamic and magnetic forces the particles tend to aggregate in linear clusters. An averaging of ten initial random seedings was performed in order to evaluate the temporal growth of the mean cluster size $\langle S(t) \rangle$. This is defined as

$$\langle S(t) \rangle = \frac{\sum_s s^2 n(s)}{\sum_s s n(s)},$$

where $n(s)$ denotes the number of clusters of size s in the suspension at the sampled time.

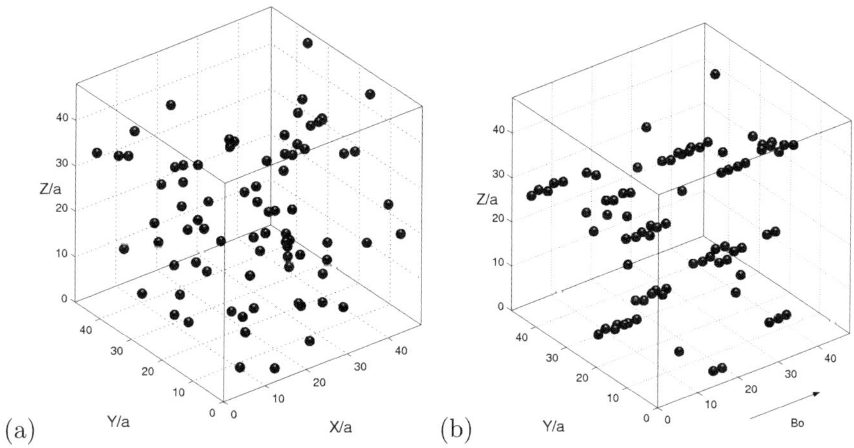

FIGURE 13.3. (a) Initial random seeding of paramagnetic beads, and (b) configuration at $t/t_B^* = 16$. Concentration is $c = 0.003$ and dipole strength $\lambda = 100$.

In (Climent et al., 2004), several different configurations were considered corresponding to a particle volume fraction c ranging from 0.3% to 3% and dipole strength ratio λ from 1 to 104. This concentration is in the range of the experiments of (Promislow et al., 1995), so that comparisons are possible. Figure 13.3(a) shows typical initial conditions with eighty paramagnetic particles seeded at random positions in a cubic domain. Figure 13.3(b) shows a typical aggregation of particles with chains of various lengths formed but with some particles still isolated. As time evolves, these particles will join together and form longer chains. Chain/chain interaction leads also to long structure formation. Most of the chains are linear, since head-to-tail aggregation of magnetic dipoles is energetically preferable. Lateral merging of chains is also possible, leading to thick clusters of particles. Such lateral merging is important for the resistance to deformation and threshold of rupture of long chains. It is directly connected to the yield stress of the suspension and has a strong impact on nanotechnology applications. In Figure 13.3(b), chain defects are observable; e.g., observe one chain divided into two branches that connect again. Careful experiments of microrheology using dual-trap optical tweezers have highlighted clearly the impact of annealing defects on the mechanical properties of chains (Furst and Gast, 2000).

Von Smoluchowski's theory (von Smoluchowski, 1916) provides a solid basis for predicting aggregation rates in very dilute solutions. It states that

- the rate of change in the number of clusters containing n particles is connected to the reaction kernel that represents the rate of coalescence of two smaller clusters.

In diffusion-limited aggregation, the numerical simulations in (Miyazima et al., 1987), based on the motion of oriented particles on periodic square

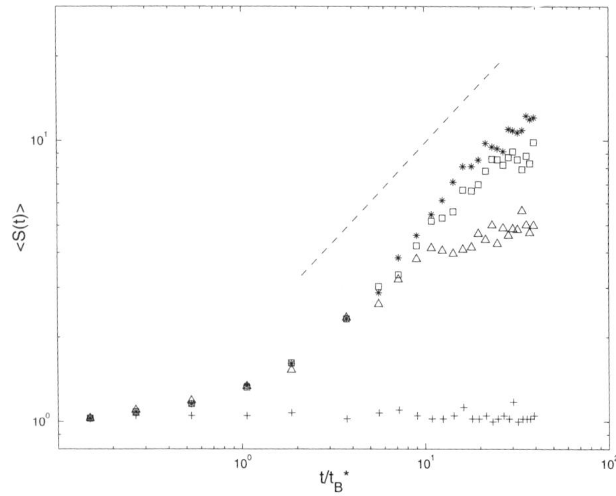

FIGURE 13.4. Evolution of the mean cluster size in terms of particle diameter. The symbols denote: plus - $\lambda = 1$; triangle - $\lambda = 10$; squares - $\lambda = 100$, and stars - $\lambda = 10{,}000$. The dashed line indicates $t^{0.7}$ dependence. $\langle S(t) \rangle$ is normalized with the particle diameter.

lattices support the scaling evolution $\langle S(t) \rangle \propto t^z$. The exponent z is related to the diffusion coefficient of a cluster of length s. Under the assumption of a long chain, it is commonly accepted that the diffusion coefficient scales like s^{-1}, which in turn leads to the relation

$$\langle S(t) \rangle \propto t^{0.5}.$$

Even though this behavior has never been strictly observed in experiments, it provides a valuable point of reference.

The determination of a characteristic time scale is a critical issue. We can expect a collapse of all the data if the time scale is properly defined, i.e., accounting for the effect of the dipole strength λ and the particle volume fraction c. A simplistic model of doublet formation is based on pure diffusion of particles in a very dilute suspension. Particles diffuse as long as collisions do not occur. When two particles are touching, a doublet is formed. The characteristic time t_B in this case is inversely proportional to diffusivity D and concentration c as it corresponds to the flux of particles toward an isolated test particle, i.e.,

$$t_B = \frac{a^2}{6Dc}.$$

Based on experimental results, a correction was introduced in (Promislow et al., 1995), that improves this scaling. The identification of mechanisms leading to chain formation is of fundamental significance for the time scale determination. In particular, the motion of a paramagnetic particle can be split into two main regimes. When particles are widely separated, the attraction force that decays quickly (r^{-4}) is not strong enough, and particles diffuse randomly. When the separation between two beads or between a particle and a chain is smaller than a critical value r_c determined by $U(r_c) \propto k_{\mathrm{B}}T$, the attraction force becomes dominant and aggregation occurs. Therefore, the flux of particles captured by a single sphere has to be augmented. The capture volume related to the surface $|U| = k_{\mathrm{B}}T$ allows the determination of a new time scale based on the envelope of the anisotropic dipolar potential. Promislow et al. proposed to use a modified time scale t_{B}^* given by

$$t_{\mathrm{B}}^* = \frac{a^2}{24[(1/3)^{1/2} - (1/3)^{3/2}]Dc\lambda}.$$

This characteristic time is inversely proportional to D, c as before, but also to the dipole strength λ. The constant prefactor is derived based on the capture volume that corresponds to the combined conditions of attraction, i.e., $U < 0$ and $r < r_c$.

In the simulations of (Climent et al., 2004), initially, the mean cluster size is equal to one particle diameter, since all the particles are seeded randomly throughout the domain respecting the nonoverlapping condition. Figure 13.4 shows results for a very dilute suspension ($c = 0.003$) that experiences a constant magnetic field characterized by a dipole strength λ ranging from 1 to 104. We notice that in the case of $\lambda = 1$, no chains are forming and the particles always diffuse randomly. The magnetic attraction is not strong enough to join the particles together, since Brownian forcing dominates the behavior of the suspension. A clear transition is observed when $\lambda \gg 1$, with $\langle S(t) \rangle$ slowly increasing during a characteristic time of order t_B^*. Subsequently, aggregation sets in as particles or chains gradually join together forming linear supraparticle structures. The linear increase in a log-log plot is evident for almost one decade. When the mean length is on the order of half of the domain width, the periodic conditions become invalid and the simulation is stopped.

In the previous example, the geometry was not taken into account, and chains became longer as a function of time, although they may bifurcate into smaller chains due to chain–chain interactions. Geometry effects are, however, very important, and they can be used cleverly to affect the pattern formation. The effect of geometric confinement on dynamic self-assembly has been simulated for the first time in (Liu, 2004), following the experimental studies of (Hayes et al., 2001). Figure 13.5 shows four snapshots of the dynamic aggregation process inside a trianglular duct. Initially, eight paramagnetic beads are randomly scattered inside the duct. Upon the application of a horizontal external magnetic field, the particles in close proximity form pairs first, then triplets, and so on. Due to the magetic dipole repulsion and attraction mechanisms, the particles tend to form a straight

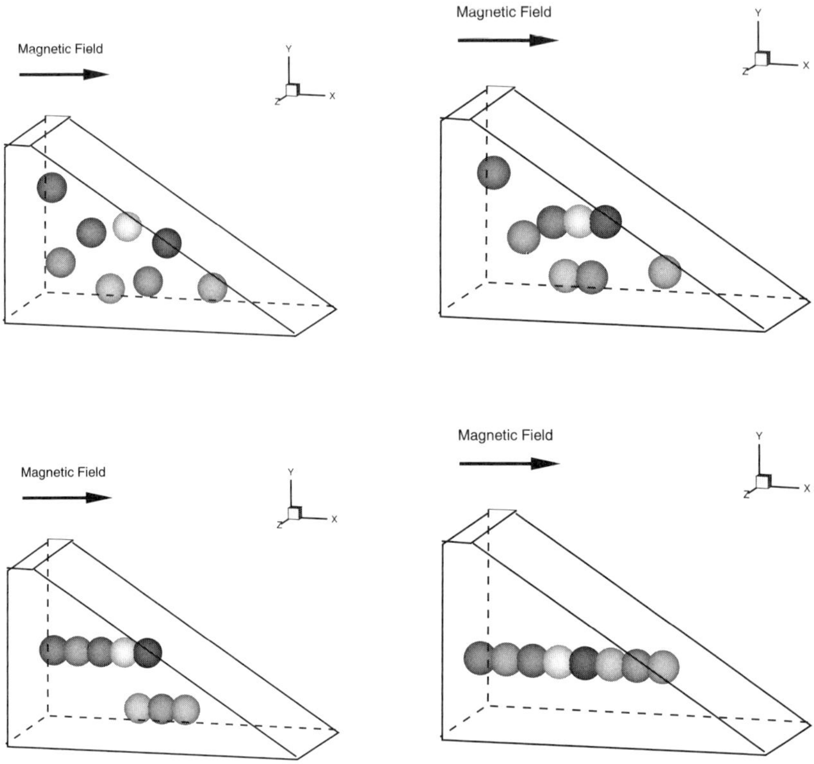

FIGURE 13.5. Aggregation of eight particles inside a trianglular duct due to magnetic dipole interactions: initial seeding (top left); shortly after initial seeding (top right); shortly before the final aggregation (bottom left); final position (bottom right).

line to minimize the magnetic potential energy in the system. However, because of the constraint imposed by the geometric boundaries, not all particles can be accommodated at that height. Therefore, a five-particle chain moves down to a proper right height, while a triplet moves upward and to the right in order to join the finally formed chain parallel to the magnetic field direction.

13.1.2 Electrophoretic Deposition

Electrophoretic deposition of colloidal particles at charged surfaces has been used as a manufacturing technique for coating metals, oxides, phosphores, inorganic and organic paints, rubber, dielectrics, superconductors, and glasses (Trau et al., 1997). In his seminal paper, (Hamaker, 1940) described the formation of a colloidal deposit by electrophoresis. Electrophoretic deposition (EPD) is essentially a two-step process that involves:

1. Application of an electric field to transport the colloidal particles in the liquid suspension toward an electrode (electrophoresis), and
2. Collection of particles on the electrode to form a coherent deposit.

Building on these early ideas, flocculation at the electrode surface was explained in terms of dipole interactions (Estrelalopis et al., 1982). Hamaker's law relates the deposit yield to the electric field strength, electrophoretic mobility, surface area of the electrode, and the particle mass concentration in the suspension, and it is given as

$$\frac{dY}{dt} = \mu ECA, \tag{13.1}$$

where Y is the yield (in kg), E is the electric field strength (in V/m), μ is the electrophoretic mobility (in $m^2/(V.s)$), A is the surface area of the electrode (in m^2), and C is the mass concentration of the colloidal particle in the suspension (in kg/m^3). For cylindrical coaxial electrodes, a closed-form expression for the yield during EPD is provided in (Avgustinik et al., 1962). For more advanced models of the EPD, the reader is referred to (Biesheuvel and Verweij, 1999).

The majority of work in EPD has been concerned with measuring the deposition rate, and maximizing film thickness and porosity of the deposited film. However, very little attention was devoted to the microscopic dynamics that give rise to the resulting morphology of the deposited layers. For a long time, the dynamics of electrophoretically deposited layers was assumed to be analogous to particle sedimentation. (Hamaker and Verwey, 1940) theorized that the primary action of the electric field was to move the particles toward the electrode and to produce a force that pressed the particles together on the electrode surface, the same way gravity acts on particle sedimentation. However, systematic experimentation has revealed new interesting physics during electrophoretic deposition (Richetti et al., 1984; Bohmer, 1996; Trau et al., 1996; Trau et al., 1997). Particle motion transverse to the applied electric field was observed close to the electrode surface, when particles do not stick to the electrode. This resulted in particle aggregation to form densely packed self-ordered planar clusters. This phenomenon is unexpected from pure electrostatic considerations, since like-charged particles repel each other. Consequently, the attractive interaction between particles must be strong enough to overcome electrostatic repulsion to facilitate the aggregation. In the following, we discuss in detail the aggregation dynamics during EPD.

Colloidal Aggregation

The colloidal particle aggregation on a charged surface under an electric field was first reported in (Richetti et al., 1984). Subsequent experiments

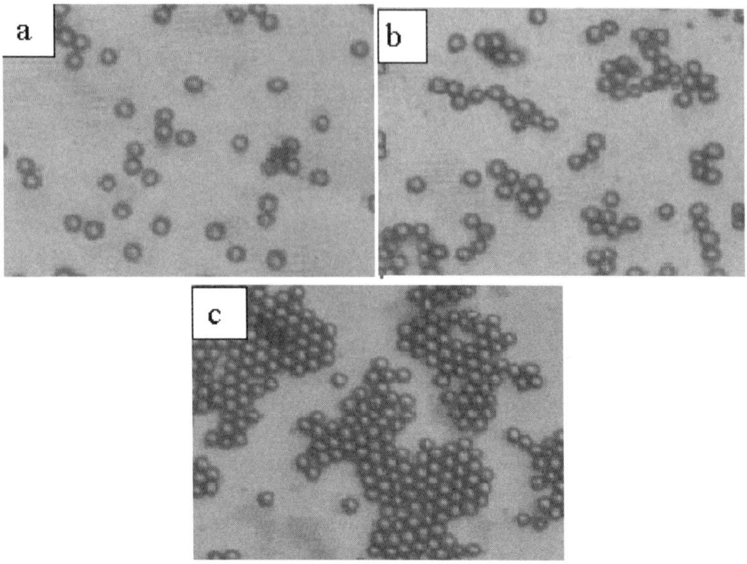

FIGURE 13.6. Colloid arrays obtained by electrodeposition on nonpatterned ITO substrates for 3.88 μm size polystyrene (PS) particles at a DC potential of 20 V. Times of electrodeposition were (a) 30 s, (b) 90 s, and (c) 210 s. (Courtesy of E. Kumacheva.)

on electrophoretic deposition have also confirmed the migration of colloidal particles toward each other over very large distances (even greater than five particle diameters) to form highly ordered two-dimensional structures. This long-range attraction between colloidal particles close to electrodes has been observed in both AC and DC electric fields (Bohmer, 1996; Trau et al., 1996; Trau et al., 1997). In addition, different two-dimensional colloidal phases (analogous to) gas, liquid, and solid (crystal structures) can be formed on the surface of the electrode by manipulating the current density, which alters the magnitude of lateral attractions between particles. In the following we review observations by (Bohmer, 1996) regarding the self-ordering of colloidal particles on electrode surfaces:

1. The aggregation occurred after the particles were close to the surface or deposited.
2. The aggregation was reversed when the polarity of the applied field was reversed.
3. Two or more clusters that formed on the surface aggregated to form a larger cluster.
4. The clusters dispersed when the field was turned off.

Figure 13.6 shows the colloidal arrays obtained by the electrophoretic deposition of 3.88 μm size polystyrene (PS) particles at a DC potential of

20 V on ITO electrodes (Golding et al., 2004). Evolution of the colloidal system from a gaseous state in Figure (a) to a liquid phase in Figure (b) and finally to a crystalline phase in Figure (c) can be seen. The colloidal assembly follows a three-stage process. In the initial nucleation stage, the colloidal particles randomly deposit on the electrode surface (Figure a). Following deposition, the colloidal particles laterally migrate on the electrode surface, forming dyad- and triad-like structures, thereby increasing the surface area (Figure b). In the final stage, small particle clusters begin to merge to form two-dimensional islands (Figure c). It is important to understand that although the Brownian motion tends to redistribute and break up the particle clusters formed after aggregation, it is a very slow process (Ristenpart et al., 2004).

The details of the aggregation mechanism are still not completely understood. Comprehensive theories that can completely describe the aggregation dynamics at the electrode surface are not available in the literature. As the particles and clusters were seen to interact over long ranges, (Bohmer, 1996) suggested that hydrodynamic effects resulting from electroosmotic flow around each particle were responsible for particle aggregation. This also ruled out the contribution of short-ranged van der Waals forces to particle aggregation. The motion of particles deposited on an electrode surface is governed by the relative interplay of electrokinetics, electrohydrodynamics, and Brownian diffusion. (Solomentsev et al., 1997) considered the electroosmotic flow around the charged particles near the electrode surface, and proposed an electrokinetic model for particle aggregation. This model was able to qualitatively explain the observations in (Bohmer, 1996). In addition, the particle trajectories predicted by the model were in good quantitative agreement with experimentally measured trajectories of three particles aggregating near the electrode surface. In a later work, (Solomentsev et al., 2000) studied the aggregation dynamics for two particles during electrophoretic deposition under steady electric fields. They proposed a convective–diffusive model based on electrokinetics to explain the mechanism behind the particle aggregation.

Figure 13.7(a) shows a schematic of two equal-sized colloidal particles electrophoretically deposited on an electrode. The two particles are assumed to be at the same height h above the electrode surface, and the electric field $\mathbf{E}$ is normal to the electrode surface. The particles attract each other due to the electroosmotic flow around each particle (Solomentsev et al., 2000). The authors solved Laplace's equation to obtain the electric field about a single particle, and then they used this result to solve for the electroosmotic flow field. The streamlines around a particle are shown in Figure 13.7(b). The electrokinetic slip velocity at the surface of the colloidal particles drives the flow. The electroosmotic flow around a single particle entrains the neighboring particle, and draws it closer. Secondary electrophoretic effects also become important, since the electric field has a component that affects the relative motion of the two particles. The

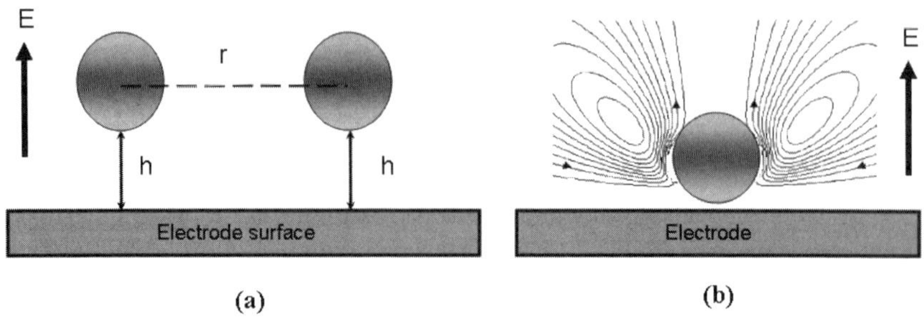

FIGURE 13.7. (a) Schematic of two equal-sized particles of diameter D deposited on an electrode by EPD. The particles remain mobile due to the Brownian motion, (b) Electroosmotic flow around a colloidal particle held stationary near an electrode. When the direction of electric field is reversed, the direction of the flow is also reversed. (Courtesy of M.A. Bevan.)

electroosmotic flow brings the particles together, while the electrophoresis moves them apart. (Solomentsev et al., 2000) have shown that at $r \approx 3D$, the magnitude of the electroosmotic velocity is about seven times greater than the electrophoretic velocity, and this facilitates the aggregation process. It has to be noted that this analysis is applicable only for DC electric fields.

(Trau et al., 1996) and (Trau et al., 1997) explained the transverse migration of colloidal particles using an electrohydrodynamic mechanism wherein fluid flow transports the particles toward each other. They theorized that the particles near the electrode alter the local electric fields, and these perturbations can result in concentration and current density gradients at the electrode surface, resulting in fluid motion. This model is applicable for both AC and DC electric fields, and it is consistent with experimental studies.

In summary, the current theories are not elaborate enough to provide quantitative estimates of particle aggregation behavior. More experiments are needed to provide an insight into the origin of the lateral convective forces. A model that can completely describe the particle aggregation dynamics should take into account the combined effects of electrokinetics, electrohydrodynamics, and Brownian diffusion, similar to the model we presented for magnetorheological fluids.

13.2 Electrolyte Transport Through Carbon Nanotubes

There is great interest in investigating fluid flow through carbon nanotubes. In Chapter 10 and Chapter 11, we presented results on transport of simple fluids and water through carbon nanotubes. The fundamental question is whether electrolytes can be transported easily through small-diameter carbon nanotubes. One of the motivations to investigate electrolyte transport through carbon nanotubes is to create nanoscale devices or design concepts to mimic biological ion channels. In this section, after an introduction to carbon nanotubes and ion channels, we address the issue of electrolyte transport through carbon nanotubes and present ideas to mimic some aspects of biological ion channels.

13.2.1 Carbon Nanotubes

Carbon nanotubes were discovered by (Iijima, 1991). Since their discovery, nanotubes have aroused great excitement because of their unique physical properties, which span a wide range, from structural to electronic. For example, nanotubes have a light weight and a record-high elastic modulus, and they are one of the strongest fibers that can be made. There are two main types of carbon nanotubes (Saito et al., 1998; Dekker, 1999; Baughman et al., 2002):

- A single-walled carbon nanotube (SWNT), made of a single atomic layer thick of graphite (called graphene) rolled into a seamless cylinder.
- A multiwalled carbon nanotube (MWNT), made of several concentric nanotube shells.

Single-walled carbon nanotubes have emerged as a very promising new class of materials for a variety of applications. The diameter of a SWNT is about 0.7 nm to 10 nm, though most of the observed single-wall nanotubes have diameters less than 2 nm. The structure of a SWNT is one-dimensional with axial symmetry. A SWNT exhibits a spiral conformation, called *chirality*. The chirality is given by a single vector called the chiral vector, $\mathbf{c_h}$. The chirality vector connects crystallographically equivalent sites on a two-dimensional graphene sheet. The diameter and helicity of a defect-free SWNT are uniquely characterized by the chirality vector

$$\mathbf{c_h} = n\mathbf{a_1} + m\mathbf{a_2} \equiv (n, m),$$

where n, m are integers ($0 \leq |m| \leq n$), and $\mathbf{a_1}$ and $\mathbf{a_2}$ are the graphene lattice vectors (see Figure 13.8). An armchair nanotube corresponds to the case of $n = m$, that is, $\mathbf{c_h} = (n, n)$, and a zigzag nanotube corresponds to the case of $m = 0$, or $\mathbf{c_h} = (n, 0)$. All other (n, m) chiral vectors correspond to chiral nanotubes. The indices (n, m) also determine the metallic or semiconducting behavior (electronic properties) of SWNT. Carbon nanotubes for which $n - m = 3i$, with i an integer, are metallic; all others are semiconducting. Armchair nanotubes are metallic. The electronic properties of MWNTs are rather similar to those of perfect SWNTs, because the coupling between the cylinders is weak in MWNTs.

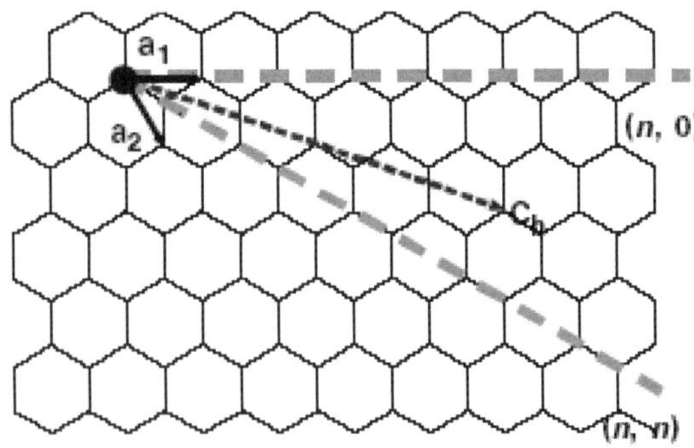

FIGURE 13.8. Schematic of a two-dimensional graphene sheet illustrating lattice vectors $\mathbf{a_1}$ and $\mathbf{a_2}$, and the chiral vector $\mathbf{c_h} = n\mathbf{a_1} + m\mathbf{a_2}$.

The diameter of SWNT, d_{cnt}, is given by L/π, where L is the circumferential length of the carbon nanotube given by

$$L = |\mathbf{c_h}| = \sqrt{\mathbf{c_h} \cdot \mathbf{c_h}} = a\sqrt{n^2 + m^2 + nm},$$

where $a = 2.49$ Å is the lattice constant of the honeycomb lattice. The *chiral angle* θ is defined as the angle between the vectors $\mathbf{c_h}$ and $\mathbf{a_1}$, with values of θ in the range $0 \leq |\theta| \leq 30°$, because of the hexagonal symmetry of the honeycomb lattice. The chiral angle θ denotes the tilt angle of the hexagons with respect to the direction of the nanotube axis, and the angle θ specifies the chiral symmetry. The indices (n, m) can also be used to compute the chiral angle, i.e.,

$$\cos\theta = \frac{2n + m}{2\sqrt{n^2 + m^2 + nm}}.$$

Zigzag and armchair nanotubes correspond to $\theta = 0°$ and $\theta = 30°$, respectively; $0 < \theta < 30°$ corresponds to chiral nanotubes.

Single-walled and multiwalled nanotubes have excellent mechanical properties, since the two-dimensional arrangement of carbon atoms in a graphene sheet allows large out-of-plane distortions. The strength of carbon–carbon in-plane bonds keeps the graphene sheet exceptionally strong against any in-plane distortion or fracture. For example, the Young's modulus of a (10, 10) SWNT is estimated to be $\propto$ 0.64 TPa. The density-normalized modulus and strength of a (10, 10) SWNT are, respectively, $\propto$ 19 and $\propto$ 56 times that of a steel wire and, respectively, $\propto$ 2.4 and $\propto$ 1.7 times that of silicon carbide nanorods. The measured room temperature thermal conductivity for an individual MWNT (>3000 W/mK) is greater than that of natural diamond. Because of their excellent structural and material characteristics, carbon nanotubes are being explored for a wide variety of applications including nanoelectronics, microscopy, sensors, and actuators, field emission, nanotube-polymer composites, and many others. It is not surprising that carbon nanotubes are also being explored for applications in nanoflows. In particular, the transport of liquids, gases, and electrolytes through carbon nanotubes have many applications including biomimetics, sensors, nanofluidic interconnects, and many others.

13.2.2 Ion Channels in Biological Membranes

Ion channels are proteins in cell membranes (lipid bilayers) that are crucial for shaping electrical signals and controlling diffusion and flow of ions and fluids across cells. They provide a high conducting, hydrophilic pathway across the hydrophobic interior of the membrane. The key features of ion channels that give rise to their extraordinary range of biological functions include variations in:

1. Their *selectivity* (what ions can pass through),
2. Their *conductance* (how rapidly can ions get through), and
3. Their *sensitivity* (how the conductance is modulated by such factors as the chemical composition of their environment, the transmembrane voltage, the membrane surface tension, and the chemical binding of the ion).

Some of the biological functions of channels include signaling and computation, triggers for cellular events, electrical power generation, energy transduction, fluid pumping and filtration, chemical sensing and mechanotransduction.

The gating property of the ion channel:

Whenever there is a conformational change in the protein structure that lets the channel open or close, allowing the ionic current to be turned "on" or "off," the channel is said to be gated, analogous to opening and closing of a gate. Ion channels can be classified according to which chemical or physical modulator controls their gating activity. Thus we have different groups of channels as summarized below:

1. Ligand gated channels: Ligand gated channels open or close in response to the binding of a small signaling molecule or "ligand." Some ion channels are gated by extracellular ligands and others by intracellular ligands.
2. Voltage gated channels: Voltage gated channels are found in neurons and muscle cells. They open or close in response to changes in the voltage drop across the plasma membrane.
3. Mechanosensitive channels: These channels open or close in response to some mechanical deformation. For example, sound waves bending

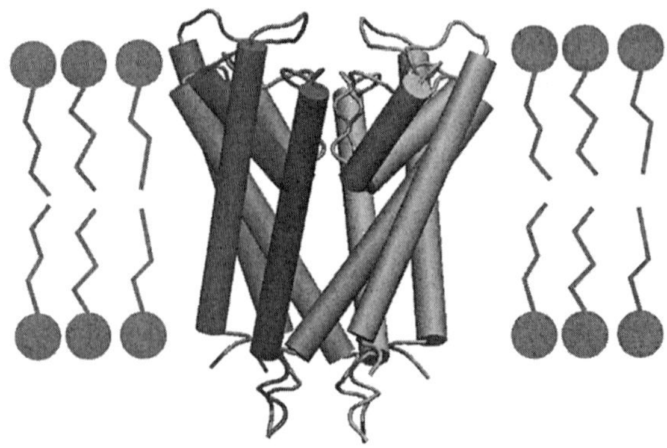

FIGURE 13.9. A potassium channel in a lipid membrane.

the cilia-like projections on the hair cells of the inner ear open up some type of ion channels leading to the creation of nerve impulses that the brain interprets as sound.

Selectivity of ion channels: Ion channels are highly specific filters, allowing only desired ions to go through the cell membrane. They can discriminate between size and charge of the permeant molecule. For example, potassium channels are selective to potassium but not sodium even though the latter is smaller in diameter. The potassium channel is shown in Figure 13.9. The KcsA K^+ channel is composed of four subunits, each with two transmembrane helical domains and a pore region. Recent structural imaging studies by (Doyle et al., 1998) show this three-dimensional organization in detail and have helped in understanding better the mechanism of channel selectivity (see Figure 13.10).

Engineered ion channels have been developed to function as a single-molecule detection system (Bayley and Cremer, 2001; Woodhouse et al., 1999). In an applied potential, an ionic current is carried by the ions that bathe both sides of the lipid bilayer. When the target molecule binds to the binding site in the pore, the current is modulated. The frequency of binding reveals the concentration of the analyte, and the duration and amplitude of the events reveal its identity. Though engineered channels have significant advantages including high sensitivity, wide dynamic range, and biocompatibility, their lack of durability makes them reliable only in a lab setting. The possibility of a nanoscale device that incorporates the functionality of ion channels into artificial nanotubes and is far less complex than a biological system has also been investigated. A review of the literature shows that promising options for such a device that can be practically realized include gold nanotubule membranes (Kang and Martin, 2001; Li et al., 2001), ion beam etched silicon nitride membranes (Martin et al., 2001), and single carbon nanotubes.

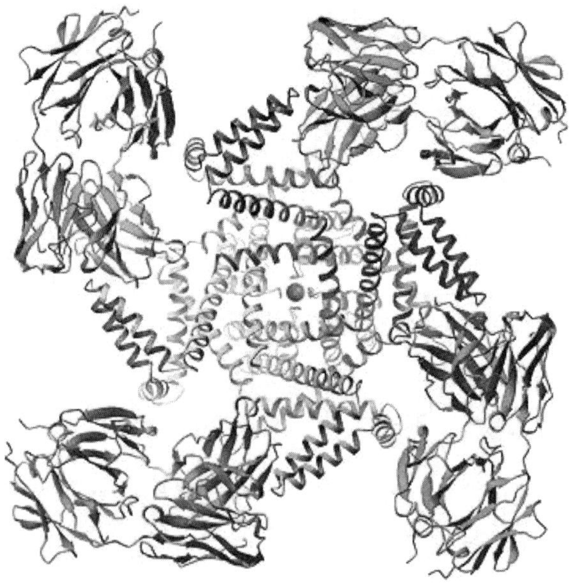

FIGURE 13.10. An ion entering the selectivity filter of the potassium channel embedded in a lipid bilayer.

13.2.3 Transport Through Unmodified Nanotubes

Quantitative information on fundamental modes of water transport and transport rates have been obtained through carbon nanotubes of diameters of about 150 nm (Sun and Crooks, 2000). As mentioned earlier, in Chapter 11, (Hummer et al., 2001) showed by molecular dynamics simulations that water molecules enter nanotubes of as small as 8.4 Å diameter even though carbon is hydrophobic. (Miller and Martin, 2002) showed experimentally that it is possible to control the rate and direction of electroosmotic flow through nanotube membranes of large diameters (120 nm). The behavior of water molecules and ions through hydrophobic carbon nanotubes is of critical importance because ion channels such as K^+ channels have their selectivity filter lined with hydrophobic residues that facilitate transmembrane movement of ions and water without sticking to the side wall. However, transport of ions in these channels is stabilized by polar interactions with surrounding proteins, which is a feature that the nanotube lacks. Therefore, fundamental questions have to be answered regarding transport of ions in an electrolytic solution through nanotubes, and in this section we summarize some results.

We have performed MD simulations using a system consisting of the carbon nanotube, water, and ions using GROMACS (Lindahl et al., 2001) at a constant pressure of 1 bar and a constant temperature of 300 K. In some examples discussed below, a slab and functional (charge) groups are also used. The slab consisted of neutral atoms that mimic the interior of a hydrophobic phospholipid bilayer. The box size varied from 33 Å to 75 Å depending on the length and diameter of the tube and was at least twice as long as the tube length. The tube was fixed at the center of the box when there was no slab, but had limited freedom of motion when inserted into a slab that was held stationary. Periodic boundary conditions were applied in all three directions. The parameters for Lennard–Jones potentials and bonded interaction were taken from a GROMACS force field with a carbon–carbon bond length of 1.42 Å. Values for the Lennard–Jones interaction parameters for the nanotube (C), and water oxygen (O) were:

- $\sigma_{\rm CC} = 3.36$ Å, $\epsilon_{\rm CC} = 0.0969$ kcal/mol,
- $\sigma_{\rm OC} = 3.23$ Å, $\epsilon_{\rm OC} = 0.129$ kcal/mol,

and were based on a cutoff of 15 Å. Bulk SPC/E water, having atomic charges of $-0.8476e$ and $0.4238e$ on oxygen and hydrogen, respectively, was equilibrated (300 K) and used to fill the nanotube interior. Electrostatic interactions used the particle-mesh Ewald method with a 10-Å real-space cutoff, a 1.5-Å reciprocal space gridding, and splines of order 4 with a 10^{-5} tolerance (see Section 16.1 for more details on MD simulations). Depending on the box size, 1033 to 2970 water molecules were used to yield a constant density of 1 g/cm^3. Concentrations of KCl varied from 1 M to 1.85 M. The slab consisted of pseudoatoms with the same Lennard–Jones parameters as those of a lipid bilayer. The partial charges for NH_3^+ and COO^- were taken from the side groups of lysine and glutamic acid respectively in the GROMACS amino acid database. The simulation time varied between 2 ns and 7 ns, and a time step of 2 fs was used in a leapfrog scheme.

Even though both water and molten ions seem to enter the tube spontaneously, when both ions and water molecules are involved, the situation could be more complex because of the possible interplay between hydrophilic interactions involving the ions and water and the hydrophobic interactions involving the water and the carbon nanotube. Four sets of simulations were performed, comprising:

1. Carbon nanotubes fixed in the center of a box in a solution of KCl,
2. Tubes fixed in water with artificial charges at the ends,
3. Asymmetrically functionalized tube in a pseudo bilayer (opposite charge groups at either ends), and
4. Symmetrically functionalized tube (same charge groups at the ends) in a pseudo bilayer.

The results of the above simulations are summarized below.

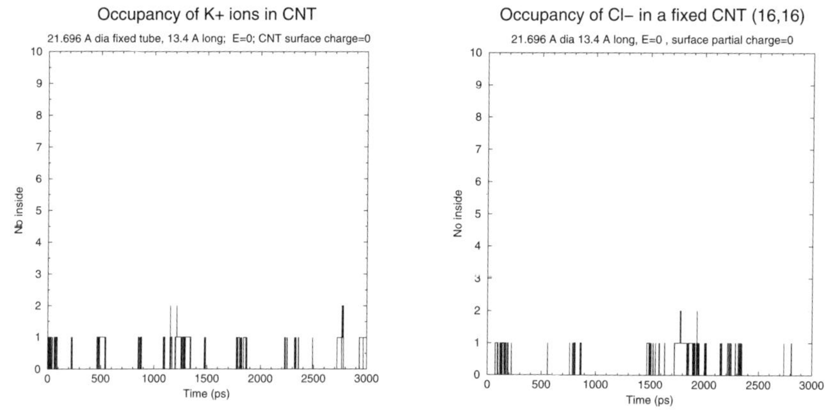

FIGURE 13.11. Very low ion occupancy in a (16, 16) carbon nanotube (13.4 Å long, 21.696 Å diameter) fixed in the center of the box with a solution of 1.85 M KCl. There is neither an external electric field nor partial charges on the rim.

The ion occupancy (or the number of ions) inside uncapped carbon nanotubes of various diameters ranging from (6, 6) to (16, 16) tubes in a solution of 1.85 M KCl was tested. The tubes were fixed in the box center and were of length 13.4 Å. For a simulation with a tube of 21.7 Å diameter, the box had 925 water molecules and 40 potassium and chloride ions each. Initially, the tube had 7 potassium and 6 chloride ions. Over the course of 3 ns, the ion occupancy was observed to be largely 0 or 1, as shown in Figure 13.11. The ions enter the tube from either end, but they do not travel across the length of the tube.

13.2.4 Transport Through Nanotubes with Charges at the Ends

To increase the occupancy of ions in the tube, partial charges of $\pm 0.38e$ were placed at atoms on the rim of the tube to create a dipole, the positive charges being on the top rim and the negative charges on the bottom rim. Nonequilibrium molecular dynamics simulations (NEMD) were performed with the tube fixed in the center of a box of length 33 Å and an applied external electric field of 0.015 V/nm, which is of comparable order to that of some ion channel membrane potentials, along the axial direction to mimic the membrane potential. It was observed that the ion occupancy is much higher compared to the ion occupancy in nanotubes with no charges (see Figure 13.12). Figure 13.13 shows how the ion occupancy varies with the diameter of the tube. An interesting observation is that the average occupancy of chloride ions is higher than that of the potassium ions with a ratio of 3:2. This could be due to the difference in water structure around the

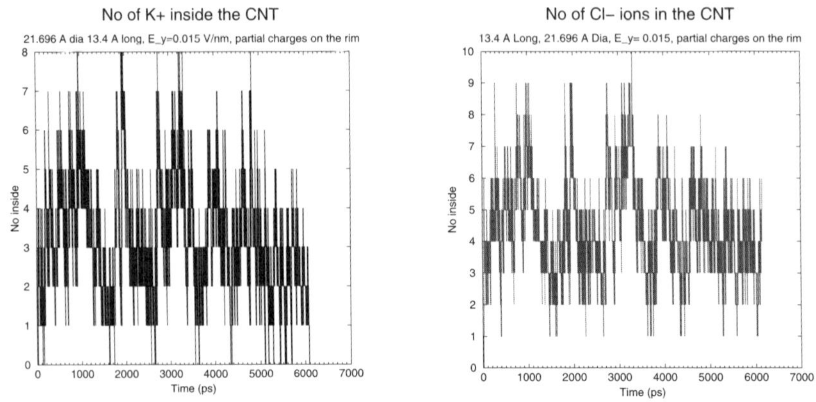

FIGURE 13.12. Ion occupancy in a (16, 16) carbon nanotube (13.4 Å long, 21.696 Å diameter) fixed in a solution of 1.85 M KCl with external electric field of $E = 0.015$ V/nm and partial charges of $\pm 0.38e$ on the rim atoms.

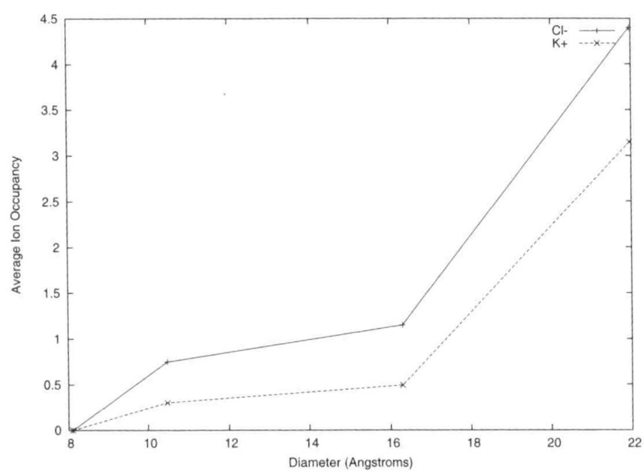

FIGURE 13.13. Variation of average ion occupancy with diameter in CNT (13.4 Å long) fixed in 1.85 M KCl with an external applied electric field of 0.015 V/nm.

two ions as well as the van der Waals interaction between the ions and the nanotube.

13.2.5 Transport Through Functionalized Nanotubes

Even though an electric field alone would drive ions into the tube, the partial charge on the rim increases the sensitivity and could be used to control the type and rate of ionic flow into the tube. This principle is the basis of tube end functionalization. Once partial charges were shown to increase occupancy, the next step was to replace them with functional groups. NH_3^+ and COO^- were used as the functional groups. Both outer wall and end wall chemical functionalization attachments of nanotubes have been successfully realized in experiments (Chen et al., 2001; Chen et al., 1998; Halicioglu and Jaffe, 2002). Functional group attachment can increase solubility and alter nanotube properties, among other things (Sinnott, 2002). In our simulations, an asymmetric functionalization of carboxylate and amino residues was used in place of the partial charges to mimic a real ion channel at either end. Even though functionalizing the inner wall would mimic some ion channels more closely, end wall functionalization is more feasible than inner wall functionalization for tubes of small diameter (Chen et al., 1998). The functionalized carbon nanotube was then placed in a membrane-mimic with properties similar of those of a lipid bilayer with a surrounding bath of 1.5 M KCL solution (see Figure 13.14). An electric field of 0.15 V/nm was used to drive the ions through the tubes, and the trajectories of K^+ and Cl^- ions are shown in Figure 13.15. Over the course of 2 ns of simulation time, the chloride current was found to be much higher than the potassium ion current because K^+ ions are bound electrostatically to the COO^- groups at the mouth, causing an energy barrier, thereby reducing the K^+ occupancy in the tube. Thus, this decorated nanotube is an example of one possible nanotube functionalization that can lead to a modulation of the ionic current via ion selectivity.

Symmetric functionalization can be used to selectively transport one species. Figure 13.16 shows the trajectories of ions in a symmetrically functionalized (16, 16) tube with COO^- on both ends and in the middle with protonated carbon atoms. The chloride occupancy is much larger than that of potassium.

13.2.6 Anomalous Behavior

An interesting aspect of electrolytic transport through charged carbon nanotubes is the observation of anomalous behavior (Qiao and Aluru, 2003d). Specifically, we have performed molecular dynamics simulations of electroosmotic transport of NaCl solution through a 5.42 nm diameter single-walled carbon nanotube with different surface charge densities. We observed that if the carbon nanotube surface is negatively charged (charge density of -0.076 C/m^2), a significant amount of Na^+ ion is contact adsorbed on the nanotube wall and immobilized, resulting in negligible electroosmotic transport. However, if the carbon nanotube surface is positively charged (charge density of 0.076 C/m^2), the contact adsorption of the Cl^- ion is moderate, and the adsorbed Cl^- ions are not immobilized, thus generating a significant electroosmotic transport and a velocity slip on the carbon nanotube surface. The flow is driven by an external electric field E_{ext}, applied along the channel in the z-direction. Because of the extremely high thermal noise, a strong electric field was applied in our simulations so that the fluid

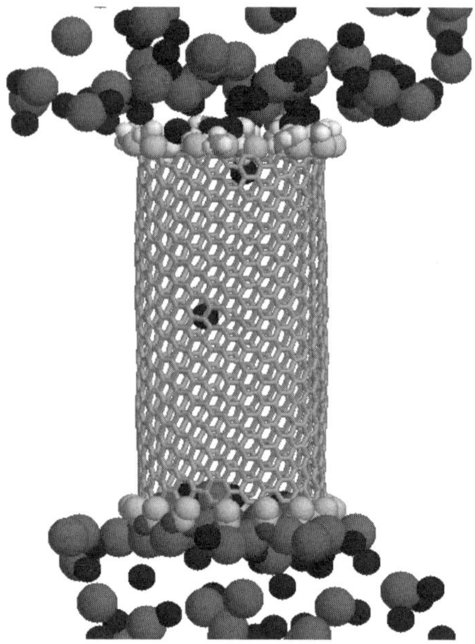

FIGURE 13.14. Functionalized carbon nanotube in a bath of 1.5 M KCl. CH_2-NH_3^+ is attached one per carbon ring to the top, and CH_2-COO^- is attached at the bottom. The tube is inserted in a slab (not shown). Chloride ions are shown in the lighter shade. Water molecules are also not shown. Chloride occupancy is higher than potassium. A buildup of potassium ions near the COO^- is also observed.

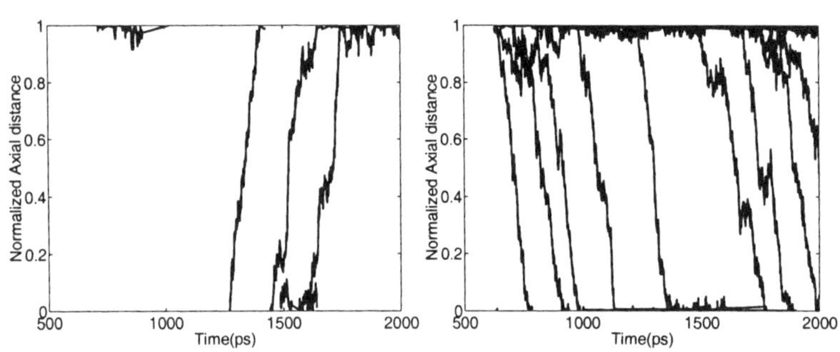

FIGURE 13.15. Positions of individual ions (K^+ on the left and Cl^- on the right) inside an end wall functionalized nanotube along the nanotube axis in a (16, 16) tube fixed in a slab in a solution of 1.5 M KCl. The axial distance is scaled, with a unit of 1 being the length of the tube. K^+ enters from the bottom and Cl^- enters from the top. There is an external electric field of 0.15 V/nm. The rate of chloride ion passage is higher than that of K^+ ion passage, indicating a selectivity of anions over cations.

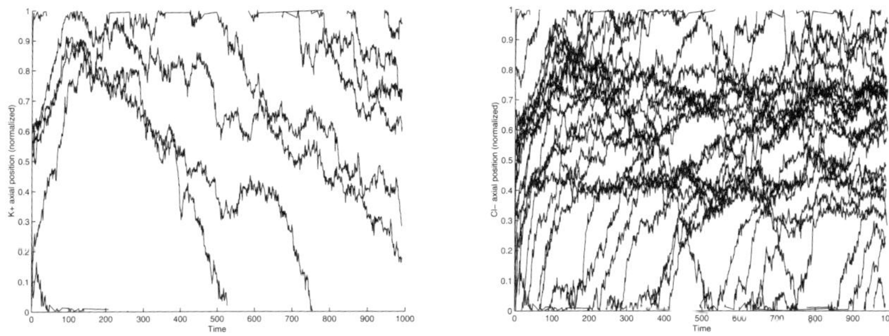

FIGURE 13.16. K^+ trajectory (left) and Cl^- trajectory (right) in a symmetrically functionalized tube with COO^- groups on both ends and protonated carbon atoms in the center of the $(16, 16)$ tube. The axial distance is scaled such that the length of the tube corresponds to a unit of 1. The chloride current is much higher than that of potassium current wherein there is asymmetric functionalization because K^+ ions are electrostatically bound to the COO^- group at the mouth of the tube. The chloride ion also has to clear a barrier created by protonated carbon atoms in the center to reach the other end.

velocity could be retrieved with reasonable accuracy. External electric field strengths of -0.2 V/nm and 0.2 V/nm are used for positive and negative surface charge densities, respectively.

Figure 13.17 shows the water and ion concentration profile across the channel for a positively charged CNT surface. A significant layering of water, as indicated by the strong fluctuation of the water concentration, is observed near the channel wall. The Cl^- ion concentration is maximum at a position very close to the channel wall ($r = 2.34$ nm), and decreases toward the channel center. However, such a decrease is not monotonic, and we observe a weak second concentration peak at $r = 2.05$ nm. It is also observed that the Na^+ ion concentration does not decrease monotonically toward the channel wall, but has a peak located at $r = 1.93$ nm. Both the second peak of Cl^- ion and the peak of Na^+ ion are very close to the second concentration valley ($r = 1.94$ nm) of the water molecules.

Figure 13.18 shows the water and ion concentration profile across the channel for a negatively charged CNT surface. The counterion (Na^+ ion) concentration distribution is significantly different from that observed in the positively charged case. Specifically, we observe that:

1. The location of the counterion adsorption peak is closer to the channel wall than the first water concentration peak, and

2. The peak concentration of the counterion is much higher than the value observed in the positively charged surface.

Both observations are caused by the smaller size of Na^+ ion compared to that of the Cl^- ion, i.e., the bare radius of a Na^+ ion is smaller than that

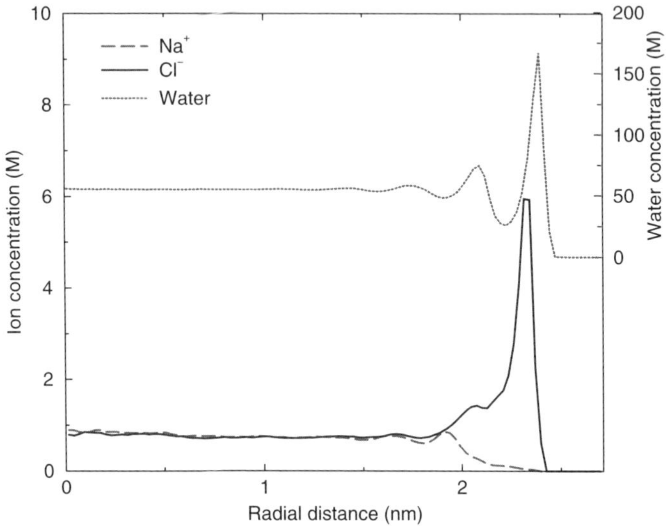

FIGURE 13.17. Water and ion concentration distribution across the channel for a positively charged CNT surface.

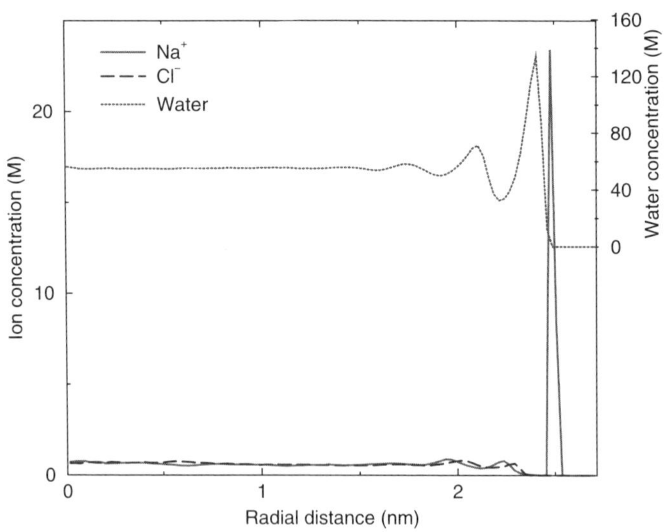

FIGURE 13.18. Water and ion concentration distribution across the channel for a negatively charged CNT surface.

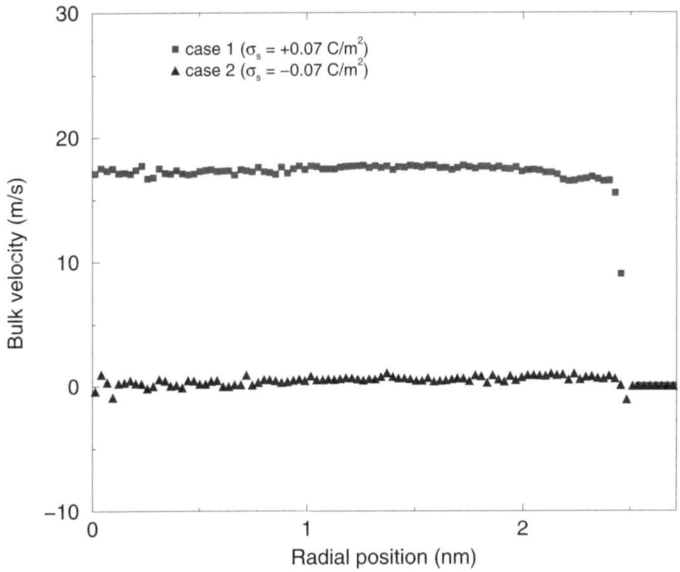

FIGURE 13.19. Bulk water velocity profile across the channel for a positively charged surface (shown as filled squares) and a negatively charged surface (shown as filled triangles).

of a Cl^- ion and a water molecule (the radii of the Na^+ ion, Cl^- ion, and a water molecule are 0.095, 0.181, and 0.14 nm, respectively); hence, the Na^+ ion can approach the channel wall more closely compared to a Cl^- ion or a water molecule.

Figure 13.19 shows the bulk water velocity for positively and negatively charged CNT surfaces. We observe that for a positively charged surface:

1. The velocity profile is flat in the central portion of the channel,

2. A significant velocity slip is observed near the channel wall, and

3. The velocity for a negatively charged surface is very small, indicating that the bulk transport is negligible in this case.

Figure 13.20 (a) shows the typical trajectory of 8 Cl^- ions during a 60 ps period. Note that the starting z-position of all 8 ions is shifted to $z = 0$ nm. We observe that the contact adsorbed Cl^- ions are not immobilized, since the Cl^- charged surface atom interactions are not very strong. Figure 13.20 (b) shows the typical trajectory of 8 Na^+ ions during a 120 ps period. We observe that the contact adsorbed Na^+ ions are immobilized because of the stronger Na^+ charged surface atom interactions. Since the movement of Na^+ ions is negligible, the driving force on the fluid is negligible. Thus,

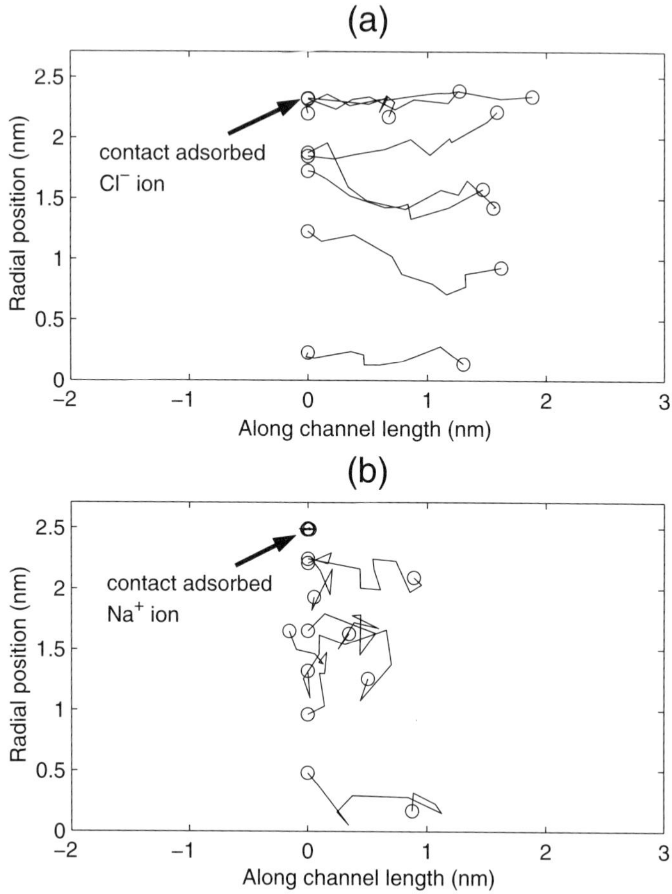

FIGURE 13.20. (a) Typical trajectory of Cl^- ions for a positively charged surface during a time period of 60 ps. (b) Typical trajectory of Na^+ ions for a negatively charged surface during a time period of 120 ps. The starting z−position of all ions are shifted to $z = 0$ nm.

the bulk fluid velocity for a negatively charged surface is much smaller than that observed for a positively charged surface.

The observed dependence of electroosmotic transport on the surface charge is significantly different from the results obtained using the conventional continuum theories, which predict that the electroosmotic transport will simply reverse its direction if the surface charge density is flipped. This anomalous dependence of electroosmotic transport on the surface charge is mainly caused by the different adsorption behavior of the counterions, which depends on the size of the counterion, the local electrostatic interactions between ion-water and ion-charged surface atoms, and on the external electric field.

In summary, by investigating the behavior of water and ions (or electrolytes) and their interactions in carbon nanotubes and functionalized carbon nanotubes and other nanopores, it is possible to create nanofluidic devices that open new opportunities for sensing, detection, and probing matter at nanoscales.

Part III: Simulation Techniques

14

Numerical Methods for Continuum Simulation

Full-system simulation of microsystems typically involves simulations of coupled electrical, mechanical, thermal, and fluid domains. Even within the fluid domain only, in applications such as multiphase microflows, different subdomains are required to handle the stationary and moving components. It is clear that in order to reduce computational complexity, the numerical discretizations employed should be both highly efficient as well as robust. The significant geometric complexity of flows in microsystems suggests that finite elements and boundary elements are more suitable than finite differences for efficient discretization. Because of the nonlinear effects, either through convection or boundary conditions, boundary element methods are also limited in their application range despite their efficiency for linear flows. However, they have been used routinely for efficient computation of the electrostatics. A particularly promising approach for microflows makes use of meshless techniques, where particles are "sprinkled" almost randomly into the flow and boundary. This approach handles the geometric complexity of microflows effectively, but the issues of accuracy and efficiency have not yet been fully resolved. As regards nonlinearities, one may argue that at such low Reynolds numbers the convection effects should be neglected, but in complex geometries with abrupt turns the convective acceleration terms may be substantial, and thus they need to be taken into account. The same is true for particulate microflows, where nonlinear effects are important, and thus Stokesian dynamics – a very effective simulation approach – may be limited for simulation of biofluidic applications, since it can simulate only zero Reynolds number flows. Also, flows in micronozzles and other aerospace applications may result in large Reynolds numbers, exceeding one thousand!

In this chapter we present three main numerical methodologies to analyze flows in microdomains:

- High-order finite-element (spectral element) methods for Navier–Stokes equations. Formulations for both incompressible and compressible flows in stationary and moving domains are presented.
- Meshless methods with random point distribution.
- The force coupling method for particulate microflows.

These methods represent three different classes of discretization philosophies. They have been used with success in diverse applications of microsystems, from microfilters, valves, and mixers to self-assembly processes. Clearly, any other discretization method based on finite-differences, finite-elements or finite-volumes can also be used.

14.1 Spectral Element Method: The μFlow Program

In this section we present the spectral element method (Karniadakis and Sherwin, 1999) implemented in the program μ**Flow**, which was used in many examples included in this book. A summary of the capabilities of this program or gas flows in different regimes is given in Table 14.1. This is just an example of how a continuum-based approach can be employed to simulate microflows, and thus the spectral element method can be replaced by finite elements, finite volumes, or finite differences. Most of the ideas we present next apply also to these other discretizations. The specific capabilities and issues we cover in this section are to be used as reference in designing a similar program using other continuum-based discretizations.

In gas microflow simulations both the incompressible and the compressible Navier–Stokes equations can be employed to compute the relative effects of compressibility and rarefaction. Strictly speaking, from the theoretical point of view, there is an inconsistency in using the incompressible form of the Navier–Stokes equations with the slip boundary condition (Aoki, 2001). In practice and for very small Reynolds numbers, the limits in Knudsen number for the incompressible models are dictated by the physics of the problem as well as by numerical stability considerations. The first incompressible version of μ**Flow** (I1 in Table 14.1) solves the Navier–Stokes equations as well as the energy equation. It employs the first-order slip (equation (2.19)) and temperature jump (equation (2.20b)) and thermal creep boundary conditions. It is general for two- and three-dimensional flows. Explicit (in time) implementation of the boundary conditions results in a Knudsen number limit of typically $\mathrm{Kn} < 0.1$. The second version of

TABLE 14.1. Gas flow models and boundary conditions implemented in μ**Flow**. This program is used in many examples in this book, and it can be replaced by other equivalent discretizations.

Flow	B.C.	2-D/3-D	V. Slip	T. Creep	T. Jump	Kn limit
I1	$\mathcal{O}$(Kn, Kn^2)	2-D/3-D	Yes	Yes	Yes	< 0.1
I2	$\mathcal{O}$(Kn^n)	2-D	Yes	assigned	No	< 0.5
C	$\mathcal{O}$(Kn^n)	2-D/Axi-Sym	Yes	Yes	Yes	< 0.5

incompressible μ**Flow** (I2 in Table 14.1) employs the slip boundary condition (equation (2.26)). It is stable for high Kn flows, and applicability is restricted with the flow geometry and the validity of the slip flow model. It does not solve the energy equation, and therefore thermal creep effects are imposed explicitly by a prescribed tangential temperature gradient. In the compressible version of μ**Flow** (C in Table 14.1) the general high-order slip boundary condition (equation (2.26)) and high-order temperature jump boundary condition (equation (2.31)) are used. Their limitations are based on the limitations of the slip flow theory. This version of the program is restricted to shock-free flows; therefore, it is used for subsonic and shock-free transonic flows.

The spatial discretization of μ**Flow** is based on the spectral element method, which is similar to the *hp* version of finite element methods (Karniadakis and Sherwin, 1999). Typical meshes for simulation in a rough microchannel with different types of roughness (presented in Section 4.1.4) are shown in Figure 14.1. The two-dimensional domain is broken up into elements similar to finite elements, but each element employs high-order interpolants based on Legendre polynomials. The N points that determine the interpolant correspond to locations of maxima of the Legendre polynomials and include the end-points. For smooth solutions, the spatial discretization error decays exponentially to zero (spectral or p-convergence). This means that if we *double* N, the error will decay by *two orders* of magnitude. This is a much faster decay than in standard second-order methods, which yield an error reduction by a factor of only four. In microflow simulations, the spectral element method can be used efficiently by exploiting the *dual path of convergence* allowed by the method. For example, in regions of geometric complexity a finite element-like discretization (low N and small-size elements) can be employed, whereas in regions of homogeneous geometry with steep gradients a spectral-like discretization (high N and large-size elements) can be employed. In particular, the boundary conditions for microflows, either for gases (Knudsen effects) or liquids (electrokinetic effects) can be resolved very accurately. On the other hand, the computational cost of the spectral element method is $\mathcal{O}(KN^{d+1})$, where $d = 2$ and $d = 3$ for two-dimensional and three-dimensional discretizations, respectively, and K is the number of elements. This is higher compared

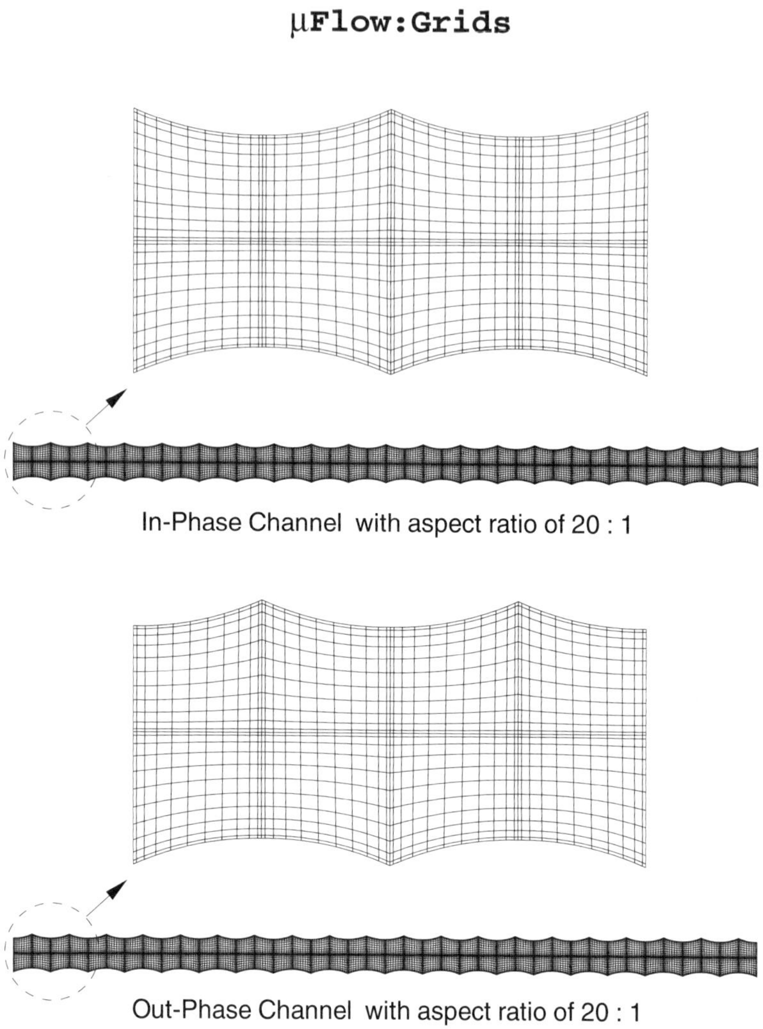

FIGURE 14.1. Two-dimensional meshes used in μ**Flow** simulations of flow in a rough microchannel of aspect ratio 20 : 1. The top plot shows *in-phase* roughness, and the lower plot shows *out-of-phase* roughness. The domain is broken up into large elements, and in each element a spectral expansion defined by Gauss–Lobatto–Legendre (GLL) points (see detail) is used to represent all fields and data (including the geometry). Here $N = 12$ GLL points are used in each direction.

to the corresponding cost for standard finite element methods, but for the same accuracy the spectral element method is more efficient.

Our objective here is to focus on the following two issues, which are very important for microflows, instead of presenting complete discretization details; many different algorithms can be found in (Karniadakis and Sherwin, 1999) for general incompressible and compressible flows. Specifically:

- For incompressible microflows, which are viscous-dominated, we present the correct **pressure boundary condition** to supplement the consistent Poisson equation for the pressure.

- For compressible flows we present the **characteristic treatment of boundary conditions**, which guarantee stability and accuracy.

The efficiency of the overall algorithm is based on time-splitting of the advection and diffusion operators, which are treated with a collocation formulation and Galerkin projection, respectively. The time-splitting is demonstrated here for the one-dimensional linear advection–diffusion equation

$$U_t + VU_x = \alpha U_{xx}, \tag{14.1}$$

where V is the constant advection velocity. Splitting the advection and the diffusion terms and discretizing the time derivative, we obtain

$$\frac{\hat{U} - U^n}{\Delta t} = -\frac{23}{12}(VU_x)^n + \frac{16}{12}(VU_x)^{n-1} - \frac{5}{12}(VU_x)^{n-2}, \tag{14.2a}$$

$$\frac{U^{n+1} - \hat{U}}{\Delta t} = \alpha U_{xx}^{n+1}, \tag{14.2b}$$

where Δt is the time step, U^n denotes the value of U at time t, and U^{n+1} denotes the value of U at time $t + \Delta t$. The intermediate (predicted) state is denoted by $\hat{U}$. Therefore, this splitting scheme corresponds to obtaining a predicted field ($\hat{U}$) first by advection, and then correcting it with diffusion. For enhanced accuracy and stability of the advection step we have used a third-order Adams–Bashforth scheme. For the diffusion step we have used the Euler backward scheme. This can easily be extended to higher-order time-accurate schemes.

This time-splitting procedure accommodates the implementation of the elemental interface connectivity conditions properly and efficiently. Specifically, the elemental interface connectivity conditions are handled by a characteristic decomposition for the advection substep, and a *direct stiffness summation* for the diffusion substep. The characteristic treatment is stable for elemental interfaces and inflow/outflow conditions for hyperbolic problems (such as the Euler equations), and the direct stiffness summation procedure is widely used for elemental interface conditions in parabolic problems.

It has been demonstrated in earlier chapters that the majority of microflows (with the exception of micronozzles, Section 6.6) are in the low Reynolds number regime ($\mathrm{Re} \leq \mathcal{O}(1)$) due to the small characteristic dimensions. In some cases, the inertial terms in the Navier–Stokes equations can be neglected, but the continuity equation for density must still be solved in the characteristic form. Thus, use of time-splitting is inevitable for simulating microflows with our algorithm. For creeping Stokes flow, with the velocity rather than the characteristic dimension being small, the Mach number becomes very small, and use of compressible algorithms becomes inefficient. This is due to the fact that the wave speeds are dominated by the speed of sound c_s, and $u \ll c_s$. In such a case we use the incompressible version of μ**Flow**.

14.1.1 Incompressible Flows

The most efficient way of solving the incompressible (unsteady) Navier–Stokes equations is based on the fractional step method; see Chapter 8 in (Karniadakis and Sherwin, 1999). It is based on the projection of the velocity field obtained from the momentum equation into a divergence-free space. The latter involves the pressure equation and corresponding boundary conditions. However, the fractional step method was first proposed for *high Reynolds number* flows, and therefore it should be corrected for strong viscous effects in microflows. To this end, a *consistent pressure boundary condition* should be employed, as we demonstrate next. We follow here the work of (Karniadakis et al., 1991), where the correct pressure boundary condition was employed, leading to a consistent Poisson equation with a proper Neumann condition for the pressure. In two substeps the time-discrete scheme is as follows: First we solve

$$\frac{\hat{\mathbf{v}} - \boldsymbol{v}^n}{\Delta t} + (\boldsymbol{v}^{n+1/2} \cdot \nabla)\boldsymbol{v}^{n+1/2} = \frac{1}{2}\nu\nabla^2(\hat{\mathbf{v}} + \boldsymbol{v}^n), \tag{14.3a}$$

$$\hat{\mathbf{v}} = 0 \quad \text{on } \partial\Omega, \tag{14.3b}$$

where

$$\boldsymbol{v}^{n+1/2} \equiv 23/12\boldsymbol{v}^n - 16/12\boldsymbol{v}^{n-1} + 5/12\boldsymbol{v}^{n-1},$$

corresponding to a third-order Adams–Bashforth scheme. In the second substep we enforce

$$\begin{aligned} \frac{\boldsymbol{v}^{n+1} - \hat{\mathbf{v}}}{\Delta t} &= -\nabla p^{n+1}, \\ \nabla \cdot \boldsymbol{v}^{n+1} &= 0, \\ \frac{\partial p^{n+1}}{\partial n} &= -\nu \boldsymbol{n} \cdot (\nabla \times \omega)^{n+1} - \boldsymbol{n} \cdot \hat{\mathbf{v}} \quad \text{on } \partial\Omega. \end{aligned} \tag{14.4}$$

Note that in the classical fractional step we have

$$\frac{\partial p^{n+1}}{\partial n} = 0,$$

which is an inviscid-type boundary condition. This can be used for high Reynolds number flows but not for microflows. We assume here that the no-slip and no-penetration conditions are valid. Otherwise, the pressure boundary condition also needs to be modified, e.g., in electroosmotic flow with slip. Assuming now Stokes flow for simplicity, the rotational form of the boundary condition for the pressure in equation (14.4) is equivalent to the Laplacian form of the boundary condition

$$\frac{\partial p^{n+1}}{\partial n} = \nu \boldsymbol{n} \cdot \nabla^2 \boldsymbol{v}^{n+1} \quad \text{on } \partial\Omega. \tag{14.5}$$

The rotational form, unlike the Laplacian form, satisfies the compatibility condition (Karniadakis et al., 1991), and it also reinforces the incompressibility condition, since

$$\nabla^2 \boldsymbol{v} = \nabla(\nabla \cdot \boldsymbol{v}) - \nabla \times \omega.$$

In addition, it leads to a stable approximation, since the magnitude of the boundary-divergence is directly controlled by the time step.

To illustrate the differences between the rotational (equation (14.4)) and Laplacian (equation (14.5)) forms of the pressure boundary condition, we consider the exact boundary condition at time step $(n+1)\Delta t$, i.e.,

$$\frac{\partial p^{n+1}}{\partial n} = \nu \left[\frac{\partial Q^{n+1}}{\partial n} - \omega_s^{n+1} \right],$$

where we have introduced $\omega_s = \boldsymbol{n} \cdot \nabla \times \omega$ and $Q = \nabla \cdot \boldsymbol{v}$. We can now expand ω_s in a Taylor series to obtain

$$\frac{\partial Q^{n+1}}{\partial n} = \frac{1}{\nu} \frac{\partial p^{n+1}}{\partial n} + \omega_s^n + \Delta t \frac{\partial \omega_s^n}{\partial t} + \cdots .$$

Inserting the Laplacian form (equation (14.5)) in the above equation, we obtain

$$\frac{\partial Q^{n+1}}{\partial n} \propto \frac{\partial Q^n}{\partial n} + \Delta t \frac{\partial \omega_s^n}{\partial t},$$

which shows an accumulation of divergence flux at the boundary at every time step and therefore the possibility for instability. In contrast, if the rotational form in equation (14.4) is used, we obtain

$$\frac{\partial Q^{n+1}}{\partial n} \propto \Delta t \frac{\partial \omega_s^n}{\partial t},$$

and therefore the magnitude of the *boundary-divergence flux* is controlled directly by the time step. We can reduce the boundary-divergence errors further by using a multistep approximation to represent the right-hand side in equation (14.4):

$$\frac{\partial p^{n+1}}{\partial n} = -\nu \boldsymbol{n} \cdot \sum_{q=0}^{J_p \quad 1} \beta_q (\nabla \times \omega)^{n-q},$$

where J_p is the number of previous steps from which information is used. Following the above argument we find that

$$\frac{\partial Q^{n+1}}{\partial n} \propto (\Delta t)^{J_p},$$

and therefore the boundary-divergence flux can be made arbitrarily small by controlling the time step Δt. Note that for the inviscid pressure boundary condition

$$\frac{\partial p^{n+1}}{\partial n} = 0,$$

the boundary-divergence flux is $\mathcal{O}(1)$, independent of the size of the time step Δt.

To relate the boundary-divergence to the overall accuracy of the velocity field we consider the equation that the divergence $Q \equiv Q^{n+1}$ satisfies, i.e.,

$$\frac{Q}{\Delta t} - \gamma_0 \nu \nabla^2 Q = 0,$$

where we set the right-hand side to zero, since the pressure satisfies a consistent Poisson equation, and the divergence at previous time steps $(Q^n, Q^{n-1}, \ldots)$ is assumed zero; γ_0 is a coefficient due to time-discretization (Karniadakis et al., 1991). It is clear that there exists a numerical boundary layer of thickness $\delta = \sqrt{\gamma_0 \nu \Delta t}$, so that $Q = Q_w e^{-s/\delta}$, and thus the boundary-divergence is $Q_w = -\delta(\partial Q/\partial n)_w$. (Here s is a general coordinate normal to the boundary). Similarly, from order of magnitude analysis we have $Q_w = \mathcal{O}(\partial v/\partial n)$, and therefore

$$v \propto Q_w \delta \propto \left(\frac{\partial Q}{\partial n}\right)_w \gamma_0 \nu \Delta t.$$

This relation shows that the time-differencing error of the velocity field is an order smaller in Δt than the corresponding error in the boundary-divergence. For the inviscid-type pressure boundary condition, we are therefore limited to first-order accuracy, since the boundary-divergence flux is $\mathcal{O}(1)$. However, in general, we obtain

$$v \propto (\Delta t)^{J_p+1}$$

if a high-order time integration scheme is used to advance the velocity field. In numerical experiments, it was found in (Karniadakis et al., 1991) that with $J_p = 2$ we obtain a third-order-accurate velocity field. Note that the boundary-divergence scales as

$$Q_w \propto \sqrt{\nu}(\Delta t)^{J_p}.$$

These heuristic arguments have been more rigorously documented in (Karniadakis et al., 1991) and confirmed by numerical results. To demonstrate the effect of the incorrect inviscid pressure boundary condition versus the correct rotational form in the boundary condition, we consider a

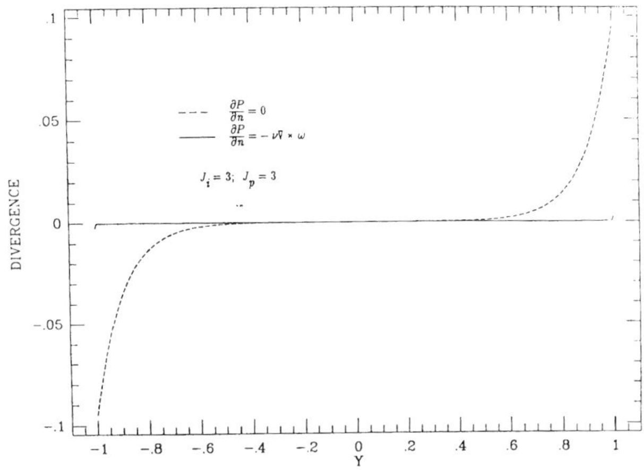

FIGURE 14.2. Divergence of velocity field across the channel for a Stokes flow. $\Delta t = 10^{-2}$. The spatial discretization is based on 20 spectral elements of order 10, which eliminates any spatial errors (Tomboulides et al., 1989). The inviscid-type boundary condition leads to large divergence errors at the boundaries.

decaying Stokes channel flow subject to compatible initial conditions (Karniadakis and Sherwin, 1999). In Figure 14.2 we plot the divergence of the velocity field across the channel. It is seen that incorporation of the rotational form of the pressure boundary condition almost eliminates the artificial boundary layer. High-order treatment produces smaller boundary divergence errors consistent with the aforementioned arguments.

Remark: The above pressure boundary condition assumes that the *no-penetration* condition at the wall holds. However, if the surface moves perpendicular to its plane with velocity $\boldsymbol{v}_w \cdot \boldsymbol{n}$, e.g., squeezed film applications in Section 6.1, then a time-dependent term is present, i.e.,

$$\frac{\partial p}{\partial n} = -\nu \boldsymbol{n} \cdot (\nabla \times \omega) - \frac{\partial \boldsymbol{v}_w \cdot \boldsymbol{n}}{\partial t} \quad \text{on } \partial\Omega.$$

If the *slip condition* is applied and the nonlinear effects are nonnegligible, then this term should also be added, appropriately, on the right-hand side of the above equation. In addition, the boundary condition in equation (14.3b) should be corrected appropriately.

14.1.2 Compressible Flows

We present here the two-dimensional formulation in order to discuss some important issues associated with the interface and boundary conditions. Both Galerkin and discontinuous Galerkin projections can be employed, but here for simplicity we present the standard Galerkin approach. For an introduction to discontinuous Galerkin methods with emphasis on compressible flow simulations the interested reader can consult (Cockburn et al., 2000).

The compressible Navier–Stokes equations in nondimensional flux form are (see Section 2.2, equation (2.16))

$$\frac{\partial}{\partial t}\begin{pmatrix}\rho\\ \rho u\\ \rho v\\ E\end{pmatrix}+\frac{\partial}{\partial x}\begin{pmatrix}\rho u\\ \rho u^2+p\\ \rho uv\\ (E+p)\cdot u\end{pmatrix}+\frac{\partial}{\partial y}\begin{pmatrix}\rho v\\ \rho vu\\ \rho v^2+p\\ (E+p)\cdot v\end{pmatrix}$$

$$=\frac{1}{\mathrm{Re}}\frac{\partial}{\partial x}\begin{pmatrix}0\\ \frac{2}{3}\mu\,(2\frac{\partial u}{\partial x}-\frac{\partial v}{\partial y})\\ \mu\,(\frac{\partial u}{\partial y}+\frac{\partial v}{\partial x})\\ \frac{2}{3}\mu\,(2\frac{\partial u}{\partial x}-\frac{\partial v}{\partial y})\cdot u+\mu\,(\frac{\partial u}{\partial y}+\frac{\partial v}{\partial x})\cdot v+\frac{k\gamma}{Pr}\cdot\frac{\partial T}{\partial x}\end{pmatrix}$$

$$+\frac{1}{\mathrm{Re}}\frac{\partial}{\partial y}\begin{pmatrix}0\\ \mu\,(\frac{\partial u}{\partial y}+\frac{\partial v}{\partial x})\\ \frac{2}{3}\mu\,(2\frac{\partial v}{\partial y}-\frac{\partial u}{\partial x})\\ \frac{2}{3}\mu\,(2\frac{\partial v}{\partial y}-\frac{\partial u}{\partial x})\cdot v+\mu\,(\frac{\partial u}{\partial y}+\frac{\partial v}{\partial x})\cdot u+\frac{k\gamma}{Pr}\cdot\frac{\partial T}{\partial y}\end{pmatrix},$$

where k is the thermal conductivity and γ is the ratio of specific heats. The unknowns are the *conservative variables*, i.e.,

$$(\rho,\ \rho u,\ \rho v,\ E).$$

The energy is defined as

$$E=\rho[T+1/2(u^2+v^2)],$$

and the pressure p is obtained from the equation of state

$$p=(\gamma-1)\rho T.$$

The nondimensionalization is done with respect to reference velocity, density, and length scales (i.e., U_0, ρ_0, l_0), and the reference temperature is chosen as $T_0=U_0^2/C_v$; here, C_v is the constant-volume specific heat. The dynamic viscosity μ is related to temperature by *Sutherland's law*

$$\frac{\mu}{\mu_0}=\left(\frac{T}{T_0}\right)^{3/2}\frac{T_0+S_1}{T+S_1},$$

where μ_0 is the viscosity at the reference temperature T_0, and S_1 is a constant, which for air is

$$S_1=110\,\mathrm{K}.$$

It is convenient to simplify this equation to a simpler power law of the form

$$\frac{\mu}{\mu_0} = \left(\frac{T}{T_0}\right)^{\omega}, \quad \text{with} \quad 0.5 < \omega < 1.$$

Sutherland's law can be approximated at high temperature values from the equation above and $\omega \to 0.5$, whereas for low temperature values $\omega \to 1$.

Interface and Boundary Conditions

Following the time-splitting algorithm of equation (14.1), we first consider the inviscid (Euler) equations in order to present a proper treatment of interface conditions. Specifically, the Euler equations are solved by a spectral (Gauss–Lobatto–Legendre) collocation formulation (Karniadakis and Sherwin, 1999). The flow domain is divided into elements, where collocation discretization is applied in each element locally. This procedure brings up the issue of elemental interface treatment. Since the Euler equations are hyperbolic, simple averaging at elemental interfaces is inappropriate. Instead, a characteristic treatment is necessary, as we explain next.

Specifically, the interface problem is solved in three main steps:

- linearization of the Euler equations,
- characteristic decomposition, and
- characteristic treatment.

Let us consider the following one-dimensional system of nonlinear hyperbolic partial differential equations in conservative form:

$$\frac{\partial W}{\partial t} + \frac{\partial F(W)}{\partial x} = 0. \tag{14.6}$$

We first define the Jacobian A by

$$A(W) = \frac{\partial F}{\partial W},$$

and then we linearize the equations around some known state o, yielding the following form:

$$\frac{\partial W}{\partial t} + A(W_o)\frac{\partial W}{\partial x} = 0. \tag{14.7}$$

Here $A(W_o)$ is a constant $N \times N$ matrix (N is the dimension of the vector W). We then transform the linearized equations into the characteristic form

$$\frac{\partial R_i}{\partial t} + \lambda_i \frac{\partial R_i}{\partial x} = 0, \tag{14.8}$$

where λ_i are the eigenvalues of $A(W_o)$, and $R_i(W, W_o)$ are the characteristic variables. For Euler equations the eigenvalues and the corresponding characteristic variables are

$$\lambda_1 = \lambda_2 = \mathbf{u}_o \cdot \mathbf{n}, \quad \lambda_3 = \mathbf{u}_o \cdot \mathbf{n} - c_s, \quad \lambda_4 = \mathbf{u}_o \cdot \mathbf{n} + c_s,$$

$$\begin{aligned}
R_1 &= \rho \frac{c_s^2}{\gamma - 1} - \left(\frac{1}{2} \rho \mathbf{u}_o \cdot \mathbf{u}_o - \mathbf{m} \cdot \mathbf{u}_o + E \right), \\
R_2 &= (\rho u - \rho u_o) n_y - (\rho v - \rho v_o) n_x, \\
R_3 &= -(\mathbf{m} - \rho \mathbf{u}_o) \cdot \mathbf{n} + \left(\frac{\gamma - 1}{c_s} \right) \left(\frac{1}{2} \rho \mathbf{u}_o \cdot \mathbf{u}_o - \mathbf{m} \cdot \mathbf{u}_o + E \right), \\
R_4 &= (\mathbf{m} - \rho \mathbf{u}_o) \cdot \mathbf{n} + \left(\frac{\gamma - 1}{c_s} \right) \left(\frac{1}{2} \rho \mathbf{u}_o \cdot \mathbf{u}_o - \mathbf{m} \cdot \mathbf{u}_o + E \right).
\end{aligned}$$

Here c_s is the local speed of sound, $\mathbf{m}$ is the momentum flux vector ($\mathbf{m} = (\rho u, \rho v)$), E is the energy ($E = \rho T + 0.5\rho(u^2 + v^2)$), $\mathbf{u}_o$ is the *Roe-averaged* velocity vector, and $\mathbf{n}$ is the surface normal, with n_x and n_y the x and y components of the surface normal, respectively. The subscript o corresponds to the Roe-averaged states obtained by averaging two states (denoted by left, L, and right, R) as follows:

$$u_o = \frac{\sqrt{\rho_L} u_L + \sqrt{\rho_R} u_R}{\sqrt{\rho_L} + \sqrt{\rho_R}}; \quad v_o = \frac{\sqrt{\rho_L} v_L + \sqrt{\rho_R} v_R}{\sqrt{\rho_L} + \sqrt{\rho_R}}.$$

The characteristic treatment is performed as follows: Each elemental interface is treated by analyzing the sign of the eigenvalues, λ_i. We define the right (R) and the left (L) states at an interface depending on the local flow direction. We linearize the equations around the Roe-averaged states obtained from the right and left values. If the sign of the eigenvalue is positive, we choose the characteristic variables from the left calculated state, and if the sign of the eigenvalue is negative, we choose the characteristic variables from the right calculated state. Once the characteristic variables are obtained, we convert them back to the conservative variables and update the elemental interface values as follows:

$$\begin{aligned}
\rho &= \left(\frac{\gamma - 1}{c_s^2} \right) \left[R_1 + \frac{c}{\gamma - 1} \left(\frac{R_3 + R_4}{2} \right) \right], \\
\rho u &= \left(\frac{R_4 - R_3}{2} n_x + R_2 n_y \right) + (\rho u_o), \\
\rho v &= \left(\frac{R_4 - R_3}{2} n_y - R_2 n_x \right) + (\rho v_o), \\
E &= \left(\frac{c_s}{\gamma - 1} \right) \left(\frac{R_3 + R_4}{2} \right) - \frac{1}{2} \rho \mathbf{u}_o \cdot \mathbf{u}_o + (\rho u) u_o + (\rho v) v_o.
\end{aligned}$$

The above elemental interface treatment is one-dimensional and can be used in multiple dimensions by directional splitting. However, it cannot be used directly at element corners. To this end, we apply the treatment suggested in (Kopriva, 1991). The problem is divided into two one-dimensional problems. The corresponding Riemann invariants are

$$R^+ = u + \frac{2}{\gamma - 1} c_s; \quad R^- = u - \frac{2}{\gamma - 1} c_s,$$
$$S^+ = v + \frac{2}{\gamma - 1} c_s; \quad S^- = v - \frac{2}{\gamma - 1} c_s.$$

We assume locally isentropic flow in the neighborhood of the corner and obtain the entropy value of the corner from the element that the flow is leaving. We also define a domain of influence, and choose the calculated values of Riemann invariants from the corresponding elements, which lie in the domain of dependence. Figure 14.3 shows the domain of influence at the corners of four elements. For this specific example, since the flow is leaving element 1, we get the entropy and Riemann invariants of R^+ and S^+ from element 1. The Riemann invariants R^- and S^- are obtained from elements 2 and 4, respectively. The flow variables are calculated as follows:

$$u = \frac{1}{2}\left(R^+ + R^-\right); \quad v = \frac{1}{2}\left(S^+ + S^-\right),$$
$$c_s = \frac{\gamma - 1}{4}\left(R^+ - R^- + S^+ - S^-\right) - c_{\text{up}}; \quad s = s_{\text{up}},$$

where the subscript "up" denotes the upstream values (i.e., values from the element that the flow is leaving). Assuming ideal gas and locally isentropic flow, the state of the gas is fixed by solving the local pressure and density using speed of sound and the entropy

$$\frac{p}{\rho^\gamma} = \text{constant}; \quad c_s^2 = \gamma p/\rho,$$

and thus the conservative variables and local temperature can be calculated.

Next, we discuss *inflow* and *outflow* boundary conditions. The Euler equations require specification of three inflow and one outflow boundary condition for subsonic flow. However, these boundary conditions are not known ahead of time. Therefore, we select predicted inlet and exit stages, and perform again a characteristic decomposition. The only difference is that we select the left state as the specified inlet state and the right state as the calculated values at the inlet. Then, we perform the characteristic decomposition. Similarly at exit, the predicted state is specified as the right state, and the calculated state is chosen as the left state.

The wall boundary conditions must be designed to reflect incident pressure waves with high-order accuracy, resulting in minimal numerical entropy creation near the boundaries. One implementation is to impose zero

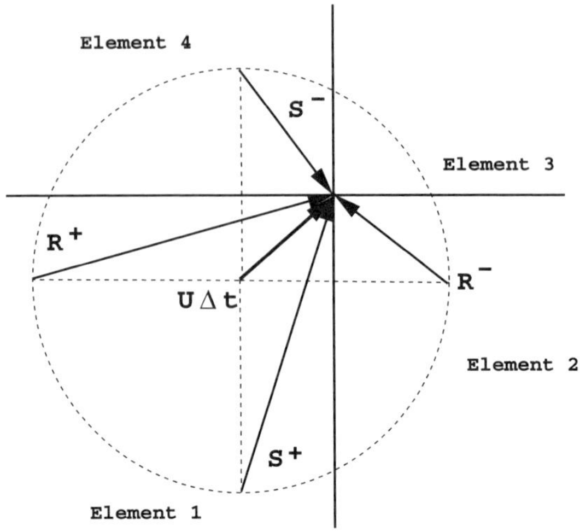

FIGURE 14.3. The domain of influence at a corner of four elements, and corresponding Riemann invariants.

normal velocity at the surface by modifying the local pressure to account for the changes in energy while the density remains unchanged. For a homentropic (i.e., constant entropy in the entire flow domain) inviscid flow, simulation with μ**Flow** resulted in change in the entropy near the wall by $\approx 10^{-6}$. This is an indication of the very low values of numerical diffusion of the spectral element discretization combined with the characteristic approach for the boundary conditions.

Turning now to the full compressible Navier–Stokes equations, we need to treat properly the *viscous fluxes* at elemental interfaces as well as the *no-slip* or *slip* condition. First, unlike the spatial discretization for the Euler equations, where we employ a spectral collocation approach, for the viscous contributions we employ a standard Galerkin projection (Karniadakis and Sherwin, 1999). Specifically, the viscous diffusion terms in the momentum equation are treated by the chain rule of differentiation, because the unknowns are the conservative variables, not the primitive variables, i.e., $(\rho u, \rho v, E)$, not (u, v, T). The Laplacian of temperature is treated by assuming that T is also an independent variable. This seems necessary; otherwise, the spatial derivatives of T must include spatial derivatives of terms obtained from the chain rule of differentiation of the energy relation $(E = \rho[T + 1/2(u^2 + v^2)])$. The elemental interface is treated by applying a *direct stiffness summation* procedure as in standard finite element methods (Karniadakis and Sherwin, 1999). Specifically, simple addition of all contributions at nodes at the interface is performed to ensure continuity of the variables.

No-slip and Dirichlet temperature boundary conditions are implemented at the walls. It is possible to specify Dirichlet boundary conditions at the inflow and outflow, which can be a function of time and space. Use of the characteristic treatment is essential for the stability of the inviscid part of the equations, and viscous boundary conditions must be designed to maintain minimal wave reflection from the outflow and inflow boundaries. At the outflow we let the pressure be infinitesimally smaller than the value calculated from the inviscid step. This treatment is usually enough to release the pressure waves with minimal reflection from the boundaries for the viscous substep.

For steady internal flows, specifying the back-pressure at the exit of the domains constitutes a difficulty, since the characteristic decomposition method is based on $\rho, (\rho u), (\rho v)$, and E. In the implementation of back-pressure we have used the calculated values of velocity and temperature at the exit of the channel. Then, the back-pressure is imposed indirectly by calculating the density corresponding to the calculated temperature and the desired back-pressure. This implementation results in uniform back-pressure with good accuracy and eliminates numerical boundary layers at the channel exit.

Implementation of Slip Boundary Conditions

The numerical implementation of equations (2.19) and (2.20b) is somewhat complicated due to the mixed-type (Robin) boundary conditions. An explicit (in time) implementation of equation (2.19) at time level $(n+1)\Delta t$ (neglecting for simplicity the temperature term) is as follows:

$$U_s^{n+1} - U_w = \sum_{i=0}^{J-1} \alpha_i \left[\frac{2-\sigma_v}{\sigma_v} (Kn) \frac{\partial U}{\partial n} \right]_s^{n-i} \quad + \quad \mathcal{O}(\Delta t^J), \qquad (14.9)$$

where α_i denotes the weights necessary to obtain the time-accuracy $\mathcal{O}(\Delta t^J)$ with Δt the time step. However, explicit treatment of boundary conditions is an extrapolation process, and thus it is numerically unstable, e.g., for relatively high values of Knudsen number.

We have determined through numerical experimentation that the overall Navier–Stokes solution with explicitly implemented velocity slip boundary conditions becomes unstable when the calculated slip amount $(U_s - U_w)$ at a certain time step is sufficiently large to cause a sudden change of the sign of wall vorticity in the next time step. This empirical finding can be readily justified by considering the following argument. For a linear Couette flow (see Chapter 3) with driving velocity U_0 and local gas velocity U_1 at a distance (Δy) away from the wall, it is possible to approximate (to first-order accuracy and for $\sigma_v = 1$) the velocity slip U_s as

$$U_s - U_0 = \mathrm{Kn}\, \frac{U_1 - U_0}{\Delta y}.$$

For no change in the sign of vorticity at the wall, we require that $(U_0 - U_1) > (U_0 - U_s) = -\mathrm{Kn}\,\frac{U_1 - U_0}{\Delta y}$; this is satisfied if $\Delta y > \mathrm{Kn}$ (in nondimensional form). This limit is a significant restriction, especially for spectral-based methods such as the one we use in our discretization, where collocation points are clustered very rapidly close to the boundaries. Therefore, spectral and other high-order methods that typically provide high-order accuracy are subject to numerical instabilities of this form.

We have found that the new boundary conditions (equations (2.26) and (2.31)) are numerically stable for relatively higher values of Knudsen number. Their applications are usually restricted by the flow geometry. For example, the limit of applicability of (2.26) and (2.31) in a channel flow is $\mathrm{Kn} = 0.5$. Since these boundary conditions obtain the slip information, one mean free path away from the surface, meaningful results are achieved when $\lambda < h/2$, where h is the channel thickness.

14.1.3 Verification Example: Resolution of the Electric Double Layer

As an example of how to verify resolution-independence with the spectral element discretization, we consider the numerical solution of the Poisson–Boltzmann equation (7.4) and the incompressible Navier–Stokes equations (7.12); see Section 7.1. The weak (variational) form of equation (7.4) is solved via a Galerkin projection. A Newton iteration strategy for a variable-coefficient Helmholtz equation is employed to treat the exponential nonlinearity in the following form:

$$\begin{aligned}\left[\nabla^2 - \alpha\beta\cosh(\alpha\,(\psi^*)^n)\right](\psi^*)^{n+1} = {} & \beta\sinh(\alpha\,(\psi^*)^n) \\ & -\alpha\beta\,(\psi^*)^n\cosh(\alpha\,(\psi^*)^n),\end{aligned}$$

where (n) denotes the iteration number. The solution from the previous iteration is used for evaluation of the nonlinear forcing function, and the resulting system is solved until the residual is reduced beyond a certain level (typically 10^{-13}).

The numerical solution of equation (7.4) is challenging due to the *exponential nonlinearity* associated with the hyperbolic sine function. In particular, for large values of α the nonlinear forcing increases rapidly for any value of β. Also, for very large values of β with $\alpha = 1$, similar difficulties exist. Accurate resolution of the problem requires high grid density within the EDL. A typical mesh for the $\alpha = 1$ and $\beta = 10{,}000$ case is presented in Figure 14.4 (right). It consists of 22 elements across the channel width, spaced in biased fashion with minimum width of $0.001h$ very near the walls. Once the mesh topology is fixed, the modal expansion order N is increased to resolve the problem further. For rectangular elements shown in Figure 14.4, $N = 2$ corresponds to a quadratic solution for ψ^*, typically employed in finite element discretizations.

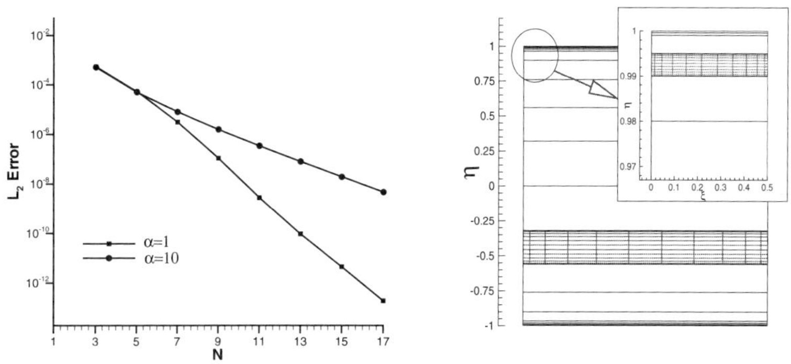

FIGURE 14.4. EDL resolution: Exponential decay of the L_2 error norm as a function of the spectral expansion order N (left). Sample grid used to resolve sharp electric double layer consists of 22 elements across the channel, and each element is discretized with Nth-order modal expansion per direction (right). The quadrature points for su-elemental discretization at select elements are also shown. Simulations are performed for $\beta = 10{,}000$.

The numerical accuracy of the results is determined by using equation (7.9). In Figure 14.4 (left) the variation of the L_2 error norm as a function of the modal expansion order N is presented. The results are obtained for the mesh topology shown on the right plot. Convergence results for $\alpha = 1$, $\beta = 10{,}000$ and $\alpha = 10$, $\beta = 10{,}000$ are shown. The L_2 error norm is defined as

$$L_2 = \frac{\left[\int_\Omega \mathcal{R}\left(\psi^*\right)^2 d\Omega\right]^{1/2}}{\int_\Omega d\Omega},$$

where Ω represents the entire flow domain. The residual of equation (7.9) is denoted by $\mathcal{R}$, and it is given by

$$\mathcal{R}(\psi^m) = \psi^m - \frac{4}{\alpha}\tanh^{-1}\left[\tanh\left(\frac{\alpha}{4}\right)\exp\left(-\sqrt{\alpha\beta}\ \ \eta^*\right)\right],$$

where the superscript m denotes the numerical results. The convergence results presented in Figure 14.4 show *exponential decay of the discretization error* with increased N, typical of the spectral/hp element methodology. This high-resolution capability enables accurate resolution of the electric double layers with substantially fewer elements compared to the low-order finite element discretizations. Figure 14.4 shows exponential convergence for both $\alpha = 1$ and $\alpha = 10$.

14.1.4 Moving Domains

The spectral element method that we described in the previous sections is a suitable method for simulations in moving domains, which are often encountered in microsystems, e.g., valves and mixers or other microactuators with moving parts. A robust treatment of the moving boundary requirement can be achieved by the Arbitrary Lagrangian Eulerian (ALE) formulation, where the arbitrary motion and acceleration of the moving domain can be handled independently of the fluid motion. The ALE method was developed in the early 1970s for fluid flow problems in arbitrarily moving domains (Hirt et al., 1974). Finite-element-based ALE formulations for incompressible viscous flows to study dynamic fluid structure interaction problems were developed by (Donea et al., 1982), and (Nomura and Hughes, 1992). Further advances in the ALE method, especially in improvement of the mesh velocities for moving boundaries, have been developed by (Lohner and Yang, 1996). The first spectral element ALE algorithm using quadrilateral spectral elements to study free surface flows was developed by (Ho, 1989). In the following we present an ALE algorithm for solving the two-dimensional incompressible Navier–Stokes and heat transfer/scalar transport equations in moving domains (Beskok and Warburton, 2001).

ALE Formulation

In this section, we consider domains that are arbitrarily moving in time. This is not a trivial generalization, because we have to discretize the time-dependent operators as well as the time-dependent fields. In the ALE formulation, the local elemental operators are formed at every time step. This is necessary in order to handle the mesh shape variations in time. The high-order ALE formulation implemented with spectral element discretizations exhibits the usual advantages of low dispersion and robustness to large deformations, as we will demonstrate below. This, in turn, implies that no remeshing is required during the simulation, which frequently dominates the computational cost.

We consider the nondimensionalized incompressible Navier–Stokes equations with a passively advected scalar field $\theta(\boldsymbol{x}, t)$. The domain is time-dependent $(\Omega(t))$, and it is moving with velocity $\mathbf{w}$. The governing equations are

$$\begin{aligned}\frac{\partial \boldsymbol{v}}{\partial t} + (\boldsymbol{v} - \mathbf{w}) \cdot \nabla \boldsymbol{v} &= -\nabla p + \frac{1}{\mathrm{Re}} \nabla^2 \boldsymbol{v} + \mathbf{f} \quad \text{in } \Omega(t),\\ \frac{\partial \theta}{\partial t} + (\boldsymbol{v} - \mathbf{w}) \cdot \nabla \theta &= \frac{1}{\mathrm{Pe}} \nabla^2 \theta \quad \text{in } \Omega(t),\\ \nabla \cdot \boldsymbol{v} &= 0 \quad \text{in } \Omega(t).\end{aligned}$$

The Peclet number Pe is the Reynolds number Re multiplied by either the Prandtl number Pr or the Schmidt number Sc for heat transfer or species transport applications, respectively. For heat transfer problems, the nondimensional temperature is given as

$$\theta = \frac{T - T_0}{\Delta T},$$

where T_0 is a reference temperature and ΔT is a predefined or desired temperature difference. For the species transport applications, θ can be identified as the concentration density normalized by a reference value.

To discretize the equations in time we use a high-order stiffly stable scheme (see (Karniadakis and Sherwin, 1999) in two passes:

Pass I

$$\hat{\boldsymbol{v}} = \sum_{q=0}^{Ji-1} \alpha_q \boldsymbol{v}^{n-q} + \Delta t \left[\sum_{q=0}^{J_e-1} \beta_q \mathbf{N}(\boldsymbol{v}^{n-q}, \mathbf{w}^{n-q}) + \mathbf{f}^{n+1} \right],$$

$$\hat{\theta} = \sum_{q=0}^{Ji-1} \alpha_q \theta^{n-q} + \Delta t(\sum_{q=0}^{J_e-1} \beta_q \tilde{\mathbf{N}}(\boldsymbol{v}^{n-q}, \mathbf{w}^{n-q}, \theta^{n-q}),$$

$$\mathbf{x}^{n+1} = \sum_{q=0}^{Ji-1} \alpha_q \mathbf{x}^{n-q} + \Delta t(\sum_{q=0}^{J_e-1} \beta_q \mathbf{w}^{n-q}),$$

$$\frac{\partial p^{n+1}}{\partial n} = \mathbf{n} \cdot \left[- \sum_{q=0}^{J_e-1} \beta_q \mathbf{N}(\boldsymbol{v}^{n-q}, \mathbf{w}^{n-q}) \right] - \mathbf{n} \cdot \left[\frac{1}{\mathrm{Re}} \sum_{q=0}^{J_e-1} \beta_q [\nabla \times (\nabla \times \boldsymbol{v}^{n-q})] \right].$$

PassII

$$\nabla^2 p^{n+1} = \nabla \cdot \left(\frac{\hat{\boldsymbol{v}}}{\Delta t} \right),$$

$$\nabla^2 \boldsymbol{v}^{n+1} - \frac{\gamma_0 \mathrm{Re}}{\Delta t} \boldsymbol{v}^{n+1} = -\frac{\mathrm{Re}}{\Delta t} (\hat{\boldsymbol{v}} - \Delta t \nabla p^{n+1}),$$

$$\nabla^2 \theta^{n+1} - \frac{\gamma_0 \mathrm{Pe}}{\Delta t} \theta^{n+1} = -\frac{\mathrm{Pe}}{\Delta t} \hat{\theta},$$

$$\nabla^2 \mathbf{w}^{n+1} = 0,$$

where $\mathbf{x}(\mathbf{X}, t)$ are the coordinates of the moving frame, relative to a fixed set of coordinates $\mathbf{X}$, and

$$\mathbf{N}(\boldsymbol{v}, \mathbf{w}) = (\boldsymbol{v} - \mathbf{w}) \cdot \nabla \boldsymbol{v},$$

$$\tilde{\mathbf{N}}(\boldsymbol{v}, \mathbf{w}, \theta) = (\boldsymbol{v} - \mathbf{w}) \cdot \nabla \theta.$$

In the first pass all steps are explicit and computed using the values of $\theta, \boldsymbol{v}, \mathbf{w}$, which are computed at the quadrature points. In the second pass all steps are computed implicitly. This scheme is of first order in time, but second-order schemes can be constructed based on staggered algorithms or predictor–corrector methods. The constants $\alpha_q, \beta_q, \gamma_0$ are integration weights (see Table 6.1 in (Karniadakis and Sherwin, 1999)). The mesh velocity is arbitrary, and can be specified explicitly or can be obtained from a Laplace equation, following (Ho, 1989). A better approach is to employ a variable coefficient in the Laplacian to provide enhanced smoothing, thereby preventing sudden distortions in the mesh (Lohner and Yang, 1996).

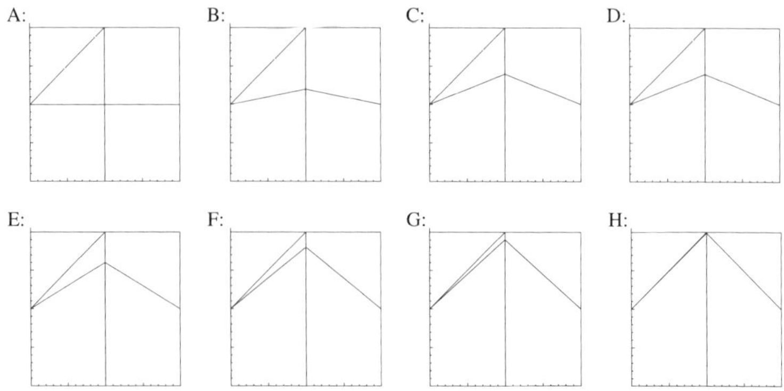

FIGURE 14.5. Meshes (A–H) consist of three quadrilaterals and two triangles, which are progressively skewed by shifting the interior vertex. Case H results in element minimum angle of 10^{-3} degrees. Convergence plots for all these cases are shown in Figure 14.6.

We now consider the effect of the skewness of the physical elements on the accuracy of the projection operator without any mesh motion. In Figure 14.5, we examine eight different meshes consisting of triangles and quadrilaterals. We start by projecting $\sin(\pi x)\sin(\pi y)$ onto a square domain covered with standard elements. Figure 14.6 shows the results for the *modal* basis, demonstrating that exponential convergence is achieved. Subsequently, we make the elements covering the domain progressively more skewed in the B-H meshes. In each case, we see that exponential convergence is achieved, even when one of the triangular elements has a minimum angle of about 10^{-3} degrees (See case H in Figure 14.5). Hence, the accuracy of the method is extremely robust for badly shaped elements. Also, we note that the similarity of the convergence curves demonstrates that the rate of exponential convergence is unaffected by the skewing for meshes A through G.

Example

Applications of the spectral element ALE approach to microfluidic mixing have been published in (Kumar and Beskok, 2002; Sert and Beskok, 2002). Here we present some results for mixing of two fluids of the same density in a microchannel by stirring the fluid with an oscillating cylinder. Although detailed mixing studies require three-dimensional simulations, the following results are intended to demonstrate the capability of this approach, and they are limited to two-dimensional analysis. Two fluids are introduced at the channel entrance from separate inlets as shown in Figure 14.7 (a). Fully developed flow conditions are assumed at the inlet with parabolic velocity distribution and identical flow rates. The top stream carries a scalar quantity with concentration of $\theta = 1$, and the stream entering from the bottom

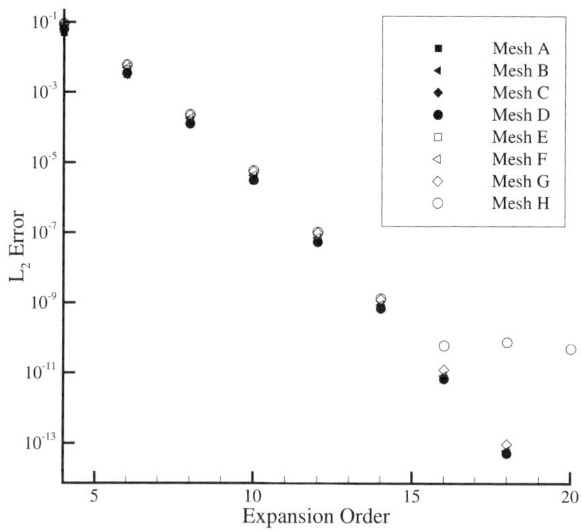

FIGURE 14.6. Convergence in the L_2 norm for *modal projection* of the function $u = \sin(\pi x)\sin(\pi y)$ on meshes A through H shown in Figure 14.5.

has a concentration of $\theta = 0$. The zero-flux boundary conditions for the species are used on the channel side-walls and on the cylinder surface. At the channel exit fully developed conditions are assumed, and the pressure is set to zero (gauge pressure) at the outflow.

The computational domain and the corresponding spectral/hp element discretization are shown in Figure 14.7. Here a 6th-order modal expansion is employed in each direction inside each element with 32 quadrilateral and 314 triangular elements. The total number of elements is fixed during a simulation. The elemental discretization is shown by the thick lines, and quadrature points are shown by the thin lines (a). Figure 14.7 (b) shows the deformed mesh at the cylinder minimum position, caused by the cylinder oscillation. The dashed–dotted line shows the center of the channel. The elements near the cylinder experience large deformations. Based on the test results in Figures 14.5 and 14.6, we expect the unstructured hp mesh to sustain high-order accuracy under large deformations. Thus, we do not have to remesh the computational domain for most practical applications.

The oscillating cylinder perturbs the two streams with concentration densities of $\theta = 1$ and $\theta = 0$, and promotes mixing. Mixing of the two streams depends on the Schmidt number of the fluid, $\mathrm{Sc} = \nu/D$. The ratio of fluid convection to mass diffusion is determined by the Peclet number (based on the mass diffusion coefficient D). In this case the Peclet number is defined as $\mathrm{Pe} = \mathrm{Re}\,\mathrm{Sc}$. Since mixing is enhanced by the oscillating cylinder, the Strouhal number $\mathrm{St} = U/\omega d$ (defined by the maximum inlet velocity U,

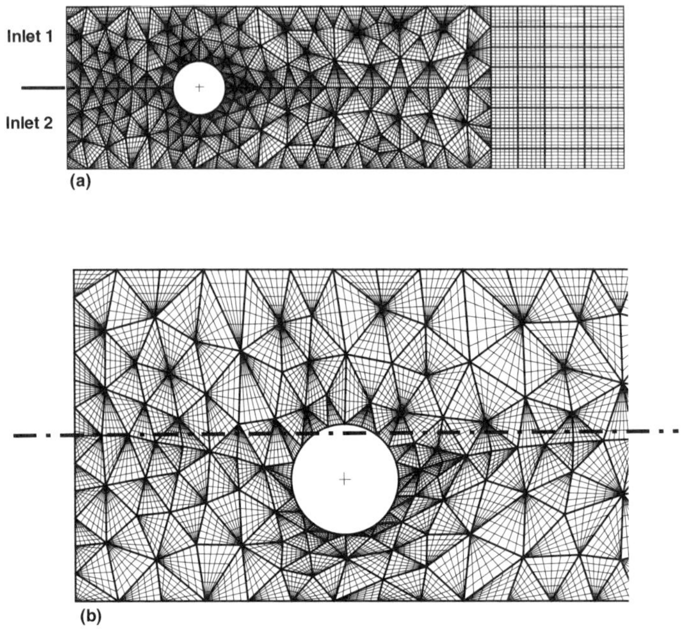

FIGURE 14.7. The mesh consists of 32 quadrilateral and 314 triangular elements. The elemental discretization is shown by thick lines, and the quadrature points for 6th-order modal expansion are shown by the thin lines. The undeformed mesh is shown in (a). The deformed mesh at the cylinder minimum position, due to the cylinder oscillation, is shown in (b).

the cylinder diameter d, and oscillation frequency ω) also becomes important in characterizing the micromixer. In this study Re $= 100$, St $= 0.6$ and Sc $= 5$. The concentration contours are shown in Figure 14.8.

Mixing simulations for large Schmidt number flows are computationally expensive for the following reasons. First, the concentration gradients at the interface increase with the Schmidt number, requiring enhanced spatial resolution. Second, mixing is an unsteady process, and it requires long-time integration. Therefore, for large Schmidt number flows we need to increase both the spatial resolution and the integration time of the simulation. For such cases, accumulation of the phase and dissipation errors can become problematic. The spectral/*hp* methods, exhibiting exponential reduction in the time rate of growth of phase and dissipation errors may be effective for such simulations. More details on mixing at microscales can be found in Chapter 9.

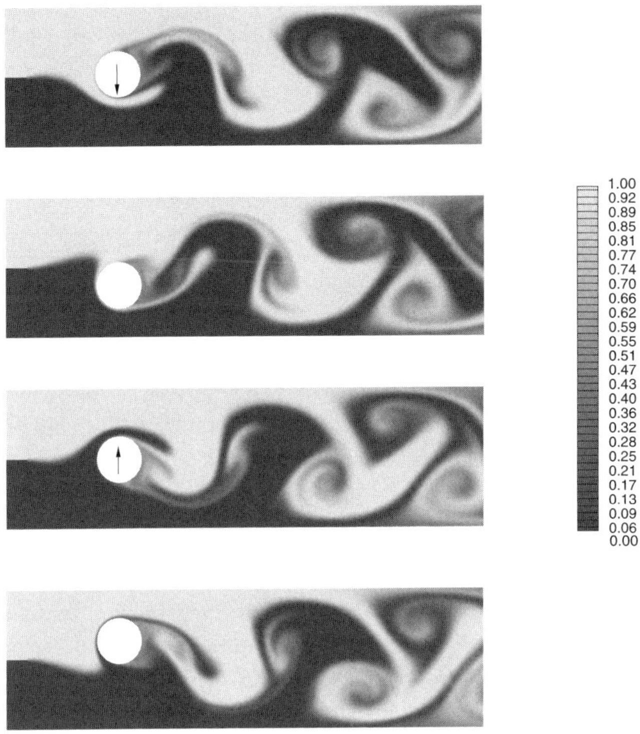

FIGURE 14.8. Four snapshots for mixing promoted by an oscillating cylinder. The concentration density contours are shown for $Re = 100, Sc = 5, St = 0.6$ A total of 346 spectral/*hp* elements with 8th order modal expansion are used in the simulations. Arrows show the direction of motion of the cylinder.

14.2 Meshless Methods

The spectral element method presented above is a typical Galerkin method, and in the limit of linear basis it reduces to the standard finite element method. The high-order equivalent of the finite volume method, which would be more appropriate for compressible high-speed flows, e.g., in micronozzles, is a discontinuous Galerkin method with a spectral basis, see (Cockburn et al., 2000), and references therein. Both classes of methods and their low-order counterparts are based on a mesh that consists of triangles, quadrilaterals in two dimensions, and tetrahedra, hexahedra, etc. in three dimensions.

A popular research topic in numerical methods has been the development of meshless methods as alternatives to the traditional finite element, finite volume, and finite difference methods. The traditional methods all require some connectivity knowledge a priori, such as the generation of a mesh, whereas the aim of meshless methods is to sprinkle only a set of points or nodes covering the computational domain, with no connectivity information required among the set of points. Multiphysics (specifically involving problems with moving domains) and multiscale analysis can be simplified by meshless techniques, since we deal with only nodes or points instead of a mesh. Meshless techniques are also appealing because of their potential in adaptive techniques, where a user can simply add more points in a particular region to obtain more accurate results.

Extensive research has been conducted in the area of meshless methods; see (Belytschko et al., 1996; Li and Liu, 2002; Atluri, 2002; Li, 2003), for an overview. Broadly defined, meshless methods contain two key steps:

- Construction of meshless approximation functions and their derivatives, and
- Meshless discretization of the governing partial differential equations.

Least-squares (Lancaster and Salkauskas, 1981), **kernel based** (Monaghan, 1992), and **radial basis function** (Hardy, 1971; Kansa, 1990a; Kansa, 1990b) approaches are three techniques that have gained considerable attention for construction of meshless approximation functions (see (Jin et al., 2001), for a detailed discussion on least-squares and kernel approximations). The meshless discretization can be categorized into three classes:

1. Cell integration (Belytschko et al., 1994),
2. Local point integration (Atluri, 2002), and
3. Point collocation (Liszka et al., 1996; Aluru, 2000).

Both interior and exterior domain problems (using a boundary-only formulation such as the boundary-integral formulation) encountered in microsystems have been solved with meshless methods. In this section we provide a brief overview of the application of meshless methods for interior (or domain) and boundary-only problems.

14.2.1 Domain Simulation

A summary of the various meshless techniques that have been developed for domain simulation is provided in Table 14.2. Here we outline the key steps in the *finite cloud method* and show its application to some examples.

The meshless finite cloud method uses a fixed kernel technique to construct the interpolation functions and a point collocation technique to discretize the governing partial differential equations. In a two-dimensional fixed kernel approach, given a scattered set of points as shown in Figure

TABLE 14.2. A summary of meshless methods developed for domain simulation. See (Belytschko et al., 1996; Li and Liu, 2002; Atluri, 2002; Li, 2003), for references to the various methods listed here.

	Point collocation	Cell integration Galerkin	Local domain integration Galerkin
Moving least-squares	Finite point method	Element-free Galerkin method, Partition of unity finite element method	Meshless local Petrov–Galerkin method, Method of finite spheres
Fixed least-squares	Generalized finite difference method, h-p meshless cloud method, Finite point method	Diffuse element method	
Reproducing kernel	Finite cloud method	Reproducing kernel particle method	
Fixed kernel	Finite cloud method		
Radial basis	Many techniques	Many techniques	

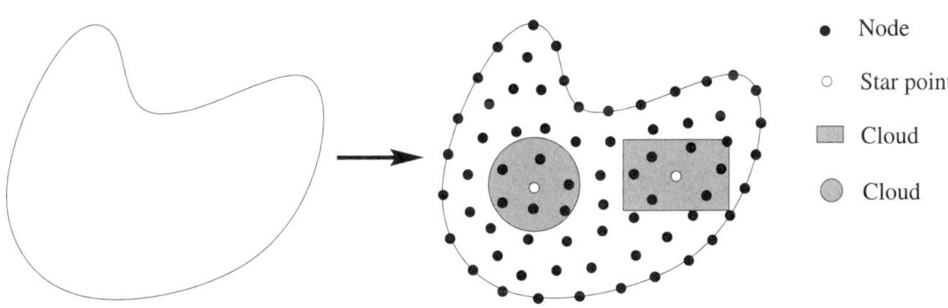

FIGURE 14.9. Distribution of random points and typical clouds in the meshless method.

14.9, an approximation $u^a(x, y)$ to an unknown $u(x, y)$ is given by

$$u^a(x,y) = \int_\Omega \mathcal{C}(x, y, x_k - s, y_k - t)\phi(x_k - s, y_k - t)u(s,t)\,ds\,dt, \quad (14.10)$$

where ϕ is the kernel function centered at (x_k, y_k), which is usually taken as a cubic spline or a Gaussian function. For the results shown here, ϕ is taken as a modified Gaussian function, i.e.,

$$\phi(x - x_I) = \frac{w(x - x_I)}{1 - w(x - x_I) + \hat{\epsilon}},$$

where $\hat{\epsilon}$ is a small number that is used to avoid the singularity of $\phi(x - x_I)$; typically, $\hat{\epsilon}$ is chosen to be 10^{-5}. Also, $w(x - x_I)$ is a normalized Gaussian function given by

$$w(z) = \begin{cases} \frac{e^{-(z/c)^2} - e^{-(d_{\mathrm{mi}}/c)^2}}{1 - e^{-(d_{\mathrm{mi}}/c)^2}}, & z \le d_{\mathrm{mi}}, \\ 0, & z > d_{\mathrm{mi}}, \end{cases}$$

where d_{mi} is the support size of a cloud (see Figure 14.9), and c is the dilation parameter, which is taken as $d_{\mathrm{mi}}/2$. In two dimensions, the kernel function is constructed as

$$\phi(x - x_I, y - y_I) = \phi(x - x_I)\phi(y - y_I).$$

Also, $\mathcal{C}(x, y, x_k - s, y_k - t)$ is the correction function, which is given by

$$\mathcal{C}(x, y, x_k - s, y_k - t) = \mathbf{p}^T(x_k - s, y_k - t)\mathbf{c}(x, y), \tag{14.11}$$

$\mathbf{p}^T = \{p_1, p_2, \ldots, p_m\}$ is an $m \times 1$ vector of basis functions. In two dimensions, a quadratic basis vector is given by

$$\mathbf{p}^T = \left[1, x_k - s, y_k - t, (x_k - s)^2, (x_k - s)(y_k - t), (y_k - t)^2\right], \quad m = 6.$$

Finally, $\mathbf{c}(x, y)$ is an $m \times 1$ vector of unknown correction function coefficients. The correction function coefficients are computed by satisfying the consistency conditions, i.e.,

$$\int_\Omega \mathbf{p}^T(x_k - s, y_k - t)\mathbf{c}(x, y)\phi(x_k - s, y_k - t)p_i(s, t)\,ds\,dt = p_i(x, y), \quad i = 1, \ldots, m.$$

, In discrete form, the above equation can be written as

$$\sum_{I=1}^{\mathrm{NP}} \mathbf{p}^T(x_k - x_I, y_k - y_I)\mathbf{c}(x, y)\phi(x_k - x_I, y_k - y_I)p_i(x_I, y_I)\Delta V_I = p_i(x, y), \quad i = 1, \ldots, m,$$

where NP is the number of points in the domain and ΔV_I is the nodal volume of node I (the nodal volumes can be simply set to unity; see (Aluru and Li, 2001; Aluru, 2000), for a more detailed discussion on nodal volumes). The above equation can be written in matrix form as

$$\mathbf{M}\mathbf{c}(x, y) = \mathbf{p}(x, y),$$

$$M_{ij} = \sum_{I=1}^{\mathrm{NP}} p_j(x_k - x_I, y_k - y_I)\phi(x_k - x_I, y_k - y_I)p_i(x_I, y_I)\Delta V_I, \quad i, j = 1, \ldots, m.$$

From the above equation, the unknown correction function coefficients are computed as

$$\mathbf{c}(x,y) = \mathbf{M}^{-1}\mathbf{p}(x,y).$$

Since $\mathbf{M}$ is a small matrix (6×6 matrix for a quadratic basis in 2-D, i.e., $m = 6$), the correction function coefficients can be computed using either a direct solver or any iterative solver. Substituting the correction function coefficients into equation (14.11) and employing a discrete approximation for equation (14.10), we obtain

$$u^a(x,y) = \sum_{I=1}^{\mathrm{NP}} N_I(x,y)\hat{u}_I,$$

where $\hat{u}_I$ is the nodal parameter for node I, and $N_I(x,y)$ is the fixed kernel meshless interpolation function defined as

$$N_I(x,y) = \mathbf{p}^T(x,y)\mathbf{M}^{-T}\mathbf{p}(x_k - x_I, y_k - y_I)\phi(x_k - x_I, y_k - y_I)\Delta V_I.$$

The interpolation functions obtained from the above equation are multivalued. A unique set of interpolation functions can be constructed by computing $N_I(x_k, y_k)$, $I = 1, 2, \ldots, \mathrm{NP}$, when the kernel is centered at (x_k, y_k) (see (Aluru and Li, 2001), for more details). The derivatives of the unknown u are approximated by

$$\frac{\partial u^a(x,y)}{\partial x} = \sum_{I=1}^{NP} \frac{\partial N_I(x,y)}{\partial x}\hat{u}_I,$$

$$\frac{\partial u^a(x,y)}{\partial y} = \sum_{I=1}^{NP} \frac{\partial N_I(x,y)}{\partial y}\hat{u}_I,$$

$$\frac{\partial^2 u^a(x,y)}{\partial x^2} = \sum_{I=1}^{NP} \frac{\partial^2 N_I(x,y)}{\partial x^2}\hat{u}_I,$$

$$\frac{\partial^2 u^a(x,y)}{\partial y^2} = \sum_{I=1}^{NP} \frac{\partial^2 N_I(x,y)}{\partial y^2}\hat{u}_I,$$

$$\frac{\partial^2 u^a(x,y)}{\partial x \partial y} = \sum_{I=1}^{NP} \frac{\partial^2 N_I(x,y)}{\partial x \partial y}\hat{u}_I.$$

The finite cloud method uses a point collocation technique (Aluru, 2000) to discretize the governing equations. Point collocation is the easiest way to discretize the governing equations. In a point collocation approach, the governing equations for a physical problem can be written in the following general form:

$$\begin{aligned} \mathcal{L}(u(x,y)) &= f(x,y) \quad && \text{in } \Omega, \\ \mathcal{G}(u(x,y)) &= g(x,y) \quad && \text{on } \Gamma_g, \\ \mathcal{H}(u(x,y)) &= h(x,y) \quad && \text{on } \Gamma_h, \end{aligned}$$

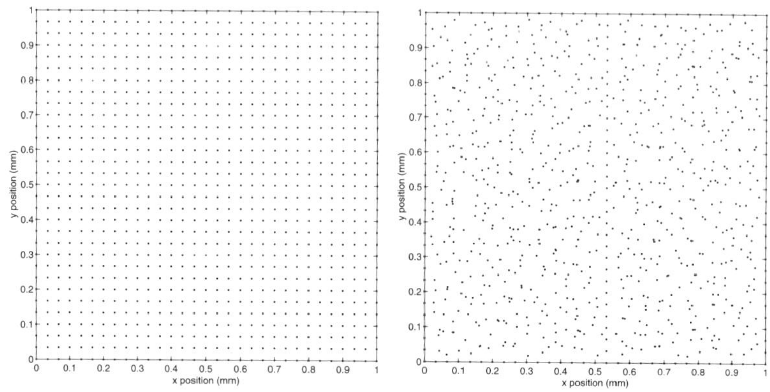

FIGURE 14.10. Meshless method: Uniform (left) and random (right) point distribution.

where Ω is the domain, Γ_g is the portion of the boundary on which Dirichlet boundary conditions are specified, Γ_h is the portion of the boundary on which Neumann boundary conditions are specified, and $\mathcal{L}$, $\mathcal{G}$, and $\mathcal{H}$ are the differential, Dirichlet, and Neumann operators, respectively. The boundary of the domain is given by $\Gamma = \Gamma_g \cup \Gamma_h$. After the meshless approximation functions are constructed, for each interior node, the point collocation technique simply substitutes the approximated unknown into the governing equations. For nodes with prescribed boundary conditions the approximate solution or the derivative of the approximate solution is substituted into the given Dirichlet or Neumann-type boundary conditions, respectively. Therefore, the discretized governing equations are given by

$$\begin{aligned} \mathcal{L}(u^a) &= f(x,y) && \text{for points in } \Omega, \\ \mathcal{G}(u^a) &= g(x,y) && \text{for points on } \Gamma_g, \\ \mathcal{H}(u^a) &= h(x,y) && \text{for points on } \Gamma_h. \end{aligned}$$

In the following we present an example of meshless simulation.

Flow in a Driven Cavity: As an application of the finite cloud method to fluids, a typical flow solution using the incompressible Navier–Stokes equations and no-slip condition is presented here. We consider a square cavity of dimensions 1×1 mm with the top wall moving at constant velocity corresponding to Re $= 3.2$. The two-dimensional Navier–Stokes equations are discretized in collocation fashion. Two different node distributions are employed as shown in Figure 14.10. In the first one, 961 points are distributed uniformly in the cavity. In the second case, 961 points are distributed randomly in the cavity. The corresponding velocity vectors are shown in Figure 14.11. Examination of velocity profiles at different locations shows negligible differences in the two solutions.

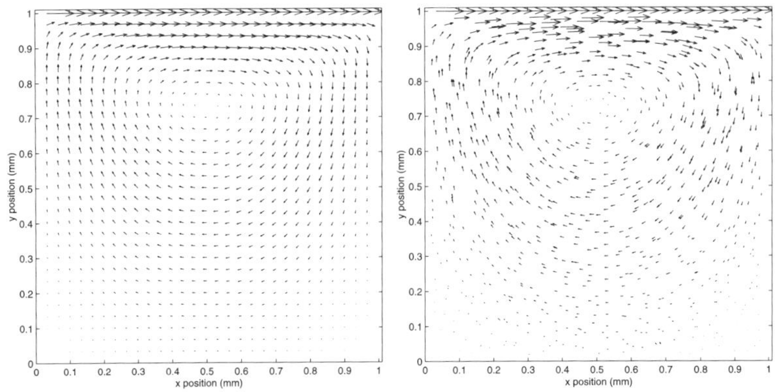

FIGURE 14.11. Velocity vectors of flow in a driven cavity: Uniform (left) and random (right) point distribution.

In summary, domain meshless methods are attractive alternatives to traditional numerical methods. The implementation of boundary conditions in Galerkin-based meshless methods has some difficulties, but a number of approaches have been suggested to overcome these difficulties (see, e.g., (Li and Liu, 2002), and references therein). The collocation meshless methods can impose the boundary conditions exactly, but the robustness of the method can be an issue for random distribution of points. This issue has been addressed in (Jin et al., 2004), but more progress is desirable. The construction of approximation functions is more expensive in meshless methods compared to the cost associated with construction of interpolation functions in the finite element method. The integration cost in Galerkin meshless methods is more expensive. Galerkin meshless methods can be a few times slower (typically about 5 times) than finite element methods. Collocation meshless methods are much faster, since no numerical integrations are involved. However, they may need more points, and as mentioned above, the robustness needs to be improved. For a quadratic basis in 2-D, the collocation meshless method has been shown to converge quadratically. Even though not much work has been published on time-stepping schemes in meshless methods, much of the published literature on time-stepping schemes for finite difference and finite element methods is applicable to meshless methods.

14.2.2 Boundary-Only Simulation

The key idea in boundary-only simulation is to discretize only the boundary of the problem. For linear and exterior problems (infinite domain problems), where there is a well-defined Green's function, boundary integral formulations are attractive, since they need only the discretization of a surface. The discretization of a surface into a mesh can, however, be quite involved for

TABLE 14.3. A summary of various boundary-only meshless methods. See (Li, 2003), for references to the various methods listed here.

	Boundary integral collocation	Dual reciprocal collocation	Boundary local domain discretization
Moving least-squares	Boundary node method		Local boundary integral equation method, Meshless local Petrov-Galerkin method, Hybrid boundary node method
Fixed least-squares	Boundary cloud method		
Boundary point interpolation	Boundary point interpolation method		
Radial basis function approximation		Boundary Knot Method	

complicated surfaces. In meshless boundary-only methods, the basic idea is to combine boundary-integral formulations with meshless approximation and discretization. A summary of the various boundary-only methods that have been developed is given in Table 14.3. In this section, we provide an overview of the boundary cloud method method and its application to solving the Stokes equations.

The various least-squares and kernel approaches that have been discussed in the context of domain meshless methods can be applied to compute the approximation functions for boundary-only meshless methods. However, instead of Cartesian coordinates one needs to use the cyclic (for 2-D problems where the boundary is one-dimensional) or curvilinear coordinates (for 3-D problems where the boundary is two-dimensional) to overcome singularity issues in the moment matrix. The *boundary cloud method* uses a varying basis approach (Li and Aluru, 2003) and Cartesian coordinates to compute the approximate functions. In a varying basis approach, the unknown $u(x, y)$ is approximated by

$$u(x, y) = \mathbf{p}_v^T(x, y)\mathbf{b}_t, \tag{14.12}$$

where $\mathbf{p}_v$ is the varying base interpolating polynomial and $\mathbf{b}_t$ is the un-

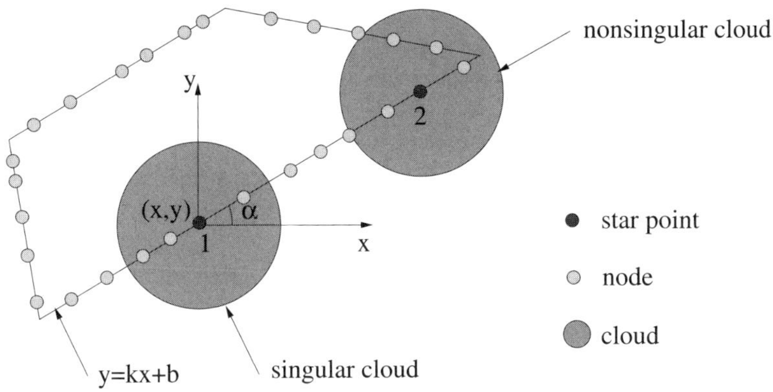

FIGURE 14.12. Definition of singular and nonsingular cloud.

known coefficient vector for point t. To construct the varying basis interpolation functions, clouds are classified into two types: singular and nonsingular. When all the points inside a cloud lie along a straight line, the cloud is defined as singular; otherwise, it is nonsingular. As shown in Figure 14.12, the cloud of point 1 is singular, and the cloud for point 2 is nonsingular, since the points do not lie along a straight line. Using a linear polynomial basis, the base interpolating polynomial is given by

$$\mathbf{p}_v^T(x,y) = \begin{cases} \left[1\ x\ y\right] & m = 3 \text{ (nonsingular cloud)}, \\ \\ \left[1\ x\right] \ or \ \left[1\ y\right] & m = 2 \text{ (singular cloud)}. \end{cases}$$

For a point t, the unknown coefficient vector $\mathbf{b}_t$ is computed by minimizing

$$J_t = \sum_{i=1}^{\mathrm{NP}} w_i(x_t, y_t) \left[\mathbf{p}_v^T(x_i, y_i)\mathbf{b}_t - \hat{u}_i\right]^2,$$

where NP is the number of nodes, $w_i(x_t, y_t)$ is the weighting function centered at (x_t, y_t) and evaluated at node i, whose coordinates are (x_i, y_i), and $\hat{u}_i$ is a nodal parameter. Once the unknown coefficient vector $(\mathbf{b}_t)$ is computed, the approximation for the unknown $u(x,y)$ in equation (14.12) is defined. Numerical integrations in the boundary cloud method are implemented using the standard cell structure and Gaussian quadrature.

The application of the boundary cloud method to the Stokes equations (see Chapter 2 for a discussion on the Stokes equations) is now discussed. The boundary integral equation of the Stokes equations without body forces can be written as (Phan et al., 2002)

$$c_{ik}(\mathbf{x})u_i(\mathbf{x}) = \int_{\partial B} [U_{ik}(\mathbf{x},\mathbf{y})\sigma_{ij}(\mathbf{y}) - W_{ijk}(\mathbf{x},\mathbf{y})u_i(\mathbf{y})]n_j(\mathbf{y})dS, \quad (14.13)$$

where $i, j, k = 1, 2$; c_{ik} is the corner tensor, n_j is the unit outward normal at ∂B; u_i is velocity; U_{ik} and W_{ijk} are the kernel tensors; $\sigma_{ij} = -p\delta_{ij} + \mu(u_{i,j} + u_{j,i})$ is the stress tensor, and $\mathbf{x}$, $\mathbf{y}$ are the source point and field point, respectively. For 2-D problems, ∂B is the boundary curve defining the body B. The kernel tensors in equation (14.13) for 2-D Stokes flow are given by

$$U_{ik} = -\frac{1}{4\pi\mu}[\delta_{ik} ln(r) - r_{,i} r_{,k}],$$
$$W_{ijk} = -\frac{1}{\pi r} r_{,i} r_{,j} r_{,k},$$

where δ_{ik} is the Kronecker delta function, $r = \|\mathbf{y} - \mathbf{x}\|$, $r_{,i} = r_i/r$, and $r_i = x_i(\mathbf{y}) - x_i(\mathbf{x})$.

In the above equations, $U_{ik}, W_{ijk} \to \infty$ as $r \to 0$, and the integral is singular. Thus, regularization of equation (14.13) is necessary. Using the linear and constant modes given in (Mukherjee, 2000), we obtain

$$\begin{aligned} 0 = &- \int_{\partial B} W_{ijk}(\mathbf{x}, \mathbf{y})[u_i(\mathbf{y}) - u_i(\mathbf{x}) - u_{i,p}(\mathbf{x})(x_p(\mathbf{y}) - x_p(\mathbf{x}))]n_j(\mathbf{y})dS \\ &+ \int_{\partial B} U_{ik}(\mathbf{x}, \mathbf{y}) [\sigma_{ij}(\mathbf{y}) - \sigma_{ij}(\mathbf{x})] n_j(\mathbf{y})dS, \end{aligned} \tag{14.14}$$

where $p = 1, 2$. If the velocity and the stress are sufficiently smooth, the singularity in equation (14.14) is removed. Therefore, the integration can be evaluated using Gaussian quadrature schemes.

For the numerical implementation, the velocity and the traction are approximated in each cloud by the varying base interpolating polynomial as described above (Li and Aluru, 2003), and can be written in a general form as

$$u_k(\mathbf{y}) = \sum_{I=1}^{\mathrm{NP}} N_I(\mathbf{y}) \hat{u}_k^I$$

and

$$\tau_k(\mathbf{y}) = \sum_{I=1}^{\mathrm{NP}} N_I(\mathbf{y}) \hat{\tau}_k^I,$$

where k = 1,2, τ_k is the traction, $\hat{u}_k^I$ and $\hat{\tau}_k^I$ are the nodal parameters, $N_I(\mathbf{y})$ is the approximation function, and NP is the number of points.

The above approximations and their derivatives are substituted into equation (14.14). In order to evaluate the integrals in equation (14.14), the boundary is decomposed into cells, and equation (14.14) is satisfied on every boundary node. By using Gaussian quadrature and looping over all the boundary nodes, equation (14.14) can be written in matrix form

$$\mathbf{A}\hat{\mathbf{u}} + \mathbf{B}\hat{\tau} = 0,$$

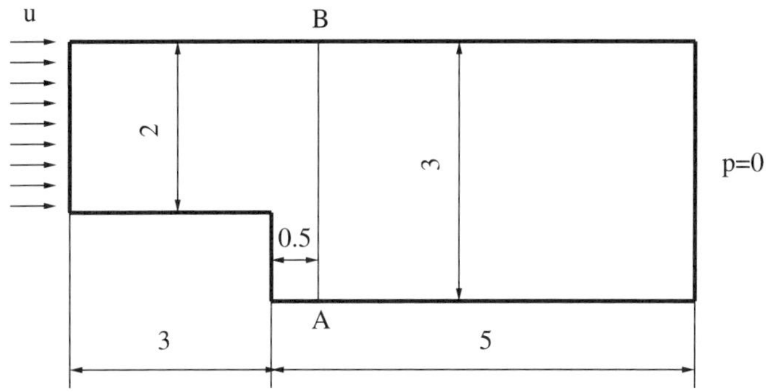

FIGURE 14.13. Sketch of Stokes flow through a step channel. Also shown is the inlet velocity.

where **A** and **B** are 2NP × 2NP matrices, and $\hat{\mathbf{u}}$ and $\hat{\tau}$ are 2NP vectors. After applying the boundary conditions, the unknowns on the boundary nodes can be calculated.

As an example, the results for a step flow are presented here. The geometry of the channel is shown in Figure 14.13. The velocity at the inlet is uniformly distributed; the value of velocity u is 1 unit/s. The point distribution on the boundary is shown in Figure 14.14. The velocity at the interior nodes is computed by using the boundary integral equations. The computed x-component of the velocity across line AB and at the outflow are shown in Figure 14.15. The results from the boundary cloud method match well with the results obtained from the finite element method.

In summary, boundary-only formulations are more efficient for *linear problems* with a known Green's function(s), since they eliminate the need to discretize the entire domain. Meshless boundary-only formulations further improve the efficiency by eliminating the need for a mesh (a scattered set of points is used to discretize the boundary) on the boundary. Typically, the meshless boundary-only formulation is a factor of two slower than the boundary element method. However, the error obtained with the meshless boundary-only formulation is lower than the error obtained with the boundary element method for the same point distribution. Meshless boundary-only formulations are relatively new approaches compared to meshless domain formulations. As a result, a complete mathematical analysis of the boundary-only formulations is not yet available. The extension of boundary element methods or meshless boundary-only methods to unsteady methods can be quite involved.

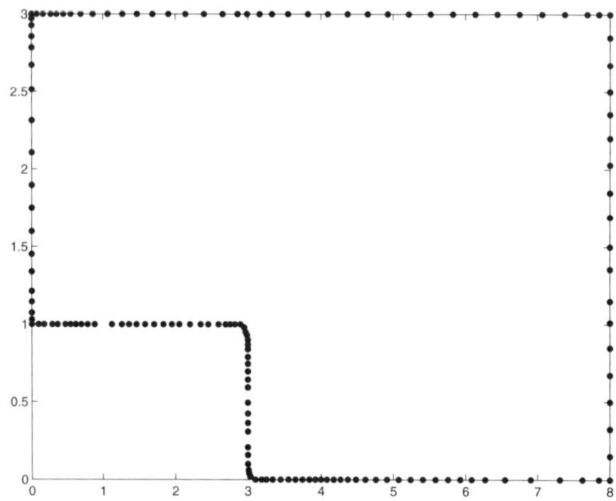

FIGURE 14.14. Boundary cloud method: Scattered point distribution along the boundary of the step channel example.

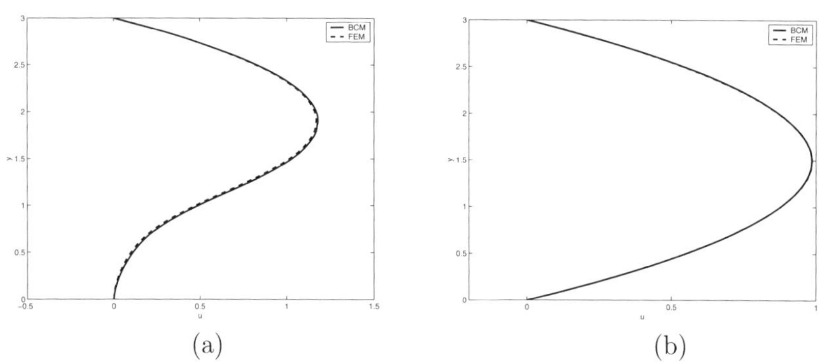

FIGURE 14.15. Comparison between the boundary cloud method (BCM) and the finite element method (FEM) for the x-velocity profiles (a) along line AB; (b) at the outflow.

14.3 Particulate Microflows

Here we consider numerical approaches for simulating particulate microflows in the applications described in chapter 1 (see Figures 1.5, 1.6) and also in self-assembly applications; see Section 13.1. There are a number of approaches in dealing with particulate flows based on direct but computationally expensive methods such as the front tracking technique of (Tryggvason et al., 1998); the ALE (Arbitrary Lagrangian Eulerian) method, see Section 14.1.4 and (Johnson and Tezduyar, 1996); and the fictitious domain method

(DLM) (Glowinsky et al., 1999). A mesoscopic method based on the lattice–Boltzmann equation has been developed by (Ladd, 1994a; Ladd, 1994b); we discuss this method in some detail in Section 15.5. Also, a stochastic molecular dynamics approach, the dissipative particle dynamics (DPD) method, can be used, especially for dense systems; See section 16.4. At low Reynolds numbers, the Stokesian dynamics approach has been developed by (Brady and Bossis, 1988), which deals effectively with the lubrication forces in particle–particle and particle–wall interactions.

The complexity of dynamics of a particulate flow depends on the *volume fraction* occupied by the particles relative to the total volume, defined as

$$\alpha = \frac{\sum_i N_i V_{p,i}}{V},$$

where V is the total volume, $V_{p,i}$ is the particle volume, and N_i, is the number of particles present in volume $V_{p,i}$. For $\alpha \leq 10^{-3}$ the particulate flow is dilute, but for $\alpha \geq 10^{-3}$ strong fluid-dynamic interactions and particle collisions occur.

To model particulate microflows accurately, the *full* Navier–Stokes equations need to be employed; otherwise, important phenomena such as the wake behind the particles are inaccurately represented. The wake influences the history of the particle's trajectory. Additionally, convective inertia influences the vorticity transport, resulting in the faster decay of the so-called *Basset history* force. In the following we first review some classical results on hydrodynamic interactions between spheres, and subsequently we focus on the force coupling method, a particularly fast and easy to implement modeling approach for particulate flows.

14.3.1 Hydrodynamic Forces on Spheres

We first discuss some classical semianalytical results for a single sphere, and subsequently we present known results for hydrodynamic interactions between spherical particles, and between particles and solid walls. We assume that the flow is either in the Stokes or in the low Reynolds number regime.

For a *steady motion of an isolated particle*, modeled as a sphere, the drag force is given by

$$F_D = \frac{1}{2} C_D \rho_f \pi a^2 |V - U_\infty| (V - U_\infty),$$

where V is the particle velocity, U_∞ is the undisturbed flow at the particle's center, C_D is the drag coefficient, ρ_f is the fluid density, and a is the particle radius. Semiempirical expressions for the drag coefficient have been given by (Clift et al., 1978) in the form

$$C_D = \phi(\mathrm{Re}_p) C_D^{SS},$$

where $C_D^{SS} = \frac{24}{\mathrm{Re}_p}$ is the steady Stokes drag coefficient. Here,

$$\mathrm{Re}_p = \frac{2a|V - U_\infty|}{\nu},$$

and the correction $\phi(\mathrm{Re}_p)$ is obtained from a best-fit to experimental data:

$$\phi(\mathrm{Re}_p) = \begin{cases} 1 + \frac{3}{16}\mathrm{Re}_p & \text{for } \mathrm{Re}_p < 10^{-2}, \\ 1 + 0.315\mathrm{Re}_p^{0.82-0.05\log \mathrm{Re}_p} & \text{for } 10^{-2} \leq \mathrm{Re}_p < 20, \\ 1 + 0.1935\mathrm{Re}_p^{0.6305} & \text{for } 20 \leq \mathrm{Re}_p < 260. \end{cases}$$

However, in an *unsteady flow* even in the limit of the Stokes flow, the above semiempirical relations are not valid. The problem of a single sphere in nonuniform unsteady flow was first solved by (Maxey and Riley, 1983) assuming a no-slip boundary condition at the particle surface. The initial slip condition was originally discussed in (Maxey, 1993), and subsequently it was introduced in the equation of motion (Michaelides, 1997); it states that

$$V_i \neq U_i^\infty + \frac{a^2}{6}\frac{\partial^2 U_i^\infty}{\partial x_j^2},$$

so the complete particle equation of motion is

$$\begin{aligned} m_p \frac{dV_i}{dt} &= (m_p - m_f)g_i + m_f \frac{DU_i^\infty}{Dt} \\ &- \frac{m_f}{2}\frac{d}{dt}\left(V_i - U_i^\infty - \frac{a^2}{10}\frac{\partial^2 U_i^\infty}{\partial x_j^2}\right) \\ &- 6\pi\mu a\left(V_i - U_i^\infty - \frac{a^2}{6}\frac{\partial^2 U_i^\infty}{\partial x_j^2}\right) \\ &- 6\pi\mu a^2 \int_0^t \frac{1}{\sqrt{\pi\nu(t-s)}}\frac{d}{ds}\left(V_i - U_i^\infty - \frac{a^2}{6}\frac{\partial^2 U_i^\infty}{\partial x_j^2}\right) ds \\ &- 6\pi a\mu\left(V_i - U_i^\infty - \frac{\partial^2 U_i^\infty}{dx_j^2}\right)\left(\frac{1}{\sqrt{\pi\nu t}}\right). \end{aligned}$$

Here m_p and m_f are the particle mass and fluid mass respectively, and the convective total derivatives are denoted by $\frac{D}{Dt}$ for the flow and $\frac{d}{dt}$ for the particle (Maxey and Riley, 1983).

The second-derivative terms are corrections due to the flow curvature. The last term is due to initial slip velocity, and the term before the last one is the Basset history force, which is associated with diffusion generated at the particle surface. A more general expression valid for small but finite particle Reynolds number was derived by (Lovalenti and Brady, 1993), who considered also the slip effects and enhanced the Basset history term with a long-time correction. This equation was later extended in (Mei and Adrian, 1992), to particle Reynolds numbers up to 100, again introducing convection to the history term associated with the vorticity transport from the particle surface onto the bulk of the fluid. It is clear from this work that convective inertia is very important even at very small scales and cannot be neglected in microflows, as the numerical and experimental results of (Lomholt, 2000) have also shown.

We now discuss **hydrodynamic interactions** between two particles in Stokes flow, which generally scale as the inverse of the distance between their centers. First, we consider a particle interacting with a wall, and subsequently we will present results for particle–particle interactions.

An analytical solution for drag force on a sphere moving *perpendicular to a plane wall* was obtained by (Goldman et al., 1967) in the form of an infinite series solution. Using regression techniques, (Bevan and Prieve, 2000) developed the following rational function approximation:

$$F_D = \frac{6\pi\mu V a}{f(h)},$$
$$f(h) = \frac{6(h/a)^2 + 2(h/a)}{6(h/a)^2 + 9(h/a) + 2},$$

where a is the particle radius, h is the particle–surface separation distance, and V is the particle velocity. In the asymptotic limits the above relation gives

$$\lim_{h\to\infty} f_e(h) \to 1, \qquad \lim_{h\to 0} f_e(h) \to h/a.$$

An analytical solution for drag force on a sphere moving *parallel to a plane wall* was obtained in (O'Neill and Majumdar, 1970). A rational function approximation for this solution is given by

$$F_D = \frac{6\pi\mu V a}{f(h)},$$
$$f(h) = \frac{368(h/a)^3 + 559(h/a)^2 + 81(h/a)}{368(h/a)^3 + 779(h/a)^2 + 250(h/a)}.$$

In the asymptotic limits the above relation gives

$$\lim_{h\to\infty} f_e(h) \to 1, \qquad \lim_{h\to 0} f_e(h) \to 0.324.$$

Hydrodynamic interaction of a particle near a flat surface also affects its diffusion characteristics. The *particle diffusion* in the *translational direction* (parallel) to the wall is given as

$$D(h) = \frac{k_B T}{6\pi\mu a} g(h),$$
$$g(h) = \frac{368(h/a)^3 + 559(h/a)^2 + 81(h/a)}{368(h/a)^3 + 779(h/a)^2 + 250(h/a)},$$

and the hindered particle diffusion in the *normal direction* to the wall is given as (Bevan and Prieve, 2000)

$$D(h) = \frac{k_B T}{6\pi\mu a} g(h),$$
$$g(h) = \frac{6(h/a)^2 + 2(h/a)}{6(h/a)^2 + 9(h/a) + 2},$$

where h is the particle–wall distance.

We now consider *particle—particle* interactions. Relative motion of two spheres toward each other, *squeezing flow*, was solved in (O'Neill and Majumdar, 1970). A rational function approximation of drag force for this case is given by

$$F_D = \frac{6\pi\mu V a}{f(h)},$$
$$f(h) = \frac{54(h/a)^3 + 71(h/a)^2 + 8(h/a)}{54(h/a)^3 + 154(h/a)^2 + 60(h/a) + 4},$$

where h is the separation distance between the two spheres. In the asymptotic limits the above relation gives

$$\lim_{h\to\infty} f_e(h) \to 1, \qquad \lim_{h\to 0} f_e(h) \to 2h/a.$$

Collective motion of two spheres with the same velocity V *parallel to their line of centers* so that there is no relative velocity between them was solved analytically in (Stimson and Jeffery, 1926). A rational function approximation of drag force for this case is given by

$$F_D = \frac{6\pi\mu V a}{f(h)},$$
$$f(h) = \frac{2(h/a)^3 + 14(h/a)^2 + 31(h/a)}{2(h/a)^3 + 11(h/a)^2 + 20(h/a) + 4},$$

where h is the separation distance between the two spheres. In the asymptotic limits of $h \to 0$ and $h \to \infty$, the above relation gives

$$\lim_{h\to\infty} f_e(h) \to 1, \qquad \lim_{h\to 0} f_e(h) \to 1.55.$$

The relative motion of two spheres with the velocity $\pm V$ *perpendicular to their line of centers* was solved by the multiple reflection method analytically (O'Neill and Majumdar, 1970). A rational function approximation of the drag force for this case is given by

$$F_D = \frac{6\pi\mu V a}{f(h)},$$
$$f(h) = \frac{3433(h/a)^2 + 4815(h/a) + 67}{3433(h/a)^2 + 2591(h/a) + 31},$$

where h is the separation distance between the two spheres. In the asymptotic limits of $h \to 0$ and $h \to \infty$, the above relation gives

$$\lim_{h\to\infty} f_e(h) \to 1, \qquad \lim_{h\to 0} f_e(h) \to 2.17.$$

Collective motion of two spheres with the same velocity V *perpendicular to their line of centers* was obtained by (O'Neill and Majumdar, 1970). A rational function approximation of drag force for this case is given by

$$F_D = \frac{6\pi\mu V a}{f(h)},$$
$$f(h) = \frac{-31(h/a)^3 + 8275(h/a)^2 + 14720(h/a) + 45}{-31(h/a)^3 + 8252(h/a)^2 + 20843(h/a) + 62},$$

where h is the separation distance between the two spheres. In the asymptotic limits the above relation gives

$$\lim_{h\to\infty} f_e(h) \to 1, \qquad \lim_{h\to 0} f_e(h) \to 0.725.$$

The rational function approximations for the drag force, presented above, are valid for very specific particle/surface configurations. In 1976, Batchelor combined all of these two-body hydrodynamic problems and specified particle interactions in a tensor form. This related the velocity and force field vectors via the mobility tensor (Batchelor, 1976). This treatment enabled development of Stokesian dynamics algorithms (Brady and Bossis, 1988).

14.3.2 The Force Coupling Method (FCM)

For a dilute limit with volume fraction much smaller than 10^{-3}, particulate microflows can be modeled by simply adding a source term of the form

$$S^p = -\sum_n^0 F_i^n \delta(x - Y^n)$$

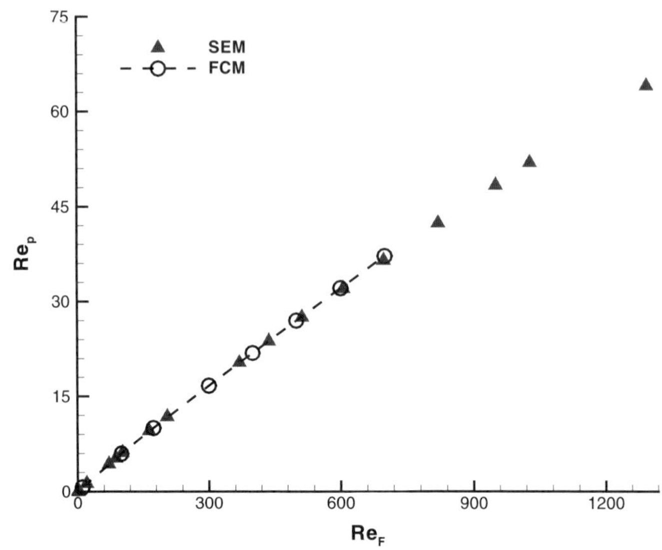

FIGURE 14.16. Comparison of FCM against results from direct numerical simulation (SEM) for flow past an array of spheres. The particle Reynolds number Re_p is plotted against the force Reynolds number, Re_f, the latter being proportional to the drag coefficient. (Courtesy of G. Dent.)

on the right-hand side of the Navier–Stokes equations, where the force on each particle is

$$F_i^n = m_p \frac{dV_i(Y^n)}{dt}$$

and Y^n is the vector coordinate associated with the particle's center of mass; also, $\delta(x - Y^n)$ is the Dirac delta function. This is the so-called point-force model, which is not resolvable in numerical simulations of Navier–Stokes equations, and the results depend on the implementation details of the numerical scheme employed. It is used formally in Stokesian dynamics, but the corresponding equations are solved analytically; it has also been used in earlier gas–solid flows, e.g., dusty gases. The point-force approach does not directly take into account the finite particle size or interactions of particles in the flow. For example, this approach cannot capture the experimentally observed phenomena associated with acceleration of particles such as the DKT (drafting–kissing–tumbling) event. This phenomenon refers to nonlinear interactions between two spheres where the trailing sphere is first drawn into the wake of the leading sphere, it touches it, and then overtakes the leading sphere by tumbling around it (Fortes et al., 1987).

For a dense particulate microflow where the size of the particle becomes comparable with the characteristic length of the domain, the above description may be of very limited use. On the other hand, computing in great detail all the pair collisions is a computationally formidable task (Johnson and Tezduyar, 1996). To this end, we describe next the *force coupling method* (FCM), which was developed by Maxey and his students (Maxey et al., 1997; Maxey and Dent, 1998; Maxey and Patel, 2001). It was used and validated for particulate microflows with great success in (Lomholt, 2000; Liu et al., 2002). In FCM, the same *stationary mesh* is used throughout the simulation, and in combination with the spectral element method we described earlier it gives very accurate results (Liu, 2004). A comparison of the overall accuracy of FCM for an array of spheres is shown in Figure 14.16. The comparison is against a direct numerical simulation using the spectral element method in computations performed by (Dent, 1999). A very good agreement is achieved even at relatively high Re_p and this holds for quite dense particulate flows, e.g., up to 20% concentration. Typically, in particulate microflows the particle Reynolds number is small, e.g., $\mathrm{Re}_p \leq 10$.

The basic idea of the FCM is to model the disturbance flow via a calibrated multipole expansion modifying the Navier–Stokes equation. The particles are *virtual*, and the slip or no-slip boundary conditions on their surface is only approximately satisfied. This alleviates the severe numerical stiffness of the problem associated with exact boundary condition enforcement on many moving surfaces. The modified Navier–Stokes equations are then

$$\begin{aligned}\frac{\partial u_i}{\partial t} + u_j \frac{\partial u_i}{\partial x_j} = & -\frac{1}{\rho_f}\nabla p + \nu \nabla^2 u_i \\ & + \frac{1}{\rho_f}\sum_n F_i^n \Delta(\mathbf{x} - \mathbf{Y}^n; \sigma) \\ & + \frac{1}{\rho_f}\sum_n F_{ij}^n \frac{\partial \Theta(\mathbf{x} - Y^n; \sigma_D)}{\partial x_j}. \end{aligned} \tag{14.15}$$

The first extra source term on the right is associated with a monopole contribution caused by a *virtual particle* that occupies a finite localized region defined by the Gaussian envelope

$$\Delta(x - Y^n, \sigma) = \frac{1}{(2\pi\sigma^2)^{3/2}} \exp\left[-\frac{(\mathbf{x} - \mathbf{Y}^n)^2}{2\sigma^2}\right].$$

The length scale σ is related to the size of the particle, i.e., radius a, as follows (Maxey and Patel, 2001):

$$\sigma = \frac{a}{\sqrt{\pi}}.$$

This choice corresponds to a particle having velocity

$$V_i(t) = \tilde{u}_i(Y, t),$$

where the latter is a volume-averaged velocity obtained from the convolution

$$\tilde{u}_i = \int u_i(\mathbf{x}, t)\Delta(\mathbf{x} - Y, \sigma)d^3\mathbf{x},$$

and $u_i(\mathbf{x}, t)$ denotes the flow field computed from the Navier–Stokes equations. Note that if more particles are present, the particle interaction is already included in $u_i(t)$ and therefore in the particle velocity $V_i(t)$. The volume-averaged velocity ensures that

- The mass of particles is conserved; see (Maxey and Dent, 1998), and
- A consistent energy balance exists between the potential energy corresponding to the settling of the particle and fluid viscous dissipation; see (Maxey and Patel, 2001).

In essence, the convolution procedure represents a *filtering process* of the small scales, which are energetically insignificant, and thus they do not affect the particle motion. The magnitude of the monopole F_i^n represents the force on the fluid by particle n, and it can be computed from the equations of the motion of the particle

$$m_p \frac{dV_i}{dt} = F_i^B + F_i^H,$$

where F_i^B includes all body forces, e.g., gravity or electromagnetic forces, and F_i^H is the force from the fluid on the particle. Since flow inertia is already included in FCM, it has to be subtracted from the particle inertia. The force monopole strength is therefore

$$F_i^n = -(m_p^n - m_f)\frac{dV_i^n}{dt} + F_i^B.$$

For example, if gravity is the only body force, then

$$F_i^B = (\rho_p - \rho_f)V_p^n g_i,$$

where V_p^n is the volume occupied by particle n.

The last term in the FCM governing equations (14.15) is associated with the dipole contribution (Kim and Karrila, 1991). It is caused by two different effects:

- an antisymmetric part, which is due to *torque* exerted on the fluid by particle n, and
- a symmetric part, the *stresslet*, which corresponds to the rate of strain tensor.

The formulation of the force dipole effect ensures that the angular momentum/kinetic budget is consistently related to the work done by any torque and that the stresslet term does not impart net work on the flow.

The envelope $\Theta(x, \sigma_D)$ employed for the dipole contribution is also a Gaussian envelope but with a different width, characterized by the length scale σ_D, i.e.,

$$\Theta(x, \sigma_D) = \Delta(x, \sigma_D).$$

This length scale is determined by matching the particle angular velocity found from the convolution

$$\tilde{\Omega}_i = \frac{1}{2} \int \omega_i(x) \Theta(\mathbf{x} - \mathbf{Y} \sigma_D) d^3x,$$

with the angular velocity of a fixed sphere in an unbounded quiescent flow of vorticity ω_i; see (Kim and Karrila, 1991), and also (Lomholt, 2000). This matching results in

$$\sigma_D = \frac{a}{\sqrt[3]{6\sqrt{\pi}}}.$$

The total force dipole strength is the sum of the contributions due to the torque and due to stresslet, and is of the form

$$F_{ij}^n = F_{ij}^{n(s)} + \frac{1}{2} \epsilon_{ijk} T_k^n,$$

where T_k^n is the torque acting on the fluid due to either an external torque on the particle or the effects of moment of intertia of the particle. Often this term is zero or negligible. Also, $F_{ij}^{n(s)}$ is the *stresslet* and is adjusted so that the condition

$$\tilde{E}_{ij}^n = \int E_{ij} \Theta(\mathbf{x} - \mathbf{Y}, \sigma_D) d^3x$$

is zero for each particle, where

$$E_{ij} = \frac{1}{2} \left(\frac{\partial u_i}{\partial x_j} + \frac{\partial u_j}{\partial x_i} \right).$$

In the context of *Stokes's flow* for an isolated particle, the torque needed to maintain the rotation of a sphere with angular velocity $\mathbf{\Omega}$ is

$$\mathbf{T} = 8\pi a^3 \mu \mathbf{\Omega}.$$

For a sphere placed in an external flow $\mathbf{U}^\infty$ that has a uniform rate of strain E_{ij}^∞, the symmetric force dipole needed to neutralize this is

$$F_{ij}^s = \frac{20}{3} \pi \mu a^3 E_{ij}^\infty.$$

Combining the above equations, we obtain

$$F_{ij}^n = \rho_f \nu V_p (3\Omega_{ij} + 5E_{ij}^*).$$

The torque term is

$$\Omega_{ij} = \epsilon_{ijk}\Omega_k^p,$$

where ϵ_{ijk} is the tensorial index and Ω_k^p is the angular velocity of particle k. The stresslet contribution is defined implicitly by assuming that the average rate of strain is zero, since the particles are not allowed to deform. To this end, we obtain E_{ij}^* from the convolution

$$E_{ij}^{*n} = \int E_{ij}^\infty \Theta(\mathbf{x} - \mathbf{Y}^n, \sigma_D) d^3\mathbf{x},$$

where

$$E_{ij}^\infty = \frac{1}{2}\left(\frac{\partial U_i^\infty}{\partial x_j} + \frac{\partial U_j^\infty}{\partial x_i}\right)$$

is the strain tensor of the undisturbed flow field.

The details of the implementation were presented by (Lomholt, 2000), who employed an iterative procedure to impose the constraint of zero average strain rate. A more direct approach would be to consider the mobility matrix, which is the inverse of an influence matrix, and can be constructed from the linear response to multiunit pulses. This, however, may be computationally expensive, but hybrid approaches can be followed that combine efficiency and accuracy. A robust and simple to implement approach is the *penalty method* developed in (Liu, 2004).

To complete FCM, we need to update the locations of the virtual particles Y^n from

$$\frac{dY_i}{dt} = V_i,$$

and following an explicit (third-order Adams–Bashforth) time-stepping scheme we obtain

$$\frac{Y_i^{k+1} - Y_i^k}{\Delta t} = \frac{23}{12}V_i^k - \frac{16}{12}V_i^{k-1} + \frac{5}{12}V_i^{k-2}.$$

Here k refers to the time level, and Δt is the time step used. Note that in FCM we have $V_i^k = \tilde{u}_i^k$.

FCM has been studied systematically by (Lomholt, 2000), who compared it with experimental results for one, two, and three particles. In Figure 14.17 we plot the results of Lomholt for a flow in a channel with two particles interacting, with gravity the body force. The DTK process revealed by the experiments of (Fortes et al., 1987) was reproduced in experiments described in (Lomholt, 2000). We see that very good agreement for the particles, trajectory is obtained with the FCM predictions for this strongly nonlinear interaction.

In the following, we compare the accuracy of FCM to the accuracy of direct numerical simulation (DNS) and to the distributed Lagrange multiplier (DLM) methods for a flow past a microsphere. All discretizations

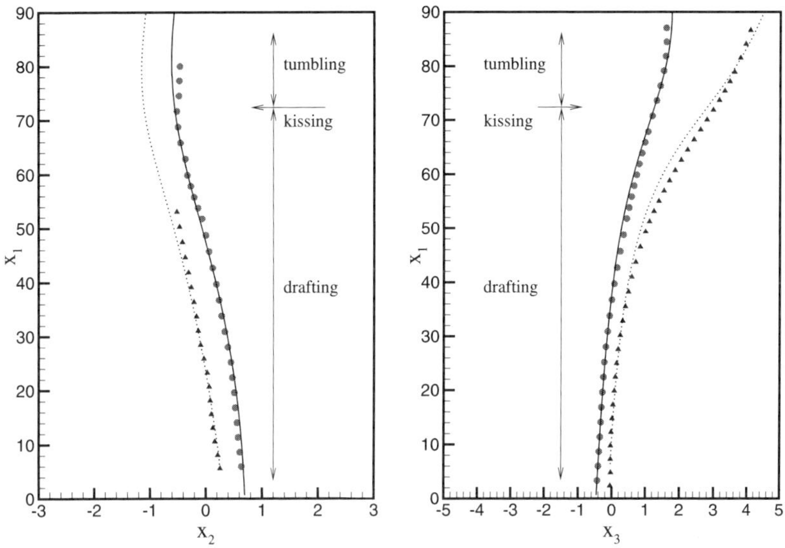

FIGURE 14.17. Particle trajectories in (left) streamwise-normal directions and (right) streamwise-spanwise directions. Comparison of FCM predictions (lines) and experimental results (symbols). The geometry is a channel of width 10 mm with the walls located at $x_2 = \pm 5$ mm. The particle Reynolds number is 1.71 for the leading particle (triangle) and 1.60 for the trailing particle (circle). All particles have the same radius of 1 mm. (Courtesy of S. Lomholt.)

are based on spectral elements. By DNS we mean that we enforce the no-slip boundary condition on the particle directly as a Dirichlet condition. We consider the prototype problem of a flow past a sphere inside a three-dimensional channel. The sphere is located near one wall of the channel, which has length 15, height 7, and width 7. The origin of the coordinate system is located at the center of the channel, so that the flow domain covers $-7.5 \leq x \leq 7.5$, $-3.5 \leq y \leq 3.5$, and $-3.5 \leq z \leq 3.5$. A sphere of radius $R = 1$ is placed near the lower channel wall with its center at $(x_0, y_0, z_0) = (0, -1.5, 0)$. The fluid density and viscosity are chosen to be $\rho_f = 1.0$ and $\nu = 1.0$, respectively. Periodic conditions are imposed in the x- and z-directions, and no-slip conditions are applied on the two walls of the channel. The flow is driven by a constant force in the x-direction, $F = \frac{2\nu}{h^2}$, where h is the channel half-width.

The spectral DLM simulations employ Fourier expansions in the x- and z-directions and a spectral element discretization in the y-direction (Dong et al., 2004). Distributed Lagrange multipliers are used to impose the zero-flow constraint inside the sphere. The "collocation points" consist of the flow grid points lying inside the sphere and intersection points between the surface of the sphere and the underlying flow grid lines. On the boundary

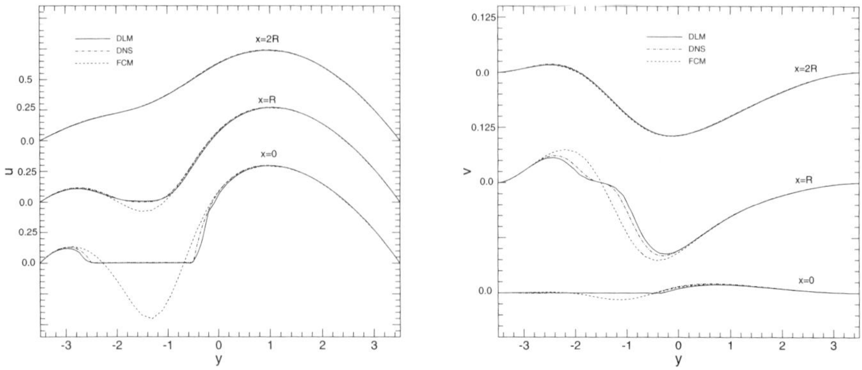

FIGURE 14.18. Sphere off center: comparison of velocity profiles between spectral DLM, DNS, and FCM at three streamwise locations at Reynolds number $\mathrm{Re} = 1.55$ (based on the channel center line maximum velocity and sphere diameter). Left: streamwise velocity. Right: wall-normal velocity.

collocation points the flow velocities are obtained via the spectral element interpolation procedure. In the FCM calculations a restoring force and torque are computed via a penalty method to keep the sphere from moving and rotating in the channel. The restoring force and torque are then used as the force monopole strength and dipole strength, respectively.

In Figure 14.18 we plot the streamwise (left figure) and wall-normal (right figure) velocity profiles at three downstream locations, $x = 0$, R, and $2R$, computed with all three methods. All three methods agree with one another quite well in the region far away from the sphere ($x = 2R$). Closer to the sphere surface and inside the sphere, the spectral DLM results are in good agreement with DNS results, while larger discrepancies between FCM and DNS results are observed in these regions. Due to asymmetry in the configuration, the flow exerts a torque and a lift force on the sphere. The coefficients for the drag and lift forces on the sphere and the torque with respect to the z-axis computed with all three methods are summarized in Table 14.4. FCM overpredicts the drag force on the sphere by about 6%. However, given the small number of elements, FCM simulations have produced lift and torque values that are in quite good agreement with DNS. FCM is a very fast method and scales favorably for a large number of particles. Even in this application with a single sphere, FCM was one to two orders of magnitude faster than DNS or DLM; the latter was as expensive as DNS. The drag and lift forces produced by spectral DLM are in good agreement with DNS results; the errors are within 2%. It is noted that poorly resolved regions on the surface of the sphere could affect the accuracy of the torque in DLM. For example, if the boundary collocation points consist only of intersection points between the sphere surface and the

	Elements/Grid	C_D	C_L	C_T
DNS	4608(5th order)	33.932	1.236	5.064
	4608(7th order)	33.929	1.236	5.066
	4608(8th order)	33.929	1.236	5.066
FCM	360(4th order)	35.122	1.274	4.872
	360(6th order)	35.175	1.274	5.251
	360(8th order)	35.306	1.276	5.251
DLM	$96 \times 60 \times 96$	33.475	1.172	5.236
	$96 \times 72 \times 96$	33.533	1.211	5.157
	$96 \times 84 \times 96$	33.724	1.210	5.175

TABLE 14.4. Drag coefficient ($C_D = \frac{\text{Drag}}{\frac{1}{2}\rho_f U_m^2 d}$, where U_m is the maximum velocity at the center line of the channel and d is the cylinder diameter), lift coefficient ($C_L = \frac{\text{lift}}{\frac{1}{2}\rho_f U_m^2 d}$), and the torque coefficient ($C_T = \frac{\text{Torque}}{\frac{1}{4}\rho_f U_m^2 d^2}$) on the sphere near one wall in a 3D channel.

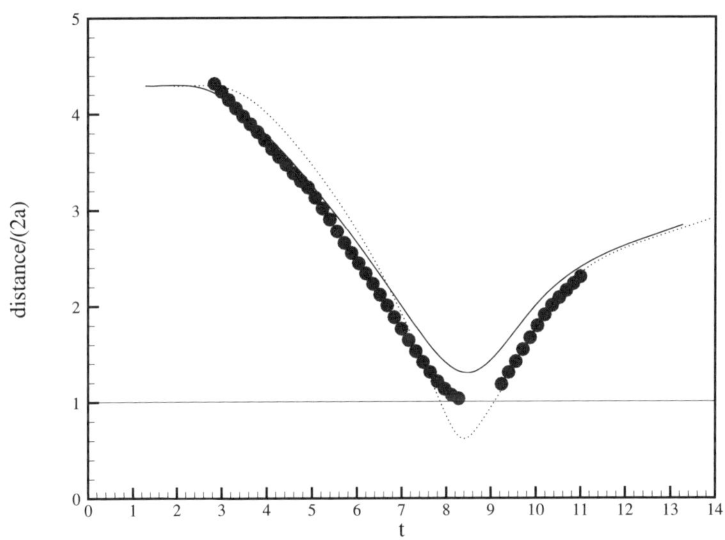

FIGURE 14.19. Relative distance between the centers of two particles in a microchannel. The circles denote experimental results, the solid line FCM predictions with monopole and dipole terms, and the dotted line FCM predictions with monopole terms only. (Courtesy of S. Lomholt.)

vertical grid lines, thc torque will be overpredicted on the grid $96 \times 84 \times 96$ (with a value 5.37).

Collisions and Contact Forces

In Section 14.3.1 we have presented several semianalytical results that can be used to analyze the hydrodynamic interactions between the particles as they approach each other, i.e., in the lubrication limit. The effect of lubrication is not explicitly included in FCM, and it has to be added via separate models (Dance and Maxey, 2003). For exact representation of both the lubrication effect and the far-field interaction, an infinite number of multipoles is required, which will make the method inefficient. However, it is interesting to note the effect of the dipole term in this context. Lomholt (2000) has studied this effect by considering the dynamics of the two particles when they touch, using FCM simulations with both the monopole and the dipole terms. A typical result is shown in Figure 14.19, where we see that the inclusion of the dipole contribution improves the results. Specifically, the trajectories of the two particles show a general agreement with the experimental results, but the numerical results, obtained with the monopole term show only erroneous overlap between the two particles. On the other hand, the inclusion of the dipole term leads to particles that touch but do not overlap; similar conclusions were reached in the simulations presented in (Liu et al., 2002). Including force dipoles helps to represent the particle–particle interactions in the flow more accurately, but they are not adequate on their own. Lubrication forces still play a role in low Reynolds number flows. At finite Reynolds numbers collisions do occur, and contact forces need to be explicitly included, e.g., elastic forces that cause particles to bounce. To this end, simple models based on the interparticle distance to mimic the elastic collision effect can be employed, as we show next.

The component of the repulsion force on particle j from particle k in the x_i direction is computed as:

$$F_i^{jk} = C_1 \frac{x_i^j - x_i^k}{2a} \left(\frac{(1.05 \times 2a)^2 - r^2}{C_2 \times 4a^2} \right)^n , \tag{14.16}$$

where $C_1 = 1.02$ and $C_2 = 0.011$ are two constants to adjust the strength of the force; a is the particle radius; $x_i^j - x_i^k$ is the difference in the x_i coordinates of particles j and k; and r is the distance between the two particles. The exponent n can take the value 2 or 3 for weak or strong repulsions, respectively. The collision force works in pairs, i.e., $\mathbf{F}^{jk} = -\mathbf{F}^{kj}$. The cut-off distance is 5% of the diameter away from the particle surface. In other words, when two particles are within distance $1.05 \times 2a$, the repulsion force will be automatically activated on each particle to push them away. For example, when $r \geq 1.05 \times 2a$, this repulsive force will remain 0. When $r = 1.025 \times 2a$, it is about 6 times the Stokes drag force. When $r = 2a$, it equals $43.8 \times 6\pi a\mu$. It is much stronger when particles start going into each other. Figure 14.20 shows the magnitude of the repulsion force versus the distance between two particles for $n = 2$ and $n = 3$.

A collision model of the same nature is developed for particles bumping

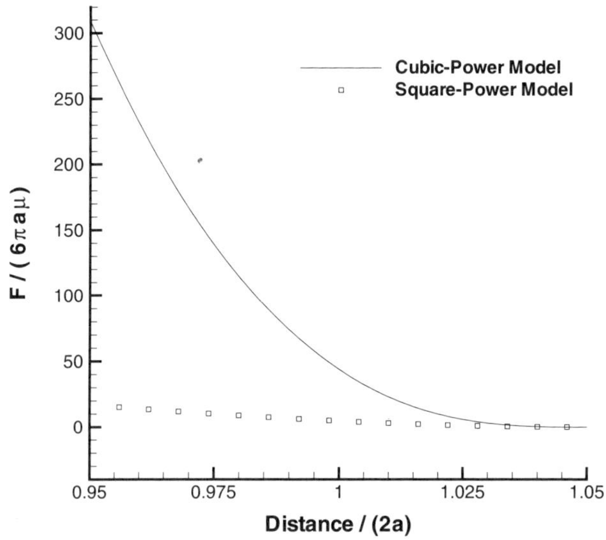

FIGURE 14.20. Normalized elastic collision force versus normalized distance.

into a wall at an angle. Here the force is no longer pairwise, but instead it acts on the particle from the wall. This force is computed as

$$F_i^j = C_1 \frac{x_i^j - Y_i}{a} \left(\frac{(1.05a)^2 - r^2}{C_2 a^2} \right)^n, \tag{14.17}$$

where Y_i is the coordinate of the contact point at a wall, and r is the distance from the center of particle j to the contact point. The same cut-off distance is used here. The contact point is found by a loop over the grid points on the wall and picking the point closest to the center of the particle.

Nonspherical Particles

The FCM as described above is valid for spherical nondeformed particles. However, extensions to ellipsoid particles are straightforward by setting the two axes of the ellipsoid thus:

$$a_1 = \sqrt{\pi}\sigma_1 \quad \text{and} \quad a_2 = \sqrt{\pi}\sigma_2.$$

In this case the Gaussian envelope is ellipsoidal with extent defined by σ_1 and σ_2. The expressions above are obtained by matching the settling velocities in FCM with the analytically obtained settling velocities for ellipsoid particles. The additional computational complexity is that equations of angular momentum have to be solved for each particle in order to obtain the required orientation information. Moreover, arbitrary shapes, e.g., particle chains or clusters, can be modeled as an assembly of ellipsoids with appropriate linkages and constraints in their motion. The important point is that the same governing equations for the microflow are valid, i.e., equation (14.15), and only the particle dynamics have to be modeled properly. Details of the implementation and validation results are presented in (Liu, 2004).

Summary

In summary, particulate microflows are difficult to compute directly, but the relatively low particle Reynolds number limit allows some simplifications. Microflows typically correspond to values of volume fraction greater than 10^{-3}, and thus nonlinear hydrodynamic interactions have to be accurately modeled. FCM is suitable for particulate microflows because it models accurately such interactions without the extra expense of special remeshing or solution of stiff algebraic systems as in the arbitrary Lagrangian Eulerian (ALE) method, the front tracking method, or the fictitious domain method (DLM). To appreciate the numerical resolution requirements, FCM typically employs approximately *5 grid points per particle* compared to 15 to 20 points in the aforementioned approaches. With regard to lubrication effects, simple models derived from Stokesian dynamics can be incorporated, when such corrections are needed as in the case of very large volume fraction. Similarly, different collision strategies can be implemented that are application-specific, as is typically done in particulate macroflows. The main drawback of FCM is that it assumes rigid particles, which may not be valid for some microflow applications. In addition, for dense suspensions the results may not be accurate.

15

Multiscale Modeling of Gas Flows

In this chapter we discuss theory and numerical methodologies for simulating gas flows at the mesoscopic and atomistic levels. Such a description is necessary for gases in the transition and free-molecular regimes. First, we present the direct simulation Monte Carlo (DSMC) method, a stochastic approach suitable for gases. We discuss limitations and errors in the steady version of DSMC and subsequently present a similar analysis for the unsteady DSMC. In order to bridge scales between the continuum and atomistic scales we present an iterative coupling algorithm used often in elliptic problems. To demonstrate convergence and accuracy of this algorithm we revisit the microfilters (see Section 6.5) and present specific details for several cases. We then give an overview of the Boltzmann equation, describing in some detail gas–surface interactions, and include benchmark solutions for validation of numerical codes and of macromodels. A main result relevant for bridging microdynamics and macrodynamics is the Boltzmann inequality, which we also discuss in the last section on lattice Boltzmann methods (LBM) as the H-theorem. These methods represent a "minimal" discrete form of the Boltzmann equation, and are applicable to both compressible and incompressible flows. An interesting new version is the entropic LBM, which enforces the H-theorem in order to guarantee Galilean invariance and numerical stability even for small viscosity values.

15.1 Direct Simulation Monte Carlo (DSMC) Method

The direct simulation Monte Carlo method was first developed by G. Bird, its inventor and pioneer (Bird, 1994). Several review articles about the DSMC method are also available (Bird, 1978; Muntz, 1989; Cheng and Emmanuel, 1995; Oran et al., 1998). Most of these articles present reviews of the DSMC method for low-pressure rarefied gas flow applications, with the exception of (Oran et al., 1998), where microfluidic applications are also addressed. The reader is referred to these books and articles for technical algorithmic details and sample DSMC programs.

Under standard conditions in a volume of $10\,\mu\text{m} \times 10\,\mu\text{m} \times 10\,\mu\text{m}$ there are about 10^{10} molecules. A molecular-based simulation model that can compute the motion and interactions of all these molecules is not possible. The typical DSMC method employs hundreds of thousands or even millions of "simulated" molecules that mimic but do not follow exactly the motion of real molecules.

DSMC is based on the splitting of the molecular motion and intermolecular collisions by choosing a time step smaller than the mean collision time, and tracking the evolution of this molecular process in space and time. For efficient numerical implementation, the space is divided into cells similar to the finite volume method. The DSMC cells are chosen proportional to the mean free path λ. In order to resolve large gradients in flow with realistic (physical) viscosity values, the average cell size Δx_c is about $\lambda/3$ (Oran et al., 1998). The time- and cell-averaged molecular quantities are presented as the macroscopic values at cell centers.

DSMC involves four main steps:

- Motion of the particles.
- Indexing and cross-referencing of particles.
- Simulation of collisions.
- Sampling of macroscopic properties of the flow field.

The basic steps of a DSMC algorithm are given in Figure 15.1 and are summarized below.

The **first step** involves motion of the simulated molecules during a time interval of Δt. Since the molecules will go through intermolecular collisions, the overall time step for simulation is chosen smaller than the mean collision time Δt_c. Once the molecules are advanced in space, some of them will have gone through wall collisions or will have left the computational domain through the outflow boundaries. Hence, the boundary conditions must be enforced at this level, and the macroscopic properties along the solid surfaces must be sampled. This is done by modeling the surface molecule interactions by application of conservation laws on individual molecules, rather

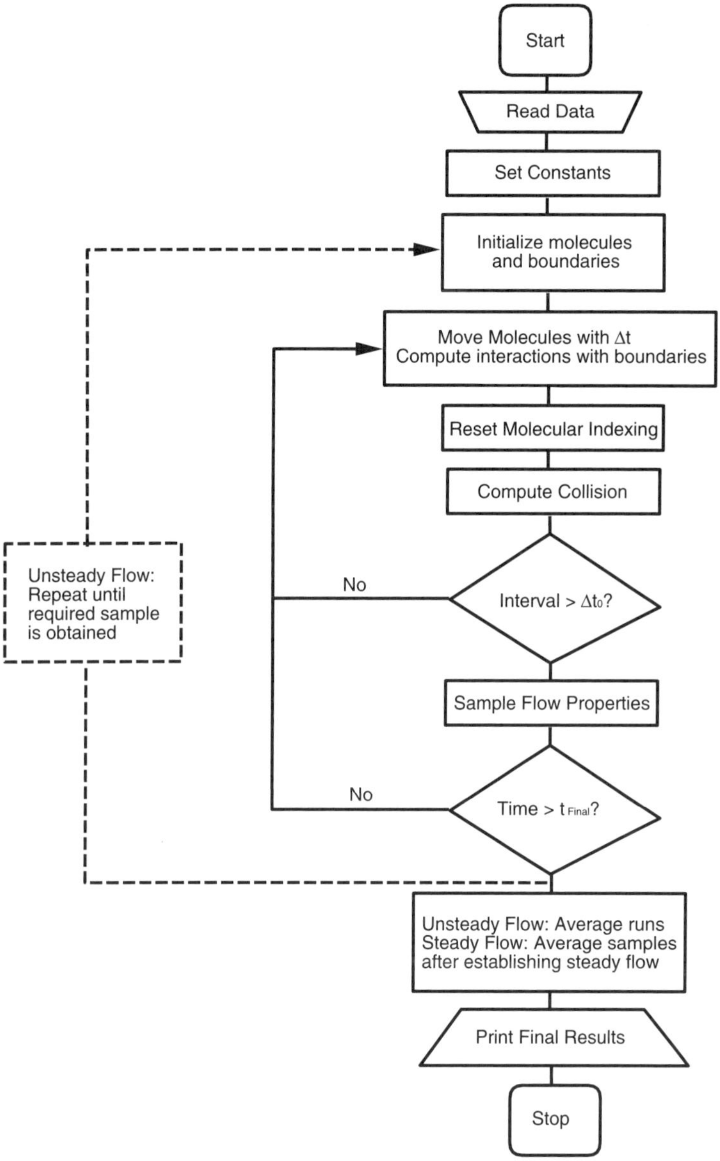

FIGURE 15.1. Typical steps for a DSMC method. (Courtesy of E. Oran.)

than using a velocity distribution function (commonly used in Boltzmann equation algorithms; see sections 15.4 and 15.5). This approach allows inclusion of many other physical processes, such as the chemical reactions, radiation effects, three-body collision, and ionized flow effects, without major modifications to the basic DSMC procedure. However, a priori knowledge of the accommodation coefficients must be used in this process (see

Section 2.2.2), and this constitutes a weakness of the DSMC method, similar to the Navier–Stokes-based slip and even Boltzmann-equation-based simulation models. This issue is discussed in detail in the following section.

The **second step** is indexing and tracking of the particles. This is necessary, since during the first stage the molecules might have moved to a new cell. The new cell locations of the molecules are indexed; hence the intermolecular collisions and flow field sampling can be handled accurately. This is a very crucial step in an efficient DSMC algorithm. The indexing, molecule tracking, and data structuring algorithms should be carefully designed for specific computing platforms. Dietrich and Boyd presented DSMC calculations for more than 100 million simulated particles on 400-node IBM SP2 computer with 90% parallel efficiency (Dietrich and Boyd, 1996). Parallel efficiency of DSMC algorithms requires very effective load balancing methods based on the number of molecules, because the computational work of a DSMC method is proportional to the number of simulated molecules.

The **third step** is simulation of collisions via a probabilistic process. Since only a small portion of the molecules is simulated, and the motion and collision processes are decoupled, the probabilistic treatment becomes necessary. A common collision model is the **no-time-counter** (NTC) technique of Bird (1994), which is used in conjunction with the subcell technique, where the collision rates are calculated on the basis of the DSMC cell, and the collision pairs are selected within the subcells. This improves the accuracy of the method by maintaining the collisions of molecules near to each other (Oran et al., 1998).

The **last step** is the sampling of macroscopic flow properties, which is done within a cell, and the appropriate macroscopic values are presented at cell centers. The macroscopic properties for unsteady flow conditions are obtained by ensemble averaging of many independent calculations. For steady flows, time-averaging can also be used.

15.1.1 Limitations and Errors in DSMC

While DSMC is a very powerful approach, and in fact the only possible numerical approach in simulating high Knudsen number rarefied flows in complex geometries, there are several possible limitations and error sources that the reader should be aware of:

1. **Finite Cell Size:** The typical DSMC cell should be about one-third of the local mean free path. Values of cell sizes larger than this may result in erroneously enhanced diffusion coefficients. In DSMC one cannot directly specify the dynamic viscosity of the fluid. The dynamic viscosity is calculated via diffusion of linear momentum. (Breuer et al., 1995) have performed one-dimensional Rayleigh flow problems in the continuum flow regime. They discovered that for cell

sizes larger than one mean free path the apparent viscosity of the fluid increases. Also, the dependence of viscosity and thermal conductivity on the cell size has been studied using the Green–Kubo theory (Alexander et al., 1998). Specifically, it was shown that the transport coefficients deviate from the Enskog values for dilute gas as the square of the cell size Δy. For example, for the viscosity coefficient it was found that

$$\mu = \frac{5}{16d^2}\sqrt{\frac{mk_{\mathrm{B}}T}{\pi}}\left[1 + \frac{16}{45\pi}\frac{\Delta y^2}{\lambda^2}\right],$$

where d is the molecular diameter. Therefore, if the cell size is equal to the mean free path, the error is on the order of 15%.

2. **Finite Time Step:** Due to the time splitting of the molecular motion and collisions, the maximum allowable time step should be smaller that the local collision time Δt_c. Values of time steps larger than Δt_c will result in traveling of molecules through several cells prior to a cell-based (even subcell-based) collision calculation. In (Hadjiconstantinou, 2000), it was found that the error in transport coefficients is proportional to the square of the time step; e.g., the viscosity has the form

$$\mu = \frac{5}{16d^2}\sqrt{\frac{mk_{\mathrm{B}}T}{\pi}}\left[1 + \frac{16}{75\pi}\frac{(v_m\Delta t)^2}{\lambda^2}\right],$$

where $v_m = \sqrt{2RT}$ is the most probable velocity.

The time-step and cell-size restrictions stated above are not a Courant Friederichs Lewy (CFL) stability restriction of a typical wave equation formulation. The DSMC method is always stable. Violation of the physical restrictions stated in 1 and 2 may result in highly diffused numerical results.

3. **Ratio of the Simulated Molecules to the Real Molecules:** Due to the vast number of molecules in simulations and limited computational resources, one always has to choose a sample of molecules to simulate. If the ratio of the actual to the simulated molecules becomes too high, the statistical scatter of the solution is increased. The details for the statistical error can be found in (Oran et al., 1998; Bird, 1994; Chen and Boyd, 1996).

4. **Boundary Condition Treatment:** Especially the inflow/outflow boundary conditions can become important in a microfluidic simulation. A subsonic microchannel flow simulation may require specification of inlet and exit pressures. Based on the apparent viscosity of the fluid, the flow will develop under this pressure gradient, and result in a certain mass flowrate. During such a simulation, specification of back-pressure for subsonic flows becomes a problem. For this case we

have implemented and recommend a DSMC algorithm mimicking the characteristic treatment of subsonic compressible flow algorithms in the following fashion: At the entrance of the channels we specified the number density, temperature, and average velocity of the molecules. At the outflow region we specified the desired number density and temperature (for a given pressure drop) as well as the corresponding average velocity of the molecules that are entering the computational domain from the outflow boundary. By this treatment we were able to significantly reduce the spurious numerical boundary layers at the inflow and outflow regions.

Also, for high Knudsen number flows (i.e., $\mathrm{Kn} > 1$) in a channel with blockage (such as a sphere in a pipe) we have observed that the location of the inflow and outflow becomes important. For example, the molecules reflected from the front of the body may reach the inflow region with very few intermolecular collisions. This creates a largely diffused flow at the front of the bluff body (Liu et al., 1998).

5. **Uncertainties in the Physical Input Parameters:** These include typically the input for molecular collision cross-section models such as the hard sphere (HS), variable hard sphere (VHS), and variable soft sphere (VSF) models; see (Oran et al., 1998; Vijayakumar et al., 1999), and the references therein. The hard sphere model is usually sufficient for monatomic gases or for cases with negligible vibrational and rotational nonequilibrium effects. To demonstrate this claim we investigate the accuracy of our DSMC results in Section 4.2 by performing higher-resolution DSMC studies. The original resolution studies for microchannels employed 6,000 cells with 80,000 simulated molecules, and the results were sampled for 2.0×10^6 time steps. In Figure 15.2 comparisons of $\mathrm{Kn} = 0.1$ and $\mathrm{Kn} = 2.0$ channel flow velocity profiles obtained by VHS and VSS models for various cell and simulated molecule resolutions are presented. The 24,000 cells with 480,000 molecule runs are time-averaged for 1 to 2×10^5 time steps. No significant differences in various DSMC results are observed. However, the higher-resolution cases converged faster. The predictions of various DSMC models agree well with the linearized Boltzmann solutions obtained by (Ohwada et al., 1989a).

Along with these possible error sources and limitations there are some disadvantages of the DSMC method applied to gas microflows. These are:

1. **Slow Convergence:** The error in the DSMC method is inversely proportional to the square root of the number of simulated molecules. Reducing the error by a factor of two requires increasing the number of simulated molecules by a factor of four! This is a very slow convergence rate compared to continuum-based simulations with spatial accuracy of second or higher order. Hence, if the continuum equations

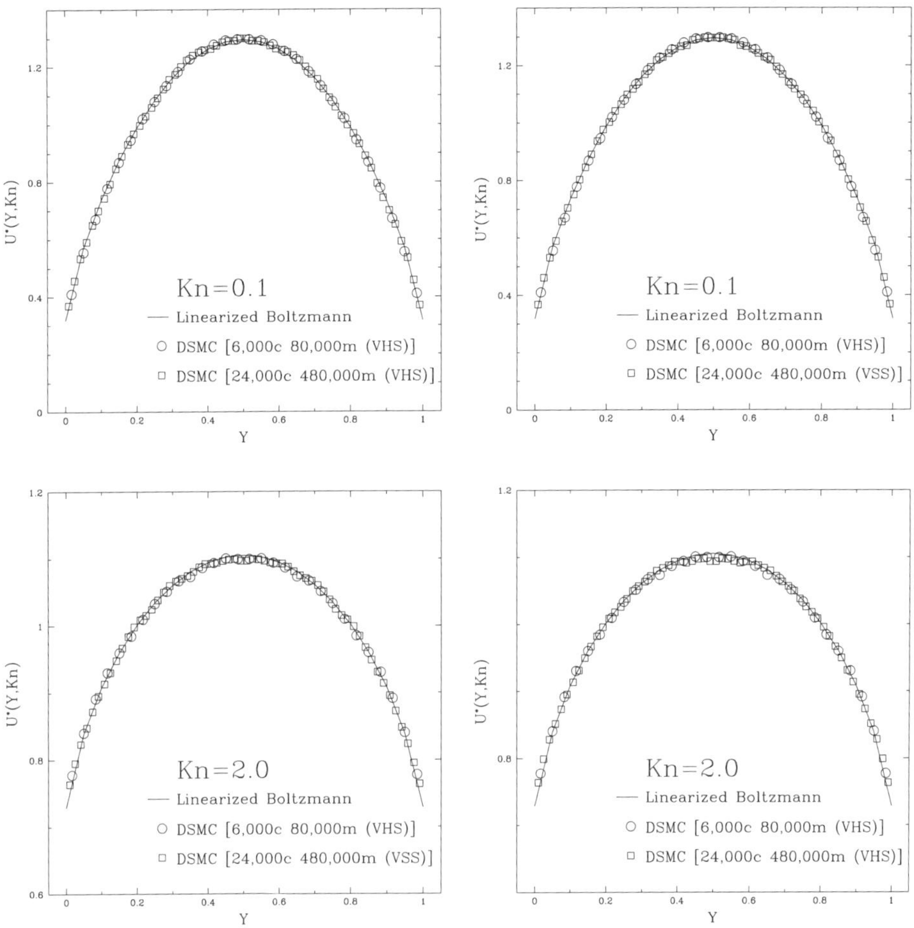

FIGURE 15.2. Nondimensionalized velocity distribution across the channel for Kn = 0.1 and Kn = 2.0 flows. Comparisons of various DSMC resolutions against the linearized Boltzmann solution of (Ohwada et al., 1989a) are presented. It is seen that the DSMC simulations using the VSS (left) and VHS (right) models are in good agreement with the linearized Boltzmann solutions.

are valid (for Kn $<$ 0.1), it is computationally more efficient to use the continuum models than the DSMC method.

2. **Large Statistical Noise:** Microflows are typically low-speed flows (1 mm/s to 1 m/s). The macroscopic fluid velocities are obtained by averaging the molecular velocities (of approximate value 500 m/s) for a long time. This five to two orders of magnitude difference between the molecular and average speeds results in large statistical noise, thus microflows require a very long time averaging for gas microflow

simulations. The statistical fluctuations decrease with the square root of the sample size. Time or ensemble averages of low-speed microflows on the order of 0.1 m/s require about 10^8 samples in order for one to be able to distinguish such small macroscopic velocities. A modified version was introduced by (Fan and Shen, 1999), who developed an information-preservation (IP) technique that enables DSMC simulation of low-speed flows at higher efficiency; see Section 15.1.3.

3. **Long Time to Reach Steady State:** For low-speed microflows the time required to reach steady state is usually dictated more by the convective time scales than the diffusive time scales. For example, gas flow in a microchannel of length 1 cm and height 1 μm with an average speed of 1 cm/s will require the DSMC simulation to reach time 1 second for the macroscopic disturbances to travel from the inflow toward the outflow of the channel. On the other hand, the viscous time scale for this problem can be calculated by $t_{\text{visc}} \approx \sqrt{h^2/\nu}$, where ν is the kinematic viscosity ($\nu_{\text{air}} \approx 10^{-6}$ m^2/s), and h is the height of the channel. For air in a 1-μm channel the viscous time scale is 10^{-3} seconds, a value three orders of magnitude less than the convective time scale. The mean collision time for air at standard conditions is on the order of 10^{-10} seconds, and the time step of the DSMC method should be smaller than the mean collision time. Hence, a DSMC algorithm must run at least 10^{10} time steps for this microchannel prior to settling into a steady-state condition. In practice, due to the simple geometry, settling into a steady state may occur much earlier. However, for a large aspect ratio channel with geometric complications common in microsystems, the above order of magnitude analysis will be valid.

4. **Extensive Number of Molecules:** If we discretize a rectangular domain of 1 mm $\times 100\,\mu$m $\times 1\,\mu$m under standard conditions for Kn $=$ 0.1 flow, we will need at least 30 cells per 1 μm length scale. This results in a total of 2.7×10^9 cells. Each of these cells should contain at least 20 simulated molecules, resulting in a total of 5.4×10^{10} particles. Combined with the number of time step restrictions given above, low-speed microfluidic simulation with DSMC easily goes beyond the capabilities of most computers.

   An alternative treatment to overcome the extensive number of simulated molecules and long integration times is utilization of the *dynamic similarity* of low-pressure rarefied gas flows to gas microflows under atmospheric conditions. The key parameters for the dynamic similarity is the geometric similarity, and matching of the flow Knudsen, Mach, and Reynolds numbers. Performing actual experiments under dynamically similar conditions may be very difficult. However, parametric studies via numerical simulations are possible. The fundamental question to answer for such an approach is whether a specific, unforeseen microscale phenomenon is missed with the dynamic similarity approach. However, all numerical simulations are inherently model-based. Unless microscale-specific models are implemented in the algorithm, we will not be able to obtain more physical information from an atomistic simulation than from a dynamically similar low-pressure simulation.

5. **Lack of Deterministic Surface Effects:** Molecule–wall interactions are specified by the accommodation coefficients (σ_v, σ_T); for diffuse reflection, $\sigma_v = 1$. Hence, the reflected molecules lose their incoming tangential velocity and are reflected with the tangential velocity of the wall. For $\sigma_v = 0$, the tangential velocity of the impinging molecules is not changed. For any other value of σ_v, a combination of these procedures can be applied. This level of wall and boundary interaction treatment is more fundamental than the slip conditions presented in Chapter 2. However, it still lacks the most fundamental way of simulating the molecule–wall interactions, which includes the molecular structure of the walls. Such approaches are obtained by molecular-dynamics simulations (see Section 16.1 and (Tehver et al., 1998)). Also, the accommodation coefficients for every surface and gas pair are not available (see Section 2.2.2 for details on recent experimental research).

15.1.2 DSMC for Unsteady Flows

Microsystems often experience unsteady or time-periodic flows, which result in dynamic variations of lift and drag forces, and torques on the device components. Computations of unsteady flows require time-dependent numerical simulations, and DSMC provides an effective tool in the transition and free molecular flow regimes (Kn ≥ 0.1). To this end, this section presents the key concepts that need to be addressed in DSMC computations of unsteady flows.

In order to make these ideas specific, in the following we discuss the concepts and conditions relevant to the computations of oscillatory Couette flows, presented in Section 3.3. Lateral oscillations in this prototype geometry require consideration of one-dimensional unsteady flows. In these computations, we utilized the hard sphere (HS) model for molecular collision, and the no-time-counter (NTC) scheme for collision pair selection (Bird, 1994). The choice of the HS model facilitates easy comparisons with the theoretical solutions of the linearized Boltzmann equation, and it also enables easier code implementation. Argon gas is simulated with a reference temperature $T_e = 273$ K. The surfaces are assumed to be fully accommodating. Hence, the particles are reflected from the surfaces according to a Maxwellian distribution with surface velocity and temperature. Simulation parameters are chosen such that the compressibility and viscous heating effects are negligible. Although the gas temperature increases with increased oscillation frequency, the maximum temperature rise in the simulations is less than 2%. More than 100 simulated particles per cell are employed. The entire domain is discretized into 40 to 100 equally spaced cells, to ensure that the cell size (Δy) is smaller than the mean free path for all simulations. A finer grid is used for high Stokes number flows to resolve the flow inside the Stokes layers, while a coarser grid is used for low Stokes number flows. With respect to selecting the time step (Δt), the following issues need to be considered:

1. Δt has to be significantly smaller than the mean collision time, Δt_c, to ensure accurate simulations.

2. The cell size Δy should be smaller than the mean free path, λ, and the molecules should not move across more than one cell between two consecutive time steps. Hence, $\Delta t \ll \Delta y / v_m$, where $v_m = \sqrt{2RT}$ is the most probable velocity.

3. The characteristic length scale of the problem, L, and kinematic viscosity, ν, in micro- and nanoflows can result in small viscous diffusion time, which scales as L^2/ν. Hence, $\Delta t \ll L^2/\nu$.

4. The time period of oscillations ($T_0 = 2\pi/\omega$) can become smaller than the mean collision time; hence $\Delta t \ll 2\pi/\omega$, where ω is the frequency of oscillations.

We must indicate that the first three requirements are also valid for steady DSMC computations, while the last requirement is specific for unsteady flows. In all simulations of Section 3.2, the time step was chosen to satisfy all four constraints. The range of the total simulation time was around three to eighteen time periods. This ensured that the transients starting from the quiescent initial conditions decay, and a time-periodic state is achieved. In unsteady DSMC, ensemble-average at each time step replaces the time-average used in steady computations. Ensemble-averaging is performed over 5000 different realizations of the stochastic process for each time step.

In all simulations, the amplitude of the oscillating wall is kept constant at $u_0 = 100$ m/s, maintaining relatively low Mach number flows so that compressibility effects are negligible. The gas number density (n_0), excitation frequency (ω), and the characteristic system length (L) are adjusted to simulate different combinations of Kn and β.

Effects of Statistical Scatter

Figure 3.8(d) aids in better visualization of statistical scatter in the DSMC results, which is insignificant for this case even near the stationary wall ($y/L \leq 0.1$). Note that the normalized amplitude does not drop below 1%

of the maximum signal for this case. In the simulations, onset of statistical fluctuations is observed when the normalized signal amplitude drops below 1% of the maximum signal. Some of the DSMC results presented in Section 3.3 exhibit statistical fluctuations. In order to explore the statistical fluctuations induced by finite sampling in the presence of thermal fluctuations, we follow (Hadjiconstantinou et al., 2003). In *equilibrium statistical mechanics*, the ratio of excitation velocity u_0 to the thermal fluctuation u' for an ideal gas is given by (Hadjiconstantinou et al., 2003)

$$\frac{u_0}{\sqrt{\langle (u')^2 \rangle}} = M\sqrt{\gamma N_0}, \tag{15.1}$$

where M is the Mach number, γ is the ratio of specific heats, and N_0 is the average number of particles per single cell. The velocity fluctuation is defined as $u' = u - \langle u \rangle$, where u is the instantaneous velocity and $\langle u \rangle$ is the average velocity. Since the unsteady DSMC algorithm uses ensemble averages over K repeating runs, the "noise-to-signal" ratio E_u can be expressed as

$$E_u = \frac{\sqrt{\langle (u')^2 \rangle}}{u_0} = \frac{1}{M\sqrt{\gamma K N_0}}. \tag{15.2}$$

Based on the above definition, and the typical simulation parameters used in Section 3.3 (u_0 = 100 m/s, K = 5000, N_0 = 100, and $\gamma = 5/3$), we obtain $E_u = 3.4 \times 10^{-3}$. Considering that the above expression is obtained for a medium in equilibrium, the noise level in unsteady computations is expected to be higher due to the presence of strong nonquilibrium effects in high Stokes number rarefied flows.

Finally, we must indicate that DSMC computations of more complex cases, such as the squeeze film damping (see Section 6.1), often require consideration of two- or three-dimensional unsteady flows. Although such computations are very challenging, ensemble averaging requires independent realizations of unsteady flows. This may enable code parallelization with perfect parallel efficiency if the stochastic processes as well as the random number generators employed in each processor are independent of each other.

15.1.3 DSMC: Information-Preservation Method

(Fan and Shen, 1999) have developed an information-preservation (IP) scheme for low-speed rarefied gas flows. Their method uses the molecular velocities of the DSMC method as well as an *information velocity* that records the collective velocity of a large number of molecules that a simulated particle represents. The information velocity is based on an inelastic collision model. The results presented for one-dimensional Couette, Poiseuille, and Rayleigh flows in the slip, transition, and free-molecular flow regimes show very good agreement with the corresponding analytical solutions. Also, Boyd and his co-workers have developed a DSMC-IP scheme for simulation of two-dimensional gas microflows (Cai et al., 2000). In (Sun and Boyd, 2002), a general two-dimensional unsteady IP formulation is described that relaxes the incompressible, isothermal assumptions of (Fan and Shen, 1999). The new IP version was shown in (Sun and Boyd, 2002), to provide good agreement with DSMC fundamental test cases including Couette flow and Rayleigh flow. This method was also used in (Sun and Boyd, 2004), to investigate fundamental aerodynamic behavior under flow conditions typical of microsystems.

The DSMC-IP method overcomes the statistical scatter in low-speed constant-density flow systems. The statistical scatter of DSMC is due to the thermal motion of particles with mean thermal speed, which is on the order of 450 m/s. However in the DSMC-IP scheme, the thermal motion causes statistical scatter only at the information level, which is smaller than the average macroscopic velocity of the system, typically on the order of 1 m/s. Therefore, the DSMC-IP method can be used for simulation of gas microflows. The information-preservation algorithm requires about 37.5% memory increase, compared to a regular DSMC algorithm. Since the DSMC-IP reduces the statistical scatter, it is possible to identify convergence to steady state by monitoring the average number density and streamwise velocity (or the kinetic energy) of the system. This approach cuts down the sample size and correspondingly decreases the CPU time required by a standard DSMC method for low-speed flows by *orders of magnitude*.

In Figure 15.3, we present a typical result from (Boyd and Sun, 2001). A comparison of density contours obtained from the DSMC, DSMC-IP, and Navier–Stokes with slip simulations for flow past an NACA 0012 airfoil is shown. The free stream flow conditions correspond to $M = 0.1$, $\mathrm{Re} = 1$, $\mathrm{Kn} = 0.013$. The computational domain consists of 9120 nonuniform structured cells, which are clustered near the airfoil. Each DSMC cell has 45 particles on average, while the DSMC-IP has 40 particles per cell. The time step is set to 5×10^{-8} s, which is smaller than the mean collision time of the particles. Both the DSMC and DSMC-IP are executed for 30,000 iterations to reach a steady state, before sampling the flow field. The sampling continued until each cell had an average of 400,000 sampled particles (corresponding to about 1000 samples per cell). Large statistical scatter is observed in the density contours obtained in the DSMC algorithm, as shown in Figure 15.3 (a). However, the density contours of the DSMC-IP are smooth, and agree well with the Navier–Stokes with slip results (plot in (c)) (Boyd and Sun, 2001). (Shen et al., 2003) utilized the DSMC and IP-DSMC methods to simulate low-speed flows in long microchannels. The IP results agree with the experimental measurements of pressure distribution and mass flow through microchannels in (Arkilic et al., 1994; Liu et al., 1993; Pong et al., 1994). The IP and DSMC also predict the Knudsen's

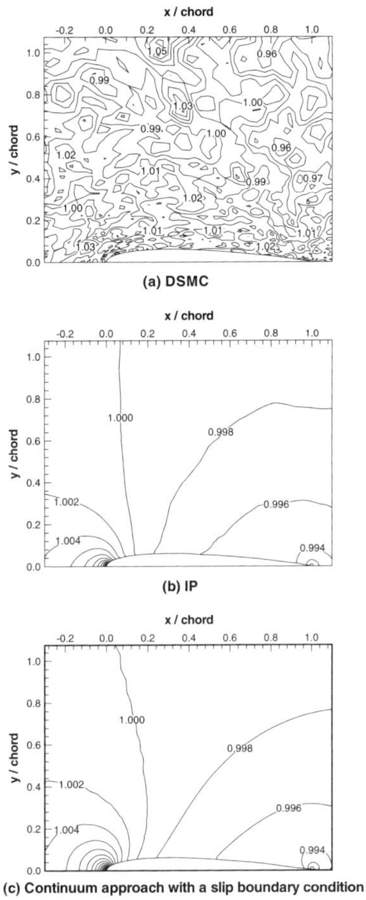

FIGURE 15.3. Density contours of flow past an NACA 0012 airfoil, obtained by the DSMC (a), DSMC-IP (b), and Navie–Stokes with slip (c) algorithms. The free stream flow conditions are $M = 0.1$, $\mathrm{Re} = 1$, $\mathrm{Kn} = 0.013$. (Courtesy of I.D. Boyd.)

minimum equally well.

The DSMC-IP method is relatively new, and further developments for nonisothermal flow conditions, which require preservation of the internal energy, are necessary. However, the result presented in Figure 15.3 is a clear indication of significant advancement of the DSMC-IP over the classical DSMC method for nearly isothermal flow conditions. Another direction in which the IP method has been developed and utilized is as an interface in hybrid methods that combine particle-based DSMC with continuum-based discretizations; see (Sun et al., 2004), and also Section 15.2.

15.2 DSM: Continuum Coupling

In this section we discuss possible procedures for coupling the DSMC method with the Navier–Stokes equations. This is important particularly for simulation of gas flows in MEMS components. If we consider the microcomb drive mechanism, the flow in most of the device can be simulated by slip continuum-based solvers. Only when the gap between the surfaces becomes submicron should the DSMC method be employed. Similarly, in microchannel flows mixed slip-transition flows could occur. Hence, it is necessary to implement multidomain DSMC/continuum solvers. Depending on the application, hybrid Euler/DSMC (Roveda et al., 1998) or DSMC/Navier-Stokes algorithms (Hash and Hassan, 1997) can be used. Such hybrid methods require compatible kinetic-split fluxes for the Navier–Stokes portion of the scheme (Lou et al., 1998) so that an efficient coupling can be achieved. An adaptive mesh and algorithm refinement (AMAR) procedure, which embeds a DSMC-based particle method within a continuum grid, has been developed in (Garcia et al., 1999); it enables molecular-based treatments even within a continuum region. Hence, the AMAR procedure can be used to deliver microscopic and macroscopic information within the same flow region. An effective coupling approach is based on the DSMC-IP technique presented earlier in this chapter. The original idea for introducing the IP method was to reduce the high level of statistical fluctuations found in DSMC calculations. This same property makes it suitable as an interface for communicating information to large scales described by continuum-based approximations (Sun et al., 2004).

To make these ideas more concrete we discuss some possible algorithms. We assume that in the highly rarefied region (Kn $\geq$ 0.1) we employ a DSMC discretization that is computationally efficient in this regime. As Kn $\rightarrow$ 0 the DSMC approach becomes increasingly inefficient, and thus the need for a continuum description in the low Kn region. Such a situation, for example, arises in gas flow through a long narrow tube or slit where the pressure drop downstream produces a high degree of rarefaction. The same is true for the expansion region of micronozzle flows we studied in Section 6.6.

The change from a microscopic to macroscopic (or vice versa) behavior is gradual, and therefore the question arises as to where the interface separating rarefied from nonrarefied behavior should be located. A possible criterion for determining this interface is the local Knudsen number defined as

$$\mathrm{Kn}_l = (\lambda/\rho)|\nabla\rho|$$

or a continuum breakdown parameter suggested by (Bird, 1994):

$$P = M(\lambda/\rho)|\nabla\rho|,$$

where M is the Mach number. Values of $P = 0.1$ and 0.01 to 0.02, respectively, have been suggested by Bird and other researchers, e.g., for rarefied

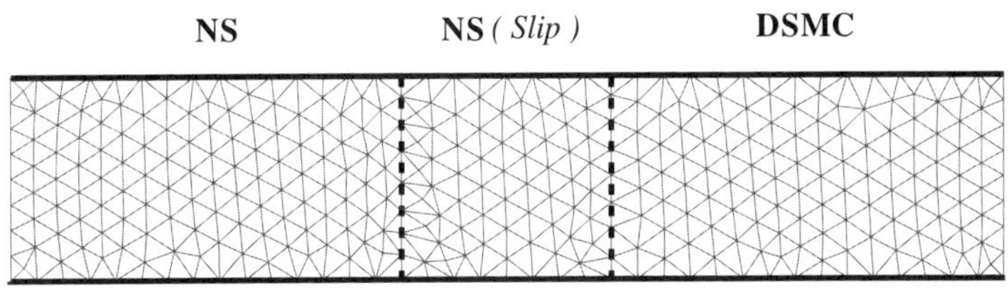

FIGURE 15.4. Domain for DSMC-continuum coupling. The interface is an overlap region where DSMC and modified Navier–Stokes with slip are both valid.

flows encountered in high-altitude applications. However, for internal flows at microscales the criterion is application-specific. For small Reynolds number viscous flows, deviation of velocity distribution from the Maxwellian may provide a more definitive metric. This interface from continuum to rarefied has been treated as a single surface in previous attempts to couple DSMC with Navier–Stokes equations starting with the work of (Wadsworth and Erwin, 1992) and in subsequent work, see (Hash and Hassan, 1997), and references therein. It is perhaps better to treat the interface as a finite zone giving an overlap between the region of validity of DSMC and the Navier–Stokes equations. A typical situation is sketched in Figure 15.4, corresponding to the aforementioned example of a long narrow channel. The proposed interface extends from Kn_1 to Kn_2, inside which the modified Navier–Stokes with the high-order slip condition (see equation (2.28)) and DSMC are both valid.

The key theoretical issue here in developing a hybrid description from the atomistic to macroscopic scale is the identification and theoretical justification of a proper set of interface conditions. In past work associated with high-altitude rarefied flows, three different procedures have been proposed and implemented with various degrees of success:

- the Marshak condition,
- extrapolating the fluxes, and
- extrapolating the properties.

The Marshak condition is an extension of a concept in radiative heat transfer, and it was first proposed by (Golse, 1989) for higher-atmosphere applications. It employs the half-fluxes at the interface, i.e., the flux of particles with velocity in the half normal velocity range. The total flux is then set to the sum of the half-flux based on the DSMC side and the half-flux based on the Navier–Stokes side. Matching the fluxes by extrapolation leads to a conservative global scheme, but the main difficulty comes from the large scatter of fluxes (i.e., momentum and heat flux) on the DSMC side that cannot match the smooth macroscopic fluxes on the Navier–Stokes side. In low Reynolds number cases, the scatter effect is more pronounced than in the high-altitude rarefied flows, and is particularly large for high-order moments, e.g., the fluxes. Finally, the extrapolation of macroscopic properties at the interface (density, velocity, temperature) does not guarantee monotonicity and is critically dependent on both smoothing and accuracy of extrapolation of these quantities. In addition, a large number of samples is required for averaging in order to obtain the macroscopic properties (an input to the Navier–Stokes solver), which renders this approach inefficient.

While these issues have been addressed to some extent in high-altitude rarefied flows, no significant progress has been made (Hash and Hassan, 1997). In addition, flows at microscales correspond to very viscous subsonic flows, whereas the previously studied flows corresponded to high-speed transonic and supersonic flows. A more appropriate approach is to develop interface conditions based on the overlap zone shown in Figure 15.4. The Navier–Stokes along with the slip conditions in the overlap zone will facilitate a smooth transition between the no-slip continuum flow and the large-slip rarefied flow. The size of the overlap is a parameter that will be selected in such a way that conservativity, solution continuity, and solution convergence are guaranteed. With regard to convergence in particular, the following acceleration scheme can be employed based on a relaxation procedure:

$$\phi_{\mathrm{MC}}^{n+1} = \theta\phi_{\mathrm{MC}}^{n} + (1-\theta)\phi_{\mathrm{NS}}^{n},$$

where θ is the acceleration parameter with $\theta \in (0, 1)$. Here the subscript (MC) denotes quantities on the DSMC side and (NS) on the Navier–Stokes side. Note that the microscopic quantities in the DSMC region will be computed based on the Chapman–Enskog distribution, which involves *gradients* of interpolated quantities, since Maxwellian distributions are certainly inappropriate for these nonequilibrium flows. The overlap region also provides a spatially homogeneous regime over which appropriate averages can be performed to reduce scatter from the DSMC solution. On the Navier–Stokes side, subsonic outflow conditions can be imposed following a characteristic decomposition that could involve fluxes (if the Marshak condition is enforced) or property interpolated quantities in analogy with the above interpolation procedure. In either case, appropriate formulations need to be developed in order to take into account the overlap zone.

As far as frequency of coupling is concerned, the time step restrictions on the Navier–Stokes side are dictated by the Courant number (CFL; $\Delta tU/\Delta x$) and most probably the diffusion number ($\Delta t\nu/\Delta x^2$) for the very low Reynolds number we consider, whereas on the DSMC side the time step is controlled by the mean collision time λ/c. This suggests that coupling should take place at almost every time step if the diffusion time step constraint dominates or a subcycling procedure should be followed, in which one Navier–Stokes step is followed by many DSMC steps if the convective time step dominates.

To understand the various algorithmic and implementation issues, we consider the example of multiscale analysis of microfilters. The specific algorithmic issues are elaborated below and in the next section, and the results are presented in the subsequent section. A high-level description of the multiscale method is presented in *Algorithm* 1. Given an arbitrary initial state and a set of boundary conditions along the overlapping interfaces, a Schwarz technique is implemented to find a self-consistent solution to the Stokes (or Navier–Stokes) and the DSMC subdomains. Self-consistency is determined by a convergence check that requires that the DSMC noise as well as the updates to the solution (for example, pressure and the velocities) be less than a specified tolerance value. Self-consistency also ensures that the boundary conditions at the interface have converged to the specified tolerance. To further reduce the DSMC noise in the solution, a postprocessing step is performed after the initial convergence check is satisfied. In the postprocessing step, several coupling iterations between Stokes (or Navier–Stokes) and DSMC subdomains are performed. The final results in the DSMC subdomain are obtained as an average of the samples collected during the postprocessing step. As a last step, the results from the DSMC subdomains are used as boundary conditions to find the solution in the continuum subdomains. The Schwarz algorithm and the interpolation between the Stokes (or Navier–Stokes) and the DSMC domains is discussed next.

15.2.1 The Schwarz Algorithm

To understand the Schwarz technique, consider two overlapping subdomains as shown in Figure 15.5(a). An alternating Schwarz method for this geometry can be summarized as follows:

$$
\begin{array}{lll}
\text{Begin} & : & n=0;\, u_2^{(0)}|_{\Gamma_1} = \text{initial condition} \\
\text{Repeat} & :\{ & n = n+1 \\
 & & \text{Solve } Lu_1^{(n)} = f_1 \text{ in } \Omega_1 \text{ with BC } u_1^{(n)} = u_2^{(n-1)} \text{ on } \Gamma_1 \\
 & & \text{Solve } Lu_2^{(n)} = f_2 \text{ in } \Omega_2 \text{ with BC } u_2^{(n)} = u_1^{(n)} \text{ on } \Gamma_2 \\
 & \} & \text{until convergence}
\end{array}
$$

where n is the iteration number, $u_i^{(n)}$ is the solution in domain Ω_i at iteration n, L is the partial differential operator describing the governing equations, and f_i are forcing functions of position in domain Ω_i. Here BC refers to the operator imposing the boundary conditions. In the alternating Schwarz method, the subdomains are overlapped, and Dirichlet-type

Algorithm 1 Description of DSMC/Stokes (or Navier-Stokes (NS)) coupling in various overlapping Schwarz methods.

Main Loop
initialization step
While *not converged** **do**
  Do *coupling_iteration*
end while

coupling_iteration
serial alternating Schwarz
for each DSMC domain D_i **do**
  Make N_{step} time steps in D_i
end for
Interpolate DSMC to Stokes/NS
for each Stokes/NS domain S_i **do**
  Solve Stokes/NS equations in S_i
end for
Interpolate Stokes/NS to DSMC

Coupling_iteration
colored alternating Schwarz
Make N_{step} time steps in all D_i
Interpolate DSMC to Stokes/NS
Solve Stokes/NS in all S_i
Interpolate Stokes/NS to DSMC

coupling_iteration
parallel alternating Schwarz
Make N_{step} time steps in all D_i
Solve Stokes/NS in all S_i
Interpolate DSMC to Stokes/NS
Interpolate Stokes/NS to DSMC

* Convergence requires the convergence of coupling iterations and the reduction of DSMC noise below a specified tolerance.

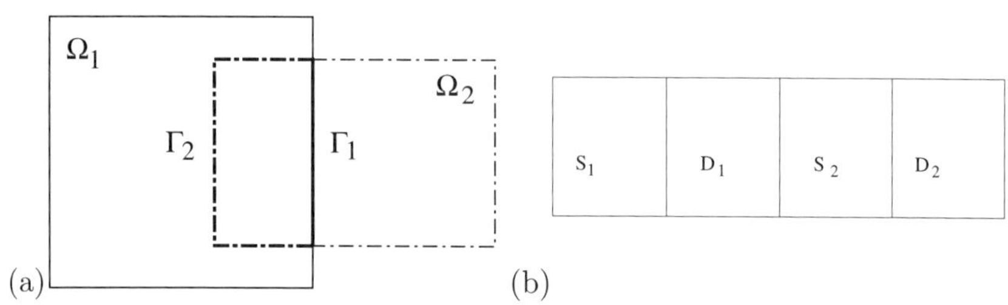

FIGURE 15.5. Decomposition of a sample geometry into (a) two overlapping subdomains, and (b) four subdomains.

boundary conditions are employed. The alternating Schwarz method can be modified to use nonoverlapping domains and Neumann-type boundary conditions, as explained in the previous section using a relaxation technique; see also (LeTallec and Mallinger, 1997), for a description on coupling time-dependent Navier–Stokes with DSMC using a nonoverlapped Schwarz coupling and Robin-type (mixed) boundary conditions. Several variations of the basic Schwarz technique for elliptic partial differential equations are presented in (Smith et al., 1996).

The alternating Schwarz method as described above is a serial technique. In Algorithm 1, the serial alternating Schwarz method and two parallel implementations of the Schwarz technique for coupling Stokes (or Navier–

Stokes) and DSMC subdomains are shown. To understand the implementation of various Schwarz techniques, consider the geometry and its decomposition shown in Figure 15.5(b). Here S_i denotes the Stokes (or the Navier–Stokes) subdomain, and D_i denotes the DSMC subdomain. In the serial alternating Schwarz method, each subdomain is solved sequentially, i.e., S_1, followed by D_1, followed by S_2, and so on. In a colored Schwarz method, the subdomains are divided into groups (i.e., colored), and the subdomains in each group are solved concurrently, while each group is solved sequentially. The optimal coloring depends on the geometry and its decomposition. For example, for the subdomains shown in Figure 15.5(b), all the D_i's are assigned one color, and all the S_i's are assigned a different color. All the D_i's are solved at once, followed by the solution of all the S_i's. In a parallel alternating Schwarz method, all the subdomains are solved at once; i.e., the subdomains S_1, S_2, D_1, and D_2 are all solved concurrently.

15.2.2 Interpolation Between Domains

In general, the position of the continuum nodes and the DSMC particles do not match in the overlapped regions. When a solution is computed in the DSMC subdomain, the solution for the continuum nodes in the DSMC subdomain can be obtained using an interpolation scheme. This interpolation scheme is referred to as the *DSMC-to-continuum interpolation*. Similarly, when a solution in the Stokes subdomain is computed, the solution for the DSMC particles (or DSMC cell centers) in the Stokes subdomain can again be computed by an interpolation scheme. This interpolation scheme will be referred to as the *continuum-to-DSMC interpolation*. Both continuum-to-DSMC and DSMC-to-continuum interpolation schemes can be implemented by using the meshless interpolation techniques described in Chapter 14. We explain the two interpolation techniques next.

DSMC-to-Continuum Interpolation

Let us denote by c_i the DSMC cells that will be used to compute the solution at the continuum boundary nodes n_j. Let x_{ci} denote the position of the DSMC cell center and x_{nj} denote the position of a continuum boundary node at which the solution needs to be interpolated. Let u_{ci} denote the nodal parameters for the solution at the DSMC cell centers (the solution that needs to be interpolated can be pressure, velocity, etc.) and $u(x_{nj})$ be the interpolated solution at the continuum boundary node n_j. To compute $u(x_{nj})$, a kernel or a weighting function is first centered at the position x_{nj}. The kernel centered at x_{nj} defines the cloud Ω_j and the number of DSMC cell centers, NC_j, that lie within the cloud Ω_j (see Figure 15.6(a) for the definition of the cloud and the cells that fall within the cloud). When the kernel is centered at x_{nj}, the interpolation functions, $N_i(x_{nj}), i = 1, 2, \ldots, \text{NC}_j$, are computed. Once the interpolation functions

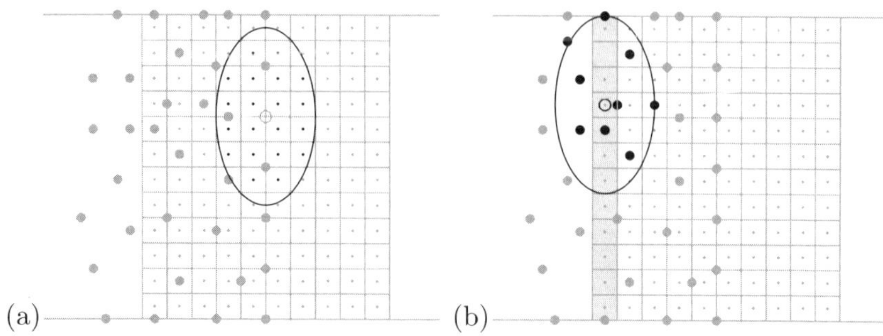

FIGURE 15.6. Placement of the kernel and the definition of the cloud for (a) DSMC-to-continuum interpolation and for (b) continuum-to-DSMC interpolation.

are computed, $u(x_{nj})$ is computed by

$$u(x_{nj}) = \sum_{i=1}^{\mathrm{NC}_j} N_i(x_{nj})u_{ci}. \tag{15.3}$$

Continuum-to-DSMC Interpolation

Let n_i denote the nodes in the continuum domain, and let c_j denote the DSMC cells. Let x_{ni} and x_{cj} denote the position of the continuum nodes and the DSMC cells, respectively. Let u_{ni} denote the nodal parameter for the solution at the continuum node with location x_{ni}, and let $u(x_{cj})$ be the interpolated solution at the DSMC cell center with location x_{cj}. To compute $u(x_{cj})$, a kernel or a weighting function is centered at the position x_{cj}. The kernel centered at x_{cj} defines the cloud Ω_j and the number of continuum nodes, NP_j, that lie within the cloud Ω_j (see Figure 15.6(b) for the definition of the cloud and the continuum nodes that fall within the cloud). When the kernel is centered at x_{cj}, the interpolation functions, $N_i(x_{cj}), i = 1, 2, \ldots, \mathrm{NP}_j$, are computed. Once the interpolations functions are computed, $u(x_{cj})$ is computed by

$$u(x_{cj}) = \sum_{i=1}^{\mathrm{NP}_j} N_i(x_{cj})u_{ni}. \tag{15.4}$$

15.3 Multiscale Analysis of Microfilters

In this section we present results on multiscale analysis of microfilters using two coupling schemes. For a discussion on the physics of microfilters, see Section 6.5. The first scheme couples DSMC with the Stokes equations and the second scheme couples DSMC with the Navier–Stokes equations. The differences in the two coupling schemes are highlighted, and the dependence of the convergence of the multiscale method on various parameters is discussed.

15.3.1 Stokes/DSMC Coupling

The coupled DSMC/Stokes approach is applied to the microfilter geometry shown in Figure 15.7. The various geometrical parameters shown in Figure 15.7 are set to be $h_f = 5$ μm, $l_c = 1$ μm, $l_{in} = 7$ μm, $l_{\text{out}} = 7$ μm and $h_c = 0.2\mu$m; see also (Aktas and Aluru, 2002), for results on $h_c = 0.8\mu$m. Figure 15.7 also shows the decomposition of the filter geometry into Stokes and DSMC subdomains. The extension of the DSMC subdomain on each side of the channel is denoted by d_{ext}. Observe that for this example, we have one DSMC subdomain and two Stokes subdomains. In order to make sure that the flow is approximately incompressible at the interface, $d_{\text{ext}} = 2$ μm is used. The overlap between DSMC and Stokes subdomains is denoted by d_{ov}. The overlap is measured from the center of the DSMC estimation cells to the continuum nodes; i.e., the generation cells are not counted in the overlap because these cells do not have valid data that can be used. An identical overlap distance, d_{ov}, is used for both the input and the output regions. The initial state and the boundary conditions for the DSMC subdomain are selected far from the expected steady-state solution in order to test the convergence characteristics of the coupled approach. The boundary conditions imposed on various surfaces of the microfilter geometry are listed in Table 15.1. The initial pressure was set to 5.0 atm, and the initial velocity was set to 0 m/s for the DSMC subdomain. For all the simulations, a DSMC time step of 10 ps was used. For the coupled DSMC/Stokes analysis, a total of 50×10^3 DSMC iterations were performed to make sure the coupling procedure has converged, and the averages were collected for 1 μs. For the DSMC simulations, a transient of 1.5 μs was simulated, before the averages were collected for 1 μs.

TABLE 15.1. A summary of boundary conditions on various surfaces of the microfilter geometry.

	Pressure	x-velocity	y-velocity
Surface A	1.3 atm	-	0
Surface B	1.0 atm	-	0
Surface C,E	$\partial P/\partial y = 0$	0	$\partial v_y/\partial y = 0$
Surface D,F	$\partial P/\partial y = 0$	0	$\partial v_y/\partial y = 0$
Surface Si,So	-	v_x=DSMC estimate	-
Surface Di,Do	p=Stokes solution	v_x=Stokes solution	v_y=Stokes solution
Surface G,H	-	diffusive	diffusive

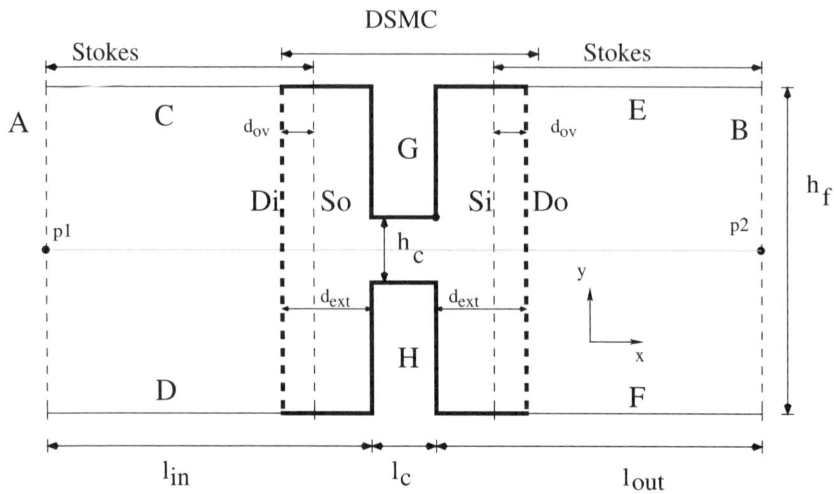

FIGURE 15.7. The geometry of the microfilter device. Also shown in the figure are the Stokes (or the Navier–Stokes) and DSMC subdomains and the overlap between the two subdomains.

To test the accuracy of the coupled DSMC/Stokes method, the geometry shown in Figure 15.7 is simulated by DSMC only (which will serve as the exact solution) and by the coupled approach. The coupled simulations use an overlap of $d_{\text{ov}} = 0.0$ μm. Figures 15.8 and 15.9 compare the pressure and x-velocity, respectively, obtained with the DSMC and the coupled simulations for $h_c = 0.2$ μm. We observe that the coupled simulations are in good agreement with the DSMC results. The good agreement between the two methods establishes the accuracy of the coupled method. From the results, we can also conclude that the multiscale approach achieves proper coupling between the DSMC and Stokes subdomains.

An important aspect of the multiscale approach is its convergence behavior. The dependence of the convergence on various parameters contained in the multiscale approach needs to be understood to guarantee the convergence of the multiscale approach. Here we investigate the convergence characteristics on the overlap size, d_{ov}, the DSMC particle weight, w_p, and the number of DSMC time steps between coupling iterations, N_{step}.

Overlap size: The overlap size in Figure 15.7, d_{ov}, describes the overlap between the Stokes and the DSMC subdomains, i.e., both DSMC and Stokes equations are solved in the overlap region. The dependence of the convergence on the overlap is investigated by considering various overlap sizes: $d_{\text{ov}} = 0.6$, 0.4, 0.2, and 0.0 μm. The overlap is increased or decreased by adjusting the distance d_{ext} (see Figure 15.7). The implementation of the interface boundary conditions for the $d_{\text{ov}} = 0.0$ μm case is not different from

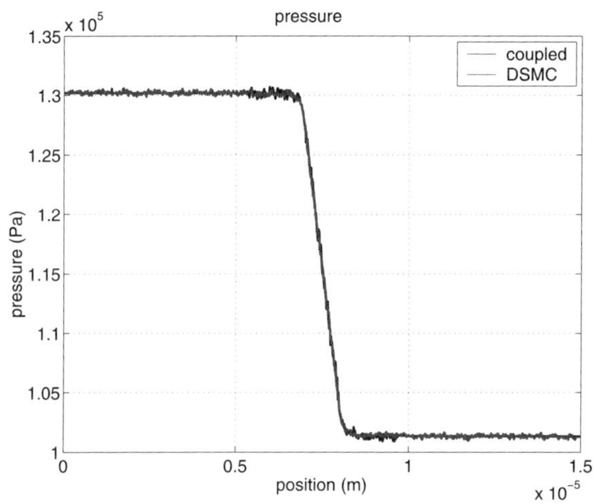

FIGURE 15.8. Comparison of pressure along the midline of the channel obtained from the DSMC only and coupled simulations ($h_c = 0.2\,\mu\text{m}$, $d_{\text{ov}} = 0\,\mu\text{m}$).

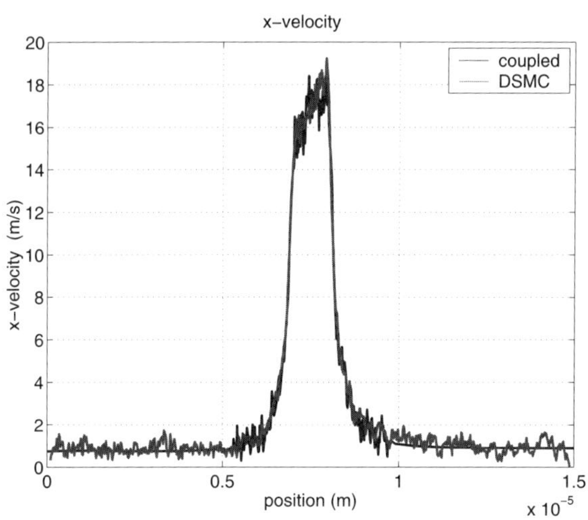

FIGURE 15.9. Comparison of velocity along the midline of the channel obtained from the DSMC only and coupled simulations ($h_c = 0.2\,\mu\text{m}$, $d_{\text{ov}} = 0\,\mu\text{m}$).

the other d_{ov} cases.

The convergence results are summarized in Figures 15.10 and 15.11. Figure 15.10 shows the convergence of the pressure boundary condition at the input section (see (Aktas and Aluru, 2002) for convergence of velocity boundary condition). In Figure 15.10, the pressure boundary condition

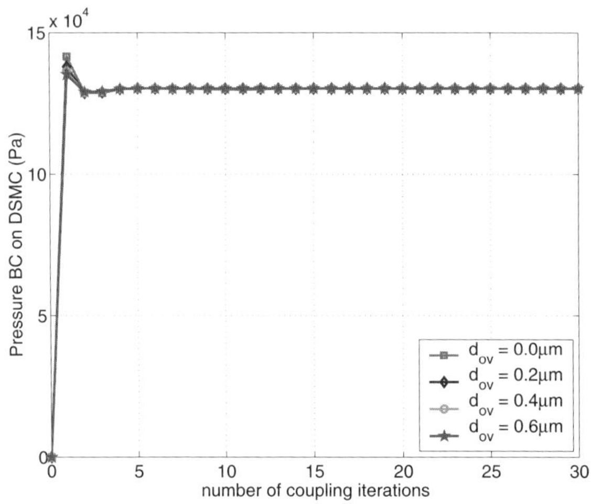

FIGURE 15.10. Convergence of pressure boundary condition transferred from the continuum side to the DSMC subdomain for different overlaps. $N_{\text{step}} = 5000$ is used.

is plotted by computing the average of the boundary condition imposed on all the buffer cells. The figure indicates that the convergence is weakly dependent on the overlap size for this problem, with larger overlap size exhibiting slightly better convergence. In order to further investigate the dependence of the convergence characteristics on the overlap size, we show the absolute error in the pressure boundary condition transferred from the Stokes subdomain to the DSMC subdomain in Figure 15.11. Absolute error is defined as the absolute deviation between the computed value and an exact value, which is determined from the DSMC-only simulation. Again, the values plotted are the average of all the DSMC buffer cells. The results in Figure 15.11 were obtained for $N_{step} = 5000$. This result shows that the number of iterations until convergence is weakly dependent on the overlap size. A mathematical analysis of the Schwarz method for elliptic problems reveals that a faster convergence should be obtained for a larger overlap (Smith et al., 1996). However, the result in Figure 15.11 indicates clearly that the number of coupling iterations is only weakly dependent on the overlap size. The observed behavior leads to the conclusion that the Stokes and the DMSC subdomains are weakly coupled for this example.

Particle weight: We have already discussed the effect of noise in DSMC; this is particularly troubling to the convergence of the multiscale approach. The noise in the DSMC estimates can be controlled by decreasing the particle weight or by increasing the number of DSMC time steps. To evaluate the dependence of convergence on the particle weight, w_p, simulations are

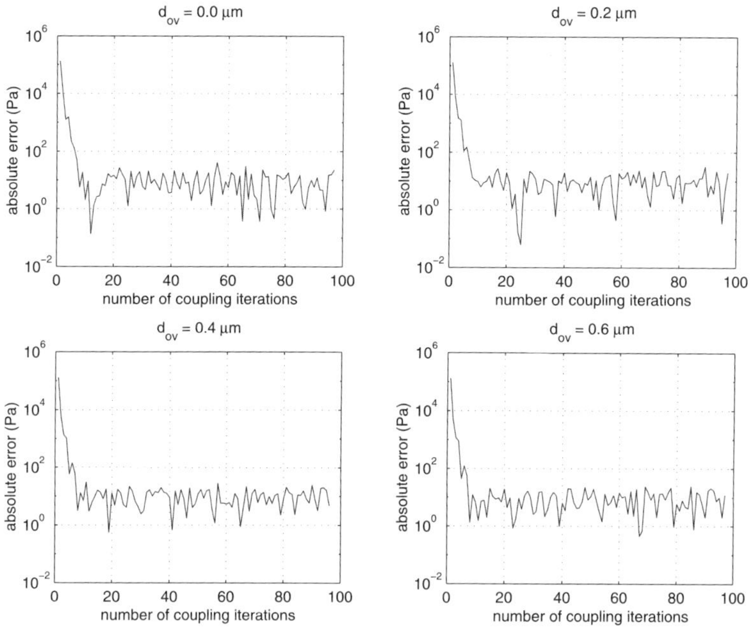

FIGURE 15.11. The absolute error in the pressure boundary condition transferred from the continuum subdomain to the DSMC subdomain for different overlaps. $N_{\text{step}} = 5000$ is used.

performed using particle weights of 25×10^4, 5×10^4, 10^4, and 10^3. We observed (see (Aktas and Aluru, 2002), for convergence plots) that a smaller particle weight, which accounts for more particles, exhibits better convergence characteristics with less noise. We also observed that the convergence is not delayed significantly because of a larger noise (due to a larger particle weight). Thus, we can conclude that the method is fairly robust. The observation that the coupled method converges when a large particle weight is used is important for application of the particle cloning method (Chen and Boyd, 1996) to speed up the DSMC simulation. The particle weight that should be selected according to the DSMC accuracy requirements is 10^3. However, a lower weight requires extensive CPU times, and the simulation time can be reduced by using a particle cloning method that starts with 16 times the desired weight of 10^3 and using four cloning steps to get the desired accuracy.

Number of time steps in DSMC: The selection of the number of DSMC time steps, N_{step}, during each coupled iteration is important for the efficiency of the coupled method. Two issues need to be considered in selecting N_{step}. The first issue deals with the noise in the DSMC solution. The noise considerations that were discussed in connection with particle weight also apply for the selection of N_{step}. In the simulations reported here, the particle weight, w_p, is reduced proportionally as N_{step} is decreased. The second issue that needs to be considered in selecting N_{step} is the time-dependent nature of the DSMC solution computed during the coupled iterations. Starting from an initial state, the multiscale coupled algorithm will take a certain number of iterations to compute a converged solution. During each iteration, the flow in the DSMC subdomain will evolve in a time-accurate manner toward a steady-state solution determined by the boundary conditions. If N_{step} is large enough during each coupled iteration, a steady-state solution can be reached. However, there is no need to compute steady-state solutions during each coupled iteration because the boundary conditions enforced on the DSMC subdomain are not necessarily steady-state boundary conditions. Since the goal is to compute a steady-state solution for the entire system (including both Stokes and DSMC subdomains), N_{step} can be selected shorter, and the boundary conditions can be updated in an iterative manner until a steady-state solution is reached.

To investigate the effect of N_{step}, the filter geometry is simulated by keeping N_{step}/w_p constant while N_{step} is changed. This keeps the noise in the DSMC estimates at the same level. N_{step} values of 200, 1000, 5000, and 25,000 are investigated. The convergence of the pressure boundary condition in the input section was investigated (see (Aktas and Aluru, 2002), for convergence plots). We observed that for $N_{\text{step}} = 200$ and 1000, a larger number of coupling iterations are needed for convergence when compared to $N_{step} = 5000$, and 25,000. A comparison of the total simulated DSMC iterations until convergence shows that for $N_{step} = 200$ and 1000, the total DSMC iterations are approximately equal. However, for the other two cases, the DSMC iterations until convergence are much larger. Thus, we can conclude that if $N_{\text{step}} \times N_{\text{cpl}}$ (where N_{cpl} is the number of coupling iterations until convergence) is longer than the number of iterations the DSMC subdomain takes to reach a steady state, convergence is determined by the properties of the coupling method, whereas if $N_{\text{step}} \times N_{\text{cpl}}$ is smaller, then the DSMC subdomain will evolve in a quasi-static manner and N_{cpl} will be increased. For an efficient implementation, $N_{\text{step}} \times N_{\text{cpl}}$ should be close to the time constant of the DSMC subdomain.

15.3.2 Navier–Stokes/DSMC Coupling

When compared to DSMC/Stokes coupling, the application of DSMC/Navier–Stokes coupling involves several issues: coupling of temperature, the presence of larger gradients at the interface and/or the overlap region, the behavior of Navier–Stokes equations in the presence of noisy boundary conditions, and the distribution function for the particles injected into the DSMC region. These and related issues are discussed below.

The coupling of the pressure and velocity is performed in the same manner as the DSMC/Stokes coupling. For the coupling of temperature, several alternatives can be implemented. Typically, coupling the temperature in the same way as velocity gives the best results. That is, the temperature estimated from DSMC is interpolated to the Navier–Stokes subdomains; and after the Navier–Stokes solution, the temperature from within the continuum subdomain is interpolated back to the DSMC boundary cells. The transfer of variables between subdomains for coupling is summarized in Table 15.2 (see Figure 15.7 for the microfilter geometry). In contrast to velocity coupling, in the absence of overlap, the temperature does not converge. For this reason, the DSMC/Navier–Stokes coupling described here uses *nonzero overlap.*

In addition to the scheme described in Table 15.2, other possibilities for the coupling of temperature can also be implemented. However, we observed that the Navier–Stokes solution does not converge if the temperature is not specified at the interfaces with the atomistic model. Also, it was observed that when the temperature at the boundary of the DSMC subdomain is updated by extrapolating the value from the neighboring cells, the method becomes unreliable, with the temperature solution differing significantly from the DSMC solution in some cases.

TABLE 15.2. A summary of boundary conditions on various surfaces of the microfilter geometry.

Surface	Pressure	x-velocity	y-velocity	temperature
A	1.3 atm	-	0	300 K
B	1.0 atm	-	0	300 K
C, E	$\partial P/\partial y = 0$	0	$\partial v_y/\partial y = 0$	$\partial t/\partial y = 0$
D, F	$\partial P/\partial y = 0$	0	$\partial v_y/\partial y = 0$	$\partial t/\partial y = 0$
Si, So	-	v_x = DSMC est.	v_y = DSMC est.	t = DSMC est.
Di, Do	p = NS sol	v_x = NS sol.	v_y = NS sol.	v_y = NS sol.
G, H	-	diffusive	diffusive	300 K

Distribution function used for DSMC boundary cells: Previous work on DSMC coupling is not conclusive about the distribution function used for injection into the DSMC domain. It was first suggested that Chapman–Enskog distribution be used whenever DSMC is being coupled to Navier–Stokes equations (Hash and Hassan, 1996). However, it was later suggested that Chapman–Enskog distribution may not be necessary in all cases (Hash and Hassan, 1997). In (Garcia et al., 1999), a dimensionless parameter B is used to analyze the validity range of the Chapman–Enskog distribution. For the examples discussed here, the dimensionless parameter B is 0.04, which is smaller than the maximum value for the range of validity of the Chapman–Enskog distribution, which indicates that the Chapman–

Enskog distribution can be utilized. For the example presented here, the Chapman–Enskog distribution was used for the particles generated in the buffer cells. Chapman–Enskog distribution was selected because Kn at the interface was in the slip flow regime, indicating the presence of nonequilibrium. The parameters that are needed for the Chapmann–Enskog distribution were taken from the Navier–Stokes simulation of the continuum subdomain and interpolated back to the DSMC subdomain.

Number of buffer cells used: An additional optimization that can be implemented for DSMC/Navier–Stokes coupling is to use two rows of buffer cells instead of one. Using two or more rows of buffer cells makes certain that the buffer region has enough depth to inject a proper number of high-velocity particles. Furthermore, as compared to using one big buffer cell, this approach helps capture some of the gradient in the buffer zone.

Problems induced by noise in the DSMC estimates: In some cases, the noise in DSMC estimates can result in wiggles in the Navier–Stokes solution. The wiggles die off further into the continuum domain. However, the presence of wiggles may cause problems with interpolation because wrong values may be interpolated. In order to help avoid problems, the flexibility provided by meshless interpolation can be utilized by adjusting the cloud sizes to filter out the effect of the wiggles in the Navier–Stokes solution.

Analysis of Peclet and local Kn numbers: The Peclet number, Reynolds number, and local Kn of the flow in the interface zone can be calculated to provide information on the characteristics of the filter problem. An analysis of the Peclet number shows that in the Navier–Stokes subdomains, the local Peclet number, defined by using point-to-point spacing, is smaller than or close to 1. The Peclet number is highest at the regions closest to the input and output of the channels. In the simulations described here, numerical stabilization is not used. However, for larger values of Peclet number stabilization will be necessary. An analysis of the local Kn shows that a local Kn between 0.4 μm and 2 μm from the channel exit is in the slip flow regime. Far from the channel exit, the global Kn is 0.011, which is at the boundary of continuum and slip flow regimes. The use of no-slip boundary conditions close to the DSMC/NS boundary may introduce some error to the coupled solution. Since the slip velocity observed from the DSMC simulation is small with respect to the maximum velocity, the error is expected to be small. The Reynolds number of the flow for the conditions studied is about 4.5. Thus, it is seen that the flow in the filter is compressible, and the use of Navier–Stokes equations is necessary in order to bring the coupling interface close to the filter channel.

Example: The coupled DSMC/Navier–Stokes approach is applied to the microfilter geometry shown in Figure 15.7. The various geometrical param-

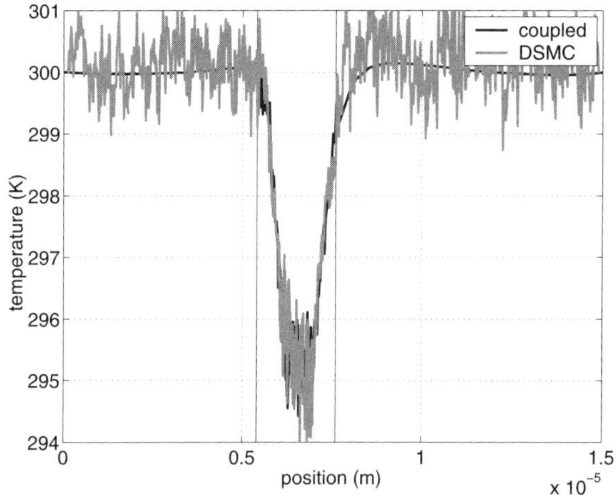

FIGURE 15.12. The DSMC and coupled results for temperature plotted along the midline for $d_{\mathrm{MC}} = 0.6$ μm, $d_{\mathrm{ov}} = 0.29$ μm. The vertical lines denote the boundaries of the Navier–Stokes subdomains.

eters shown in Figure 15.7 are set to be $h_f = 5$ μm, $l_c = 1$ μm, $l_{\mathrm{in}} = 6$ μm, $l_{\mathrm{out}} = 8$ μm, and $h_c = 0.8$ μm. Figure 15.7 also shows the decomposition of the filter geometry into Navier–Stokes and DSMC subdomains. The extension of the DSMC subdomain on each side of the channel is denoted by d_{ext}. The region simulated by DSMC on each side of the channel is denoted by $d_{\mathrm{MC}} = d_{\mathrm{ext}} - d_{\mathrm{ov}}$. The overlap between DSMC and Navier–Stokes subdomains is denoted by d_{ov}. Again, the overlap is measured from the center of the DSMC estimation cells to the continuum nodes; i.e., the generation cells are not counted in the overlap. An identical overlap distance, d_{ov}, is used for both the input and the output regions. The boundary conditions imposed on various surfaces of the microfilter geometry are listed in Table 15.2. For all the simulations, a DSMC time step of 10 ps was used. For the coupled DSMC/NS analysis, a total of 80×10^3 DSMC iterations were performed to make certain the coupling procedure converged, and the averages were collected for at least 2 μs. For the DSMC simulations, a transient of 1.5 μs was simulated, and averages were collected for 1 μs. The parameters of the fluid for Navier–Stokes simulation were set at the values corresponding to the DSMC values. The filter geometry shown in Figure 15.7 was simulated with the coupled method using $d_{\mathrm{MC}} = 0.6$ μm and $d_{\mathrm{ov}} = 0.29$ μm. The results for temperature are compared with DSMC solution in Figure 15.12 (see (Aktas et al., 2003), for results on pressure and velocity). The good agreement between the DSMC and multiscale results demonstrates the accuracy of the coupled method.

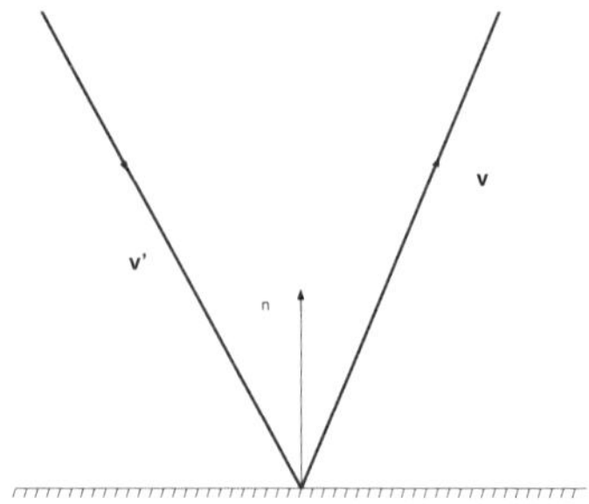

FIGURE 15.13. Gas–surface interaction.

15.4 The Boltzmann Equation

The Boltzmann equation can be derived rigorously from Newton's laws at least for the low-density limit, but it is also used for the dense limit, although there is no totally rigorous theoretical basis for the latter (Cercignani et al., 1994). Research efforts in the 1990s focused on obtaining most of the fluid-dynamics limits of the Boltzmann equation including the incompressible limit; see (Bardos et al., 1991; Bardos et al., 1993), and (Desjardins et al., 1999; DiPerna and Lions, 1991), and references therein.

Here we assume that microscale fluidic and thermal gas transport in the entire Knudsen regime ($0 \leq \text{Kn} < \infty$) is governed by the Boltzmann equation (BE). The Boltzmann equation describes the evolution of a velocity distribution function by molecular transport and binary intermolecular collisions. An analysis of microflows based on the Boltzmann equation has been presented in series of papers that study silicon accelerometer motion and squeezed film damping as a function of the Knudsen number and the time-periodic motion of the surfaces (Veijola et al., 1995a; Veijola et al., 1998; Fukui and Kaneko, 1988; Fukui and Kaneko, 1990); see Section 6.1. More applications using solutions of the Boltzmann equation are presented in (Aoki, 2001).

To illustrate some of the complexities in performing simulations based on the Boltzmann equation, we provide a brief review next. Let us consider monoatomic gases, the state of which can be described by a velocity distribution function $f(t, \boldsymbol{x}, \boldsymbol{v})$, where $\boldsymbol{x}$ denotes the position and $\boldsymbol{v}$ denotes the velocity of the molecules. The distribution function represents the number of particles in the *six-dimensional phase space* $d\boldsymbol{x}\, d\boldsymbol{v}$ at time t. This distribution function obeys the Boltzmann equation (Sone, 2002; Cercignani, 1988; Bird, 1994)

$$\frac{\partial f}{\partial t} + \boldsymbol{v} \cdot \frac{\partial f}{\partial \boldsymbol{x}} + \mathbf{F} \cdot \frac{\partial f}{\partial \boldsymbol{v}} = Q(f, f_*), \tag{15.5}$$

with $\boldsymbol{x} \in \Omega$ and $\boldsymbol{v} \in R^3$. Here $\mathbf{F}$ is an external body force, and the term on the right-hand side represents molecule collisions; it is given by

$$Q(f, f_*) = \int_{R^3} \int_{S^+} |\mathbf{V} \cdot \boldsymbol{n}|[f(\boldsymbol{x}, \boldsymbol{v}'_*) f(\boldsymbol{x}, \boldsymbol{v}') - f(\boldsymbol{x}, \boldsymbol{v}_*) f(\boldsymbol{x}, \boldsymbol{v})] d\boldsymbol{n}\, d\boldsymbol{v}_*. \tag{15.6}$$

This represents collisions of two molecules with postcollision velocities $\boldsymbol{v}$ and $\boldsymbol{v}_*$, and corresponding precollision velocities denoted in addition by primes (see Figure 15.13). Here, the integration is taken over the three-dimensional velocity space R^3 and the hemisphere S^+, which includes the particles moving away from each other after the collision. Also, we have the definitions

$$\mathbf{V} = \boldsymbol{v} - \boldsymbol{v}_*; \quad \boldsymbol{v}' = \boldsymbol{v} - \boldsymbol{n}(\boldsymbol{n} \cdot \mathbf{V}); \quad \boldsymbol{v}'_* = \boldsymbol{v}_* + \boldsymbol{n}(\boldsymbol{n} \cdot \mathbf{V}),$$

where $\boldsymbol{n}$ is the unit vector along $(\boldsymbol{v} - \boldsymbol{v}')$.

The definition of the rest of the terms in equation (15.5) is as follows: The first term is the rate of change of the number of class $\boldsymbol{v}$ molecules in the phase space. The second term shows convection of molecules across a fluid volume by molecular velocity $\boldsymbol{v}$. The third term is convection of molecules across the velocity space as a result of the external force $\mathbf{F}$.

Let us now define the quadratic function

$$\phi(\boldsymbol{v}) \equiv a + \mathbf{b} \cdot \boldsymbol{v} + c|\boldsymbol{v}|^2;$$

then the collision term satisfies the following relations:

$$\int_{R^3} \phi(\boldsymbol{v}) Q(f, f_*) d\boldsymbol{v} = 0, \tag{15.7a}$$

$$\int (\ln f)\, Q(f, f_*) \leq 0. \tag{15.7b}$$

The first one represents conservation of mass, momentum, and energy for the $a, \mathbf{b}$, and c terms, respectively. In the second one, known as the *Boltzmann inequality*, the equal sign applies if $(\ln f)$ is collision invariant. This leads to the solution

$$f = \exp(a + \mathbf{b} \cdot \boldsymbol{v} + c|bv|^2),$$

which is known as the *Maxwellian distribution* and represents an *equilibrium state* for number density n_0 and temperature T_0. It can be rewritten in the familiar form

$$f_0 = \pi^{-3/2} n_0 \beta_0^3 \exp(-\beta_0^2 v^2), \tag{15.8}$$

where we have defined

$$\beta_0^2 = \frac{m}{2k_{\mathrm{B}} T_0}$$

with $k_B = 1.3805 \times 10^{-23}$ J/K the Boltzmann constant, and m the molecular mass.

All the **macroscopic quantities** are defined in terms of the distribution function: for example,

- Density
$$\rho(\boldsymbol{x}, t) = m \int f(\boldsymbol{x}, \boldsymbol{v}, t) d\boldsymbol{v}$$

- Bulk velocity
$$\rho \mathbf{u}(\boldsymbol{x}, t) = m \int \boldsymbol{v} f(\boldsymbol{x}, \boldsymbol{v}, t) d\boldsymbol{v}$$

- Temperature
$$T(\boldsymbol{x}, t) = \frac{m}{3nk_B} \int c^2 f(\boldsymbol{x}, \boldsymbol{v}, t) d\boldsymbol{v},$$
where $\mathbf{c} \equiv \boldsymbol{v} - \mathbf{u}$ is the *peculiar velocity.*

The **boundary conditions** take into account the wall type via the nonnegative *scattering kernel*, representing a probability density

$$R(\boldsymbol{v}' \to \boldsymbol{v}; \boldsymbol{x}; t)$$

that molecules hitting the wall with velocity between $\boldsymbol{v}'$ and $\boldsymbol{v}' + d\boldsymbol{v}'$ at location $\boldsymbol{x}$ at time instance t will be reflected with velocity between $\boldsymbol{v}$ and $\boldsymbol{v} + d\boldsymbol{v}$. If R is known, then we can obtain the boundary condition for the distribution function as

$$f(\boldsymbol{x}, \boldsymbol{v}, t)\, v_n = \int H(-v_n')|v_n'| R(\boldsymbol{v}' \to \boldsymbol{v}; x, t) f(\boldsymbol{x}, \boldsymbol{v}', t) d\boldsymbol{v}', \tag{15.9}$$

where $H(x)$ is the Heaviside step function, and $v_n = \boldsymbol{v} \cdot \boldsymbol{n}$ is the velocity normal to the surface. If there is no adsorption on the wall surface, then

$$\int H(v_n) R(\boldsymbol{v}' \to \boldsymbol{v}; x, t) d\boldsymbol{v}' = 1,$$

and also the scattering kernel satisfies a reciprocity condition; see (Sone, 2002; Cercignani, 1988).

Different types of scattering kernels express different gas–surface interactions and define the accommodation coefficient introduced earlier in Section 2.2.2. Assuming, as before, that σ_v molecules are reflected diffusely and $(1 - \sigma_v)$ are reflected specularly, then the popular Maxwell's scattering kernel, used exclusively before the 1960s, has the form

$$R(\boldsymbol{v}' \to \boldsymbol{v}; x) = (1 - \sigma_v)\delta(\boldsymbol{v}' - \boldsymbol{v} + 2\boldsymbol{n} v_n) + \frac{2\sigma_v \beta_w^4}{\pi} \exp(-\beta_w^2 v^2), \tag{15.10}$$

where β_w involves the wall temperature T_w. For $\sigma_v = 1$ we obtain solely diffuse scattering, which physically means that we have *perfect accommodation*, in the sense that the molecules "forget their past" and reemerge after the wall collision with a Maxwellian distribution function.

Other scattering kernels have been proposed by many authors, but the most popular one is the **Cercignani–Lampis model** (Cercignani and Lampis, 1971). It was obtained by other methods using Brownian motion and through an analogy with electromagnetic scattering. It introduces two accommodation coefficients: the tangential accommodation coefficient, $0 \leq \sigma_t \leq 2$, and the normal accommodation coefficient, $0 \leq \sigma_n \leq 1$. It has the form

$$\begin{aligned} R(\boldsymbol{v}' \to \boldsymbol{v}; x) = {} & \frac{2\sigma_n \sigma_t (2 - \sigma_t)\beta_w^4}{\pi} \\ & \times \exp\left(-\beta_w^2 \frac{v_n^2 + (1 - \sigma_n)(v_n')^2}{\sigma_n} - \beta_w^2 \frac{(v_t - (1 - \sigma_t)v_t')^2}{\sigma_t(2 - \sigma_t)} \right) \\ & \times I_0 \left(\beta_w^2 \frac{2\sqrt{1 - \sigma_t} v_n v_n'}{\sigma_n} \right), \end{aligned}$$

where v_n and v_t are the normal and tangential components of the molecular velocities, and I_0 is the zeroth-order modified Bessel function of the first kind.

For the continuum description presented in Chapter 2, we did not need the scattering kernel but rather an equivalent *lumped* accommodation coefficient, which was determined empirically depending on the gas and the wall type. For example, for light gases, such as helium and neon, the tangential accommodation coefficient may be much less than unity, but for heavy gases, e.g., xenon and krypton, the tangential accommodation coefficient is close to one. For typical surfaces in microsystems and argon or nitrogen the value obtained in (Arkilic, 1997), is around 0.8.

- In general, the cleaner the surface, the smaller the value of the tangential accommodation coefficient.

Also, the position-dependent accommodation coefficient is a matrix, the elements of which depend on the distribution function of the impinging molecules.

A comparison of the Maxwell scattering kernel and the Cercignani–Lampis kernel for micronozzle flows was presented in (Ketsdever et al., 2000b). In particular, the free molecule microresistojet (FMMR) discussed in Section 6.6 was considered with argon as propellant. Two forms of the Cercignani–Lampis–Lord (CLL) model were employed in the simulations (Lord, 1991; Lord, 1995): the original model as well as a generalized form permitting diffuse reflection at a surface with *incomplete* energy accommodation. A comparison of the normalized specific impulse at the exit of the FMMR is shown in Figure 15.14. For the same value of the accommodation

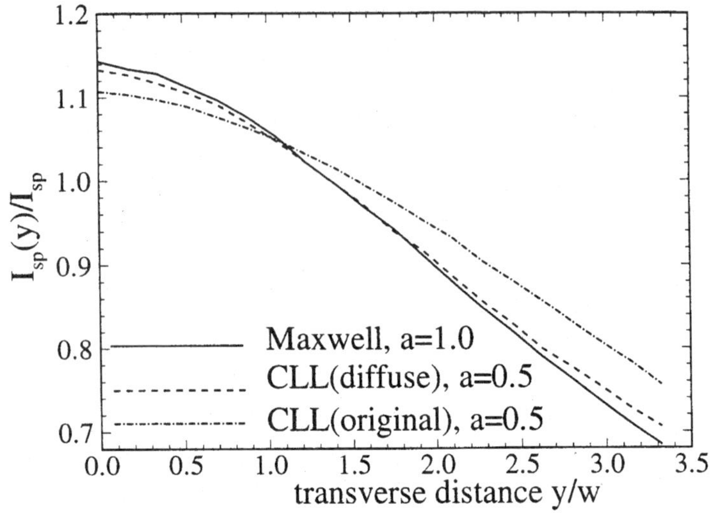

FIGURE 15.14. Gas–surface interaction in micronozzle flows. The normalized specific impulse is plotted across the expansion slot of a free molecule microresistojet (Ketsdever et al., 2000b). (a denotes accommodation coefficient, same as σ_v in this book). (Courtesy of A. Ketsdever.)

coefficient (symbol a in the figure) the models differ by about 5%, whereas varying the accommodation coefficient from specular ($\sigma_v = a = 0$) to diffuse ($\sigma_v = a = 1$) leads to about 20% variation in the value of the predicted specific impulse I_{sp}. It was shown in (Ketsdever et al., 2000b), that other quantities, such as axial velocity distribution functions, are more sensitive to the scattering model assumption.

15.4.1 Classical Solutions

We review here some of the most popular solution methods for the Boltzmann equation; more details can be found in (Cercignani, 1988; Cercignani, 2000; Cercignani et al., 1994), and in the comprehensive review articles (Sharipov and Sleznev, 1998; Aoki, 2001). The degree of success in deriving semianalytical solutions depends on the Knudsen number and the geometry of the flow. More specifically, there are different methods for

- the hydrodynamic limit, where $\text{Kn} \to 0$;
- the free-molecular limit, where $\text{Kn} \to \infty$, and
- the transition limit, which is the most difficult regime.

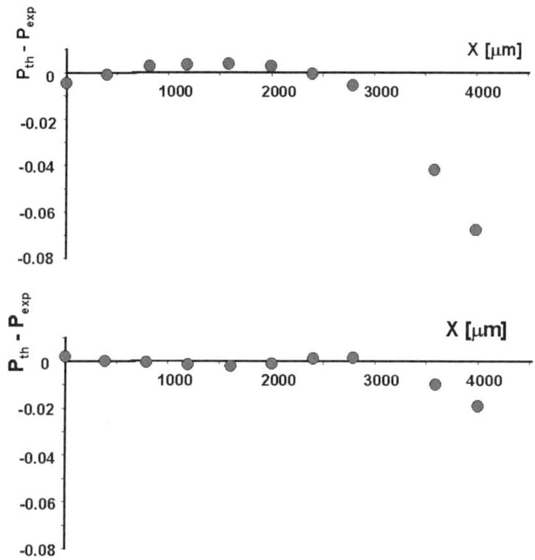

FIGURE 15.15. Measured pressure deviation from theoretical prediction in helium microchannel flow. Top: No-slip boundary condition is assumed. Bottom: Grad's 13-moment method with slip is applied. Kn = 0.15. (Courtesy of C.-M. Ho.)

Hydrodynamic Regime: The Grad and Chapman–Enskog Methods

One of the earliest solution approaches for the Boltzmann equation is the **method of moments** proposed by (Grad, 1949), where the distribution function is represented by a series

$$f(\boldsymbol{x}, \boldsymbol{v}, t) = f_{\text{loc}}\left[a^{(0)}H^{(0)} + a^{(1)}H^{(1)} + \cdots + \frac{1}{N!}a^{(N)}H^{(N)}\right],$$

where f_{loc} is the local Maxwellian distribution obtained from the equilibrium Maxwellian of equation (15.8) by replacing the equilibrium quantities with the local number density $n(\boldsymbol{x}, t)$ and local temperature $T(\boldsymbol{x}, t)$, and $H^{(N)}$ are orthogonal Hermite polynomials. The coefficients $a^{(N)}$ are expressed in terms of the N moments of the distribution function. Similarly, the boundary conditions are handled using projections with the Hermite polynomials, but a closure condition is also needed, which is based on some physical condition.

This method has been used by C.-M. Ho and his colleagues using $N = 13$ moments to predict the pressure distribution in helium flow in a channel with Kn = 0.15 at the exit. The results are shown in Figure 15.15. On the first plot, the difference in pressure between the measurements and Grad's method with no-slip boundary conditions is shown. On the second plot, the pressure difference is shown again but with the theoretical result obtained with Grad's 13-moment method and slip boundary condition. The deviations start at about 2/3 of the channel, where Kn increases from 0.1 to 0.15. It is clear that the Grad method captures accurately the deviation due to velocity slip at the wall. For Grad's method no information about the accommodation coefficient is needed, and this is one of its great advantages.

Another approach is the **Chapman–Enskog method**, where the distribution function is expanded into a perturbation series with the Knudsen number being the small parameter (see Chapter 2):

$$f = f^0 + \mathrm{Kn}\, f^{(1)} + \mathrm{Kn}^2\, f^{(2)} + \cdots .$$

Here, the first term corresponds to the local Maxwellian distribution, i.e.,

$$f^0 = f_{\mathrm{loc}}.$$

Successive high-order equations are obtained by substituting this expansion into the Boltzmann equation (15.5).

From the equation for $f^{(1)}$ we obtain the Newtonian Navier–Stokes equations and Fourier's law of conduction. Furthermore, assuming a model for the molecular interaction we can obtain explicit expressions for the transport coefficients of the momentum and energy equations. For example, for the hard sphere molecules model the dynamic viscosity is

$$\mu = \frac{5}{16} \frac{\sqrt{\pi m k_{\mathrm{B}} T}}{\pi d^2},$$

with d the molecular diameter.

Approximate Forms of Boltzmann Equation

Approximate formulations of the Boltzmann equation can be obtained by simplifying the collision integral. In the limit of very high Knudsen number, i.e., the free-molecular flow, the collision integral is zero, but for arbitrary rarefaction, simplifications are needed to make the Boltzmann equation computationally tractable.

One such approach is the method of Bhatnagar, Gross, and Krook (Bhatnagar et al., 1954), the so-called **BGK model**. In this method, the collision integral is approximated as

$$Q_{\mathrm{BGK}}(f, f_*) = \nu_*(f_{\mathrm{loc}} - f),$$

where ν_* is the collision frequency, which is assumed to be independent of the molecular velocity $\boldsymbol{v}$, but it is a function of spatial coordinates and time. We can obtain a relation of the collision frequency from the mean thermal velocity $\bar{v} = \sqrt{8k_{\mathrm{B}}T/(\pi m)}$ and the mean free path $\lambda = \sqrt{2}/(2n\pi d^2)$, i.e.,

$$\nu_* = \frac{\bar{v}}{\lambda} = \frac{4}{\pi} \frac{p(\boldsymbol{x}, t)}{\mu(T)},$$

where $p(\boldsymbol{x}, t)$ is the pressure obtained from the distribution function

$$p(\boldsymbol{x}, t) = \frac{m}{3} \int c^2 f(\boldsymbol{x}, \boldsymbol{v}, t) d\boldsymbol{v}.$$

However, this expression for the collision frequency leads to expressions for the dynamic viscosity (also for the thermal conductivity) that are inconsistent with those derived using the Chapman–Enskog method to obtain solutions of the BGK model equation. Specifically, the Chapman–Enskog solution leads to

$$\nu_* = \frac{p(\boldsymbol{x}, t)}{\mu(T)},$$

which is the same equation as the one corresponding to solutions of the Boltzmann equations with the *full* collision term. Other models have been proposed for the collision frequency, including corrections that allow dependence on the molecular velocity, since full simulations suggest this to be the case; the interested reader can find appropriate references in (Sharipov and Sleznev, 1998).

- In general, numerical evidence from full solutions of the Boltzmann equation suggests that the BGK model is an accurate method for isothermal flows. However, for nonisothermal flows, corrections for the Prandtl number (and collision frequency) need to be introduced.

Another approximation that allows computationally tractable models is **linearization** of the Boltzmann equation. The distribution function is then written as

$$f(\boldsymbol{x}, \boldsymbol{v}, t) = f_0(n_0, T_0)[1 + h(\boldsymbol{x}, \boldsymbol{v}, t)],$$

where f_0 is the *absolute Maxwellian distribution* corresponding to equilibrium state (n_0, T_0), and $h(\boldsymbol{x}, \boldsymbol{v}, t)$ is the perturbation distribution function. The **linearized Boltzmann equation** is then

$$\frac{\partial h}{\partial t} + \boldsymbol{v} \cdot \frac{\partial h}{\partial t} - \tilde{Q}h = 0,$$

where the linearized collision term is

$$\tilde{Q}h = \int_{R^3} \int_{S^+} f_0(\boldsymbol{v}_*)(h' + h'_* - h - h_*)\, d\boldsymbol{n}\, d\boldsymbol{v}_*.$$

All macroscopic (continuum) parameters can then be written in terms of the perturbation function, i.e.,

$$n = n_0 + (1, h); \quad n_0 \mathbf{u} = (\boldsymbol{v}, h); \quad T = T_0 + \frac{m}{3k_B n_0}(v^2, h) - \frac{T_0}{n_0}(1, h),$$

where $(\cdot, \cdot)$ denotes a weighted inner product with f_0 as the weight function.

A different linearized Boltzmann equation can be obtained by linearizing around the *local Maxwellian distribution*, i.e.,

$$f(\boldsymbol{x}, \boldsymbol{v}, t) = f_{\text{loc}}(n, T)[1 + h(\boldsymbol{x}, \boldsymbol{v}, t)],$$

where the macroscopic local quantities $n(\boldsymbol{x}, t)$ and $T(\boldsymbol{x}, t)$ are involved. In this case the perturbation function satisfies an inhomogeneous equation:

$$\frac{\partial h}{\partial t} + \boldsymbol{v} \cdot \frac{\partial h}{\partial t} - \tilde{Q}h = -\boldsymbol{v} \cdot \left[\frac{1}{n}\frac{\partial n}{\partial \boldsymbol{x}} + \left(\frac{mv^2}{2k_BT} - \frac{3}{2}\right)\frac{1}{T}\frac{\partial T}{\partial \boldsymbol{x}}\right].$$

Further approximations are required for both the boundary conditions and the linearized collision operator $\tilde{Q}$ to bring the Boltzmann equation to a computationally friendly form; see (Cercignani, 1988; Cercignani et al., 1994; Sharipov and Sleznev, 1998).

15.4.2 Sone's Asymptotic Theory

When the Knudsen number is small, the contribution of the collision term in the Boltzmann equation becomes large, and the velocity distribution function approaches a local Maxwellian. Then, the behavior of the gas may be described as continuum. In this section, we present the work of Sone and collaborators (Sone, 2002) for the small Knudsen number limit, which is pertinent to microflows. In particular, we will consider small and large deviations from equilibrium. This is measured by the Mach number, and we recall that the Knudsen number, the Mach number, and the Reynolds number are related (see Chapter 2):

$$M \propto \text{Kn} \cdot \text{Re}.$$

We first introduce the nondimensional variables, which are chosen in such a way that they express a perturbation from a Maxwellian distribution with $v_i = 0$. Let L, t_0, T_0, and p_0 be the reference length, time, temperature, and pressure, and

$$\rho_0 = p_0(RT_0)^{-1}, \tag{15.11}$$

where the reference state is the Maxwellian distribution with $v_i = 0$, $p = p_0$, and $T = T_0$:

$$f_0 = \frac{\rho_0}{(2\pi RT_0)^{3/2}} \exp\left(-\frac{\xi_i^2}{2RT_0}\right). \tag{15.12}$$

Then, the nondimensional variables are defined as follows:

$$\begin{array}{lll} x_i = X_i/L, & \bar{t} = t/t_0, & \zeta_i = \xi_i/(2RT_0)^{1/2}, \\ \phi = f/f_0 - 1, & \omega = \rho/\rho_0 - 1, & u_i = v_i/(2RT_0)^{1/2}, \\ \tau = T/T_0 - 1, & P = p/p_0 - 1, & P_{ij} = p_{ij}/p_0 - \delta_{ij}, \\ Q_i = q_i/p_0(2RT_0)^{1/2}, & u_{wi} = v_{wi}/(2RT_0)^{1/2}, & \tau_w = T_w/T_0 - 1, \\ \omega_w = \rho_w/\rho_0 - 1, & P_w = p_w/p_0 - 1, & (P_w = \omega_w + \tau_w + \omega_w\tau_w). \end{array} \tag{15.13}$$

We also redefine the Knudsen number to be consistent with Sone's notation as follows:

$$k = \frac{\sqrt{2RT_0}}{\rho_0 A_c L} = \frac{\sqrt{\pi}}{2}\frac{\lambda_0}{L}. \tag{15.14}$$

Case with $Re \ll 1$

Following (Sone, 2002), we first analyze the small Reynolds number regime, for which we have

$$\text{Re} \ll \text{Kn} \ll 1,$$

and thus small deviations from equilibrium, i.e., $M \ll 1$. The steady-state Boltzmann equation in nondimensional abstract form is

$$\zeta_i \frac{\partial \phi}{\partial x_i} = \frac{1}{k}\mathcal{L}(\phi), \tag{15.15}$$

where the right-hand side denotes the collision operator. The boundary condition expressed also in abstract form is

$$\phi = \phi_w \quad (\zeta_i n_i > 0) \tag{15.16}$$

on the boundary, where ϕ_w may depend on ϕ ($\zeta_i n_i < 0$) linearly.

We are looking for the asymptotic behavior of ϕ for small k following the method of analysis by (Sone, 1969; Sone, 1971) and (Sone and Aoki, 1994). In particular, expansions of the Grad–Hilbert form are sought for a moderately varying solution of equation (15.15), whose length scale of variation is on the order of the characteristic length L of the system $[\partial\phi/\partial x_i = O(\phi)]$, in a power series of k:

$$\phi = \phi + \phi k + \phi k^2 + \cdots. \tag{15.17}$$

Corresponding to this expansion, the macroscopic variables ω, u_i, τ, etc. are also expanded in k:

$$h = h + hk + hk^2 + \cdots, \tag{15.18}$$

where $h = \omega, u_i, \ldots$.

The asymptotic solution obtained by Sone is summarized below: A solvability condition yields the zeroth-order equation for the pressure, i.e.,

$$\frac{\partial P_0}{\partial x_i} = 0, \tag{15.19}$$

and the following governing equations for different orders (m) of the expansion:

$$\frac{\partial u_{im}}{\partial x_i} = 0, \tag{15.20a}$$

$$\frac{\partial P_{m+1}}{\partial x_i} = \gamma_1 \frac{\partial^2 u_{im}}{\partial x_j^2}, \tag{15.20b}$$

$$\frac{\partial^2 \tau_m}{\partial x_j^2} = 0, \tag{15.20c}$$

$$(m = 0, 1, 2, \ldots),$$

where

$$P_m = \omega_m + \tau_m, \tag{15.20d}$$

and γ_1 is a constant related to the collision integral.

The corresponding stress tensor and temperature gradient vector of the Grad–Hilbert solution are

$$\left.\begin{aligned} &P_{ij0} = P_0\delta_{ij}, \quad P_{ij1} = P_1\delta_{ij} + \gamma_1 S_{ij0}, \\ &P_{ij2} = P_2\delta_{ij} + \gamma_1 S_{ij1} + \gamma_3 \frac{\partial^2 \tau_0}{\partial x_i \partial x_j}, \\ &P_{ij3} = P_3\delta_{ij} + \gamma_1 S_{ij2} + \gamma_3 \frac{\partial^2 \tau_1}{\partial x_i \partial x_j} - \frac{2\gamma_6}{\gamma_1}\frac{\partial^2 P_1}{\partial x_i \partial x_j}, \end{aligned}\right\} \tag{15.21a}$$

$$\left.\begin{aligned} &Q_{i0} = 0, && Q_{i1} = \tfrac{5}{4}\gamma_2 G_{i0}, \\ &Q_{i2} = \tfrac{5}{4}\gamma_2 G_{i1} + \frac{\gamma_3}{2\gamma_1}\frac{\partial P_1}{\partial x_i}, && Q_{i3} = \tfrac{5}{4}\gamma_2 G_{i2} + \frac{\gamma_3}{2\gamma_1}\frac{\partial P_2}{\partial x_i}. \end{aligned}\right\} \tag{15.21b}$$

Here we have defined the strain tensor and the heat flux as

$$S_{ijm} = -\left(\frac{\partial u_{im}}{\partial x_j} + \frac{\partial u_{jm}}{\partial x_i}\right), \quad G_{im} = -\frac{\partial \tau_m}{\partial x_i}.$$

The constants are as follows:

- For a *hard sphere molecular gas* (Ohwada and Sone, 1992):

$$\left.\begin{aligned} &\gamma_1 = 1.270042427, \quad \gamma_2 = 1.922284066, \\ &\gamma_3 = 1.947906335, \quad \gamma_6 = 1.419423836. \end{aligned}\right\} \tag{15.22a}$$

- For the *BKG model*,

$$\gamma_n = 1. \tag{15.22b}$$

Remark: The governing equations are the steady-state Stokes and heat conduction equation at various orders with appropriately defined stress tensor and heat flux. In particular, in equation (15.21a) the term proportional to S_{ijm} corresponds to *viscous stress* in the classical fluid dynamics, the higher-order term proportional to $\partial^2 \tau_m / \partial x_i \partial x_j$ is called *thermal stress*, and the term proportional to $\partial^2 P_1 / \partial x_i \partial x_j$ is the *pressure stress.* In the second and higher orders, the heat flux vector Q_{im} depends on pressure gradient as well as on temperature gradient.

Boundary Conditions: We will ignore here the Knudsen correction, which is also derived asymptotically by (Sone, 2002). Instead, we will focus our attention to consistent boundary conditions, which have been presented in Chapter 2 from a different perspective.

The boundary conditions up to zeroth order are

$$u_{i0} - u_{wi0} = 0, \tag{15.23a}$$

$$\tau_0 - \tau_{w0} = 0. \tag{15.23b}$$

The boundary conditions up to *first order* are

$$(u_{i1} - u_{wi1})t_i = k_0 S_{ij0} n_i t_j + K_1 G_{i0} t_i \tag{15.24a}$$

$$u_{i1} n_i = 0, \tag{15.24b}$$

$$\tau_1 - \tau_{w1} = -d_1 G_{i0} n_i. \tag{15.24c}$$

The boundary conditions up to *second order* are

$$\begin{aligned}(u_{i2} - u_{wi2})t_i = k_0 S_{ij1} n_i t_j &+ a_1 \frac{\partial S_{ij0}}{\partial x_r} n_j n_r t_i + a_2 \bar{\kappa} S_{ij0} n_i t_j + a_3 \kappa_{ij} S_{jr0} n_r t_i \\ &+ a_4 \frac{\partial G_{i0}}{\partial x_j} n_j t_i + a_5 \bar{\kappa} G_{i0} t_i + a_6 \kappa_{ij} G_{j0} t_i \\ &- K_1 \frac{\partial \tau_{w1}}{\partial x_i} t_i,\end{aligned} \tag{15.25a}$$

$$u_{i2} n_i = b_1 \frac{\partial S_{ij0}}{\partial x_r} n_i n_j n_r + b_2 \left(\frac{\partial G_{i0}}{\partial x_j} n_i n_j + 2\bar{\kappa} G_{i0} n_i \right), \tag{15.25b}$$

$$\begin{aligned}\tau_2 - \tau_{w2} = -d_1 G_{i1} n_i - d_4 \frac{\partial S_{ij0}}{\partial x_r} n_i n_j n_r - d_3 \frac{\partial G_{i0}}{\partial x_j} n_i n_j \\ -d_5 \bar{\kappa} G_{i0} n_i,\end{aligned} \tag{15.25c}$$

where

$$\left.\begin{aligned} \bar{\kappa} &= \tfrac{1}{2}(\kappa_1 + \kappa_2), & \kappa_{ij} &= \kappa_1 \ell_i \ell_j + \kappa_2 m_i m_j, \\ u_{wi} &= u_{wi0} + u_{wi1} k + \cdots, & \tau_w &= \tau_{w0} + \tau_{w1} k + \cdots . \end{aligned}\right\} \tag{15.25d}$$

The terms κ_1/L and κ_2/L are the principal curvatures of the boundary, where κ_1 or κ_2 is taken negative when the corresponding center of curvature lies on the side of the gas; the ℓ_i and m_i are the direction cosines of the principal directions corresponding to κ_1 and κ_2, respectively. Also, $\sqrt{2RT_0}u_{wi}$ ($u_{wi}n_i = 0$) and $T_0(1+\tau_w)$ are the velocity and temperature of the boundary, respectively; $k_0, K_1, a_1, \ldots, a_6, b_1, b_2$, and d_1, d_3, d_4, and d_5 are constants called slip coefficients, determined by the molecular model (e.g., hard sphere, BKG) and the reflection law on the boundary (e.g., diffuse reflection). The slip coefficients depend on the parameter $(U_0/k_B T_0)$ except for a hard sphere molecular gas and the BKG model.

- For a *hard sphere* molecular gas under diffuse reflection ($\sigma_v = 1$), the slip coefficients are

$$\left.\begin{array}{lll} k_0 = -1.2540, & K_1 = -0.6463, & d_1 = 2.4001, \\ a_4 = 0.0330, & b_1 = 0.1068, & b_2 = 0.4776. \end{array}\right\} \tag{15.26}$$

- For a BGK gas the slip coefficients are

$$\left.\begin{array}{lll} k_0 = -1.01619, & K_1 = -0.38316, & d_1 = 1.30272, \\ a_1 = 0.76632, & a_2 = 0.50000, & a_3 = -0.26632, \\ a_4 = 0.27922, & a_5 = 0.26693, & a_6 = -0.76644, \\ b_1 = 0.11684, & b_2 = 0.26693, & d_3 = 0, \\ d_4 = 0.11169, & d_5 = 1.82181. & \end{array}\right\} \tag{15.27}$$

In order to obtain dimensional quantities we need the viscosity and conductivity, which are given by

$$\mu = \frac{\sqrt{\pi}}{2}\gamma_1 p_0 \frac{\lambda_0}{\sqrt{2RT_0}}; \quad k = \frac{5\sqrt{\pi}}{4}\gamma_2 R p_0 \frac{\lambda_0}{\sqrt{2RT_0}}.$$

Remark 1: The second term on the right-hand side of the slip condition of equation (15.24a) shows that a flow is induced over a wall with a temperature gradient along it. This is the thermal creep flow, as we have discussed in Section 5.1. The fifth term on the right-hand side of equation (15.25a) shows the existence of another type of flow, called thermal stress slip flow, which we have also discussed in Section 5.1.

Remark 2: The above equations demonstrate rigorously that in order to simulate steady-state microflows when $\mathrm{Re} \ll \mathrm{Kn} \ll 1$, we simply need to solve the Stokes equations under the slip boundary condition. The effect of gas rarefaction for the macroscopic variables, such as density, flow velocity, and temperature, enters only through the slip boundary condition.

Case with Re ≈ 1 and Small ΔT

The next step in Sone's theory is to extend the asymptotic theory to the case the Reynolds number takes a finite value. In that case, the Mach number is of the same order of magnitude as the Knudsen number. Correspondingly, the deviation of the velocity distribution function from a uniform equilibrium state at rest is $\mathcal{O}(\mathrm{Kn})$. In addition, the temperature variation should be $\mathcal{O}(\mathrm{Kn})$ for the theory to be valid. The solutions are obtained from the steady-state nonlinear Boltzmann equation. Specifically, the nonlinear terms of the (perturbed) velocity distribution function ϕ cannot be neglected when powers of k are considered, since $\phi = O(k)$ by assumption.

Sone uses the so-called S expansion to expand ϕ in the form

$$\phi = \phi_1 k + \phi_2 k^2 + \cdots , \tag{15.28}$$

looking for a moderately varying solution whose length scale of variation is on order of the characteristic length L of the system $[\partial\phi/\partial x_i = O(\phi)]$. Here the series starts from the first-order term of k, since $\phi = O(k)$, and $\phi_m = O(1)$, in contrast to the previous Grad–Hilbert expansion. The macroscopic variables ω, u_i, τ, etc. are also expanded in k in a similar fashion, i.e.,

$$h = h_1 k + h_2 k^2 + \ldots, \tag{15.29}$$

where $h = \omega,\ u_i,\ \ldots$.

The governing equations obtained from the expansion at various orders (m) are:

First, the *solvability* pressure condition:

$$\frac{\partial P_1}{\partial x_i} = 0. \tag{15.30}$$

The *first-order* conservation equations are

$$\frac{\partial u_{i1}}{\partial x_i} = 0, \tag{15.31a}$$

$$u_{j1}\frac{\partial u_{i1}}{\partial x_j} = -\frac{1}{2}\frac{\partial P_2}{\partial x_i} + \frac{1}{2}\gamma_1\frac{\partial^2 u_{i1}}{\partial x_j^2}, \tag{15.31b}$$

$$u_{j1}\frac{\partial \tau_1}{\partial x_j} = \frac{1}{2}\gamma_2\frac{\partial^2 \tau_1}{\partial x_j^2}. \tag{15.31c}$$

The *second-order* conservation equations are

$$\frac{\partial u_{j2}}{\partial x_j} = -u_{j1}\frac{\partial \omega_1}{\partial x_j}, \tag{15.32}$$

$$u_{j1}\frac{\partial u_{i2}}{\partial x_j}+(\omega_1 u_{j1}+u_{j2})\frac{\partial u_{i1}}{\partial x_j}$$
$$=-\frac{1}{2}\frac{\partial}{\partial x_i}\left[P_3-\frac{1}{6}(\gamma_1\gamma_2-4\gamma_3)\frac{\partial^2\tau_1}{\partial x_j^2}\right]$$
$$+\frac{1}{2}\gamma_1\frac{\partial^2 u_{i2}}{\partial x_j^2}+\frac{1}{2}\gamma_4\frac{\partial}{\partial x_j}\left[\tau_1\left(\frac{\partial u_{i1}}{\partial x_j}+\frac{\partial u_{j1}}{\partial x_i}\right)\right], \tag{15.33a}$$
$$u_{j1}\frac{\partial \tau_2}{\partial x_j}+(\omega_1 u_{j1}+u_{j2})\frac{\partial \tau_1}{\partial x_j}-\frac{2}{5}u_{j1}\frac{\partial P_2}{\partial x_j}$$
$$=\frac{1}{5}\gamma_1\left(\frac{\partial u_{i1}}{\partial x_j}+\frac{\partial u_{j1}}{\partial x_i}\right)^2+\frac{1}{2}\frac{\partial^2}{\partial x_j^2}\left(\gamma_2\tau_2+\frac{1}{2}\gamma_5{\tau_1}^2\right), \tag{15.33b}$$

where γ_n are constants related to the collision integral.

The coefficient functions of the S expansion of the stress tensor P_{ij} and heat flux vector Q_i are given as follows:

$$\left.\begin{aligned}
P_{ij1}&=P_1\delta_{ij},\\
P_{ij2}&=P_2\delta_{ij}-\gamma_1\left(\frac{\partial u_{i1}}{\partial x_j}+\frac{\partial u_{j1}}{\partial x_i}\right),\\
P_{ij3}&=P_3\delta_{ij}-\gamma_1\left(\frac{\partial u_{i2}}{\partial x_j}+\frac{\partial u_{j2}}{\partial x_i}-\frac{2}{3}\frac{\partial u_{k2}}{\partial x_k}\delta_{ij}\right)\\
&\quad-\gamma_4\tau_1\left(\frac{\partial u_{i1}}{\partial x_j}+\frac{\partial u_{j1}}{\partial x_i}\right)+\gamma_3\left(\frac{\partial^2\tau_1}{\partial x_i\partial x_j}-\frac{1}{3}\frac{\partial^2\tau_1}{\partial x_k^2}\delta_{ij}\right),
\end{aligned}\right\} \tag{15.34}$$

$$\left.\begin{aligned}
Q_{i1}&=0,\\
Q_{i2}&=-\frac{5}{4}\gamma_2\frac{\partial\tau_1}{\partial x_i},\\
Q_{i3}&=-\frac{5}{4}\gamma_2\frac{\partial\tau_2}{\partial x_i}-\frac{5}{4}\gamma_5\tau_1\frac{\partial\tau_1}{\partial x_i}+\frac{1}{2}\gamma_3\frac{\partial^2 u_{i1}}{\partial x_j^2},
\end{aligned}\right\} \tag{15.35}$$

where γ_4 and γ_5 are defined as follows.

- For a hard sphere molecular gas

$$\gamma_4=0.635021,\quad \gamma_5=0.961142, \tag{15.36}$$

- and for the BKG model

$$\gamma_4=\gamma_5=1. \tag{15.37}$$

Remark 1: The last terms of P_{ij3} and Q_{i3}, i.e.,

$$\gamma_3\left[\partial^2\tau_1/\partial x_i\partial x_j-(1/3)(\partial^2\tau_1/\partial x_k^2)\delta_{ij}\right]$$

and

$$(\gamma_3/2)\partial^2 u_{i1}/\partial x_j^2,$$

are not present in the Newton and Fourier laws, respectively. The former is called thermal stress and its effect has been discussed in Section 5.1. The terms before these terms (terms with γ_4 and γ_5) are due to the dependence of the viscosity and thermal conductivity on the temperature of the gas.

Boundary Conditions: On the wall boundary the following boundary expansions at different orders are valid. For expansions of *first order* we have

$$u_{i1} - u_{wi1} = 0, \tag{15.38a}$$

$$\tau_1 - \tau_{w1} = 0. \tag{15.38b}$$

For terms of *second order* we have

$$(u_{i2} - u_{wi2})t_i = k_0 S_{ij1} n_i t_j + K_1 G_{i1} t_i \tag{15.39a}$$

$$u_{i2} n_i = 0, \tag{15.39b}$$

$$\tau_2 - \tau_{w2} = -d_1 G_{i1} n_i, \tag{15.39c}$$

where the slip coefficient k_0, K_1, and d_1 are the same as those in the linear theory.

Remark 2: The equations derived from the *first order* expansion are the incompressible Navier–Stokes equations with no slip. The next-order equations are the compressible Navier–Stokes equations valid for a slightly compressible fluid, but there is a difference. Specifically, the corresponding Navier–Stokes equations have $\gamma_3 = 0$. This difference is due to the thermal stress in P_{ij3} in equation (15.34). Following Sone's analysis and introducing a new variable P_3^*, we obtain

$$P_{S3}^* = P_{S3} + \frac{2}{3}\gamma_3 \frac{\partial^2 \tau_{S1}}{\partial x_i^2}. \tag{15.40}$$

Then, we can incorporate the γ_3 term in the pressure term. So in the case of small temperature differences, we recover exactly the compressible Navier–Stokes equations by absorbing the thermal stress in the pressure term. This is not, however, true for large Reynolds number or for large temperature variations, as we will see below.

Remark 3: The above analysis illustrates that the slip boundary condition should be used in conjunction with the compressible Navier–Stokes equations. The combination of slip boundary conditions and the incompressible Navier–Stokes equations, which is often used because of convenience, is theoretically *inconsistent* (Aoki, 2001).

Remark 4: Unlike the Chapman–Enskog expansion, Sone's asymptotic theory leads to a set of equations whose degree of differentiation does not increase with the order of approximation. In addition, Sone's theory proves directly that the velocity distribution function depends on the spatial variables only through the five macroscopic variables and their derivatives and is not used as an assumption as done in the classical theory.

Case with $M \approx 1$

This case corresponds to a finite Mach number, which for a small Knudsen number that we study here implies that the Reynolds number can be very large, since $\mathrm{Re} \propto \mathcal{O}(1/\,\mathrm{Kn}) \gg 1$. The method of analysis of Sone is to first obtain the solution $[\partial\phi/\partial x_i = O(\phi)]$ describing the overall behavior of the gas without limiting the size of ϕ. To this end, the Hilbert expansion in the Knudsen number is introduced in the form

$$\phi = \phi_0 + \phi_1 k + \cdots . \tag{15.41}$$

The macroscopic variables ω, u_i, τ, etc. are also expanded in a similar fashion. The leading term ϕ_0 of the expansion is the local Maxwellian, characterized by the leading terms ω_0, u_{i0}, and τ_0 of the five macroscopic variables, i.e., density, flow velocity, and temperature. The variables ω_0, u_{i0}, and τ_0 are governed by the Euler set of equations:

$$\frac{\partial \widetilde{\omega}_0 u_{i0}}{\partial x_i} = 0, \tag{15.42a}$$

$$\widetilde{\omega}_0 u_{j0} \frac{\partial u_{i0}}{\partial x_j} + \frac{1}{2}\frac{\partial P_0}{\partial x_i} = 0, \tag{15.42b}$$

$$\widetilde{\omega}_0 u_{j0} \frac{\partial}{\partial x_j}(u_{i0}^2 + \frac{5}{2}\widetilde{\tau}_0) = 0, \tag{15.42c}$$

where

$$\widetilde{\omega}_0 = 1 + \omega_0, \quad \widetilde{\tau}_0 = 1 + \tau_0, \quad P_0 = \omega_0 + \tau_0 + \omega_0\tau_0. \tag{15.43}$$

The higher-order macroscopic variables ω_m, u_{im}, and τ_m $(m \geq 1)$ are governed by inhomogeneous linear Euler-type equations.

In addition, still in leading order (Kn^0) a correction is required in order to make the solution satisfy the kinetic boundary condition. This introduces boundary layer terms with the *no-slip* condition. To this end, the solution is expanded in powers of $\sqrt{\mathrm{Kn}}$ (Sone et al., 2000). Thus, the next-order term is $\mathcal{O}(\sqrt{Kn})$ rather than $\mathcal{O}(\mathrm{Kn})$. The governing equations at this order are also perturbed Euler equations with boundary layer corrections but with *slip* boundary conditions representing velocity jump and thermal creep. In addition, at this order a Knudsen layer correction is required. The technical details of this approach can be found in (Sone et al., 2000), where some earlier erroneous published results are rectified.

Case with $Re \approx 1$ and Large ΔT

For a hard sphere molecular gas, the continuum equations governing the macroscopic variables at their leading-order term are given as follows (Sone, 2002):

$$\frac{\partial P_0}{\partial x_i} = 0, \quad \frac{\partial P_1}{\partial x_i} = 0, \tag{15.44a}$$

$$\frac{\partial \widetilde{\omega}_0 u_{j1}}{\partial x_j} = 0, \tag{15.44b}$$

$$\begin{aligned}
\widetilde{\omega}_0 u_{j1}\frac{\partial u_{i1}}{\partial x_j} = & -\frac{1}{2}\frac{\partial P_2}{\partial x_i} + \frac{1}{2}\gamma_1 \frac{\partial}{\partial x_j}\left[\widetilde{\tau}_0^{1/2}\left(\frac{\partial u_{i1}}{\partial x_j} + \frac{\partial u_{j1}}{\partial x_i} - \frac{2}{3}\frac{\partial u_{k1}}{\partial x_k}\delta_{ij}\right)\right] \\
& -\frac{1}{2}\gamma_7 \frac{1}{\widetilde{P}_0}\frac{\partial}{\partial x_j}\left[\frac{\partial \tau_0}{\partial x_i}\frac{\partial \tau_0}{\partial x_j} - \frac{1}{3}\left(\frac{\partial \tau_0}{\partial x_k}\right)^2 \delta_{ij}\right] \\
& -\frac{1}{2}\gamma_3 \frac{1}{\widetilde{P}_0}\frac{\partial}{\partial x_j}\left[\widetilde{\tau}_0\left(\frac{\partial^2 \tau_0}{\partial x_i \partial x_j} - \frac{1}{3}\frac{\partial^2 \tau_0}{\partial x_k^2}\delta_{ij}\right)\right],
\end{aligned} \tag{15.44c}$$

$$\widetilde{\omega}_0 u_{j1}\frac{\partial \tau_0}{\partial x_j} = \frac{1}{2}\gamma_2 \frac{\partial}{\partial x_j}\left(\widetilde{\tau}_0^{1/2}\frac{\partial \tau_0}{\partial x_j}\right), \tag{15.44d}$$

where

$$\widetilde{\omega}_0 = 1 + \omega_0, \quad \widetilde{\tau}_0 = 1 + \tau_0, \quad \widetilde{P}_0 = 1 + P_0 = \widetilde{\omega}_0 \widetilde{\tau}_0,$$
$$\gamma_1 = 1.270042427, \gamma_2 = 1.922284066, \gamma_3 = 1.947906335, \gamma_7 = 0.188106.$$

In the present expansion, u_{i1} is the leading term of the flow velocity, since the case with $u_{i0} \equiv 0$ is considered.

The *boundary conditions* are

$$\tau_0 = \tau_{w0}, \quad u_{i1} = -\frac{\widetilde{\tau}_0^{1/2}}{\widetilde{P}_0}\frac{\partial \tau_0}{\partial x_j}(\delta_{ij} - n_i n_j), \tag{15.45}$$

for the temperature and velocity, respectively.

Remark: The third and fourth terms on the right-hand side of equation (15.44c) are due to thermal stress. The thermal stress contribution is not in the form of a gradient of some function, and thus it cannot be incorporated in the pressure term, in contrast to the small temperature variation presented earlier. Therefore, the solution $u_{i1} = 0$ is possible under a special temperature field. The condition for u_{i1} to be zero is

$$\varepsilon_{ijk}\frac{\partial \tau_0}{\partial x_j}\frac{\partial}{\partial x_k}\left(\frac{\partial \tau_0}{\partial x_l}\right)^2 = 0, \tag{15.46}$$

which is obtained by putting $u_{i1} = 0$ in equation (15.44c). When the distance between isothermal lines or surfaces is constant along such contours, the condition is satisfied. Thus, even when the temperature of the boundary is uniform and the thermal creep flow is absent, a flow may be induced in the gas. This flow is called *nonlinear thermal stress flow*, and it was presented in Section 5.1.

15.4.3 Numerical Solutions

Numerical simulations based on the Boltzmann equation are computationally very expensive, and they have been obtained mostly for simple geometries, such as pipes and channels. In particular, a number of investigators have considered numerical solutions of the linearized BGK and exact Boltzmann equations, valid for flows with small pressure and temperature gradients (Huang et al., 1997; Sone, 1989; Ohwada et al., 1989a; Loyalka and Hamoodi, 1990; Aoki, 1989). These studies used hard sphere and Maxwellian molecular models.

In the following, we summarize some benchmark solutions that can be used for validating numerical simulations and experimental results for microchannels and micropipes.

Rarefied flow in a channel has been studied extensively using both the full Boltzmann equation and different versions of the BGK model. A comparison of different numerical solutions in (Sharipov and Sleznev, 1998), reveals a small difference of about 2% in most of the published solutions for the normalized flowrate. This quantity is defined as

$$\dot{M}_c = \frac{\dot{M}}{HP_i\beta_i},$$

where $\dot{M}$ is the mass flow rate, H is the channel height, P_i is the pressure at the inlet, and the parameter β_i is defined as

$$\beta_i = \sqrt{\frac{m}{2k_B T_i}},$$

where T_i is the temperature at the inlet. In Tables 15.3 and 15.4 (adopted and modified from (Sharipov and Sleznev, 1998)) we list values of the normalized flowrate as a function of the rarefaction parameter δ defined as

$$\delta = \frac{\sqrt{\pi}}{2}\frac{1}{\mathrm{Kn}},$$

where the Knudsen number is based on the channel's height H (δ is the same as the parameter D defined in Section 6.1). These results are obtained for different accommodation coefficients employing Maxwell's scattering kernel. More specifically, the fully diffuse scattering case ($\sigma_v = 1$)

TABLE 15.3. Pressure-driven flow in a channel: Normalized flowrate $\dot{M}_c$ vs. δ (rarefaction parameter) and σ_v (accommodation coefficient) $(0 < \delta \leq 1)$.

$\dot{M}_c$							
Loyalka (1975)[a]						Loyalka and Hickey (1991)[b]	
δ	$\sigma = 1$	$\sigma = 0.92$	$\sigma = 0.88$	$\sigma = 0.84$	$\sigma = 0.80$	$\sigma = 0.75$	$\sigma = 0.50$
0.01	3.0489	3.2417	3.6697	3.9085	4.1695	...	...
0.02	2.7107	3.0548	3.2463	3.4530	3.6771	...	...
0.03	2.5234	3.0381	3.0131	3.2021	3.4070	...	...
0.04	2.3964	2.6915	2.8556	3.0328	3.2249	...	...
0.05	2.3016	2.5823	2.7383	2.9069	3.0897	...	...
0.07	2.1655	2.4259	2.5706	2.7270	2.8967	...	...
0.09	2.0698	2.3163	2.4532	2.6011	2.7618	...	...
0.1	2.0314	2.2723	2.4060	2.5507	2.7077	2.7860	4.3628
0.3	1.7092	1.8937	2.0011	2.1176	2.2448	...	...
0.5	1.6017	1.7776	1.8766	1.9844	2.1023	2.2128	3.4748
0.7	1.5591	1.7272	1.8220	1.9254	2.0388	...	...
0.8	1.5482	1.7052	1.7976	1.8986	2.0092	...	...
0.9	1.5416	...	...	...	...	2.1269	3.3392
1.0	1.5389	1.7005	1.7921	1.8921	2.0019	2.1204	3.3270

[a] Solutions of BGK model equation.
[b] Solutions of Boltzmann equation.

corresponds to BGK solutions of (Cercignani et al., 1994), while the hybrid specular-diffuse scattering cases correspond to solutions of BGK model equation by (Loyalka, 1975) and solutions of the full Boltzmann equation by (Loyalka and Hickey, 1991).

The good approximation of solutions to Boltzmann's equation by the

TABLE 15.4. Pressure-driven flow in a channel: Normalized flowrate $\dot{M}_c$ vs. δ (rarefaction parameter) and σ_v (accommodation coefficient) $(1 < \delta \leq 10)$.

$\dot{M}_c$							
Loyalka (1975)[a]						Loyalka and Hickey (1991)[b]	
δ	$\sigma = 1$	$\sigma = 0.92$	$\sigma = 0.88$	$\sigma = 0.84$	$\sigma = 0.80$	$\sigma = 0.75$	$\sigma = 0.50$
1.1	1.5379	...	...	...	...	2.1171	3.3192
1.2	1.5394	...	...	...	...	2.1164	3.3149
1.3	1.5427	...	...	...	...	2.1178	3.3136
1.4	1.5473	...	...	...	...	2.1209	3.3144
1.5	1.5530	1.7107	1.7999	1.8974	2.0046	2.1254	3.3171
2.0	1.5942	1.7503	1.8386	1.9352	2.0414	2.1625	3.3491
2.5	1.6480	1.8039	1.8918	1.9881	2.0939	...	...
3.0	1.7092	1.8653	1.9531	2.0493	2.1551	2.2748	3.4618
3.5	1.7751	1.9316	2.0196	2.1158	2.2217	...	...
4.0	1.8440	2.0013	2.0894	2.1858	2.2918	...	...
5.0	1.9883	2.1472	2.2356	2.3324	2.4388	2.5555	3.7496
6.0	2.1381	2.2988	2.3876	2.4848	2.5916	...	...
7.0	2.2914	2.4541	2.5433	2.6408	2.7480	2.8625	4.0633
9.0	2.6048	2.7722	2.8620	2.9601	3.0679	...	...
10.0	2.7638	2.9340	3.0241	3.1225	3.2305	3.3407	4.5490

[a] Solutions of BGK model equation.
[b] Solutions of Boltzmann equation.

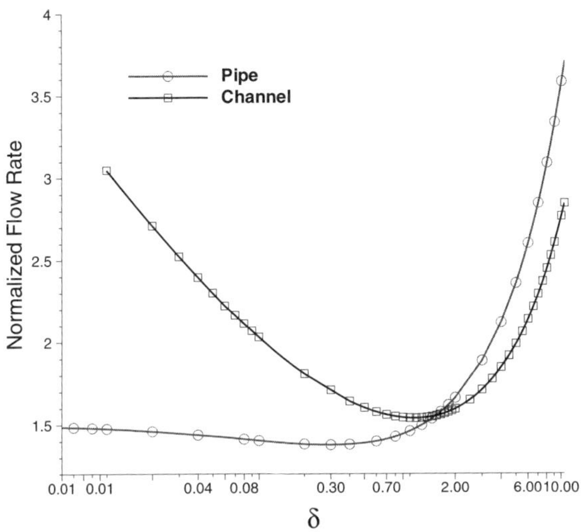

FIGURE 15.16. Normalized flowrate versus the rarefaction parameter $\delta = \sqrt{\pi}/(2Kn)$ for channel and pipe.

BGK kinetic model is also valid for **rarefied pipe flow**, and the discrepancy in the mass flowrate is about 2%. Here the normalized flowrate is defined as

$$\dot{M}_p = \frac{\dot{M}}{\pi R^2 P_i \beta_i},$$

where R is the pipe radius. In Table 15.5 (adopted and modified from (Sharipov and Sleznev, 1998) we present values of the normalized flowrate as a function of the rarefaction parameter δ and the accommodation coefficient σ_v. The data for the diffuse scattering ($\sigma_v = 1$) are from (Cercignani and Sernagiotto, 1966), and they are solutions to the BGK model equations. The data for the hybrid diffuse-specular scattering is due to Porodnov et al. (Porodnov et al., 1978; Porodnov and Tuchvetov, 1979), and they are also solutions of the BGK equation.

In Figure 15.16 we plot the normalized flowrate versus the rarefaction parameter δ for both the channel and pipe; the data are from Tables 15.3, 15.4, and 15.5. We observe the familiar Knudsen's minimum discussed earlier in Section 4.2, with the asymptote for a pipe only as $\delta \to 0$, which is equivalent to $\mathrm{Kn} \to \infty$.

TABLE 15.5. Pressure-driven flow in a pipe: Normalized flowrate $\dot{M}_p$ vs. rarefaction parameter (δ) and accommodation coefficient (σ_v). Solutions of BGK equations by Porodnov and Tuchvetov (1979).

$\dot{M}_p$						
δ	$\sigma = 1$	$\sigma = 0.94$	$\sigma = 0.90$	$\sigma = 0.84$	$\sigma = 0.80$	$\sigma = 0.6$
0.01	1.4768	1.657	1.791	2.026	2.187	3.374
0.02	1.4608	1.635	1.764	1.983	2.144	3.255
0.04	1.4391	1.605	1.728	1.933	2.085	3.137
0.08	1.4131	1.569	1.685	1.873	2.014	...
0.1	1.4043	1.556	1.668	1.853	1.992	2.944
0.2	1.3820	1.523	1.627	1.806	1.931	...
0.4	1.3796	1.510	1.615	1.768	1.888	2.720
0.6	1.3982	1.523	1.621	1.772	1.888	2.691
0.8	1.4261	1.547	1.638	1.791	1.904	...
1.0	1.4594	1.578	1.668	1.818	1.930	2.706
2.0	1.6608	1.773	1.861	2.007	2.116	2.879
3.0	1.8850	1.994	2.081	2.227	2.336	3.096
4.0	2.1188	2.225	2.312	2.458	2.567	3.372
5.0	2.3578	2.461	2.548	2.694	2.803	3.565
6.0	2.5999	2.700	2.787	2.934	3.003	...
7.0	2.8440	2.942	3.029	3.167	3.285	...
8.0	3.0894	3.185	3.272	3.420	3.529	4.293
9.0	3.3355	3,430	3.517	3.664	3.778	...
10.0	3.5821	3.675	3.761	3.910	4.019	4.785

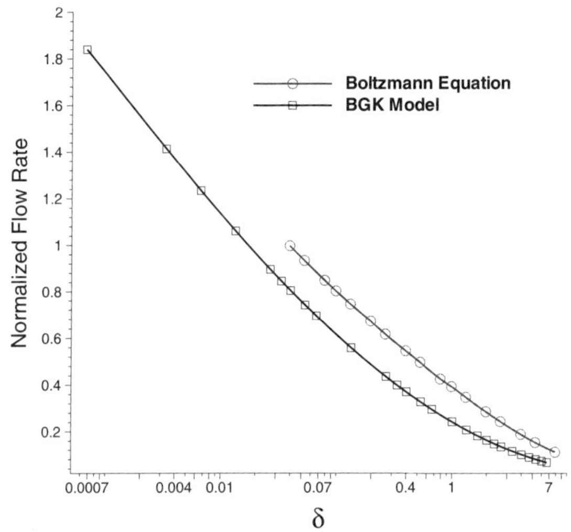

FIGURE 15.17. Thermal creep in a channel: Normalized flowrate versus the rarefaction parameter $\delta = \sqrt{\pi}/(2\,\mathrm{Kn})$.

15.4.4 Nonisothermal Flows

Unlike isothermal flows, where the BGK model equations result in accurate approximations to the Boltzmann equation, in the case of **heat transfer** the BGK model is inadequate, since it gives the wrong Prandtl number. This is particularly evident in simulations of thermal creep, where the difference in solutions of the BGK and Boltzmann equations may be as high as 30%. This is shown in the Figure 15.17, where we plot the normalized flowrate in a channel due to thermal creep. The BGK solution, due to (Loyalka, 1974), differs significantly from the solution of the Boltzmann equation obtained by (Ohwada et al., 1989a). The BGK solution can be improved if the assumption that states that the collision frequency is independent of the molecular velocity is relaxed.

- In fact, a simple modification where the nondimensional collision frequency is taken as $(2/3)\delta$ instead of exactly equal to δ, which corresponds to the standard BGK model, improves the BGK predictions significantly.

15.5 Lattice–Boltzmann Method (LBM)

The lattice-Boltzmann method (LBM) offers potentially great advantages over conventional methods for simulating microflows. It represents a "min-

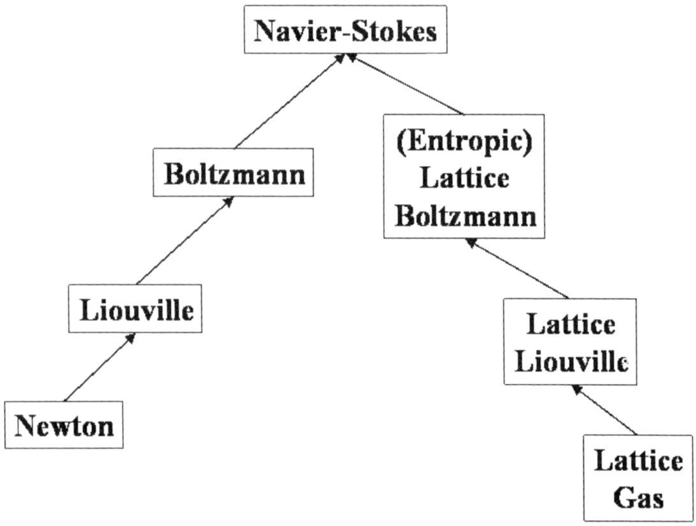

FIGURE 15.18. Analogue between the BBGKY hierarchy and its lattice counterpart. Adopted from (Succi, 2001).

imal" form of the Boltzmann equation and can be used for gas or liquid as well as for particulate microflows. It can handle arbitrarily complex geometries, even random geometries, in a fairly straightforward way, and it seems to be particularly effective in the regime in which microdevices operate. This method solves a simplified Boltzmann equation on a discrete lattice. Because of its intrinsic kinetic nature it can also handle the high Knudsen number regime, and it is very effective for problems where both mesoscopic dynamics and microscopic statistics are important. However, initially there was only limited use of the method in microflows, and this may have to do with its originally intended use in simulating high Reynolds number flows. In the following, we review its origin and basic idea, we compare it with Navier–Stokes solutions, and finally we present flow simulation examples, including microflows.

There are three main theoretical developments: The first one took place in the mid 1980s leading to the lattice gas methods. The second one started in the early 1990s leading to the lattice Boltzmann equation. Finally, the third main development took place in the early 2000s leading to the entropic lattice Boltmann method. A schematic representation of these methods is shown in Figure 15.18, adopted from the book by (Succi, 2001). The left column shows the classical BBGKY (Bogoliubov–Born–Green–Kirwood–Yvon) hierarchy leading from atomistic to continuum flow equations. The right column shows the corresponding approximations in the framework of lattice methods. At the atomistic level we have a description by Newton's law that uses molecular positions and molecular velocities. This is the basis of the molecular dynamics (MD) (see Section 16.1) but also for the lattice gas method. This is followed by the many-body kinetic model of Liouville, which employs distribution functions $f_N(\boldsymbol{x}_1, \boldsymbol{v}_1, \ldots, \boldsymbol{x}_N, \boldsymbol{v}_N, t)$ that satisfy the Liouville equation

$$\left[\frac{\partial}{\partial t} + \sum_{i=1}^{N} \boldsymbol{v}_i \cdot \frac{\partial}{\partial \boldsymbol{x}_i} + \mathbf{a}_i \cdot \frac{\partial}{\partial \boldsymbol{v}_i}\right] f_N = 0, \tag{15.47}$$

where $\mathbf{a}_i$ are the molecular accelerations. This is a $6N$-dimensional continuity equation; it can be simplified by coarse-graining, i.e., averaging over single-particle coordinates to obtain a distribution function of reduced order

$$f_M = f_{12\ldots M<N} = \int f_{12\ldots N} d\mathbf{z}_{M+1} \ldots d\mathbf{z}_N,$$

where $d\mathbf{z}_k = d\boldsymbol{x}_k \, d\boldsymbol{v}_k$, $k = M+1, \ldots, N$. This averaging procedure results in the *BBGKY hierarchy* expressed by the equation

$$\left[\frac{\partial}{\partial t} + \sum_{i=1}^{M} \boldsymbol{v}_i \cdot \frac{\partial}{\partial \boldsymbol{x}_i} + \mathbf{a}_i \cdot \frac{\partial}{\partial \boldsymbol{v}_i}\right] f_M = C_M, \tag{15.48}$$

where the right-hand-side term C_M contains effects of intermolecular collisions represented by $f_{M+1}, \ldots, f_{M+B}$, where B denotes the number of bodies involved in the interaction. The Boltzmann equation corresponds to $B = 1$; also, to obtain the Navier–Stokes equations we keep only the two lower levels, i.e., $M = 1, 2$ in the BBGKY hierarchy.

Next, we present the main results of all three aforementioned lattice approaches.

In the last half of the 1980s, a new class of numerical approaches was developed for solving the Navier–Stokes equations indirectly. These new algorithms were based on discrete lattice models of interacting "particles," whose continuum description could govern the continuum fluid flow equations. The most interesting of these methods was the cellular automaton model of Frisch, Hasslacher, and Pomeau (hereafter called FHP) (Frisch et al., 1986). The basic idea of lattice gas (or cellular automata (CA)) methods is to represent the fluid as an *ensemble* of interacting low-order-bit computers situated at regularly spaced lattice sites. In the FHP model of two-dimensional hydrodynamics, the underlying lattice is a close-packed equilateral triangular lattice with sites at triangle vertices. Each site has a seven-bit state with the first six bits specifying the presence or absence of a particle traveling at angle $\mathbf{e}_j = j\, 60^\circ$ $(0 \leq j < 6)$ along the edges of the triangular lattice and the last bit specifying the existence or nonexistence of a particle at rest at the lattice site. Each particle (except a rest particle) moves one lattice distance in one fundamental time interval. After the particles propagate they then interact according to certain collision rules (Frisch et al., 1986).

Soon it was discovered, however, that there were some difficulties with these CA methods. First, their work requirements increase rapidly with Reynolds number, and in fact, for some cases CA-based computations could be more expensive than corresponding computations with the Navier–Stokes equations (Orszag and Yakhot, 1986). Also, the methods represent the correct incompressible hydrodynamics only in the limit of small Mach number. Finally, CA methods are statistical in nature and suffer from significant noise. The velocity field is computed as the average velocity over a large number N of CA sites, so that there is an error of order $1/\sqrt{N}$ in the evaluation of this velocity field. The first two arguments are not so critical for microflows, where both the Reynolds number and the Mach number are typically low, except for micronozzles (see Section 6.6). However, the third one, which is similar to the problem in DSMC, is related to the efficiency of these methods, and new theoretical developments have extended CA methods, improving their accuracy considerably.

An effective way to avoid the difficulties with noisiness of CA systems is to use a lattice Boltzmann approach (McNamara and Zanetti, 1988; Higuera and Jiménez, 1989); also see (Succi, 2001; Chen and Doolen, 1998) and references therein. This approach seems to be more efficient than Monte Carlo CA methods for moderate to low Reynolds numbers. The idea is to integrate a kinetic equation for the CA system; here the kinetic equation is for average particle distribution functions along each of the discrete allowed particle velocities at each lattice site. For a two-dimensional CA system there are seven distribution functions (corresponding to the seven bits) at each lattice site. These functions are smooth nonrandom functions governed by nonlinear partial differential equations that are integrated in space-time to obtain the flow description. Velocities are determined as averages over a number of lattice sites of the LB system. The method extends easily to three dimensions in which the lattice is a 24-bit or 25-bit projection of a four-dimensional FCHC lattice onto three-dimensional space.

A more recent version developed by Chen and collaborators (Chen and Doolen, 1998) employs a square lattice in two dimensions with three speeds and nine velocities. Specifically, they have eight nonzero velocities for moving along the edges of the square and one zero velocity for the rest particle as follows:

$$(\pm 1, 0),\ (0, \pm 1),\ (\pm 1, \pm 1),\ (0, 0).$$

The LB equations can be thought of as discrete analogues of the continuous Boltzmann equation we presented in Section 15.4 but in an *incomplete velocity space* (phase space); more rigorous work has in fact proved this analogy (Abe, 1997). Let us denote by $f_i(\boldsymbol{x}, t)$ the distribution function at $\boldsymbol{x}, t$ with velocity $\boldsymbol{c}_i$, and assume that the collision operator can be described by the BGK approximation we described earlier (Section 15.4). Then, the

LBM-BGK equation of motion is given by

$$f_i(\boldsymbol{x} + \boldsymbol{c}_i \Delta t, t + \Delta t) - f_i(\boldsymbol{x}, t) = \frac{1}{\tau}(f_i - f_i^{\text{eq}}), \tag{15.49}$$

where $f_i^{\text{eq}}, (i = 0, 1, \ldots, 8)$ is the equilibrium distribution function and τ is the relaxation time. An equilibrium distribution that approximates the Maxwellian–Boltzmann equilibrium distribution up to second–order was derived in (Qian et al., 1992), and is given by

$$f_i^{\text{eq}} = w_i \rho \left[1 + \frac{c_{i\alpha} u_\alpha}{c_s^2} + \frac{(c_{i\alpha} c_{i\beta} - c_s^2 \delta_{\alpha\beta})}{2c_s^4} u_\alpha u_\beta \right],$$

where $i = 0, 1, \ldots, 8$, while α and β are the two Cartesian directions. All fluid velocities are normalized by $\sqrt{3RT}$, and thus the speed of sound is $c_s = 1/\sqrt{3}$; also, w_i are weights. The density and velocity are obtained from formulas similar to the Boltzmann equation, where the integral is replaced by summations, i.e.,

$$\rho = \sum_i f_i \quad \text{and} \quad \rho \boldsymbol{v} = \sum_i \boldsymbol{c}_i f_i.$$

In simulating incompressible isothermal flows, the relaxation time τ is taken constant in the BGK mode, but as we have discussed already in Section 15.4, this is not valid for nonisothermal flows or flows with variable density, as in gas microflows. To this end, it has been proposed by Chen and collaborators (Nie et al., 1998) to modify the relaxation parameter as follows:

$$\tau' = \frac{1}{2} + \frac{\rho_0}{\rho}\left(\tau - \frac{1}{2}\right),$$

where $\rho_0 = 1$. The viscosity ν and mean free path are defined based on the relaxation parameter from

$$\nu = c_s^2 \Delta t(\tau' - 1/2) \quad \text{and} \quad \lambda = \frac{\rho_0}{\rho}(\tau' - 1/2). \tag{15.50}$$

In order to arrive at the Navier–Stokes equation from the LBM we also need to employ the limit of long wavelength and low Mach number, and using the Chapman–Enskog multiscale expansion the resulting equations are

$$\frac{\partial \rho}{\partial t} + \nabla_\alpha \cdot (\rho \boldsymbol{v}_\alpha) = 0,$$

$$\frac{\partial (\rho \boldsymbol{v}_\alpha)}{\partial t} + \nabla_\beta \cdot (\rho \boldsymbol{v}_\alpha \boldsymbol{v}_\beta) = -\nabla_\alpha P + \nabla_\beta \cdot [\nu(\nabla_\alpha(\rho \boldsymbol{v}_\beta) + \nabla_\beta(\rho \boldsymbol{v}_\alpha)].$$

The pressure is given by $P = c_s^2 \rho$. If the density variations are small, we recover the more familiar form of Navier–Stokes from the above equations.

The BGK version of LBM is typically more accurate for values of $\tau < 1$. A systematic truncation error analysis of LBM is performed in (Holdych et al., 2004). The method is second-order accurate in time for fixed lattice spacing, but is first-order accurate in time and second-order accurate in space for a constant value of τ. If the lattice spacing is reduced for a constant ratio $\Delta x/\Delta t$, then LBM does not converge. It is also recognized that LBM has the same type of inconsistency as the classical Dufort–Frankel scheme for diffusion, and thus consistency is satisfied only for $\Delta t/\Delta x^2$ remaining constant during refinement. For constant values of τ, LBM is consistent in the classical sense.

While successful for simple isothermal fluid flows, the above LBM formulation is not Galilean invariant for nonisothermal and multiphase flows; for isothermal flows it is Galilean invariant up to order $\mathcal{O}(M^4)$, where M is the Mach number. It has also been shown to be unstable for small values of the viscosity. Indeed, in the above formulation we see that the viscosity is proportional to $(\tau - \frac{1}{2})$, and for values of $\tau \in (\frac{1}{2}, 1]$ instabilities may arise. Such considerations have led to the development of the *entropic* lattice Boltzmann method in (Karlin et al., 1999; Boghosian et al., 2003). At the heart of this formulation is the use of the H-theorem of Boltzmann, which measures irreversibility. This was presented as the Boltzmann inequality in equation (15.7b). Specifically, H is the Boltzmann function based on which the entropy S is computed as $S = -k_B H$, where

$$H = \int f(\boldsymbol{x}, \boldsymbol{v}, t) \ln f(\boldsymbol{x}, \boldsymbol{v}, t)\, d\boldsymbol{v}\, d\boldsymbol{x}.$$

Regardless of the form of the collision operator, the H-theorem states that $dH/dt \leq 0$, which in general is not satisfied in LB methods globally, although it may be satisfied locally in some versions. Specifically, it has been shown rigorously by (Yong and Luo, 2003) that the H-theorem does not exist for the lattice Boltzmann equation with *polynomial* equilibria.

Enforcing the H-theorem in the lattice Boltzmann formulation guarantees an asymptotically homogeneous spatial distribution of particles as $t \to \infty$, which in turn translates into numerical stability. In the continuum case, the Boltzmann inequality produces the Maxwellian distribution (used as equality in equilibrium). This is obtained as a minimizer of the H function defined above subject to the five conservation laws (as constraints) of mass, momentm (3), and energy. The local Maxwellian can also be interpreted as the zero point of the collision term, and also the zero point of the entropy production (Succi et al., 2002). In the discrete case the Maxwellian distribution is not necessarily the minimizer of the disrete H function (irrespective of the lattice), and this results in the aforementioned problems. To this end, in entropic lattice Boltzmann, use of different convex H functions is made that can be minimized relatively easily. The requirements are that Galilean invariance be preserved, as well as realizability $(0 < f^{\text{eq}} < 1)$

and solvability. The latter condition implies that local equilibrium can be expressed in terms of the density and velocity, i.e., the continuum variables.

In particular, the following form of the H function was introduced by (Karlin et al., 1999):

$$H = \sum_{i=1}^{N} f_i \ln(f_i/W_i),$$

where W_i are velocity-dependent weights with $W_i = f_i^{\mathrm{eq}}$ at zero flow. An example in two dimensions for the nine-velocity lattice (2D9V) is

$$H = H_B + \ln(3/8) f_0 + \ln(3/2) \sum_{i=1}^{4} f_{1i} + \ln(6) \sum_{i=1}^{4} f_{2i},$$

where $H_B = \sum_{k=1}^{2} \sum_{l=1}^{4} f_{kl}(\ln f_{kl} - 1)$, and at equilibrium,

$$f_0^{\mathrm{eq}} = \frac{8}{3}\exp[a], \quad f_{1l}^{\mathrm{eq}} = \frac{2}{3}\exp[a + \lambda_\alpha c_{1l\alpha}],$$
$$f_{2l}^{\mathrm{eq}} = \frac{1}{6}\exp[a + \lambda_\alpha c_{2l\alpha}].$$

Also, $a = \ln(\rho/6) - u^2/(2c_s^2)$, and $\lambda_\alpha = u_\alpha/c_s^2$. Constructing the corresponding local equilibrium from these expressions, (Karlin et al., 1999) recovered the equilibrium distribution used in (Qian et al., 1992), on the 2D9V lattice. Such H functions enforce Galilean invariance up to $\mathcal{O}(M^4)$, where M is the Mach number. Given, however, that the lattice Boltzmann equation is an approximation to the Navier–Stokes equations within $\mathcal{O}(M^2)$, such H functions lead to very accurate results for the local equilibrium distributions.

A second-order $\mathcal{O}(M^2)$ H function of different form was employed in (Boghosian et al., 2003), where uniform contributions to the H function from the lattice sites are assumed of the form

$$H = \sum_{i=1}^{N} h(f_i).$$

This H function is also minimized subject to the usual hydrodynamic constraints. In this case h is not a Boltzmann entropy, but its form depends on the space dimension. For example, in three dimensions it follows the Tsallis form, which is typically associated with lack of ergodicity (Boghosian et al., 2003). Many other entropy functions are possible depending on the specific kinetic model, i.e., single- or multispeed. For example, in (Boghosian et al., 2004), $h(f_i)$ is obtained from a certain functional differential equation, which admits power-law solutions for certain types of lattices.

In summary, this new entropic LB method offers many possibilities and potentially can overcome the aforementioned shortcomings of the more traditional LBM. However, its extensions to nonisothermal flows as well as complex fluid flows has not been established rigorously yet. We also note that the limit of LBM to the Navier–Stokes equations has been obtained so far only for incompressible flows but not for compressible flows. In a sense, we have a quasi-compressible formulation that can be used for both incompressible and compressible flows. In this limit the Mach number scales with the Knudsen number, and the fluctuation of density around its mean value scales with the Knudsen number squared (Boghosian et al., 2003).

15.5.1 Boundary Conditions

With regard to boundary conditions, the so-called *bounce-back* scheme with its origin in CA has been used to simulate wall boundary conditions. In the bounce-back scheme, when a particle distribution streams to a wall node, the particle distribution scatters back to the node it came from (Chen and Doolen, 1998). However, this approach leads to relatively low-order accuracy, and more recent work has attempted to include corrections in the distribution function by including velocity gradients at the wall nodes (Skordos, 1993), explicitly imposing a pressure constraint (Noble et al., 1995) or employing extrapolation techniques on staggered meshes (nodes at midpoints of lattice), similarly to the classical finite difference discretizations (Chen et al., 1996). An analysis of the accuracy of the boundary conditions as a function of the relaxation parameter τ was performed in (Holdych et al., 2004).

A robust boundary condition was analyzed in (Wagner and Pagonabarraga, 2002), for LBM, the so-called Lees–Edwards periodic boundary conditions. These are appropriate for simulating flows with simple boundaries, like Couette flows, subject to severe shear, which could be problematic using other types of boundary conditions. One issue is the interpolation scheme employed, since it may introduce artificial dissipation. Finally, new boundary conditions with error estimates have been formulated in (Ansumali and Karlin, 2002), and have been applied to slip flows as well as to no-slip flows.

15.5.2 Comparison with Navier–Stokes Solutions

A validation of the traditional LBM with direct solutions of the Navier–Stokes equations was presented in (Karniadakis et al., 1993) (LBM computations were performed by G. Zanetti). Specifically, the spectral element method described earlier in Section 14.1 was employed, and the incompressible Navier–Stokes equations were solved. The version of LBM used was similar to the method of (Higuera and Succi, 1989), which was made Galilean invariant by using a different choice of equilibrium distribution function. The resulting numerical scheme has some of the flavor of a finite-volume technique. It is, however, more microscopic in nature because it still involves the integration of a Boltzmann equation but in a limited phase

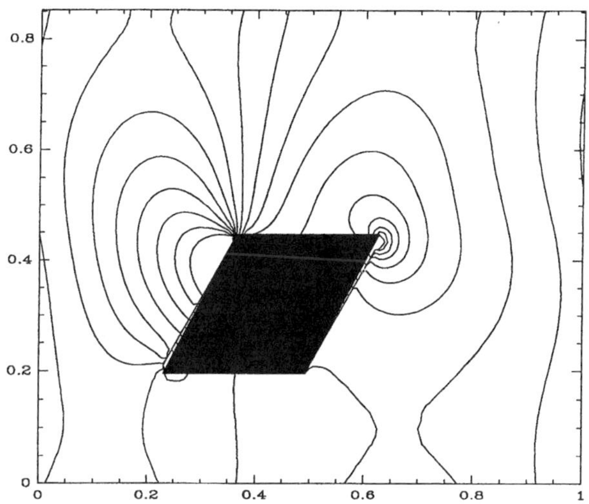

FIGURE 15.19. Pressure contours for the shear-driven flow past the shaded body. The top wall is moving, while the lower wall is stationary; Re = 110 (Karniadakis et al., 1993).

space with only a finite number of possible particle velocities. This new model successfully passed a series of consistency tests: tests of rotational symmetry, tests of the stress tensor, tests of Galilean invariance, sound, and shear waves. Specifically, we present the simulation of a sheared flow in a channel with a bluff body. This flow has simple boundary conditions: periodic in the flow direction and no-slip on the bottom wall and on the body. The flow is driven by a moving top wall. Moreover, the flow is steady in the range of Reynolds number considered.

The domain was mapped onto a parallelogram of 128×128 lattice sites, with a top wall velocity of $U = 0.042$, and the reduced density, ρ, of the LBE fluid was chosen to be $\rho = 0.3$; the kinematic viscosity was adjusted so that the macroscopic Reynolds number defined using V_0, the height of the channel, and the kinematic viscosity of the fluid was Re = 110. The same simulation was repeated using a higher resolution, i.e., 192×192 lattice sites, and a small discrepancy was found, less than 2% between the two runs. The results of the LBM simulations were rescaled appropriately for the comparison with the spectral element results. It is worth mentioning that there are no free parameters in this conversion process, except for an arbitrary constant added to the pressure. In the reference solution computed using the spectral element method the flow domain was subdivided into 28 elements, and to test for convergence, the same problem was run with three different spectral resolutions, 5×5, 7×7, and 9×9 modes per

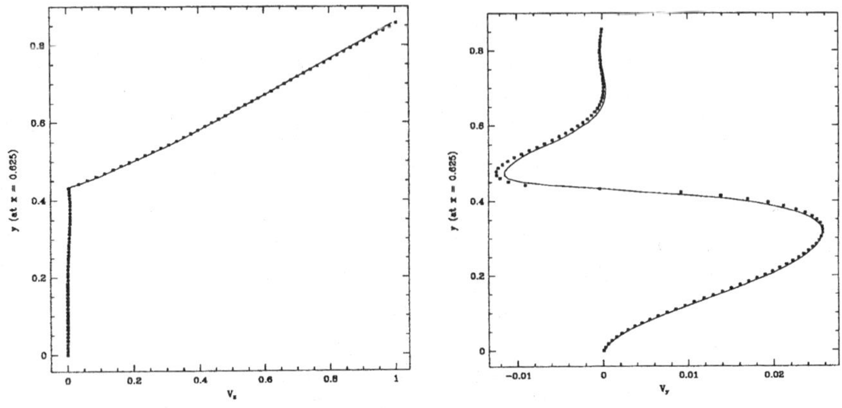

FIGURE 15.20. Streamwise and normal velocity profiles. The LBM data are indicated with symbols, and the direct Navier–Stokes solutions with the solid line (Karniadakis et al., 1993).

element. The results between LBM and spectral element simulations were in very good agreement. The domain and pressure contours are shown in Figure 15.19. A typical comparison for the velocity components v_x and v_y as a function of y downstream of the bluff body at $x = 0.625$ is shown in Figure 15.20. The discrepancies between the two solution methods seem to be particularly prominent, on the order of 10%, where the local velocity is small, typically less than 1% or 2% of the wall velocity $V_0 = 1$. This small discrepancy between the two methods is physical in origin, i.e., due to weak compressibility effects in LBM, what we described earlier as quasi-compressible fluid. In fact, the amplitude of compressibility corrections to the local density scales as the square of Mach number, and the latter varies widely between different regions of the flow.

15.5.3 LBM Simulation of Microflows

A discrete Boltzmann equation for microfluidics has been developed by (Li and Kwok, 2003) based on the aforementioned BGK single-relaxation-time collision model. A statistical-mechanical approach was employed to derive an equivalent external acceleration force acting on the lattice particles. The potentials used accounted for electrostatic interactions as well as intermolecular interactions between fluid–fluid and fluid–substrates. For Poiseuille microflow, the slip velocity U_s is

$$U_s = \frac{2\tau(\tau - 1)}{3\rho\nu} F \delta_x^2, \tag{15.51}$$

where τ is the relaxation factor in the BGK model. Also, F is the external force due to the mean-field potential, and δ_x is the lattice constant or grid spacing. The validity of this model and in particular equation (15.51) was questioned by (Luo, 2004), who interpreted δ_x as the grid spacing. In that case, for fixed fluid properties, the slip velocity vanishes as $\delta_x \rightarrow 0$. It is also not clear what the dependence of the slip velocity U_s on the single relaxation time constant τ means physically. Li and Kwok argued that δ_x is a lattice spacing whose value depends on the physical properties of the microfluidic system, but they did not specify how. The main problem with such models is how rigorously the boundary condition is imposed at the wall; for example, the bounce-back boundary condition can create any amount of slip if not properly implemented.

While more work is required to fully resolve such issues at the fundamental level, an evaluation of LB models in the context of microflows is instructive. Next, we first present some simulations of the standard LBM method, i.e., the BGK version we described above, and subsequently we present an example obtained using the entropic LBM. We have already presented LBM simulations of a microcavity flow in Chapter 3 as well as simulations using the entropic LBM in Chapter 5. Here we first present results for microchannels obtained by (Nie et al., 1998) using a square lattice and the equations described above. The microchannel has length $L = 1000$ and height $H = 10$ in lattice units. A pressure boundary condition was used at the inlet and the outlet with ratio $\Pi = P_i/P_o = 2$. The parameter ρ_0 in equation (15.50) was set by comparing the flowrate from the simulation to experimental results presented in Figure 4.2. The slip velocity and mass flowrate at the outlet of the microchannel are plotted in Figure 15.21. The slip velocity at the outlet was obtained from

$$U(Y) = U_0(Y - Y^2) + V_s,$$

where $Y = y/H$. From a least-squares fit to the data of Figure 15.21, Nie et al. found that

$$V_s = 8.7\,\mathrm{Kn}^2,$$

and based on this equation, they obtained an analytical formula for the normalized flowrate (with respect to continuum flowrate):

$$M_f = 1 + 12V_s(\mathrm{Kn})\frac{\ln(\Pi)}{\Pi^2 - 1}.$$

This formula agrees well with the numerical data shown in Figure 15.21 if we set $\Pi = 2$. With regard to nonlinear distribution of pressure, for $\mathrm{Kn} \leq 0.2$ the LBM results agree with the DSMC results presented in Chapter 4. However, for higher Knudsen number values a change in the curvature of the pressure distribution was observed, indicating a slower than linear pressure drop. This latter result, however, has not been verified by DSMC approaches, and it deserves further investigation.

In the following we present an application of the *entropic LBM*. Specifically, the slip boundary condition is examined for a simple setup, where

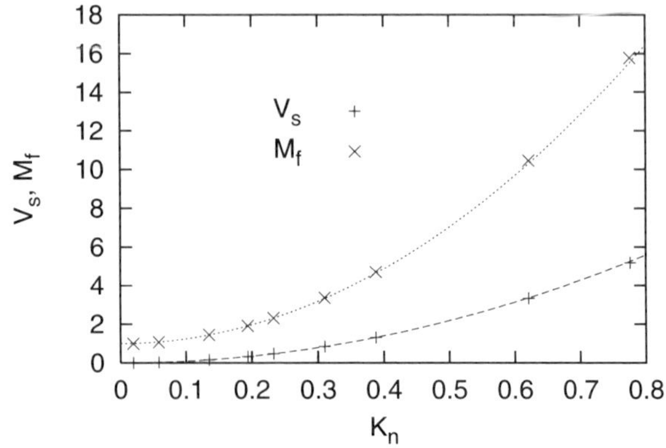

FIGURE 15.21. Slip velocity and normalized mass flowrate at theThe lines correspond to equations described in the text. (Courtesy of X. Nie, G.D. Doolen, and S. Chen.)

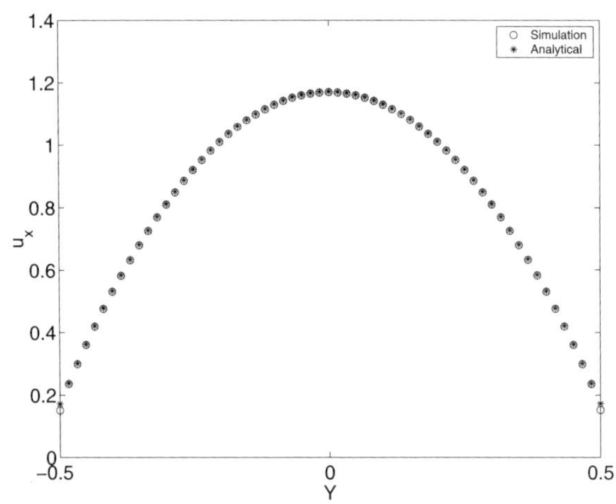

FIGURE 15.22. Velocity profile in a 2-D body force driven Poiseuille flow at Kn = 0.035 and $Ma = 0.01$. Comparison between the analytical solution of the BGK equation and a simulation of the entropic lattice Boltzmann method with the diffusive boundary condition. (Courtesy of I.V. Karlin and S. Ansumali.)

asymptotic analysis of the continuous Boltzmann BGK equation is possible, using the assumption of the isothermal condition. In Figure 15.22 we compare the analytical solution (Cercignani, 1975) with the simulation using the isothermal entropic lattice Boltzmann model (Karlin et al., 1999; Ansumali et al., 2003) and the diffusive boundary condition we presented earlier (Ansumali and Karlin, 2002). Good agreement is obtained for this benchmark problem without the use of any adjusted parameters. Another application of entropic LBM is presented in Section 5.4.2, where the assumption of isothermal conditions is relaxed and temperature variations across the microchannel are allowed. DSMC simulations for this case reported in (Zheng et al., 2002), show that the assumption of the isothermal flow is not valid. The quasi-continuum approaches, e.g., the Navier–Stokes–Fourier description based on the slip boundary conditions or Burnett description, fail to capture even qualitatively the temperature profile. This is especially true at values of Knudsen number in the transition regime, where a minimum in the temperature profile appears in the middle of the channel. Other hybrid extensions of LBM have been developed to address the difficulties encountered in simulating such temperature effects with the standard LBM. In (Lallemand and Luo, 2003), the mass and momentum equations are solved using a multiple-relaxation-time model (instead of the single-relaxation-time BGK model), whereas the energy equation is solved using finite differences.

16

Multiscale Modeling of Liquid Flows

In this chapter we discuss theory and numerical methodologies for simulating liquid flows at the atomistic and mesoscopic scales. The atomistic description is necessary for liquids contained in domains with dimension of fewer than ten molecules. First, we present the molecular dynamics (MD) method, a deterministic approach suitable for liquids. We explain details of the algorithm and focus on the various potentials and thermostats that can be used. This selection is crucial for reliable simulation of liquids at the nanoscale. In the next section we consider various approaches in coupling atomistic with mesoscopic and continuum levels. Such coupling is quite difficult, and no fully satisfactory coupling algorithms have been developed yet, although significant progress has been made. An alternative method is to embed an MD simulation in a continuum simulation. This is demonstrated in the next section in the context of electroosmotic flow in a nanochannel, where examples for various parameters in Poisson–Boltzmann and Navier–Stokes applications are included. In the last section we discuss a new method, developed in the late 1990s primarily in Europe: the dissipative particle dynamics (DPD) method. It has features of both LBM and MD algorithms and can be thought of as a coarse-grained version of MD. It employs stochastic forces to account for the eliminated degrees of freedom and thus new integration techniques need to be used. We present different such methods and ways of implementing boundary conditions.

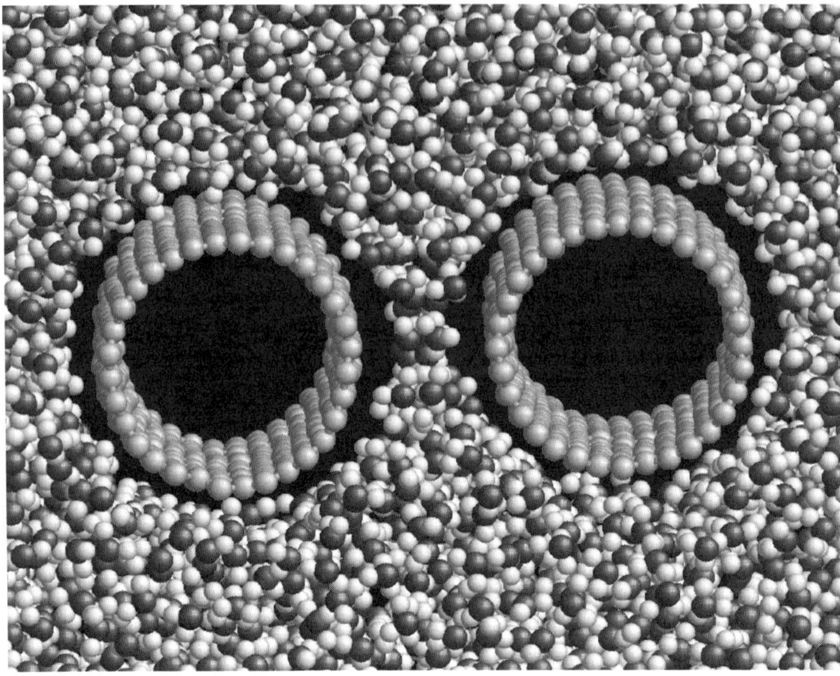

FIGURE 16.1. MD simulation of hydrophobic hydration of two $(16, 0)$ carbon nanotubes of 5 nm diameter. The white color represents hydrogen, the dark oxygen, and the grey shows molecules of the single-wall carbon nanotubes. (Courtesy of P. Koumoutsakos.)

16.1 Molecular Dynamics (MD) Method

The molecular dynamics (MD) method is suitable for simulating very small volumes of liquid flow, with linear dimensions on the order of 100 nm or less and for time intervals of several tens of nanoseconds. It can deal effectively with nanodomains and is perhaps the only accurate approach in simulating flows involving very high shear where the continuum or the Newtonian hypothesis may not be valid. For dimensions less than approximately ten molecules, the continuum hypothesis breaks down even for liquids (see Chapters 10 and 12), and MD should be employed to simulate the atomistic behavior of such a system. MD is, however, inefficient for simulating gas microflows due to the large intermolecular distances that require relatively large domains. Gas microflows are simulated more efficiently using the DSMC method that we describe in Section 15.1.

Another emerging application of MD simulation is investigation of the fluid–thermal behavior of carbon nanotubes, such as the one shown in Figure 1.23, from first principles. Carbon nanotubes have very interesting hydrophobic and hydrophilic behavior, as discussed in Section 13.2.1.

In Figure 16.1 we show the results of a constant-temperature (300 K) MD simulation of hydrophobic hydration around two carbon nanotubes of 5 nm diameter (Walther et al., 2001b). The objective is to quantify the behavior of water in the presence of single-wall carbon nanotubes and obtain the wetting angles for different systems.

Molecular dynamics (MD) computes the trajectories of particles that model the atoms of the system, since they result from relatively simplified interaction force fields. The MD simulations generate a sequence of points in phase space as a function of time; these points belong to the same ensemble, and they correspond to the different conformations of the system and their respective momenta. An ensemble is a collection of points in phase space satisfying the conditions of a particular thermodynamic state. Several ensembles, with different constraints on the thermodynamic state of the system, are commonly used in MD. For example:

- The NPT ensemble (also known as the isobaric–isothermal ensemble) is characterized by a fixed number of atoms, N, a fixed pressure, P, and a fixed temperature, T.

- The microcanonical ensemble (NVE) has a thermodynamic state characterized by a fixed number of atoms, N, a fixed volume, V, and a fixed energy, E. This corresponds to an isolated system.

- The canonical ensemble (NVT) is a collection of all systems whose thermodynamic state is characterized by a fixed number of atoms, N, a fixed volume, V, and a fixed temperature, T.

- The grand canonical ensemble (μVT) has a thermodynamic state characterized by a fixed chemical potential, μ, a fixed volume, V, and a fixed temperature, T.

Molecular-dynamics simulations can be classified into:

- Equilibrium MD (EMD) simulations, and

- Nonequilibrium MD (NEMD) simulations.

The properties of the fluid that are not in equilibrium can be described by nonequilibrium statistical mechanics (Sadus, 1999) and calculated from NEMD simulations. Typically, NEMD involves applying a perturbation to the usual equations of motion. The perturbation can be constant throughout the simulation, it can evolve with time, or alternatively, a sinusoidally oscillating perturbation can be used.

The motion of an ensemble of atoms in MD simulations is governed by interatomic forces arising from the interaction of electrons and nuclei. Thus, the results obtained from MD simulations are linked with the ability of the potential energy function to represent the underlying system. In a classical

MD simulation, first, a model system consisting of N particles is selected and Newton's equations of motion of the form

$$m_i \frac{d^2\mathbf{r}_i}{dt^2} = \mathbf{F}_i, \tag{16.1}$$

where m_i is the mass of atom i, $\mathbf{a_i} = d^2\mathbf{r}_i/dt^2$ its acceleration, and $\mathbf{F}_i$ the force acting on atom i, are solved until the properties of the system no longer change with time. Once a steady state is reached, the required measurements are performed. The key steps in MD simulation are:

1. *Initialization:* Before starting the simulation, initial positions and velocities are assigned to all particles in the system. The particle positions should be chosen compatible with the structure being simulated.

2. *Force calculation:* At each step the interactions between the particles are examined and the forces due to these interactions are calculated. This is the most time-consuming step in typical MD simulations (Frenkel and Smit, 2002). The various intermolecular potentials employed in MD and their calculation is discussed in Sections 16.1.1 and 16.1.2.

3. *Integration of the equations of motion:* The equations of motion are integrated using time integration algorithms that are based on finite difference methods. The most commonly used time integration algorithm is the Verlet integration rule

$$\mathbf{r}^{n+1} = 2\mathbf{r}^n - \mathbf{r}^{n-1} + \Delta t^2 \mathbf{a}(t) + \mathcal{O}(\Delta t^4).$$

   Variants of the Verlet algorithm, such as the leap-frog scheme and the velocity Verlet algorithm, are popularly employed for time integration.

4. *Data storage and analysis:* After the equations of motion are integrated, the relevant properties of the system (e.g., temperature, pressure, volume) are calculated and stored.

In the rest of this section, we provide details on some key steps in MD. Since there are many excellent textbooks on MD (see, e.g., (Allen and Tildesley, 1994)), only the important details are highlighted.

16.1.1 Intermolecular Potentials

The definition of accurate intermolecular potentials is key to any atomistic simulation, and here we provide an overview of some of the intermolecular potentials developed in the past. In general, the potential energy (V) of a system consisting of N interacting particles can be expressed as:

$$V = \sum_i V_1(\boldsymbol{r}_i) + \sum_i \sum_{j>i} V_2(\boldsymbol{r}_i, \boldsymbol{r}_j) + \sum_i \sum_{j>i} \sum_{k>j>i} V_3(\boldsymbol{r}_i, \boldsymbol{r}_j, \boldsymbol{r}_k) + \cdots, \tag{16.2}$$

where $\boldsymbol{r}_i$ is the position of particle i. The first term on the right-hand side (V_1) is the potential energy due to the external fields, and the remaining terms, which are modeled by intermolecular potentials, represent the particle interactions (e.g., V_2 is the potential between pairs of particles and V_3 is the potential between particle triplets and so on (Sadus, 1999)). Typically, equation (16.2) is truncated after the second term; i.e., the three-body and higher-order interactions are neglected. The intermolecular interaction potentials have been discussed in detail in (Maitland et al., 1981), and (Stone, 1996). The many-body effects on the intermolecular interactions have been reviewed in (Elrod and Saykally, 1994).

Pairwise Intermolecular Potentials

In many atomistic simulations, it is sufficient to use the simplest models to represent the essential physics, and many pairwise (two-body) potentials have been proposed (Maitland et al., 1981). Typically, these potentials are parameterized such that the simulations based on them can reproduce certain experimental measurements, e.g., the second virial coefficient, diffusion coefficient, and viscosity of a certain material (Sadus, 1999). Here we outline some of the most commonly used pairwise intermolecular potentials. Unless mentioned otherwise, r denotes the distance between the two particles interacting via the pairwise intermolecular potentials.

1. *Square-Well Potential.* The square-well potentials is one of the simplest intermolecular potentials that is capable of representing the properties of liquids (Sadus, 1999). It is given by

$$V(r) = \begin{cases} \infty, & r \leq \sigma, \\ -\epsilon, & \sigma < r \leq \lambda\sigma, \\ 0, & r > \lambda\sigma, \end{cases} \tag{16.3}$$

where λ is some multiple of the hard-sphere diameter (σ) and ϵ is a measure of the attractive interaction. This potential belongs to the genre of "hard-sphere + attractive term" potentials. In the "hard-sphere" potential (Allen and Tildesley, 1994) the atoms are approximated as impenetrable hard spheres. The properties of the square-well fluid have been investigated widely (Haile, 1992), and it remains a useful starting point for the development of liquid state theories (Yuste and Santos, 1994).

2. *Yukawa Potential.* Considerable interest has been demonstrated in simulating the properties of atoms interacting via the hard-sphere Yukawa potential (Frenkel and Smit, 2002; Rudisill and Cummings, 1989; Rosenfeld, 1993; Kalyuzhnyi and Cummings, 1996), which is described by

$$V(r) = \begin{cases} \infty, & r \leq \sigma, \\ -\epsilon\sigma/re^{-z(r/\sigma-1)}, & r > \sigma, \end{cases} \quad (16.4)$$

where ϵ is an attractive term, σ is the hard-sphere diameter, and z is an adjustable parameter. The inverse power dependence of this potential means that it can be applied to ionic systems (Rowlinson, 1989). It has been shown that when $z = 1.8$, the potential behaves very similarly to the Lennard–Jones (12-6) potential (Duh and Mier-Y-Teran, 1997).

3. *Lennard–Jones Potential.* Lennard–Jones potential is one of the most widely used potentials for nonpolar molecules. The generalized form of a Lennard–Jones potential is

$$V(r) = \epsilon \left[\frac{m}{n-m} x^{-n} - \frac{n}{n-m} x^{-m} \right], \quad (16.5)$$

where n and m are constants, $x = r/r_m$, and r_m is the separation corresponding to minimum potential energy. The "hard-sphere" diameter is related to the energy-minimum separation r_m by

$$\sigma = r_m \left(\frac{m}{n} \right)^{\frac{1}{n-m}}. \quad (16.6)$$

The most common form of the Lennard–Jones potential is obtained when $n = 12$ and $m = 6$, i.e.,

$$V(r) = 4\epsilon \left[\left(\frac{\sigma}{r} \right)^{12} - \left(\frac{\sigma}{r} \right)^{6} \right]. \quad (16.7)$$

The first term in equation (16.7) represents a short-range repulsive force, which prevents overlap of the atoms, while the second term represents an attractive interaction. Figure 16.2 shows the variation of the Lennard–Jones potential and the corresponding force ($F = -\nabla_r V(r)$). The advantage of the Lennard–Jones potential is that it combines a realistic description of the intermolecular interaction with computational simplicity.

4. *WCA Potential.* The WCA (Weeks–Chandler–Andersen) potential is a modification of the Lennard–Jones potential, where the atoms

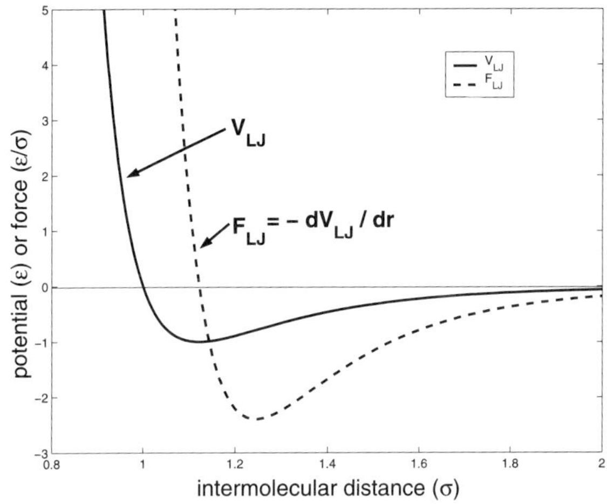

FIGURE 16.2. The variation of the Lennard–Jones potential (in units of ϵ) and the corresponding force (in units of ϵ/σ) as a function of the intermolecular distance.

interact via a cut and shifted Lennard–Jones interatomic potential function defined by

$$V(r) = \begin{cases} 4\epsilon\left[\left(\frac{\sigma}{r}\right)^{12} - \left(\frac{\sigma}{r}\right)^{6}\right] - V_{LJ}(r_c), & r \leq r_c, \\ 0, & r > r_c, \end{cases} \tag{16.8}$$

where r_c $(= \sigma/2^{1/6})$ is the truncation distance, $V_{\text{LJ}}(r_c)$ is the value of the Lennard–Jones potential at the point of truncation, and σ and ϵ are the Lennard–Jones distance and energy parameters. We note that in this case, the interactions between the atoms are purely repulsive. Therefore, this potential is used in those cases in which we want the atoms to purely repel each other.

5. *Buckingham Potential.* The Buckingham potential is given by

$$V(r) = A\exp(-Br) - \frac{C_6}{r^6}, \tag{16.9}$$

where A, B, and C_6 are empirical constants. The major difference from the Lennard–Jones potential is that the repulsion term now has an exponential dependence on the distance, which is shown to be more realistic than the Lennard–Jones potential (Born and Mayer, 1932). However, compared to the Lennard–Jones potential, the Buckingham potential is much more expensive to evaluate.

6. *Coulomb Potential.* The Coulomb potential accounts for the electrostatic interactions between particles when charges are present, e.g., for ions or polyatomic molecules with partial charges. The Coulomb potential between two particles is given by

$$V(r) = \frac{1}{4\pi\epsilon_r\epsilon_0}\frac{q_1 q_2}{r}, \tag{16.10}$$

where q_1, q_2 are the charges of the two particles, and ϵ_0 is the vacuum permittivity.

Table 16.1 gives a summary of the pairwise (i.e., two-body) intermolecular interaction potential schemes that are commonly used in MD simulation.

Many-Body Intermolecular Potentials

Though the pairwise potentials have been fairly successful in describing the intermolecular interactions, there is evidence that the three-body interactions (or even higher-order interactions) can be important in some cases (Bobetic and Barker, 1970; Monson et al., 1983; Rittger, 1990b; Rittger, 1990a; Rittger, 1990c). Here we introduce the Tersoff potential, which is a three-body potential.

The Tersoff potential is based on the concept of bond order; i.e., the strength of a bond between two atoms is not constant, but depends on the local environment. The Tersoff potential has the form

$$V(r) = \frac{1}{2}\sum_{ij}\phi_R(r_{ij}) + \frac{1}{2}\sum_{ij}B_{ij}\phi_A(r_{ij}) + \cdots, \tag{16.11}$$

where R and A mean "repulsive" and "attractive." The Tersoff potential is not a pair potential because B_{ij} is not a constant. In fact, it is the *bond order* for the bond joining atoms i and j, and is a decreasing function of the "coordination" G_{ij} assigned to the bond; i.e., $B_{ij} = B(G_{ij})$. G_{ij} is defined as

$$G_{ij} = \sum_k f_c(r_{ik})g(\theta_{jik})f(r_{ij} - r_{ik}),$$

where $f_c(r)$, $f(r)$, and $g(\theta)$ are empirical functions. The basic idea is that the bond ij is weakened by the presence of other bonds ik involving atom i. The amount of weakening is determined by the location of the other bonds. The angular terms are introduced to help construct a realistic model. Tersoff potential has been calibrated for silicon (Tersoff, 1988b) and carbon (Tersoff, 1988a). In practice, there are two major problems with this potential. First, since the potential involves a large number of parameters to be calibrated, finding a good parameterization for a given material is rather difficult. Second, the calculation of the potential and the associated force is very expensive. However, because simulations based on it can reproduce many important solid material properties, e.g., the lattice constant

TABLE 16.1. Summary of the commonly employed intermolecular potentials.

Potential	Expression	Parameters
Square-well	$V(r) = \begin{cases} \infty, & r \leq \sigma, \\ -\epsilon, & \sigma < r \leq \lambda\sigma, \\ 0, & r > \lambda\sigma \end{cases}$	$\sigma,\ \epsilon$
Yukawa	$V(r) = \begin{cases} \infty, & r \leq \sigma, \\ -\frac{\epsilon\sigma}{r} \exp[-z(\frac{r}{\sigma} - 1)], & r > \sigma, \end{cases}$	$\sigma,\ \epsilon, z$
Lennard–Jones	$V(r) = 4\epsilon \left[(\frac{\sigma}{r})^{12} - (\frac{\sigma}{r})^{6} \right]$	$\sigma,\ \epsilon$
Buckingham	$V(r) = A \exp(-Br) - \frac{C_6}{r^6}$	$A,\ B, C_6$
Coulomb	$V(r) = \frac{1}{4\pi\epsilon_r\epsilon_0} \frac{q_1 q_2}{r}$	$q_1,\ q_2, \epsilon_r$
WCA	$V(r) = \begin{cases} 4\epsilon \left[(\frac{\sigma}{r})^{12} - (\frac{\sigma}{r})^{6} \right] - V_{LJ}(r_c), & r \leq r_c, \\ 0, & r > r_c \end{cases}$	$\sigma,\ \epsilon, r_c$

and bulk modulus of diamond, it is widely used in atomistic simulation of solids.

16.1.2 Calculation of the Potential Function

The calculation of the potential function and the associated force accounts for most of the computational cost in an atomistic simulation. Therefore, efficient algorithms for the calculation of the potential and force are essential to any atomistic simulation program. Here we give a brief overview of the potential calculation, and the force can usually be calculated in a similar way. Depending on the nature of the potential, the algorithm for potential calculation can be quite different. It is useful to divide the potentials into two categories, i.e., those for the short-range interactions and those for the long-range interactions. For the short-range interactions between particles, the potential energy usually decreases to essentially zero for an intermolecular distance of 1 nm to 2 nm or even smaller. Most intermolecular potentials (e.g., the Lennard–Jones potential) belong to this category. For the long-range interactions between particles, the potential energy decreases very slowly and is typically not negligible even at a very large distance, e.g., tens of nanometers. The Coulomb potential belongs to this category.

Calculation of Short-Range Interactions

To compute the potential due to the short-range interactions, a "cutoff" method is usually used (Allen and Tildesley, 1994). The idea is to compute the potential only for the particle pair (or particle triplet) that are within certain cutoff distance r_{cutoff}. This value is usually chosen such that the potential energy between particles whose distance is larger than r_{cutoff} is negligible. For Lennard–Jones potential (see equation (16.7)), $r_{\text{cutoff}} = 2.5\sigma$ is usually used. In practice, a neighbor list is maintained for each particle in the system, and one computes the potential energy between a particle pair only when one particle is within another particle's neighbor list. Because the particles move, the neighbor list needs to be updated during the simulation. Many algorithms have been developed to construct and update the neighbor list, and they are discussed in detail in (Sadus, 1999).

Calculation of Long-Range Interactions

The calculation of potential due to the long-range interactions (long-range potential) is much more difficult than to the calculation of potential due to the short-range interactions. For long-range potentials, using a cutoff method with a small r_{cutoff} typically gives rise to significant artifacts in the simulations, while using a large r_{cutoff} is computationally expensive. Thus, in general, the cutoff method is not preferred in the calculation of long-range potentials. In this section, we give an overview of the algorithms developed to compute the Coulomb potential due to the electrostatic interactions. There are many algorithms, e.g., Ewald summation (Frenkel and Smit, 2002), particle–mesh Ewald (Darden et al., 1993), the fast multipole method (Greengard, 1987), the particle–particle particle-mesh method (PPPM) (Hockney and Eastwood, 1981; Darden et al., 1993; Luty et al., 1995), and the reaction field method (Sadus, 1999); here we will focus on the first three methods. We note that many of these algorithms are developed in the context of periodic boundary conditions, which are commonly used in atomistic simulations. Because of the periodic boundary conditions, a particle i in a system consisting of N particles interacts not only with the $N-1$ particles, but also with an infinite number of images of these particles.

1. *The Ewald Summation.* The Ewald summation is a popular technique to compute electrostatic interactions (Frenkel and Smit, 2002). The total electrostatic energy of N particles with charges q_i in a cubic simulation box (box length: L) and their periodic images is given by

$$V = \frac{1}{2}\sum_{\mathbf{n}=\mathbf{0}}^{\infty\,\prime}\sum_{i=1}^{N}\sum_{j=1}^{N}\frac{q_i q_j}{\mid \mathbf{r_{ij}} + \mathbf{n}\mid}, \tag{16.12}$$

where the summation over $\mathbf{n}$ is taken over all periodic images, $\mathbf{n} = (n_x L,\, n_y L,\, n_z L)$ with n_x, n_y, n_z integers. The prime indicates the omission of $i = j$ for $\mathbf{n} = 0$. Note that the prefactor $1/4\pi\epsilon_0$ is omitted for simplicity. Direct implementation of equation (16.12) is difficult because the summation is conditionally convergent, and converges very slowly. In the Ewald summation, equation (16.12) is converted into two series terms each of which converges more rapidly (De Leeuw et al., 1980; Heyes, 1981). The first step is to impose a neutralizing charge distribution (typically a Gaussian distribution) of equal magnitude but of opposite sign to each charge. Then, the summation over point charges becomes a summation of the interaction between charges plus the neutralizing distributions. The new summation is often referred to as the "real space" summation. The real space energy is given by

$$V_{\text{real}} = \frac{1}{2}\sum_{\mathbf{n}=\mathbf{0}}^{\infty\,\prime}\sum_{i=1}^{N}\sum_{j=1}^{N}\frac{q_i q_j \operatorname{erfc}(\kappa|\mathbf{r}_{ij} + \mathbf{n}|)}{|\mathbf{r_{ij}} + \mathbf{n}|}. \tag{16.13}$$

The value of κ is chosen such that only terms corresponding to $\mathbf{n} = 0$ (i.e., interactions involving charges in the central box only) make a contribution. Because of this, V_{real} can be computed using the usual cutoff method. In the second step, another charge distribution is imposed to counteract exactly the neutralizing distribution. The contribution of this charge distribution to the potential energy is called a "reciprocal space" energy,

$$V_{\text{recip}} = \frac{1}{2} \sum_{k \neq 0} \sum_{i=1}^{N} \sum_{j=1}^{N} \frac{4\pi q_i q_j}{k^2 L^3} \exp\left(-\frac{k^2}{4\kappa^2}\right) \cos(\mathbf{k} \cdot \mathbf{r_{ij}}), \tag{16.14}$$

where $\mathbf{k}= 2\pi \boldsymbol{n}/L^2$ are the reciprocal vectors. The summation of Gaussian functions in real space also includes the interaction of each Gaussian with itself. This interaction energy is given by

$$V_{\text{Gauss}} = -\frac{\kappa}{\pi^{0.5}} \sum_{k=1}^{N} q_i^2. \tag{16.15}$$

The medium surrounding the sphere of simulation boxes must also be considered in the calculation, since the sphere can interact with its surroundings (Sadus, 1999). No correction is required if the surrounding medium is a good conductor. However, if the surrounding medium is a vacuum, the following correction applies:

$$V_{\text{corr}} = \frac{2\pi}{3L^3} \left| \sum_{i=1}^{N} q_i \mathbf{r_i} \right|^2. \tag{16.16}$$

Consequently, the final expression for the *total* potential energy is

$$V = V_{\text{real}} + V_{\text{recip}} + V_{\text{Gauss}} + V_{\text{corr}}. \tag{16.17}$$

The computational cost of the reciprocal-space energy (equation (16.14)) scales as $N^{3/2}$ (Sadus, 1999). Thus, the Ewald summation approach can still be very expensive for large systems.

2. *Particle–Mesh Ewald (PME).* PME is a method proposed by (Darden et al., 1993) and (Essmann et al., 1995) to improve the performance of the reciprocal summation in the Ewald method. Instead of directly summing reciprocal vectors, the charges are assigned to a grid using cardinal B-spline interpolation. This grid is then Fourier transformed with a 3D FFT algorithm, and the reciprocal energy term is calculated by a single sum over the grid in k-space. The potential at the grid points is calculated by inverse transformation, and by using the interpolation factors, the forces on each atom can be calculated. The PME algorithm scales as $N \log(N)$, and is substantially faster than the ordinary Ewald summation for medium to large systems. For very small systems it might still be better to use Ewald summation to avoid the overhead in setting up grids and performing transformations.

3. *The Fast Multipole Method (FMM):* The traditional particle–particle, particle–mesh, or a combination of these methods commonly benefit

from potential truncation and neighbor list strategies (Sadus, 1999). The philosophy behind these computational strategies is to identify and distinguish between neighboring molecules that make only small contributions. This distinction between near and far interactions can be handled efficiently by ordering the molecules in a hierarchical tree structure. The tree-based methods provide large gains in computational efficiency, particularly for calculation of long-range interactions. For example, the fast multipole method (FMM) is of order N compared with N^2 (or $N^{3/2}$) for a traditional particle–particle calculation.

The fast multipole method was developed using the hierarchical tree concept (Greengard, 1987; Greengard and Rokhlin, 1987; Carrier et al., 1988; Schmidt and Lee, 1991). The FMM algorithm involves multipole expansion for boxes at the lowest level of the tree. These expansions are combined and shifted as they are passed up and down the tree. Particles are assigned to the cells at the finest level of the tree. In the FMM algorithm some cells may be empty, while some cells may have several particles. The multipole expansion of the particle configuration on the finest level is formed about the center of the box. Each "child" box communicates this information to the "parent" box on the next level. Aggregate information about distant particles comes back down the low-level boxes. In the FMM algorithm, the simulation box of length L is subdivided into a box of length $L/2^r$, where r is an integer representing the level of refinement. The division is equivalent to forming 8^r equal-sized subvolumes. This is done for every single box to a maximum level of refinement R irrespective of the number of particles that they contain. The maximum level of refinement R is approximately equal to the number of particles (N), i.e., $R = \log_8 N$. At the maximum level there is on average one particle per box.

The comparison of the efficiency of the FMM algorithm to that of the particle–particle algorithm depends on the number of particles in the simulation, the maximum level of refinement, and the number of multipoles (Sadus, 1999). (Schmidt and Lee, 1991) analyzed the different scenarios. Typically, the FMM algorithm is advantageous for simulations involving tens of thousands of molecules. Simple particle–particle algorithms assisted by time-saving concepts such as a neighbor list are adequate for atoms or simple polyatomic molecules. However, the rigorous evaluation of the properties of real macromolecules by molecular simulation requires thousands of interaction sites per molecule. Alternative computational strategies are required to deal with this increased level of complexity. It is in this context that hierarchical tree algorithms (e.g., FMM) are likely to play an increasingly important role in molecular simulation.

16.1.3 Thermostats

For several reasons (e.g., drift during equilibration, drift as a result of force truncation and integration errors, heating due to external or frictional forces), it is necessary to control the temperature of the system in MD simulations. In a canonical ensemble of finite systems, the instantaneous kinetic temperature fluctuates (Frenkel and Smit, 2002). In fact, if the average kinetic energy per particle is kept constant (as is done in the isokinetic MD scheme (Evans and Morriss, 1990)), then the true constant-temperature ensemble would not be simulated. In practice, the difference between isokinetic and canonical schemes is often negligible (Frenkel and Smit, 2002). In this section we discuss some of the most commonly employed temperature coupling schemes.

The Berendsen Thermostat

The Berendsen algorithm simulates weak coupling with first-order kinetics to an external heat bath with a given temperature T_0. According to this algorithm the deviation of the system temperature from T_0 is slowly corrected according to the following equation (van der Spoel et al., 2004):

$$\frac{dT}{dt} = \frac{T_0 - T}{\tau}. \tag{16.18}$$

As equation (16.18) shows, the temperature deviation decays exponentially with a time constant τ. This method of coupling has the advantage that the strength of the coupling can be varied and adapted to the user's requirement. At each step the velocity of each particle is rescaled by a factor λ, given by

$$\lambda = \left[1 + \frac{\Delta t}{\tau_T}\left(\frac{T_0}{T} - 1\right)\right]^{1/2}. \tag{16.19}$$

The parameter τ_T (in equation (16.19)) is close to, but not exactly equal to, the time constant τ of the temperature coupling (equation (16.18)) (van der Spoel et al., 2004):

$$\tau = \frac{2C_V\tau_T}{N_{\mathrm{df}}k_{\mathrm{B}}},$$

where C_V is the total heat capacity of the system, k_{B} is Boltzmann's constant, and N_{df} is the total number of degrees of freedom. Here τ is not equal to τ_T, because the kinetic energy change caused by scaling the velocities is partly redistributed between kinetic and potential energy, and hence the change in temperature is less than the scaling energy (Berendsen et al., 1984).

The Nose–Hoover Thermostat

The Berendsen thermostat is very efficient in relaxing a system to the target temperature. However, it does not generate states in a canonical ensemble, even though the deviation is small. To enable canonical ensemble simulations, one may use the extended-ensemble approach first proposed by Nose (Nose, 1984) and later modified by Hoover (Hoover, 1985). The system Hamiltonian is extended by introducing a thermal reservoir and a frictional term in the equations of motion. The frictional force is proportional to the product of each particle's velocity and a friction parameter ξ. This friction parameter (or "heat bath" variable) is a fully dynamic quantity with its own equation of motion; the time derivative is calculated from the difference between the current kinetic energy and the reference temperature.

In Hoover's formulation, the particles' equations of motion are given by

$$\frac{d^2\mathbf{r_i}}{dt^2} = \frac{\mathbf{F_i}}{m_i} - \xi\frac{d\mathbf{r_i}}{dt},$$

where the equation of motion for the heat bath parameter ξ is

$$\frac{d\xi}{dt} = \frac{1}{Q}(T - T_0).$$

The reference temperature is denoted by T_0, while T is the current instantaneous temperature of the system. The strength of the coupling is determined by the constant Q (usually called the "mass parameter" of the reservoir) in combination with the reference temperature. Since the mass parameter is dependent on the reference temperature, it is an awkward way of describing the coupling strength.

An important difference between the weak coupling scheme and the Nose–Hoover algorithm is that using a weak coupling one gets a strongly damped exponential relation, while the Nose–Hoover approach produces an *oscillatory* relaxation. For further discussion and some implementation issues of the Nose–Hoover algorithm, we refer the reader to (van der Spoel et al., 2004).

The Andersen Thermostat

In the constant-temperature method proposed by (Andersen, 1980), the system is coupled to a heat bath that imposes the desired temperature. The coupling to a heat bath is represented by stochastic impulsive forces that act occasionally on randomly selected particles. These stochastic collisions with the heat bath can be considered as Monte Carlo moves that transport the system from one constant-energy shell to another (Frenkel and Smit, 2002). Between stochastic collisions, the system evolves at constant energy according to the normal Newtonian laws of motion.

Before starting such a constant-temperature simulation, the strength of the coupling to the heat bath should be selected. This coupling strength is determined by the frequency of stochastic collisions. Let us denote this frequency by Γ. If successive collisions are uncorrelated, then the distribution of time intervals between two successive stochastic collisions, $\mathcal{P}(t;\Gamma)$, is of the Poisson form

$$\mathcal{P}(t;\Gamma) = \Gamma exp[-\Gamma t],$$

where $\mathcal{P}(t;\Gamma)dt$ is the probability that the next collision will take place in the interval $[t, t + dt]$.

A constant-temperature simulation employing Andersen thermostat consists of the following steps:

1. Start with an initial set of positions and momenta and integrate the equations of motion for a time interval of Δt.

2. A number of particles are selected to undergo a collision with the heat bath. The probability that a particle is selected in a time step of length Δt is $\Gamma \Delta t$.

3. If particle i has been selected to undergo a collision, its new velocity will be drawn from a Maxwell-Boltzmann distribution corresponding to the desired temperature T_0. All other particles are unaffected by this collision.

It should be noted that rigorously, the dynamics generated by the Andersen scheme are unphysical (Frenkel and Smit, 2002). Therefore, it is risky to use the Andersen method when studying dynamical properties. The key properties of the Andersen thermostat can be summarized as follows:

1. It relies on stochastic collisions with heat reservoirs to control the temperature.

2. It produces a canonical NVT distribution, but probabilistically, rather than deterministically.

3. The algorithm is very similar to MD, but each particle undergoes a stochastic collision with probability $\Gamma \Delta t$ after every time step.

The Andersen thermostat has been used very effectively in dissipative particle dynamics (DPD) methods; see Section 16.4 in this chapter. Table 16.2 gives a comparison of the temperature coupling schemes that are commonly used in MD simulation.

16.1.4 Data Analysis

MD simulation generates the trajectories of all the particles in the system, but to obtain a deeper insight into the system being studied, we need to analyze the trajectories obtained during the MD simulation. In this section, we summarize some of the most commonly performed data analyses in the simulation of fluid transport.

TABLE 16.2. Summary of the commonly employed thermostats.

Thermostat	Key concept	Suitability/Application
Berendsen	First-order-kinetics-based weak coupling	Easy implementation and computationally inexpensive.
Nose–Hoover	Extended Lagrangian	Most rigorous implementation of the NVT ensemble.
Andersen	Stochastic collision	Suitable for thermal coupling of atomistic and continuum domains; also for DPD.

Density Profiles

To investigate nanoflows in channels and pores, where the fluid density is inhomogeneous (see Chapters 10 and 11), it is useful to compute the spatial distribution of fluid density, e.g., density profile along the radial direction of a nanopore. This is usually performed using the "binning method" (Allen and Tildesley, 1994). In this scheme the relevant spatial domain (i.e., the domain where the density distribution of the species needs to be computed) is partitioned into a number of cells, which are identified as the "bins." The number of atoms in each bin is computed from the knowledge of the positions of the atoms. In order to obtain a better statistical analysis of the number density in a bin, we add the number of atoms in the bin for a number of steps and then divide the total number of atoms in the bin by the number of steps and the volume of the bin. Thus, the number density, c_i, of the ith bin, averaged over s steps, is given by

$$c_i = \langle n_i/\text{VOL}_i \rangle_s,$$

where n_i is the total number of atoms in the ith bin during each step and VOL_i is the volume of the ith bin.

Velocity Profiles

The velocity profile is one of the most important measurables for fluid transport, and can be computed in a similar manner as the density profile. Usually, the simulation system is partitioned into n bins, and statistics of the fluid velocity are gathered separately in each bin (Tysanner and Garcia, 2004). Assuming that during an s-step simulation, at each step k, there are $n_{k,i}$ particles in the ith bin, and the velocity of each of these particles (denoted by j) is given by $\boldsymbol{v}_{k,i}^j$, then the average fluid velocity $\boldsymbol{u}_i$ in the ith bin can be computed by

$$\boldsymbol{u}_i = \frac{\sum_{k=1}^{s} \sum_{j=1}^{n_{k,i}} \boldsymbol{v}_{k,i}^{j}}{\sum_{k=1}^{s} n_{k,i}}. \tag{16.20}$$

Equation (16.20) is used to compute the steady-state velocity profile. If one is interested in the transient behavior of the velocity profile, an ensemble of simulations will need to be performed. In this case, the velocity profile can still be analyzed using equation (16.20), and the only difference is that the parameter s now denotes the different simulations rather than different time steps.

Diffusion Coefficient

The diffusion coefficient can be calculated using the Einstein relationship

$$D = \frac{1}{6} \lim_{t \to \infty} \frac{\langle [\boldsymbol{r}(t_0 + t) - \boldsymbol{r}(t_0)]^2 \rangle}{t},$$

where $\boldsymbol{r}$ is the atom position, which can be obtained from the trajectories generated by the MD simulation. See Chapter 10 for a discussion of the calculation of diffusion coefficients of simple fluids.

Stress Tensor

The stress or pressure tensor of an atomic fluid, denoted by $\mathbf{P}$, is often defined as the infinitesimal force $d\mathbf{F}$ acting on an infinitesimal area $d\mathbf{A}$, which moves with the local streaming velocity $\mathbf{u}(r, t)$ of the fluid (Todd et al., 1995):

$$d\mathbf{F} = -d\mathbf{A} \cdot \mathbf{P}. \tag{16.21}$$

The pressure tensor can be written as a linear sum of a kinetic component, $\mathbf{P}^k$, and a potential component, $\mathbf{P}^u$. In equation (16.21), the kinetic component is deemed to be across the surface $d\boldsymbol{A}$ if at a time t a particle moves through (or across) the surface. The potential component $\mathbf{P}^u$, due to intermolecular forces, is, however, not as easily defined (Todd et al., 1995). An interatomic force between two atoms is often said to be "across" the surface if the line between the centers of mass of the two atoms cuts through (or across) the surface defined by $d\boldsymbol{A}$. This is known as the Irving–Kirkwood convention (Todd et al., 1995).

However, there is really no unambiguous definition of "across" for either the kinetic or the potential contribution to the pressure tensor. For example, there are obvious difficulties in handling many-body force contributions to the potential part of the pressure tensor. The ambiguities in both components of the pressure tensor are best illustrated by the fact that the predictions of hydrodynamics are unaltered if the curl of an arbitrary vector field is added to the pressure tensor. In hydrodynamics it is only the gradient of the pressure tensor that appears in the equations of motion.

The Irving–Kirkwood (IK) expression for the pressure tensor at time t (Todd et al., 1995) is

$$\mathbf{P}(\mathbf{r},t) = \frac{1}{\mathrm{VOL}}\left[\sum_i m_i[\mathbf{v}_i(t) - \mathbf{u}(\mathbf{r}_i,t)][\mathbf{v}_i(t) - \mathbf{u}(\mathbf{r}_i,t)] + \frac{1}{2}\sum_{ij} \mathbf{r}_{ij}(t) O_{ij}(t) \mathbf{F}_{ij}(t)\big|_{\mathbf{r}_i(t)=\mathbf{r}}\right],$$

where $\mathbf{v}_i$ is the total particle velocity, $\mathbf{u}$ is the streaming velocity of the fluid, VOL is the volume of the system, $\mathbf{F}_{ij}$ is the force on atom i due to atom j, and O_{ij} is the differential operator:

$$O_{ij} = 1 - \frac{1}{2!}\mathbf{r}_{ij}\cdot\frac{\partial}{\partial \mathbf{r}} + \cdots + \frac{1}{n!}\left[-\mathbf{r}_{ij}\cdot\frac{\partial}{\partial \mathbf{r}}\right]^{n-1} + \cdots . \tag{16.22}$$

Thus, from the knowledge of the positions of the atoms, velocities of the atoms, and forces acting on the atoms obtained from a typical "MD run," the stress (or pressure) tensor in the fluid medium can be computed.

Shear Viscosity

Each molecular-dynamics method for calculating the shear viscosity of a fluid falls into one of two main categories: equilibrium molecular dynamics (EMD) or nonequilibrium molecular dynamics (NEMD) techniques. The EMD techniques involve either the calculation of time correlation functions by measuring the decay of near-equilibrium fluctuations in properties of the fluid (Green–Kubo methods) or by accumulating displacements in properties over time (Einstein methods). For example, the Green-Kubo relation for shear viscosity, η, is given by (Arya et al., 2000)

$$\eta = \frac{\mathrm{VOL}}{k_{\mathrm{B}}T}\int_0^\infty \langle P_{xz}(0) P_{xz}(t)\rangle dt,$$

where P_{xz} is the xz component of the pressure tensor $\boldsymbol{P}$ given by (Arya et al., 2000)

$$\mathbf{P} = \frac{1}{\mathrm{VOL}}\left(\sum_{i=1}^{N}\frac{\mathbf{p}_i\mathbf{p}_i}{m_i} + \sum_{i=1}^{N}\sum_{j>i}^{N}\mathbf{r}_{ij}\mathbf{F}_{ij}\right), \tag{16.23}$$

where $\mathbf{p}_i$ is the momentum vector for atom i, and $\mathbf{r}_{ij} = \mathbf{r}_i - \mathbf{r}_j$ is the vector joining the centers of molecules i and j. A weakness of these EMD methods is that the shear viscosity suffers from substantial nonmonotonic system size dependence (Hess, 2002).

The NEMD techniques usually involve measuring the macroscopic steady-state response of the system to a perturbing field and relating the linear response to a transport coefficient. One of the earliest NEMD techniques, which maintains conventional periodic boundary conditions, involves imposing a spatially periodic external force on the molecules to generate an oscillatory velocity profile (Arya et al., 2000). The amplitude of this velocity profile at steady state is inversely related to the shear viscosity, and hence the viscosity can be calculated. The shear viscosity is wavelength dependent, however, and the Newtonian shear viscosity is obtained only in the long wavelength limit, i.e., in the limit $k \to 0$, where k is the wave vector of the oscillatory perturbation. This means that a very large simulation box is required to get reasonably accurate values of shear viscosity, which limits the usefulness of this technique.

The more successful NEMD techniques involve imposing a planar Couette flow velocity profile (i.e., zero wave vector techniques). One of the most efficient NEMD algorithms for shear viscosity is the Sllod algorithm (Arya et al., 2000). The Sllod algorithm has been used by several authors, and has been shown to be exact for arbitrarily large shear rates $\dot{\gamma}$, and is therefore appropriate for studying non-Newtonian regimes. The modified equations of motion for Sllod algorithm are (Arya et al., 2000)

$$\begin{aligned} \frac{d\mathbf{r}_i}{dt} &= \frac{\mathbf{p}_i}{m_i} + \mathbf{r}_i \cdot \nabla \mathbf{u}, \\ \frac{d\mathbf{p}_i}{dt} &= \mathbf{F}_i - \mathbf{p}_i \cdot \nabla \mathbf{u} - \alpha \mathbf{p}_i, \end{aligned}$$

where $\mathbf{F}_i$ is the force on molecule i, and α is the thermostating multiplier. The strain-rate-dependent shear viscosity is obtained from the constitutive equation

$$\eta(\dot{\gamma}) = -\frac{\langle P_{xz} \rangle}{\dot{\gamma}}.$$

The Newtonian shear viscosity is estimated by extrapolating the shear viscosities to zero shear rate. Both EMD and NEMD methods give similar values for the Newtonian shear viscosities. However, an advantage of the NEMD method is that the shear rate dependence of the viscosity is obtained directly from NEMD, while EMD provides the zero shear rate value only. One of the drawbacks of the NEMD method is that there is no generally accepted theoretical model for the shear rate dependence of the shear viscosity. The resulting Newtonian viscosity obtained from an NEMD simulation depends on the model used in the extrapolation procedure. To overcome this limitation, NEMD simulations at very small shear rates may be performed. However, this defeats the purpose of the NEMD method, since these low shear rate simulations require nearly as much computation time as the EMD methods. Although the NEMD runs can be parallelized for different shear rates, the computational time required to obtain the viscosity is limited by those long simulation runs at low shear rates. Refinements to the traditional NEMD methods have been developed that reduce the computational cost by improving the signal-to-noise ratio at small fields (Arya et al., 2000; Hess, 2002); however, viscosity calculation is still quite demanding.

Error Estimation

Sources of systematic error in MD include system-size dependence, possible effects of random number generators, and poor equilibration (Allen and Tildesley, 1994). These should be estimated and eliminated wherever possible. It is also essential to obtain an estimate of the statistical significance of the results. Simulation averages are taken over runs of finite length, and this is the main cause of statistical imprecision in the mean values so obtained.

1. *Errors in Equilibrium Averages:* Suppose that we are analyzing some simulation results that contain a total of τ_{run} time steps, or configurations. The average of some property $\mathcal{A}$ is (Allen and Tildesley, 1994)

$$\langle \mathcal{A} \rangle_{run} = \frac{1}{\tau_{\text{run}}} \sum_{\tau=1}^{\tau_{\text{run}}} \mathcal{A}(\tau).$$

If it is assumed that each quantity $\mathcal{A}(\tau)$ is statistically independent of the others, then the *variance* in the mean would simply be given by

$$\sigma^2(\langle \mathcal{A} \rangle_{\text{run}}) = \sigma^2(\mathcal{A})/\tau_{\text{run}}, \tag{16.24}$$

where

$$\sigma^2(\mathcal{A}) = \langle \delta \mathcal{A}^2 \rangle_{\text{run}} = \frac{1}{\tau_{\text{run}}} \sum_{\tau=1}^{\tau_{\text{run}}} (\mathcal{A}(\tau) - \langle \mathcal{A} \rangle_{\text{run}})^2.$$

The estimated error in the mean is given by $\sigma(\langle \mathcal{A} \rangle_{\text{run}})$. However, the data points are not usually independent, because configurations are often stored sufficiently frequently, so that those are highly correlated with each other. In those cases, the number of steps for which these correlations persist must be built into equation (16.24) (Allen and Tildesley, 1994). For example, suppose that the τ_{run} configurations actually consist of blocks, each containing $2\tau_{\mathcal{A}}$ identical configurations. For large $\tau_{\mathcal{A}}$, this corresponds to a correlation "time" $\tau_{\mathcal{A}}$. Then

$$\sigma^2(\langle \mathcal{A} \rangle_{\text{run}}) = 2\tau_{\mathcal{A}}\sigma^2(\mathcal{A})/\tau_{\text{run}}.$$

2. *Errors in Fluctuations:* Errors in the estimate of fluctuation averages of the type $\langle \delta \mathcal{A}^2 \rangle$ may be estimated simply on the assumption that the process $\mathcal{A}(t)$ obeys Gaussian statistics. The resulting formula is (Allen and Tildesley, 1994)

$$\sigma^2(\langle \delta \mathcal{A}^2 \rangle_{\text{run}}) = 2t'_{\mathcal{A}} \langle \delta \mathcal{A}^2 \rangle^2 / t_{\text{run}},$$

where a slightly different correlation time appears:

$$t'_{\mathcal{A}} = 2 \int_0^\infty dt \langle \delta \mathcal{A}(t) \delta \mathcal{A} \rangle^2 / \langle \delta \mathcal{A}^2 \rangle^2.$$

For an exponentially decaying correlation function, $t'_{\mathcal{A}} = t_{\mathcal{A}}$ is the usual correlation time.

16.1.5 Practical Guidelines

In this section we outline some of the issues that are faced in setting up and running an MD simulation. If one is using an MD package or an in-house program, it is necessary to choose the various parameters properly. This section discusses how to make a prudent choice of the important parameters and how various choices affect the accuracy and the speed of the simulation.

1. *Size of the Time Step.* We would like to use as large a time step as possible so that we can explore more of the phase space of the system. However, since we truncate the Taylor's series expansions, the time step needs to be small enough so that the expansions can provide a reliable estimate of the atomic positions and velocities at the end of the time step. For typical time-marching algorithms with a time accuracy of order three, one uses a time step that is a fraction of the period of the highest-frequency motion in the system. A good way of checking whether the time step is small enough is to run an equilibrium simulation without temperature coupling. If the fluctuation in the total energy is less than 0.5% of the total energy of the system, the time step is typically acceptable. For a typical simulation of water transport, where the O–H bond length is fixed, a time step size of 1.0 to 2.0 fs is commonly used.

2. *Cut-Off Scheme.* The most expensive part of energy and force calculations is the nonbonded interactions, since there are $N(N-1)/2$ such interactions to calculate in an N-atom system. To save computer time, it is usual practice to neglect nonbonded interactions that occur between atoms separated by more than a given distance (say 10 Å). This use of a cut-off scheme in potential/force calculation is justified for potentials like the Lennard–Jones potential. However,

for charged atoms, the use of a cut-off scheme is not easily justified: charge–charge interactions are of much longer range (because of the $1/r$ dependence of the interaction energy). In this case, one usually needs to use the more complicated method for the calculation of the interaction energy, e.g., the particle mesh Ewald method (PME) or the fast multipole method (FMM), as discussed in Section 16.1.2.

3. *Boundary Conditions:* The most commonly employed boundary conditions in MD are the periodic boundary conditions. The atoms of the system to be simulated are put into a space-filling box, which is surrounded by translated copies of itself. Thus there are no boundaries of the system; the artifact caused by unwanted boundaries in an isolated cluster is now replaced by the artifact of periodic conditions. If a crystal is simulated, such boundary conditions are desired (although motions are naturally restricted to periodic motions with wavelengths fitting into the box). If one wishes to simulate nonperiodic systems, as liquids or solutions, the periodicity by itself causes errors. The errors can be evaluated by comparing various system sizes; they are expected to be less severe than the errors resulting from an unnatural boundary with vacuum (van der Spoel et al., 2004). Many packages (e.g., GROMACS (van der Spoel et al., 2004)) use periodic boundary conditions, combined with the minimum image convention, where only one, the nearest, image of each particle is considered for short-range nonbonded interaction terms. For long-range electrostatic interactions this is not always accurate enough, and therefore other techniques for the calculation of the interaction energy, e.g., the particle mesh Ewald method (PME) or the fast multipole method (FMM) (see Section 16.1.2) are necessary.

4. *Freezing a Group of Atoms.* Atoms that are supposed to remain stationary can be frozen to reduce the computational time. This is also useful during equilibration, e.g., to avoid badly placed solvent molecules from interacting unreasonably with other species (say protein atoms), although the same effect can also be obtained by putting a restraining potential on the atoms that must be protected (van der Spoel et al., 2004). The freeze option can be used, if desired, on just one or two coordinates of an atom, thereby freezing the atoms in a plane or in a line (van der Spoel et al., 2004).

5. *Storing the Simulation Results.* MD simulation generates the trajectories of each atom in the simulation system. The properties of the system can be obtained by analyzing these trajectories. The common practice is to save the system trajectories at certain intervals during the simulation, and analyze the trajectories when the simulation is finished. For most measurables, e.g., density, this is sufficient. Because the trajectories of neighboring time steps are usually highly

correlated, it is usually not necessary to store the trajectory at each time step. However, for the analysis of certain variables, e.g., streaming velocity, which has huge statistical fluctuations, one has to store the data at each step to obtain a reasonable estimation of the variable, which will result in prohibitive disk space consumption. In this case, the data analysis is usually performed in an "on-the-fly" manner.

6. *Number of Bins.* In the binning method (described in Section 16.1.4), the system is partitioned into cells. The total number of cells or the size of the individual cell is an important choice. Since we want to estimate a certain variable, which varies continuously in the space, a larger number of bins means a finer grid and thus a better estimation (small systematic error). However, because of the statistical nature of data analysis in MD simulations, a smaller-sized bin or cell means a smaller number of particles in the bin and thus a larger statistical error. Thus there is a trade-off between systematic error and statistical error. It is to be noted that since the computation time for data analysis using the binning method is negligible compared to the other computational steps of MD simulation, coarser or finer partitioning (for the purpose of binning) does not affect the speed of the simulation significantly. The choice of the number of bins also depends on the parameter that needs to be extracted.

16.1.6 MD Software

A number of MD programs are available on the Web for free download. Some of the MD programs that are popularly used by the computational chemistry and biology community are listed in Table 16.3.

16.2 MD-Continuum Coupling

MD simulations can be employed only for a short time and very small length scales due to their large computational requirements compared to the computational complexity of continuum discretizations. Multiscale approaches both in time and space can overcome this difficulty. The coupling of MD to Navier–Stokes equations would extend the range of applicability of both approaches and provide a unifying description of liquid flows from nanoscales to larger scales. The *incompressible* Navier–Stokes equations are involved in the coupling, although some authors have attempted to couple MD to compressible Navier-Stokes equations; see, for example, (Flekkoy et al., 2000). In the following we will present the main ideas of four different coupling strategies:

- The relaxation method of (O'Connell and Thompson, 1995).

TABLE 16.3. A list of some commonly used MD packages.

MD Package	*Website*
GROMACS	*http://www.gromacs.org*
AMBER	*http://amber.scripps.edu/*
CHARMM	*http://yuri.harvard.edu/*
NAMD	*http://www.ks.uiuc.edu/Research/namd/*
Chemical	*http://www.uku.fi/ thassine/chemical/*
Insight II	*http://www.accelrys.com/insight/*
ORAC	*http://www.chim.unifi.it:8080/orac*

- The Maxwell demon method of (Hadjiconstantinou and Patera, 1997).
- The flux-exchange method of (Flekkoy et al., 2000), and
- A hybrid method of (Nie et al., 2004).

Coupling of heterogeneous representations was first attempted in solid mechanics problems (Kohlhof et al., 1991). Similar coupling was extended to liquids in the works of (O'Connell and Thompson, 1995; Hadjiconstantinou, 1999; Flekkoy et al., 2000; Nie et al., 2004). In the MD-continuum coupling the use of an overlap region is necessary, since MD induces local

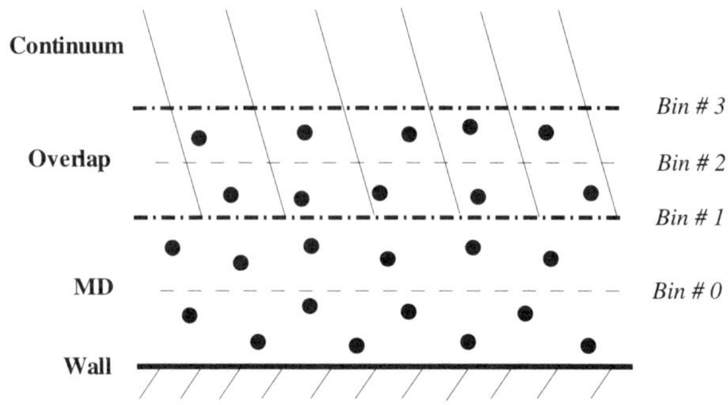

FIGURE 16.3. Domain for MD-Continuum coupling.

structure in the fluid at interfaces and especially for density (Koplik and Banavar, 1995). The two-domain coupling would therefore have the form shown in the sketch of Figure 16.3, consisting of a region (MD) where the molecular dynamics simulation is performed, the region where the Navier–Stokes equations will be solved, and an overlap region, where both descriptions are valid. However, in order to terminate the MD region, in addition to standard particle motion in the overlap region a constraint should be imposed of the form

$$\sum_{n=1}^{N_i} p_n - M_i v_i = 0,$$

where N_i is the total number of particles and M_i is the mass of the continuum fluid element in the i-bin, and p_n is the momentum of the nth particle in the v-direction. This constraint can be integrated into standard Lagrange's equations governing the motion of the rest of the molecules in the MD region. This approach was successfully implemented by (O'Connell and Thompson, 1995), who used an overlap region of 14σ. A free parameter in this approach is the strength of the constraint ξ in relation to the extent of the overlap region. Specifically, the equations of motion in the i-bin are

$$\dot{\mathbf{x}}_i = \frac{p_i}{m} + \xi \left[\frac{M_i}{mN_i} - \frac{1}{N_i} \sum_{n=1}^{N_i} \frac{p_n}{m} \right], \tag{16.25}$$

$$\dot{\mathbf{p}}_i = -\frac{\partial V}{\partial \mathbf{x}}, \tag{16.26}$$

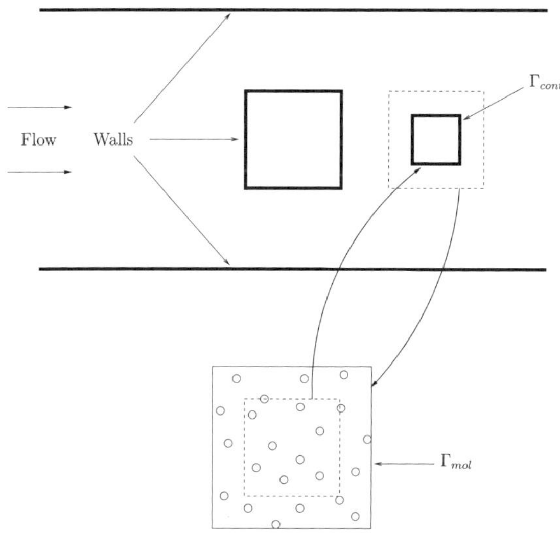

FIGURE 16.4. Schematic of one Schwarz iteration cycle (Courtesy of N.G. Hadjiconstantinou.)

where m is the particle mass. Small values of the parameter will provide an inadequate coupling between MD and Navier–Stokes, while large values will lead to excessive damping of particle fluctuations, which in turn will lead to divergence in the solution. In (O'Connell and Thompson, 1995), a value of $\xi = 0.01$ was used for simulating a slow-startup Couette flow. It was shown that the best choice of ξ is to have $\Delta t_{\text{MD}}/\xi$ greater than the autocorrelation time t_{vv}, where Δ_{MD} is the time step in the MD simulation. This relaxation approach does not handle correctly the mass flux at the MD-continuum interface, and this is an important limitation of the method.

A more effective approach would be to use the constraint dynamics to control the density fluctuations at the end of the interface (last bin), while at the same time resorting to a relaxation iterative procedure similar to the DSMC-continuum coupling (see Section 15.1). In the context of liquid flows, we cannot provide a modification of the Navier–Stokes equations to account for the slip, and thus the constraint, on the MD side, will enforce the no-slip condition as interface to the continuum description. In other words, we interpolate quantities as

$$\phi_{\text{MD}}^{n+1} = \theta \phi_{\text{MD}}^{n} + (1 - \theta)\phi_{\text{NS}}^{n},$$

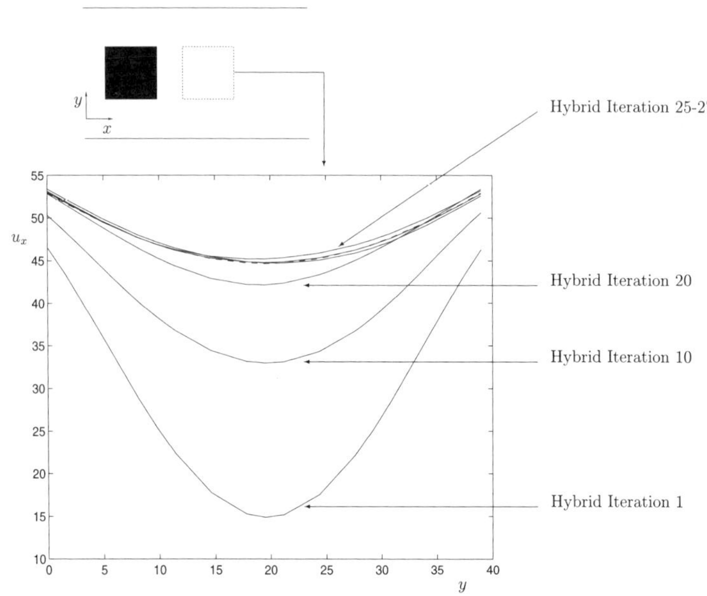

FIGURE 16.5. Convergence history for the obstructed channel problem of Figure 16.4. The velocity profile at different iterations is shown. (Courtesy of N.G. Hadjiconstantinou.)

with θ the relaxation parameter. For $\theta = 0$ we obtain the alternating Schwarz algorithm used in the work of (Hadjiconstantinou, 1999), which is explained schematically in the plot of Figure 16.4; see also Section 15.2.1. The example is flow in an obstructed channel in which a square region behind the solid block is treated by MD. The continuum iterate receives Dirichlet data on Γ_{cont} from the molecular solution obtained in the previous iteration cycle (top); the continuum solution on Γ_{mol} is subsequently used as Dirichlet data on the molecular simulation (bottom). A typical result of the convergence history of the iterative process is shown in Figure 16.5. The hybrid solution is compared to the full continuum solution (denoted by a dashed line); the continuum solution is taken as "exact," since the molecular region is representing bulk fluid far from any boundaries. An initial guess of zero velocity in the molecular region was used. The oscillations around the "exact" solution are due to the statistical fluctuations in the molecular description.

This method employs a *particle reservoir* to satisfy the mass flux across the hybrid interface; this reservoir is taken as the outer ring of the MD domain. The particles do not drift away, because of the periodic boundary conditions imposed on the MD domain. The particle velocity in a bin of the overlap region is drawn from a Maxwellian distribution at each time step, adopting the so-called *Maxwell demon* procedure. In this approach the particle nonequilibrium distributions are not included. In addition, there are sudden changes of the particle velocities from one step to another.

In general, this type of iterative coupling may depend on the relative size of the domains and could lead to possible divergence of solutions for large disparity in domain sizes, which is typical in microfluidic and nanofluidic applications. The relaxation procedure for $\theta \neq 0$ has been shown to weaken the strong dependence on the domain size (Henderson and Karniadakis, 1991). This is typical of algorithms based on the classical Schwarz algorithm, and for coupling of elliptic problems with heterogeneous discretization, convergence of the θ-relaxation iterative procedure is independent of the domain sizes. In addition to the iterative coupling and the dynamic constraint in the overlap region, convergence of the coupled solution is accomplished only if there is compatibility between transport coefficients on the two sides and the scatter of quantities on the microscopic side is minimized using spatial averages, for example, by exploiting homogeneity in planar slices about 2σ thick.

The *flux-exchange* method developed by (Flekkoy et al., 2000) is *conservative*, since it relies in the matching of fluxes of mass and momentum between the MD and Navier–Stokes domains. In particular, the mass flux continuity is enforced by the equation

$$m\, s(\boldsymbol{x}, t) = A\rho \boldsymbol{v} \cdot \boldsymbol{n},$$

where A is the area and $\boldsymbol{n}$ the unit normal vector. Here $s(\boldsymbol{x}, t)$ is the number of particles that need to be added to (if s is positive) or removed (if s is negative) the top bin of the overlap domain. Similarly, the momentum flux continuity is enforced by the equation

$$ms(\boldsymbol{x}, t)\langle \mathbf{u}' \rangle + \sum_i \mathbf{F}_i = A\mathbf{\Pi} \cdot \boldsymbol{n},$$

where $\mathbf{u}'$ is the velocity of the added or removed particles, and $\mathbf{F}_i$ is an external force acting on particle i in the region of flux-exchange from continuum to particles ($P \leftarrow C$). Flekkoy et al. employed the *compressible* Navier–Stokes equations in the coupling, and thus the momentum flux tensor $\mathbf{\Pi}$ has the form

$$\mathbf{\Pi} = \rho \boldsymbol{v}\boldsymbol{v} + p - \mu(\nabla \boldsymbol{v} + \nabla \boldsymbol{v}^T - \nabla \cdot \boldsymbol{v}) - (\mu/3)\nabla \cdot \boldsymbol{v}.$$

Combining the two equations, we observe that the momentum equation is satisfied if the mass equation is satisfied, but in addition we need to enforce

$$\langle \mathbf{u}' \rangle = \boldsymbol{v} \quad \text{and} \quad \sum_i \mathbf{F}_i = A\,(\mathbf{\Pi} - \rho \boldsymbol{v}\boldsymbol{v}) \cdot \boldsymbol{n}.$$

In order to avoid drifting of particles, a weight function $g(x)$ was introduced. In particular, this function obeys $g(x) = g'(x) = 0$ for $x \leq 0$ and diverges as

$$g(x \to L/2) \propto \frac{1}{L/2 - x}$$

at the edge of the region $(P \leftarrow C)$. The coordinate x runs parallel to $\boldsymbol{n}$, and $x = 0$ is in the middle of the region $(P \rightarrow C)$ where flux-exchange from continuum to particles take place. Also, L is the size of the bins in the $\boldsymbol{n}$ direction. In addition, in order to maintain thermal equilibrium it was found necessary to thermalize the particles in the subdomain $P \leftarrow C$ using Langevin dynamics. Specifically, a force of the form

$$\mathbf{F}_{Li} = -\gamma(\mathbf{u}_i - \boldsymbol{v}) + \tilde{\mathbf{F}}, \quad \langle \tilde{\mathbf{F}}(t)\tilde{\mathbf{F}}(t') \rangle = 2k_{\mathrm{B}}T\gamma\delta(t - t')$$

was added to the force $\mathbf{F}_i$; here γ is a measure of dissipation.

This scheme was tested for steady Couette and Poiseuille flow using a shifted Lennard–Jones potential, with time steps $\Delta t_{\mathrm{MD}} = 0.0017\tau$ on the MD side and $\Delta t = 100\Delta_{\mathrm{MD}}$ on the Navier–Stokes side.

In the *hybrid method* of (Nie et al., 2004) the relaxation procedure of O'Connel and Thompson is abandoned, and new motion equations for the MD are derived. In the overall region, the average continuum velocity $\boldsymbol{v}_J$ in each cell J is computed by averaging the velocities on the cell's grid points where the Navier–Stokes equations are discretized. Continuity of the mean velocity is imposed by requiring that the averaged particle velocity in this cell be equal to $\boldsymbol{v}_J$, i.e.,

$$\frac{1}{N_J} \sum_i \mathbf{u}_i = \boldsymbol{v}_J(t),$$

where N_J is the number of particles in cell J. Taking the Lagrangian derivative of the above, we have

$$\frac{1}{N_J} \sum_i \ddot{\boldsymbol{x}}_i = \frac{D\boldsymbol{v}_J(t)}{Dt},$$

which is a constraint in the equations of motion $\ddot{\boldsymbol{x}} = \mathbf{F}_i/m$. A general solution that satisfies the constraint has the form

$$\ddot{\boldsymbol{x}} = \frac{D\boldsymbol{v}_J(t)}{Dt} + \zeta_i, \quad \sum_i \zeta_i = 0.$$

By choosing

$$\zeta_i = \frac{\mathbf{F}_i}{m} - \frac{1}{mN_J} \sum_{i=1}^{N_J} \mathbf{F}_i$$

we obtain the generalized equations of motion for the ith particle,

$$\ddot{\boldsymbol{x}} = \frac{\mathbf{F}_i}{m} - \frac{1}{mN_J}\sum_{i=1}^{N_J}\mathbf{F}_i + \frac{D\boldsymbol{v}_J(t)}{Dt} + \zeta_i, \quad \sum_i \zeta_i = 0. \tag{16.27}$$

The discretization of the above equation gives

$$\frac{\boldsymbol{x}(t+\Delta t_{MD}) - 2\boldsymbol{x}(t) + \boldsymbol{x}(t-\Delta t_{MD})}{\Delta t_{MD}^2}$$
$$= \frac{\mathbf{F}_i}{m} - \frac{1}{mN_J}\sum_{i=1}^{N_J}\mathbf{F}_i - \frac{1}{\Delta t_{MD}}\left[\frac{1}{N_J}\sum_{i=1}^{N_J}\dot{\boldsymbol{x}}(t)_i - \boldsymbol{v}_J(t+\Delta t_{MD})\right].$$

The motion equations employed in the relaxation method in (O'Connell and Thompson, 1995), are similar to the equations (16.27) but with the last two terms on the right-hand side premultiplied by the relaxation parameter $\xi \ll 1$. This means that the solution relaxes to a converged solution after time on the order of $\Delta t_{\mathrm{MD}}/\xi$. This delay may be necessary in order to prevent the constraint from canceling intrinsic thermal fluctuations on time scales less than the autocorrelation time t_{vv}. However, as arugued in (Nie et al., 2004), it has the undesirable effect that the particle velocities always *lag* the continuum solution, which is incorrect in an accelerating flow.

A weight function similar to the flux-exchange method was also introduced in (Nie et al., 2004), in order to prevent drifting of molecules away from the MD domain but also to minimize density oscillations. In order to ensure mass continuity at the MD–continuum interface, the number of particles in each cell is modified by the net flux over a time step of the Navier–Stokes equations,

$$n' = -A\Delta t \rho \boldsymbol{v} \cdot \boldsymbol{n}/m.$$

In general, this method has features of the relaxation method and the flux-exchange method, but momentum flux continuity is not directly imposed. It has been used successfully in simulating accelerating Couette flow, cavity flow, and flow over an obstacle. Typical values were 15.5σ for the overlap region in a domain 52.1σ wide. The Navier–Stokes time step was $\Delta t = 50\Delta t_{\mathrm{MD}}$ with $\Delta t_{\mathrm{MD}} = 0.005\tau$, and $\tau = \sqrt{m\sigma^2/\epsilon}$ the characteristic time of the Lennard–Jones potential.

In summary, some of the key ingredients for effective MD–continuum coupling are:

- Use of a constraint to minimize density variations at the interface.
- Modification of the potential at the interface to eliminate local artificial structure.

- Use of a relaxation procedure to accelerate convergence of the coupled algorithm.
- Incorporation of constrained dynamics.
- Mass and momentum flux exchange to maintain conservativity.

In the following section we present another approach to multiscale modeling in the context of electroosmotic flow in a nanochannel.

16.3 Embedding Multiscale Methods

In Chapter 12 (Sections 12.2 and 12.3), we observed that near the channel wall various atomistic characteristics (e.g., finite size of the ions, layering of water molecules) that were neglected in the classical continuum theory for the electroosmotic flow become important. In order to predict the electroosmotic flow in the entire channel accurately, one has to capture these atomistic details in the near-wall region. Multiscale simulation can be very helpful in such a scenario. In this section, we discuss a multiscale simulation method that is based on the embedding technique. Figure 16.6 shows a schematic of the embedding technique. The central ideas in the embedding technique are:

1. To simulate a wider (or coarser) length scale problem (see Figure 16.6(a)), we first set up an auxiliary smaller (or finer) length scale problem (see Figure 16.6(b)) using similar input conditions (e.g., wall surface charge density) such that the near-wall noncontinuum behavior is captured.
2. An MD simulation is performed on the finer length scale problem.
3. The MD results from the finer scale channel are embedded into the continuum simulation of the coarser length scale problem.

For this approach to be *accurate*, the size of the fine-scale problem must be large enough such that all the critical regions in the coarser length scale problem, where atomistic details are important, are included in the auxiliary problem. For this approach to be *efficient*, the size of the fine-scale problem must be much smaller than the size of the original system, i.e., $W_1 \ll W_0$. In practice, the size of the auxiliary problem is chosen as a compromise between these two objectives.

In the rest of this section, we describe two embedding multiscale examples to compute the ion concentration and the velocity distribution in a nanochannel electroosmotic flow.

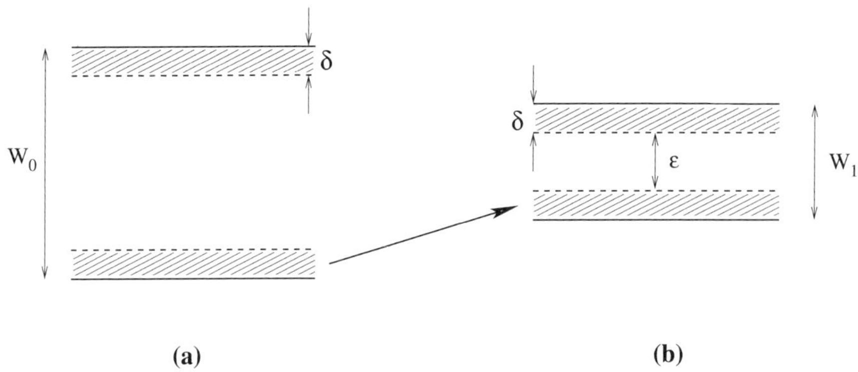

FIGURE 16.6. Schematic of the embedding technique for multiscale simulation. (a) represents the original coarser length scale problem and (b) represents auxiliary fine scale problem set up to solve the coarser length scale problem. W_0 and W_1 are the characteristic length scales of the two systems. The shaded areas of width δ denote the critical regions where atomistic details are important in determining the system behavior. The region of width ϵ in panel (b) is a buffer region.

16.3.1 Application to the Poisson–Boltzmann Equation

In Section 12.2.1, we presented a modified Poisson–Boltzmann equation to account for the wall–ion, water–ion, and ion–ion interactions in a more accurate manner. The key issue in the implementation of the modified Poisson–Boltzmann equation is to compute the excess chemical potential for an ion, i, denoted by $\phi_{\text{ex},i}$ (see Section 12.2.1 for details). In the embedding multiscale approach, we extract the electrochemical potential correction term from the ion concentration profile obtained from MD simulation of a smaller width channel using equation (12.4). Once the electrochemical potential correction term is obtained, one can use it in the modified Poisson–Boltzmann equation (12.5b) to simulate the ion distribution in a bigger channel with the same wall structure and similar surface charge density. In such an approach, one circumvents the difficulty of obtaining a closed form expression for the electrochemical potential correction term by utilizing the MD simulation results.

The accuracy of this approach depends on how the fine-scale problem is set up and on how significantly the electrochemical potential correction term differs in the two problems (i.e., in the coarse-scale and in the fine-scale problems). In setting up the fine-scale problem, one needs to include the near-wall region, where the electrochemical potential correction is nonzero. By setting up the fine-scale problem using similar operating conditions (e.g., wall charge density, wall structure, and bulk concentration) as in the original system, the electrochemical potential correction will not differ significantly in the two problems. This is because:

1. The wall–ion interaction included in the electrochemical potential correction term is the Lennard–Jones potential, which depends only on the wall structure and the Lennard–Jones parameters, and thus will not change when the channel width is increased.

2. The water–ion interactions depend primarily on the water concentration (i.e., how closely the water molecules are packed). MD simulation results of water concentration profile in channels of different width but with the same surface charge density indicate that the water concentration profile near the channel wall is independent of the channel width.

In summary, the electrochemical potential correction term is primarily due to the wall effects (e.g., ion–wall interactions and wall-induced water layering). Since these interactions are short-ranged, further addition of water layers in the bulk (corresponding to a wider channel) would not affect the electrochemical potential correction term significantly. Hence, the use of the same electrochemical potential correction term for wider channels can produce reasonably accurate results.

The efficiency of this approach depends on whether the length scale of the fine-scale problem can be significantly smaller compared to the original problem. This can be achieved by choosing a small ϵ in Figure 16.6. However, if ϵ is too small in the fine-scale problem, the system behavior in one critical region may be influenced by the system behavior in another critical region (e.g., the ion distribution near the upper channel may be influenced by that near the lower channel wall), which may not exist in the original coarse-scale problem. As an example, for a 6-nm-wide coarse-scale channel, typically a 2-nm-wide fine-scale channel (note that this can depend on a number of parameters such as the surface charge density, and Debye length) is used. MD simulation of a 2-nm-wide channel requires much smaller computational time than the MD simulation of the 6-nm-wide channel. Another good example where the embedding multiscale approach can be efficient is the nanofluidic system studied by (Kemery et al., 1998), where nanochannels are connected by microchannels. In this case, the MD simulation of the nanochannels is possible, but the MD simulation for the entire system is impossible.

Figure 16.7 (a) shows the MD concentration profile of Cl^- ion across the channel for case 4 (see Table 12.2 for details), where the channel width is 2.22 nm and the surface charge density on the channel wall is 0.120 C/m^2. Figure 16.7 (b) shows the electrochemical potential correction term extracted by using equation (12.4). Note that the electrochemical potential correction term is close to zero at about 0.8 nm away from the channel wall. ϕ_{ex} reaches a minimum at about 0.34 nm away from the channel wall, and this roughly corresponds to the position of the minimum of the potential energy due to the Lennard–Jones potential between the wall and the Cl^- ion (see Figure 12.3 (b)). This indicates that the electrochemical potential correction term at this position is primarily due to the Lennard–Jones potential between the wall and the Cl^- ion. Using the electrochemical potential correction term shown in Figure 16.7 (b), the ion distribution in various channels with different widths and similar surface charge densities were calculated. Figure 16.7 (c) shows the comparison of Cl^- concentration in a 3.49-nm channel (case 1, charge density: 0.120 C/m^2 (see Table 12.2)) predicted by MD simulation and by the modified Poisson-Boltzmann equation. Figure 16.7 (d) shows the comparison of Cl^- concentration in a 10.0 nm channel (case 6, charge density: 0.124 C/m^2) predicted by MD simulation and by the modified Poisson–Boltzmann equation. The results in Figures 16.7 (c) and 16.7 (d) suggest that the extraction of the electrochemical potential correction using a fine-scale channel and employing it in the modified Poisson–Boltzmann equation to predict the variation of the ion concentration in the coarser channel works very well.

16.3.2 Application to Navier–Stokes Equations

The results in Section 12.3 indicate that the continuum flow theory is not valid for electroosmotic flow in a 0.95-nm-wide channel. However, continuum theory can be used to describe flow in channels wider than 2.22 nm, provided that viscosity variation near the channel wall is taken into account. It is, however, very difficult to obtain a closed-form expression for viscosity variation near the channel wall. To simulate electroosmotic flow in wide channels, where MD simulation can be very expensive, one possible way is to first perform an MD simulation in a smaller channel under similar conditions (e.g., using the same wall structure and charge density as in a wider channel) and then extract the viscosity from the MD data. The extracted viscosity can then be used in continuum theory to model flow in a wider channel. In this approach, one assumes that the viscosity near the channel wall would not change appreciably when the channel width increases. This assumption typically holds, since viscosity depends on the fluid properties and ion concentrations near the channel wall, and these parameters would not change significantly when the channel width changes, provided that other operating conditions (e.g., wall structure and wall charge density) do not change significantly or remain the same. The evaluation of viscosity from molecular dynamics data can be difficult, since one needs to compute the derivative of the velocity obtained from MD data. Since the velocity obtained from the MD simulation is usually very noisy, unless the simulation is carried out for a very long time, the derivative of the velocity would be even noisier, leading to significant noise in the extracted viscosity. It is possible to smooth the velocity data using a filter, but this may introduce additional errors into the viscosity estimation. An alternative approach is to use the embedding multiscale approach, in

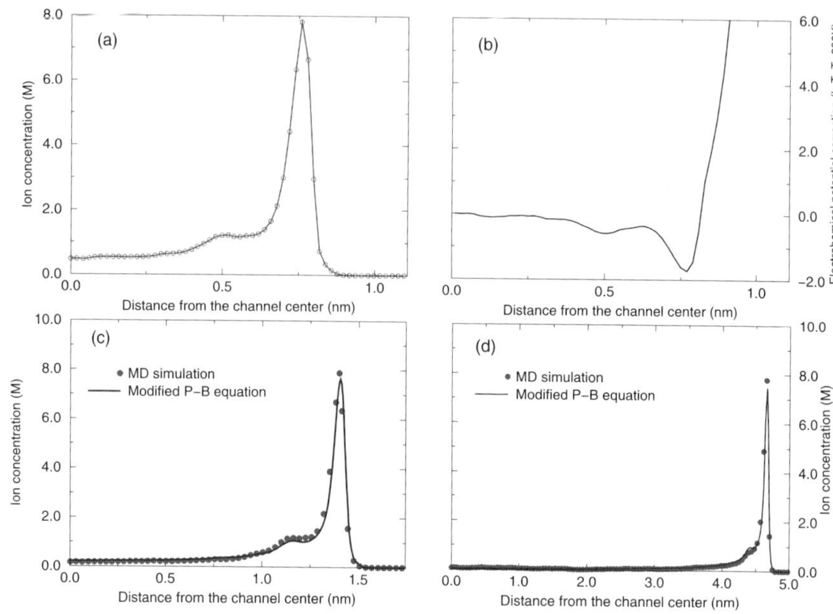

FIGURE 16.7. (a) Cl^- ion concentration across the channel for case 4 ($W = 2.22$ nm, $\sigma_s = +0.120$ C/m^2). (b) The electrochemical potential correction term extracted from the ion distribution shown in (a). (c) Comparison of Cl^- ion concentration across the channel for case 1 ($W = 3.49$ nm, $\sigma_s = +0.120$ C/m^2) as predicted by the MD simulation and modified P-B equation. (d) Comparison of Cl^- ion concentration across the channel for case 6 ($W = 10.0$ nm, $\sigma_s = +0.124$ C/m^2) as predicted by the MD simulation and modified Poisson–Boltzmann (P-B) equation.

which the velocity near the wall, obtained from the fine-scale simulation, is used in the continuum modeling of a coarse-scale channel. This approach is described below.

Figure 16.8 presents details on the simulation of electroosmotic flow in a large channel using the velocity obtained from MD simulation of electroosmotic flow in a small channel. For any position within δ' from the no-slip plane, the velocity in the large channel is obtained by embedding the MD velocity obtained for the electroosmotic flow in a small channel. Once the velocity at $z = \delta'$ is obtained, it is used as the boundary condition for the continuum flow modeling in the central portion of the large channel using a constant viscosity. To embed the small channel MD velocity u within δ' from the no-slip plane into the simulation of flow in a large channel, we first integrate the momentum equation from the channel center (c'' is the center of the small channel and c' is the center of the large channel) to a

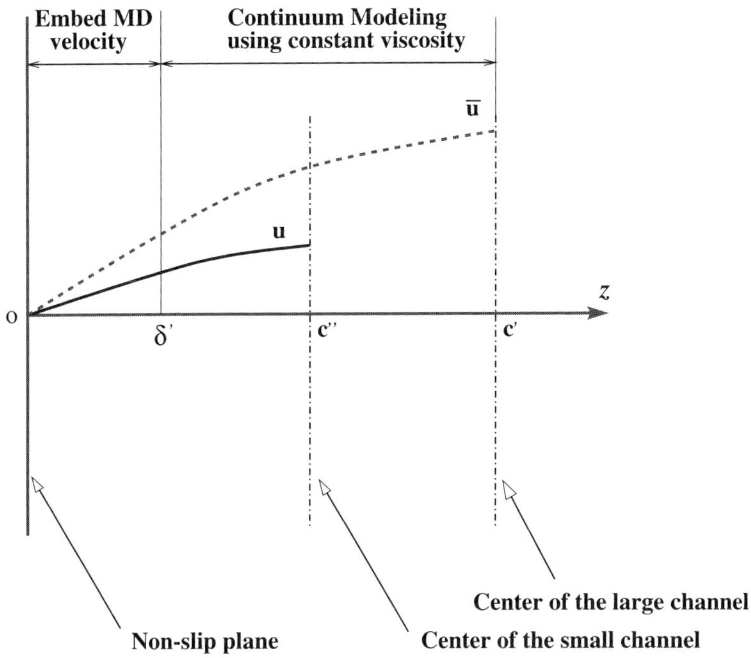

FIGURE 16.8. Continuum simulation of electroosmotic flow in a large channel using the velocity embedding technique. The channel is partitioned into two regions: a near wall region and a channel center region. The velocity near the wall is computed by embedding the velocity obtained from MD simulation of electroosmotic flow in a smaller channel. The velocity in the channel center region is computed using the continuum flow theory based on a constant viscosity.

position z:

$$\mu \left. \frac{du}{dz} \right|_{s=c''}^{z} = \int_{c''}^{z} -\tilde{z}qc(s)E_{\text{ext}}\, ds, \tag{16.28}$$

$$\mu \left. \frac{d\bar{u}}{dz} \right|_{s=c'}^{z} = \int_{c'}^{z} -\tilde{z}q\bar{c}(s)\bar{E}_{\text{ext}}\, ds, \tag{16.29}$$

where u, c, and E_{ext} are the velocity, the ion concentration, and the external electric field at any position z in the small channel, respectively. Furthermore, $\bar{u}$, $\bar{c}$, and $\bar{E}_{\text{ext}}$ are the velocity, the ion concentration, and the external electric field at any position z in the big channel, respectively. Using the symmetry of the velocity profile with respect to the channel center,

$$\left. \frac{du}{dz} \right|_{s=c''} = \left. \frac{d\bar{u}}{dz} \right|_{s=c'} = 0, \tag{16.30}$$

dividing equation (16.29) by equation (16.28), and applying equation (16.30) gives

$$\left.\frac{d\bar{u}}{dz}\right|_z = \frac{\int_{c'}^{z} -\tilde{z}q\bar{c}(s)\bar{E}_{\text{ext}}\,ds}{\int_{c''}^{z} -\tilde{z}qc(s)E_{\text{ext}}\,ds}\left.\frac{du}{dz}\right|_z = F(z)\left.\frac{du}{dz}\right|_z, \tag{16.31}$$

where $F(z)$ is defined by

$$F(z) = \frac{\int_{c'}^{z} -\tilde{z}q\bar{c}(s)\bar{E}_{\text{ext}}\,ds}{\int_{c''}^{z} -\tilde{z}qc(s)E_{\text{ext}}\,ds}. \tag{16.32}$$

Integrating equation (16.31) from the no-slip plane (i.e., $z = 0$) to any position z, and using the fact that the velocity is zero at the no-slip plane, we obtain

$$\bar{u}(z) = \int_{s=0}^{z} F(s)\frac{du}{ds}\,ds = F(z)u(z) - \int_0^z \frac{dF(s)}{ds}u(s)\,ds. \tag{16.33}$$

Equation (16.33) can be used to compute the velocity near the channel wall in large channels. Note that no derivatives of the MD velocity in the small channel are needed. Instead, one needs to calculate the derivative of the function $F(z)$, which is obtained by integrating the ion concentration, and it is much easier to obtain good statistics for ion concentrations in MD simulations. In principle, equation (16.33) can be applied in the region from the no-slip plane to the center of the small channel (i.e., point c'' in Figure 16.8). However, equation (16.33) is used only in the region within δ' from the channel wall. There are two reasons for this:

1. Evaluation of the function $F(z)$ is difficult as we approach the center of the small channel because the integration term in the denominator is close to zero.

2. Since the viscosity variation is important only near the channel wall, we can use a constant viscosity in the region δ' away from the no-slip plane instead of embedding the MD velocity.

In the simulations presented here, δ' is taken to be 0.64 nm, since MD simulations indicate that the viscosity variation beyond this length scale is small. As mentioned earlier, the no-slip plane is typically located at 0.14 nm from the channel wall. Hence, the region in which the velocity is obtained from the embedding technique is $\delta = 0.78$ nm from the channel wall for the larger channel.

Figure 16.9 shows the velocity profile across a 10.0-nm channel (case 6; see Table 12.2 for details) obtained by using the embedding multiscale approach. The velocity within the δ distance from the channel wall is embedded using equation (16.33) and the MD velocity of a 2.22-nm channel. The velocity in the central portion of the channel is computed using a constant viscosity of 0.743 mPa·s. Although there is considerable noise in

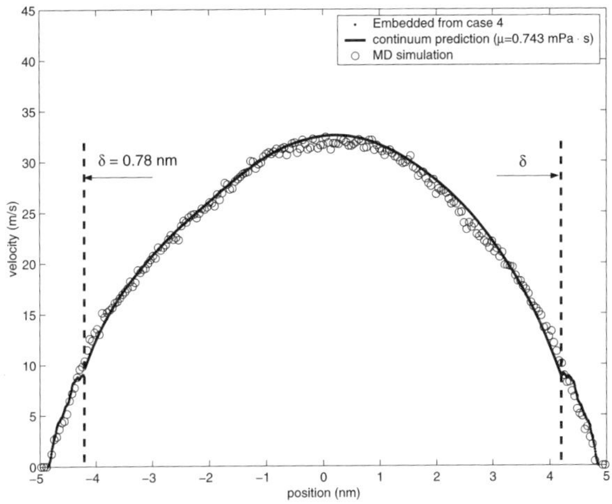

FIGURE 16.9. Velocity profile across the channel for case 6 (W = 10.00 nm, σ_s = +0.124 C/m^2). The velocity in the region within δ from the channel wall is obtained using equation (16.33) and the MD velocity within the same region from case 4 (W = 2.22 nm, σ_s = +0.120 C/m^2). The velocity in the central portion of the channel is computed using a constant viscosity of 0.743 mPa·s.

the MD velocity profile for case 4 (i.e., in the MD velocity of a 2.22-nm channel), the velocity obtained using equation (16.33) matches the MD simulation results reasonably well.

16.4 Dissipative Particle Dynamics (DPD)

So far, we have discussed the molecular dynamics (MD) method and the lattice Boltzmann method (LBM) for simulations in the atomistic and mesoscopic regimes, respectively. A potentially very powerful alternative to both methods has more recently emerged: the dissipative particle dynamics method (DPD), which combines features from both MD and LBM. The initial model was proposed by Hoogerburgge and Koelman as a simulation method to avoid the artifacts associated with traditional LBM simulations while capturing spatiotemporal hydrodynamic scales much larger than those achievable with MD (Hoogerburgge and Koelman, 1992).

The DPD model consists of particles that correspond to *coarse-grained* entities, thus representing molecular clusters rather than individual atoms. The particles move off-lattice, interacting with each other through a set of prescribed and velocity-dependent forces (Hoogerburgge and Koelman, 1992; Espanol and Warren, 1995). Specifically, there are three types of forces acting on each dissipative particle:

- A purely repulsive conservative force,
- A dissipative force that reduces velocity differences between the particles, and
- A stochastic force directed along the line connecting the center of the particles.

These forces can be interpreted as follows: The conservative forces cause the fluid particles to be as evenly distributed in space as possible as a result of certain "pressures" among them. The frictional forces represent viscous resistances between different parts of the fluid. Finally, the stochastic forces represent degrees of freedom that have been eliminated during the coarse-graining process. The last two forces effectively implement a thermostat, so that thermal equilibrium is achieved. Correspondingly, the amplitude of these forces is dictated by the fluctuation–dissipation theorem (Espanol and Warren, 1995), which ensures that in thermodynamic equilibrium the system will have a canonical distribution. All three forces are modulated by a weight function that specifies the range of interaction between the particles and renders the interaction local.

The distinguishing feature of the DPD forces is that they conserve momentum, and therefore, the DPD model satisfies mass and momentum conservation, which are responsible for the hydrodynamic behavior of a fluid at large scales (Espanol, 1995). Also, by changing the conservative interactions between the fluid particles, one can easily construct "complex" fluids, such as polymers, colloids, amphiphiles, and mixtures.

In summary, the DPD method is characterized by the following conditions:

- Position and velocity variables are continuous, as in MD, but the time step is updated in discrete steps, as in LBM.
- The conservative forces between DPD particles are soft-repulsive, which makes it possible to extend the simulations to longer time scales compared to MD.
- Hydrodynamic behavior is expected at much smaller particle numbers than in classical MD.
- DPD has the advantages of standard LBM but avoids the disadvantages of lacking Galilean invariance and introducing spurious conservation laws.
- Mass and momentum are locally conserved, which results in hydrodynamic flow effects on the macroscopic scale.
- The characteristic kinetic time in DPD is large compared to MD time scales.

A conceptual picture then of DPD is that of soft microspheres randomly moving around but following a preferred direction dictated by the conservative forces. DPD can be interpreted as a Lagrangian discretization of the equations of fluctuating hydrodynamics as the particles follow the classical hydrodynamic flow but they display thermal fluctuations. These fluctuations are included *consistently* based on the principles of statistical mechanics.

In the following we will present the governing equations, and we will discuss in some detail the numerical integration of these stochastic differential equations. Subsequently, we will discuss how to implement boundary conditions and we will present some examples.

16.4.1 Governing Equations

We consider a system consisting of N particles having equal mass (for simplicity in the presentation) m, position, $\mathbf{r}_i$, and velocities $\boldsymbol{v}_i$. The aforementioned three types of forces exerted on a particle i by particle j are given by

$$\mathbf{F}_{ij}^C = \mathcal{F}_{ij}^C(r_{ij})\hat{\mathbf{r}}_{ij}, \tag{16.34a}$$

$$\mathbf{F}_{ij}^D = -\gamma\omega^D(r_{ij})(\boldsymbol{v}_{ij}\cdot\hat{\mathbf{r}}_{ij})\hat{\mathbf{r}}_{ij}, \tag{16.34b}$$

$$\mathbf{F}_{ij}^R = \sigma\omega^R(r_{ij})\xi_{ij}\hat{\mathbf{r}}_{ij}, \tag{16.34c}$$

where $\mathbf{r}_{ij} \equiv \mathbf{r}_i - \mathbf{r}_j$, $r_{ij} \equiv |\mathbf{r}_{ij}|$, $\hat{\mathbf{r}}_{ij} \equiv \mathbf{r}_{ij}/r_{ij}$, and $\boldsymbol{v}_{ij} \equiv \mathbf{v}_i - \mathbf{v}_j$. The variables γ and σ determine the strengths of the dissipative and random forces, respectively. Also, ξ_{ij} are symmetric Gaussian random variables with zero mean and unit variance, and are independent for different pairs of particles and at different times; $\xi_{ij} = \xi_{ji}$ is enforced in order to achieve momentum conservation.

The conservative force $\mathbf{F}_{ij}^C$ is similar to the MD formulation, and it can include any pressure-drop terms, electrostatic or magnetic interactions, as well as van der Waals forces. This force as well as the other two forces are acting within a sphere of radius r_c, which is the length scale of the system. It corresponds to a soft interaction potential, similar to what has been proposed independently by (Forrest and Suter, 1995) in simulation studies of polymers. By averaging systematically Lennard–Jones potentials or the corresponding molecular field over the *rapidly* fluctuating motions of atoms during short time intervals, an effective average potential is obtained of the form shown in Figure 16.10. An approximation of this can be given as (Groot and Warren, 1997)

$$\mathbf{F}_{ij}^C = \begin{cases} a_{ij}(1-r_{ij})\hat{\mathbf{r}}_{ij} & \text{for } r_{ij} \le r_c = 1, \\ 0 & \text{for } r_{ij} > r_c = 1, \end{cases}$$

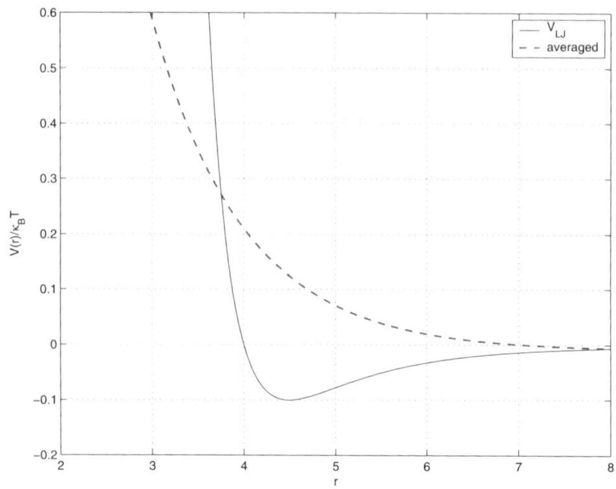

FIGURE 16.10. Lennard–Jones potential and its averaged soft potential.

where a_{ij} is a maximum repulsion between particle i and particle j. Unlike the hard Lennard–Jones potential, which is unbounded at $r = 0$, the soft potential employed in DPD has a finite value at $r = 0$ equal to a_{ij}. This reflects the fact that there is a finite probability that there will be no atoms at $r = 0$ for some realizations.

For water in the nondimensional units set here we have that $a_{ij} = 25$, while for other liquids we have that $a_{ij} = 75k_{\rm B}T/\rho$. These values were obtained in (Groot and Warren, 1997), by enforcing the proper compressibility of the system defined by

$$\kappa^{-1} = \frac{1}{k_{\rm B}T}\frac{\partial p}{\partial n}\bigg|_T,$$

where n is the number density. For water at room temperature (300 K) this dimensionless compressibility is $\kappa^{-1} = 15.9835$. The pressure is obtained as a function of the density, e.g.,

$$p = \rho k_{\rm B}T + \frac{1}{3V}\left\langle \sum_{j>i} (\mathbf{r}_i - \mathbf{r}_j)\cdot \mathbf{F}_{ij}^C \right\rangle,$$

where V is the volume. Also, the density is a free parameter, and for $\rho \geq 3$ the fluid behaves as liquid. However, since the number of interactions between particles scales linearly with density, the required computational time per unit volume increases with the square of the density. Therefore, in the simulations it suffices to work with the lowest value of $\rho = 3$ for computational efficiency (Groot and Warren, 1997).

The requirement of canonical distribution sets two conditions on the weight functions and the amplitudes of the dissipative and random forces.

Specifically, we have that

$$\omega^D(r_{ij}) = \left[\omega^R(r_{ij})\right]^2 \tag{16.35a}$$

and

$$\sigma^2 = 2\gamma k_{\mathrm{B}} T, \tag{16.36}$$

where T is the system temperature and k_{B} the Boltzmann constant. The weight function takes the form

$$\omega^R(r_{ij}) = \begin{cases} 1 - r_{ij}/r_c & \text{for } r_{ij} \leq r_c, \\ 0 & \text{for } r_{ij} > r_c, \end{cases}$$

where r_c is the cutoff radius. This is simply a convenient model to localize the interactions.

In the initial formulation of DPD by (Hoogerburgge and Koelman, 1992), the above conditions were not satisfied and energy was not conserved. This was corrected by (Espanol and Warren, 1995), who employed solutions of the Fokker–Planck equation. Let us consider the distribution function $f(\mathbf{r}_i, \mathbf{p}_i, t)$, which describes the probability of finding the system in a state with particles located at $\mathbf{r}_i$ having momenta $\mathbf{p}_i$ at time t. The time evolution of this distribution is expressed by the Fokker–Planck equation

$$\frac{\partial f}{\partial t} = \mathcal{L}^C f + \mathcal{L}^D f,$$

where $\mathcal{L}^C$ denotes the Liouville operator of the Hamiltonian system interacting with conserved forces $\mathbf{F}^C$; also, $\mathcal{L}^D$ represents the dissipative and random terms. If the last term is set to zero, we obtain a Hamiltonian system, which admits the canonical Gibbs–Boltzmann distribution as a solution. That is, $f^{\mathrm{eq}}(\mathbf{r}_i, \mathbf{p}_i) = \exp(-\sum_i \mathbf{p}_i^2/2mk_{\mathrm{B}}T - U/k_{\mathrm{B}}T)$ is a solution of

$$\frac{\partial f^{\mathrm{eq}}}{\partial t} = \mathcal{L}^C f^{\mathrm{eq}} = 0.$$

However, in the presence of the extra two nonconservative forces the equilibrium distribution will be different from the above unless the condition

$$\mathcal{L}^D f^{\mathrm{eq}} = 0$$

is satisfied. This condition is satisfied if the amplitude of the random and dissipative forces and also the weight functions are related as expressed by equations (16.35a) and (16.36).

The time evolution of DPD particles is described by Newton's law:

$$d\mathbf{r}_i = \boldsymbol{v}_i dt, \tag{16.37}$$

$$d\boldsymbol{v}_i = \frac{1}{m}\left(\mathbf{F}_i^C dt + \mathbf{F}_i^D dt + \mathbf{F}_i^R \sqrt{dt}\right). \tag{16.38}$$

Here $\mathbf{F}_i^C = \sum_{i \neq j} \mathbf{F}_{ij}^C$ is the total conservative force acting on particle i; $\mathbf{F}_i^D$ and $\mathbf{F}_i^R$ are defined similarly. The velocity increment due to the random force has a factor $\sqrt{dt}$ instead of dt. This term represents Brownian motion, which is described readily by a standard Wiener process with a covariance kernel given by

$$R_{FF}(t_i, t_j) = e^{-\gamma |t_i - t_j|}.$$

We see therefore that $(1/\gamma)$ is the *correlation length* in time for this stochastic process.

16.4.2 Numerical Integration

Unlike the MD equations, the DPD equations are stochastic, and this represents an extra degree of difficulty. In addition, the dissipative force depends on the velocity, which in turn depends on the force, so there is nonlinear coupling. In the following we represent a few schemes that have been used to integrate the DPD equations with various degrees of success. A detailed comparison of the performance of these integrators is given in (Nikunen et al., 2003).

Explicit Euler Forward (EF)

The explicit Euler integrator is never used, but it helps to set the notation for the more complicated algorithms and also as starter for multistep integrators of higher order. We assume that we know the solution at time step n and we want to obtain the solution at time step $(n+1)$. We update both the velocities and the positions of all particles in a straightforward manner. All three forces contribute within a region refined by the cutoff radius r_c.

- At time t^{n+1} we have

$$\begin{aligned} \boldsymbol{v}_i^{n+1} &= \boldsymbol{v}_i^n + \frac{1}{m_i} \left(\mathbf{F}_i^C \Delta t + \mathbf{F}_i^D \Delta t + \mathbf{F}_i^R \sqrt{\Delta t} \right), \\ \mathbf{r}_i^{n+1} &= \mathbf{r}_i^n + \boldsymbol{v}_i^n \Delta t. \end{aligned}$$

- Compute physical quantities.

Velocity–Verlet Method (DPD-VV)

The velocity–Verlet method is based on the classical MD velocity–Verlet algorithm, which we discuss in Section 16.1, and is time-reversible. One difference, however, with the classical algorithm is that the dissipative force is computed twice at the end of each time step to account for the implicit relationship to velocity mentioned earlier.

- At time t^{n+1} we have

$$
\begin{aligned}
\hat{\boldsymbol{v}}_i &= \boldsymbol{v}_i + \frac{1}{2}\frac{1}{m_i}\left(\mathbf{F}_i^C \Delta t + \mathbf{F}_i^D \Delta t + \mathbf{F}_i^R \sqrt{\Delta t}\right), \\
\mathbf{r}_i^{n+1} &= \mathbf{r}_i^n + \hat{\boldsymbol{v}}_i \Delta t, \\
\boldsymbol{v}_i^{n+1} &= \hat{\boldsymbol{v}}_i + \frac{1}{2}\frac{1}{m_i}\left(\mathbf{F}_i^C \Delta t + \mathbf{F}_i^D \Delta t + \mathbf{F}_i^R \sqrt{\Delta t}\right).
\end{aligned}
$$

- Compute $\mathbf{F}_i^D$.

- Compute physical quantities.

Here the forces at the last step are reevaluated with $\hat{\boldsymbol{v}}_i$ and $\mathbf{r}_i^{n+1}$.

Shardlow's Splitting Method

Another approach, which is often used in classical computational fluid dynamics, is operator splitting. In (Shardlow, 2003), the conservative forces were separated from the dissipative and random forces. First, the conservative components are computed using the classical MD approach, and subsequently the dissipative and random contributions are handled using a Langevin formulation. Two splitting schemes were developed (S1 and S2) corresponding to first-order and second-order accuracy, respectively. S1 is more efficient and is presented below.

- At time t^{n+1} we have

$$\hat{\boldsymbol{v}}_i = \boldsymbol{v}_i^n - \frac{1}{2}\frac{1}{m}\gamma\omega^2(r_{ij})(\boldsymbol{v}_{ij}^n \cdot \hat{\mathbf{r}}_{ij})\hat{\mathbf{r}}_{ij}\Delta t + \frac{1}{2}\frac{1}{m}\sigma\omega(r_{ij})\boldsymbol{\xi}_{ij}\hat{\mathbf{r}}_{ij}\sqrt{\Delta t},$$

$$\hat{\boldsymbol{v}}_j = \boldsymbol{v}_j^n + \frac{1}{2}\frac{1}{m}\gamma\omega^2(r_{ij})(\boldsymbol{v}_{ij}^n \cdot \hat{\mathbf{r}}_{ij})\hat{\mathbf{r}}_{ij}\Delta t - \frac{1}{2}\frac{1}{m}\sigma\omega(\mathbf{r}_{ij})\boldsymbol{\xi}_{ij}\hat{\mathbf{r}}_{ij}\sqrt{\Delta t},$$

$$\begin{aligned}\hat{\hat{\boldsymbol{v}}}_i = \hat{\boldsymbol{v}}_i &+ \frac{1}{2}\frac{1}{m}\sigma\omega(r_{ij})\boldsymbol{\xi}_{ij}\hat{\mathbf{r}}_{ij}\sqrt{\Delta t} \\ &- \frac{1}{2}\frac{1}{m}\frac{\gamma\omega^2(r_{ij})\Delta t}{1+\gamma\omega^2(r_{ij})\Delta t}\left[(\hat{\boldsymbol{v}}_{ij}\cdot\hat{\mathbf{r}}_{ij})\hat{\mathbf{r}}_{ij} + \sigma\omega(r_{ij})\boldsymbol{\xi}_{ij}\hat{\mathbf{r}}_{ij}\sqrt{\Delta t}\right],\end{aligned}$$

$$\begin{aligned}\hat{\hat{\boldsymbol{v}}}_j = \hat{\boldsymbol{v}}_j &- \frac{1}{2}\frac{1}{m}\sigma\omega(r_{ij})\boldsymbol{\xi}_{ij}\hat{\mathbf{r}}_{ij}\sqrt{\Delta t} \\ &+ \frac{1}{2}\frac{1}{m}\frac{\gamma\omega^2(r_{ij})\Delta t}{1+\gamma\omega^2(r_{ij})\Delta t}\left[(\hat{\boldsymbol{v}}_{ij}\cdot\hat{\mathbf{r}}_{ij})\hat{\mathbf{r}}_{ij} + \sigma\omega(r_{ij})\boldsymbol{\xi}_{ij}\hat{\mathbf{r}}_{ij}\sqrt{\Delta t}\right],\end{aligned}$$

$$\tilde{\boldsymbol{v}}_i = \hat{\hat{\boldsymbol{v}}}_i + \frac{1}{2}\frac{1}{m_i}\mathbf{F}_i^C\Delta t,$$

$$\mathbf{r}_i^{n+1} = \mathbf{r}_i^n + \tilde{\boldsymbol{v}}_i\Delta t,$$

$$\boldsymbol{v}_i^{n+1} = \tilde{\boldsymbol{v}}_i + \frac{1}{2}\frac{1}{m_i}F_i^C(\mathbf{r}_j^{n+1})\Delta t.$$

- Compute physical quantities.

Lowe's Alternative Method

The approach by Lowe differs from all other integration approaches of the DPD scheme. Here, the integration of the dissipative and random forces is bypassed. Specifically, Newton's equation of motion are integrated first, and then the system is *thermalized* as follows:

- For all particles within a sphere defined by r_c we assign a probability ($0 < \Gamma\Delta t \leq 1$) to predict an updated relative velocity from a Maxwelian distribution. So for a specific pair (ij) we first generate the relative velocity $\boldsymbol{v}_{ij}^0 \cdot \mathbf{e}_{ij}$ from a distribution $\boldsymbol{\xi}_{ij}\sqrt{2k_{\mathrm{B}}T/m}$, where $\boldsymbol{\xi}_{ij}$ is a Gaussian random variable with zero mean and with variance as before.

This procedure effectively sets up an Andersen thermostat; see Section 16.1.3. In particular, for $\Gamma\Delta t = 0$ the system is not coupled to the thermostat, whereas for $\Gamma\Delta t = 1$ the particle velocities are thermalized at every time step.

Two particularly attractive features of this method are:

- energy conservation even at large time steps, and
- the tracer diffusion properties of the fluid can match those of a real liquid, in conrast to other DPD versions.

- At time t^{n+1} we have

$$\hat{\boldsymbol{v}}_i = \boldsymbol{v}_i^n + \frac{1}{2}\frac{1}{m}\mathbf{F}_i^C \Delta t,$$

$$\mathbf{r}_i = \mathbf{r}_i^n + \hat{\boldsymbol{v}}_i^n \Delta t,$$

$$\hat{\hat{\boldsymbol{v}}}_i = \hat{\boldsymbol{v}}_i + \frac{1}{2}\frac{1}{m}\mathbf{F}_i^C(r_j)\Delta t.$$

- For all pairs $r_{ij} < r_c$, select probability $\Gamma\Delta t$. Obtain $\boldsymbol{v}_{ij}^0 \cdot \hat{\mathbf{r}}_{ij}$:

$$\Delta_{ij} = \frac{1}{2}\hat{\mathbf{r}}_{ij} \cdot (\boldsymbol{v}_{ij}^0 - \hat{\hat{\boldsymbol{v}}}_{ij}) \cdot \hat{\mathbf{r}}_{ij},$$

$$\boldsymbol{v}_i^{n+1} = \boldsymbol{v}_i^n + \Delta_{ij},$$

$$\boldsymbol{v}_j^{n+1} = \boldsymbol{v}_j^{n+1} - \Delta_{ij}.$$

- Compute physical quantities.

Example

Conservation is one of the key conditions for accurate simulations of the canonical ensemble. To this end, we study how a DPD fluid reaches equilibrium using various discretization schemes. In particular, we set to zero all conservative forces, and we simulate in time the effective thermostat expressed by the balance of dissipative and random forces. The simulation is conducted in a 3D periodic box of size $10 \times 10 \times 10$, where the length scale is defined by $r_c = 1$, and a particle number density is set to $\rho = 4$; thus $N = 4{,}000$ particles. The random force strength is chosen as $\sigma = 3$ in units of $k_B T$ and the dissipative force amplitude is $\gamma = 4.5$.

The physical quantities considered here are the average kinetic temperature

$$\langle k_B T \rangle = \frac{m}{3N - 3}\left\langle \sum_{i=1}^{N} \boldsymbol{v}_i^2 \right\rangle, \tag{16.39}$$

and the normal component of the pressure tensor defined by

$$p_{xx} = \rho \sum_{i=1}^{N} \boldsymbol{v}_{ix}^2, \tag{16.40}$$

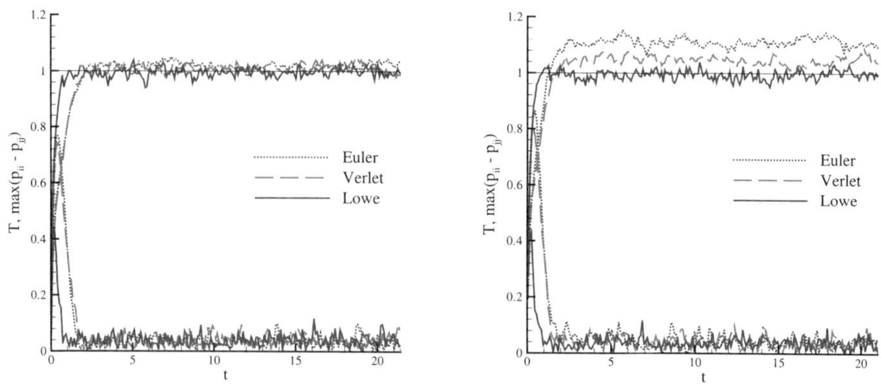

FIGURE 16.11. Time evolution of average kinetic temperature (approaching unity) and normal component of pressure (approaching zero) for fluid in equilibrium. Left: $\Delta t = 0.01$; Right: $\Delta t = 0.05$.

where a constant mass for all particles is assumed. When the system reaches equilibrium, the average kinetic temperature reaches unit value and the pressure approaches zero.

In Figure 16.11, we plot both quantities as a function of time for two different time steps. We use three different methods for integration, namely, the Euler, the Verlet, and the Lowe methods. We see that at $\Delta t = 0.01$ all methods converge to the correct value after some initial transients, whereas for $\Delta t = 0.05$ only Lowe's method gives the correct value of temperature. In all methods, however, a stationary state is achieved, which implies that the simulated thermostat is effective in all cases.

In general, the splitting method gives similar results to those of Lowe's method. (Nikunen et al., 2003) recommend the use of the splitting method and Lowe's method, since they are more accurate and more efficient than several other methods that they tested, including the DPD-VV scheme. Their tests include problems with zero conservative forces as well as polymer flows where hard Lennard–Jones potentials combined with proper springs are employed to represent the polymer; the solvent is represented by soft potentials. Lowe's method, in particular, which does not deal with the computations of dissipative or random forces, seems to be the most efficient. It is based on normally distributed random numbers, while other integration schemes employ random numbers, from a uniform distribution. Another important consideration is that with Lowe's approach, one can achieve realistic values of the Schmidt number Sc $\propto$ 1000 for liquids, whereas with the other approaches the Schmidt number is of order one (Sc $\propto$ 1), which is clearly incorrect, since this corresponds to gases.

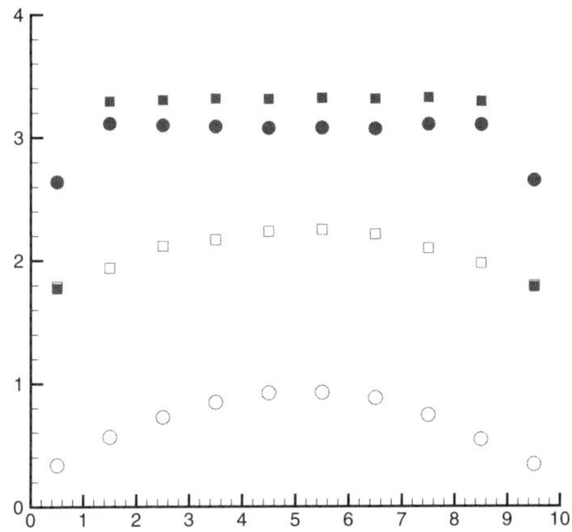

FIGURE 16.12. Comparison of boundary conditions for the two cases described in the text: case 1 (high wall density): squares; case 2 (equal wall density): circles. Plotted are velocity profiles (open symbols) and density profiles (solid symbols) across a channel of height $10r_c$.

16.4.3 Boundary Conditions

The example we presented in the previous section involved only periodic boundaries, and hence there was no need to specify boundary conditions. In a confined geometry, however, the effect of the no-slip or slip boundary condition with the wetted surface has to be modeled carefully. To this end, the boundary conditions that have been used in DPD are based on ideas implemented in both LBM and MD formulations.

There are three main methods to impose boundary conditions in DPD (Revenga et al., 1999):

1. The Lees–Edwards method, also used in LBM, which is a way to avoid directly modeling the physical boundary.

2. Freezing regions of the fluid to create a rigid wall or a rigid body, e.g., in particulate microflows.

3. Combining freezing with proper reflections, namely, specular reflection or bounce-back reflection, or Maxwellian reflection.

Let us first consider how to implement the Lees–Edwards (Lees and Edwards, 1972; Wagner and Pagonabarraga, 2002) boundary conditions for a shear flow (e.g., Couette flow) with the upper wall moving with velocity $-U_x/2$ and the lower wall moving at $U_x/2$. Let us assume that we have a system of particles with positions (r_x, r_y, r_z) and velocities (v_x, v_y, v_z)

within a box of dimensions (L_x, L_y, L_z). We can describe the boundary conditions by providing the new positions r' and velocities v' of the particles after the particles have been moved, as follows:

$$r'_x = \begin{cases} (r_x + d_x) \mod L_x, \, r_y \geq L_y, \\ r_x \mod L_x, \qquad\qquad 0 \leq r_y < L_y, \\ (r_x - d_x) \mod L_x, \, r_y < 0, \end{cases}$$
$$r'_y = r_y \mod L_y,$$
$$r'_z = r_z \mod L_z,$$
$$v'_x = \begin{cases} (v_x + U_x), \, r_y \geq L_y, \\ v_x, \qquad\qquad 0 \leq r_y \leq L_y, \\ (v_x - U_x), \, r_y < 0, \end{cases}$$
$$v'_y = v_y \quad \text{and} \quad v'_z = v_z.$$

Here $d_x = U_x \Delta$ is the time-dependent offeset. Therefore, we avoid providing explicitly any new boundary conditions in this case.

Next, we examine how to incorporate reflections at the wall. We have already described the aforementioned types of reflection earlier in this book, but we briefly review them here as well. In specular reflections the velocity component that is tangential to the wall does not change, while the normal component is reversed. In the bounce-back reflection both components are reversed. A Maxwelian reflection involves particles that are introduced back into the flow with a velocity following a Maxwelian distribution centered around the wall velocity.

In the continuum limit, it is interesting to investigate which one of these boundary conditions honors the no-slip condition. We focus the discussion on the third boundary condition from the above list, since it is the most general approach. In order to parameterize the DPD flow system, following the work of Espanol and his group (Revenga et al., 1999), we identify the key nondimensional parameter that affects the slip velocity. Specifically, there are five governing parameters in the DPD fluid system: m (the mass of particles); γ (the friction coefficient); r_c (the cutoff radius); $k_B T$ (temperature); and $\lambda = N^{-1/d}$ (the average distance between particles, where d is the space dimension). Based on these we can define the dimensionless friction coefficient

$$\tau \equiv \frac{\gamma \lambda}{d v_T},$$

where $v_T = \sqrt{k_B T / m}$ is the thermal velocity. Large values of τ mean that the particles move very little in the time scale associated with the velocity decaying due to thermal fluctuations. Also, we define a measure of overlapping between particles by

$$s \equiv \frac{r_c}{\lambda},$$

which represents the number of particles that interact within the cutoff radius r_c. These two nondimensional parameters define the dynamics of the DPD fluid system. For example, the kinematic viscosity can be computed in terms of these two parameters following a kinetic theory description, i.e.

$$\nu = \frac{1}{2}\lambda v_T \left[\frac{1}{\tau} + \frac{3}{40}s^2\tau\right].$$

Espanol and other researchers have used the plane Couette flow to evaluate the above boundary conditions. The Lees–Edwards boundary conditions work well for this model, as we discussed above, but the objective is to see what type of reflections are appropriate with the freeze-and-reflect approach. In (Revenga et al., 1999), it was shown that for large values of τ all three reflections satisfy the no-slip condition. However, for small values of τ the specular and Maxwellian reflections produce an excessive slip velocity at the wall, while the bounce-back approach satisfies the no-slip condition. An anomaly, however, has been presented for the temperature profile very close to the wall at small values of τ even with the bounce-back boundary conditions. This was explained as large axial velocity dispersion for small values of τ. For $\tau \approx 0$ the particles move along almost straight lines, while the bounce-back condition causes a continuous acceleration of the particles after each collision with the wall. However, a closer inspection of the definition employed for the temperature contributions in (Revenga et al., 1999), shows that this behavior for small values of τ is not really an anomaly!

There are many possibilities for an effective implementation of boundary conditions. Next we compare two different approaches to implementing the no-slip boundary condition for a Poiseuille flow, following the studies of (Pivkin, 2005); see Figure 16.12. Similar procedures have been proposed in (Willemsen et al., 2000), for no-slip conditions. We consider a box with dimensions $10 \times 10 \times 10$ (where $r_c = 1$), $N = 3{,}000$, $\sigma = 3.0$, $\rho = 3$, and $\gamma = 4.5$. The walls are composed of three layers of DPD particles and are four times denser than the fluid (squares) in the first case and equal to the wall density (circles) in the second case. The repulsion force coefficient is $a_{ij} = 25$ for the particles in the flow and also for the wall particles. The results of this comparison are summarized in Figure 16.12 for the velocity profiles and for the corresponding density profiles. In both cases the bounce-back condition is implemented, and the only difference is the density at the wall.

17
Reduced-Order Modeling

In this chapter, we introduce several reduced-order modeling techniques for analyzing microsystems following the discussion of Section 1.7. Specifically, techniques such as generalized Kirchhoffian networks, black box models, and Galerkin methods are described in detail. In generalized Kirchhoffian networks, a complex microsystem is decomposed into lumped elements that interact with each other as constituent parts of a Kirchhoffian network. Techniques such as equivalent circuit representations and description-language-based approaches are described under generalized Kirchhoffian networks. In black box models, detailed results from simulations are used to construct simplified and more abstract models. Methods such as nonlinear static models and linear and nonlinear dynamic models are described under the framework of black box models. Finally, Galerkin methods, where the basic idea is to create a set of coupled ordinary differential equations, are described. Both linear and nonlinear Galerkin methods are discussed. The advantages and limitations of the various techniques are highlighted.

17.1 Classification

Several techniques have been developed for reduced-order modeling or macromodeling of microsystems. Each technique has its own advantages and disadvantages, and the selection of a technique for a particular problem depends on a number of parameters such as the desired accuracy and nonlinearity. Many of the macromodels are created directly from physical-level simulations and often require human input at some stage of the process; i.e.,

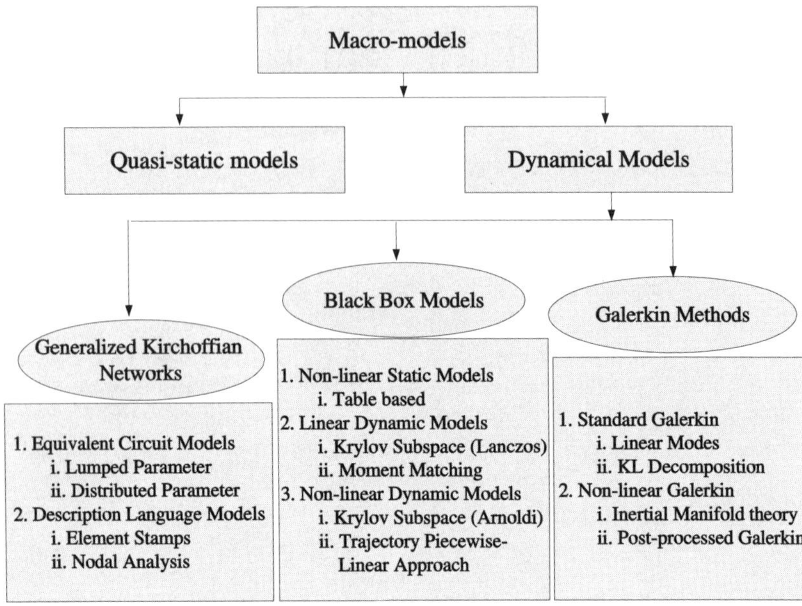

FIGURE 17.1. Classification of macro-models used in microsystem design.

there exists no systematic procedure to extract them automatically from the physical simulations. To identify macromodel extraction steps that can be automated in these cases is an important research topic in the field of microsystem simulation. In this section, we introduce the different macromodels and classify them into several broad categories. Figure 17.1 shows the classification of the various types of reduced-order models.

17.1.1 Quasi-Static Reduced-Order Modeling

Quasi-static macromodels are particularly useful for conservative systems with no dissipative terms. The distinction between energy domains in which the energy is strictly conserved (such as ideal elasticity, electromagnetic fields in linear lossless media, and inviscid flows) and those that have intrinsic dissipation mechanisms (fluidic viscosity, friction, heat flow, viscoelasticity, and internal loss mechanisms such as domain–wall motion that can lead to hysteresis (Senturia, 1998a)) is important, since dynamical behavior in a conservative domain can be derived from quasi-static behavior. All forces can be expressed as gradients of suitable potential-energy functions. If only conservative mechanisms are involved, one can use quasi-static simulations together with the mass distribution to fully characterize the dynamical behavior. Quasi-static macromodels are appropriate for cases in which a steady-state behavior is a reasonable assumption. In many cases, such as the squeeze film damping in a moving MEMS structure (see Chapter 18), such an assumption may be incorrect, in which case a dynamical macromodel is needed.

This procedure is fairly accurate only for conservative energy domains. For a more accurate analysis, the inertia and damping terms must be considered. Forces are expressed as appropriate gradients of suitably constructed potential energy or coenergy functions, and these functions are calculated quasi-statically. If one has knowledge of mass distribution, one can assess accelerations and kinetic energy in response to these forces and hence can construct complete dynamic models of the device using only quasi-static simulations in the potential energy domain.

The steps followed in the quasi-static reduced-order modeling are as follows (Senturia et al., 1997):

1. Select an idealized structure that is close to the desired model.
2. Model the idealized problem analytically, either by solving the governing differential equation, or by approximating the solution with Rayleigh–Ritz energy minimization methods.
3. Identify a set of nondimensionalized numerical constants that can be varied within the analytical form of the solution.
4. Perform meshed numerical simulations of the desired structure over the design space of interest, and adjust the nondimensionalized numerical quantities in the macromodel for agreement with the numerical simulations.

The method has some advantages: (i) simple to use and easy to implement; (ii) reasonably accurate for conservative energy systems when mass distribution is known; (iii) can be used to determine material constants; (iv) even if analytical functions exist for nonlinear behavior, in most cases nonlinearities can be taken care of by a simple fit function. However, the major disadvantage of this method is that it cannot be used in a nonconservative energy system, i.e., when dissipation is involved.

17.1.2 Dynamical Reduced-Order Modeling

Explicit dynamical formulation of microsystems can be very time-consuming, and computationally expensive to insert in a system-level simulator. As a result, it is difficult for the designer to use it in an iterative design cycle or to probe sensitivities to variations in the geometry and material constants by repeated simulations. This demands the development of dynamical reduced-order models for projecting the results of the fully meshed analysis onto physically meaningful reduced variables, containing algebraic dependencies on structural dimensions and material constants. Dynamical reduced-order modeling is much more challenging than the quasi-static reduced-order modeling, since the design space involves large motions and nonlinear forces. The various reduced-order modeling methods that fall under dynamical methods are shown in Figure 17.1. These methods can be broadly classified into three categories: (1) generalized Kirchhoffian networks, (2) black box models, and (3) Galerkin methods. In the following sections, we will look into the different methods that fall under these three categories in detail.

17.2 Generalized Kirchhoffian Networks

In this method, a complex microsystem is decomposed into components (or lumped elements) that interact with each other as constituent parts of a Kirchhoffian network (Voigt and Wachutka, 1997). Compact models with very few degrees of freedom are formulated for each of the components. All the system components are given a mathematical description in terms of conjugate thermodynamic state variables and the pertinent currents (fluxes or through quantities) and the driving forces (affinities or across quantities) such as mass flow and pressure gradient, electrical current, and voltage drop. A system component is called a "block" and is characterized by the number and nature of its terminals, which allow for the exchange of flux quantities across subsystem boundaries. The components can be decomposed further by either (a) interconnection of basic components (structural modeling) or (b) description by a set of algebraic equations (behavioral modeling). The models of reusable components are stored in a block library.

Behavioral description of devices and subsystems when supported by hardware description languages (HDL) leads to the approach of macromodeling based on HDLs. Some examples of HDLs are HDL-A[1] (Mentor Graphics, 1995), MAST[2] (Mantooth and Vlach, 1992), and SpectreHDL[3] (Cadence, Ltd., 1997). The features of these HDLs are (i) Multidomain description, (ii) clear distinction between interface and algorithmic kernel of the model, (iii) interface to embedded C programs, and (iv) mechanisms for handling nonlinear ODEs to be solved by the internal algorithm.

[1]Hardware description language for analog and mixed signal applications: Mentor Graphics.

[2]Analog hardware description language used in the SABER simulator from Synopsys Inc.

[3]Analog Hardware Description Language: Cadence Ltd.

17.2.1 Equivalent Circuit Representation

In the equivalent circuits approach, the microsystem comprising the mechanical, electrical, and fluidic components is represented by their electrical equivalents (Tilmans, 1996). The approach is based on the mathematical analogy between electrical and mechanical systems; specifically, the formal similarities between the integrodifferential equations governing the behavior of electrical and mechanical systems is the basis of the analogy. Newton's second law of motion relates the force F acting on a body of mass m by the relation

$$F = m\frac{\partial u}{\partial t} = m\frac{\partial^2 x}{\partial t^2},$$

where u and x are the velocity and displacement of the mass. This is analogous to the constitutive equation of an electrical inductor

$$V = L\frac{\partial i}{\partial t} = L\frac{\partial^2 q}{\partial t^2},$$

where V is voltage, i is current, q is charge, and L is the inductance in the electrical circuit. Here F plays the same role as the voltage V, the velocity u as the current i, and the displacement u as the charge q. The mass m in mechanical systems represents the inductance L in an electrical circuit. However, while force is a "through variable" in mechanical systems, voltage (its representative in the electrical system) is an "across variable" in electrical systems. In general, it can be seen that a through variable in the mechanical system is represented as an across variable in an electrical system and vice versa. Hence, in the equivalent circuit representation, series connections in mechanical systems are represented by parallel connections in electrical systems and vice versa.

Once all the mechanical systems are converted to electrical counterparts, a single representation of a system operating in more than one energy domain is finally obtained. Kirchhoffian conservation laws are then applied to solve the system. Commercial packages like SPICE[4] (Quarles et al., 1987) can be used for such purposes. This method gives a better understanding and visualization of the system and facilitates further analysis of the system in order to investigate the effects of the connecting subsystems or modifications to the system.

Lumped Parameter System

The lumped parameter approach is a reduced-order modeling method based on the equivalent circuit representation of microsystems. The basic idea in this method is to concentrate or lump the physical properties of thc system, such as mass, stiffness, capacitance, and inductance, into single physical elements (Tilmans, 1996). The elements representing the mass are assumed

[4]Integrated Circuit Simulator: UC Berkeley

TABLE 17.1. Direct electromechanical analogies for lumped translational systems.

Mechanical Quantity	Electrical Quantity
Force: F	Voltage: V
Velocity: u	Current: i
Displacement: x	Charge: q
Mass: m	Inductance: L
Compliance: $1/k$	Capacitance: C
Viscous resistance: c	Resistance: R

to be perfectly rigid, and conversely, elastic elements have no mass. Lumped parameter modeling is typically valid as long as the wavelength of the signal is greater than all the dimensions of the device or the system under consideration. The device is finally represented as a set of lumped electrical network elements using the electrical-mechanical analogies. Once the circuit representation is constructed, commercially available circuit simulators such as SPICE can be used. Alternatively, dynamical state equations can be obtained from the network and simulated using standard math packages like MATLAB. There are two major issues in creating lumped-element macromodels: the first issue is to partition the continuum device into a network of lumped elements, and the second issue is to determine the parameter values for each element.

The partitioning problem can be particularly troublesome, because unlike purely electric circuits, general mechanical structures do not offer a clean mapping between geometry and the corresponding network topology. For example, when considering the electrostatic pull-in of the beam, the moving beam simultaneously serves as the moving boundary of a capacitor used for actuation and sets a moving boundary condition at the bounding surface of a gas film. The lumped parameter values for the physical elements are typically determined from a combination of analysis, numerical simulation, and constitutive properties obtained from test structures. An alternative approach to determine the lumped parameters is to use energy methods. In energy methods, a reasonable shape function with one or more undetermined parameters is assumed, the total stored energy is calculated with that shape function, and the stored energy is then minimized with respect to the parameters. The accuracy of energy methods depends on the quality of the shape function employed and on how well the stored energy

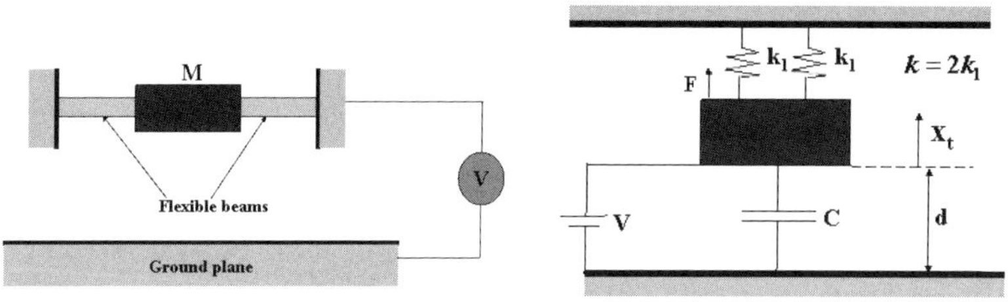

(a) (b)

FIGURE 17.2. (a) A parallel plate electrostatic actuator. (b) Schematic representation of the parallel plate electrostatic actuator.

calculation is implemented. For a first-order analysis of device behavior, accounting for the dependence of device behavior on the geometry and material properties, energy methods are unmatched in terms of simplicity and speed. A generalized procedure for lumped parameter modeling is as follows:

1. The device is decomposed into a combination of rigid bodies, ideal springs, and ideal variable capacitors. The damping is considered external to the transducer.

2. The governing equations are linearized about an equilibrium signal.

3. The characteristic equations, which relate the effort variables as a function of state variables, and the transfer matrix, which relates the effort-flow variables at the electrical port directly to those at the mechanical port, are derived.

4. The transfer matrix is used to obtain the equivalent circuit representation. Typically, the equivalent circuit representation may not be unique. So a practical situation is chosen.

We now illustrate the equivalent circuit representation of a parallel plate electrostatic actuator (in the absence of air-damping) using the above procedure. Figure 17.2(a) shows the parallel plate electrostatic actuator consisting of a rigid mass suspended by two flexible beams. A potential difference is applied between the ground plane and the mass, giving rise to attractive electrostatic forces. The mass moves down, and the beams bend due to these forces. The schematic representation of the device is given in Figure 17.2(b). Since the mass is rigid, the suspension structure can be modeled as a mass–spring system. The two identical beams represent the two springs. The spring constant ($k = 2k_1$) can be derived from the beam flexure formula. We use the energy method as described earlier to generate

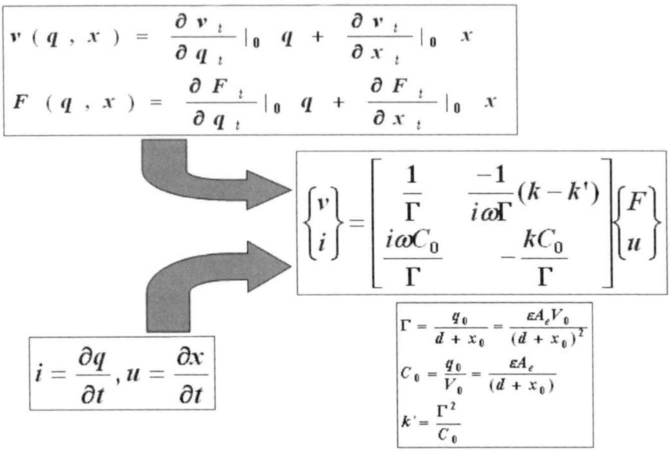

FIGURE 17.3. Transfer matrix computation for the parallel plate actuator using lumped parameters.

the lumped parameters. The total energy W of the system, consisting of electrical (W_e) and mechanical (W_m) energies, is given by

$$W = W_e(q_t, x_t) + W_m(q_t, x_t) = \frac{q_t^2}{2C(x_t)} + \frac{1}{2}k(x_t - x_r)^2, \tag{17.1}$$

where q_t and x_t are the charge and displacement at the time instant t, and x_r is the equilibrium position of the mass; C is the capacitance of the system at the time instan, which is a function of x_t and is given as $C(x_t) = \epsilon_0 A/(d+x_t)$; ϵ_0 is the permittivity of vacuum and A denotes area. Taking the total differential of the energy represented by equation (17.1), we obtain

$$\delta W = \frac{\partial W}{\partial q_t}\delta q_t + \frac{\partial W}{\partial x_t}\delta x_t = v_t dq_t + F_t dx_t, \tag{17.2}$$

where v_t is the voltage between the plates and F_t is the mechanical force acting on the movable plate. Using equations (17.1), (17.2) and the expression for capacitance, we obtain

$$v_t(q_t, x_t) = \frac{\partial W}{\partial q_t} = \frac{q_t(d + x_t)}{\epsilon_0 A} \quad \text{and} \quad F_t(q_t, x_t) = \frac{\partial W}{\partial x_t} = \frac{q_t^2}{2\epsilon_0 A} + k(x_t - x_r).$$

Since the equations are nonlinear, they are linearized using the Taylor's series expansion around some bias point (x_0, q_0). The constitutive equation, describing the linear relations between the incremental or small signal effort variables and the state variables, for voltage at the bias point (x_0, q_0) is

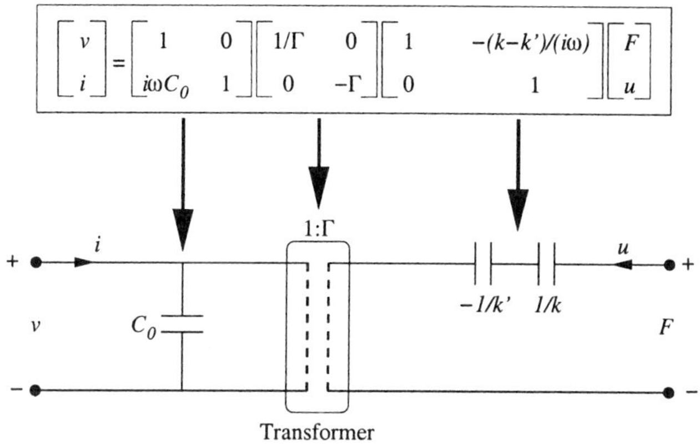

FIGURE 17.4. Decomposition of the transfer matrix into elemental matrices for circuit representation.

given by

$$v(q,x) = \left.\frac{\partial v_t}{\partial q_t}\right|_0 q + \left.\frac{\partial v_t}{\partial x_t}\right|_0 x = \frac{(d+x_0)}{\epsilon_0 A} q + \frac{q_0}{\epsilon_0 A} x = \frac{q}{C_0} + \frac{v_0}{(d+x_0)} x. \quad (17.3)$$

Using the constitutive equation for force and the expression for F_t, we obtain the final expression for the force, i.e.,

$$F(q,x) = \left.\frac{\partial F_t}{\partial q_t}\right|_0 q + \left.\frac{\partial F_t}{\partial x_t}\right|_0 x = \frac{q_0}{\epsilon_0 A} q + kx = \frac{v_0}{(d+x_0)} q + kx. \quad (17.4)$$

The constitutive equations and the final expressions for v and F as given by equations (17.3) and (17.4), respectively, can be used to construct the transfer matrix as shown in Figure 17.3. The transfer matrix can be decomposed into elemental matrices in several ways, giving rise to many feasible circuit representations of the device. Figure 17.4 shows one decomposition of the transfer matrix and the corresponding circuit representation. There are several other circuit representations possible for the same device (see (Tilmans, 1996), for details) including some with pure capacitive circuits. Typically, the designer chooses the most appropriate circuit representation based on the application.

The advantages of the lumped parameter method are as follows: (i) It is easy to use and can be easily incorporated into system simulators. (ii) Equivalent lumped resistors treated external to the system can be used to model dissipation. (iii) The equivalent circuit representation can be used to analyze complex structures and coupled subsystems with several electrical and mechanical ports. The disadvantages of the method are these: (i) Unlike pure electric circuits, mechanical structures do not offer a clean

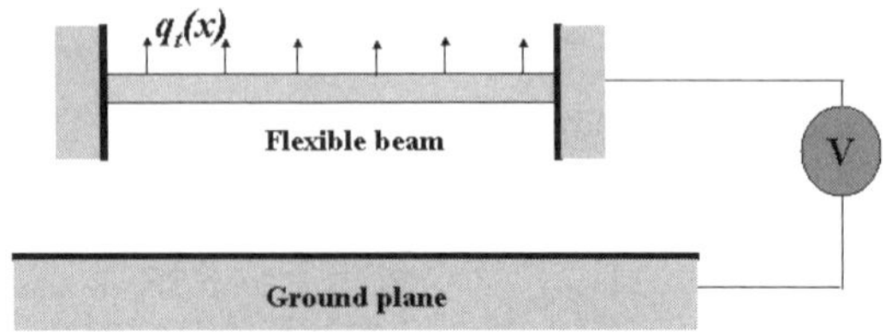

FIGURE 17.5. Continuously distributed system with infinite degrees of freedom.

mapping between the geometry and the corresponding network analogy. (ii) No CAD tools are currently available that can automatically construct an energetically correct lumped-element topology directly for an arbitrary device geometry. (iii) Large-signal and nonlinear analysis is cumbersome, difficult, and error-prone. (iv) In most cases the conservative and dissipative energy domains are to be modeled separately.

Distributed Parameter System

In a distributed parameter system (see (Tilmans, 1997)), the mass, compliance, capacitance, etc., are not easily identifiable as lumped elements at individual points. These elements are, instead, continuously distributed throughout the system. Figure 17.5 shows another parallel plate actuator, but in this case, in contrast to the lumped parameter case, the mass is a continuously deformable beam with a uniformly distributed load on it. The electrical and mechanical domains are coupled either through the boundary of the flexible beam or throughout the entire system as in the case of transducers employing piezoelectric materials. In such cases, it is difficult to distinguish between the mechanical and electrical forces, and the lumped parameter system cannot be easily used to extract the circuit parameters. Instead, the distributed parameter approach needs to be employed. The fundamental difference between lumped parameter models and distributed parameter systems is that while the former method has a finite number of degrees of freedom, the latter has an infinite number of degrees of freedom. The lumped parameter models and the distributed parameter models are just two distinct mathematical models of the same physical system and the distributed parameter approach can be considered as a more general approach compared to the lumped parameter approach (Tilmans, 1997).

The procedure for distributed parameter modeling is given by:

1. A quasi-one-dimensional modeling of the system is first performed, where only the displacements associated with the dominant modes are considered, while the ones in the other directions are neglected. This reduces the dependence of the system behavior to a single coordinate.

2. The characteristic equations are derived using modal analysis, which is described in detail later in this section.

3. Modal analysis techniques are used to find the solution to the governing equations in terms of normalized mode shapes and generalized coordinates by the use of the mode-superposition principle.

4. A Galerkin-like approach is then used to generate an infinite set of uncoupled ODEs that represent the system by means of an infinite number of single-degree-of-freedom lumped-parameter systems.

5. The characteristic equations of the system that describe the linear relations between incremental variations of the port variables around a stable bias point are derived. The equations are then linearized around the bias point.

6. Using the possible characteristic equations and the numerous equivalent circuit representations, a circuit representation that represents the practical situation in the most appropriate way is selected.

The steps in the distributed parameter approach are similar to those in the lumped parameter approach, except that the continuous system is modeled using modal analysis in the distributed parameter approach to reduce the degrees of freedom. We consider the actuator shown in Figure 17.5 to illustrate the distributed parameter approach. The electrical and mechanical domains are coupled through the surface of the flexible beam, which can be thought of as an infinite number of localized individual electrostatic transducers. The electrical energy stored in each such element of infinitesimal area δA is given by

$$U\delta A = \frac{1}{2}\frac{(\sigma_t(x,y)\delta A)^2}{\epsilon_0 \delta A/(d+w_t(x))} = \frac{1}{2}\frac{\sigma_t(x,y)^2}{\epsilon_0/(d+w_t(x))}\delta A, \tag{17.5}$$

where $\sigma_t(x,y)$ denotes the surface electric charge density as a function of position. The total differential of equation (17.5) is given by

$$\delta U = \frac{\partial U}{\partial \sigma_t}\delta\sigma_t + \frac{\partial U}{\partial w_t}\delta w_t.$$

The voltage at time instant t, v_t (see the discussion leading to equation (17.2)), can be expressed as

$$v_t = \frac{\partial U}{\partial \sigma_t} = \frac{\sigma_t(x,y)}{\epsilon_0/(d+w_t(x))}.$$

Similarly, the mechanical pressure acting along the surface p_t can be expressed as

$$p_t = \frac{\partial U}{\partial w_t} = -\frac{\sigma_t(x,y)^2}{2\epsilon_0}.$$

Similar to the derivation of equations (17.3), (17.4), the expressions for $v \equiv \Delta v_t$ and $p \equiv \Delta p_t$ in terms of $w \equiv \Delta w_t$ and $\sigma \equiv \Delta \sigma_t$ are given by

$$v = \frac{d + w_0}{\epsilon_0}\sigma + \frac{v_0}{d + w_0}w \quad \text{and} \quad p = -\frac{v_0}{d + w_0}\sigma + 0 \cdot w.$$

The operating point is indicated by the subscript 0, and the coefficient "0" arises because the stiffness properties of the beam are considered to be external to the transducer (Woodson and Melcher, 1968). Rewriting the above equations, the expressions for σ and p are given by

$$\sigma = \frac{\epsilon_0}{d + w_0}v - \frac{\epsilon_0 v_0}{(d + w_0)^2}w \quad \text{and} \quad p = -\frac{\epsilon_0 v_0}{(d + w_0)^2}v + \frac{\epsilon_0 v_0^2}{(d + w_0)^3}w.$$

In the absence of elastic stiffness, the exterior mechanical pressure p is completely counterbalanced by the electrostatic pressure p_e, and hence

$$p_e = -p = \frac{\epsilon_0 v_0}{(d + w_0)^2}v - \frac{\epsilon_0 v_0^2}{(d + w_0)^3}w.$$

Next, the elastic properties of the beam are taken into account. Employing energy methods (Shames and Dym, 1985) and excluding dynamic terms, the differential equation of motion is

$$L\left[w(x)\right] = -q_e(x) + q(x), \tag{17.6}$$

where $q_e(x)$ and $q(x)$ are transverse forces per unit length of electrical and mechanical origin, and L is the differential operator given by

$$L = EI\frac{\partial^4}{\partial x^4} - N\frac{\partial^2}{\partial x^2}, \tag{17.7}$$

where E and I are the Young's modulus and the second moment of inertia of the beam, respectively, and N is the applied axial force. Rewriting equation (17.6), we have

$$q(x) = \frac{b\epsilon_0 v_0}{(d + w_0(x))^2}v + L_e\left[w(x)\right], \tag{17.8}$$

where

$$L_e \equiv L - \frac{b\epsilon_0 v_0^2}{(d + w_0(x))^3} = L - k_e(x).$$

The eigenvalues and the mode shapes (eigenvectors) of the governing equation (17.8) are now computed by solving a standard eigenvalue problem. Using the mode superposition principle, an infinite number of ordinary differential equations are formed (this topic is discussed in more detail in Section 17.4). Typically, a few mode shapes contain most of the mechanical energy, and these few modes can satisfactorily capture the mechanical deformation, thereby reducing the order of the problem.

In summary, we have a few ODEs to describe the parallel plate actuator, which can now be used to construct the equivalent circuit of the system in the same way as described earlier for the lumped parameter modeling. The advantages of the distributed parameter approach are the following: (i) It can be used to model continuous systems where most other methods fail. (ii) It can be incorporated into system simulators. (iii) Distributed parameter electrical devices can be coupled to the mechanical and the electrical terminal pairs as done in the lumped networks case. In a general case, the system has one electrical port characterized by the voltage v and the current i, and an infinite number of mechanical ports characterized by a generalized load and a generalized velocity. The disadvantages of the method are these: (i) It needs designer input, and test structures are required to verify whether the modeling results are correct. (ii) In most cases the conservative and dissipative energy domains have to be modeled separately. (iii) Since there is no unique representation possible, macromodel generation cannot be automated easily.

17.2.2 Description Languages

Even though the equivalent circuit approach is popular, there are several drawbacks to using an equivalent circuit representation, the most important one being that not all microdevices can be represented by equivalent circuits, and even if an equivalent circuit representation exists, its construction may not be trivial. Besides, the physical meaning of the problem gets complicated due to representation of nonelectrical quantities such as force and velocity in terms of electrical quantities such as current and voltage. As a result, it may not be easy to understand how well the model captures the physics of the device.

Several other methods have also been developed which are based directly on the algebraic-differential equations that describe the device behavior. Suitable languages are used to describe the equations, with hardware description language being one of them. Simulations using description languages are, however, slower. The speed can be increased by using built-in libraries (stamps) for some standard structures or devices. Nodal analysis is another method, where the differential equations are solved directly. These two techniques are summarized below.

Element Stamps

One way of building coupled systems of equations in the electrical and mechanical domain is through the use of element stamps. Element stamps are the building blocks of conventional circuit simulators. They are derived from lumped-constant models of individual microdevices, and built into the circuit simulator (Casinovi, 2002). The use of element stamps for the constitutive elements allows one to simulate a system in a faster and more efficient way compared to models written in hardware description languages. Lumped constant models are sufficiently accurate for many applications. Ordinary circuit simulators use lumped constant models for electronic devices. This approach enables the simulation of complex mixed technology systems starting from their constitutive elements. Like Kirchhoff's laws for electrical components, the equations governing the dynamics of constrained rigid bodies are additive with respect to the number of elements in the system. As a consequence, models of microdevices can be represented by stamps, which contain all the terms that contribute to the global system of equations.

The Modified Nodal Analysis (MNA) technique for circuit simulation is based on Kirchhoff's current law, which states that the sum of all outgoing currents at each node is equal to zero, i.e.,

$$\sum_k i_k = 0.$$

Kirchhoff's current law, when considered at each node, generates a set of algebraic differential equations, which can be transformed into a set of algebraic equations by applying a suitable numerical integration scheme. The nonlinear algebraic equations can be solved by numerical methods (e.g., a Newton's method), which require repeated solution of a set of linear equations of the form

$$\mathbf{A}\mathbf{v} = \mathbf{b}.$$

The observation that each element in the circuit contributes to the above equation leads to the element stamp concept. Figure 17.6 shows a simple resistor (of conductance G) connected between nodes t and o. The branch current, i, given by $i = G(v_t - v_0)$, appears with a positive sign in the current equation at node t and with a negative sign at node o. Hence, the resistor contributes the term $+G(v_t - v_0)$ to the tth equation and the term $-G(v_t - v_0)$ to the oth equation. An element stamp (which is the coefficient matrix $\mathbf{A}$ having rows corresponding to each node, and columns corresponding to each variable) for the resistor is shown in Figure 17.6. If v_t and v_o are the tth and the oth elements of the voltage vector, $\mathbf{v}$, respectively, the resistor's contribution to the coefficient matrix adds a quantity $+G$ to positions (t, t) and (o, o) and a $-G$ to the positions (t, o) and (o, t). The lumped-constant models for MEMS is based on the fact that all the MEMS structures are built from a common set of basic (or atomic) elements, such as beams, anchors, and plates. Though these are less accurate than distributed constant models, they reduce the computational effort by a great deal. For many applications the accuracy obtained from a lumped-constant model could be sufficient. Just as the conventional circuit simulation relies on the lumped-constant models of the electronic devices, microsystem simulation can be achieved by extending the concept of element stamps to microdevices and their constitutive elements. In this

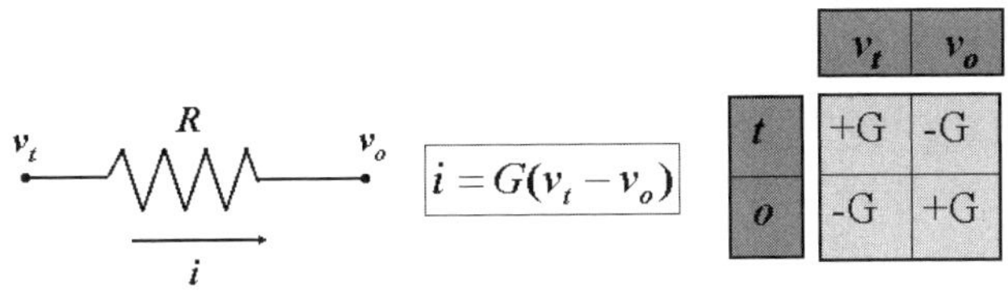

FIGURE 17.6. Element stamp for a resistor.

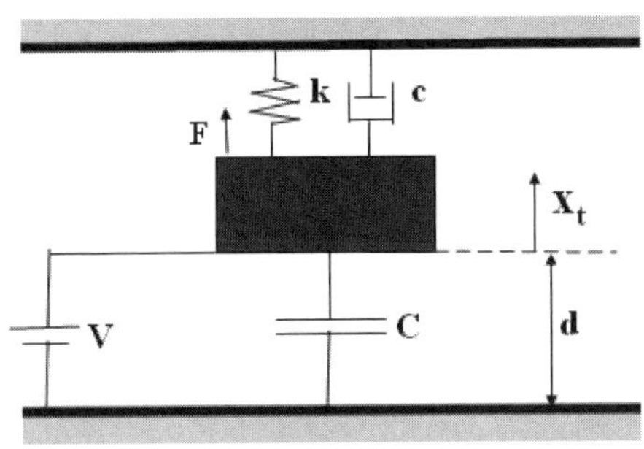

FIGURE 17.7. Schematic diagram of the parallel plate actuator with damping.

case, the Kirchhoff's circuit laws are replaced by the Newtonian equations of motion. The equations governing the dynamics of rigid bodies are given by

$$\sum_k F_k = 0 \quad \text{and} \quad \sum_k T_k = 0,$$

where F and T stand for forces and torques, respectively. The similarity between Kirchhoffian current laws and these equations is apparent. However, one difference is that since microsystems involve both mechanical and electrical domains, there are both mechanical and electrical variables and equations contained in element stamps for microsystems. We illustrate the development of an element stamp for the parallel plate actuator shown in Figure 17.2. External (air) damping is also considered and is represented by the external damper (c) as shown in the schematic diagram in Figure 17.7. The governing equations for the system are given by

$$m\frac{\partial^2 x}{\partial t^2}+c\frac{\partial x}{\partial t}+kx = F_e = -\frac{\epsilon_0 A}{2}\left[\frac{V^2}{(d-x)^2}\right] \quad \text{and} \quad i = \frac{\partial q}{\partial t} = \frac{\partial}{\partial t}\left[\frac{\epsilon_0 AV}{d+x}\right].$$

Trapezoidal integration and linearization of the governing equations gives

Stamp

	v_t	v_b	x	u_x
t	C(x)	-C(x)	-B	0
b	-C(x)	C(x)	B	0
x	0	0	1	-h/2
u_x	-E	E	D	(m+hb/2)

Right Hand Side

$-Q_C$	Q_C	0	0

$$B = \frac{\varepsilon_0 A}{(d+x)^2}(v_t - v_b) \qquad E = \frac{h\varepsilon_0 A}{2(d-x)^2}(v_t - v_b)$$

$$D = \frac{h}{2}\left(k + \frac{\varepsilon_0 A}{(d-x)^3}(v_t - v_b)^2\right)$$

FIGURE 17.8. Element stamp for the parallel plate actuator.

rise to the following simple algebraic equation:

$$mu_x + \frac{h}{2}cu_x + k\frac{h}{2}x + \frac{h}{2}\left[\frac{\partial F_e}{\partial v_t}v_t + \frac{\partial F_e}{\partial v_b}v_b\right] = 0,$$

where $u_x = \partial x/\partial t$, h is the time step of integration, and v_t and v_b are the voltages on the beam and the ground element, respectively. Using the node-wise analysis as described above for the resistor, the element stamp for this MEM actuator is shown in Figure 17.8. The stamp has four rows, two for the electrical equations and two for the mechanical-dynamical equations. The four columns correspond to the four variables, namely, v_t, v_b, x, and u_x. This method takes much less time to simulate in a circuit simulator. It provides a compact and efficient way of adding the contribution of a particular element to the overall system. Since built-in models are used, this procedure can handle only those devices that can be described by the built-in models. This might not be a disadvantag, since all MEMS devices can be described in terms of a set of basic elements. The computational effort required is much less compared to general purpose simulators like SABER[5] (Mantooth and Vlach, 1992) or MATLAB that rely on user-provided HDL models.

Nodal Analysis

Nodal analysis has been widely used for formulating system equations in circuit analysis tools such as SPICE. The circuit is decomposed into N-terminal devices, and each device is modeled by ordinary differential equations (ODEs) with coefficients parameterized by device geometry and material properties (Zhou et al., 1998). The devices are linked together at their terminals or nodes, and the resulting coupled differential equations can be solved as a system of nonlinear ODEs. This approach is fast, reasonably accurate, flexible, and can be used for a higher-level simulation of

[5] Mixed mode circuit simulators from Synopsys Inc.

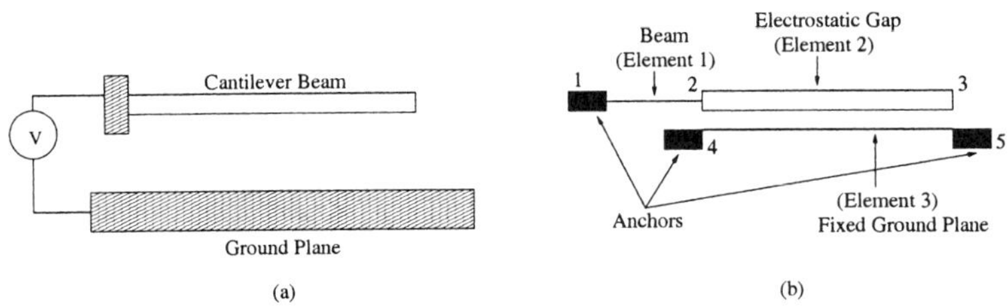

FIGURE 17.9. (a) A cantilever-beam-based microdevice. (b) Nodal representation of the microdevice.

microsystems.

In nodal analysis, the microdevice is represented using atomic elements like anchors, gaps, and beams (Zhou et al., 1998). Figure 17.9 shows a microdevice and its nodal representation. The nodal representation contains three anchor elements, one beam element, and an electrostatic gap element. Each atomic element has a lumped behavioral model with geometric parameters that can be specified individually. This simplifies the evaluation of changes in size on the device performance in each design iteration. The system matrices formed are much smaller than those in finite element analysis, and the models are implemented in analog HDLs supporting mixed physical domain simulations. The total system is formulated by formulating each individual element first. For the beam element defined between nodes 1 and 2, we have

$$f_n^1 = f_n^1(q_1, q_2), \qquad n = 1, 2,$$

and for the gap element (nodes 2, 3, 4, 5) we have

$$f_n^2 = f_n^2(q_2, q_3, q_4, q_5), \qquad n = 2, 3, 4, 5,$$

where f_n, represents the internal forces (the forces in the x-direction, y-direction, and the moment) acting at node n, and q_n represents the node displacements (the displacements in the x- and y-directions and the angle of rotation). The superscript and the subscript denote the element number and the node number, respectively. Each node has three degrees of freedom in 2D: the displacements in the x- and y-directions and the rotation. The sum of all the internal forces (f_n) acting at a given node is equal to the external load P acting at the node, which in this case is the electrostatic force. The assembled equations for each node are given by

$$\begin{aligned}
P_1 &= f_1^1(q_1, q_2),\\
P_2 &= f_2^1(q_1, q_2) + f_2^2(q_2, q_3, q_4, q_5),\\
P_3 &= f_3^2(q_2, q_3, q_4, q_5),\\
P_4 &= f_4^2(q_2, q_3, q_4, q_5),\\
P_5 &= f_5^2(q_2, q_3, q_4, q_5).
\end{aligned}$$

The displacements and the rotations associated with nodes 1, 4, and 5 are zero, so they are removed. The final system of equations is given by

$$\begin{aligned}
P_2 &= f_2^1(q_2) + f_2^2(q_2, q_3),\\
P_3 &= f_3^2(q_2, q_3).
\end{aligned}$$

This system of equations can be solved by standard numerical methods.

NODAS (Fedder and Jing, 1998) is a circuit-level behavioral simulation tool that uses the concept of nodal analysis. Design with NODAS starts from schematic entry, where microsystem elements (such as beams and fluidic channels) and circuit elements (such as transistors) can be wired together. A composite net list for the entire system is generated and sent to the circuit simulator. In the schematic generation phase, terminals of element instances are represented by groups of pins. Each pin has an associated discipline determining its physical nature. Since schematic assembly consumes a lot of effort and is prone to error, if pins were used for each degree of freedom, buses are used in digital circuit schematics for compactness of schematic representation. Similarly, for the same reason analog buses are used in NODAS; however, since existing limitations in analog HDLs (hardware description languages) allow only pins of the same discipline to be grouped as one bus, they result in three buses per terminal: translational, rotational, and electrical. This compact terminal representation reduces wiring effort and wiring errors. Splitters are the behavioral blocks used to convert scalar wires to bus wires. They also apply stimuli and monitor simulation results at the individual degree of freedom. Global acceleration and rotational rate pins are used and shared by all elements in combination with the hierarchical schematic for each model to reduce clutter in the schematic. These pins take care of the external dynamics influence. The "through" and "across" variables are chosen in accordance with Kirchhoff's laws. The across variables are chosen depending on the output required. Since Kirchhoff's network laws are applied in the chip's reference frame, coordinate transformation matrices are used to transform from one coordinate system to another. Some of the basic lumped models used are the linear beam model, nonlinear beam model, and nonlinear gap model. SUGAR (Zhou et al., 1998) uses a similar approach by modeling the MEMS structures in terms of three basic elements (i.e., beams, gaps, and anchors) and builds the ODE models for each kind. The system equations are then formulated according to node connectivity information given as an input file and solved using nodal analysis.

The advantages of the nodal analysis method for microsystem design are as follows: (i) It can solve coupled nonlinear differential equations. (ii) It is fast, reasonably accurate, flexible, and can be used for higher-level simulation of microsystems. (iii) It can perform DC, steady-state, and transient analysis. The disadvantages of the nodal analysis technique are these: (i) The approach can still be expensive for complex systems. (ii) It cannot account for all the nonlinear behavior encountered in microsystems. More details on nodal analysis as well as more examples can be found in (Fedder and Jing, 1998; Vandemeer et al., 1998; Mukherjee and Fedder, 1998; Baidya and Mukherjee, 2002).

17.3 Black Box Models

Black box models stem from basic ideas in system and control theory. Black box models are based on measured input-output behavior, hence the name "black box models." Detailed results from simulations are used to construct simplified and more abstract models. The various models that fall under this category can be broadly classified into:

1. Nonlinear static models: These models use mathematical optimization, approximation, and interpolation methods for curve fitting and parameter adaptation. Table-based numerical reduced-order modeling falls in this category.

2. Linear dynamic models: These are usually formulated in the Laplace domain. The system is simulated in the time or frequency domain. Algorithms from control and system theory are used to calculate the transfer function. The response function is calculated using the convolution integral principles on the impulse function and the actual input function. If the system is complicated, random input functions may be needed to simulate the system instead of step or impulse functions. Krylov subspace techniques and moment matching methods fall under this category.

3. Nonlinear dynamic models: In these models, the modeling is done based on assumptions about the internal structure composed of basic functional blocks. A few control theory approaches are also available. Krylov subspace methods fall under this category.

17.3.1 Nonlinear Static Models

The method that falls in this category is table-based reduced-order modeling.

Table-Based Reduced-Order Modeling

Table-based reduced-order models use tables of numerical data to describe the relationship between variables (Wu and Carley, 2001). The table-based models are built directly from data obtained from measurements or device simulations without detailed knowledge of underlying physics. Hence, the difficulties and errors associated with extracting analytical models are eliminated and the process can be automated. For behavioral-level simulations, cubic spline interpolation is used to evaluate the functions. In addition to the nonlinear behavioral numerical model, a set of linear numerical models is constructed to assist in solving for operating points, to perform AC analysis, and to design closed-loop feedback systems. The procedure for table-based macromodeling can be described as:

1. The device is described by a set of ODEs, which are solved using standard numerical methods.
2. The functions that describe the relationships between the variables are represented in numerical tables, obtained from device simulations.
3. During simulations, the models are evaluated by interpolating the data in the tables.

The microdevice shown in Figure 17.5 is modeled using the steps outlined above. The governing equation for the dynamics of the device is given by

$$m\frac{\partial^2 x}{\partial t^2} + c\frac{\partial x}{\partial t} + f_s = f_e,$$

where m is the effective mass, c is the damping factor, f_s is the spring force, and f_e is the electrostatic actuation force. First, a full simulation of the device is performed to generate the table of data. The mechanical part (or the spring force) is computed using any standard finite element solver, and the electrostatic force is computed using the capacitance solver based on the boundary element method. The functions that represent the relationship between the variables, in this case f_s and f_e, are generated from tables of numerical data obtained from device simulations. During the simulations, the tabl-based models are evaluated by interpolating the data; typically, a cubic spline interpolation is used. As an example, suppose y depends on x, and for some value of x between x_i and x_{i+1}, y can be computed by

$$y = a_1 y_i + a_2 y_{i+1} + a_3\frac{\partial^2 y_i}{\partial x^2} + a_4\frac{\partial^2 y_{i+1}}{\partial x^2},$$

$$a_1 = \frac{x_{i+1} - x}{x_{i+1} - x_i}, \qquad a_2 = \frac{x - x_i}{x_{i+1} - x_i},$$

$$a_3 = \frac{1}{6}(a_1^3 - a_1)(x_{i+1} - x_i)^2, \qquad a_4 = \frac{1}{6}(a_2^3 - a_2)(x_{i+1} - x_i)^2.$$

Since second derivatives are required, typically, a second table comprising second derivatives is also constructed.

The merits of table-based reduced-order modeling are: (i) The procedure eliminates the difficulties and errors associated with analytical model extraction. (ii) The method can be easily automated once the order and state variables of the ODE are decided. The problems with table-based macromodeling are: (i) The method does not preserve the physical meaning of the system. (ii) It is not easy for the designer to propose modifications and expect the system to behave as desired. (iii) At least some expensive full-scale physical simulations are required to generate the table(s) of data.

17.3.2 Linear Dynamic Models

The reduced-order modeling methods that fall in this category are the Krylov subspace technique and the moment matching technique, which are discussed in this section.

Krylov Subspace Technique Based on the Lanczos Method

The governing equation for a continuous time-invariant multi-input multi-output (MIMO) system (e.g., a comb-drive microresonator) is of the form (Srinivasan et al., 2001; Bai, 2002)

$$\mathbf{C}\dot{\mathbf{x}}(t) + \mathbf{G}\mathbf{x}(t) = \mathbf{B}\mathbf{u}(t), \qquad \mathbf{y}(t) = \mathbf{L}^T\mathbf{x}(t), \tag{17.9}$$

where t is the time variable, $\mathbf{x}(t) \in \Re^N$ is a state vector, $\mathbf{u}(t) \in \Re^m$ is the input excitation vector, and $\mathbf{y}(t) \in \Re^p$ is the output vector. Here $\mathbf{C}, \mathbf{G} \in \Re^{N\times N}$ are system matrices, $\mathbf{B} \in \Re^{N\times m}$ and $\mathbf{L} \in \Re^{N\times p}$ are the input and output distribution arrays, respectively, N is the state space dimension, m and p are much smaller than N, and $m \geq p$.

A variety of analyses can be performed for the linear dynamical system given in equation (17.9). For example:

1. A static analysis to compute the equilibrium condition.
2. A steady-state analysis, also called the frequency response analysis, to determine the frequency responses of the system to external steady-state oscillatory (e.g., sinusoidal) excitation.
3. A transient analysis to compute the output behavior $\mathbf{y}(t)$ subject to time varying excitation $\mathbf{u}(\mathbf{t})$.
4. A sensitivity analysis to determine the proportional changes in the time response $\mathbf{y}(t)$ and/or steady-state response to a proportional change in system parameters.

Some of these analyses can be very expensive, especially if performed using equation (17.9). If a reduced-order system to equation (17.9) can be developed, the analysis can be performed quickly. The reduced-order system should have the following desirable attributes:

1. The reduced system should have a much smaller state-space dimension compared to the state-space dimension of the full-order system.

2. The error between the full-order and the reduced-order models should be as small as possible.

3. The reduced-order model should preserve the essential properties of the full-order system.

Therefore, the reduced-order linear system should be of the form

$$\mathbf{C}_n\dot{\mathbf{z}}(t) + \mathbf{G}_n\mathbf{z}(t) = \mathbf{B}_n\mathbf{u}(t), \qquad \mathbf{y}(t) = \mathbf{L}_n^T\mathbf{z}(t), \tag{17.10}$$

where $\mathbf{z}(t) \in \Re^n$ is a state vector, $\mathbf{u}(t) \in \Re^m$ is the input excitation vector, and $\mathbf{y}(t) \in \Re^p$ is the output vector. Here $\mathbf{C}_n, \mathbf{G}_n \in \Re^{n\times n}$ are system matrices, $\mathbf{B}_n \in \Re^{n\times m}$ and $\mathbf{L}_n \in \Re^{n\times p}$ are the input and output distribution arrays, respectively; n is the state space dimension, which should be much smaller than N. Assuming a single-input single-output (SISO) system for simplicity, $p = m = 1$. In this case, we use $\mathbf{b}$ and $\mathbf{l}$ for input and output vectors, respectively. The MIMO system can be dealt with in a similar manner.

The Krylov subspace technique (Srinivasan et al., 2001) reduces the original system (equation (17.9)) to the reduced system (equation (17.10)). Before we discuss the reduction method, it is important to understand the concept of Krylov subspaces. A Krylov subspace is a subspace spanned by a sequence of vectors generated by a given matrix and a vector as follows. Given a matrix $\mathbf{A}$ and a starting vector $\mathbf{r}$, the nth Krylov subspace $K_n(\mathbf{A}, \mathbf{r})$ is spanned by a sequence of n column vectors:

$$K_n(\mathbf{A}, \mathbf{r}) = \text{span}[\mathbf{r}, \mathbf{A}\mathbf{r}, \mathbf{A}^2\mathbf{r}, \mathbf{A}^3\mathbf{r}, \ldots, \mathbf{A}^{n-1}\mathbf{r}].$$

This is called the right Krylov subspace. When $\mathbf{A}$ is asymmetric, there exists a left Krylov subspace generated by $\mathbf{A}^T$ and a starting vector $\mathbf{l}$ defined by

$$K_n(\mathbf{A}^T, \mathbf{l}) = \text{span}[\mathbf{l}, \mathbf{A}^T\mathbf{l}, (\mathbf{A}^T)^2\mathbf{l}, \ldots, (\mathbf{A}^T)^{n-1}\mathbf{l}].$$

Next, we need to define a set of basis functions such that they span the desired Krylov subspaces. Let $\mathbf{V} = [v_1, v_2, v_3, \ldots, v_n]$ and $\mathbf{W} = [w_1, w_2, w_3, \ldots, w_n]$ be basis vectors such that

$$K_n(\mathbf{A}, \mathbf{r}) = \text{span}[v_1, v_2, v_3, \ldots, v_n], \qquad K_n(\mathbf{A}^T, \mathbf{l}) = \text{span}[w_1, w_2, w_3, \ldots, w_n].$$

The Lanczos process is an elegant way of generating such basis vectors. The Lanczos vectors $\mathbf{V}$ and $\mathbf{W}$ are constructed to be biorthogonal, i.e.,

$$w_j^T v_k = 0 \qquad\qquad \forall \quad j \neq k.$$

The Lanczos algorithm to generate $\mathbf{V}$ and $\mathbf{W}$ from $\mathbf{A}$, $\mathbf{r}$, and $\mathbf{l}$ can be found in (Freund, 1999).

The reduced system, using Krylov subspaces, can be generated using the following steps (see also Figure 17.10):

1. Taking s_0 as the expansion point of equation (17.9), $\mathbf{A}$ and $\mathbf{r}$ are defined as

$$\mathbf{A} = -(\mathbf{G} + s_0\mathbf{C})^{-1}\mathbf{C} \quad \text{and} \quad \mathbf{r} = (\mathbf{G} + s_0\mathbf{C})^{-1}\mathbf{b}.$$

2. Using the Lanczos method, $\mathbf{V}$ and $\mathbf{W}$ are computed as

$$\text{span}[\mathbf{V}] = K_n(\mathbf{A}, \mathbf{r}), \qquad \text{span}[\mathbf{W}] = K_n(\mathbf{A}^T, \mathbf{l}).$$

3. The reduced matrices $\mathbf{C_n}, \mathbf{G_n}, \mathbf{b_n}, \mathbf{l_n}$ are computed as follows: For double-sided projection

$$\mathbf{C}_n = \mathbf{V}^T\mathbf{C}\mathbf{W}, \quad \mathbf{G}_n = \mathbf{V}^T\mathbf{G}\mathbf{W}, \quad \mathbf{b}_n = \mathbf{W}^T\mathbf{b}, \quad \mathbf{l}_n = \mathbf{V}^T\mathbf{l},$$

and for single-sided projection,

$$\mathbf{C}_n = \mathbf{V}^T\mathbf{C}\mathbf{V}, \quad \mathbf{G}_n = \mathbf{V}^T\mathbf{G}\mathbf{V}, \quad \mathbf{b}_n = \mathbf{V}^T\mathbf{b}, \quad \mathbf{l}_n = \mathbf{V}^T\mathbf{l}.$$

The double-sided projection formula does not always guarantee a stable reduced-order model except for certain trivial cases (e.g., RC networks), whereas single-sided projection onto $\mathbf{V}$ guarantees an unconditionally stable reduced-order system. However, double-sided projection generally gives more accurate results than single-sided projection. The order of the subspace is chosen according to the frequency range where matching is required. For matching of q resonant peaks, n has to be at least $2q$.

Often, second-order systems are encountered in microsystems (Bai, 2002; Ramaswamy and White, 2001), of the form

$$\mathbf{M}\ddot{\mathbf{q}}(t) + \mathbf{D}\dot{\mathbf{q}}(t) + \mathbf{K}\mathbf{q}(t) = \mathbf{P}\mathbf{u}(t), \qquad \mathbf{y}(t) = \mathbf{Q}^T\mathbf{q} + \mathbf{R}^T\dot{\mathbf{q}}(t),$$

where $\mathbf{Q}$ and $\mathbf{R}$ are chosen depending on the output variable of interest. The second-order system can be formulated into an equivalent linear system of the form given in equation (17.10) such that the symmetry of the original

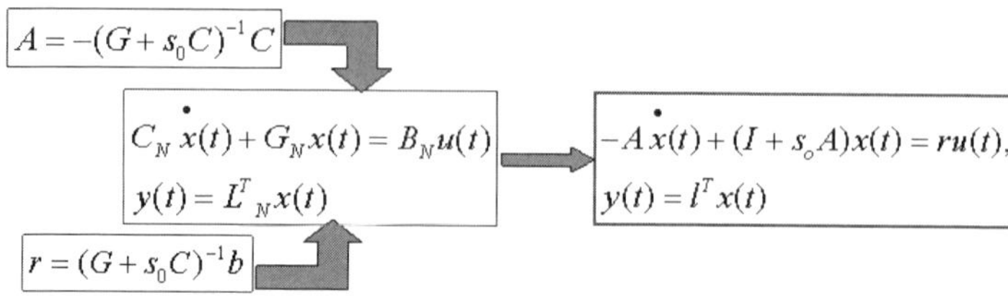

FIGURE 17.10. Basic transformations in the Krylov subspace method for macromodeling.

system is preserved, i.e.,

$$\mathbf{x}(t) = \begin{bmatrix} \mathbf{q}(t) \\ \dot{\mathbf{q}}(t) \end{bmatrix}, \qquad \mathbf{C} = \begin{bmatrix} \mathbf{D}\,\mathbf{M} \\ \mathbf{F}\,0 \end{bmatrix},$$

$$\mathbf{G} = \begin{bmatrix} \mathbf{K}\,0 \\ 0\ -\mathbf{F} \end{bmatrix}, \qquad \mathbf{B} = \begin{bmatrix} \mathbf{p} \\ 0 \end{bmatrix}, \qquad \mathbf{l} = \begin{bmatrix} \mathbf{Q} \\ \mathbf{R} \end{bmatrix},$$

where $\mathbf{F}$ can be any $N \times N$ nonsingular matrix. Generally, $\mathbf{F}$ is chosen to be the identity matrix, $\mathbf{I}$, while $\mathbf{F} = \mathbf{M}$ is also a reasonable choice if $\mathbf{M}$, $\mathbf{D}$, and $\mathbf{K}$ are symmetric. The advantages of the Krylov subspace method are: (i) It is fairly accurate for linear systems and can be automated. (ii) It is computationally very effective. The disadvantages are: (i) It is not very accurate for highly nonlinear systems. (ii) It does not preserve the physical meaning of the original system.

Moment Matching Techniques

The main idea behind the moment matching technique (Ismail, 2002) is to construct the transfer function directly from the system equations using Laplace transformation, and then to approximate the transfer function by some rational function. Consider again equation (17.9). Taking the Laplace transform of this equation, the frequency domain formulation is given by

$$s\mathbf{C}\dot{\mathbf{X}}(s) + \mathbf{G}\mathbf{X}(s) = \mathbf{B}\mathbf{U}(s), \qquad \mathbf{Y}(s) = \mathbf{L}^T\mathbf{X}(s). \tag{17.11}$$

Eliminating the variable $\mathbf{X}(s)$ from equation (17.11), the input and output are related by a $p \times m$ matrix-valued rational function

$$\mathbf{H}(s) = \mathbf{L}^T(\mathbf{G} + s\mathbf{C})^{-1}\mathbf{B},$$

where $\mathbf{H}(s)$ is known as the transfer function of the linear system, and the state-space dimension of the system is N. The Taylor series expansion of the scalar transfer function $H(s)$ about s_0 is given by

$$\begin{aligned} H(s) = \mathbf{l}^T(\mathbf{I}-(s-s_0)\mathbf{A})^{-1}\mathbf{r} &= \mathbf{l}^T\mathbf{r}+(\mathbf{l}^T\mathbf{A}\mathbf{r})(s-s_0)+(\mathbf{l}^T\mathbf{A}^2\mathbf{r})(s-s_0)^2+\cdots \\ &= m_0 + m_1(s-s_0) + m_2(s-s_0)^2 + \cdots , \end{aligned}$$

where m_j are the moments about s_0. The objective is to approximate $H(s)$ by a rational function $H_q(s) \in \Re^{q-1,q}$ over the range of frequencies of interest, where $q \leq N$. One choice is the Padé approximation (Bultheel and Barvel, 1986). A function $H_q(s) \in \Re^{q-1,q}$ is said to be a qth Padé approximant of $H(s)$ about an expansion point s_0 if it matches with moments of $\mathbf{H}(s)$ as far as possible. It is required that (Bai, 2002)

$$H(s) = H_q(s) + O((s-s_0)^{2q}). \tag{17.12}$$

Note that we have $2q$ conditions on the $2q$ degrees of freedom that describe the approximation function. Specifically, let

$$H_q(s) = \frac{P_{q-1}(s)}{Q_q(s)} = \frac{a_0 + a_1 s + a_2 s^2 + \cdots + a_{q-1}s^{q-1}}{1 + b_1 s + b_2 s^2 \cdots + b_q s^q}.$$

The coefficients $[a_i]$ and $[b_i]$ of the polynomials $P_{q-1}(s)$ and $Q_q(s)$, and also the moments can be computed by multiplying both sides of equation (17.12) by $Q_q(s)$ and comparing the first $q(s-s_0)^k$ terms for $k = 0, 1, 2, \ldots, n-1$. A system of $2q$ nonlinear equations is solved to find the $2q$ unknowns. This takes into account the dominant q poles, while the poles larger than this value are neglected.

The advantages of this method are: (i) Reduction in computational effort. (ii) A wide variety of physical phenomena encountered in microsystems, including dissipation, can be modeled. (iii) The accuracy can be improved by taking more moments at each node. (iv) Static, steady-state, and transient analysis can be performed. The disadvantages of this approach are: (i) It is applicable for linear dynamical systems only. (ii) It is inefficient if the number of inputs is large. (iii) It is not stable for higher-order approximations. (iv) It is computationally expensive for each expansion point.

The multinode moment matching (MMM) method (Ismail, 2002) is an extension of the single point moment matching (SMM) method and has much better efficiency than to the SMM technique. The MMM technique simultaneously matches the moments at several nodes of a circuit using explicit moment matching around $s = 0$. MMM requires a smaller computational effort, since only $(q+1)$ moments are required (see (Ismail, 2002) for details). MMM is numerically stable, as the higher powers are not used in the expansion, avoiding truncation errors.

17.3.3 Nonlinear Dynamic Models

Nonlinear dynamic models are frequently encountered in microsystems. Linearizing the nonlinear equations and using reduced-order methods like the linear Krylov subspace method based on a Lanczos process or other linear basis function techniques may not be sufficient to capture the nonlinear behavior of the system. Arnoldi-based Krylov subspace methods (Chen and White, 2000) and the trajectory piecewise linear approach (Rewienski and White, 2001) and its modifications are found to work better for nonlinear systems. These techniques are summarized in this section.

Krylov Subspace Technique Based on the Arnoldi Method

Consider a nonlinear system of the form

$$\dot{\mathbf{x}} = \mathbf{f}(\mathbf{x}) + \mathbf{b}\mathbf{u}(t), \qquad \mathbf{y} = \mathbf{l}^T\mathbf{x}, \tag{17.13}$$

where $\mathbf{x}$ is a vector of length n, $\mathbf{f}$ is a nonlinear vector function, $\mathbf{u}(t)$ is the input of the system, and $\mathbf{y}(t)$ is the output. Taylor series expansion of the function $\mathbf{f}$ about the origin (the equilibrium point) to second order yields a quadratic approximation of the form

$$\dot{\mathbf{x}} = \mathbf{J_f}\mathbf{x} + \mathbf{x}^T\mathbf{W}\mathbf{x} + \mathbf{b}\mathbf{u}(t), \qquad \mathbf{y} = \mathbf{l}^T\mathbf{x}, \tag{17.14}$$

where $\mathbf{J_f}$ is the Jacobian of $\mathbf{f}$ evaluated at the origin and $\mathbf{W}$ is an $N \times N \times N$ Hessian tensor. The matrices $\mathbf{J_f}$ and $\mathbf{W}$ are given by

$$\mathbf{J_f}_{i,j} = \frac{\partial f_i}{\partial x_j} \quad \text{and} \quad \mathbf{W}_{i,j,k} = \frac{\partial^2 f_i}{\partial x_j \partial x_k}.$$

We assume that $\mathbf{J_f}$ is nonsingular. Let $\mathbf{A} = \mathbf{J_f}^{-1}$ be the inverse of the Jacobian. Multiplying equation (17.14) by $\mathbf{A}$ yields

$$\mathbf{A}\dot{\mathbf{x}} = \mathbf{x} + \mathbf{A}\mathbf{x}^T\mathbf{W}\mathbf{x} + \mathbf{A}\mathbf{b}\mathbf{u}(t), \qquad \mathbf{y} = \mathbf{l}^T\mathbf{x}. \tag{17.15}$$

The orthogonal basis for the Krylov subspace span$[\mathbf{A}\mathbf{b}, \mathbf{A}^2\mathbf{b}, \ldots, \mathbf{A}^q b]$, where $q \ll N$, is the size of the reduced system that will be generated by using the Arnoldi method (Chen and White, 2000) for numerical stability. The Arnoldi process generates $\mathbf{V}$, an $n \times q$ orthonormal matrix whose columns span the Krylov subspace. Using the change of variables $\mathbf{x} = \mathbf{V}\mathbf{z}$ in equation (17.15), we have

$$\mathbf{A}\mathbf{V}\dot{\mathbf{z}} = \mathbf{V}\mathbf{z} + \mathbf{A}\mathbf{z}^T\mathbf{V}^T\mathbf{W}\mathbf{V}\mathbf{z} + \mathbf{A}\mathbf{b}\mathbf{u}(t), \qquad \mathbf{y} = \mathbf{l}^T\mathbf{V}\mathbf{z}.$$

Left-multiplying by $\mathbf{V}^T$, and defining $\mathbf{H} = \mathbf{V}^T\mathbf{A}\mathbf{V}$, we have

$$\mathbf{H}\dot{\mathbf{z}} = \mathbf{z} + \mathbf{V}^T\mathbf{A}\mathbf{z}^T\mathbf{V}^T\mathbf{W}\mathbf{V}\mathbf{z} + \mathbf{V}^T\mathbf{A}\mathbf{b}\mathbf{u}(t), \qquad \mathbf{y} = \mathbf{l}^T\mathbf{V}\mathbf{z}.$$

The system can be expressed in the original form (equation (17.13)) by left-multiplying by $\mathbf{H}^{-1}$ to obtain

$$\dot{\mathbf{z}} = \mathbf{H}^{-1}\mathbf{z} + \mathbf{H}^{-1}\mathbf{V}^T\mathbf{A}\mathbf{z}^T\mathbf{V}^T\mathbf{W}\mathbf{V}\mathbf{z} + \mathbf{H}^{-1}\mathbf{V}^T\mathbf{A}\mathbf{b}\mathbf{u}(t), \qquad \mathbf{y} = \mathbf{l}^T\mathbf{V}\mathbf{z}. \tag{17.16}$$

Setting

$$\mathbf{J}' = \mathbf{H}^{-1}, \qquad \mathbf{b}' = \mathbf{H}^{-1}\mathbf{V}^T\mathbf{b}, \qquad \mathbf{l}' = \mathbf{V}^T\mathbf{l},$$

where $\mathbf{H}^{-1}\mathbf{V}^T\mathbf{A}\mathbf{z}^T\mathbf{V}^T\mathbf{W}\mathbf{V}\mathbf{z}$ is quadratic in $\mathbf{z}$ and can be written in the form $\mathbf{z}^T\mathbf{W}'\mathbf{z}$ for some $\mathbf{W}'$. Then, equation (17.16) can be reduced to a quadratic system of the form

$$\dot{\mathbf{z}} = \mathbf{J}'\mathbf{z} + \mathbf{z}^T\mathbf{W}'\mathbf{z} + \mathbf{b}'\mathbf{u}(t), \qquad \mathbf{y}' = \mathbf{l}'^T\mathbf{z},$$

where $\mathbf{y}'$ is an approximation to $\mathbf{y}$. The key step in this approach is the use of Arnoldi projection to reduce the large quadratic tensor to a small quadratic tensor.

The merits of this approach are: (i) This method is much more accurate than the linearized models and can be automated. (ii) It is computationally very effective. The problem with this approach is that it is not very accurate for highly nonlinear systems even though it has a quadratic nonlinear term. If higher-order terms are included, the cost of the reduced order model increases as $O(n^4)$, and the number of coefficients to be evaluated is very large.

Trajectory Piecewise-Linear Approach

The key idea in the trajectory piecewise-linear approach is based on representing the nonlinear system with a piecewise-linear system and then reducing each of these pieces with Krylov subspace projection methods (Rewienski and White, 2001). Instead of approximating the individual components as piecewise-linear and then composing hundreds of components to make a system with exponentially many different linear regions, a small set of linearizations is generated about the state trajectory, which is the response to a "training input." Introducing the change of variables $\mathbf{x} = \mathbf{V}\mathbf{z}$ in equation (17.13) and multiplying the resulting equation by $\mathbf{V}^T$ yields

$$\dot{\mathbf{z}} = \mathbf{V}^T\mathbf{f}(\mathbf{V}\mathbf{z}) + \mathbf{V}^T\mathbf{b}\mathbf{u}(t) \quad \text{and} \quad \mathbf{y} = \mathbf{l}^T\mathbf{V}\mathbf{z}.$$

The two key issues here are first, selecting a reduced basis $\mathbf{V}$ such that the system provides a good approximation of the original system. This has already been addressed in the previous sections. The second issue, which makes the Taylor-series-expansion-based reduced-models inefficient, is the computation of the term $\mathbf{V}^T\mathbf{f}(\mathbf{V}\mathbf{z})$. For linear and quadratic reduced-order models, the linear and the quadratic terms from the Taylor expansion about an equilibrium point $\mathbf{x}_0$ are considered, and all higher-order terms are neglected, i.e.,

$$\mathbf{f}(\mathbf{x}) \approx \mathbf{f}(\mathbf{x}_0) + \mathbf{A}_0(\mathbf{x} - \mathbf{x}_0) + \frac{1}{2}\mathbf{W}_0(\mathbf{x} - \mathbf{x}_0) \otimes (\mathbf{x} - \mathbf{x}_0),$$

where $\otimes$ is the Kronecker product and $\mathbf{A}_0$ and $\mathbf{W}_0$ are the Jacobian and the Hessian of $\mathbf{f}(\cdot)$. For the linear case, the reduced-order model becomes

$$\dot{\mathbf{z}} = \mathbf{V}^T\mathbf{f}(\mathbf{x}_0) + \mathbf{A}_{0r}\mathbf{z} + \mathbf{V}^T\mathbf{b}\mathbf{u}(t) \quad \text{and} \quad \mathbf{y} = \mathbf{l}^T\mathbf{V}\mathbf{z},$$

while for the quadratic case, the reduced order model becomes

$$\dot{\mathbf{z}} = \mathbf{V}^T\mathbf{f}(\mathbf{x}_0) + \mathbf{A}_{0r}\mathbf{z} + \frac{1}{2}\mathbf{W}_{0r}(\mathbf{z}\otimes\mathbf{z}) + \mathbf{V}^T\mathbf{b}\mathbf{u}(t) \quad \text{and} \quad \mathbf{y} = \mathbf{l}^T\mathbf{V}\mathbf{z},$$

where $\mathbf{A}_{0r} = \mathbf{V}^T\mathbf{A}_0\mathbf{V}$ and $\mathbf{W}_{0r} = \mathbf{V}^T\mathbf{W}_0(\mathbf{V}\otimes\mathbf{V})$ are $q\times q$ and $q\times q^2$ matrices, respectively, which are typically dense and must be represented explicitly. As a result, the cost of computing $\mathbf{V}^T\mathbf{f}(\mathbf{V}\mathbf{z})$ and the cost of storing the matrices $\mathbf{A}_{0r}$ ($\mathbf{A}_{0r}$ and $\mathbf{W}_{0r}$ for the quadratic case) are $O(q^2)$ in the linear case and $O(q^3)$ in the quadratic case. Hence, although the method based on Taylor expansion may be extended to higher orders of nonlinearities, this approach is limited in practice to cubic expansions due to exponentially growing memory and computational costs.

In the piecewise-linear approach, s linearized models of the nonlinear system with expansion points $\mathbf{x}_0,\ldots,\mathbf{x}_{s-1}$ are considered, i.e.,

$$\dot{\mathbf{x}} = \mathbf{f}(\mathbf{x}_i) + \mathbf{A}_i(\mathbf{x}-\mathbf{x}_i) + \mathbf{b}\mathbf{u}(t),$$

where $\mathbf{x}_0$ is the initial state of the system and $\mathbf{A}_i$ are the Jacobians of $\mathbf{f}(\cdot)$ evaluated at the states $\mathbf{x}_i$. Considering a weighted combination form,

$$\dot{\mathbf{x}} = \sum_{i=0}^{s-1}\left[w_i(\mathbf{x})\mathbf{f}(\mathbf{x}_i) + w_i(\mathbf{x})\mathbf{A}_i(\mathbf{x}-\mathbf{x}_i)\right] + \mathbf{b}\mathbf{u}(t),$$

where $w_i(\mathbf{x})$ are weights that sum to 1. Assuming that a qth-order basis $\mathbf{V}$ has already been generated, the following reduced-order model is obtained:

$$\dot{\mathbf{z}} = (\mathbf{A}_r\cdot\mathbf{w}'(\mathbf{z})^T)\mathbf{z} + \gamma\cdot\mathbf{w}'(\mathbf{z})^T + \mathbf{b}_r\mathbf{u}(t), \qquad \mathbf{y} = \mathbf{l}_r\mathbf{z},$$

where $\gamma = \left[\mathbf{V}^T(\mathbf{f}(\mathbf{x}_0)-\mathbf{A}_0\mathbf{x}_0),\ldots,\mathbf{V}^T(\mathbf{f}(\mathbf{x}_{s-1})-\mathbf{A}_{s-1}\mathbf{x}_{s-1})\right]$, $\mathbf{b}_r = \mathbf{V}^T\mathbf{b}$, $\mathbf{l}_r = \mathbf{l}^T\mathbf{V}$, $\mathbf{A}_r = \left[\mathbf{A}_{0r}\mathbf{A}_{1r}\ldots\mathbf{A}_{(s-1)r}\right]$, $\mathbf{w}' = [w_0, w_1,\ldots,w_{s-1}]$ is a vector of weights, and $\mathbf{A}_{ir} = \mathbf{V}^T\mathbf{A}_i\mathbf{V}$. The weights are computed in the following manner.

1. For $i = 0,\ldots,(s-1)$ compute $d_i = \|\mathbf{z}-\mathbf{z}_i\|_2$.
2. Compute $m = \min[d_i : i = 1,\ldots,(s-1)]$.
3. For $i = 0,\ldots,(s-1)$ compute $w_i = (\exp(d_i)/m))^{-25}$.
4. Normalize w_i.

This implies that the linearized point that is closest to the current position gets the maximum weight. Instead of finding linearized models covering the entire N-dimensional state space, a collection of models is generated along a single trajectory of the system. This is done by simulating the system at the initial point and moving ahead by a very small interval from that point to get a new point, and the process is repeated for each point. However, this method requires performing simulation of the initial nonlinear system, which may be very costly due to the initial size of the problem. This problem is avoided by using the Arnoldi-based Krylov subspace method, instead of the full simulation, to simulate the nonlinear system and obtain approximate trajectory and linearization points, making the process faster.

Further improvements to the method have been reported in (Rewienski and White, 2002), where a richer aggregated reduced basis is obtained by applying the Arnoldi method at each linearization point instead of only once. This results in improved accuracy and consequently reduces the order of the reduced model further. In the original implementation, the projection matrix $\mathbf{V}$ was constructed using a Krylov subspace based on a linearization about the initial state x_0. In the new implementation, the above approach has been replaced by a three-step procedure. First, at each of the linearization points $\mathbf{x}_i$, a reduced-order basis is generated in a suitable Krylov space corresponding to a linearized model generated at $\mathbf{x}_i$. Second, the union of all the bases is formed, and third, the set is reduced using singular value decomposition. The method for basis generation was replaced from the Arnoldi-based Krylov subspace method to the TBR (Truncated Balanced Realization) algorithm in (Vasilyev et al., 2003), and a hybrid method using both TBR and Krylov subspace has been implemented. It was found that the TBR-based methods gave better accuracy than the method in which the Krylov subspace was used solely.

17.4 Galerkin Methods

Galerkin methods are popular techniques for reduced-order modeling. In this section, we summarize both linear and nonlinear Galerkin methods for reduced-order modeling.

17.4.1 Linear Galerkin Methods

The objective is to create a set of coupled ordinary differential equations that give an accurate representation of the dynamical behavior of the device. The approach is to formulate the dynamical behavior in terms of a finite set of orthonormal spatial basis functions, each with a time-dependent coefficient. Though this method is not typically analytical, it still forms a very important tool in the reduced-order modeling of microsystems that cannot be represented as lumped elements. Because of the relatively small number of state variables, the models can be quickly evaluated, and integrated over time. Such models can be readily inserted into circuit simulators for behavioral representation at the system level, including feedback effects around nonlinear devices. For completely numerical sets of ODEs, automatic model-order reduction can be implemented, at least for linear problems, and for nonlinear problems by using a combination of methods like the Krylov subspace techniques. The two most popular methods that fall under this category are the linear modes of vibration and Karhunen–Loève decomposition.

Linear Modes of Vibration

The basic idea in this method is to represent the physical variable, e.g., the deformed shape of the microdevice, as a summation of the linear normal mode shapes. This results in the transformation from the nodal coordinates to the time-dependent coefficients of the mode shapes, called modal coordinates (Ananthasuresh et al., 1996). This approach also eliminates the coupling between the inertia and stiffness matrices of the governing equations. Assuming that higher modes of vibration have negligible effect on the system's response, a reduced-order model is obtained by using only the first few modes. Instead of the original system of N coupled equations, N being the total number of degrees of freedom, only n equations need to be solved in the reduced-order model, where n is the number of modes considered. The number of modes considered determines both the accuracy and the computation time of the system. A general procedure for this method is given as (Ananthasuresh et al., 1996):

1. Derive basis functions from an initially meshed structure by solving for the small-amplitude (linear) modes of a structure.
2. Form a basis set that is orthonormal in the state space.
3. Consider the first few modes to represent the physical variable(s) of interest (e.g., structural deformation).
4. Represent the solution to the system as a linear combination of the modes with time-dependent coefficients.

The undamped dynamical behavior of a fully meshed structure is given by

$$\mathbf{M}\ddot{\mathbf{x}} + \mathbf{K}\mathbf{x} = \mathbf{F}(\mathbf{x}, t), \tag{17.17}$$

where $\mathbf{M}$ is the global inertial matrix, $\mathbf{K}$ is the global stiffness matrix, and $\mathbf{F}(\mathbf{x}, t)$ is the nonlinear external force. Let $\mathbf{S}$ be the modal matrix, i.e., an $N \times N$ matrix whose columns are the mode shape vectors. The generalized inertia matrix $\mathbf{M_G}$ and the generalized stiffness matrix $\mathbf{K_G}$ are defined as

$$\mathbf{M}_G = \mathbf{S}^T\mathbf{M}\mathbf{S} \quad \text{and} \quad \mathbf{K}_G = \mathbf{S}^T\mathbf{K}\mathbf{S},$$

where both $\mathbf{M}_G$ and $\mathbf{K}_G$ are diagonal. Substituting $\mathbf{x} = \mathbf{Sq}$ in equation (17.17) and premultiplying by $\mathbf{S}^\mathbf{T}$, we get

$$\mathbf{M}_G\ddot{\mathbf{q}} + \mathbf{K}_G\mathbf{q} = \mathbf{S}^T\mathbf{F}(\mathbf{Sq}, t).$$

Since $\mathbf{M}_G$ and $\mathbf{K}_G$ are diagonal matrices, the coupling of the $\mathbf{q}$'s is through the nonlinear force term. In practice, only a few modes are sufficient to describe the deformation. So the N equations are reduced to a smaller number of m equations, where m is the number of mode shapes considered.

If damping properties are to be included, the damping force is added as an additional term on the right-hand side of the equation, or a new set of geometric basis functions is generated by including the space external to the structure where the damping is present. If an additional force term is added, it contains a dependence on velocity. Another approach to account for damping is to assume Rayleigh damping, in which case the linear modes obtained from $\mathbf{M}$ and $\mathbf{K}$ would be sufficient to capture the full behavior. However, in general, to include the effect of damping, one may have to solve the quadratic eigenvalue problem $(\lambda^2\mathbf{M} + \lambda\mathbf{D} + \mathbf{K})\mathbf{s} = 0$ for the desired modal matrix $\mathbf{S}$.

If the structure undergoes large-amplitude deformation, then in an ideal case, the stiffness matrix needs to be recomputed as the amplitude changes. An alternative approach is to retain the original stiffness matrix and add an extra force term to account for the large-amplitude effects, such as stress stiffening of the structure. It is convenient to express the right-hand side in terms of modal coordinates instead of the meshed coordinates. The energy method (Senturia, 1998b) can be used for this purpose, and this procedure is summarized below:

1. Find the linear modes for the elastic problem assuming the no-load condition.

2. Perform quasi-static simulation over a design space that includes the deformations described by a superposition of p modes and develop a suitable potential energy function for other conservative forces (e.g., electrostatic) and large-amplitude elastic effects (e.g., for stress stiffening). Create analytical expressions for the variation of potential energy as a function of the selected mode set (this function is nonlinear and must include products of modal amplitudes, etc.).

3. Replace the right-hand side of the modal dynamic formulation with suitable derivatives of the potential energy functions with respect to modal amplitudes. The net result is a small coupled set of $2p$ (2 state variables per mode) ODEs that can be easily integrated forward in time, without requiring any conversions to and from the original meshed space.

The advantage with modal methods is that they break open the coupled-domain problem. The original modal basis functions are obtained from a single energy domain, e.g., elasticity, together with the associated mass distribution, and the nonlinear potential energy functions can be computed one energy domain at a time. Therefore, it is not necessary to perform complex self-consistent coupled-domain simulations. This approach requires many (single-energy domain) simulations combined with fitting parameters to obtain analytical functions. So it is difficult to do it manually. It is possible to automate the procedure for nonlinear conservative problems. There are limits to the basis-function approach. Thus far, it has been difficult to calculate accurately the stress stiffening of an elastic body undergoing large-amplitude deformation using superposition of modal coordinates. Additionally, when the device undergoes nonlinear motion, such as contact, modal approaches fail. However, the class of microsystems that can now be handled with the automated basis-function approach is large enough to be interesting. More examples using modal basis functions for MEMS simulations are given in (Gabbay and Senturia, 1998; Varghese et al., 1999).

Karhunen–Loève Decomposition Method

The basic idea in the Karhunen-Loève (KL) decomposition method is similar to the basic idea in the linear modes method, i.e., to develop a few global basis functions to represent the entire system by a reduced-order model. In the case of the linear modes method, the linear modes of the system obtained through the solution of the generalized eigenvalue problem form the set of basis functions. Karhunen–Loève decomposition is another method to generate global basis functions, and the advantage with the KL decomposition is that it works better than the linear modes technique for nonlinear cases. The procedure for the extraction of the basis functions in the KL decomposition method is summarized below:

1. Simulate the entire system dynamics first by using a time-stepping scheme that is stable and known to give accurate results.

2. The spatial distributions of each state variable $\mathbf{u}(x, t)$ are sampled at a series of t_n different time instants during the simulations. These sampled distributions are stored as a series of vectors $\mathbf{u}_i$, and each vector represents a "snapshot" in time.

3. The basis functions are determined using either the singular value decomposition (SVD) or the KL approach.

In the SVD approach, which is mathematically equivalent to the KL decomposition technique, n orthogonal basis functions $[\mathbf{a}_1, \ldots, \mathbf{a}_n]$ are determined by minimizing the following expression:

$$\sum_{i=1}^{t_n} \mid \mathbf{u}_i - \text{proj}(\mathbf{u}_i, \text{span}[\mathbf{a_1}, \ldots, \mathbf{a_n}]) \mid^2 . \tag{17.18}$$

This is accomplished by taking the singular value decomposition (SVD) of the matrix $\mathbf{U}$, whose columns are $\mathbf{u}_i$. The SVD of $\mathbf{U}$ is given by

$$\mathbf{U} = \mathbf{V}\boldsymbol{\Sigma}\mathbf{W}^T,$$

where $\mathbf{V}$ contains the eigenvector as columns and $\boldsymbol{\Sigma}$ contains the eigenvalues in the diagonal. By taking $\mathbf{a}_i = \mathbf{v}_i$ for $i = 1, 2, \ldots, n$, the minimization of the expression in equation (17.18) is accomplished.

The KL approach (Sirovich, 1987) is a procedure for extracting an empirical basis for a modal decomposition from an ensemble of signals. Assuming that the signals are an ensemble of the functions $\mathbf{u}_i$, the objective is to find a single deterministic function that is most similar to members of $\mathbf{u}_i$ on average. In other words, one needs to find a function that maximizes the inner product with the field $\mathbf{u}_i$. That is, one needs to maximize

$$\lambda = \frac{\langle(\phi, \mathbf{u}_i)^2\rangle}{(\phi, \phi)},$$

where $\langle\rangle$ is the averaging operator, which can be a time, space, or ensemble average, and $(\phi, \mathbf{u}_i) = \int_\Omega \phi(x)\mathbf{u}_i(x)d\Omega$ is the inner product defined in the function space Ω. It turns out that this condition is met when $\phi(x)$ is an eigenfunction of the two-point tensor given by

$$K\phi = \int_\Omega K(x, x')\phi(x')dx' = \lambda\phi(x),$$

where $K(x, x')$ is a nonnegative Hermitian operator given by

$$K(x, x') = \langle \mathbf{u}_i(x)\mathbf{u}_i(x')\rangle = \frac{1}{t_n}\sum_{i=1}^{t_n} \mathbf{u}_i(x)\mathbf{u}_i^T(x').$$

The above equation can be solved by the direct method or by the method of "snapshots" (Sirovich, 1987). In the method of "snapshots" the eigenfunction can be represented as the summation of snapshots $\mathbf{u}_i$, i.e.,

$$\phi(x) = \sum_k \alpha_k \mathbf{u}_k(x). \tag{17.19}$$

Substituting this into the two-point tensor equation yields a matrix eigenvalue problem that determines the eigenvalues and eigenvectors α. Substituting this set of eigenvectors in equation (17.19) yields a set of eigenfunctions $\phi(x)$, which is the set of global basis functions. It is important to note that the eigenfunction corresponding to the largest eigenvalue corresponds to the most energetic of the snapshots ensemble followed by the eigenfunction corresponding to the next-largest eigenvalue, and so on. After the basis functions are generated by using either the SVD or the KL method, the Galerkin procedure is employed with the basis functions to

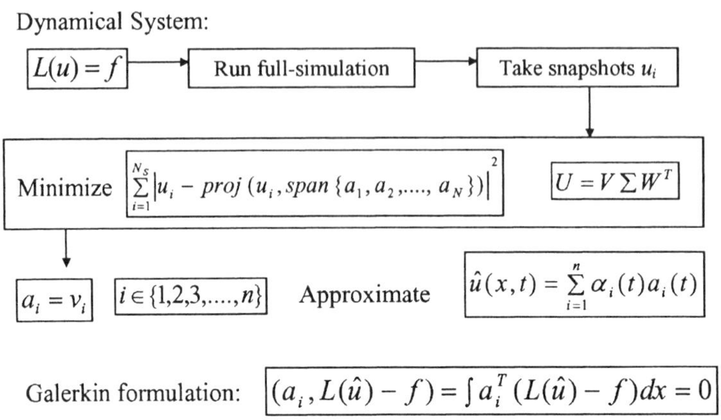

FIGURE 17.11. Basic steps in the Karhunen–Loève decomposition method.

convert the original nonlinear governing PDE to a set of coupled ODEs much smaller in size. Figure 17.11 explains the various steps for implementing the Karhunen–Loève decomposition method.

The advantages of this method are: (i) In general, very few basis functions are required. (ii) Nonlinear behavior is efficiently captured. (iii) The same basis functions can be employed to simulate different input parameters, even though regeneration of the basis functions for the new input parameters can provide more accurate results. The disadvantages of the method are: (i) Using a combined state vector instead of independent basis functions can result in a reduced number of basis functions, but this can distort the physical meaning of the problem. (ii) Problems can still occur, since the basis functions chosen may not capture the entire state space. This is usually the case when there are multiple attractors, rare intermittent fluctuations, or bifurcations in the parameter space. (iii) Complications can result in systems exhibiting intermittent chaotic behavior. (iv) The overhead with the initial full-scale simulation is quite high. Low-dimensional flow dynamical systems may converge to erroneous states after long-time integration and can be observed in reduced-order models constructed from the Karhunen–Loève decomposition method. A dissipative model based on a spectral viscosity diffusion convolution operator has been used in (Sirisup and Karniadakis, 2004), for improving the asymptotic behavior of KL predictions.

In spite of the existing voids, the KL decomposition technique is promising for nonlinear dynamic simulation, especially as the complexity increases. As coupled domain simulation tools improve, this approach can be a powerful tool for microsystem simulation. The standard KL decomposition has been modified under several circumstances to generate more efficient macromodels. Nonlinear responses have been well captured by an arc-length-based KL decomposition presented in (Chen and Kang, 2001b). This method is motivated by the fact that while rapidly varying events can occur in a very short time period, they typically traverse a relatively large interval in the phase space. Hence, considering an ensemble average based not only on time but also on arc length in phase space can lead to better macromodels. This has been shown in (Chen and Kang, 2001b), for a capacitive pressure sensor, where the arc-length-based approach is found to capture the rapidly changing dynamics of the device better than the standard KL approach. Further analysis of this approach shows that the arc-length-based ensemble average is a weighted time average, with the weighting function equal to the magnitude of the vector field, thereby stressing the event of rapid change.

Weighted Karhunen–Loève Decomposition Method

Another modification to the standard KL decomposition is the use of a weighting function (Qiao and Aluru, 2003c; Graham and Kevrekidis, 1996; Zhang et al., 2003). The basic idea is that instead of trying to minimize equation (17.18), we assign different weights to different snapshots and try to minimize the weighted residual, i.e.,

$$\sum_{i=1}^{N_s} |\ w_i\mathbf{u}_i - \text{proj}\left(w_i\mathbf{u}_i, \text{span}\{\boldsymbol{a}_1, \ldots, \boldsymbol{a}_N\}\right)|^2. \tag{17.20}$$

Observe the difference between equations (17.18) and (17.20): w_i is the weighting assigned to snapshot $\mathbf{u}_i$. In the weighted KL approach, instead of minimizing a least-squares measure of "error" between the linear subspace spanned by the basis functions and the observation space, we minimize the weighted "error" between these two spaces.

By using the fact that the SVD of a snapshot ensemble gives the basis that minimizes equation (17.18), it is easy to show that the basis that minimizes equation (17.20) is the column vector of matrix $\boldsymbol{L_2}$:

$$\tilde{\boldsymbol{U}}\boldsymbol{W} = \boldsymbol{L_2}\boldsymbol{\Sigma_2}\boldsymbol{R_2}^T, \tag{17.21}$$

where $\boldsymbol{W}$ is an $N_s \times N_s$ diagonal matrix whose diagonal elements are the weighting coefficients for each snapshot, i.e., $[\boldsymbol{W}]_{i,i} = w_i$.

Remarks:

1. If the weighting function matrix is the identity, i.e., $w_i = 1$ ($i = 1, 2, \ldots, N_s$), then the weighted KL technique and the standard KL technique produce identical bases.

2. In the implementation of the weighted KL technique, once the snapshots are obtained, a weight is assigned to each snapshot.

3. Since the snapshot ensemble matrix is multiplied by a diagonal matrix, the computational cost of a weighted KL decomposition based on equation (17.21) is almost the same as the computational cost of the classical KL decomposition.

Significance of Weighting: The concept of assigning different weights to different snapshots is useful when the transient behavior of certain variables (for example, velocity or pressure) changes significantly with time. For example, in the case of electroosmotic transport, the flow gets to steady state at different times for different locations in the channel (see the discussion in (Qiao and Aluru, 2003c), for more details). If a higher weighting is assigned to those snapshots taken during the fast-changing transient, then the basis obtained with SVD will, according to equation (17.20), be able to produce more accurate results. In other words, the new basis obtained with weighted snapshots will be able to represent the system behavior much better than the basis functions obtained with the classical KL decomposition technique. If the transient behavior of the system is gradual, then the use of the weighting function is limited, and both weighted and classical KL decomposition techniques can be expected to produce comparable accuracy results.

A feasible approach for rapidly varying transient solutions is to obtain more snapshots during the time when the solution is changing rapidly and to compute the basis using the classical KL decomposition technique. However, there are several situations in which obtaining snapshots is not straight-forward. For example, when snapshots are obtained from experiments, repeating the experiment to obtain more snapshots can be very expensive. Similarly, if the snapshots are obtained from numerical simulations and if a rapidly varying transient is represented by a few snapshots, repeating the simulation to get more snapshots with a smaller time step can be very expensive. A good compromise in such cases is to use weighted snapshots to get better basis functions, instead of repeating the experiments or the numerical simulations. Many times it is difficult to foresee the various time scales encountered in the system. The concept of weighting in a KL decomposition technique introduces more flexibility and accuracy to represent multiple time scales in a dynamical system.

Discussion on Weighted Basis versus Error in the Solution: It is important to note that weighting is a concept introduced to improve the accuracy over certain time scales or periods rather than a technique that can be used to improve the accuracy over the *entire* time period. In fact, reduced-order modeling using a weighted basis usually exhibits slightly higher error in the time period that is less significantly weighted. Typically, reduced-order modeling exhibits a very nonuniform error in the whole time domain, i.e., it might behave very well in certain time periods but not be able to capture the basic characteristics in certain other time periods. By using a weighted KL basis, it is possible to achieve a more uniform reduction in error in the solution.

A second question of significant interest is, how does a weighted basis compare with other bases in capturing the system transient? There is no easy answer without a detailed mathematical analysis. However, we do know that (1) increasing the number of basis functions used in approximating, for example, the velocities and pressure generally improves the accuracy of the simulation; (2) different methods generate different bases, and the number of significant basis functions that need to be included in each method is different. The accuracy of the solution is influenced by both the number of basis functions and the quality of the bases. Typically, the number of basis functions that need to be included in a weighted approach is less than the number of basis functions that need to be included in the classical KL approach for comparable accuracy.

Example: Transient Analysis of Electroosmotic Flow

Electroosmotic flow (see Chapter 7 for governing equations and other details) in a cross channel, shown in Figure 17.12, is used as an example to demonstrate the KL and the weighted KL techniques. The cross channel is an interesting problem, since the flow in the intersection exhibits many interesting characteristics. In the case of balanced applied potentials, the net flow into the side channel is negligible but the fluid velocity does not vanish in the side channel. A good reduced-order model should capture both the flow in the main channel and flow within the intersection of the cross channel. The flow in the cross channel exhibits multiple time scales, i.e., the flow within the intersection reaches steady state much more quickly than the flow in the main channel. In addition, the velocity profile within the intersection is more complex than the velocity profile in the main channel. To capture the multiple time scales encountered in the cross channel example, a weighted KL decomposition is used to generate the basis functions for the reduced-order model.

Sixty snapshots are used to generate a reduced-order model. The snapshots are equispaced in time with a time period of 8.85 μs between snapshots. Figure 17.13(a) shows the weighting function employed to generate the weighted KL basis. The weighting coefficient, $w(i)$, for the ith snapshot is calculated by

$$w(i) = r + (1-r)\frac{e^{-(i/c)^2} - e^{-(N_s/c)^2}}{1 - e^{-(N_s/c)^2}}, \tag{17.22}$$

where N_s is the total number of snapshots, r is the minimal weighting for all snapshots, and c is a parameter controlling the steepness of the weighting function. In this calculation, since the first few snapshots contain the information of how V-velocity near the intersection reaches steady state, they are weighted more heavily than the other snapshots. The snapshots closer to the steady-state value are not critical, so they are assigned a lower

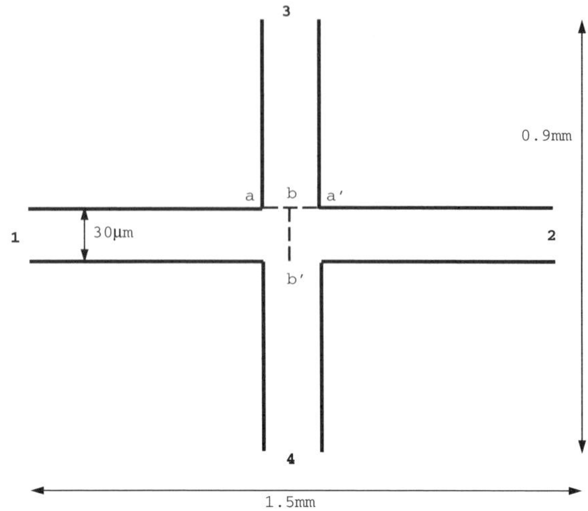

FIGURE 17.12. Geometry of a cross channel used in transient analysis.

weighting. The minimal weighting is set to be 1/6. The steepness parameter is chosen to be 8.5 in this case.

Figure 17.13(b) shows the U-velocity prediction in the main channel, and both methods (standard and weighted KL techniques) produce almost identical results at steady state, though the reduced-order model using weighted KL basis gives slightly better accuracy at the beginning of the simulation. In Figure 17.14(a), we compare the performance of weighted and standard KL techniques by fixing the number of snapshots and investigating the number of basis functions required with each technique to get comparable accuracy. We use 60 snapshots in each approach. For the weighted KL technique we use 4 basis functions for the x-component of the velocity, 4 basis functions for the y-component of the velocity, and 3 basis functions for the pressure (referred to as (4U+4V+3P) in Figure 17.14(a)). To reproduce the results obtained by the weighted KL technique, 6 basis functions had to be used for the x-velocity, 6 basis functions for the y-velocity, and 6 basis functions for the pressure (denoted by (6U+6V+6P)). For comparable accuracy, we need almost *twice* the number of basis functions in a standard KL approach than in the weighted KL technique. This result indicates that when the number of snapshots is fixed, a weighted KL technique needs fewer basis functions than a standard KL technique to reproduce full simulation results. The use of fewer basis functions leads to a computationally more efficient approach. We also observed that a technique with (4U+4V+3P) basis functions is almost twice as fast as the technique with (6U+6V+6P) basis functions; i.e., a reduced-order model based on the weighted KL technique can be twice as fast as the reduced-order model based on the standard KL technique while achieving essentially the same

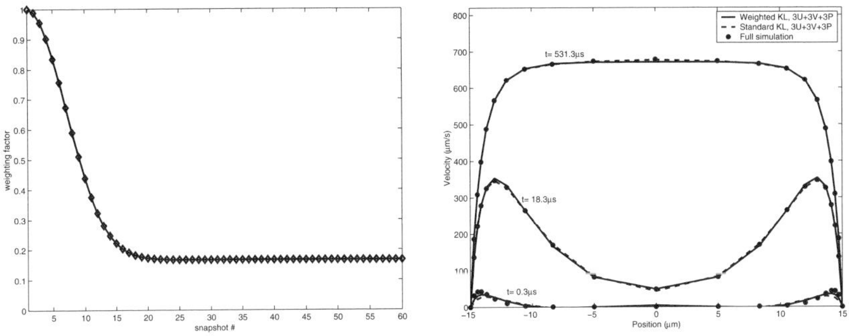

(a) (b)

FIGURE 17.13. (a) A weighting function that assigns different weighting coefficients for different snapshots. (b) Comparison of U-velocity in the main channel far upstream of the intersection. 60 snapshots are used and 3U+3V+3P basis functions are used in both methods.

accuracy.

In Figure 17.14(b), we compare the performance of weighted and standard KL techniques by fixing the number of snapshots and basis functions. The number of snapshots is fixed to 20, and the number of basis functions for U, V, and P is fixed to 3, i.e., (3U+3V+3P). The snapshots are equispaced in time with a time period of 26.6 μs between snapshots. The weighting coefficients for the weighted KL technique are computed using equation (17.22). The minimal weighting r is 1/6, and the steepness parameter is set to be 9. Figure 17.14(b) compares the weighted and standard KL techniques with the full transient simulation. The results indicate that the weighted KL basis is able to capture the velocity profile during the initial transient much more effectively than the standard KL approach. The steady-state solution predicted by both techniques is almost the same and compares well with the full transient simulation. From this we can conclude that with the same number of snapshots and basis functions, the weighted KL approach can offer better accuracy in resolving multiple time scales than the standard KL approach. In Figure 17.15, we compare the performance of weighted and standard KL techniques by fixing the number of basis functions (3U+3V+3P) and using different snapshots with each approach. The weighted KL method uses 22 snapshots to generate the basis functions (the weighting coefficients are again selected by the approach described in the previous paragraph), and the standard KL method uses 66 snapshots to generate basis functions. The result in Figure 17.15 indicates that the weighted KL technique offers better accuracy during the initial transient than the standard KL method, while both methods produce comparable accuracy at steady state. From this result, we can conclude that for a fixed number of basis functions, a weighted KL technique using fewer snapshots

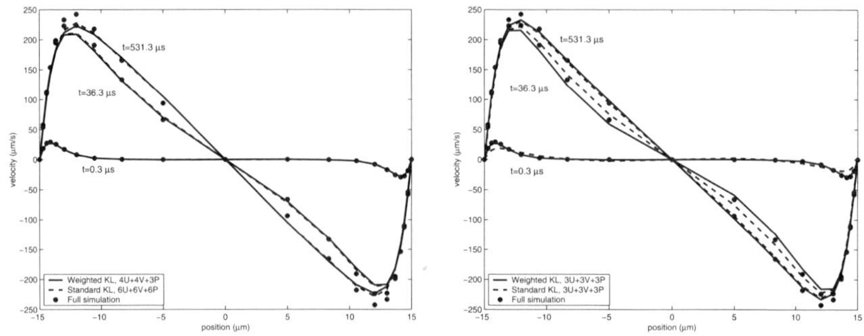

(a) (b)

FIGURE 17.14. (a) Comparison of V-velocity at position $a - a'$ of Figure 17.12. 60 snapshots are used in both methods. (b) Comparison of V-velocity at position $a - a'$ of Figure 17.12. 20 snapshots and 3U+3V+3P basis functions are used in both reduced-order modeling techniques.

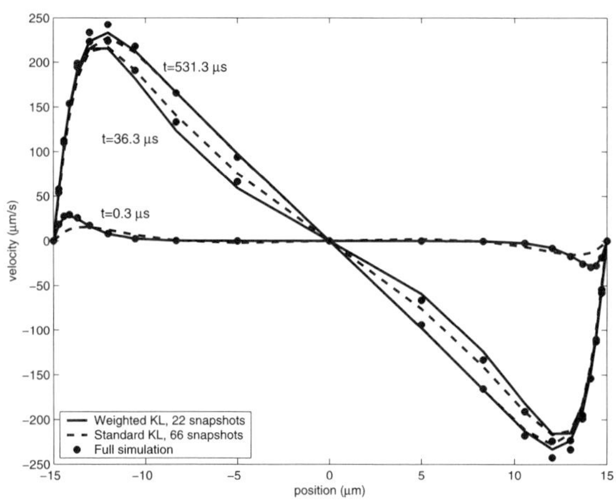

FIGURE 17.15. Comparison of V-velocity at position $a - a'$ of Figure 17.12. 3U+3V+3P basis functions are used in both reduced-order modeling techniques. 22 snapshots are used in the weighted KL method and 66 snapshots are used in the standard KL method to generate basis functions.

can produce better accuracy than a standard KL technique.

17.4.2 Nonlinear Galerkin Methods

A dynamical system can be represented by a differential equation of the form

$$\dot{\mathbf{v}} = \mathbf{G}(\mathbf{v}, \lambda), \tag{17.23}$$

where in the general case $\mathbf{v}$ is an element of a Hilbert space $\mathbf{E}$ and $\mathbf{G}(\mathbf{v}, \lambda)$ is a smooth nonlinear operator. We investigate the loss of stability of an equilibrium $\mathbf{v}_e$ of equation (17.23) under quasi-static variation of a distinguished system parameter λ. Equation (17.23) can be rewritten in the form

$$\dot{\mathbf{u}} = \mathbf{L}(\lambda)\mathbf{u} + \mathbf{g}(\mathbf{u}, \lambda), \tag{17.24}$$

where $\mathbf{L} = \mathbf{G}_v(\mathbf{v}_e)$ is the linearization of the operator $\mathbf{G}$ at $\mathbf{v}_e$, the equilibrium position; $\mathbf{g}$ is a smooth nonlinear operator, and $\mathbf{u} = \mathbf{v} - \mathbf{v}_e$ is the deviation from $\mathbf{v}_e$. From the point of stability we assume that equation (17.24) has an asymptotically stable equilibrium position $\mathbf{u}_e = 0$ for a range of parameter values λ. Now λ is varied quasi-statically, and it is assumed that for $\lambda = \lambda_c$ a loss of stability occurs at $\mathbf{u}_e = 0$. Then, two cases exist for which a proper dimension reduction can be performed ($\epsilon \ll 1$):

1. For $\mid \lambda - \lambda_c \mid = O(\epsilon)$, the center manifold theory applies.
2. For $\mid \lambda - \lambda_c \mid = O(1)$, the Galerkin methods apply.

According to the center manifold theory (Troger and Steindl, 1991), the field variable can be decomposed into a form:

$$\mathbf{u}(\mathbf{x}, t) = \mathbf{u}_c(\mathbf{x}, t) + \mathbf{u}_s(\mathbf{x}, t) = \sum_{i=1}^{n_c} q_i(t)\chi_i(\mathbf{x}) + \mathbf{U}(q_i(t), \mathbf{x}), \tag{17.25}$$

where $\chi_i(\mathbf{x})$ are the active spatial modes, obtained from the solution of the eigenvalue problem related to the linear system

$$\dot{\mathbf{u}} = \mathbf{L}(\lambda_c)\mathbf{u}.$$

Also, $q_i(t)$ are the time-dependent amplitudes, and $\mathbf{u}_s(\mathbf{x}, t)$ can be represented by an infinite sum. The key point is that the influence sum of the higher modes contained in $\mathbf{u}_s(\mathbf{x}, t)$ can be expressed in terms of the lower-order modes by the function $\mathbf{U}(q_i(t), \mathbf{x})$.

In applying the Galerkin methods, the field variable $\mathbf{u}(\mathbf{x}, t)$ is expressed in the form

$$\mathbf{u}(\mathbf{x}, t) = \sum_{j=1}^{m} q_j(t)\phi_j(\mathbf{x})$$

by a set of m comparison vectors $\phi_j(\mathbf{x})$ called the Galerkin basis, which satisfies the geometric and natural boundary conditions. Two subdivisions in the Galerkin methods exist, namely, the standard (linear) Galerkin method and the nonlinear Galerkin method. In the linear Galerkin method one neglects $\mathbf{u}_s(\mathbf{x}, t)$ in equation (17.25). Therefore, the fast dynamics taken into account by the center manifold theory are completely ignored in the reduction process. Nonlinear Galerkin methods take into consideration the influence of higher modes and are also known by the name inertial manifolds in the mathematical literature. Two nonlinear Galerkin methods that have gained importance are (Steindl and Troger, 2001):

1. Approximate inertial manifold theory.

2. Postprocessed Galerkin method.

Application of Karhunen–Loève decomposition in nonlinear Galerkin methods has been presented in (Bangia et al., 1997). The dynamics of incompressible Navier–Stokes flow in a spatially periodic array of cylinders in a channel (for a mixing application) have been investigated using this method.

Approximate Inertial Manifold Theory

Equation (17.24) can be rewritten in the form (Steindl and Troger, 2001)

$$\dot{\mathbf{u}}_c = P\mathbf{L}\mathbf{u}_c + P\mathbf{g}(\mathbf{u}_c + \mathbf{u}_s), \tag{17.26}$$

$$\dot{\mathbf{u}}_s = Q\mathbf{L}\mathbf{u}_s + Q\mathbf{g}(\mathbf{u}_c + \mathbf{u}_s), \tag{17.27}$$

by decomposing $\mathbf{E} = \mathbf{E}_c \oplus \mathbf{E}_s$, where $\mathbf{E}_c$ is finite-dimensional and $\mathbf{E}_s$ is closed. This decomposition is achieved by defining the projection P onto $\mathbf{E}_c$ along $\mathbf{E}_s$, giving $\mathbf{u}_c = P\mathbf{u} \in \mathbf{E}_c$ and $\mathbf{u}_s = Q\mathbf{u} \in \mathbf{E}_s$, where $Q = I - P$ (see (Troger and Steindl, 1991), for details). In the standard Galerkin approximation of equation (17.24), from the eigenfunctions of $\mathbf{L}$, m modes are selected, equation (17.27) is completely ignored, and $\mathbf{u}_s$ is set to be zero in equation (17.26) to obtain

$$\dot{\mathbf{u}}_{ml} = P\mathbf{L}\mathbf{u}_{ml} + P\mathbf{g}(\mathbf{u}_{ml}),$$

where the index l denotes linear approximation. The influence of fast dynamics on the slow (essential) dynamics is completely ignored. Sometimes, a much better approximation is obtained if it is assumed that equation (17.24) has an inertial manifold of dimension m that can be realized as the graph of a function $\mathbf{h} : P\mathbf{E} \to Q\mathbf{E}$, or in other words, $\mathbf{u}_s = \mathbf{h}(\mathbf{u}_{mn})$. The projection of the inertial form onto $P\mathbf{E}$ is then given by

$$\dot{\mathbf{u}}_{mn} = P\mathbf{L}\mathbf{u}_{mn} + P\mathbf{g}(\mathbf{u}_{mn} + \mathbf{h}(\mathbf{u}_{mn})). \tag{17.28}$$

Now the approximation of $\mathbf{u}$ is given by $\mathbf{u}_{\text{approx}} = \mathbf{u}_{mn} + \mathbf{h}(\mathbf{u}_{mn})$, which is analogous to equation (17.25). In the actual process, first, one makes a standard Galerkin approximation using n nodes. Then the m-dimensional approximation of the inertial manifold, $\mathbf{h}$, is calculated (see (Brown et al., 1990), for details) and used in equation (17.28) and in the expression for $\mathbf{u}_{\text{approx}}$. This method can capture nonlinear behavior better than standard Galerkin methods, but it involves extra cost, since the inertial manifold needs to be computed at every integration step.

Postprocessed Galerkin Method

This method is computationally more efficient than the approximate inertial manifold theory (Garcia-Archilla et al., 1998). In this method, first the standard Galerkin method is used, and at time (t) only when some output is required, the variables are approximated by the inertial manifold. That is, the solution $\mathbf{q}_m = \mathbf{q}_{ml}$ is calculated from

$$\dot{\mathbf{q}}_{ml} = P\mathbf{L}\mathbf{q}_{ml} + P\mathbf{g}(\mathbf{q}_{ml}),$$

which requires less effort than computing $\mathbf{q}_m = \mathbf{q}_{mn}$ from

$$\dot{\mathbf{q}}_{mn} = P\mathbf{L}\mathbf{q}_{mn} + P\mathbf{g}(\mathbf{q}_{mn} + \mathbf{h}_{\text{approx}}(\mathbf{q}_{mn})).$$

The final approximate solution for $\mathbf{u}$, $\mathbf{u}_{\text{approx}}$, is computed by $\mathbf{u}_{\text{approx}} = \mathbf{u}_{ml} + \mathbf{h}_{\text{approx}}(\mathbf{u}_{ml})$, where $\mathbf{h}_{\text{approx}}$ is the approximate inertial manifold. The computational cost is reduced greatly as a result of this simplification.

The concept of dynamic postprocessing has been introduced in (Margolin et al., 2003), for highly oscillatory systems. For a variety of systems, the normal postprocessed Galerkin method has been found to be a very efficient technique for improving the accuracy of ordinary Galerkin/nonlinear Galerkin methods with very little extra computational cost. The normal postprocessed Galerkin methods are based on truncation analysis using asymptotic (in time) estimates for the low and high mode components, which hold only when the solutions are on or near an attractor. As a result, these estimates may not hold for nonautonomous systems with highly oscillatory (in time) forcing, long transients, etc. Dynamic postprocessing can handle such situations by integration along transients (see (Margolin et al., 2003), for more details).

18

Reduced-Order Simulation

In chapter 17 we discussed various techniques for reduced-order modeling of microsystems. In this chapter, we discuss the application of these techniques to several examples in microflows. First, we present circuit and device models and their application to lab-on-a-chip systems. Then, we discuss macromodeling of squeeze film damping by applying equivalent circuit, Galerkin, mixed-level, and black box models. Next, we present a compact model for electrowetting. Finally, we summarize some of the software packages that are available for reduced-order simulation.

18.1 Circuit and Device Models for Lab-on-a-Chip Systems

The concept of a micro-total analysis system (μ-TAS) or a lab-on-a-chip for integrated chemical and biochemical analysis has grown considerably in scope since its introduction (Manz et al., 1990; Reyes et al., 2002). μ-TAS involves the miniaturization of all the functions found in chemical analysis, including fluidic transport, mixing, reaction, and separation (Greenwood and Greenway, 2002), so that the entire chemical measurement laboratory could be miniaturized onto a device of a few square centimeters. For example, the system shown in Figure 18.1 incorporates the essential processes (fluidic transport, mixing, reaction, and separation) involved in a μ-TAS. One of the critical elements of any microfluidic system or μ-TAS is its fluidic transport system. For the example considered in Figure 18.1, the fluid

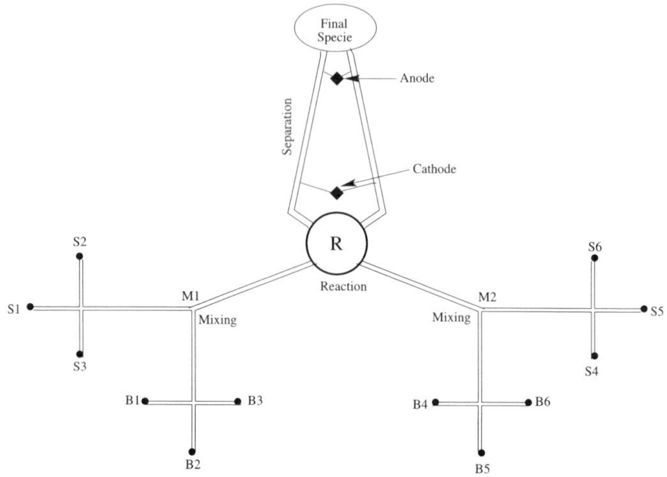

FIGURE 18.1. A prototype chemical analysis system. The system incorporates fluidic transport, mixing, reaction, and separation.

is transported from the ends of the cross-shaped segments to the reservoir marked as the "Final Species." Most microfluidic chips transport the fluid electrokinetically (see Chapter 7) and/or by pressure. Electrokinetic transport and control of fluids has the advantage that it eliminates the need for mechanically moving parts, such as valves and pumps, which have thus far been difficult to construct and interface to microchip systems (Weigl et al., 2003).

An important element of the μ-TAS is the reaction chamber. As shown in Figure 18.1, chemical/biological species are transported to the reaction chambers, where chemical reactions take place leading to the formation of a product. The rate of formation of the product is dependent on the flux of the reactant, the proportion of the various reactants in the solution, the order of the reaction, and the reaction kinetics. The solution from the reaction chamber is sometimes tapped for detection. The detection of the product is typically easier than the detection of the reacting species. In such a case, the presence of the product and the concentration of the product can give quantitative information about the reacting species. Thus, often reaction and detection schemes are intrinsically linked together, and both of these form an integral part of the μ-TAS.

Another important functionality in μ-TAS is the separation of biomolecules and biochemical species. Electrophoresis and isoelectric focusing (see Chapter 7 for details) are the most commonly employed methods of separation. In Figure 18.1, for example, the separation is based on electrophoresis. Higher field intensity is generally tolerable for electrophoretic separation in microchannels (Ehrfeld, 2003). Smaller characteristic dimensions in combination with higher field intensities lead to a shorter time scale of separation, which is a fundamental advantage in μ-TAS compared to macroscopic devices.

When designing integrated microfluidic systems of the type shown in Figure 18.1, some important objectives are to:

1. Increase the throughput.
2. Improve the homogeneity of the mixture.
3. Obtain higher separation efficiency.
4. Perform detection faster.

However, it may not be possible to attain all these objectives, and there can be a trade-off leading to an optimized design. In this section, using the techniques discussed in Chapter 17, "easy-to-use" circuit and device models are presented, which can be used to explore the design space and select an optimal design for integrated microfluidic systems to perform various functions. The model development is illustrated using the example shown in Figure 18.1. The models are, however, general enough that they can be applied or extended to other microfluidic systems. The development of a compact/circuit model for fluid flow due to a combined pressure and electrical potential gradient is first discussed. The compact model is described in two parts, namely, the electrical model and the fluidic model.

18.1.1 Electrical Model

For microfluidic devices that rely on the electrokinetic force as the driving force, the electric field must be computed first. In the case of electroosmotic flow (see Chapter 7 for details; here we restate only the essential equations to derive the circuit models), the potential field due to an applied potential can be computed by solving the Laplace equation:

$$\nabla^2 \phi = 0, \tag{18.1}$$

where ϕ is the electrical potential. Since equation (18.1) predicts a linear potential drop for simple straight channels, the potential variation can be represented by linear electrical resistances. In order to develop a complete circuit that takes into account the charge stored in the electrical double layer (EDL), capacitive elements also need to be included in modeling the electrical domain. The EDL can be decomposed into the stern layer and the diffuse layer. As the stern layer and the diffuse layer store charge, the capacitance associated with these layers is important. In addition, the capacitance of the channel wall, which arises due to a potential difference across the channel wall, needs to be taken into account. The electrical resistance of the EDL can be safely neglected, since the effective resistance of the EDL is much higher than the resistance of the channel filled with buffer (Hayes and Ewing, 2000). Figure 18.2(a) and Figure 18.2(b) illustrate a typical cross-shaped channel segment (this is similar to the cross shapes formed by S1, S2, M1, S3 or B1, B2, B3, M1 or B4, B5, B6, M2 or S4, S5, S6, M2 in Figure 18.1) in a microfluidic system and its circuit representation, respectively.

The electrical resistance of a solution-filled simple straight channel is given by the expression

$$R_{ch,i} = \frac{\rho_{\text{sol},i} L_i}{A_{c,i}},$$

where $\rho_{\text{sol},i}$ is the electrical resistivity of the solution in the ith channel, $i = 1, 2, \ldots 4$ (see Figure 18.2(b)), L_i is the length of the ith channel, $A_{c,i}$ is the cross-sectional area of the ith channel, and $R_{\text{ch},i}$ is the electrical resistance of the ith channel.

The expression for the *effective capacitance*, shown in Figure 18.2(b), is given by

$$(C_{\text{eff},i})^{-1} = (C_{\text{st},i})^{-1} + (C_{\text{dl},i})^{-1} + (C_{\text{wall},i})^{-1},$$

where $C_{\text{st},i}$ is the capacitance of the stern layer of the ith channel, $C_{\text{dl},i}$ is the capacitance of the diffuse layer of the ith channel, and $C_{\text{wall},i}$ is the capacitance of the ith channel wall; $C_{\text{st},i}$ is given by the expression (Oldham and Myland, 1994)

$$C_{\text{st},i} = \frac{\epsilon A_{s,i}}{x_{H,i}},$$

where ϵ is the permittivity of the fluid in the channel, $A_{s,i}$ is the inner surface area of the ith channel, and $x_{H,i}$ is the thickness of the stern layer. The *capacitance of the diffuse layer*, $C_{\text{dl},i}$, is given by the expression (Davies and Rideal, 1966)

$$C_{\text{dl},i} = \frac{\sigma_{T,i} A_{s,i}}{\left\{ \left(\frac{2k_{\text{B}}T}{ze}\right) \sinh^{-1}\left(\frac{\sigma_{T,i}}{c^{\frac{1}{2}}} \left[\frac{500\pi}{\epsilon RT}\right]^{\frac{1}{2}} \right) \right\}},$$

where $\sigma_{T,i}$ is the intrinsic surface charge density on the channel wall, k_{B} is Boltzman's constant, T is the temperature, z is the valence of the counterion, e is the charge of an electron, c is the concentration of the counterion in the bulk solution, and R is the universal gas constant. The *capacitance of the wall* for a cylindrical channel, $C_{\text{wall},i}$, is given by the expression

$$C_{\text{wall},i} = \frac{\epsilon A_{s,i}}{r_i \ln(\frac{r_o}{r_i})},$$

where r_i is the inner radius of the channel and r_o is the outer radius of the channel.

When no potential difference is applied across the channel wall, no charge is induced in the channel wall. As a result, the capacitance of the channel wall can be neglected in the computation of the effective capacitance.

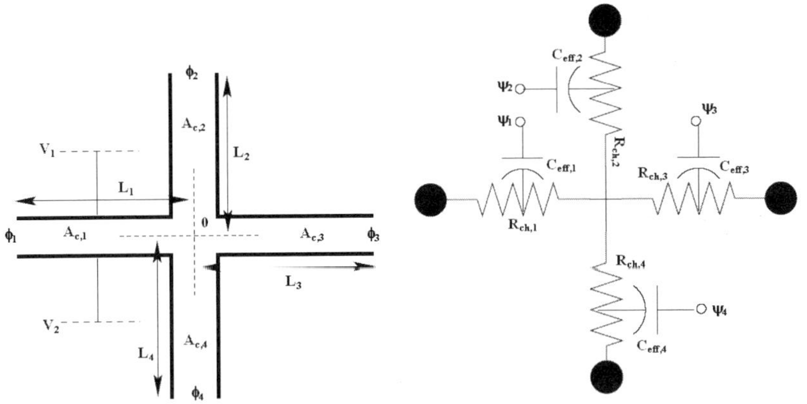

FIGURE 18.2. (a) A typical cross-shaped channel segment of a microfluidic system. The electrical potentials, ϕ_{1-4}, are given. V_1 and V_2 are the transverse applied potentials. (b) The electrical network representation for the cross-shaped channel. $R_{\mathrm{ch},1-4}$ are the electrical resistances, ψ_{1-4} are the surface potentials of the channel walls, and $C_{\mathrm{eff},1-4}$ are the capacitances of the EDLs.

For example, there is no wall capacitance for $i = 2, 3, 4$, since there is no applied voltage across the channel, as shown in Figure 18.2(a). Typically, the capacitance of the stern layer is much higher than the capacitance of the diffuse layer (Oldham and Myland, 1994). Also, when capacitances are connected in series (as in this case), the capacitance with the smaller value dominates. Therefore, in most cases the effective capacitance, C_{eff}, can be approximated by the diffuse layer capacitance, C_{dl}. The effective capacitance can be related to the surface potential by the expression

$$C_{\mathrm{eff},i}\psi_{0,i} = q_{st,i} = \sigma_{T,i}A_{s,i},$$

or

$$\psi_{0,i} = \frac{\sigma_{T,i}A_{s,i}}{C_{\mathrm{eff},i}}, \tag{18.2}$$

where $\psi_{0,i}$ is the surface potential on the ith channel and $q_{\mathrm{st},i}$ is the total charge stored in the EDL of the ith channel.

18.1.2 Fluidic Model

For the fluidic transport driven by an electrical field and/or a pressure gradient, the "through quantities" are the flow rates through the channels, while the "across quantities" are the electrical potential differences and the pressure differences imposed on the fluidic channels. In this section we present a derivation of the constitutive equation relating the "through quantities" to the "across quantities" making use of the continuity equation and the steady-state momentum equation for electroosmotic flows (see Chapter 7 for details).

Slip Case

The slip case model can be used when the thickness of the EDL is insignificant compared to the depth or diameter of the channel. The body force, $\mathbf{F} = \rho_e \mathbf{E}$ (see Chapter 7), is nonzero only within a few Debye lengths from the channel wall, since the potential induced by the zeta potential drops to zero very quickly near the channel wall (Mitchell et al., 2000). In the development of the compact model for the slip flow case, we will assume that the flow is fully developed and the thickness of the EDL is insignificant compared to the thickness or diameter of the channel (this assumption usually holds good for channels larger than 200 nm; see Chapter 7 for details). As a result, the effect of the electrokinetic force can be represented by a slip velocity at the wall given by the Helmholtz–Smoluchowski equation (see section 7.3)

$$u_p = -\frac{\epsilon\zeta}{\mu}\nabla\phi, \tag{18.3}$$

where $\nabla\phi$ is the potential gradient across the fluidic channel and ζ is the zeta potential of the fluidic channel. The Poisson–Boltzmann equation, which is used for the full-scale simulation of electroosmotic flow, can be linearized for low values of surface charge density. Then, the Debye–Hückel theory predicts the following relationship between the zeta potential, ζ, and the surface potential, ψ_0 :

$$\zeta = \psi_0 \exp(-\kappa\chi),$$

where κ is the inverse of the Debye length and χ is the radius of the counterion. The surface potential can be computed from equation (18.2) using the capacitance model. Thus, from knowing the surface potential, the zeta potential of the channel wall can be computed. The velocity profile across a capillary slit is a function of only the slip velocity and the pressure gradient, i.e.,

$$u = -\frac{1}{2\mu}\frac{dp}{dx}\left(y^2 - \frac{h^2}{4}\right) + u_p, \tag{18.4}$$

where x denotes the stream direction of the channel, y denotes the transverse direction of the channel, and h is the channel depth. Since u_p is given by equation (18.3), solving for the velocity in equation (18.4) is reduced to computing the pressure distribution in the fluidic network. By taking divergence of the momentum equation and applying the continuity condition, we get the expression

$$\nabla^2 p = \nabla\cdot\mathbf{F} - \nabla\cdot(\rho_f(\mathbf{u}\cdot\nabla)\mathbf{u}).$$

In the regions where the flow is fully developed, the convection term $(\mathbf{u}\cdot\nabla)\mathbf{u}$ is zero. Thus, $\nabla \cdot (\rho_f(\mathbf{u}\cdot\nabla)\mathbf{u})$ vanishes. The term corresponding to the divergence of the force must be zero in the fully developed flow regions; otherwise, the flow would not be fully developed due to the nonuniform body force. Hence, for the region where the flow is fully developed, the pressure calculation is reduced to a Laplace equation,

$$\nabla^2 p = 0. \tag{18.5}$$

Thus, equation (18.5) decouples the solution of pressure from the solution of velocity.

Integrating the velocity profile given in equation (18.4) across the cross-section of the capillary slit and using equations (18.1), (18.3), and (18.5), we get the following expression for the flowrate per unit width:

$$Q = \left(\frac{h^3}{12\mu L}\right)\Delta p + \left(\frac{\epsilon\zeta h}{\mu L}\right)\Delta\phi. \tag{18.6}$$

For the ith channel in an array of channels, equation (18.6) can be rewritten as:

$$Q_i = H_i \Delta p_i + E_i \Delta\phi_i, \tag{18.7}$$

where H_i is the hydraulic conductance of the ith channel, E_i is the electrohydraulic conductance of the ith channel, Δp_i is the pressure drop in the ith channel, and $\Delta\phi_i$ is the electrical potential drop in the ith channel. The expressions for H_i and E_i for the capillary slit are given in equation (18.6). For a cylindrical channel, the hydraulic conductance and the electro-hydraulic conductance are given by

$$H_i = \frac{\pi {r_i}^4}{8\mu_i L_i} \quad \text{and} \quad E_i = \frac{\epsilon\zeta_i\pi {r_i}^4}{\mu_i L_i},$$

where r_i is the inner radius of the ith cylindrical channel. Equation (18.7) is the constitutive relationship, which relates the "through quantity" to the "across quantities" (a combined pressure and electrical potential drop). If the flow is driven by only a pressure gradient, then the second term in equation (18.6) can be neglected. Similarly, if the flow is driven by only an electric field, then the first term on the right-hand side of equation (18.6) can be neglected. Figure 18.4 shows the circuit representations of the fluidic domain for the cross-shaped channel segment shown in Figure 18.3. It is to be noted that the total flow is the sum of the electrokinetically driven flow and the pressure-driven flow.

No-Slip Case

The slip velocity model discussed above can be employed when the Debye length is thin compared to the channel width. However, when the Debye

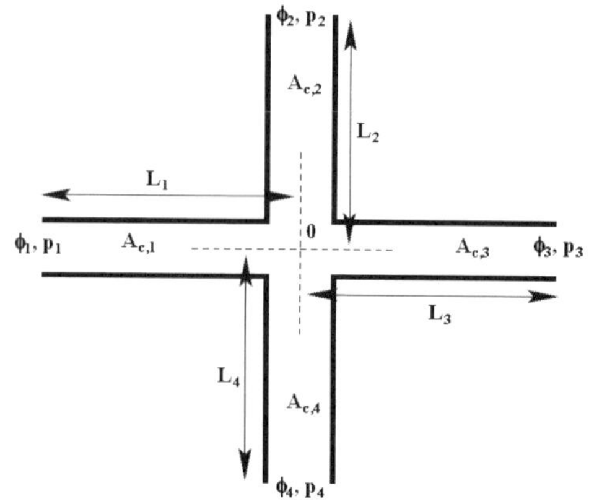

FIGURE 18.3. A cross-shaped channel with a combined pressure and electrical potential gradients. The electrical potentials, ϕ_{1-4}, and pressures, p_{1-4}, are given.

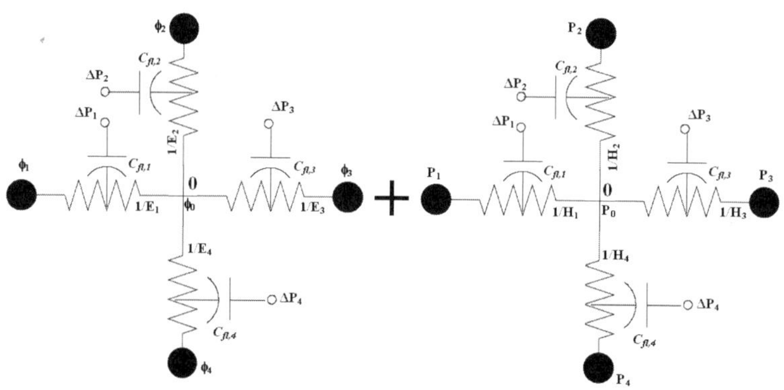

FIGURE 18.4. Circuit representation for the electrokinetically driven flow is on the left. E_{1-4} are the electrohydraulic conductances of the channels and $(C_{\text{fl},1-4})$ are the fluidic capacitances. Circuit representation for the pressure-driven flow is given on the right. H_{1-4} are the hydraulic conductances of the channels. The plus sign between the two figures indicates that the total flow is the sum of the electrokinetically driven flow *and* the pressure-driven flow.

length is comparable to the channel width, the slip velocity model may not be accurate. For a capillary slit, the velocity profile is given by the expression (Patankar and Hu, 1998; Keh and Tseng, 2001)

$$u_{(y)} = -\frac{1}{2\mu}\frac{dp}{dx}\left(y^2 - \frac{h^2}{4}\right) - \frac{\epsilon}{\mu}\nabla\phi\left(\psi_0 - \psi_{(y)}\right), \tag{18.8}$$

where

$$\psi_{(y)} = \frac{\psi_0 \cosh(\frac{y}{\lambda_\mathrm{D}})}{\cosh(\frac{h}{\lambda_\mathrm{D}})}, \tag{18.9}$$

and λ_D is the Debye length; see equation (7.1). Integrating the velocity profile (given in equation (18.8)) across the cross-section and using equation (18.9), we get the following expressions for the hydraulic conductance and the electrohydraulic conductance of the ith channel:

$$H_i = \frac{{h_i}^3}{12\mu_i L_i} \quad \text{and} \quad E_i = \frac{\epsilon}{\mu_i L_i}\psi_0 \left(2h_i - 2\lambda_D \frac{\sinh\left(\frac{h_i}{\lambda_\mathrm{D}}\right)}{\cosh\left(\frac{h_i}{\lambda_\mathrm{D}}\right)} \right).$$

Fluidic Channels with Elastic Membranes

In the case of channels with integrated elastic parts in them (e.g., a flexible membrane) a capacitive element needs to be included in the circuit model of the fluidic domain as shown in Figure 18.4. The fluidic capacitor can be modeled as

$$C_\mathrm{fl} = \frac{\iint_\Gamma w(x,y) d\Gamma}{p}, \tag{18.10}$$

where C_fl is the fluidic capacitance, w is the deflection, Γ is the total surface area of the flexible membrane, and p is the pressure difference across the channel wall. For a rectangular membrane of dimensions $a \times b$, the fluidic capacitance from equation (18.10) is given by

$$C_{fl} = \frac{4a}{\pi^5 D_r} \sum_{m=1,3,5,\ldots}^{\infty} \frac{(-1)^{\frac{m-1}{2}}}{m^5} \frac{\sin(\frac{m\pi}{2})}{\frac{m\pi}{a}} \left\{ \frac{b}{2} + \frac{a}{2m\pi}[\alpha_m - \tanh(\alpha_m)(3 + \alpha_m \tanh(\alpha_m))] \right\},$$

where

$$\alpha_m = \frac{m\pi b}{2a},$$

and D_r is the rigidity of the membrane given by

$$D_r = \frac{E_\mathrm{mod} h_m^3}{12(1-\nu^2)},$$

where h_m is the thickness of the membrane, E_mod is the elastic modulus of the membrane, and ν is the Poisson ratio of the membrane.

The implementation of the electrical model and the fluidic model is carried out using the *modified nodal analysis* technique (Ogrodzki, 1994). Once the variations of ϕ and p are known, the flowrate in each channel can be computed using the constitutive relationship given in equation (18.7).

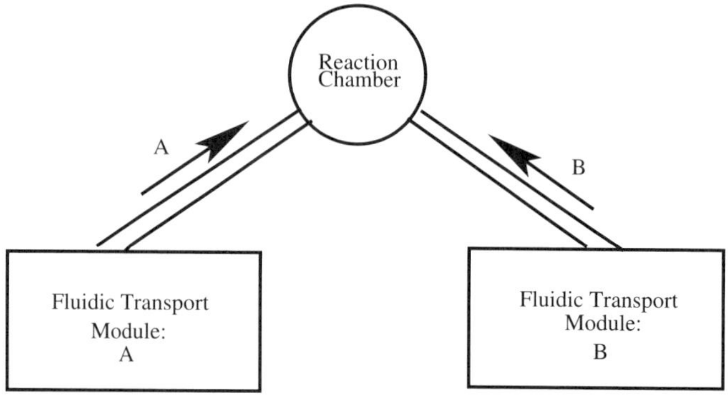

FIGURE 18.5. Chemical species A and B are transported to the reaction chamber, where they undergo a second-order reversible reaction process.

18.1.3 Chemical Reactions: Device Models

Consider a scheme (shown in Figure 18.5) in which the chemical species A and B are transported to the reaction chamber, where they undergo a second-order reversible reaction process to produce species C. The governing equations for this reaction process are given by

$$\begin{aligned} A + B &\underset{k_2}{\overset{k_1}{\rightleftharpoons}} C, \\ \frac{\partial m_A}{\partial t} &= Q_A C_A - k_1(m_A)(m_B) + k_2(m_C), \\ \frac{\partial m_B}{\partial t} &= Q_B C_B - k_1(m_A)(m_B) + k_2(m_C), \\ \frac{\partial m_C}{\partial t} &= k_1(m_A)(m_B) - k_2(m_C), \end{aligned}$$

where Q_i is the flowrate of the ith species, which is computed from the fluidic transport model (or known from the design specifications), C_i is the concentration of the ith species, m_i is the number of moles of the ith species present in the reaction chamber, k_1 is the forward reaction rate, and k_2 is the backward reaction rate. A trapezoidal scheme is used to discretize the ODEs given above. The discretized equations are given by

$$\begin{aligned} \frac{(m_A^{n+1} - m_A^n)}{\Delta t} &= Q_A C_A - \frac{k_1}{4}(m_A^{n+1} + m_A^n)(m_B^{n+1} + m_B^n) + \frac{k_2}{2}(m_C^{n+1} + m_C^n), \\ \frac{(m_B^{n+1} - m_B^n)}{\Delta t} &= Q_B C_B - \frac{k_1}{4}(m_A^{n+1} + m_A^n)(m_B^{n+1} + m_B^n) + \frac{k_2}{2}(m_C^{n+1} + m_C^n), \\ \frac{(m_C^{n+1} - m_C^n)}{\Delta t} &= \frac{k_1}{4}(m_A^{n+1} + m_A^n)(m_B^{n+1} + m_B^n) - \frac{k_2}{2}(m_C^{n+1} + m_C^n). \end{aligned}$$

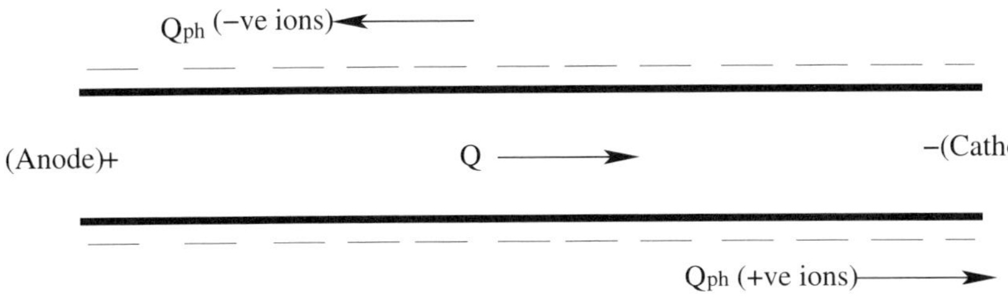

FIGURE 18.6. A basic separation unit, which can separate species that are oppositely charged, have different valences or different electrophoretic mobilities.

The nonlinear equations given above are solved by employing a Newton–Raphson scheme to compute m_A^{n+1}, m_B^{n+1} and m_C^{n+1} at the $(n+1)$th time step given m_A^n, m_B^n, and m_C^n at the nth time step. These equations constitute the device model for the reaction module.

18.1.4 Separation: Device Model

Figure 18.6 shows a simple separation mechanism, which is repeated as the basic unit in the circular separation device reported in (Kutter, 2000). The separation unit can separate species that are either oppositely charged or have different valences or different electrophoretic mobilities. The total flux of a given species through a channel is given by the following expression:

$$J_t = \left[D_{\text{diff}} \frac{\partial c}{\partial x} + \frac{z F_c D_{\text{diff}}}{RT} (\nabla \phi) c + v_{\text{conv}} c \right] A_c, \tag{18.11}$$

where J_t is the total flux, D_{diff} is the diffusion coefficient of the species, c is the concentration of the species, F_c is Faraday's constant, z is the valence of the ion, R is the universal gas constant, T is the temperature, A_c is the cross-sectional area of the fluidic channel, and v_{conv} is the convective velocity of the flow that arises due to the bulk flowrate, Q, given in equation (18.6):

$$v_{\text{conv}} = \frac{Q}{A_c}.$$

From equation (18.11), the total flux is the sum of the diffusive flux (given by the first term), the electrophoretic flux (given by the second term and it is zero for uncharged species), and the convective flux (given by the last term), which arises due to the bulk flow in the channel. Typically, the separation unit is designed in such a way that the convective flux and the electrophoretic flux (for charged species) dominate over the diffusive flux (Fletcher et al., 1999). Thus, assuming that the diffusive flux is negligible, the expression for the total flux is given by

$$J_t = \left[\frac{zF_c D_{\text{diff}}}{RT}(\nabla\phi)\, c + v_{\text{conv}} c\right] A_c$$

or

$$J_t = (Q_{\text{ph}} + Q)c,$$

where Q is the convective flowrate, which is computed using equation (18.6), and Q_{ph} is the electrophoretic flow rate, which is given by the expression (Fletcher et al., 1999)

$$Q_{\text{ph}} = \left(\frac{zF_c D_{\text{diff}}}{RT}\right) A_c \nabla\phi.$$

Thus, the constitutive equation, which relates the "through quantity" (electrophoretic flowrate) to the "across quantity" (electrical potential difference), in the case of electrophoretic flow, is given by

$$Q_{ph} = \left(\frac{zF_c D_{\text{diff}} A_c}{RTL}\right) \Delta\phi = \mathcal{F}\Delta\phi,$$

where $\mathcal{F}$ is the electrophoretic conductance of the fluidic channel.

Consider an example, where two species A and B are present in the separation channel shown in Figure 18.6. Assume that species A is unit-positively charged and species B is unit-negatively charged, while the surface of the channel has a negative fixed charge. Therefore, the electroosmotic flow through the channel would be from left to right (i.e., from the anode side to the cathode side) as shown in Figure 18.6. The electrophoretic flow for A would be from left to right, but that for B would be in the opposite direction. This is due to the difference in the electrophoretic velocities of these two species. Thus, the ratio of the rate of molar increment at the outlet of the separation channel for the two species is given by the expression

$$\textit{Separation Ratio} = \frac{(Q + \text{sign}(z_A)\times \mid Q_{\text{ph}} \mid_A) c_A^{\text{in}}}{(Q + \text{sign}(z_B)\times \mid Q_{\text{ph}} \mid_B) c_B^{\text{in}}} = \frac{(Q + \mid Q_{\text{ph}}\mid_A) c_A^{\text{in}}}{(Q - \mid Q_{\text{ph}}\mid_B) c_B^{\text{in}}},$$

where c_A^{in} is the concentration of species A at the inlet, and c_B^{in} is the concentration of species B at the inlet. Considering that the bulk flow is due to electrical potential gradient only (i.e., pressure-driven flow is absent), the separation ratio of the species can be expressed in terms of the electrophoretic conductance, electrohydraulic conductance, and the inlet concentration of the species, i.e.,

$$\textit{Separation Ratio} = \frac{(H + \mathcal{F}_A) c_A^{\text{in}}}{(H + \mathcal{F}_B) c_B^{\text{in}}}. \tag{18.12}$$

Thus, the knowledge of the electrophoretic conductance and the electrohydraulic conductance can be used to compute the separation ratio using equation (18.12), which can be considered as the device model for the separation module.

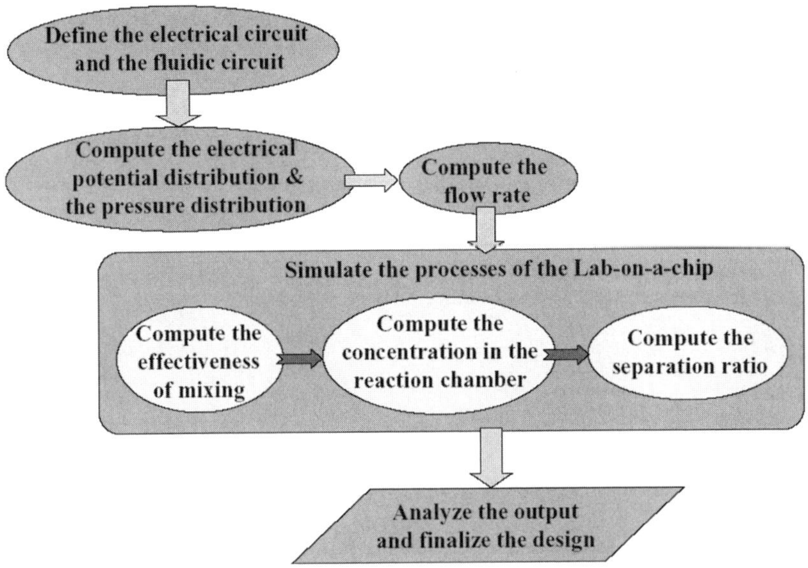

FIGURE 18.7. A block diagram for combined circuit/device analysis of the lab-on-a-chip system shown in Figure 18.1.

18.1.5 Integration of the Models

Figure 18.7 summarizes the integration of the circuit and device models for the prototype integrated microfluidic system shown in Figure 18.1. The circuit-based electrical model is first employed to compute the electrical potential distribution in the entire microfluidic system. Using the electrical potential distribution as an input, the fluidic circuit model is used to compute the flow variables (the pressure distribution, flowrate, etc.) in the entire system. The flowrates through various channels are then used to compute the mixing ratio/efficiency, reactions and the separation ratio. Even though Figure 18.7 is specific to the microfluidic system shown in Figure 18.1, it can be generalized to various other microfluidic systems by appropriately combining the electrical, fluidic, mixing, reaction/detection, and separation modules.

18.1.6 Examples

In this section, we demonstrate the application of the models and the implementation using several examples. In the first example (Figure 18.8, (Jacobson et al., 1999)), we consider microfluidic devices, which can be used for electrokinetically driven parallel and serial mixing. In the second

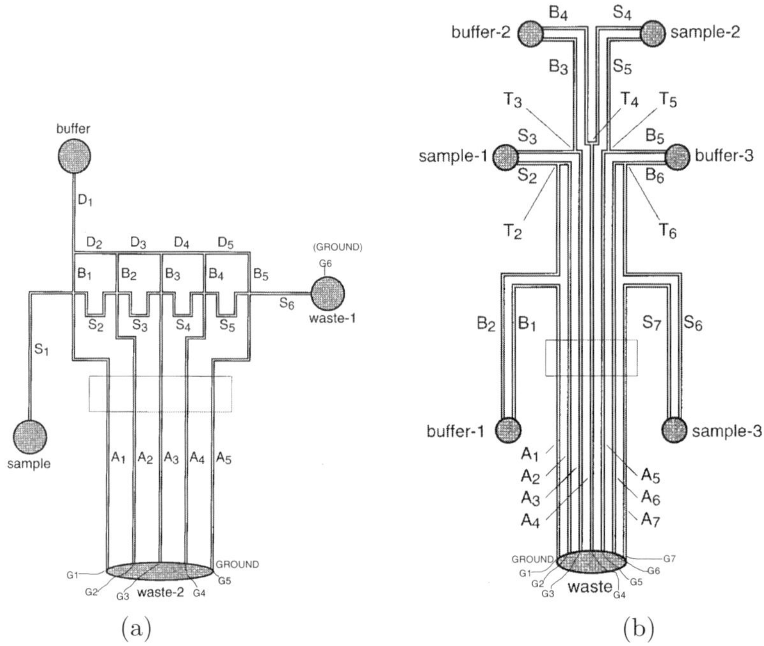

FIGURE 18.8. Schematics of the microchips for parallel (a) and serial (b) electrokinetic mixing. The circles depict sample, buffer, and waste reservoirs. The sample, buffer, and analysis channels are labeled "S," "B," and "A," respectively. The T intersections are the basic units for the parallel mixing device, while the cross intersections are the basic units for the serial mixing device (Jacobson et al., 1999).

example, we demonstrate a circuit-model-based analysis of a pneumatically controlled fluidic transport system, which has been used in a high-density microfluidic chip by (Thorsen et al., 2002). In the final example, we consider an integrated system, and a complete simulation-based analysis of the lab-on-a-chip.

Electrokinetically Driven Mixing

Microfluidic devices for parallel and serial mixing have been experimentally demonstrated (Jacobson et al., 1999). The parallel mixing device (Figure 18.8a) is designed with a series of independent T-intersections, and the serial mixing device (Figure 18.8b) is based on an array of cross-intersections. Figures 18.9(a) and 18.9(b) show the circuit representation of the mixing devices. Since the channels do not contain any flexible walls, the fluidic capacitances are neglected. The parameters (e.g., channel dimensions and applied potential) used in the simulation are the same as those used in the experiments reported in (Jacobson et al., 1999). The zeta potential of the channel walls for this example is computed from the capacitor model and has been verified with the experimental results given in (Jacobson et al., 1999). The expressions that have been used to compute the sample fraction are the same as those given in (Jacobson et al., 1999). For the parallel mixing device, the sample fraction in the jth analysis channel is computed by the expression

$$(\text{S.F.})_{A_j} = (n)_{A_j} = \frac{(Q)_{S_j}}{(Q)_{A_j}},$$

where S.F. is the sample fraction, $(Q)_{S_j}$ is the flowrate of the sample in the jth analysis channel, and $(Q)_{A_j}$ is the flowrate of the total solution in the jth analysis channel. For the serial mixing device, the sample fraction in the $(m+1)$th analysis channel is computed by the expression

$$(\text{S.F.})_{A_{m+1}} = (n)_{A_{m+1}} = \prod_{k=1}^{m} \left[1 - \frac{(Q)_{B_k}}{(Q)_{S_{k+1}}}\right],$$

where $(Q)_{B_k}$ is the flowrate of the buffer in the kth channel and $(Q)_{S_{k+1}}$ is the sample flowrate in the $(k+1)$th channel. Table 18.1 gives a comparison of the simulated and experimental results for the parallel and serial mixing devices. The simulation results show very good agreement with the experimental results. The CPU times to compute the electrical variables and the fluidic variables for the systems shown in Figure 18.8 (i.e., the mixing devices) were of order 1 second on a 800-MHz PC. Figures 18.10a and 18.10b show the variation in the sample fraction that can be obtained by controlling the electrical potential at the buffer and the sample reservoirs. These results demonstrate the advantage of the circuit model for designing microfluidic systems. It is practically impossible to get the variation of the output parameter with the input parameter varying over such a large range using experimental techniques or full-scale simulation methods.

The depth of the channels considered for parallel and serial mixing are 10 μm and 5.5 μm, respectively. For such large depths, the slip flow circuit model presented in Section 18.1.2 gives accurate results. Even if a no-slip flow circuit model is employed, the results would match exactly with the slip flow circuit model. However, as the depth of the channel gets smaller, the no-slip model can produce more accurate results than the slip-flow model. Shown in Figure 18.11 is a comparison of the relative error between the full simulation results and the slip and no-slip models for channel depths of 50 nm, 100 nm, and 200 nm. The Debye length is 10 nm in all cases. For both models, the error grows as the depth of the channel decreases. However, the error is much smaller with the no-slip model than with the slip model. Also, the rate of growth of the error is smaller with the no-slip model than with the slip model.

TABLE 18.1. A comparison of the simulated and experimental results for different types of mixing.

CHANNEL (PARALLEL MIXING)	SAMPLE FRACTION (EXPERIMENT)	SAMPLE FRACTION (SIMULATION)
A_1	0	0
A_2	0.84	0.833
A_3	0.67	0.675
A_4	0.51	0.522
A_5	0.36	0.340
A_6	0.19	0.165
A_7	1.0	1.0
CHANNEL (SERIAL MIXING)	SAMPLE FRACTION (EXPERIMENT)	SAMPLE FRACTION (SIMULATION)
A_1	1.0	1.0
A_2	0.36	0.37
A_3	0.21	0.22
A_4	0.12	0.12
A_5	0.059	0.053

Large-Scale Integration

In Chapter 1, we discussed a large-scale-integration-based microfluidic chip in Figure 1.31. Here we revisit the problem and show some results obtained using the circuit models discussed in the previous chapter. In the example shown in Figure 1.31, the fluidic transport system consists of two layers (Figure 18.12): the "control" layer, which contains all channels required to actuate the valves, is situated on top of the "flow" layer, and the "flow" layer contains the network of the channels being controlled (Unger et al., 2000). All biological assays and fluid manipulations are performed in the

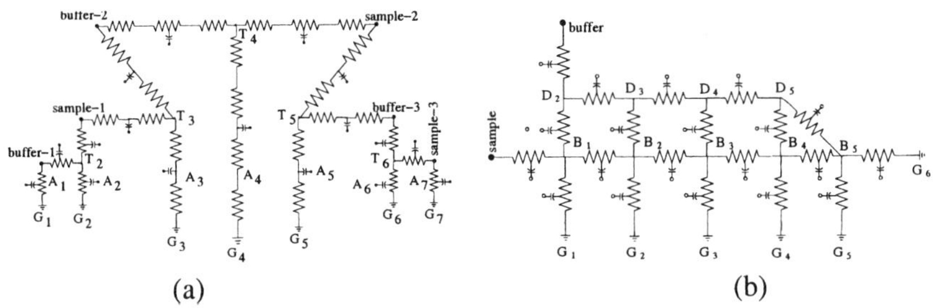

FIGURE 18.9. (a) The circuit (both fluidic and electrical) representation of the parallel mixing device. Since the flow is electrokinetically driven, the fluidic resistance of the channel is the inverse of the electrohydraulic conductance. (b) The circuit representation of the serial mixing device.

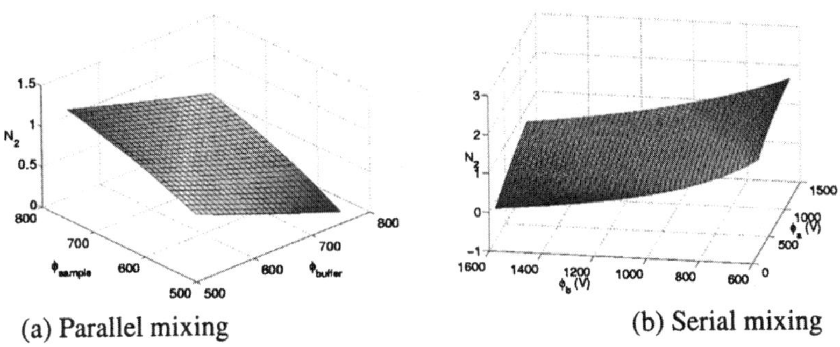

FIGURE 18.10. Variation in the sample fraction (denoted by N_2) of the second analysis channel when the applied potential (in "volts") in the sample reservoir and the buffer reservoir is changed. The plots for the other analysis channels (in both cases) have the same pattern.

flow layer. A valve is created whenever a control channel crosses a flow channel (Figure 18.12). The resulting thin membrane at the junction between the two channels can be deflected by fluidic actuation (Thorsen et al., 2002; Unger et al., 2000).

The schematic A of Figure 18.12 shows the orientation of the control layer and the flow layer. The schematic B of Figure 18.12 shows the valve closing for rectangular and rounded channels. The dotted lines indicate the contour of the top of the channel for a rectangular (left) and a rounded (right) channel as pressure is increased. In the example shown in Figure 18.13(a), rectangular channels are considered. Making multiple, independently actuated valves in a device requires independent control of the pressure applied to each control line. Figure 18.13(a) shows an example of such a device. From the "top view," the black channels oriented from west to east are

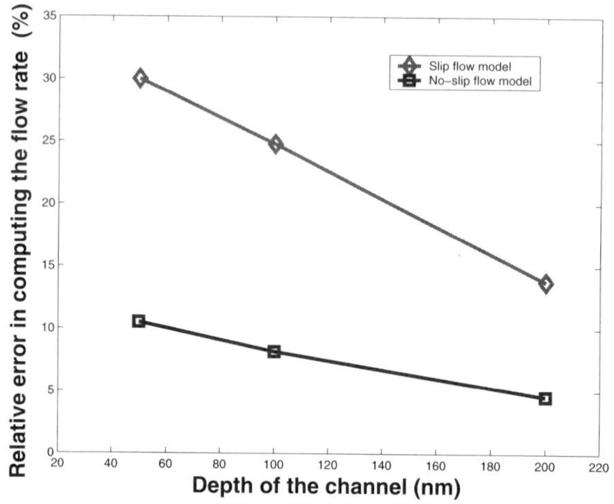

FIGURE 18.11. A comparison of the percentage relative error in the bulk flowrate Q between the slip flow model and the no-slip flow model, when compared with full-scale simulation.

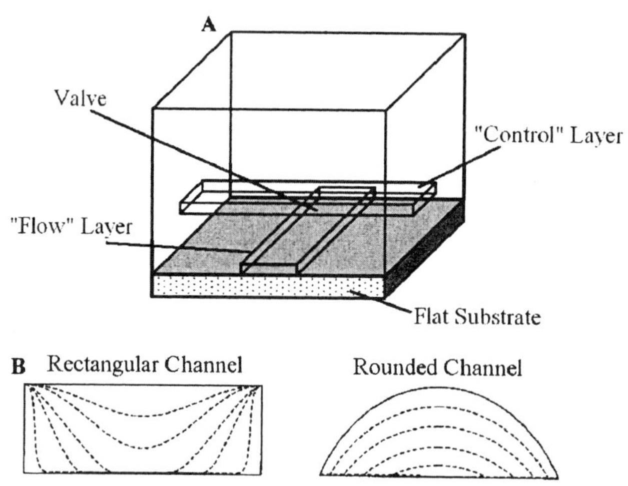

FIGURE 18.12. (A) Schematic of the arrangement of the control layer and the flow layer used for attaining pneumatic control. (B) Schematic of the valve closing for rectangular and rounded channel (Unger et al., 2000).

the control channels, and the gray channels oriented from north to south are flow channels. The control layer is on top of the flow layer. The flow channels are numbered from 0 to 7, and the control channels are named in alphabetic order from A to F. A valve at the intersection of flow channel

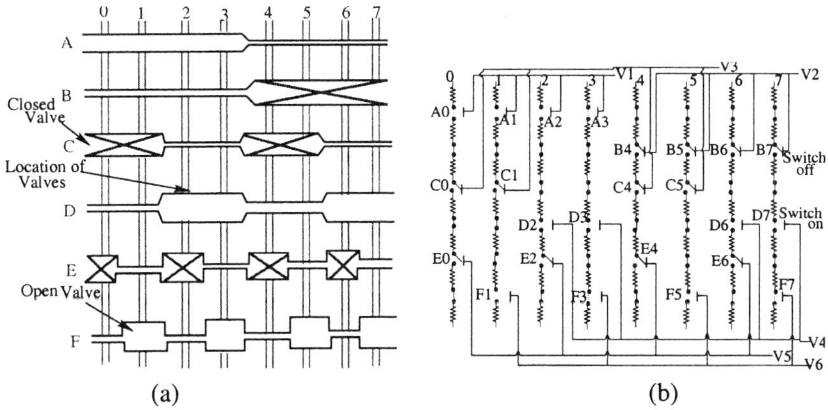

FIGURE 18.13. (a) Microfluidic system consisting of "control channels" (left to right) and "flow channels" (top to bottom). Intersections with wider control channels denote valves or switches. A cross indicates a closed valve. For further details refer to (Thorsen et al., 2002). (b) The fluidic circuit representation of the system. The valves are modeled as electrical switches.

0 and control channel A is designated as "A0." Such a designation is later used to explain the circuit representation of the system. The configuration shown in Figure 18.13(a) consists of simple "on–off" valves, which can be considered as fluidic switches to control the flow in the "flow" channels. Each control line can actuate multiple valves simultaneously. Since the dimension of the control line can be varied, it is possible to have a control line pass over multiple flow channels to actuate multiple valves. The active element is the roof of the flow channel, and the intersections that act as valves or fluidic switche, are denoted by a wider width of the control channel. The intersections that are marked by a cross (Figure 18.13(a)) indicate a closed (or off) position, and the intersections that are not marked by any cross indicate an open (or on) position.

The circuit representation for the microfluidic system shown in Figure 18.13(a) (Thorsen et al., 2002) is depicted in Figure 18.13(b). Since the flow is pressure-driven, only the fluidic circuit needs to be considered. The fluidic circuit represents the flow layer, and the intersections with valves are shown as electrical switches. The resistances (or conductances) in the fluidic circuit of Figure 18.13(b) are the fluidic resistances of the channels in the flow layer. The "on–off" position of the valves depends on the gauge pressure in the control channel compared to the pressure in the flow channel. Thus, the control layer is represented in the fluidic circuit through its gauge pressure. In Figure 18.13(b) the pressure difference of the ith control channel is represented by "Vi". The notation "V" is used because of the analogy between electrical voltage and pressure. The "on" position of a switch (in Figure 18.13(b)) is represented by a vertical dash connecting two consecutive resistances (e.g., "A0"), and the "off" position of a switch

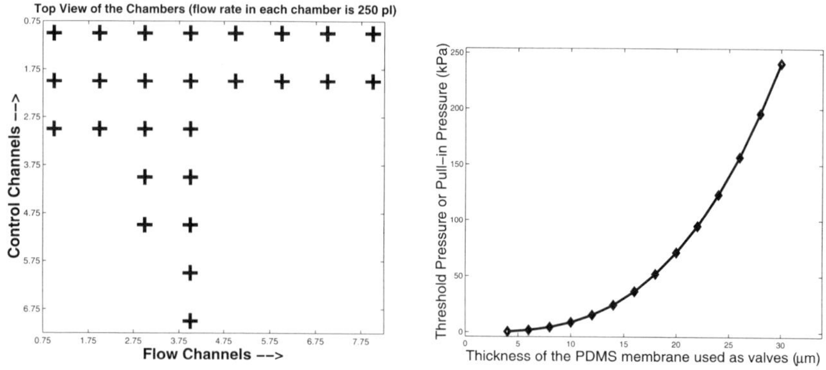

(a) (b)

FIGURE 18.14. (a) Simulation of fluid flow through the microfluidic system shown in Figure 18.13(a). The plus signs indicate presence of flow. (b) Variation of the threshold pressure with the thickness of the membrane.

is represented by a slanted dash causing a break between two consecutive resistances (e.g., "B4"). The hydraulic conductances (or hydraulic resistances) can be modeled using the approach explained in Section 18.1.2. The pressure-actuated control valves can be modeled as switches, which are considered "off" if the pressure in the control channel is above the "threshold pressure" and are considered "on" if the pressure in the control channel is below the threshold pressure. The threshold pressure can be computed from the expression (Timoshenko and Woionowsky-Krieger, 1959)

$$P_{\text{threshold}} = \frac{\left(\frac{ha}{2}\right)}{\left(\frac{4a^4}{\pi^5 D_r} \sum_{m=1,3,5\ldots}^{\infty} \frac{(-1)^{\frac{m-1}{2}}}{m^5} \frac{\sin(\frac{m\pi}{2})}{\frac{m\pi}{a}} \{1 - \frac{\alpha_m \tanh(\alpha_m)+2}{2\cosh(\alpha_m)}\}\right)},$$

where h is the height of the flow channel, a and b are the dimensions of the rectangular membrane acting as the valve, D_r and α_m are as defined in Section 18.1.2. Figure 18.14(a) shows the simulated flow distribution in the flow layer of the microfluidic circuit shown in Figure 18.13(b). A plus sign corresponding to a given flow channel indicates that the flow is "on;" otherwise, the flow is "off." A cell associated with a given flow channel will receive fluid only if the flow is on. Figure 18.14(b) shows the nonlinear variation of the threshold pressure with the thickness of the membrane, and Figure 18.15(a) shows the nonlinear variation of the threshold pressure with the dimension of the square membrane. Thus, for a specified threshold pressure one can choose the thickness of the membrane from Figure 18.14(b) and the dimension of the membrane from Figure 18.15(a). The CPU time to simulate the flow distribution for the system shown in Figure 18.13(a) was 16 seconds. Figure 18.15(b) shows the simulation result for an array

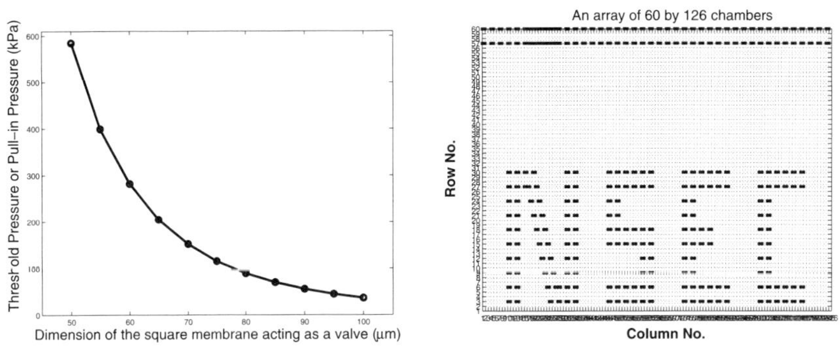

(a) (b)

FIGURE 18.15. (a) Variation of the threshold pressure with the width of the square membrane. (b) An example of large-scale integration, where the fluid is stored in a desired pattern in a microfluidic chip containing 60×126 chambers.

of 60×126 chambers. Fluid is stored in the chambers based on the filling mechanism described in Figure 18.13(a) (Thorsen et al., 2002). This result demonstrates that the fluid can be stored in any arbitrary pattern using large-scale integration of micro/nanochannels.

Lab-on-a-Chip

In the final example we consider a lab-on-a-chip system (Figure 18.16), which is designed based on the "nanochip" reported in (Becker and Locascio, 2002). The various chemical species are transported to the different modules on the chip from their sources by electrokinetic transport. One-third of the channels (marked as set A1 in Figure 18.16) perform the dual role of fluid transport and passive mixing. Each channel in the set marked as A1 is designed as shown in Figure 18.17(a) (Kutter, 2000). In this design, the characteristic dimension at a given level is half of that at the previous level. As a result, in the case of diffusion-dominated mixing, the equilibration time for mixing decreases at every level, since the equilibration time for homogeneous mixing is proportional to the square of the characteristic dimension; see Chapter 9. Thus, the homogeneity of the sample being transported increases. Figure 18.17(b) shows the circuit model, where the number of split levels used is three. In the simulations presented here, the number of splitting levels is considered as a design parameter. Figures 18.18(a) and 18.18(b) show the dependence of the flowrate and the homogeneity of the mixture, e_{mix}, on the number of split levels. The mixing

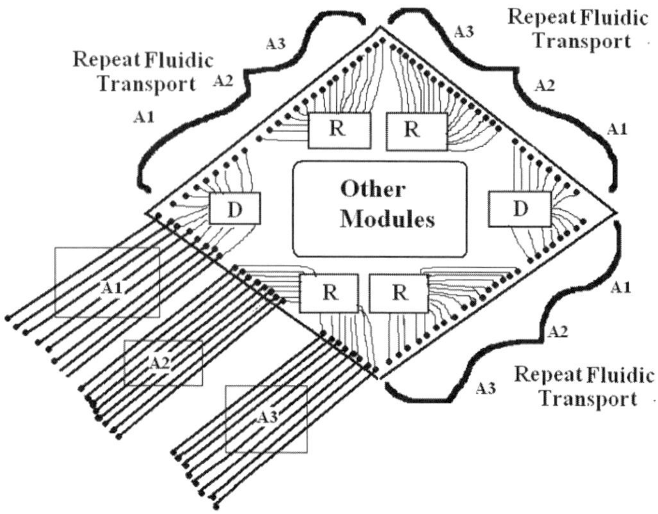

FIGURE 18.16. The schematics of the microfluidic chip considered in the lab-on-a-chip example. The fluidic transport system represented on the south-west side of the chip is duplicated on all the other sides.

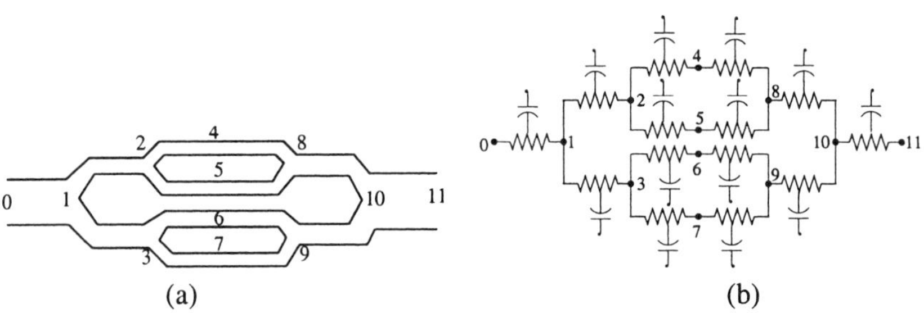

FIGURE 18.17. (a) The split channel design used for fluid transport in set A1 of the microfluidic chip. This type of channel serves a dual purpose of transporting and mixing. (b) The circuit (both fluidic and electrical) representation for the split channel design.

effectiveness is defined as (see Section 9.4)

$$e_{\text{mix}} = 1 - \frac{\sqrt{\frac{1}{N}\sum_{i=1}^{N}\left(c_i - c_i^{\text{PM}}\right)^2}}{\sqrt{\frac{1}{N}\sum_{i=1}^{N}\left(c_i^0 - c_i^{\text{PM}}\right)^2}},$$

where c_i is the concentration at the ith node, c_i^{PM} is the concentration at the ith node if the two streams (i.e., the sample and the buffer streams)

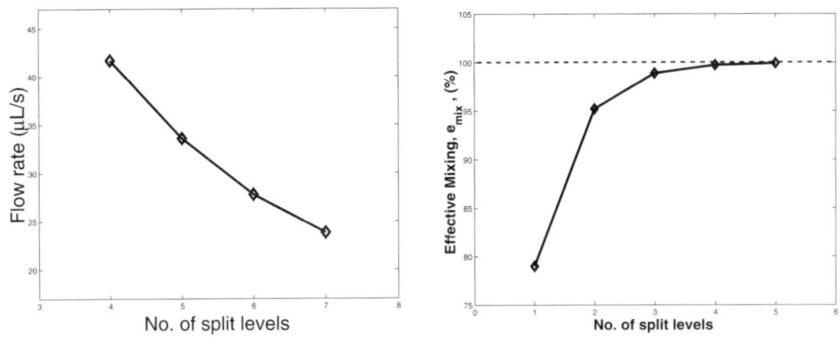

(a) (b)

FIGURE 18.18. (a) The dependence of the flowrate on the number of split levels used. (b) The dependence of the effectiveness of the mixing (i.e., homogeneity of the mixture) on the number of split levels used.

are perfectly mixed, and c_i^0 is the concentration at the ith node if the two streams do not mix at all. The analytical solution of the diffusion equation, obtained by the method of separation of variables, provides the concentration variation in the transverse direction. The mixing effectiveness, e_{mix}, ranges from 0 to 1, with 1 indicating complete mixing and 0 indicating no mixing.

The following parameters have been used for the results shown in Figures 18.18(a–b): $\Delta\phi = 100$ V (the potential difference applied between the start and the end of the channel, e.g., in Figure 18.17(a) it is applied between 0 and 11); length of each level $= 200\ \mu$m; height of the initial channel $= 16\ \mu$m; $\sigma_T = 6.2 \times 10^{-3}$ C/m^2; $\mu = 10^{-3}$ kg/m.s; $\epsilon = 6.95 \times 10^{-10}$. The concentration of species A at the inlet of the transport system is considered to be 0.1 mM. The results in Figures 18.18(a–b) indicate that there is a trade-off between the throughput and mixture homogeneity. However, if one uses a high value for the fluidic resistance of the initial channel, then that dominates the total fluidic resistance. As a result, the throughput is standardized and does not depend strongly on the number of splitting levels. Therefore, the device designer can control the effectiveness of the mixing process by varying the number of splitting levels.

Electrophoretic separation and electrokinetic transport are the governing mechanisms through the set of channels marked as A2 in Figure 18.16, while electrokinetic transport is the governing mechanism through the set of channels marked as A3. The species in set A1 (say A) is transported to the detection module (D), where it reacts with species B (already present in the detection chamber) to produce species C which can be used for off-chip detection. The reaction model given in Section 18.1.3 has been used to simulate the reaction between species A and B to produce species C. The initial condition corresponds to zero moles of A and C and one

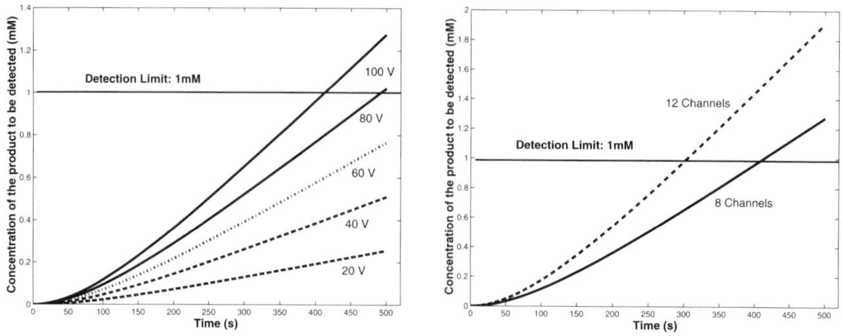

(a) (b)

FIGURE 18.19. (a) Concentration (of species C) versus time for various applied potentials. (b) Concentration (of species C) versus time for two different numbers of input ports per side of the microfluidic chip.

mole of species B in the reaction chamber (D). A second-order forward reaction is considered for this reaction chamber (i.e., D). Therefore, the backward reaction rate is considered to be zero. A forward reaction rate of 10^{-2} $(\text{mM.s})^{-1}$ has been considered. Figures 18.19(a), 18.19(b), and 18.20(a) show the variation in the rate of formation of species C with time for different design parameters (e.g., applied potential, number of input ports per side of the chip, channel length). If the minimum concentration of species C required for detection is known (say 1 mM as considered in this case), then one can predict the detection time from the simulation results or one can design the chip to meet a specific detection time. The chemical species (G and H) transported through the channels A2 and A3 are transported to the reactor module (R in Figure 18.16), where they undergo a second-order reversible chemical reaction to produce another chemical species, F. The reaction model given in Section 18.1.3 has been used for simulating the reaction between species G and H to produce species F. The initial condition corresponds to zero moles of G, H and F in the reaction chamber (R). The following parameters have been used for this phase: total length of a single channel = 2 mm; height of a single channel = 1 μm; $\sigma_T = 2 \times 10^{-1}$ C/m^2; forward reaction rate = 0.1 $(\text{mM.s})^{-1}$; backward reaction rate = 0.01 s^{-1}. The concentration of species G at the inlet of the transport system is considered to be 20 mM and the concentration of species H is considered to be 50 mM. Figures 18.20(b) and 18.21(a) show the effect of various design parameters on the variation of the concentration of F with time. Figure 18.21(b) shows the dependence of the separation ratio (taking place in the set "A2") on the ratio of the electrophoretic mobility of the species being separated. The applied potential difference is 100 V for this case. A time step of 0.1 second has been used for this case. The CPU time to do a transient analysis of the complete system (shown in Figure 18.16) for

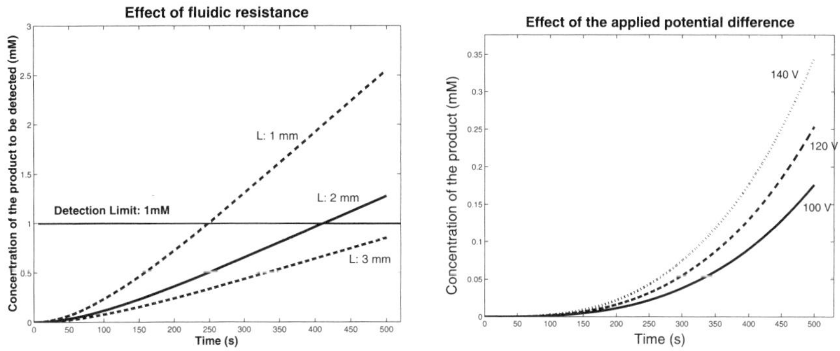

(a) (b)

FIGURE 18.20. (a) Concentration (of species C) versus time for different lengths (total) of the microfluidic channel. (b) Concentration (of species F) versus time for various applied potentials.

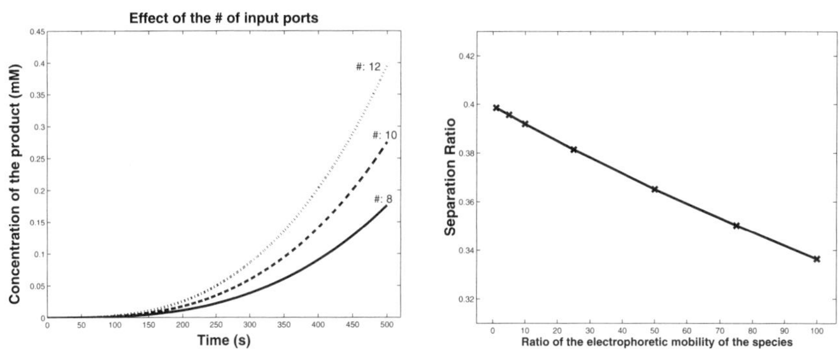

(a) (b)

FIGURE 18.21. (a) Concentration (of species F) versus time for different numbers of input ports per side of the microfluidic chip. (b) The dependence of separation ratio (taking place on set "A2" in Figure 18.16) on the ratio of the electrophoretic mobility of the species being separated.

500 seconds was on the order of 10 minutes on a PC of modest capability.

18.2 Macromodeling of Squeezed Film Damping

The dynamical behavior of microsystem components is often strongly affected by viscous air damping effects. They have to be carefully taken into account during the design and optimization process in order to get a realistic and reliable description of the device operation. The damping effects can be treated in two ways. The first is as a damping coefficient in the simulation model at the descriptional level (e.g., mechanical FEM device model or system-level compact model) as a fit parameter. The other method is to accurately treat it using a physical model (on the continuous-field level by solving the Navier–Stokes equations), which implies a large computational effort. The first method is easy. However, it lacks physical transparency, whereas the second method becomes prohibitive in the case of complex device geometry and/or coupling with other physical energy domains. However, for a large class of MEM devices, we can use a simplified form of the Navier–Stokes equations, e.g., the Reynolds squeezed film equation, which requires considerably less computational cost. Reynolds squeezed film equation (see Section 6.1) is typically applicable when a small gap between the two plates/structures opens and closes with time. This assumption holds for structures where the seismic mass moves perpendicular to a fixed wall, for plates with tilt around horizontal axes, and for clamped beams where the flexible part moves against a fixed wall. Some examples of MEM devices where the Reynolds equation is valid are fixed–fixed beams, cantilever beams, and micromirrors.

The nonlinear isothermal compressible Reynolds squeezed film equation for air damping with slip flow is (see Section 6.1)

$$\nabla \cdot \left[(1 + 6\,\mathrm{Kn})h^3 p \nabla p\right] = 12\mu \frac{\partial (ph)}{\partial t},$$

where $h(x, y, t)$ is the variable gap between the movable part and the ground electrode of the MEM device, $p(x, y, t)$ is the air pressure under the beam, $\mathrm{Kn}(x, y, t) = \lambda/h$ is the Knudsen number, where λ is the mean free path of air. Figure 18.22 shows a typical MEM device, a deformable beam/plate at a height $h(x, y, t)$ over a ground plane, which in the undeformed state is the initial gap between beam and the ground plane. The shaded region (on the xy plane) is the domain where the Reynolds equation is solved with the boundaries indicated in Figure 18.22. Depending on the example considered, the boundary conditions can change. For a mirror, the fluid system is assumed to be open (ambient pressure) on all sides, whereas for a fixed–fixed beam, the fluid system is open along the sides of the beam and closed (no flow) at the ends of the beam. Squeezed film damping in MEMS is a coupled phenomenon (mechanics, electrostatics, and fluidics). In order to obtain a self-consistent solution at any time instant, an iterative scheme has to be followed (e.g., a relaxation scheme) among the three domains. Considerable amount of work has been done in the reduced-order modeling of squeezed film damping in MEMS. They fall into the categories already discussed in Chapter 17, namely, equivalent circuit models, Galerkin methods, description language models/mixed-level simulation, and black box models.

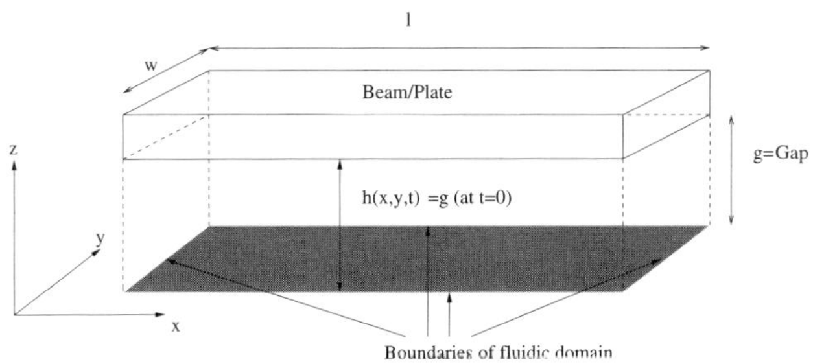

FIGURE 18.22. Domain for solving squeezed film damping equation in a MEM device.

18.2.1 Equivalent Circuit Models

Compact models for squeezed film damping based on equivalent circuit representation have the advantage of being incorporated into standard circuit simulators. The forces created by a squeezed gas film between vertically moving planar surfaces can be divided into spring and damping forces, which can be realized with frequency-dependent resistors and inductors (Veijola et al., 1995a). The second-order nonlinear Reynolds equation can be linearized when the motion of the plate(s) is small, the two plates are substantially parallel, and the motion is perpendicular to the surface of the plates. The linearized equation is written as

$$\frac{p_0 g^2}{12\mu_e}\nabla^2\left(\frac{p}{p_0}\right) - \frac{\partial}{\partial t}\left(\frac{p}{p_0}\right) = \frac{\partial}{\partial t}\left(\frac{x}{g}\right), \tag{18.13}$$

where p is a small pressure change of the static pressure p_0. The variation of the plate spacing x is also assumed to be small compared with the static gap g; μ_e is the effective viscosity of the gas given by (Veijola et al., 1995a)

$$\mu_e = \frac{\mu}{1 + 9.638\mathrm{Kn}^{1.159}}.$$

The linearized Reynolds equation (18.13) has two principal components in its solution, one in phase with the plate movement and the other out of phase, i.e., the spring term and the damping term, respectively. The force components can be calculated by integrating over the plate area, which for a rectangular plate is given (Blech, 1983) as an infinite series expansion

$$\frac{F_0}{x} = \frac{64 S p_0 A}{\pi^6 g} \sum_{m,n(\mathrm{odd})} \frac{m^2 + c^2 n^2}{(mn)^2[(m^2 + c^2 n^2) + S^2/\pi^4]}, \tag{18.14}$$

$$\frac{F_1}{x} = \frac{64 S^2 p_0 A}{\pi^8 g} \sum_{m,n(\mathrm{odd})} \frac{1}{(mn)^2[(m^2 + c^2 n^2) + S^2/\pi^4]}, \tag{18.15}$$

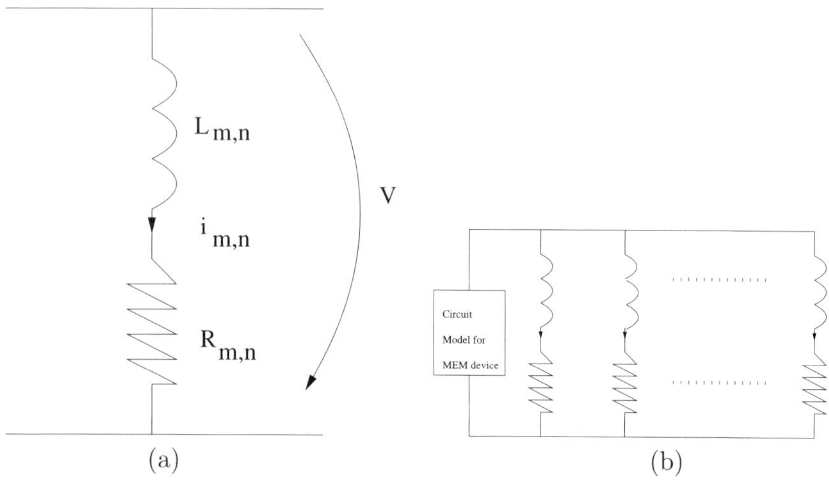

FIGURE 18.23. (a) Equivalent circuit element for squeezed film damping. (b) Equivalent circuit for squeezed film damping in the entire gap.

where m and n are odd integers, $A = WL$ is the plate area, and $c = W/L$. Here W and L are the width and length of the mass planar surfaces as shown in Figure 18.22; S is the squeeze number given by

$$S = \frac{12\mu_e W^2}{p_0 g^2}\omega,$$

and ω is the angular frequency. The forces in equations (18.14), (18.15) can be represented by the electrical equivalent circuit using a combination of an inductance (equals the spring behavior of the gas) and a resistance (acts as a damping element) in series, as shown in Figure 18.23(a). Figure 18.23(a) shows a single element for a corresponding m and n. The actual equivalent circuit for squeezed film damping would consist of a parallel combination of such elements (for various m and n) connected in parallel to the MEM device circuit as shown in Figure 18.23(b). The governing differential equation for the circuit shown in Figure 18.23(a) is

$$L_{m,n}\frac{\partial i_{m,n}}{\partial t} + R_{m,n} i_{m,n} = V = \frac{\partial u}{\partial t}, \tag{18.16}$$

where V is the voltage drop through the element (analogous to velocity in mechanics and represented by the flux term $\frac{\partial u}{\partial t}$). At steady state (all signals are sinusoidal having a single angular frequency ω), the current and velocity (flux term) can be expressed as

$$i_{m,n} = I_{m,n}\exp(j\omega t), \qquad u = U\exp(j\omega t),$$

where I and U are complex coefficients. Putting this in equation (18.16), we get

$$I_{m,n} = \frac{j\omega}{R_{m,n} + j\omega L_{m,n}} U,$$

and hence the total current in the squeezed film equivalent circuit that corresponds to the total force is the sum of all currents of the parallel sections

$$I_s = \sum_{m,n(\text{odd})} I_{m,n} = U \sum_{m,n(\text{odd})} \frac{j\omega}{R_{m,n} + j\omega L_{m,n}}.$$

The imaginary and the real parts of the ratio I_s/U are

$$Im\left(\frac{I_s}{U}\right) = \sum_{m,n(\text{odd})} \frac{R_{m,n}\omega}{R_{m,n}^2 + \omega^2 L_{m,n}^2}, \qquad \text{Re}\left(\frac{I_s}{U}\right) = \sum_{m,n(\text{odd})} \frac{L_{m,n}\omega^2}{R_{m,n}^2 + \omega^2 L_{m,n}^2},$$

which satisfy the frequency dependency specified in equations (18.14), (18.15), respectively. This requires that $Im\left(\frac{I_s}{U}\right) = F_0/x$ and $\text{Re}\left(\frac{I_s}{U}\right) = F_1/x$. This gives

$$L_{m,n} = (mn)^2 \frac{\pi^4 g}{64Ap_0}, \qquad R_{m,n} = (mn)^2(m^2 + c^2 n^2)\frac{\pi^6 g^3}{768AW^2\mu_e}.$$

The components $L_{m,n}$ and $R_{m,n}$ depend on the distance g (the static gap). If the displacement is large, the component values will also vary with the displacement and hence are nonlinear in nature. The equivalent circuit of squeezed film damping is connected in parallel with the MEM device circuit, and the whole system can be solved using any standard circuit simulator. For more details on circuit modeling of squeezed film damping, see (Veijola, 2001; Turowski et al., 1998).

18.2.2 Galerkin Methods

Linear modes of vibration of a system have been used for reduced-order modeling of MEM dynamics as discussed in Chapter 17. In this section, squeezed film damping has been considered in such a model-order reduction. The fluid (film) in between the plates typically undergoes Stokes flow (low Reynolds number, hence negligible inertia effects) and thus does not have any "normal modes" of its own that could be used for basis functions in combination with the elastic modes. One obvious approach is to linearize the dissipative effect under an assumption of small motion. Once linearized, frequency-domain analysis can be used, converting the time-dependent dissipation problem in the time domain into a time-independent frequency-domain calculation of amplitude and phase response. This approach was the basis for the early squeezed film damping work involving rigid body motion and has been widely used in the MEMS field. Even when the moving body is flexible, it is possible to use the modal amplitude to create a moving boundary condition for the fluidic system and calculate the reaction force. This has been done for small-amplitude damped resonant motions of flexible microbeams and resonators. When the amplitudes are large, such as for the electrostatic pull-in of a beam, linearized modal solutions are not accurate. In (Mehner et al., 2003), an approach has been presented to add dissipative effects of squeezed film damping (Reynolds equation) in the transient and harmonic analysis of MEMS. The macromodels are automatically generated by a modal projection technique based on the harmonic transfer functions of the fluidic domain. The transfer functions are either obtained at the initial position (small signal case) or at various deflection states (large deflection case). In this method, an equivalent damping and stiffness matrix that captures the true dependency between structural velocities and fluid pressure is computed in the modal coordinates. The damping $C_{ji}\dot{q}_i$ and the stiffness coefficients $K_{ji}q_i$ of such a matrix representation can be obtained from the following modal force balance equation:

$$C_{ji}\dot{q}_i + K_{ji}q_i = \phi_j^T \int N^T p(\phi_i \dot{q}_i) dA,$$

where q_i is the modal coordinate, ϕ_i is the ith eigenvector (mode), and N^T is the vector of finite element shape functions. Here C_{ji} and K_{ji} state the dependency between structural wall velocities caused by mode i and the reacting fluid forces that act on mode j. The damping and the squeeze coefficients of each mode are the main diagonal terms. Off-diagonal terms represent the fluidic crosstalk among modes, which happens in case of asymmetric gap separation. The following steps are performed to obtain the coefficients of C and K:

1. The squeezed film model is excited by wall velocities that are equal to the values of the first eigenvector (mode).

2. A harmonic response analysis is performed to compute the pressure response in the entire frequency range.

3. The real and the imaginary parts of the element pressure are integrated and the complex nodal force vector computed for each frequency.

4. The scalar product of all eigenvectors and the nodal force vector of step 3 is computed. The resulting numbers are modal forces, which indicate how much of the pressure distribution acts on each mode.

5. The damping and the stiffness coefficients are extracted from the real and the imaginary parts of the modal forces.

6. Steps 1 to 5 are repeated for each eigenmode.

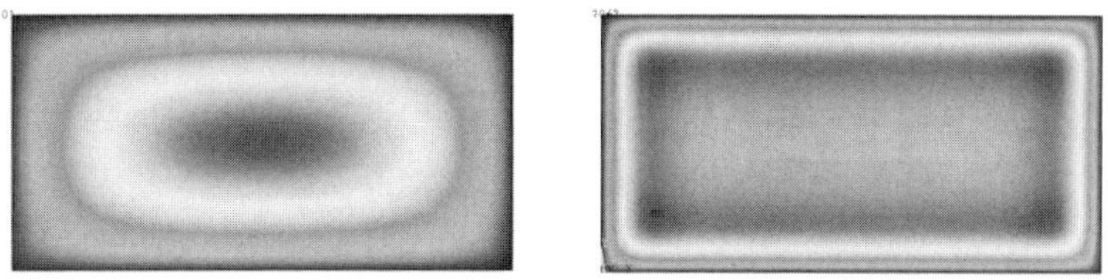

(a) (b)

FIGURE 18.24. (a) Pressure distribution under an oscillating flat plate at low frequency. (b) Pressure distribution under an oscillating flat plate at high frequency. (*http : //www.ansys.com/ansys/mems/mems_downloads/thermal_analogy_damping.pdf*)

A modal decomposition of damping effects is acceptable, since the Reynolds squeezed film equation is linear. More examples using modal basis functions for MEMS simulations are given in (Gabbay and Senturia, 1998; Varghese et al., 1999). Figure 18.24(a) shows the pressure distribution under a flat plate for a low frequency of oscillation, while Figure 18.24(b) shows the pressure distribution under a flat plate for a high frequency of oscillation. At high frequency, the fluid cannot easily escape from the sides, giving roughly uniform and high pressure under the plate.

The Karhunen–Loève decomposition method for model order reduction has also been effectively used for squeezed film damping in (Hung and Senturia, 1999). The method used is basically the same as that in the absence of damping. From a few full-simulation runs, snapshots are taken for the fluid pressure distribution and a set of pressure basis functions formed using an SVD analysis. These basis functions are then used in the dynamic simulation of the device, thereby reducing the order of the system. The nonlinear Reynolds equation can be reduced efficiently using this method with no linearization involved as in the case of the equivalent circuit representation. The trajectory piecewise-linear approach has also been used for modeling squeezed film damping in MEM devices; see (Rewienski and White, 2001), for details.

18.2.3 Mixed-Level Simulation

A mixed-level formulation uses a hardware description language for modeling squeezed film damping (Schrag and Wachutka, 2002). From an FEM model of the microstructure built by any standard FEM tool, a netlist for finite network (FN) simulation is constructed utilizing the grid and the geometric information from the FEM model. The governing equations (the Reynolds equation and the mass continuity equation) are discretized and coded in VHDL-AMS (Schrag et al., 2001; Sattler et al., 2003). The limitations of the Reynolds equation due to edge effects and perforations on the

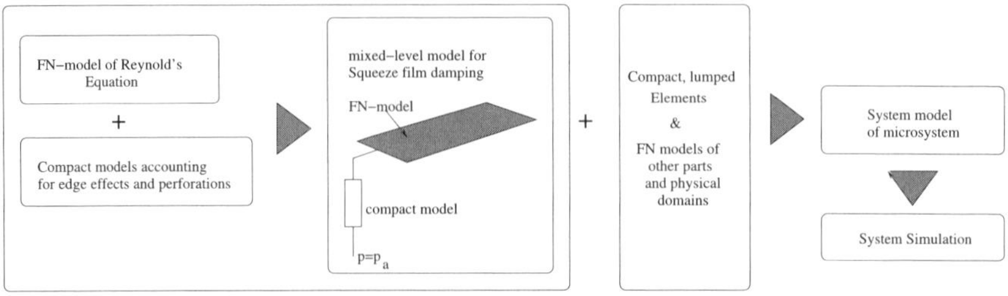

FIGURE 18.25. Mixed-level approach for modeling squeezed film damping in MEMS (see (Schrag and Wachutka, 2002), for details).

plate are rectified using error-compensating compact models. These compact models are in the form of lumped circuit elements such as resistances or constants, which can be determined from a few FEM simulations. The FN model in the sense of the Kirchhoffian network theory describes the squeezed film damping by two conjugate variables, namely, the pressure difference p_{ik} between two adjacent nodes ("across variables") and the corresponding mass flow rate Q_{ik} ("through variable") (Schrag and Wachutka, 2002). The mass balance equation is satisfied automatically as a result of the Kirchhoffian laws. However, correct formulation of the mass flowrate at each node must be done separately. The FN model can be implemented into a general-purpose system simulator and applied to arbitrary device geometries. The flowchart for the method is shown in Figure 18.25.

18.2.4 Black Box Models

The Arnoldi method has been used in (Chen and Kang, 2001a), to solve a MEMS micromirror device for both small and large deflections in the presence of fluid damping. The fluid damping equation (nonlinear isothermal Reynolds equation) has been linearized using Taylors series, and the Arnoldi method has been used to construct a reduced-order linear model for small angular deflections. For large angular deflections, both the linear and the second-order nonlinear terms from the fluid equation were retained in the Taylors expansion, and the Arnoldi method has been applied to construct a weakly nonlinear model. The accuracy can be increased by considering higher-order terms in the Taylors expansion, but the computational cost goes up, restricting the use of the process. The trajectory piecewise-linear approach overcomes some of these difficulties, as described in Chapter 17.

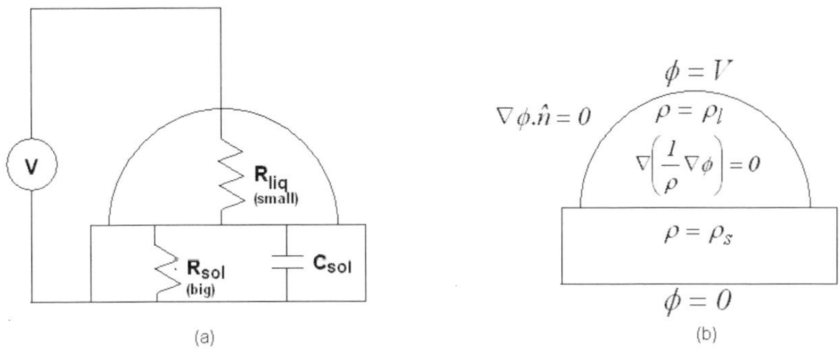

FIGURE 18.26. (a) A bulk circuit diagram for a liquid with a small amount of electrical resistance R_{liq} atop a dielectric solid with capacitance C_{sol} and a large electrical resistance R_{sol}. (b) The corresponding steady-state PDE with boundary conditions.

18.3 Compact Model for Electrowetting

In Chapter 8, we discussed electrowetting and the associated physical phenomena. Here we revisit electrowetting and discuss a compact model. Figure 18.26(a) shows the equivalent circuit diagram for the liquid-drop-dielectric solid system. Here ϕ is the electrical potential inside the drop, and ρ_l and ρ_s are the resistivities of the liquid and solid, respectively. The total impedence for the circuit diagram shown in Figure 18.26 is (Shapiro et al., 2003a)

$$\frac{V(s)}{I(s)} = z(s) = \frac{1 + \frac{R_{\text{liq}}}{R_{\text{sol}}} + sR_{\text{liq}}C_{\text{sol}}}{sC_{\text{sol}} + \frac{1}{R_{\text{sol}}}},$$

where s is the Laplace variable. For a sinusoidal signal $V(t) = \tilde{V}\cos(\omega t)$ of frequency ω, s is taken as $s = i\omega$. The voltage drop across the solid $V_{\text{sol}}(s)$ is given by

$$V_{\text{sol}}(s) = z_{\text{sol}}(s)I(s) = \frac{z_{\text{sol}}(s)}{z(s)}V(s) = \left(\frac{1}{1 + \frac{R_{\text{liq}}}{R_{\text{sol}}} + sR_{\text{liq}}C_{\text{sol}}}\right)V(s).$$

In the steady state (i.e., $s = i\omega \to 0$), the voltage and energy stored in the dielectric are

$$\tilde{V}_{\text{sol}} = \left(\frac{1}{1 + \frac{R_{\text{liq}}}{R_{\text{sol}}}}\right)\tilde{V} \quad \text{and} \quad E_{de} = \frac{1}{2}C_{\text{sol}}\left(\frac{1}{1 + \frac{R_{\text{liq}}}{R_{\text{sol}}}}\right)^2 \tilde{V}^2, \tag{18.17}$$

where $\tilde{V}$ is the applied DC voltage. The dependence shown in equation (18.17) is similar to the energy in the perfectly insulating solid; $E_{\text{de}}(R, \theta) =$

$\frac{1}{2}\frac{\epsilon_s V^2}{h}\pi R^2 \sin^2\theta$, except for the new $R_{\text{liq}}/R_{\text{sol}}$ term. Hence, the resistance of the liquid drop R_{liq} is shape-dependent. This dependence of resistance on the droplet shape gives rise to contact angle saturation in this model, see (Shapiro et al., 2003a; Shapiro et al., 2003b), for more details. The corresponding PDE and boundary conditions for the equivalent circuit diagram are shown in Figure 18.26(b).

18.4 Software

Several software packages have been developed, both commercially and by universities, for the simulation of microsystems (including microfluidic systems) using macromodels. Table 18.2 presents some of the available software packages.

TABLE 18.2. Software packages for reduced-order modeling of MEMS and microfluidics.

Name	Manufacturer	Description
HDL-A	Mentor Graphics Inc.	Analog and mixed signal analysis (equivalent circuits for MEMS/microfluidics).
MAST/SABER	Synopsys Inc.	Analog and mixed signal analysis (equivalent circuits for MEMS/microfluidics).
SpectreHDL	Cadence Design Systems	Analog and mixed signal analysis (equivalent circuits for MEMS/microfluidics).
NODAS	Carnegie Mellon Univ.	A library of parameterized components for using SABER (Synopsys Inc.) nodal simulator to simulate MEMS devices.
SPICE	U of C, Berkeley	A general-purpose circuit simulation program for nonlinear dc, nonlinear transient, and linear ac analyses (equivalent circuits for MEMS/microfluidics).
SUGAR	U of C, Berkeley	An open source simulation tool for MEMS based on nodal analysis techniques.
ANSYS Multiphysics	ANSY Inc.	Modal analysis, reduced-order models for fluid damping, system-level simulation.
MEMS Xplorer	MEMSCAP Inc.	System level simulation (Equivalent circuits for MEMS).
CoventorWare	Coventor Inc.	Macromodels and system-level models for MEMS and microfluidics.

Bibliography

Abe, T. (1997). Derivation of the lattice Boltzmann method by means of the discrete ordinate method for the Boltzmann equation. *J. Comp. Phys.*, 131:241–246.

Adamczyk, Z. and Warszynski, P. (1996). Role of electrostatic interactions in particle adsorption. *Adv. Colloid Interface Sci.*, 63:41–149.

Aerosoft Inc., Blacksburg Virginia (1996). *GASP Version 3 Users Manual.*

Agarwal, R. K. and Yun, K. Y. (2002). Burnett equations for simulation of transitional flows. *Appl. Mech. Rev.*, 55(3):219–240.

Agarwal, R. K., Yun, K. Y., and Balakrishnan, R. (2001). Beyond Navier-Stokes: Burnett equations for flows in the continuum-transition regime. *Phys. Fluids*, 13(10):3061–2085.

Ahmed, I. and Beskok, A. (2002). Rarefaction, compressibility, and viscous heating in gas microfilters. *AIAA J. Thermophys. Heat Transfer*, 16(2):161–170.

Aktas, O. and Aluru, N. R. (2002). A combined continuum/DSMC technique for multiscale analysis of microfluidic filters. *J. Comp. Phys.*, 178(2):342–372.

Aktas, O., Aluru, N. R., and Ravaioli, U. (2001). Application of a parallel DSMC technique to predict flow characteristics in microfluidic filters. *J. Microelectromech. Sys.*, 10(4):538–549.

Aktas, O., Ravaioli, U., and Aluru, N. R. (2003). A coupled DSMC/Navier-Stokes method for multiscale analysis of gas flow in microfluidic systems. *AIP Conference Proceedings*, 663(1):824–830.

Alexander, F., Garcia, A. L., and Alder, B. J. (1994). Direct Simulation Monte Carlo for thin-film bearings. *Phys. Fluids*, vol. 6 (12):3854–3860.

Alexander, F. J., Garcia, A. L., and Alder, B. J. (1998). Cell size dependence of transport coefficients in stochastic particle algorithms. *Phys. Fluids*, 10 (6):1540–1542.

Allen, M. P. and Tildesley, D. J. (1994). *Computer Simulation of Liquids*. Clarendon Press, Oxford.

Allen, R., Hansen, J. P., and Meichionna, S. (2003). Molecular dynamics investigation of water permeation through nanopores. *J. Chem. Phys.*, 119:3905–3919.

Allen, R., Melchionna, S., and Hansen, J. P. (2002). Intermittent permeation of cylindrical nanopors by water. *Phys. Rev. Lett.*, 89:175502.

Allen, T. W., Kuyucak, S., and Chung, S.-H. (1999). The effect of hydrophobic and hydrophilic channel walls on the structure and diffusion of water and ions. *J. Chem. Phys.*, 111:7985–7999.

Aluru, N. R. (2000). A point collocation method based on reproducing kernel approximations. *Int. J. Num. Meth. Eng.*, 47:1083–1121.

Aluru, N. R. and Li, G. (2001). Finite cloud method: A true meshless technique based on a fixed reproducing kernel approximation. *Int. J. Num. Meth. Eng.*, 50(10):2373–2410.

Ameel, T. A., Wang, X., Barron, R. F., and Warrington, R. O. (1997). Laminar forced convection in a circular tube with constant heat flux and slip flow. *J. Heat Transfer*, 124:346–355.

Ananthasuresh, G. K., Gupta, R. K., and Senturia, S. D. (1996). An approach to macromodeling of MEMS for nonlinear dynamic simulation. In *ASME-MEMS, DSC-Vol. 59*, pages 401–407.

Andersen, H. C. (1980). Molecular dynamics at constant pressure and/or temperature. *J. Chem. Phys.*, 72:2384–2393.

Andrienko, D., Dunweg, B., and Vinogradova, O. I. (2003). Boundary slip as a result of prewetting transition. *J. Chem. Phys.*, 119(24):13106–13112.

Ansumali, S. (2004). *Minimal kinetic modeling of hydrodynamics*. PhD thesis, Swiss Federal Institute of Technology, Zurich.

Ansumali, S. and Karlin, I. V. (2002). Kinetic boundary conditions in the lattice Boltzmann equation. *Phys. Rev. E*, 66:026311.

Ansumali, S., Karlin, I. V., and Ottinger, H. C. (2003). Minimal entropic kinetic models for hydrodynamics. *Europhys. Lett.*, 63:798–804.

Aoki, K. (1989). Numerical analysis of rarefied gas flows by finite-difference method. In Muntz, E., Weaver, D., and Campbell, D. H., editors, *Rarefied Gas Dynamics: Theoretical and Computational Techniques*, pages 297–322.

Aoki, K. (2001). Dynamics of rarefied flows: Asymptotic and numerical analyses of the Boltzmann equation. In *Paper AIAA 2001-0874, 39th Aerospace Sciences Meeting & Exhibit, January 8–11, Reno, NV*.

Aoki, K., Sone, Y., and Masukawa, N. (1995). A rarefied gas flow induced by a temperature field. In Harvey, J. and Lord, G., editors, *Proceedings of the Nineteenth International Symposium on Rarefied Gas Dynamics*, Oxford University Press, Oxford U.K.

Aref, H. (1984). Stirring by chaotic advection. *J. Fluid Mech.*, 143:1–21.

Arkilic, E. (1997). *Measurement of the mass flow and tangential momentum accommodation coefficient in silicon micro-machined channels.* PhD thesis, Massachusetts Institute of Technology.

Arkilic, E. and Breuer, K. (1993). Gaseous flow in small channels. In *AIAA Shear Flow Conference July 6-9, 1992, Orlando FL, AIAA 93-3270.*

Arkilic, E., Breuer, K., and Schmidt, M. A. (2001). Mass flow tangential momentum accommodation in silicon micromachined channels. *J. Fluid Mech.*, 437:29–43.

Arkilic, E., Breuer, K., and Schmidt, M. A. (ASME 1994). Gaseous flow in microchannels. In *FED-Vol. 197, Application of Microfabrication to Fluid Mechanics*, pages 57–66.

Arkilic, E., Schmidt, M. A., and Breuer, K. (1997). Gaseous slip flow in long microchannels. *J. Microelectromech. Sys.*, 6(2):167–178.

Arya, G., Maginn, E. J., and Chang, H.-C. (2000). Efficient viscosity estimation from molecular dynamics simulation via momentum impulse relaxation. *J. Chem. Phys.*, 113:2079–2087.

Atluri, S. N. (2002). *The Meshless Local Petrov-Galerkin (MLPG) Method.* Tech Science Press.

Avgustinik, A. I., Vigdergauz, V. S., and Zharavlev, G. I. (1962). Electrophoretic deposition of ceramic masses from suspension and calculation of deposit yields. *J. Appl. Chem. USSR.*, 35:2175–2180.

Bahukudumbi, P. and Beskok, A. (2003). A phenomenological lubrication model for the entire Knudsen regime. *J. Micromech. Microeng.*, 13 (6):873–884.

Bahukudumbi, P., Park, J. H., and Beskok, A. (2003). A unified engineering model for steady and quasi-steady shear driven gas micro flows. *Microscale Thermophys. Eng.*, 7,(4):291–315.

Bai, Z. (2002). Krylov subspace techniques for reduced-order modeling of large-scale dynamical systems. *Applied Numerical Mathematics*, 43:9–44.

Baidya, B. and Mukherjee, T. (2002). Extraction and LVS for mixed-domain integrated MEMS layouts. In *Proceedings of ICCAD*, pages 361–366.

Balakrishnan, R. (2004). An approach to entropy consistency in second-order hydrodynamic equations. *J. Fluid Mech.*, 503:201–245.

Balakrishnan, R., Agarwal, R. K., and Yun, K. Y. (1999). BGK-Burnett equations for flows in the continuum-transition regime. *J. Thermophys. Heat Transfer*, 13:(4):397–410.

Ball, K. S., Sirovich, L., and Keefe, L. R. (1991). Dynamical eigenfunction decomposition of turbulent channel flow. *Int. J. Num. Meth. Fluids*, 12:585.

Bangia, A. K., Batcho, P. F., Kevrekidis, I. G., and Karniadakis, G. E. (1997). Unsteady two-dimensional flows in complex geometries: Comparative bifurcation studies with global eigenfunction expansions. *SIAM J. Sci. Comput.*, 18:775–805.

Barber, R. W. and Emerson, D. R. (2002). Numerical simulation of low Reynolds number slip flow past a confined microsphere. In *Proceedings of the Twenty-Third International Symposium on Rarefied Gas Dynamics.*

Barber, R. W. and Emerson, D. R. (2004). Simulation of low Knudsen number isothermal flow past a confined spherical particle in a micro-pipe. In *Microchannels and Minichannels, ASME.*

Bardos, C., Golse, F., and Levermore, C. D. (1991). Fluid dynamic limit of kinetic equations. I. Formal derivations. *J. Statist. Phys.*, 63(1-2):323–344.

Bardos, C., Golse, F., and Levermore, C. D. (1993). Fluid dynamic limit of kinetic equations. II. Convergence proofs for the Boltzmann equation. *Comm. Pure Appl. Math.*, 46(5):667–753.

Barker, S. L., Ross, D., Tarlov, M. J., Gaitan, M., and Locasio, L. E. (2000). Control of flow direction in microfluidic devices with polyelectrolyte multilayers. *Anal. Chem.*, 72(24):5925–5929.

Barragan, V. M. and Bauza, C. R. (2000). Electroosmosis through a cation-exchange membrane: Effect of an AC perturbation on the electroosmotic flow. *J. Colloid Interface Sci.*, 230:359–366.

Barrat, J.-L. and Bocquet, L. (1999). Large slip effect at a nonwetting fluid-solid interface. *Phys. Rev. Lett.*, 82(23):4671–4674.

Batchelor, G. (1976). Brownian diffusion of particles with hydrodynamic interaction. *J. Fluid Mech.*, 74:1–29.

Batchelor, G. (1998). *An Introduction to Fluid Dynamics.* Cambridge University Press, New York.

Bau, H. H. and Pfahler, J. N. (2001). Experimental observations of liquid flow in micro conduits. In *Paper AIAA 2001-0722, 39th Aerospace Sciences Meeting & Exhibit, January 8–11, Reno, NV.*

Bau, H. H., Zhong, J., and Yi, M. (2001). A minute magneto hydro dynamic (MHD) mixer. *Sens. Actuators, B*, 79:207–215.

Baudry, J., Charlaix, E., Tonck, A., and Mazuyer, D. (2001). Experimental evidence for a large slip effect at a nonwetting fluid-solid interface. *Langmuir*, 17:5232–5236.

Baughman, R. H., Zakhidov, A. A., and de Heer, W. A. (2002). Carbon nanotubes: the route toward applications. *Science*, 297:787–792.

Baygents, J. C., Schwarz, B. C., Deshmukh, R. R., and Brier, M. (1997). Recycling electrophoretic separations: modeling of isotachophoresis and isoelectric focusing. *J. Chromatogr. A*, pages 165–183.

Bayley, H. and Cremer, P. S. (2001). Stochastic sensors inspired by biology. *Nature*, 413:226–230.

Bayt, R. L. (1999). *Analysis, fabrication and testing of a MEMS-based micropropulsion system.* PhD thesis, Massachusetts Institute of Technology.

Bayt, R. L. and Breuer, K. (2000). *Fabrication and testing of micron-sized cold gas thrusters*, volume 187 of *Progress in Astronautics and Aeronautics*, pages 381–398. AIAA.

Becker, H. and Locascio, L. E. (2002). Polymer microfluidic devices. *Talanta*, 56:267–287.

Beckstein, O., Biggin, P. C., and Sansom, M. S. (2001). A hydrophobic gating mechanism for nanopores. *J. Phys. Chem. B*, 105:12902–12905.

Beckstein, O. and Sansom, M. S. (2003). Liquid-vapor oscillations of water in hydrophobic nanopores. *Proc. Natl. Acad. Sci.*, 100:7063–7068.

Belytschko, T., Krongauz, Y., Organ, D., Fleming, M., and Krysl, P. (1996). Meshless methods: An overview and recent developments. *Comput. Methods Appl. Mech. Engrg.*, 139:3 47.

Belytschko, T., Lu, Y. Y., and Gu, L. (1994). Element free Galerkin methods. *Int. J. Num. Meth. Eng.*, 37:229–256.

Beni, G., Hackwood, S., and Jackel, J. L. (1982). Continuous electrowetting effect. *Appl. Phys. Lett.*, 40(10):912–914.

Berendsen, H. J., Grigera, J. R., and Straatsma, T. P. (1987). The missing term in effective pair potentials. *J. Phys. Chem.*, 91:6269–6271.

Berendsen, H. J. C., Postma, J. P. M., van Gunsteren, W. F., DiNola, A., and Haak, J. R. (1984). Molecular dynamics with coupling to an external bath. *J. Chem. Phys.*, 81:3684–3690.

Berezhkovskii, A. and Hummer, G. (2002). Single-file transport of water molecules through a carbon nanotube. *Phys. Rev. Lett.*, 89:064503.

Berkooz, G., Holmes, P., and Lumley, J. L. (1993). The proper orthogonal decomposition in the analysis of turbulent flows. *Ann. Rev. Fluid Mech.*, 25:537.

Beskok, A. (2001). Validation of a new slip model for separated gas micro-flows. *Num. Heat Transfer: Fundamentals*, 40(6):451–471.

Beskok, A. and Karniadakis, G. E. (1992). Simulation of slip-flows in complex micro-geometries. *ASME*, DSC-Vol. 40:pp. 355–370.

Beskok, A. and Karniadakis, G. E. (1994). Simulation of heat and momentum transfer in complex microgeometries. *AIAA J. Thermophys. Heat Transfer*, 8 (4):647–655.

Beskok, A. and Karniadakis, G. E. (1999). A model for flows in channels, pipes and ducts at micro and nano scales. *Microscale Thermophys. Eng.*, 3(1):43–77.

Beskok, A., Trimmer, W., and Karniadakis, G. E. (1995). Rarefaction compressibility and thermal creep effects in gas microflows. *IMECE 95, Proceedings of the ASME Dynamic Systems and Control Division*, DSC-Vol. 57-2:877–892.

Beskok, A. and Warburton, T. (2001). An unstructured h/p finite element scheme for fluid flow and heat transfer in moving domains. *J. of Comp. Phys.*, 174:492–509.

Bevan, M. A. and Prieve, D. C. (2000). Hindered diffusion of colloidal particles very near to a wall: Revisited. *J. Chem. Phys.*, 113(3):1228–1236.

Bharadwaj, R., Santiago, J. G., and Mohammadi, B. (2002). Design and optimization of on-chip capillary electrophoresis. *Electrophoresis*, 23:2729–2744.

Bhatnagar, P. L., Gross, E. P., and Krook, K. (1954). A model for collision processes in gasses. *Phys. Rev.*, 94:511–524.

Bhatt, K. H. and Velev, O. D. (2004). Control and modeling of the dielectrophoretic assembly of on-chip nanoparticle wires. *Langmuir*, 20:467–476.

Bianchi, F., Ferrigno, R., and Girault, H. H. (2000). Finite element simulation of an electroosmotic-driven flow division at a T-junction of microscale dimension. *Anal. Chem.*, 72(9):1987–1993.

Biesheuvel, P. M. and Verweij, H. (1999). Theory of cast formation in electrophoretic deposition. *Journal of the American Ceramic Society*, 82:1451–1455.

Bird, G. (1978). Monte Carlo simulation of gas flows. *Ann. Rev. Fluid Mech.*, 10:11–31.

Bird, G. (1994). *Molecular Gas Dynamics and Direct Simulation of Gas Flows.* Oxford University Press, Oxford.

Bitsanis, I., Magda, J. J., Tirell, M., and Davis, H. T. (1987). Molecular dynamics of flow in micropores. *J. Chem. Phys.*, 87(3):1733–1750.

Bitsanis, I., Somers, S. A., Davis, H. T., and Tirell, M. (1990). Microscopic dynamics of flow in molecular narrow pores. *J. Chem. Phys.*, 93(5):3427–3431.

Bitsanis, I., Vanderlick, T. K., Tirell, M., and Davis, H. T. (1988). A tractable molecular theory of flow in strongly inhomogeneous fluids. *J. Chem. Phys.*, 89(5):3152–3162.

Blake, T. D. (1990). Slip between a liquid and a solid: D.M. Tolstoi's (1952) theory reconsidered. *Colloids Surf.*, 47:135–145.

Blech, J. J. (1983). On isothermal squeeze films. *J. Lubrication Technol.*, 105:615–620.

Bobetic, M. V. and Barker, J. A. (1970). Lattice dynamics with three-body forces: argon. *Phys. Rev. B*, 2:4169–4175.

Bocquet, L. and Barrat, J.-L. (1993). Hydrodynamic boundary-conditions and correlation-functions of confined fluids. *Phys. Rev. Lett.*, 70(18).

Bocquet, L. and Barrat, J.-L. (1994). Hydrodynamic boundary-conditions, correlation-functions and Kubo relations for confined fluids. *Phys. Rev. E*, 49(4).

Boghosian, B. M., Love, P. J., Coveney, P. V., Karlin, I. V., Succi, S., and Yepez, J. (2003). Galilean-invariant lattice- Boltzmann models with H-theorem. *Phys. Rev. E Rapid Communications*, 68(2):025103.

Boghosian, B. M., Love, P. J., Yepez, J., and Coveney, P. (2004). Galilean-invariant multi-speed entropic lattice Boltzmann models. *Phys. D*, 193:169–181.

Bohmer, M. (1996). In situ observation of 2-dimensional clustering during electrophoretic deposition. *Langmuir*, 12:5747–5750.

Bonaccurso, E., Butt, H.-J., and Craig, V. (2003). Surface roughness and hydrodynamic boundary slip of a Newtonian fluid in a completely wetting system. *Phys. Rev. Lett.*, 90(14):144501.

Bonaccurso, E., Kappl, M., and Butt, H.-J. (2002). Hydrodynamic force measurements: Boundary slip of water on hydrophilic surfaces and electrokinetic effects. *Phys. Rev. Lett.*, 88(7):076103.

Born, M. and Mayer, J. E. (1932). Zur Gittertheorie der Ionenkristalle. *Z. Physik*, 75:1–18.

Borukhov, I., Andelman, D., and Orland, H. (2000). Adsorption of large ions from an electrolyte solution: a modified Poisson-Boltzmann equation. *Electrochimi. Acta*, 46:221–229.

Bottausci, F., Mezic, I., Meinhart, C. D., and Cardonne, C. (2004). Mixing in the shear superposition micromixer: three-dimensional analysis. *Phil. Trans. R. Soc. London, Ser. A*, 362:1001–1018.

Boyd, I. D. and Sun, Q. (2001). Particle simulation of micro-scale gas flows. In *AIAA 2001-0876, 39th Aerospace Sciences Meeting & Exhibit, January 8-11, Reno, NV.*

Brady, J. F. and Bossis, G. (1988). Stokesian Dynamics. *Ann. Rev. Fluid Mech.*, 20:111–157.

Brakke, K. (1992). The surface evolver. *Exp. Math.*, 1(2):141–165.

Bretherton, F. P. (1961). The motion of long bubbles in tubes. *J. Fluid Mech.*, 10(2):166–188.

Breuer, K., Piekos, E. S., and Gonzales, A. D. (1995). DSMC simulations of continuum flows. In *AIAA Thermophysics Conference, June 19–22 1995, San Diego, CA, AIAA 95-2088.*

Brodman, C. (1891). Untersuchungen ueber Reibungkoeffizienten zu Fluessigkeiten. *Dissertation, Goettingen.*

Brovchenko, I. V., Geiger, A., and Paschek, D. (2001). Simulation of confined water in equilibrium with a bulk reservoir. *Fluid Phase Equilibria*, 183:331–339.

Brown, H. S., Jolly, M. S., Kevrekidis, I. G., and Titi, E. S. (1990). Use of approximate inertial manifolds in bifurcation calculations. In D. Roose, et al., editor, *Continuation and Bifurcations: Numerical Techniques and Applications.* Kluwer Academic Press.

Bryzek, J., Petersen, K., and McCulley, W. (1994). Micromachines on the march. *IEEE Spectrum*, 31(5):20–31.

Buch, J. S., Wang, P. C., DeVoe, D. L., and Lee, S. L. (2001). Field-effect flow control in a polydimethylsiloxane-based microfluidic system. *Electrophoresis*, 22:3902–3907.

Bultheel, A. and Barvel, M. V. (1986). Padé techniques for model reduction in linear system theory: A survey. *J. Comput. Appl. Math. Phys.*, 14:401–438.

Bungay, P. M. and Brenner, H. (1973). The motion of a closely-fitting sphere in a fluid-filled tube. *Int. J. Multiphase Flow*, 1:25–56.

Burak, Y. and Andelman, D. (2001). Discrete aqueous solvent effects and possible attractive forces. *J. Chem. Phys.*, 114(7):3271–3283.

Burgdorfer, A. (1959). The influence of the molecular mean free path on the performance of hydrodynamic gas lubricated bearings. *J. Basic Eng. Trans. ASME*, 81:94–99.

Burgreen, D. and Nakache, F. R. (1964). Electrokinetic flow in ultrafine capillary silts. *J. Phys. Chem.*, 68(5):1084–1091.

Cabrera, C. R., Finlayson, B., and Yager, P. (2001). Formation of natural pH gradients in a microfluidic device under flow conditions: Model and experimental validation. *Anal. Chem.*, 73(3):658–666.

Cadence, Ltd. (1997). *SpectreHDL Reference Manual.*

Cai, C., Boyd, I. D., and Fan, J. (2000). Direct simulation methods for low-speed microchannel flows. *J. Thermophys. Heat Transfer*, 14(3):368–378.

Carrier, J., Greengard, L., and Rokhlin, V. (1988). A fast adaptive multipole algorithm for particle simulations. *SIAM J. Sci. Stat. Comput.*, 9:669–686.

Casinovi, G. (2002). Coupled-energy-domain macromodeling of MEMS devices. In *Modeling and Simulation of Microsystems*, pages 158–161.

Cercignani, C. (1963). Plane Poiseuille flow and Knudsen minimum effect. In Laurmann, J. A., editor, *Proceedings of the Third International Symposium on Rarefied Gas Dynamics*, volume 1, pages 92–101, New York. Academic Press.

Cercignani, C. (1975). *Theory and Application of the Boltzmann Equation.* Scottish Academic Press.

Cercignani, C. (1988). *The Boltzmann Equation and Its Applications.* Springer, New York.

Cercignani, C. (2000). *Rarefied Gas Dynamics: From Basic Concepts to Actual Calculations.* Cambridge University Press, New York.

Cercignani, C. and Daneri, A. (1963). Flow of a rarefied gas between two parallel plates. *J. Appl. Phys.*, 43.

Cercignani, C., Illner, R., and Pulvirenti, M. (1994). *The Mathematical Theory of Dilute Gases.* Springer-Verlag, New York.

Cercignani, C. and Lampis, M. (1971). Kinetic models for gas-surface interactions. *Transport Theory and Stat. Phys.*, 1:101–114.

Cercignani, C. and Sernagiotto, F. (1966). Cylindrical Poiseuille flow of a rarefied gas. *Phys. Fluids*, 9 (1):40.

Chan, D. Y. and Horn, R. G. (1985). The drainage of thin liquid films between solid surfaces. *J. Chem. Phys.*, 83(10):5311–5324.

Chan, H. B., Aksyuk, V. A., Kleiman, R. N., Bishop, D. J., and Capasso, F. (2001). Quantum mechanical actuation of microelectromechanical systems by the Casimir force. *Science*, 291:1941–1944.

Chan, W. K. and Sun, Y. (2003). Analytic modeling of ultra-thin-film bearings. *J. Micromech. Microeng.*, 13:463–473.

Chaplin, M. (2004). http://www.lsbu.ac.uk/water/.

Chapman, S. and Cowling, T. G. (1970). *The Mathematical Theory of Non-Uniform Gases; An account of the kinetic theory of viscosity, thermal conduction and diffusion in gases.* Cambridge University Press.

Chen, C. H., Lin, H., Lele, S. K., and Santiago, J. G. (2003). Electrokinetic microflow instability with conductivity gradients. In *μTAS2003, Squaw Valley, CA*.

Chen, C. H. and Santiago, J. G. (2002). Electrokinetic instability in high concentration gradient microflows. In *IMECE-2002, New Orleans, LA*, pages CD Vol.1, No 33563.

Chen, C. S. and Kuo, W. J. (2003). Squeeze and viscous dampings in micro electrostatic comb drives. *Sens. Actuators, A*, 107:193–203.

Chen, G. and Boyd, I. (1996). Statistical error analysis for the direct simulation Monte Carlo technique. *J. Comp. Phys.*, 126:434–448.

Chen, J. and Kang, S. M. (2001a). Dynamic macromodeling of MEMS mirror devices. In *Proceedings of IEDM*, pages 925–928.

Chen, J. and Kang, S. M. (2001b). Model-order reduction of nonlinear MEMS devices through arc-length-based Karhunen-Loève decomposition. In *Proceedings of ISCAS*, pages 457–460.

Chen, R., Zhang, Y., Wang, D., and Dai, H. (2001). Noncovalent sidewall functionalization of single-walled carbon nanotubes for protein immobilization. *J. Am. Chem. Soc.*, 123(16):3838–3839.

Chen, R., Zitter, R. N., and Tao, R. (1992). Laser-diffraction determination of the crystalline structure of an electrorheological fluid. *Phys. Rev. Lett.*, 68:2555.

Chen, S. and Doolen, G. D. (1998). Lattice Boltzmann method for fluid flows. *Ann. Rev. Fluid Mech.*, 30:329.

Chen, S., Martinez, D., and Mei, R. (1996). On boundary conditions in lattice Boltzmann methods. *Phys. Fluids*, 8:2527–2536.

Chen, X. and Xu, B. (2002). The Nusselt number and friction factor at the entrance of a micro-duct. *Int. Comm. Heat and Mass Transfer*, 29(6):763–772.

Chen, Y., Haddon, R., Fang, S., Rao, A., Eklund, P., Lee, W., Dickey, E., Grulke, E., Pendergrass, J., Chavan, A., Haley, B., and Smalley, R. (1998). Chemical attachment of organic functional groups to single-walled carbon nanotube material. *J. Mater. Res.*, 13(9):2423–2431.

Chen, Y. and White, J. (2000). A quadratic method for nonlinear model order reduction. In *Modeling and Simulation of Microsystems*, pages 477–480.

Cheng, H. K. and Emmanuel, G. (1995). Perspectives on hypersonic nonequilibrium flow. *AIAA J.*, 33:385–400.

Cheng, J., Sheldon, E., Wu, L., Uribe, A., Gerrue, L., Carrino, J. Heller, M., and O'Connell, J. (1998). Preparation and hybridization analysis of DNA/RNA from E. *coli* on microfabricated bioelectronic chips. *Nat. Biotechnol.*, 16:541–546.

Cheng, J.-T. and Giordano, N. (2002). Fluid flow through nanometer-scale channels. *Phys. Rev. E*, 65:031206.

Cheng, L., Fenter, P., Nagy, K. L., Schlegel, M. L., and Sturchio, N. C. (2001). Molecular-scale density oscillations in water adjacent to a mica surface. *Phys. Rev. Lett.*, 87(15):156103.

Chiou, P. Y., Moon, H., Toshiyoshi, H., Kim, C.-J., and Wu, M. C. (2003). Light actuation of liquid by optoelectrowetting. *Sens. Actuators, A*, 104:222–228.

Cho, S. K., Moon, H., and Kim, C.-J. (2003). Creating, transporting, cutting, and merging liquid droplets by electrowetting-based actuation for digital microfluidic circuits. *J. Microelectromech. Sys.*, 12(1):70–79.

Choi, C.-H., Westin, J., and Breuer, K. (2003). Apparent slip flows in hydrophilic and hydrophobic microchannels. *Phys. Fluids*, 15(10):2897–2902.

Chu, W. H., Chin, R. C., and Huen, T. (1999). Silicon membrane nanofilters from sacrificial oxide removal. *J. Microelectromech. Sys.*, 8(1):34–42.

Chuarev, N. V., Sobolev, V. D., and Somov, A. N. (1984). Slippage of liquids over lyophobic solid surfaces. *J. Colloid Interface Sci.*, 97:574–581.

Cieplak, M., Koplik, J., and Banavar, J. R. (2001). Boundary conditions at fluid-solid interface. *Phys. Rev. Lett.*, 86(5):803–806.

Clift, R., Grace, J. R., and Weber, M. E. (1978). *Bubbles, Drops, and Particles.* Academic Press, New York.

Climent, E., Maxey, M. R., and Karniadakis, G. E. (2004). Dynamics of self-assembled chaining in magnetorheological fluids. *Langmuir*, 20:507–513.

Cockburn, B., Karniadakis, G. E., and Shu, C.-W. (2000). *Discontinuous Galerkin methods: Theory, Computation and Applications.* Springer-Verlag, Heidelberg.

Colin, S., LaLonde, P., and Caen, R. (2004). Validation of a second-order slip flow model in rectangular microchannels. *Heat Transfer Engineering*, 25(3):23–30.

Conlisk, A. T., McFerran, J., Zheng, Z., and Hansford, D. (2002). Mass transfer and flow in electrically charged micro- and nanochannels. *Anal. Chem.*, 74:2139–2150.

Craig, V., Neto, C., and Williams, D. (2001). Shear-dependent boundary slip in an aqueous Newtonian liquid. *Phys. Rev. Lett.*, 87(5):054504.

Craighead, H. G. (2000). Nanoelectromechanical systems. *Science*, 290:1532–1535.

Crocker, J. C. and Grier, D. G. (1996). Methods of digital video microscopy for colloidal studies. *J. Colloid Interface Sci.*, 179:298–310.

Culbertson, C. T., Ramsey, R. S., and Ramsey, J. M. (2000). Electroosmotically induced hydraulic pumping on microchips: Differential ion transport. *Anal. Chem.*, 72(10):2285–2291.

Cummings, E. B. (2001). Ideal electrokinesis and dielectrophoresis in arrays of insulating posts. In *AIAA, AIAA2001-1163*, pages 1–10.

Cummings, E. B., Griffiths, S. K., and Nilson, R. H. (1999). Irrotationality of uniform electroosmosis. In *Proc. SPIE Microfluidic Devices and Systems II*, pages 180–189.

Cummings, E. B., Griffiths, S. K., Nilson, R. H., and Paul, P. H. (2000). Conditions for similitude between the fluid velocity and electric field in electroosmotic flow. *Anal. Chem.*, 72(11):2526–2532.

Cummings, E. B. and Singh, A. K. (2000). Dielectrophoretic trapping without embedded electrodes. In *Proc. of SPIE Conferences on Micromachining and Microfabrication*, pages 164–173.

da Silva, M. G., Greiner, K., Liu, C. H., Kenny, T. W., and Gilbert, J. R. (1999). Squeeze film damping effects in a micromachined tunneling accelerometer. In *Proceedings of 1999 ASME IMECE Meeting MEMS-Vol 1*, pages 87–92, Nashville, TN.

Dance, S. and Maxey, M. R. (2003). Incorporation of lubrication effects into the force-coupling method for particulate two-phase flow. *J. Comp. Phys.*, 189:212–238.

Darden, T., York, D., and Pedersen, L. (1993). Particle mesh Ewald: An N-log(N) method for Ewald sums in large systems. *J. Chem. Phys.*, 98:10089–10092.

Darhuber, A. A., Davis, J. M., Troian, S. M., and Reisner, W. (2003). Thermocapillary actuation of liquid flow on chemically patterned surfaces. *Phys. of Fluids*, 15(5):1295–1304.

Darhuber, A. A., Troian, S. M., Miller, S. M., and Wagner, S. (2000). Morphology of liquid microstructures on chemically patterned surfaces. *J. Appl. Phys.*, 87(11):7768–7775.

Darhuber, A. A., Troian, S. M., and Reisner, W. W. (2001). Dynamics of capillary spreading along hydrophilic microstipes. *Phys. Rev. E*, 64:031603–1 031603–8.

Davies, J. T. and Rideal, E. K. (1966). *Interfacial Phenomena.* Academic Press Inc.: London.

Davis, J. M. and Troian, S. M. (2003). Influence of attractive van der Waals interactions on the optimal excitations in thermocapilary-driven spreading. *Phys. Rev. E*, 67:016308–1 016308–9.

De Leeuw, S. W., Perram, J. W., and Smith, E. R. (1980). Simulation of electrostatic systems in periodic boundary conditions. I. Lattice sums and dielectric constant. *Proc. R. Soc. Lond.*, A373:27–56.

Deane, A., Kevrekidis, I. G., Karniadakis, G. E., and Orszag, S. A. (1991). Low-dimensional models for complex geometry flows: Application to grooved channels and circular cylinders. *Phys. Fluids*, 3(10):2337.

Debye, P. and Cleland, R. L. (1959). Flow of liquid hydrocarbons in porous Vycor. *J. Appl. Phys.*, 30:843–849.

DeGasperis, G., Yang, J., Becker, F. F., and Gascoyne, P. R. (1999). Microfluidic cell separation by two-dimensional dielectrophoresis. *Biomed. Microdevices*, 2:1:41–49.

deGennes, P. (1985). Wetting: Statics and dynamics. *Rev. Mod. Phys.*, 57:827–863.

deGennes, P. (2002). On fluid/wall slippage. *Langmuir*, 18:3413–3414.

deGennes, P., Brochard-Wyart, F., and Quere, D. (2004). *Capillarity and Wetting Phenomena.* Springer, New York.

Deissler, R. G. (1964). An analysis of second-order slip flow and temperature jump boundary conditions for rarefied gases. *Int. J. Heat Mass Transfer*, 7:681–694.

Dekker, C. (1999). Carbon nanotubes as molecular quantum wires. *Physics Today*, 52(5):22–28.

Dent, G. L. (1999). *Aspects of particle sedimentation in dilute flows at finite Reynolds numbers*. PhD thesis, Brown University.

Derbunovich, G. I., Zamskaya, A. S., Repik, Y. U., and Sosedko, Y. P. (1998). Hydraulic drag on perforated plates. *Fluid Mech.-Soviet Research*, 13:111–116.

Derjaguin, B. V. (1970). Superdense Water. *Scientific American*, 223 (5):52.

Derjaguin, B. V., Popovskif, Y. M., and Altoiz, B. A. (1983). Liquid-crystalline state of the wall-adjacent layers of some polar liquids. *J. Colloid Interface Sci.*, 96:492–503.

Deshmukh, A. A., Liepmann, D., and Pisano, A. P. (2000). Continuous micromixer with pulsatile micropumps. In *Solid-State Sensor and Actuator Workshop, Hilton Head Island, SC, USA, 4-8 June*, volume 73-6.

Desjardins, B., Grenier, E., Lions, P.-L., and Masmoudi, N. (1999). Incompressible limit for solutions of the isentropic Navier-Stokes equations with Dirichlet boundary conditions. *J. Math. Pures Appl.*, 78(5):461–471.

Deval, J., Tabeling, P., and Ho, C. H. (2002). A dielectrophoretic chaotic mixer. In *MEMS 2002, IEEE International Conference, Las Vegas, January 20-24*.

Dietrich, S. and Boyd, I. D. (1996). Scalar and parallel optimized implementation of the direct simulation Monte Carlo method. *J. Comp. Phys.*, 126:328–342.

Diez, J. A., Kondic, L., and Bertozzi, A. (2000). Global models for moving contact lines. *Phys. Rev. E*, 63:011208–1 011208–13.

Digilov, R. (2000). Charge-induced modification of contact angle: The secondary electrocapillary effect. *Langmuir*, 16:6719–6723.

Dinsmore, A. D., Hsu, M. F., Nicolaides, M. G., Marquez, M., Bausch, A. R., and Weitz, D. A. (2002). Colloidosomes: Selectively permeable capsules composed of colloidal particles. *Science*, 298:1006–1009.

DiPerna, R. J. and Lions, P.-L. (1991). Global solutions of Boltzmann's equation and the entropy inequality. *Arch. Rational Mech. Anal.*, 114(1):47–55.

Donea, J., Giuliani, S., and Halleux, J. P. (1982). An arbitrary Lagrangian-Eulerian finite element method for transient dynamic fluid-structure interactions. *Comput. Methods Appl. Mech. Eng.*, 33:689.

Dong, S., Liu, D., Maxey, M. R., and Karniadakis, G. E. (2004). Spectral distributed multiplier (DLM) method: Algorithm and benchmark tests. *J. Comp. Phys.*, 195:695–717.

D'Orazio, F., Bhattacharja, S., Halperin, W. P., and Gerhardt, R. (1989). Enhances self-diffusion of water in restricted geometry. *Phys. Rev. Lett.*, 63:43–46.

Dose, E. V. and Guiochon, G. (1993). Time scales of transient processes in capillary electrophoresis. *J. Chromatogr.*, 652:263–275.

Doyle, D., Cabral, J., Pfuetzner, R., an J. Gulbis, A. K., Cohen, S., Chait, B., and MacKinnon, R. (1998). The structure of the potassium channel. Molecular basis of K-conduction and selectivity. *Science*, 280:69–77.

Doyle, P. S., Bibette, J., Bancaud, A., and Viory, J.-L. (2002). Self-assembled magnetic matrices for DNA separation chips. *Science*, 295:2237.

Dozier, W. D., Drake, J. M., and Klafter, J. (1986). Self-diffusion of a molecule in porous Vycor glass. *Phys. Rev. Lett.*, 56:197–200.

Drazer, G., Koplik, J., Acrivos, A., and Khusid, B. (2002). Adsorption phenomena in the transport of a colloidal particle through a nanochannel containing a partially wetting fluid. *Phys. Rev. Lett.*, 89:244501.

Duh, D. and Mier-Y-Teran, L. (1997). An analytical equation of state for the hard-core Yukawa fluid. *Mol. Phys.*, 90:373–379.

Dussan, E. (1979). On the spreading of liquids on solid surfaces: static and dynamic contact lines. *Ann. Rev. Fluid Mech.*, 11:371–400.

Dutta, D. and Leighton, D. T. (2001). Dispersion reduction in pressure-driven flow through microetched channels. *Anal. Chem.*, 73:504–513.

Dutta, D. and Leighton, D. T. (2002). A low dispersion geometry for microchip separation devices. *Anal. Chem.*, 74:1007–1016.

Dutta, D. and Leighton, D. T. (2003). Dispersion in large aspect ratio microchannels for open-channel liquid chromatography. *Anal. Chem.*, 75:57–70.

Dutta, P. and Beskok, A. (2001a). Analytical solution of combined electroosmotic/pressure driven flows in two-dimensional straight channels: Finite Debye layer effects. *Anal. Chem.*, 73(9):1979–1986.

Dutta, P. and Beskok, A. (2001b). Analytical solution of time periodic electroosmotic flows: Analogies to Stokes' second problem. *Anal. Chem.*, 73(21):5097–5102.

Dutta, P., Beskok, A., and Warburton, T. (2002a). Electroosmotic flow control in complex microgeometries. *J. Microelectromech. Sys.*, 11(1):36–44.

Dutta, P., Beskok, A., and Warburton, T. (2002b). Numerical simulation of mixed electroosmotic/pressure driven microflows. *Numer. Heat Transfer, Part-A: Applications*, 41(2):131–148.

Ebert, H. and Albrand, K. R. (1963). The applicability of gas flows in the molecular range. *Vacuum*, 13:563.

Egorov, A. and Brodskaya, E. (2003). The effect of ions on solid-liquid phase transition in small water clusters. A molecular dynamics study. *J. Chem. Phys.*, 118:6380–6386.

Ehrfeld, W. (2003). Electrochemistry and microsystems. *Electrochimica Acta*, 48:2857–2868.

Eikerling, M. and Kornyshev, A. (2001). Proton transfer in a single pore of a polymer electrolyte membrane. *J. Electroana. Chem.*, 502.

Elrod, M. J. and Saykally, R. (1994). Many body effects in intermolecular forces. *Chem. Rev.*, 94:1975–1997.

Engl, W. L., Laur, R., and Dirks, H. K. (1982). MEDUSA–A simulator for modular circuits. *IEEE Transactions on Computer-Aided Design of Integrated Circuits*, 1:85–93.

Erickson, D. and Li, D. (2002). Influence of surface heterogeneity on electrokinetically driven microfluidic mixing. *Langmuir*, 18(5):1883–1893.

Ermakov, S. V., Bello, M. S., and Righetti, P. G. (1994). Numerical algorithms for capillary electrophoresis. *J. Chromatogr. A*, 661:265–278.

Ermakov, S. V., Jacobson, S. C., and Ramsey, J. M. (1998). Computer simulation of electrokinetic transport in micro fabricated channel structures. *Anal. Chem.*, 70(21):4494–4504.

Ermakov, S. V., Jacobson, S. C., and Ramsey, J. M. (1999). Computer simulation of electrokinetic mass transport in microfabricated fluidic devices. In *Proc. of Modeling and Simulation of Microsystems*, pages 534–537.

Ermakov, S. V., Jacobson, S. C., and Ramsey, J. M. (2000). Computer simulations of electrokinetic injection techniques in microfluidic devices. *Anal. Chem.*, 72(15):3512–3517.

Ermakov, S. V., Mazhorova, O. S., and Popov, Y. P. (1992). Mathematical modeling of the electrophoretic separation of biological mixtures. *Translated from Differentsial'nye Uravneniya*, 12:2129–2137.

Espanol, P. (1995). Hydrodynamics for dissipative particle dynamics. *Phys. Rev. E*, 52:1734.

Espanol, P. and Warren, P. (1995). Statistical mechanics of dissipative particle dynamics. *Europhys. Lett.*, 30(4):191–196.

Essmann, U., Perera, L., Berkowitz, M., Darden, T., Lee, H., and Pedersen, L. (1995). A smooth particle mesh Ewald potential. *J. Chem. Phys.*, 103:8577–8592.

Estrelalopis, V. R., Ulberg, Z. R., and Koniashvili, S. A. (1982). Mechanism of electrophoretic deposition of aqueous dispersions in a pulsed field. *Colloid Journal of the USSR*, 44:74–79.

Evans, D. J. and Morriss, G. P. (1990). *Statistical Mechanics of Non-Equilibrium Liquids*. Academic Press, London.

Evans, J. D., Lipemann, D., and Pisano, A. P. (January 26-30, 1997). Planar laminar mixer. In *MEMS-97, The Tenth Annual International Workshop on MEMS*.

Fan, J. and Shen, C. (1999). Statistical simulation of low-speed unidirectional flows in transition regime. In Brun et al., editor, *Rarefied Gas Dynamics*, volume 2, pages 245–252.

Fedder, G. K. and Jing, Q. (1998). Nodas 1.3: Nodal design of actuators and sensors. In *IEEE/VIUF Int. Workshop on Behavioral Modeling and Simulation*.

Feynman, R. P., Leighton, R. B., and Sands, M. (1977). *The Feynman Lectures on Physics*. Addison-Wesley, Massachusetts.

Fiedler, S., Shirley, S. Schnelle, T., and Fuhr, G. (1998). Dielectophoretic sorting of particles and cells in a microsystem. *Anal. Chem.*, 70:1909–1915.

Fischer, J. and Methfessel, M. (1980). Born-Green-Yvon approach to the local densities of a fluid at interfaces. *Phys. Rev. A*, 22(6):2836–2843.

Flekkoy, E. G., Wagner, G., and Feder, J. (2000). Hybrid model for combined particle and continuum dynamics. *Europhys. Lett.*, 52(3):271–276.

Fletcher, P. D., Haswell, S. J., and Paunov, V. N. (1999). Theoretical considerations of chemical reactions in micro-reactors under electroosmotic and electrophoretic control. *The Analyst*, 124:1273–1282.

Forrest, B. M. and Suter, U. W. (1995). Accelerated equilibration of polymer melts by time-coarse-graining. *J. Chem. Phys.*, 102(18):7256–7266.

Fortes, A. F., Joseph, D. D., and Lundgren, T. S. (1987). Nonlinear mechanics of fluidization of beds of spherical particles. *J. Fluid Mech.*, 177:467–483.

Freeman, D. M., Aranyosi, A. J., Gordon, M. J., and Hong, S. S. (1998). Multidimensional motion analysis of MEMS using computer microvision. In *Solid-State Sensor and Actuator Workshop, Hilton Head Is., S.C.*, pages 150–155.

Fremerey, J. K. (1982). Spinning rotor vacuum gauges. *Vacuum*, 32:685–690.

Fremerey, J. K. (1985). The spinning rotor gauge. *J. Vacuum Science Technology A*, 3(3):1715–1720.

Frenkel, D. and Smit, B. (2002). *Understanding Molecular Simulation.* Academic Press.

Freund, J. (2002). Electro-osmosis in a nanometer-scale channel studied by atomistic simulation. *J. Chem. Phys.*, 116(5):2194–2200.

Freund, R. W. (1999). Reduced order modeling techniques based on Krylov subspaces and their use in circuit simulations. *Applied and Computational Control, Signals and Circuits*, 1:435–498.

Frisch, U., Hasslacher, B., and Pomeau, Y. (1986). Lattice Gas Automaton for the Navier-Stokes equation. *Phys. Rev. Lett.*, 56:1505.

Fukui, S. and Kaneko, R. (1988). Analysis of ultra thin gas film lubrication based on linearized Boltzmann equation: First report: derivation of a generalized lubrication equation including thermal creep flow. *J. Tribology*, 110:253–262.

Fukui, S. and Kaneko, R. (1990). A database for interpolation of Poiseuille flow rates for high Knudsen number lubrication problems. *J. Tribology*, 112:78–83.

Furst, E. M. and Gast, A. P. (2000). Micromechanics of magnetorheological suspensions. *Phys. Rev. E*, 61(6):6732–6739.

Furst, E. M., Suzuki, C., Fermigier, M., and Gast, A. P. (1998). Aggregation kinetics of paramagnetic coloidal particles. *Langmuir*, 14:7334–7336.

Gabbay, L. and Senturia, S. D. (1998). Automatic generation of dynamic macromodels using quasi-static simulations in combination with modal analysis. In *Technical Digest: Solid State Sensors and Actuators Workshop*, pages 197–200.

Gad-el-Hak, M. (1999). The fluid mechanics of microdevices. *J. Fluids Eng.*, 12(1):5–33.

Gad-El-Hak, M. (2001). *Handbook of MEMS.* CRC Press.

Gaede, W. (1913). Die äussere Reibung der Gase. *Annalen der Physik*, 41:289.

Galle, J. and Vortler, H. (1999). Monte Carlo simulation of primitive water in slit-like pores: Networks and clusters. *Surf. Sci.*, 421:33–43.

Galle, J. and Vortler, H. (2001). Monte Carlo simulation of primitive water in planar slits with molecularly rough walls: Surface self-organization. *Surf. Sci.*, 481:39–53.

Gallo, P., Rapinesi, M., and Rovere, M. (2002a). Confined water in low hydration regime. *J. Chem. Phys.*, 117:369–375.

Gallo, P., Ricci, M. A., and Rovere, M. (2002b). Layer analysis of the structure of water confined in vycor glass. *J. Chem. Phys.*, 116:342–346.

Gallo, P., Rovere, M., Ricc, M. R., Harting, C., and Sphor, E. (1999). Evidence of glassy behavior of water molecules in confined states. *Philosophical Magazine*, 79:1923–1930.

Gallo, P., Rovere, M., and Spohr, E. (2000). Glass transition and layering effects in confined water: A computer simulation study. *J. Chem. Phys.*, 113:11324–11334.

Gampert, B. (1976). Inlet flow with slip. In *Proceedings of the Tenth International Symposium on Rarefied Gas Dynamics*, pages 225–235.

Gampert, B. (1978). Low Reynolds number TAC-slip flow past a circular cylinder. In *Proceedings of the Eleventh International Symposium on Rarefied Gas Dynamics*, pages 255–267.

Garcia, A. L., Bell, J. B., Crutchfield, W. Y., and Alder, B. J. (1999). Adaptive mesh and algorithm refinement using direct simulation Monte Carlo. *J. Comp. Phys.*, 154(1):134–155.

Garcia-Archilla, B., Novo, J., and Titi, E. S. (1998). Postprocessing the Galerkin method: A novel approach to approximate inertial manifold. *SIAM J. Numer. Anal.*, 35:941–972.

Gascoyne, P., Huang, Y., Pethig, R., Vykuokal, J., and Becker, F. (1992). Dielectrophoretic separation of mammalian-cells studied by computerized image-analysis. *Meas. Sci. and Tech*, 3:439–445.

Gascoyne, P. and Wang, X. (1997). Dielectrophoretic separation of cancer cells from blood. *IEEE Transactions on Industry Application*, 33(3):670–678.

Gast, A. P. and Zukoski, C. F. (1989). Electrorheological fluids as colloidal suspensions. *Adv. Col. Inter. Sci.*, 30:153.

Gatsonis, N. A., Nanson, R. A., and Le Beau, G. J. (2000). Simulations of cold-gas nozzle and plume flows and flight data comparisons. *Journal of Spacecraft and Rockets*, 37(1):39–48.

Gau, H., Herminghaus, S., Lenz, P., and Lipowsky, R. (1999). Liquid morphologies on structured surfaces: From microchannels to microchips. *Science*, 283:46–49.

Ghosal, S. (2002a). Band broadening in a microcapillary with a stepwise change in the zeta-potential. *Anal. Chem.*, 74 (16):4198–4203.

Ghosal, S. (2002b). Effect of analyte adsorption on the electroosmotic flow in microfluidic channels. *Anal. Chem.*, 74 (4):771–775.

Ghosal, S. (2002c). Lubrication theory for electro-osmotic flow in a microfluidic channel of slowly varying cross-section and wall charge. *J. Fluid Mech.*, 459:103–128.

Ghosal, S. (2003). The effect of wall interactions in capillary-zone electrophoresis. *J. Fluid Mech.*, 491:285–300.

Giridharan, M. G. and Krishnan, A. (1998). An implicit numerical model for electrophoretic systems. In *Proc. ASME International Mechanical Engineering Congress and Exposition, Micro-Electro-Mechanical Systems (MEMS)*, pages 61–68.

Glowinsky, R., Pan, T.-W., Hesla, T. I., and Joseph, D. D. (1999). A distributed Lagrange multiplier/fictitious domain method for particulate flows. *Int. J. Multiphase Flow*, 25:755–794.

Golding, R. K., Lewis, P. C., Kumacheva, E., Allard, M., and Sargent, E. H. (2004). In situ study of colloid crystallization in constrained geometry. *Langmuir*, 20:1414–1419.

Goldman, A. J., Cox, R. G., and Brenner, H. (1967). Slow viscous motion of a sphere parallel to a plane wall: I. motion through a quiescent fluid. *Chem. Eng. Sci.*, 22:637–651.

Golse, F. (1989). Applications of the Boltzmann equation within the context of upper-atmosphere vehicle aerodynamics. *Comp. Meth. Appl. Mech. Eng.*, 75:299.

Gonnet, P., Zimmerli, U., Walther, J. H., Werder, T., and Koumoutsakos, P. (2004). Wetting and hydrophobicity of nanoscale systems with impurities. *Proceedings of the ICCN,NSTI-Nanotech 2004, www.nsti.org*, 3:69–72.

Gonzalez, A., Ramos, A., Green, N., Castellanos, A., and Morgan, H. (2000). Fluid flow induced by nonuniform AC electric fields in electrolytes on microelectrodes: II. a linear double layer analysis. *Phys. Rev. E*, 61(4):4019–4028.

Gordillo, M. and Martí, J. (2003). Water on the outside of carbon nanotube bundles. *Phys. Rev. E*, 67:205425.

Grad, H. (1949). On the kinetic theory of rarefied gases. *Comm. Pure Appl. Math.*, 2:331–407.

Graham, M. D. and Kevrekidis, I. G. (1996). Alternate approaches to the Karhunen-Loève decomposition for model reduction and data analysis. *Computer Chem. Engng.*, 20:495–506.

Grateful, T. M. and Lightfoot, E. N. (1992). Finite difference modelling of continuous-flow electrophoresis. *J. Chromatogr.*, 594:341–349.

Green, M. and Lu, J. (1997). Simulation of water in a small pore: Effect of electric field and density. *J. Phys. Chem. B*, 101:6512–6524.

Green, N. and Morgan, H. (1997). Rapid Communication: Dielectrophoretic separation of nano-particles. *J. Phys. D: Allp. Phys.*, 30:L41–L44.

Green, N. and Morgan, H. (1998). Separation of submicrometer particles using a combination of dielectrophoretic and electrohydrodynamic forces. *J. Phys. D: Appl. Phys.*, 3:25–30.

Green, N., Ramos, A., Gonzalez, A., Morgan, H., and Castellanos, A. (2000). Fluid flow induced by nonuniform AC electric fields in electrolytes on microelectrodes: I. Experimental measurements. *Phys. Rev. E*, 61(4):4011–4018.

Green, N., Ramos, A., Gonzalez, A., Morgan, H., and Castellanos, A. (2002). Fluid flow induced by nonuniform AC electric fields in electrolytes on microelectrodes: III. Observation of streamlines and numerical simulation. *Phys. Rev. E*, 66:Art. No. 026305.

Greengard, L. (1987). *The Rapid Evaluation of Potential Fields in Particle Systems*. MIT Press, Cambridge, MA.

Greengard, L. and Rokhlin, V. (1987). Lattice-sum methods for calculating electrostatic interactions in molecular simulations. *J. Comput. Phys.*, 73:325–348.

Greenwood, P. A. and Greenway, G. M. (2002). Sample manipulation in micro total analytical systems. *Trends in Analytical Chemistry*, 21:726–740.

Griffiths, S. K. and Nilson, R. H. (1999). Hydrodynamic dispersion of a neutral nonreacting solute in electroosmotic flow. *Anal. Chem.*, 71:5522–5529.

Griffiths, S. K. and Nilson, R. H. (2000). Electroosmotic fluid motion and late-time solute transport for large zeta potentials. *Anal. Chem.*, 72:4776–4777.

Grisnik, T. A., Smith, S. P., and Saltz, L. E. (1987). Experimental study of low Reynolds number nozzles. In *Paper AIAA 87-0992, 19th AIAA/DGLR/JSASS International Electric Propulsion Conference.*

Groot, R. D. and Warren, P. B. (1997). Dissipative particle dynamics: Bridging the gap between atomistic and mesoscopic simulation. *J. Chem. Phys.*, 107(11):4423–4435.

Gross, E. P. and Ziering, S. (1958). Kinetic theory of linear shear flow. *Phys. Fluids A*, 1:215–224.

Gubskaya, A. V. and Kusalik, P. G. (2002). The total molecular dipole moment for liquid water. *J. Chem. Phys.*, 117:5290–5302.

Gulczinski, F. S., Dulligan, M. J., Lake, J. P., and Spanjer, G. G. (2000). Micropropulsion research at AFRL. In *AIAA-2000-3255*.

Haberman, W. L. and Sayre, R. M. (1958). Motion of rigid and fluid spheres in stationary and moving liquids inside cylindrical tubes. David Taylor Model Basin Report 1143, U.S. Navy Dept., Washington D.C.

Hadjiconstantinou, N. G. (1999). Hybrid atomistic-continuum formulations and the moving contact-line problem. *J. Comp. Phys.*, 154:245–265.

Hadjiconstantinou, N. G. (2000). Analysis of discretization in the DSMC. *Phys. Fluids*, 12:2634.

Hadjiconstantinou, N. G. (2003a). Comment on Cercignani's second-order slip coefficient. *Phys. Fluids*, 15(8):2352–2354.

Hadjiconstantinou, N. G. (2003b). Dissipation in small scale gasous flows. *J. Heat Transfer*, 125:944–947.

Hadjiconstantinou, N. G., Garcia, A., Bazant, M., and He, G. (2003). Statistical error in particle simulations of hydrodynamic phenomena. *J. Comp. Phys.*, 187:274–297.

Hadjiconstantinou, N. G. and Patera, A. T. (1997). Heterogeneous atomistic-continuum representations for dense fluid systems. *I. J. Mod. Phys. C*, 8:967–976.

Haile, J. M. (1992). *Molecular Dynamics Simulation. Elementary Methods.* John Wiley & Sons, New York.

Halicioglu, T. and Jaffe, R. L. (2002). Solvent effect on functional groups attached to edges of carbon nanotubes. *Nano Lett.*, 2(6):573–575.

Halsey, T. C. and Toor, W. (1990). Structure of electrorheological fluids. *Phys. Rev. Lett.*, 65:2820.

Hamaker, H. C. (1940). Formation of a deposit by electrophoresis. *Trans. Faraday Soc.*, 36:279–283.

Hamaker, H. C. and Verwey, E. J. (1940). The role of the forces between the particles in electrodeposition and other phenomena. *Trans. Faraday Soc.*, 36:180–185.

Hardy, R. L. (1971). Multiquadric equations for topography and other irregular surfaces. *J. Geophys. Res.*, 176:1905–1915.

Harley, C. J. (1993). *Compressible Gas Flows in Microchannels and Microjets.* PhD thesis, University of Pennsylvania.

Harley, J. C., Huang, Y., Bau, H. H., and Zemel, J. N. (1995). Gas flow in micro-channels. *J. Fluid Mech.*, 284:257–274.

Hartnig, C., Witschel, W., and Sphor, E. (1998). Molecular dynamics study of the structure and dynamics of water in cylindrical pores. *J. Phys. Chem. B*, 102:1241–1249.

Hash, D. and Hassan, H. (1996). Assessment of schemes for coupling Monte Carlo and Navier-Stokes solution methods. *J. Thermophys. Heat Transfer*, 10(2):242–249.

Hash, D. and Hassan, H. (1997). Two-dimensional coupling issues of hybrid DSMC/Navier-Stokes solvers. *Proceedings of AIAA Conference*, pages AIAA 97–2507.

Hayes, M. A. and Ewing, A. J. (2000). Electroosmotic flow control and monitoring with an applied radial voltage for capillary zone electrophoresis. *Anal. Chem.*, 64:56–62.

Hayes, M. A., Polson, N. A., and Garcia, A. A. (2001). Active control of dynamic supraparticle structures in microchannels. *Langmuir*, 17:2866–2871.

Hayward, R. C., Saville, D. A., and Aksay, I. A. (2000). Colloidal crystals with tunable micropatterns: Current density modulation with uv-light during electrophoretic deposition. *Nature*, 404:56–59.

Heinbuch, U. and Fischer, J. (1989). Liquid flow in pores: Slip, no-slip, or multilayer sticking. *Phys. Rev. A*, 40(2):1144–1146.

Henderson, D., Busath, D. D., and Rowley, R. (2001). Fluids near surfaces and in pores and membrane channels. *Prog. Surf. Sci.*, 65:279–295.

Henderson, R. D. and Karniadakis, G. E. (1991). Hybrid spectral element-low order methods for incompressible flows. *J. Sci. Comp.*, 6(2):79.

Hermanson, K. D., Lumsdon, S. O., Williams, J. P., Kaler, E. W., and Velev, O. D. (2001). Dielectrophoretic assembly of electrically functional microwires from nanoparticle suspensions. *Science*, 294:1082–1086.

Herr, A., Molho, J., Santiago, J. G., Mungal, M. G., Kenny, T. W., and Garguilo, M. G. (2000). Electroosmotic capillary flow with non uniform zeta potential. *Anal. Chem.*, 72:1053–1057.

Heslot, F., Fraysse, N., and Cazabat, A. M. (1989). Molecular layering in the spreading of wetting liquid-drops. *Nature*, 338:640–642.

Hess, B. (2002). Determining the shear viscosity of model liquids from molecular dynamics simulations. *J. Chem. Phys.*, 116:209–217.

Heyes, D. M. (1981). Electrostatic potentials and fields in infinite point charge lattices. *J. Chem. Phys.*, 74:1924–1929.

Higuera, F. J. and Jiménez, J. (1989). Boltzmann approach to lattice gas simulations. *Europhys. Lett.*, 9:663–668.

Higuera, F. J. and Succi, S. (1989). Simulating the flow around a cylinder with a lattice Boltzmann equation. *Europhys. Lett.*, 8:517–521.

Hille, B. (2001). *Ion Channels of Excitable Membranes.* Sinauer Associates, Inc.

Hirt, C. W., Amsden, A. A., and Cook, H. K. (1974). An arbitrary Lagrangian-Eulerian computing method for all flow speeds. *J. Comp. Phys.*, 14:27.

Ho, C. M. (January 21-25, 2001). Fluidics – The link between micro and nano science and technologies. In *MEMS-2001, Interlaken, Switzerland.*

Ho, C. M. and Tai, Y. C. (1998). Micro-electro-mechanical systems (MEMS) and fluid flows. *Ann. Rev. Fluid Mech.*, 30:579–612.

Ho, L.-W. (1989). *A Legendre Spectral Element Method for Simulation of Incompressible Unsteady Free-Surface Flows.* PhD thesis, Massachusetts Institute of Technology.

Hockney, R. W. and Eastwood, J. W. (1981). *Computer Simulation Using Particles.* McGraw- Hill, New York.

Holdych, D. J., Noble, D. R., Georgiadis, J. G., and Buckius, R. O. (2004). Truncation error analysis of lattice Boltzmann methods. *J. Comp. Phys.*, 193:595–619.

Hoogerburgge, P. J. and Koelman, J. M. (1992). Simulating microscopic hydrodynamic phenomena with dissipative particle dynamics. *Europhy. Lett.*, 19(3):155–160.

Hoover, W. G. (1985). Canonical dynamics: equilibrium phase-space distributions. *Phys. Rev. A*, 31:1695–1697.

Horiuchi, K., Dutta, P., Chi, H., and Ivory, C. F. (2003). High resolution separation of proteins in a polymeric micro-fluidic chip. In *2003 ASME International Mechanical Engineering Congress*, pages IMECE2003–41206, Washington, D.C., November 15–12, 2003.

Horvath, J. and Dolnik, V. (2001). Polymer wall coatings for capillary electrophoresis. *Electrophoresis*, 22:644–655.

Hou, S., Zou, Q., Chen, S., Doolen, G. D., and Cogley, A. C. (1995). Simulation of incompressible Navier-Stokes fluid flows using a lattice Boltzann method. *J. Comp. Phys.*, 118:329.

Howe, R. T., Muller, R. S., Gabriel, K. J., and Trimmer, W. (1990). Silicon micromechanics: sensors and actuators on a chip,. *IEEE Spectrum*, 27(7):29–35.

Hsia, Y. T. and Domoto, G. A. (1983). An experimental investigation of molecular rarefaction effects in gas lubricated bearings at ultra low clearances. *Journal of Lubrication Technology*, 105:120–130.

Huang, A. B. and Stoy, R. L. (1966). Rarefied gas channel flow for three molecular models. *Phys. Fluids*, 9 (12):2327–2336.

Huang, W. D., Bogy, D. B., and Garcia, A. L. (1997). Three-dimensional direct simulation Monte Carlo method for slider airbearings. *Phys. Fluids*, 9(6):1764–1769.

Hummer, G., Rasaiah, J. C., and Noworyta, J. P. (2001). Water conduction through the hydrophobic channel of a carbon nanotube. *Nature*, 414:188–190.

Hung, E. S. and Senturia, S. D. (1999). Generating efficient dynamical models for microelectromechanical systems from a few finite-element simulation runs. *J. Microelectromech. Sys.*, 8:280–289.

Hunter, R. J. (1981). *Zeta Potential in Colloid Science: Principles and Applications.* Academic Press Inc, New York.

Ichikawa, K., Kameda, Y., Yamaguchi, T., Wakita, H., and Misawa, M. (1991). Neutron-diffraction investigation of the intramolecular structure of a water molecule in the liquid-phase at high-temperatures. *Mol. Phys.*, 73:79–86.

Ichiki, K. and Brady, J. F. (2001). Many-body effects and matrix inversion in low-Reynolds number hydrodynamics. *Phys. Fluids*, 13:350–353.

Ichimura, K., Oh, S.-K., and Nakawa, M. (2000). Light-driven motion of liquids on a photoresponsive surface. *Science*, 288:1624–1626.

Iijima, S. (1991). Helical microtubules of graphitic carbon. *Nature*, 354:56–58.

Inman, R. M. (1964). Laminar slip flow heat transfer in a parallel plate channel or a round tube with uniform wall heating. Technical report, NASA, Report D-2393.

Isenberg, C. (1992). *The Science of Soap Films and Soap Bubbles.* Dover, New York.

Ismail, Y. I. (2002). Efficient model order reduction via multi-node moment matching. In *Proceedings of ICCAD*, pages 767–774.

Israelachvili, J. (1992a). *Intermolecular and Surface Forces.* Academic Press, New York.

Israelachvili, J. N. (1992b). Adhesion forces between surfaces in liquids and condensable vapours. *Surf. Sci. Rep.*, 14:109–158.

Ito, T., Sun, L., and Crooks, R. M. (2003). Simultaneous determination of the size and surface charge of individual nanoparticles using a carbon nanotube-based Coulter Counter. *Anal. Chcm.*, 75(10):2399–2406.

Ivanov, M. S., Markelov, G. N., Ketsdever, A. D., and Wadsworth, D. C. (1999). Numerical study of cold gas micronozzle flows. In *Paper AIAA 99-0166, 37th Aerospace Sciences Meeting & Exhibit, January 11–14, Reno, NV.*

Jacobs, H. O., Tao, A. R., Schwartz, A., Gracias, D. H., and Whitesides, G. M. (2002). Fabrication of a cylindrical display by patterned assembly. *Science*, 296(323-325):20–31.

Jacobson, S. C., McKnight, T. E., and Ramsey, J. M. (1999). Microfluidic devices for electrokinetically driven parallel and serial mixing. *Anal. Chem.*, 71:4455–4459.

Janos, P. (1999). Role of chemical equilibria in the capillary electrophoresis of inorganic substances. *J. Chromatogr. A*, 834:3–20.

Janson, S. W. and Helvajian, H. (1997). Batch-fabricated microthrusters: Initial results. In *Paper AIAA 96-2988, 32nd AIAA/ASME/SAE/ASEE Joint Propulsion Conference & Exhibit,*.

Jeffrey, D. J. (1982). Low Reynolds number flow between converging spheres. *Mathematika*, 29:58–66.

Jin, W., Koplik, J., and Banavar, J. R. (1997). Wetting hysteresis at the molecular scale. *Phys. Rev. Lett.*, 78:1520–1523.

Jin, X., Li, G., and Aluru, N. R. (2001). On the equivalence between least-squares and kernel approximation in meshless methods. *CMES: Computer Modeling in Engineering and Sciences*, 2(4):447–462.

Jin, X., Li, G., and Aluru, N. R. (2004). Positivity conditions in meshless collocation methods. *Comput. Methods Appl. Mech. Eng.*, 193(12-14):1171–1202.

Johnson, A. A. and Tezduyar, T. E. (1996). Simulation of multiple spheres falling in a liquid-filled tube. *Comput. Meth. Appl. Mech. Eng.*, 134:351–373.

Jones, S. W. (1991). The enhancement of mixing by chaotic advection. *Phys. Fluids A*, 3(5):1081–1086.

Jones, S. W., Thomas, O. M., and Aref, H. (1989). Chaotic advection by laminar flow in a twisted pipe. *J. Fluid Mech.*, 209:335–357.

Jones, T. B. (2001). Liquid dielectrophoresis on the microscale. *J. of Electrostatics*, 51-52:290–299.

Jones, T. B., Fowler, J. D., Chang, Y. S., and Kim, C.-J. (2003). Frequency-based relationship of electrowetting and dielectrophoretic liquid microactuation. *Langmuir*, 19:7646–7651.

Jones, T. B., Gunji, M., and Feldman, H. J. (2001). Dielectrophoretic liquid actuation and nanodroplet formation. *J. Apply. Phys.*, 89:1441–1448.

Kaiktsis, L., Karniadakis, G. E., and Orszag, S. A. (1996). Unsteadiness and convective instabilities in flow over a backward-facing step. *J. Fluid Mech.*, 321:157.

Kalko, S. G., Hernandez, J. A., Grigera, J. R., and Fischbarg, J. (1995). Osmotic permeability in molecular dynamics simulation of water transport through a single-occupancy pore. *Biochimica et Biophysica Acta*, 1240:159–166.

Kalyuzhnyi, Y. V. and Cummings, P. T. (1996). Phase diagram for the Lennard-Jones fluid modelled by the hard-core Yukawa fluid. *Mol. Phys.*, 87:1459–1462.

Kang, M. S. and Martin, C. R. (2001). Investigations of potential-dependent fluxes of ionic permeates in gold nanotubule membranes prepared via the template method. *Langmuir*, 17(9):2753–2759.

Kaniansky, D., Masar, M., and Bielcikova, J. (1997). Electroosmotic flow separation for capillary zone electrophoresis in a hydrodynamically closed separation system. *J. Chromatogr. A*, 792:483–494.

Kansa, E. J. (1990a). Multiquadrics – a scattered data approximation scheme with applications to computational fluid dynamics – I, surface approximations and partial derivative estimates. *Comp. Math. Appl.*, 19:127–145.

Kansa, E. J. (1990b). Multiquadrics – a scattered data approximation scheme with applications to computational fluid dynamics – II, solutions to parabolic, hyperbolic and elliptic partial differential equations. *Comp. Math. Appl.*, 19:147–161.

Karlin, I. V., Ferrante, A., and Ottinger, H. C. (1999). Perfect entropy functions of the lattice Boltzmann method. *Europhys. Lett.*, 47:182–188.

Karniadakis, G. E., Israeli, M., and Orszag, S. A. (1991). High-order splitting methods for incompressible Navier-Stokes equations. *J. Comp. Phys.*, 97:414.

Karniadakis, G. E., Orszag, S. A., Ronquist, E. M., and Patera, A. T. (1993). Spectral element and lattice gas methods for incompressible fluid dynamics. In Gunzburger, M. and Nicolaides, R., editors, *Incompressible Computational Fluid Dynamics*. Cambridge University Press, New York.

Karniadakis, G. E. and Sherwin, S. J. (1999). *Spectral/hp Element Methods for CFD*. Oxford University Press.

Kasianowicz, J. J., Brandin, E., Branton, D., and Beamer, D. W. (1996). Characterization of individual polynucleotide molecules using a membrane channel. *Proc. Natl. Acad. Sci. (USA)*, 93.

Kataoka, D. E. and Troian, S. M. (1999). Patterning liquid flow on the microscopic scale. *Nature*, 402:794–797.

Keh, H. and Liu, Y. C. (1995). Electrokinetic flow in a circular capillary with a surface charge layer. *J. Colloids and Interface Surfaces*, 172:222–229.

Keh, H. and Tseng, H. C. (2001). Transient electrokinetic flow in fine capillaries. *J. Colloid Interface Sci.*, 242:450–459.

Keh, H. J. and Anderson, J. L. (1985). Boundary effects on electrophoretic motion of colloidal spheres. *J. Fluid Mech.*, 153:417–439.

Kemery, P. J., Steehler, J. K., and Bohn, P. W. (1998). Electric field mediated transport in nanometer diameter channels. *Langmuir*, 14:2884–2889.

Kennard, E. H. (1938). *Kinetic Theory of Gasses*. McGraw-Hill Book Co. Inc., New York.

Ketsdever, A. D., Green, A. A., Muntz, E. P., Wadsworth, D., and Vargo, S. E. (2000a). Heat transfer measurements and calculations for a MEMS fabricated resistojet: Initial Results. In *34th AIAA Thermophysics Conference, June 19-22, 1995, Denver, CO, AIAA 2000-2505.*

Ketsdever, A. D., Wadsworth, D., and Muntz, E. P. (2000b). Influence of gas-surface interaction models on predicted performance of a micro-resistojet. In *34th AIAA Thermophysics Conference, June 19-22, 1995, Denver, CO, AIAA 2000-2430.*

Ketsdever, A. D., Wadsworth, D. C., Vargo, S., and Muntz, E. P. (1998). The free molecule micro-resistojet: An interesting alternative to nozzle expansion. In *Paper AIAA 98-3918, 34th AIAA/ASME/SAE/ASEE Joint Propulsion Conference & Exhibit, July 13–15, Cleveland, OH.*

Kihm, K. D., Banarjee, A., Choi, C. K., and Takagi, T. (2004). Near-wall hindered Brownian diffusion of nanoparticles examined by three dimensional ratiometric total internal reflection fluorescence microscopy (3D R-TIRFM). *Experiments in Fluids*, 37:811–824.

Kim, E., Xia, Y., and Whitesides, G. M. (1997). Two- and three-dimensional crystallization of polymeric microspheres by micromolding capillaries. *Advanced Materials*, 8:245–247.

Kim, H.-J. (2004). Quantification of chaotic mixing in microfluidic systems. Master's thesis, Texas A&M University, College Station, TX.

Kim, M. J., Kihm, K. D., and Beskok, A. (2002). A comparative study of μ-PIV measurements and numerical simulations of electroosmotic flows in various micro-channel configurations. *Exp. Fluids*, 33:170–180.

Kim, S. and Karrila, S. J. (1991). *Microhydrodynamics: Principles and Selected Applications.* Butterworth-Heinemann.

Kittilsland, G., Steme, G., and Nordoen, B. A. (1990). Submicron particle filter in silicon. *Sens. Actuators, A: Physical*, 23:904–907.

Kleparnik, K., Mala, Z., and Bocek, P. (2001). Fast separation of DNA sequencing fragments in highly alkaline solutions of linear polyacrylamide using electrophoresis in bare silica capillaries. *Electrophoresis*, 22:783–788.

Knudsen, M. (1909). Die Gesetze der Molekularströmung und der inneren Reibungsströmung der Gase durch Röhren. *Annalen der Physik*, 28:75–130.

Knudsen, M. (1910a). Eine Revision der Gleichgewichtsbedingung der Gase. Thermische Molekularströmung. *Ann. Phys.*, 31:205–229.

Knudsen, M. (1910b). Thermischer Molekulardruck der Gase in Rohren. *Ann. Phys.*, 33:1435–1448.

Kogan, M. (1969). *Rarefied Gas Dynamics.* Plenum Press, New York.

Kogan, M. N., Galkin, V. S., and Fridlender, O. G. (1976). Stresses produced in gases by temperature and concentration inhomogeneities. New type of free convection. *Sov. Phys. Usp.*, 19:420–438.

Kohlhof, S., Gumbsch, P., and Fischmeister, H. F. (1991). Crack propagation in B.C.C. crystals studied with a combined finite-element and atomistic model. *Philosophical Magazine A*, 64:851.

Koplik, J. and Banavar, J. R. (1995). Continuum deductions from molecular hydrodynamics. *Ann. Rev. Fluid Mech.*, 27:257–292.

Koplik, J. and Banavar, J. R. (2000). Molecular simulations of dewetting. *Phys. Rev. Lett.*, 84:4401–4404.

Koplik, J., Banavar, J. R., and Willemsen, J. F. (1987). Molecular dynamics of Poiseuille flow and moving contact lines. *Phys. Rev. Lett.*, 60(13):1282–1285.

Koplik, J., Banavar, J. R., and Willemsen, J. F. (1989). Molecular dynamics of fluid flow at solid surfaces. *Phys. Fluids A*, 1:781–794.

Kopriva, D. A. (1991). Multidomain spectral solutions of the Euler gas-dynamics equations. *J. Comp. Phys.*, 96:428.

Krishnamoorthy, S. and Giridharan, M. G. (2000). Analysis of sample injection and band-broadening in capillary electrophoresis microchips. In *2000 Int'l. Conference on Modeling and Simulation of Microsystems (MSM).*

Kruger, A. J., Larmit, A. J. F., and Wit, M. H. D. (1970). Flow of a rarefied gas through a long capillary (cylindrical or bounded by planes) generated by a temperature gradient. In *Proceedings of the Seventh International Symposium on Rarefied Gas Dynamics*, pages 1015–1022.

Kubo, R., Tudo, M., and Hashitsume, N. (1991). *Statistical Physics II.* Berlin:Springer.

Kumar, S. and Beskok, A. (2002). Heat and mass transfer in a peristaltic micro mixer. *ASME J. Heat Transfer*, 125(4):548.

Kuo, T. C., Cannon, D. M., Chen, Y., Tulock, J. J., Shannon, M. A., Sweedler, J. V., and Bohn, P. W. (2003a). Gateable nanofluidic interconnects for multilayered microfluidic separation systems. *Anal. Chem.*, 75:1861–1876.

Kuo, T. C., Cannon, D. M., Shannon, M. A., Bohn, P. W., and Sweedler, J. V. (2003b). Hybrid three-dimensional nanofluidic/microfluidic devices using molecular gates. *Sens. Actuators, A*, 102:223–233.

Kuo, T. C., Sloan, L. A., Sweedler, J. V., and Bohn, P. W. (2001). Manipulating molecular transport through nanoporous membranes by control of electrokinetic flow: Effect of surface charge density and Debye length. *Langmuir*, 17:6298–6303.

Kusalik, P. G., Liden, F., and Svishchev, I. M. (1995). Calculation of the third virial coefficient for water. *J. Chem. Phys.*, 103:10169–10175.

Kutter, J. P. (2000). Current developments in electrophoretic and chromatographic separation methods on microfabricated devices. *Trends in Analytical Chemistry*, 19:352–363.

Kwon, S. and Lee, L. P. (2001). Focal length control by microfabricated planar electrodes-based liquid lens(uPELL). *Proc. 11th International Conference on Solid-State Sensors and Actuators, Eurosensors XV, Transducers 2001*, pages 1348–1351.

Ladd, A. J. (1994a). Numerical simulations of particulate suspensions via a discretized lattice Boltzmann equation. Part 1. Theoretical foundation. *J. Fluid Mech.*, 271:285–309.

Ladd, A. J. (1994b). Numerical simulations of particulate suspensions via a discretized lattice Boltzmann equation. Part 2. Numerical results. *J. Fluid Mech.*, 271:311–339.

Lallemand, P. and Luo, L.-S. (2003). Theory of the lattice Boltzmann method: Acoustic and thermal properties in two and three dimensions. *Phys. Rev. E*, 68:036706.

Lancaster, P. and Salkauskas, K. (1981). Surface generated by moving least squares methods. *Mathematics of Computation*, 37:141–158.

Larrodé, F. E., Housiadas, C., and Drossinos, Y. (2000). Slip-flow heat transfer in circular tubes. *Int. J. Heat Mass Transfer*, 43:2669–2680.

Lauga, E. and Stone, H. A. (2003). Effective slip in pressure-driven Stokes flow. *J. Fluid Mech.*, 489:55–77.

Lee, J. and Kim, C.-J. (2000). Surface tension-driven microactuation based on continuous electrowetting. *J. Microelectromech. Sys.*, 9(2):171–180.

Lee, J., Moon, H., Fowler, J., Schoellhammer, T., and Kim, C.-J. (2002). Electrowetting and electrowetting-on-dielectric for microscale liquid handling. *Sens. Actuators, A*, 95:259–268.

Lee, W. Y., Lee, Y. K., Wong, M., and Zohar, Y. (2000). Microchannels in series with gradual contraction/expansion. In *ASME, MEMS-Vol. 2, Micro-Electro-Mechanical Systems*, pages 467–472.

Lee, Y. K. (2002). Lyapunov exponents in a micro chaotic mixer. *Int. J. Nonlinear Sci. Num. Sim.*, 3:561–564.

Lees, A. W. and Edwards, S. F. (1972). The computer study of transport processes under extreme conditions. *J. Phys. C*, 5:1921.

Leo, J. and Maranon, J. (2003). Confined water in nanotube. *J. Mol. Struct. (Theochem)*, 623:159–166.

LeTallec, P. and Mallinger, F. (1997). Coupling Boltzmann and Navier-Stokes equations by half fluxes. *J. Comp. Phys.*, 136(1):51–67.

Levesque, D. and Verlet, L. (1970). Computer experiments on classical fluids. III. time-dependent self-correlation functions. *Phys. Rev. A*, 2(4):2514–2522.

Levich, V. G. (1962). *Physicochemical Hydrodynamics.* Prentice-Hall, Inc.

Levinger, N. E. (2002). Water in confinement. *Science*, 29:1722–1723.

Li, B. and Kwok, D. Y. (2003). Discrete Boltzmann equation for microfluidics. *Phys. Rev. Lett.*, 90(12):124502.

Li, G. (2003). *Efficient Mixed-Domain Analysis of Electrostatic Microelectromechanical Systems (MEMS).* PhD thesis, University of Illinois at Urbana-Champaign.

Li, G. and Aluru, N. R. (2003). A boundary cloud method with a cloud-by-cloud polynomial basis. *Eng. Anal. Boundary Elem.*, 27(1):57–71.

Li, J., Stein, D., McMullan, C., Branton, D., Aziz, M. J., and Golovchenko, J. A. (2001). Ion-beam sculpting at nanometer length scales. *Nature*, 412.

Li, S. and Liu, W. K. (2002). Meshfree and particle methods and their applications. *Applied Mechanics Review*, 55:1–34.

Li, W. L., Lin, J. W., Lee, S. C., and Chen, M. D. (2002). Effects of roughness on rarefied gas flow in long microtubes. *J. Micromech. Microeng.*, 12:149–156.

Libera, J. and Gogotsi, Y. (2001). Hydrothermal synthesis of graphite tubes using Ni catalyst. *Carbon*, 39:1307.

Liepmann, H. W. and Roshko, A. (1957). *Elements of Gas Dynamics.* John Wiley and Sons, Inc.

Lin, H., Storey, B. D., Oddy, M. H., Chen, C. H., and Santiago, J. G. (2004). Instability of electrokinetic microchannel flows with conductivity gradients. *Phys. Fluids*, 16(6):1922–1935.

Lindahl, E., Hess, B., and van der Spoel, D. (2001). GROMACS 3.0: a package for molecular simulation and trajectory analysis. *J. Mol. Mod.*, 7(8):306–317.

Link, D., Anna, S. L., Weitz, D., and Stone, H. A. (2004). Geometrically mediated breakup of drops in microfluidic devices. *Phys. Rev. Lett.*, 92(5):054503–1–054503–4.

Lippmann, M. G. (1875). Relations entre les phenomenes electriques et capillaires. *Ann. Chim. Phys.*, 5(11):494–549.

Lishchuk, S. V. and Fischer, J. (2001). Velocity distribution of water molecules in pore under microwave electric field. *Int. J. Therm. Sci.*, 40:717–723.

Liszka, T. J., Duarte, C. A., and Tworzydlo, W. W. (1996). hp-meshless cloud method. *Comput. Methods Appl. Mech. Eng.*, 139:263–288.

Liu, D. (2004). *Spectral Element/Force Coupling Method: Application to Colloidal Microdevices and Self-Assembled Structures in 3D Domains.* PhD thesis, Brown University.

Liu, D., Maxey, M. R., and Karniadakis, G. E. (2002). A fast method for particulate microflows. *J. Microelectromech. Sys.*, 11(6):691–702.

Liu, H. F., Gatsonis, N., Beskok, A., and Karniadakis, G. E. (1998). Simulation models for rarefied sphere in a pipe. In *Proceedings of the Twenty-First International Symposium on Rarefied Gas Dynamics.*

Liu, J., Lawrence, E. M., Wu, A., Ivey, M. L., Flores, G., Javier, K., Bibette, J., and Richard, J. (1995). Field-induced structures in ferrofluid emulsions. *Phys. Rev. Lett.*, 74:2828–2831.

Liu, J. Q., Tai, Y. C., Pong, K. C., and Ho, C. M. (1993). Micromachined channel/pressure sensor systems for micro flow studies. *In The 7th Int. Conf. on Solid-State Sensors and Actuators*, pages 995–998.

Liu, M., Muzzio, F. J., and Peskin, R. L. (1994). Quantification of mixing in aperiodic chaotic flows. *Chaos, Solitons and Fractals*, 4(6):869–893.

Liu, N. and Ng, E. Y. (2001). The posture effects of a slider air bearing on its performance with a direct simulation Monte Carlo method. *J. of Micromech. Microeng.*, 11:463–473.

Liu, Y., Fanguy, J. C., Bledsoe, J. M., and Henry, C. S. (2000). Dynamic coating using polyelectrolyte multilayers for chemical control of electroosmotic flow in capillary electrophoresis microchips. *Anal. Chem.*, 72(24):5939–5944.

Liu, Y. and Ichiye, T. (1996). Soft sticky dipole potential for liquid water: A new model. *J. Phys. Chem.*, 100:2723–2730.

Lo, W. Y. and Chan, K. (1994). Poisson-Boltzmann calculations of ions in charged capillaries. *J. Chem. Phys.*, 143:339–353.

Lockerby, D. A. and Reese, J. M. (2003). High resolution Burnett simulations of micro Couette flow and heat transfer. *J. Comp. Phys.*, 188:333–347.

Loeb, L. B. (1961). *Kinetic Theory of Gases.* Dover Publications Inc., New York.

Lohner, R. and Yang, C. (1996). Improved ALE mesh velocities for moving bodies. *Comm. Num. Meth. Eng. Phys.*, 12:599–608.

Lomholt, S. (2000). *Numerical Investigations of Macroscopic Particle Dynamics in Microflows.* PhD thesis, Riso National Laboratory, Roskilde, Denmark.

London, A. P. (1996). A systems study of propulsion technologies for orbit and attitude control of microspacecraft. Master's thesis, Massachusetts Institute of Technology.

Lord, R. G. (1976). Tangential momentum coefficients of rare gases on polycrystalline surfaces. In *Proceedings of the Tenth International Symposium on Rarefied Gas Dynamics*, pages 531–538.

Lord, R. G. (1991). Some extensions of the Cercignani-Lampis gas-surface scattering kernel. *Phys. Fluids A*, 3(4):706–710.

Lord, R. G. (1995). Some further extensions of the Cercignani-Lampis gas-surface scattering kernel. *Phys. Fluids A*, 7(5):1159–1161.

Lou, T., Dahlby, D. C., and Baganoff, D. (1998). A numerical study comparing kinetic flux vector splitting for the Navier-Stokes equations with a particle method. *J. Comp. Phys.*, 145:489–510.

Lovalenti, P. M. and Brady, J. F. (1993). The hydrodynamic force on a rigid particle undergoing arbitrary time-dependent motion at small Reynolds number. *J. Fluid Mech.*, 256:561–605.

Loyalka, S. and Hamoodi, S. (1990). Poiseuille flow of a rarefied gas in a cylindrical tube: Solution of linearized Boltzmann equation. *Phys. Fluids A*, 2 (11):2061–2065.

Loyalka, S. K. (1974). Comments on "Poiseuille flow and thermal creep of a rarefied gas between parallel plates". *Phys. Fluids*, 17:1053.

Loyalka, S. K. (1975). Kinetic theory of thermal transpiration and mechanocaloric effect. II. *J. Chem. Phys.*, 63:4054.

Loyalka, S. K. and Hickey, K. A. (1991). Kinetic theory of thermal transpiration and the mechanocaloric effect – planar flow of a rigid sphere gas with arbitrary accommodation at the surface. *J. Vac. Sci. Technol. A*, 9 (1):158.

Lue, L., Zoeller, N., and Blankschtein, D. (1999). Incorporation of nonelectrostatic interactions in the Poisson-Boltzmann equation. *Langmuir*, 15:3726–3730.

Lum, K., Chandler, D., and Weeks, J. D. (1999). Hydrophobicity at small and large length scales. *J. Phys. Chem. B*, 103:4570–4577.

Lum, K. and Luzar, A. (1997). Pathway to surface-induced phase transition of a confined fluid. *Phys. Rev. E*, 56:R6283–R6286.

Luo, L.-S. (2004). Comment on "Discrete Boltzmann equation for microfluidics". *Phys. Rev. Lett.*, 92(13):139401.

Luty, B. A., Tironi, I. G., and van Gunsteren, W. F. (1995). Lattice-sum methods for calculating electrostatic interactions in molecular simulations. *J. Chem. Phys.*, 103:3014–3021.

Lyklema, J., Rovillard, S., and Coninck, J. (1998). Electrokinetics: The properties of the stagnant layer unraveled. *Langmuir*, 14(20):5659–5663.

Lynden-Bell, R. M. and Rasaiah, J. C. (1996). Mobility and solvation of ions in channels. *J. Chem. Phys.*, 105:9266–9280.

Macounova, K., Cabrera, C. R., Holl, M. R., and Yager, P. (2000). Generation of natural pH gradients in microfluidic channels for use in isoelectric focusing. *Anal. Chem.*, 72(16):3745–3751.

Macounova, K., Cabrera, C. R., and Yager, P. (2001). Concentration and separation of proteins in microfluidic channels on the basis of transverse IEF. *Anal. Chem.*, 73(7):1627–1633.

Magda, J. J., Tirrell, M., and Davis, H. T. (1985). Molecular dynamics of narrow, liquid-filled pores. *J. Chem. Phys.*, 83(4):1888–1901.

Mahoney, M. W. and Jorgensen, W. L. (2000). A five-site model for liquid water and the reproduction of the density anomaly by rigid, nonpolarizable potential functions. *J. Chem. Phys.*, 112:8910–8922.

Maitland, G. C., Rigby, M., Smith, E., and Wakeham, W. (1981). *Intermolecular Forces: Their Origin and Determination.* Clarendon Press, Oxford.

Mann, D. J. and Halls, M. D. (2003). Water alignment and proton conduction inside carbon nanotubes. *Phys. Rev. Lett.*, 90:195503.

Mantooth, H. A. and Vlach, M. (1992). Beyond SPICE with SABER and MAST. *Proceedings of IEEE International Symposium on Circuits and Systems*, 1:77–80.

Manz, A., Graber, N., and Widmer, H. M. (1990). Miniaturized total chemical analysis systems: A novel concept for chemical sensing. *Sens. Actuators, B*, 1:244–248.

Mao, Z. and Sinnott, S. B. (2000). A computational study of molecular diffusion and dynamic flow through carbon nanotubes. *J. Phys. Chem. B*, 104(19):4618–4624.

Margolin, L., Titi, E. S., and Wynne, S. (2003). The postprocessing Galerkin and nonlinear Galerkin methods – a truncation analysis point of view. *SIAM, J. Numer. Anal.*, 41:695–714.

Markelov, G. N. and Ivanov, M. S. (2001). Numerical study of 2D/3D micronozzle flows. In Bartel, T. and Galiss, M., editors, *Rarefied Gas Dynamics*, volume AIP, Melville.

Markx, G., Dyda, P., and Pethig, R. (1996). Dielectrophoretic separation of bacteria using a conductivity gradient. *J. Biotechnol.*, 51:175–180.

Markx, G. and Pethig, R. (1995). Dielectrophoretic separation of cells: Continuous separation. *Biotechnol. Bioeng.*, 45:337–343.

Marques, W., Kremer, G. M., and Sharipov, F. M. (2000). Couette flow with slip and jump boundary conditions. *Continuum Mech. Thermodyn.*, 12:379–386.

Marry, V., Dufreche, J.-F., Jardat, M., and Turq, P. (2003). Equilibrium and electrokinetic phenomena in charged porous media from microscopic and mesoscopic models: electro-osmosis in montmorillonite. *Mol. Phys.*, 101(20):3111–3119.

Martí, J., Guardia, E., and Gordillo, M. C. (2002). Reorientational motions in sub- and supercritical water under extreme confinement. *Chem. Phys. Lett.*, 365:536–541.

Martin, C. R., Nishizawa, M., Jirage, K., and Kang, M. (2001). Investigations of the transport properties of gold nanotubule membranes. *J. Phys. Chem. B*, 105(10):1925–1934.

Martin, J. E., Odinek, J., and Halsey, T. C. (1992). Evolution of structure in quiescent electrorheological fluids. *Phys. Rev. Lett.*, 69:1524.

Mashl, R. J., Joseph, S., Aluru, N. R., and Jakobsson, E. (2003). Anomalously immobilized water: a new water phase induced by confinement in nanotubes. *Nano Lett.*, 3:589–592.

Matsumoto, H. and Colgate, J. E. (1990). Preliminary investigation of micropumping based on electrical control of interfacial tension. *Proc. IEEE on MEMS*, Napa Valley, CA:105–110.

Maurer, J., Tabeling, P., Joseph, P., and Willaime, H. (2003). Second-order slip laws in microchannels for helium and nitrogen. *Phys. Fluids*, 15(9):2613–2621.

Maxey, M. R. (1993). The equation of motion for a small rigid sphere in a nonuniform or unsteady flow. In *Proccedings of Gas-Solid Flows, The Fluids Engineering Conference, Washington, D.C., June 20–24.*

Maxey, M. R. and Dent, G. L. (1998). Some features of particle sedimentation at finite Reynolds numbers. In *Third International Conference on Multiphase Flow, ICMF98, June 8–12, Lyon, France.*

Maxey, M. R. and Patel, B. K. (2001). Localized force representations for particles sedimenting in Stokes flow. *Int. J. Multiphase Flow*, 27(9):1603–1626.

Maxey, M. R., Patel, B. K., Chang, E. J., and Wang, L.-P. (1997). Simulations of dispersed turbulent multiphase flows. *Fluid Dynamics Research*, 20:143–156.

Maxey, M. R. and Riley, J. J. (1983). Equation of motion for a small rigid sphere in a non-uniform flow. *Phys. Fluids*, 26:883–889.

Mayaram, K., Chern, J., and Yang, P. (1993). Algorithms for transient three-dimensional mixed-level circuit and device simulation. *IEEE Transactions on Computer-Aided Design of Integrated Circuits*, 12:1726–1733.

Mayaram, K. and Pederson, D. O. (1987). CODECS: an object-oriented mixed-level circuit and device simulator. In IEEE, editor, *1987 IEEE International Symposium on Circuits and Systems, Dunfey City Line Hotel, Philadelphia, PA, May 4–7, 1987*, pages 604–607. IEEE Computer Society Press.

McClain, M. A., Culbertson, C. T., Jacobson, S. C., and Ramsey, J. M. (2001). Flow cytometry of Escherichia coli on microfluidic devices. *Anal. Chem.*, 73(21):5334–5338.

McNamara, G. R. and Zanetti, G. (1988). Use of the Boltzmann equation to simulate lattice-gas automata. *Phys. Rev. Lett.*, 61:2332–2335.

McQuarrie, A. (1973). *Statistical Mechanics.* Harper and Row, New York.

Megaridis, C. M., Guvenc, A., Libera, J. A., and Gogotsi, Y. (2002). Attoliter fluid experiments in individual closed-end carbon nanotubes: Liquid film and fluid interface dynamics. *Phys. Fluids*, 14(2):L5–L8.

Mehner, J. E., Doetzel, W., Schauwecker, B., and Ostergaard, D. (2003). Reduced-order modeling of fluid structural interactions in MEMS based on modal projection techniques. In *Transducers*, pages 1840–1843.

Mehregany, M., Nagarkar, P., Senturia, S., and Lang, J. H. (Napa Valley, CA., February 1990). Operation of microfabricated harmonic and ordinary side-drive motor. *IEEE Micro Electro Mechanical System Workshop.*

Mei, R. and Adrian, R. J. (1992). Flow past a sphere with an oscillation in the free-stream velocity and unsteady drag at finite Reynolds numbers. *J. Fluid Mech.*, 237:323–341.

Meinhart, C. D., Wereley, S. T., and Santiago, J. G. (1999). PIV measurements of a microchannel flow. *Exp. in Fluids*, 27:414–419.

Meinhart, C. D., Wereley, S. T., and Santiago, J. G. (2000). A PIV algorithm for estimating time-averaged velocity fields. *J. Fluids Eng.*, 122:285–289.

Melvin, M. (1987). *Electrophoresis.* John Wiley and Sons Inc, London.

Mentor Graphics (1995). *HDL-A Reference Manual.*

Meyerovich, A. E. and Stepaniants, S. (1994). Transport phenomena at rough boundaries. *Phys. Rev. Lett.*, 73:316.

Meyerovich, A. E. and Stepaniants, S. (1997). Ballistic transport in ultrathin channels and films with rough surfaces. *J. Phys. Cond. Matt.*, 9:4157–4173.

Michaelides, E. E. (1997). Review – The transient equation of motion for particles, bubbles, and droplets. *J. Fluids Eng.*, 119:233–247.

Miller, S. A. and Martin, C. R. (2002). Controlling the rate and direction of electroosmotic flow in template-prepared carbon nanotube membranes. *J. Electroanal. Chem.*, 522(1):66–69.

Mitchell, M. J., Qiao, R., and Aluru, N. R. (2000). Meshless analysis of steady-state electro-osmotic transport. *J. Microelectromech. Sys.*, 9 (4):435–449.

Miyazima, S., Meakin, P., and Family, F. (1987). Aggregation of oriented anisotropic particles. *Phys. Rev. A*, 36:1421–1427.

Mo, G. and Rosenberger, F. (1990). Molecular dynamics simulation of flow in a two dimensional channel with atomically rough walls. *Phys. Rev. A*, 42:4688–4692.

Molho, J., Herr, A., Mosier, B. P., Santiago, J. G., Kenny, T. W., Brennen, R. A., Gordon, G. B., and Mohammadi, B. (2001). Optimization of turn geometries for microchip electrophoresis. *Anal. Chem.*, 73:1350–1360.

Molho, J., Herr, A. E., Desphande, M., Gilbert, J. R., Garguilo, M. G., Paul, P. H., John, P. M., Woudenberg, T. M., and Connel, C. (1998). Fluid transport mechanisms in micro fluidic devices. In *ASME International Mechanical Engineering Congress and Exposition*, pages 69–75.

Monaghan, J. J. (1992). Smoothed particle hydrodynamics. *Annu. Rev. Astron. Astrophys*, 30:543–574.

Monson, A. P., Rigby, M., and Steele, W. A. (1983). Non-additive energy effects in molecular liquids. *Mol. Phys.*, 49:893–898.

Moon, H., Cho, S., Garrell, R. L., and Kim, C.-J. (2002). Low voltage electrowetting-on-dielectric. *J. Appl. Phys.*, 92:4080–4087.

Morgan, H. and Green, N. G. (2003). *AC Electrokinetics: Colloids and Nanoparticles.* Research Studies Press Ltd, Hertfordshire, UK.

Morgan, H., Hughes, M. P., and Green, N. G. (1999). Separation of submicron bioparticles by dielectrophoresis. *Biophysical Journal*, 77:516–525.

Moroney, R. M., White, R. M., and White, R. T. (1991). Ultrasonically induced microtransport. In *Proccedings of the IEEE Workshop on MEMS, Amsterdam, pp. 277-282.*

Mosher, R. A. and Thormann, W. (2002). High-resolution computer simulation of the dynamics of isoelectric focusing using carrier ampholytes: The post-separation stabilizing phase revisited. *Electrophoresis*, 23:1803–1814.

Mott, D. R., Oran, E. S., and Kaplan, C. R. (2001). Microfilter simulations and scaling laws. *AIAA J. Thermophys. Heat Transfer*, 15(4):473–477.

Mueller, J. (1997). Thruster options for microspacecraft: A review and evaluation of existing hardware and emerging technoloies. In *Paper AIAA 97-3058, 33rd AIAA/ASME/SAE/ASEE Joint Propulsion Conference & Exhibit.*

Mukherjee, S. (2000). Finite parts of singular and hypersingular integrals with irregular boundary source points. *Eng. Anal. Boundary Elem.*, 24:767–776.

Mukherjee, T. and Fedder, G. K. (1998). Design methodology for a mixed-domain systems-on-a-chip. In *IEEE Computer Society Workshop on VLSI*, pages 96–101.

Muller, E. A. and Bubbins, K. E. (1998). Molecular simulation study of hydrophilic and hydrophobic behavior of activated carbon surfaces. *Carbon*, 36:1433–1438.

Muntz, E. P. (1989). Rarefied gas dynamics. *Ann. Rev. Fluid Mech.*, 21:387–417.

Myong, R. S. (1998). Thermodynamically consistent hydrodynamic computational models for high Knudsen number gas flows. *Phys. Fluids*, 11 (9):2468–2474.

Myong, R. S. (2004). Gaseous slip models based on the Langmuir adsorption isotherm. *Phys. Fluids*, 16(1):104–117.

Nada, H. and van der Eerden, J. P. (2003). An intermolecular potential model for the simulation of ice and water near the melting point: A six-site model of water. *J. Chem. Phys.*, 118:7401–7413.

Nagel, L. W. and Pederson, D. O. (1973). Simulation Program with Integrated Circuit Emphasis. In *Proceedings of the 16th Midwest Symposium Circuit Theory*, Waterloo, Canada.

Naguib, N., Ye, H., Gogotsi, Y., Yazicioglu, A. G., Megaridis, C. M., and Yoshimura, M. (2004). Observation of water confined in nanometer channels of closed carbon nanotubes. *Nano Letters*, in press.

Nakane, J. J., Akeson, M., and Marziali, A. (2003). Nanopores sensors for nucleic acid analysis. *J. Phys.: Condens. Matter*, 15:R1365–R1393.

Nanbu, K. (1983). Analysis of Couette flow by means of the new direct simulation method. *J. Phys. Soc. Japan*, 5:52.

Nguyen, N. T. and Wereley, S. (2003). *Fundamentals and Applications of Microfluidics*. Artech House, Microelectromechanical Systems Library.

Nie, X., Doolen, G. D., and Chen, S. (1998). Lattice-Boltzmann simulations of fluid flows in MEMS. Technical report, Los Alamos Report, unpublished.

Nie, X. B., Chen, S. Y., E, W. N., and Robbins, M. O. (2004). A continuum and molecular hybrid method for micro- and nano-fluid flow. *J. Fluid Mech.*, 500:55–64.

Nikunen, P., Karttunen, M., and Vattulainen, I. (2003). How would you integrate the equations of motion in dissipative particle dynamic simulations? *Computer Physics Communications*, 153:407–423.

Niu, X. Z., Tabeling, P., and Lee, Y. K. (2003). Finite time Lyapunov exponent for micro chaotic mixer design. In *Proccedings of IMECE'03, ASME, Washington, D.C., November 16-21.*

Noble, D. R., Chen, S., Georgiadis, J. G., and Buckius, R. O. (1995). A consistent hydrodynamic boundary condition for the lattice Boltzmann method. *Phys. Fluids*, 7:203–209.

Nomura, T. and Hughes, T. J. (1992). An arbitrary Lagrangian-Eulerian finite element method for interaction of fluid and a rigid body. *Comput. Methods Appl. Mech. Eng.*, 95:115.

Nose, S. (1984). A molecular dynamics method for simulations in the canonical ensemble. *Mol. Phys.*, 52:255–268.

O'Brien, R. W. and White, L. R. (1978). Electrophoretic mobility of a spherical colloidal particle. *J. Chem. Soc. Faraday Trans. 2*, 74:1607–1626.

O'Connell, S. T. and Thompson, P. A. (1995). Molecular dynamics-continuum hybrid computations: A tool for studying complex fluid flows. *Phys. Rev. E*, 52:R5792.

Oddy, M. H., Santiago, J. G., and Mikkelsen, J. C. (2001). Electrokinetic instability micromixing. *Anal. Chem.*, 73(24):5822–5832.

Ogrodzki, J. (1994). *Circuit Simulation Methods and Algorithms*. Boca Raton, Florida, CRC Press.

Ohshima, H. and Kondo, T. (1990). Electrokinetic flow between two parallel plates with surface charge layers: Electroosmosis and streaming potential. *J. Colloids and Interface Science*, 135(2):443–448.

Ohwada, T. and Sone, Y. (1992). Analysis of thermal stress slip flow and negative thermophoresis using the Boltzmann equation for hard-sphere molecules. *Eur. J. Mech., B/Fluids*, 11:389–414.

Ohwada, T., Sone, Y., and Aoki, K. (1989a). Numerical analysis of the Poiseuille and thermal transpiration flows between two parallel plates on the basis of the Boltzmann equation for hard sphere molecules. *Phys. Fluids A*, 1(12):2042–2049.

Ohwada, T., Sone, Y., and Aoki, K. (1989b). Numerical analysis of the shear and thermal creep flows of a rarefied gas over a plane wall on the basis of the linearized Boltzmann equation for hard sphere molecules. *Phys. Fluids*, 1:1588–1599.

Oldham, K. B. and Myland, J. C. (1994). *Fundamentals of Electrochemical Science*. Academic Press Inc: San Diego.

Omar, M. P., Mehregany, M., and Mullen, R. L. (1992). Electric and fluid field analysis of side-drive micromotors. *J. Microelectromech. Sys.*, 1(3):130–140.

O'Neill, M. E. and Majumdar, S. R. (1970). Asymmetrical slow viscous fluid motions caused by the translation or rotation of two spheres. Part i: The determination of exact solutions for any values of the ratio of radii and separation parameters. *Z. Angew Math. Phys*, 21(2):164–179.

Oran, E. S., Oh, C. K., and Cybyk, B. Z. (1998). Direct Simulation Monte Carlo: Recent advances and applications. *Ann. Rev. Fluid Mech.*, 30:403–441.

Oron, A. (2001). *Physics of thin films*. Chapter 10, The MEMS Handbook, edited by M. Gad-el-Hak, CRC Press.

Oron, A. and Bankoff, S. G. (1999). Dewetting of a heated surface by an evaporating liquid film under conjoining/disjoining pressures. *J. Colloid Interface Sci.*, 218:152–166.

Oron, A., Davis, S. H., and Bankoff, S. G. (1997). Long-scale evolution of thin liquid films. *Rev. Modern Phys.*, 68:931–980.

Oron, A. and Peles, Y. (1998). Stabilization of thin liquid films by internal heat generation. *Phys. Fluids A*, 10:537–539.

Orszag, S. A. and Yakhot, V. (1986). Reynolds number scaling of Cellular Automaton hydrodynamics. *Phys. Rev. Lett.*, 56:1691.

Ottino, J. M. (1997). *The Kinematics of Mixing: Stretching, Chaos, and Transport*. Cambridge University Press.

Ottino, J. M. and Wiggins, S. (2004). Introduction: Mixing in microfluidics. *Phil. Trans. R. Soc. London, Ser. A*, 362:923–935.

Overbeek, J. T. (1952). *Colloid Science*, volume 1. Kruyt, H.R, Elsevier Publishing Company, Armsterdam.

Palusinski, O. A., Graham, A., Mosher, R. A., Bier, M., and Saville, D. A. (1986). Theory of electrophoretic separations, part II: Construction of a numerical simulation scheme and its applications. *AICHE Journal*, 32(2):215–223.

Pan, F., Kubby, J., Peeters, E., Tran, A. T., and Mukherjee, S. (1998). Squeeze film damping effect on the dynamic response of a MEMS torsion mirror. *J. Micromech. Microeng.*, 8:200–208.

Panton, R. L. (1984). *Incompressible Fluid Flow*. Wiley, 2nd ed. New York.

Paranjpe, R. S. and Elrod, H. G. (1986). Stability of chains of permeable spherical beads in an applied magnetic field. *J. Appl. Phys.*, 60:418–422.

Park, J. H., Bahukudumbi, P., and Beskok, A. (2004). Rarefaction effects on shear driven oscillatory gas flows: A DSMC study in the entire Knudsen regime. *Phys. Fluids*, 16(2):317–330.

Patankar, N. and Hu, H. H. (1998). Numerical simulation of electroosmotic flow. *Anal. Chem.*, 70(9):1870–1881.

Paul, P. H., Graguilo, M. G., and Rakestraw, D. J. (1998). Imaging of pressure and electrokinetically driven flows through open capillaries. *Anal. Chem.*, 70(13):2459–2467.

Paul, S. and Chandra, A. (2003). Dynamics of water molecules at liquid-vapour interfaces of aqueous ionic solutions: Effects of ion concentration. *Chem. Phys. Lett.*, 373:87–93.

Pfahler, J. N., Harley, J. C., Bau, H. H., and Zemel, J. N. (1991). Gas and liquid flow in small channels. *ASME*, 32:49–59.

Pham-van-Diep, G., Keeley, P., Muntz, E. P., and Weaver, D. (1995). A micromechanical Knudsen compressor. In Harvey, J. and Lord, G., editors, *Proceedings of the Ninenteenth International Symposium on Rarefied Gas Dynamics*, pages 715–721, Oxford University Press, Oxford U.K.

Phan, A. V., Gray, L. J., Kaplan, T., and Phan, T. N. (2002). The boundary contour method for two-dimensional Stokes flow and incompressible elastic materials. *Comp. Mech.*, 28:425–433.

Philip, J. R. (1972). Flows satisfying mixed no-slip and no-shear conditions. *J. Appl. Math. Phys.*, 23:353–372.

Piekos, E. S. and Breuer, K. (1995). DSMC modeling of micromechanical devices. In *AIAA Thermophysics Conference June 19-22 1995, San Diego, CA, AIAA 95-2089.*

Pit, R., Hervet, H., and Leger, L. (2000). Direct experimental evidence of slip at hexadecane: Solid interfaces. *Phys. Rev. Lett.*, 85(5):980–983.

Pivkin, I. (2005). *Continum and Atomistic Methods for Biological Flows.* PhD thesis, Brown University.

Pohl, H. A. (1978). *Dielectrophoresis.* Cambridge University Press, Cambridge.

Poiseuille, J. M. L. (1846). Experimental investigations upon the flow of liquids in tubes of very small diameters. *Sciences Mathématiques et Physiques*, 9:433–545.

Polard, W. G. and Present, R. D. (April 1948). On gaseous self-diffusion in long capillary tubes. *Phys. Rev.*, 73 (7):762–774.

Polson, N. A. and Hayes, M. A. (2000). Electroosmotic flow control of fluids on a capillary electrophoresis microdevice using an applied external voltage. *Anal. Chem.*, 72(5):1088–1092.

Pong, K. C., Ho, C. M., Liu, J., and Tai, Y. C. (1994). Non-linear pressure distribution in uniform microchannels. In *ASME, FED-Vol. 197, Application of Microfabrication to Fluid Mechanics*, pages 51–56.

Ponomarev, I. V. and Meyerovich, A. E. (2003). Surface roughness and effective stick-slip motion. *Phys. Rev. E*, 67:026302.

Poppe, H., Cifuentes, A., and Kok, W. T. (1996). Theoretical description of the influence of external radial fields on the electroosmotic flow in capillary electrophoresis. *Anal. Chem.*, 68:888–893.

Porodnov, B. T., Kulev, A. N., and Tuchvetov, F. T. (1978). Thermal transpiration in a circular capillary with a small temperature difference. *J. Fluid Mech.*, 88:609.

Porodnov, B. T. and Tuchvetov, F. T. (1979). Theoretical investigation of non-isothermal flow of a rarefied gas in a cylindrical capillary. *J. Eng. Phys.*, 36:61.

Pozhar, L. A. (1994). *Transport Theory of Inhomogenous Fluids.* World Scientific.

Pozhar, L. A. (2000). Structure and dynamics of nanofluids: Theory and simulations to calculate viscosity. *Phys. Rev. E*, 61(2):1432.

Pozhar, L. A. (2001). Virtual nanofabrication of electronic materials. *Computational Nanoscience,2001*, 3(9):188.

Priezjev, N. V. and Troian, S. M. (2004). Molecular origin and dynamic behavior of slip in sheared polymer flows. *Phys. Rev. Lett.*, 92(1):018302–1–018302–4.

Probstein, R. F. (1994). *Physiochemical Hydrodynamics: An Introduction.* John Wiley & Sons, Inc.

Promislow, J. H., Gast, A. P., and Fermigier, M. (1995). Aggregation kinetics of paramagnetic colloidal particles. *J. Chem. Phys.*, 102(13):5492–5498.

Puibasset, J. and Pellenq, R. (2003). Water adsorption on hydrophilic mesoporous and plane silica substrates: a grand canonical Monte-Carlo simulation study. *J. Chem. Phys.*, 118:5613–5622.

Qian, S. and Bau, H. H. (2002). A chaotic electroosmotic stirrer. *Anal. Chem.*, 74:3616–3625.

Qian, Y. H., D'Humières, D., and Lallemand, P. (1992). Lattice BGK models for Navier-Stokes equation. *Europhys. Lett.*, 17:479–482.

Qiao, R. and Aluru, N. R. (2002). A compact model for electroosmotic flows in microfluidic devices. *J. Micromech. Microeng.*, 12(5):625–635.

Qiao, R. and Aluru, N. R. (2003a). Dispersion control in nano-channel systems by localized zeta-potential variations. *Sens. Actuators, A-Physical*, 104(3):268–274.

Qiao, R. and Aluru, N. R. (2003b). Ion concentration and velocity in nanochannel electroosmotic flows. *J. Chem. Phys.*, 118(10):4692–4701.

Qiao, R. and Aluru, N. R. (2003c). Transient analysis of electro-osmotic transport by a reduced-order modeling approach. *Int. J. Numer. Methods Eng.*, 56:1023–1050.

Qiao, R. and Aluru, N. R. (2003d). A typical dependence of electroosmotic transport on surface charge in a single-wall carbon nanotube. *Nano Lett.*, 3(8):1013–1017.

Qiao, R. and Aluru, N. R. (2004). Charge inversion and flow reversal in a nanochannel electro-osmotic flow. *Phys. Rev. Lett.*, 92(19).

Quarles, T. (1989). The SPICE3 implementation guide. Technical Report No. UCB/ERL M89/44, University of California, Berkeley.

Quarles, T., Newton, A. R., Pederson, D. O., and Sangiovanni-Vincentelli, A. (1987). Spice 3b1 user's guide. *Univ. of California, Berkeley.*

Ramaswamy, D. and White, J. (2001). Automatic generation of small-signal dynamic macromodels from 3-D simulation. In *Modeling and Simulation of Microsystems*, pages 27–30.

Rapaport, D. C. (1997). *The Art of Molecular Dynamics Simulation.* Cambridge University Press, New York.

Ratulowski, J. and Chang, H.-C. (1989). Transport of air bubbles in capillaries. *Phys. Fluids A*, 1(10):1642–1655.

Ratulowski, J. and Chang, H.-C. (1990). Marangoni effects of trace impurities on the motion of long gas bubbles in capillaries. *J. Fluid Mech.*, 210:303–328.

Reed, L. (October 1993). Near term applications of microelectromechanical systems. *4th International Symposium on Micro Machine and Human Science, Nagoya, Japan*, page 77.

Reed, L. D. and Morrison, F. A. (1976). Hydrodynamic interactions in electrophoresis. *J. Colloid Interface Sci.*, 54(1):117–133.

Reuss, F. (1809). Sur un nouvel effet de l'électricité galvanique. *Mémoires de la Société Impériale de Naturalistes de Moscou*, 2:327–337.

Revenga, M., Espanol, P., and Zuniga, I. (1999). Boundary conditions in dissipative particle dynamics. *Computer Physics Communications*, 121-122:309–311.

Rewienski, M. and White, J. (2001). A trajectory piecewise-linear approach to model order reduction and fast simulation of nonlinear circuits and micromachined devices. In *Proceedings of ICCAD*, pages 252–257.

Rewienski, M. and White, J. (2002). Improving trajectory piecewise-linear approach to nonlinear model order reduction for micromachined devices using an aggregated projection basis. In *Modeling and Simulation of Microsystems*, pages 128–131.

Reyes, D. R., Lossifidis, D., Auroux, P., and Manz, A. (2002). Micro total analysis systems. 1. Introduction, theory, and technology. *Anal. Chem.*, 74:2623–2636.

Reynolds, O. (1879). On certain dimensional properties of matter in the gaseous state. Part I. Experimental researches on thermal transpiration of gases through porous plates and on the laws of transpiration and impulsion, including an experimental proof that gas is not a continuous plenum. Part II. On an extension of the dynamical theory of gas, which includes the stresses, tangential and normal, caused by a varying condition of gas, and affords an explanation of the phenomena of transpiration and impulsion. *Phil. Trans. Royal Society London*, B 170:727–845.

Rice, C. L. and Whitehead, R. (1965). Electro kinetic flow in a narrow cylindrical capillary. *J. Phys. Chem.*, 69(11):4017–4023.

Richardson, S. (1973). On the no-slip boundary condition. *J. Fluid Mech.*, 59:707.

Richetti, P., Prost, J., and Barois, P. (1984). Two-dimensional aggregation and crystallization of a colloidal suspension of latex spheres. *J. Phys. Lett.*, 45:1137–1143.

Righetti, P., Gelfi, C., Verzola, B., and Castelletti, L. (2001). The state of art of dynamic coatings. *Electrophoresis*, 22:603–611.

Righetti, P. G. (1983). *Isoclectric Focussing: Theory, Methodology and Applications.* Elsevier Biomedical, Amsterdam.

Ristenpart, W. D., Aksay, I. A., and Saville, D. A. (2004). Assembly of colloidal aggregates by electrohydrodynamic flow: Kinetic experiments and scaling analysis. *Phys. Rev. E*, 69(021405).

Rittger, E. (1990a). Can three-atom potentials be determined from thermodynamic data? *Mol. Phys.*, 69:867–894.

Rittger, E. (1990b). The chemical potential of liquid xenon by computer simulation. *Mol. Phys.*, 69:853–865.

Rittger, E. (1990c). An empirical three-atom potential for xenon. *Mol. Phys.*, 71:79–96.

Romero, L. A. and Yost, F. G. (1996). Flow in an open channel capillary. *J. Fluid Mech.*, 322:109–129.

Ronceray, M. P. (1911). Recherches sur lecoulement dans les tubes capillaires. *Annales de Chimie et de Physique*, 22:107–125.

Rosenfeld, Y. (1993). Free energy model for inhomogeneous fluid mixtures; Yukawa charged hard spheres, general interactions, and plasmas. *J. Chem. Phys.*, 98:8126–8148.

Rosenweig, R. E. (1987). Magnetic fluids. *Ann. Rev Fluid Mech.*, 19:437–463.

Rothe, D. (1971). Electron-beam studies of viscous flow in supersonic nozzles. *AIAA Journal*, 9(5):804–811.

Roveda, R., Goldstein, D. B., and Varghese, P. L. (1998). Hybrid Euler/Particle approach for continuum/rarefied flows. *J. Spacecraft and Rockets*, 35 (3):258–265.

Rovere, M. and Gallo, P. (2003). Strong layering effects and anomalous dynamical behaviour in confined water at low hydration. *J. Phys. Condens. Matter*, 15:S145–S150.

Rovere, M., Ricci, M. A., and Bruni, F. (1998). A molecular dynamics simulation of water confined in a cylindrical SiO_2 pore. *J. Chem. Phys.*, 108:9859–9867.

Rowlinson, J. S. (1989). The Yukawa potential. *Physica A*, 156:15–34.

Ruckenstein, E. and Rajora, P. (1983). On the no-slip condition of hydrodynamics. *J. Colloid Interface Sci.*, 96:488–491.

Rudisill, E. N. and Cummings, P. T. (1989). Gibbs ensemble simulation of phase equilibrium in the hard core two-Yukawa fluid model for the Lennard-Jones fluid. *Mol. Phys.*, 68:629–635.

Russel, W. B., Saville, D. A., and Scholwalter, W. R. (1989). *Colloidal Dispersions.* Cambridge Univ. Press.

Sadr, R., Yoda, M., Zheng, Z., and Conlisk, A. T. (2004). An experimental study of electro-osmotic flow in rectangular microchannels. *J. Fluid Mech.*, 506:357–367.

Sadus, R. J. (1999). *Molecular Simulation of Fluids: Theory, Algorithms and Object-Orientation*. Elsevier.

Saito, R., Dresselhaus, C., and Dresselhaus, M. S. (1998). *Physical Properties of Carbon Nanotubes*. Imperial College Press.

Sammarco, T. A. and Burns, M. A. (1999). Thermocapillary pumping of discrete drops in microfabricated analysis devices. *AIChE J.*, 45(2):350–366.

Santiago, J. G. (2001). Electroosmotic flows in microchannels with finite inertial and pressure forces. *Anal. Chem.*, 73(10):2353–2365.

Santiago, J. G., Werely, S. T., Meinhart, C. D., Beebe, D., and Adrian, R. (1998). A PIV system for microfluidics. *Exp. Fluids*, 25:316–319.

Sattler, R., Schrag, G., and Wachutka, G. (2003). Mixed-level model for highly perforated torsional actuators coupling the mechanical, the electrostatic and the fluidic domain. In *Nanotech 2003*, pages 284–287.

Saur, M., Angerer, B., Ankenbauer, W., Foldes-Papp, Z., Gobel, F., Han, K. T., Rigler, R., Schulz, A., Wolfrum, J., and Zander, C. (2001). Single molecule DNA sequencing in submicrometer channels: State of the art and future prospects. *J. Biotechnol.*, 86(3):181–201.

Schaaf, S. A. and Chambre, P. L. (1961). *Flow of Rarefied Gases*. Princeton University Press, Princeton.

Schamberg, R. (1947). *The Fundamental Differential Equations and the Boundary Conditions for High Speed Slip-Flow, and Their Application to Several Specific Problems*. PhD thesis, California Institute of Technology.

Schasfoort, R. B., Schlautmann, S., Hendrikse, J., and Berg, A. V. (2001). Electroosmotic flow in a microcapillary with one solution displacing another solution. *J. Colloid Interface Sci.*, 242:264–271.

Schmidt, K. E. and Lee, M. A. (1991). Implementing the fast multipole method in three dimensions. *J. Stat. Phys.*, 63:1223–1235.

Schnell, E. (1956). Slippage of water over nonwettable surfaces. *J. Appl. Phys.*, 27(1):1149–1152.

Schrag, G., Voigt, P., and Wachutka, G. (2001). Squeeze film damping in arbitrary shaped microdevices modeled by an accurate mixed level scheme. In *Modeling and Simulation of Microsystems*, pages 92–95.

Schrag, G. and Wachutka, G. (2002). Physically based modeling of squeeze film damping by mixed-level system simulation. *Sens. Actuators, A*, 97-98:193–200.

Seidl, M. and Steinheil, E. (1974). Measurement of momentum accommodation coefficients on surfaces characterized by auger spectroscopy, SIMS and LEED. In *Proceedings of the Nineth International Symposium on Rarefied Gas Dynamics*, pages E9.1–E9.2.

Senapati, S. and Chandra, A. (2001). Dielectric constant of water confined in a nanocavity. *J. Phys. Chem. B*, 105:5106–5109.

Senturia, S., Aluru, N. R., and White, J. (1997). Simulating the behavior of MEMS devices: Computational methods and needs. *IEEE Computational Science & Engineering*, 4 (January–March):30–43.

Senturia, S. D. (1998a). Simulation and design of microsystems: a 10-year perspective. *Sens. Actuators, A*, 70:1–7.

Senturia, S. D. (1998b). CAD challenges for microsensors, microactuators and microsystems. *IEEE Proceedings*, pages 1611–1626.

Sert, C. and Beskok, A. (2002). Time-periodic forced convection in micro heat spreaders. *Numerical Heat Transfer: A*, 42(7):685 – 705.

Shames, I. H. and Dym, C. L. (1985). *Energy and finite element methods in structural dynamics*. McGraw-Hill, New York.

Shanahan, M. E. R. (1988). Statics and dynamics of wetting on thin solids. *Rev. Phys. Appl.*, 33:1031–1037.

Shapiro, B., Moon, H., Garrell, R. L., and Kim, C.-J. (2003a). Equilibrium behavior of sessile drops under surface tension, applied external fields, and material variations. *J. Appl. Phys.*, 93:5794–5811.

Shapiro, B., Moon, H., Garrell, R. L., and Kim, C.-J. (2003b). Modeling of electrowetted surface tension for addressable microfluidic systems: Dominant physical effects, material dependencies, and limiting phenomena. In *IEEE The Sixteenth Annual International Conference on MEMS*, pages 201–205.

Shardlow, T. (2003). Splitting for dissipative particle dynamics. *SIAM J. Sci. Comput.*, 24(4):1267–1282.

Sharipov, F. and Sleznev, V. (1998). Data on internal rarefied gas flows. *J. Phys. Chem. Ref. Data*, 27, No 3:657–706.

Shaw, D. J. (1980). *Introduction to Colloid and Surface Chemistry*. Academic Press, New York.

Shen, C., Fan, J., and Xie, C. (2003). Statistical simulation of rarefied gas flows in micro-channels. *J. Comp. Phys.*, 189:512–526.

Sherman, F. S. (1990). *Viscous Flow*. McGraw-Hill, New York.

Shi, F. (1995). *Simulation and Analysis of MEMS with Applications of Sensitivity Analysis and Optimization*. PhD thesis, Cornell University.

Siewert, C. E. (2000). Poiseuille and thermal-creep flow in a cylindrical tube. *J. Comp. Phys.*, 160:470–480.

Silvaco International (1995). *ATLAS User's Manual: MixedMode*.

Simek, O. and Hadjiconstantinou, N. G. (2002). Constant-wall-temperature Nusselt number in micro and nano-channels. *J. Heat Transfer*, 124:356–364.

Singh, A., Cummings, E. B., and Throckmorton, D. J. (2001). Fluorescent liposome flow markers for microscale particle-image velocimetry. *Anal. Chem.*, 73(5):1057–1061.

Sinnott, S. B. (2002). Chemical functionalization of carbon nanotubes. *Journal of Nanoscience and Nanotechnology*, 2(2):113–123.

Sinton, D. and Li, D. (2003). Electroosmotic velocity profiles in microchannels. *Colloids Surf., A: Physicochem. Eng. Aspects*, 222:273–283.

Sirisup, S. and Karniadakis, G. E. (2004). A spectral viscosity method for correcting the long-term behavior of pod models. *J. Comp. Phys.*, 194:92–116.

Sirovich, L. (1987). Turbulence and the dynamics of coherent structures. *Quart. Appl. Math.*, 45:561–571.

Skordos, P. A. (1993). Initial and boundary conditions for the lattice Boltzmann method. *Phys. Rev. E*, 48:4823–42.

Smith, B., Petter, B., and Gropp, W. (1996). *Domain Decomposition.* Cambridge University Press, Cambridge.

Soderman, O. and Jonsson, B. (1996). Electroosmosis: Velocity profiles in different geometries with both temporal and spatial resolution. *J. Chem. Phys.*, 105(23):10300–10311.

Sokhan, V. P., Nicholson, D., and Quirke, N. (2001). Fluid flow in nanopores: An examination of hydrodynamic boundary conditions. *J. Chem. Phys.*, 115(8):3878–3887.

Solomentsev, Y., Bohmer, M., and Anderson, J. L. (1997). Particle clustering and pattern formation during electrophoretic deposition: A hydrodynamic model. *Langmuir*, 13:6058–6068.

Solomentsev, Y., Guelcher, S. A., Bevan, M., and Anderson, J. L. (2000). Aggregation dynamics for two particles during electrophoretic deposition under steady fields. *Langmuir*, 16:9208–9216.

Somers, S. A. and Davis, H. T. (1992). Microscopic dynamics of fluids confined between smooth and atomically structured solid surfaces. *J. Chem. Phys.*, 96(7):5389–5407.

Sone, Y. (1964). Kinetic theory analysis of linearized Rayleigh problem. *J. Phys. Soc. Japan*, 19(8):1463–1473.

Sone, Y. (1965). Effect of sudden change of wall temperature in rarefied gas. *J. Phys. Soc. Japan*, 20(2):222–229.

Sone, Y. (1969). Asymptotic theory of flow of a rarefied gas over a smooth boundary, I. In Trilling, L. and Wachman, H., editors, *Rarefied Gas Dynamics*, volume I, pages 243–253, Academic Press, New York.

Sone, Y. (1971). Asymptotic theory of flow of a rarefied gas over a smooth boundary, II. In Dini, D., editor, *Rarefied Gas Dynamics*, volume II, pages 737–749, Editrice Tecnico Scientifica, Pisa.

Sone, Y. (1972). Flow induced by thermal stress in rarefied gas. *Phys. Fluids*, 15:1418–1423.

Sone, Y. (1989). Analytical and numerical studies of rarefied gas flows on the basis of the Boltzmann equation for hard sphere molecules. *Phys. Fluids A*, 1(12):2042–2049.

Sone, Y. (2000). Flows induced by temperature fields in a rarefied gas and their ghost effect on the behavior of a gas in the continuum limit. *Annu. Rev. Fluid Mech.*, 32:779–811.

Sone, Y. (2002). *Kinetic Theory and Fluid Dynamics.* Birkhäuser.

Sone, Y. and Aoki, K. (1994). *Molecular Gas Dynamics.* Asakura, Tokyo (in Japanese).

Sone, Y., Aoki, K., Takata, S., Sugimoto, H., and Bobylev, A. V. (1996a). Inappropriateness of the heat-conduction equation for description of a temperature field of a stationary gas in the continuum limit: Examination by asymptotic analysis and numerical computation of the Boltzmann equation. *Phys. Fluids*, 8:628–638. Erratum, 1996, 8:841.

Sone, Y., Bardos, C., and Sugimoto, H. (2000). Asymptotic theory of the Boltzmann system for a steady flow of a slightly rarefied gas with a finite Mach number: General theory. *Eur. J. Mech., B/Fluids*, 19:325.

Sone, Y., Fukuda, T., Hokazono, T., and Sugimoto, H. (2001). Experiment on a one-way flow of a rarefied gas through a straight circular pipe without average temperature and pressure gradients. In *Rarefied Gas Dynamics, eds. T.J. Bartel & M. Galiss*, volume AIP, Melville.

Sone, Y. and Hasegawa, M. (1987). Poiseuille and thermal transpiration flows of a rarefied gas through a rectangular pipe. *J. Vac. Soc. of Japan (in Japanese)*, 30 (5):425–428.

Sone, Y., Takata, S., and Ohwada, T. (1990). Numerical analysis of the plane Couette flow of a rarefied gas on the basis of the linearized Boltzmann equation for hard-sphere molecules. *Eur. J. Mech., B/Fluids*, 9:273–288.

Sone, Y., Waniguchi, Y., and Aoki, K. (1996b). One-way flow of a rarefied gas induced in a channel with periodic temperature distribution. *Phys. Fluids*, 8:2227.

Sone, Y. and Yoshimoto, M. (1997). Demonstration of a rarefied gas flow induced near the edge of a uniformly heated plate. *Phys. Fluids*, 9:3530–3534.

Sparrow, E. M., Lundgren, T. S., and Lin, S. H. (1962). Slip flow in the entrance region of a parallel channel. In *Proc. Heat Transfer and Fluid Mechanics Institute, Stanford*, pages 223–238.

Spikes, H. and Granick, S. (2003). Equation for slip of simple liquids at smooth solid surfaces. *Langmuir*, 19:5065–5071.

Spoel, D., Buuren, A., Apol, E., Meulenhoff, P., Tieleman, D., Sijbers, A., Hess, B., Feenstra, K., Lindahl, E., Drunen, R., and Berendsen, H. (2001). *Gromacs User Manual version 3.0.* Nijenborgh 4, 9747 AG Groningen, the Netherlands. Internet: http://www.gromacs.org.

Spohr, E., Hartnig, C., Gallo, P., and Rovere, M. (1999). Water in porous glasses. a computer simulation study. *J. Mol. Liquids*, 80:165–178.

Spohr, E., Trokhymchuk, A., and Henderson, D. (1998). Adsorption of water molecules in slit pore. *J. Electroanal. Chem.*, 450:281–287.

Sreekanth, A. K. (1969). Slip flow through long circular tubes. In Trilling, L. and Wachman, H., editors, *Proceedings of the Sixth International Symposium on Rarefied Gas Dynamics*, volume 1, pages 667–680. Academic Press.

Srinivasan, V., Jog, A., and Fair, R. B. (2001). Scalable macromodels for microelectromechanical systems. In *Modeling and Simulation of Microsystems*, pages 72–75.

Steindl, A. and Troger, H. (2001). Methods for dimension reduction and their application in nonlinear dynamics. *Intl. Journal of Solids and Structures*, 38:2131–2147.

Stevens, M. J., Mondello, M., Grest, G. S., Cui, S. T., Cochran, H. D., and Cummings, P. T. (1997). Comparison of shear flow of hexadecane in a confined geometry. *J. Chem. Phys.*, 106:7303–7314.

Stillinger, F. H. (1973). Structure in aqueous solutions of nonpolar solutes from the standpoint of scaled-particle theory. *J. Solut. Chem.*, 2:141–158.

Stillinger, F. H. and Rahman, A. (1974). Improved simulation of liquid water by molecular dynamics. *J. Chem. Phys.*, 60:1545–1557.

Stimson, M. and Jeffery, G. B. (1926). The motion of two spheres in a viscous fluid. *Proc. Roy. Soc.*, 111:110 116.

Stone, A. J. (1996). *Theory of Intermolecular Forces.* Clarendon Press, Oxford.

Stone, H. A., Stroock, A. D., and Ajdari, A. (2004). Engineering flows in small devices: Microfluidics toward a lab-on-a-chip. *Ann. Rev. Fluid Mech.*, 34:381–411.

Stratton, J. A. (1941). *Electromagnetic Theory.* McGraw Hill, New York.

Stroock, A. D., Dertinger, S. K., Ajdari, A., Mezic, I., Stone, H. A., and Whitesides, G. M. (2002). Chaotic mixer for microchannels. *Science*, 295:647–651.

Succi, S. (2001). *The Lattice Boltzmann Equation: For Fluid Dynamics and Beyond.* Oxford University Press.

Succi, S., Karlin, I. V., and Chen, H. (2002). Colloquium: Role of the H theorem in lattice Boltzmann hydrodynamic simulations. *Rev. Mod. Phys.*, 74:1203–1220.

Sui, H., Han, B., Lee, J. K., Walian, P., and Jap, B. K. (2001). Structural basis of water-specific transport through the AQP1 water channel. *Nature*, 414:872–878.

Sun, L. and Crooks, R. M. (2000). Single carbon nanotube membranes: A well-defined model for studying mass transport through nanoporous materials. *J. Am. Chem. Soc.*, 122(49):2340–12345.

Sun, Q. and Boyd, I. D. (2002). A direct simulation method for subsonic, microscale gas flows. *J. Comp. Phys.*, 179:400–425.

Sun, Q. and Boyd, I. D. (2004). Flat-plate aerodynamics at very low Reynolds number. *J. Fluid Mech.*, 502:199–206.

Sun, Q., Boyd, I. D., and Candler, G. V. (2004). A hybrid continuum/particle approach for modeling subsonic, rarefied gas flows. *J. Comp. Phys.*, 194:256–277.

Svishchev, I. M., Kusalik, P. G., Wang, J., and Boyd, R. (1996). Polarizable point-charge model for water: Results under normal and extreme conditions. *J. Chem. Phys.*, 105:4742–4750.

Swaminathan, T. N. and Hu, H. H. (2004). Particle interactions in electrophoresis due to inertia. *J. Colloid Interface Sci.*, 273(1):324–330.

Szeri, A. Z. (1998). *Fluid Film Lubrication: Theory and Design.* Cambridge University Press, New York.

Tabor, D. (1991). *Gases, Liquids and Solids.* Cambridge Univ. Press, 3rd edition.

Tagawa, N. (1993). State of the art for flying head slider mechanisms in magnetic recording disk. *Wear*, 168:43.

Tai, Y. C., Fan, L. S., and Muller, R. S. (Salt Lake City, UT, February 1989). IC-processed micro-motors: design, technology, and testing. *IEEE Micro Electro Mechanical System Workshop.*

Takhistov, P., Indeikina, A., and Chang, H.-C. (2002). Electrokinetic displacement of air bubbles in microchannels. *Phys. Fluids*, 14(1):1–14.

Talanquer, V. (1997). A new phenomenological approach to gas-liquid nucleation based on the scaling properties of the critical nucleus. *J. Chem. Phys.*, 106(23):9957–9960.

Tan, M.-L., Fischer, J. T., Chandra, A., Brooks, B. R., and Ichiye, T. (2003). A temperature of maximum density in soft sticky dipole water. *Chem. Phys. Lett.*, 376:646–652.

Tang, C. W., Nguyen, T. H., and Howe, T. R. (1989). Laterally driven polysilicon resonant microstructures. *Sens. Actuators*, 20:25–32.

Tang, G. Y., Yang, C., Chai, J. C., and Gong, H. Q. (2004a). Joule heating effect on electroosmotic flow and mass species transport in a microcapillary. *Int. J. Heat Mass Transfer*, 47:215–227.

Tang, G. Y., Yang, C., Chai, J. C., and Gong, H. Q. (2004b). Numerical analysis of the thermal effect on electroosmotic flow and electrokinetic mass transport in microchannels. *Analytica Chimica Acta*, 507:27–37.

Tang, X. Z. and Boozer, A. H. (1996). Finite time Lyapunov exponent and advection-diffusion equation. *Phys. D*, 95:283–305.

Tanner, L. H. (1979). The spreading of silicone oil drops on horizontal surfaces. *J. Phys. D*, 12:1473–1484.

Tao, R. and Sun, J. M. (1991). Three-dimensional structure of induced electrorheological solid. *Phys. Rev. Lett.*, 67:398.

Technology Modeling Associates (1997). *MEDICI User's Manual: Circuit Analysis.*

Tehver, R., Toigo, F., Koplik, J., and Banavar, J. R. (1998). Thermal walls in computer simulations. *Phys. Rev. E*, 57 (1):R17–R20.

Tell, J. L. and Maris, H. J. (1983). Specific heats of hydrogen, deuterium, and neon in porous Vycor glass. *Phys. Rev. B*, 28:5122–5125.

Telleman, P., Larsen, U. D., Philip, J., Blankenstein, G., and Wolff, A. (1998). Cell sorting in microfluidic systems. In van den Berg, H. ., editor, *Micro Total Analysis Systems '98*, pages 39–44. Kluwer.

Terray, A., Oakey, J., and Marr, D. (2002). Microfluidic control using colloidal devices. *Science*, 296:1841–1843.

Tersoff, J. (1988a). Empirical interatomic potential for carbon, with applications to amorphous carbon. *Phys. Rev. Lett.*, 61:2879–2882.

Tersoff, J. (1988b). New empirical approach for the structure and energy of covalent systems. *Phys. Rev. B*, 37:6991–7000.

Thompson, P. and Troian, S. (1997). A general boundary condition for liquid flow at solid surfaces. *Nature*, 389:360–362.

Thompson, P. A. (1988). *Compressible Fluid Dynamics.* McGraw Hill, New York.

Thompson, P. A. and Robbins, M. O. (1990). Shear flow near solids: Epitaxial order and flow boundary conditions. *Phys. Rev. A*, 41(12):6830–6837.

Thompson, S. L. and Owens, W. R. (1975). A survey of flow at low pressures. *Vacuum*, 25:151–156.

Thorsen, T., Maerkl, S. J., and Quake, S. R. (2002). Microfluidic large-scale integration. *Science*, 298:580–584.

Tilmans, H. A. (1996). Equivalent circuit representation of electromechanical transducers: I. Lumped-parameter systems. *J. Micromech. Microeng.*, 6:157–176.

Tilmans, H. A. (1997). Equivalent circuit representation of electromechanical transducers: II. Distributed-parameter systems. *J. Micromech. Microeng.*, 7:285–309.

Timoshenko, S. and Woionowsky-Krieger, S. (1959). *Theory of Plates and Shells.* McGraw-Hill Book Company: London.

Tison, S. A. (1993). Experimental data and theoretical modeling of gas flows through metal capillary leaks. *Vacuum*, 44:1171–1175.

Tison, S. A. (1995). Private communications. *NIST.*

Todd, B. D., Evans, D. J., and Davis, P. J. (1995). Pressure tensor for inhomogeneous fluids. *Phys. Rev. E*, 52:1627–1638.

Tomboulides, A. G., Israeli, M., and Karniadakis, G. E. (1989). Efficient removal of boundary-divergence errors in time-splitting methods. *J. Sci. Comp.*, 4:291.

Toney, M. F., Howard, J. N., Richer, J., Borges, G. L., Gordon, J. G., Melroy, O. R., Wiesler, D. G., Yee, D., and Sorensen, L. B. (1994). Voltage-dependent ordering of water molecules at an electrode/electrolyte interface. *Nature*, 368:444–446.

Toxvaerd, S. (1981). The structure and thermodynamics of a solid-fluid interface. *J. Chem. Phys.*, 74(3):1998–2005.

Trau, M., Saville, D. A., and Aksay, I. A. (1996). Field-induced layering of colloidal crystals. *Science*, 272:706–709.

Trau, M., Saville, D. A., and Aksay, I. A. (1997). Assembly of colloidal crystals at electrode interfaces. *Langmuir*, 13:6375–6381.

Traube, J. and Whang, S.-H. (1928). Über Reibungskonstante und Wandschicht. *Z. Physikal. Chem. A*, 138:102–122.

Travis, K. P. and Evans, D. J. (1996). Molecular spin in a fluid undergoing Poiseuille flow. *Phys. Rev. E*, 55(2):1566–1572.

Travis, K. P. and Gubbins, K. E. (2000). Poiseuille flow of Lennard-Jones fluids in narrow slit pores. *J. Chem. Phys.*, 112(4):1984–1994.

Travis, K. P., Todd, B. D., and Evans, D. J. (1997). Departure from Navier Stokes hydrodynamics in confined liquids. *Phys. Rev. E*, 55(4):4288–4295.

Tretheway, D. C. and Meinhart, C. D. (2002). Apparent fluid slip at hydrophobic microchannel walls. *Phys. Fluids*, 14(3):L9–L12.

Trimmer, W. (1997). *Micromechanics and MEMS, Classic and Seminal Papers to 1990*. IEEE Press.

Troger, H. and Steindl, A. (1991). *Non-Linear Stability and Bifurcation Theory: An Introduction for Engineers and Applied Scientists*. Springer, New York.

Tryggvason, G., Bunner, B., Ebrat, O., and Tauber, W. (1998). Computations of multiphase flows by a finite difference/front tracking method. In *29th Computational Fluid Dynamics, Lecture Series, von Karman Institute*.

Tseng, W. L. and Chang, H. T. (2001). A new stategy for optimizing sensitivity, speed, and resolution in capillary electrophoretic separation of DNA. *Electrophoresis*, 22:763–770.

Tunc, G. and Bayazitoglu, Y. (2002). Heat transfer in rectangular microchannels. *Int. J. Heat Mass Transfer*, 45(4):765–773.

Turowski, M., Chen, Z., and Przekwas, A. (1998). Squeeze film behavior in MEMS for large amplitude motion: 3D simulations and non-linear circuit/behavioral models. In *IEEE/VIUF Intl. Conf. BMAS'98*.

Tyrrell, J. and Attard, P. (2002). Atomic force microscope images of nanobubbles on a hydrophobic surface and corresponding force-separation data. *Langmuir*, 18:160–167.

Tysanner, M. W. and Garcia, A. L. (2004). Measurement bias of fluid velocity in molecular simulations. *J. Chem. Phys.*, 196:173–183.

Unger, M. A., Chou, H.-P., Thorsen, T., Scherer, A., and Quake, S. R. (2000). Monolithic microfabricated valves and pumps by multilayer soft lithography. *Science*, 288:113–116.

Urbanek, W., Zemel, J. N., and Bau, H. H. (1993). An investigation of the temperature dependence of Poiseuille numbers in microchannel flow. *J. Micromech. Microeng.*, 3:206–208.

van der Spoel, D., Lindahl, E., Hess, B., van Buuren, A. R., Apol, E., Meulenhoff, P. J., Tieleman, D. P., Sijbers, A. L. T. M., Feenstra, K. A., van Drunen, R., and Berendsen, H. J. C. (2004). *Gromacs User Manual version 3.2*. www.gromacs.org.

van der Spoel, D., van Maaren, P. J., and Berendsen, H. J. (1998). A systematic study of water models for molecular simulation: derivation of water models optimized for use with a reaction field. *J. Chem. Phys.*, 108(24):10220–10230.

van Kampen, R. P. (1995). *Bulk-Micromachinend Capacitive Servo-Accelerometer*. PhD thesis, Technical University of Delft, Delft.

Vandemeer, J. E., Kranz, M. S., and Fedder, G. K. (1998). Hierarchical representation and simulation of micromachined inertial sensors. In *Modeling and Simulation of Microsystems*.

Varghese, M., Rabinovich, V. L., and Senturia, S. D. (1999). Reduced-order modeling of Lorentz force actuation with modal basis functions. In *Modeling and Simulation of Microsystems*, pages 155–158.

Vargo, S. E. and Muntz, E. P. (1996). A simple micromechanical compressor and vacuum pump for flow control and other distributed applications. In *Thirty-Fourth Aerospace Sciences Meeting and Exhibit January 15–18 1996, Reno, NV, AIAA 96-0310.*

Vargo, S. E., Muntz, E. P., Shiflett, G. R., and Tang, W. C. (1998). Knudsen compressor as a micro- and macroscale vacuum pump without moving parts or fluids. *J. Vac. Sci. Technol. A-Vacuum Surfaces and Films Part 2*, 17(4):2308–2313.

Vasilyev, D., Rewienski, M., and White, J. (2003). A TBR-based trajectory piecewise-linear algorithm for generating accurate low-order models for nonlinear analog circuits and MEMS. In *Proceedings of Design Automation Conference*, pages 490–495.

Veijola, T. (1999). Equivalent circuit models for micromechanical inertial sensors. Technical report, Helsinki University of Technology, Circuit Theory Laboratory, CT-39.

Veijola, T. (2000). Compact damping models for lateral structures including gas rarefaction effects. In *Proceedings of MSM 2000*, pages 162–165, San Diego.

Veijola, T. (2001). Acoustic impedence elements modeling oscillating gas flow in micro channels. In *Modeling and Simulation of Microsystems*, pages 96–99.

Veijola, T., Kuisma, H., and Lahdenperä, J. (1998). The influence of gas-surface on gas film damping in a silicon accelerometer. *Sens. Actuators, A*, 66:83–92.

Veijola, T., Kuisma, H., Lahdenperä, J., and Ryhänen, T. (1995a). Equivalent-circuit model of the squeezed gas film in a silicon accelerometer. *Sens. Actuators, A*, 48:239–248.

Veijola, T., Ryhänen, T., Kuisma, H., and Lahdenperä, J. (1995b). Circuit simulation model gas damping in microstructures with nontrivial geometries. In *Proceedings of Transducers'95 & Eurosensors IX, Stockholm, Sweden, June 25–29*, pages 36–39.

Veijola, T. and Turowski, M. (2001). Compact damping models for laterally moving microstructures with gas-rarefaction effects. *J. Microelectromech. Sys.*, 10:263–273.

Velev, O. D., Prevo, B. G., and Bhatt, K. H. (2003). On-chip manipulation of free droplets. *Nature*, 426:515–516.

Vergeles, M., Keblinski, P., Koplik, J., and Banavar, J. R. (1996). Stokes drag and lubrication flows: A molecular dynamics study. *Phys. Rev. E*, 53(5):4852–4864.

Verheijen, H. J. and Prins, M. W. (1999). Reversible electrowetting and trapping of charge: model and experiments. *Langmuir*, 15:6616–6620.

Vijayakumar, P., Sun, Q., and Boyd, I. D. (1999). Vibrational-translational energy exchange models for the direct simulation Monte Carlo method. *Phys. Fluids*, 11 (8):2117–2126.

Vincenti, W. G. and Kruger, C. H. (1977). *Introduction to Physical Gas Dynamics*. Robert E. Krieger Publishing Company, Huntington, New York.

Vinogradova, O. I. (1995). Drainage of a thin liquid film confined between hydrophobic surfaces. *Langmuir*, 11:2213–2220.

Vinogradova, O. I. (1996). Hydrodynamic interaction of curved bodies allowing slip on their surfaces. *Langmuir*, 12:5963–5968.

Vinogradova, O. I. (1999). Slippage of water over hydrophobic surfaces. *Int. J. Miner. Process.*, 56:31–60.

Voigt, P. and Wachutka, G. (1997). Electrofluidic microsystem modeling based on Kirchhoffian network theory. In *Transducers 1997*, pages 1019–1022.

Volpert, M., Mezic, I., Meinhart, C. D., and Dahleh, M. (1999). An actively controlled micromixer. In *Proccedings of the ASME, MEMS, pp. 483-487, Nashville, TN.*

von Smoluchowski, M. (1916). Drei vorträge über Diffusion, Brownsche Molekularbewegung und Koagulation von Kolloidteilchen. *Z. Phys.*, 17:557–585.

Wadsworth, D. C. and Erwin, D. A. (1992). Two-dimensional hybrid continuum/particle approach for rarefied flows. *AIAA*, 92-2975.

Waghe, A., Rasaiah, J. C., and Hummer, G. (2002). Filling and emptying kinetics of carbon nanotubes in water. *J. Chem. Phys.*, 117:10789–10795.

Wagner, A. J. and Pagonabarraga, I. (2002). Lees-Edwards boundary conditions for Lattice Boltzmann. *J. Stat. Phys.*, 107:521–537.

Walther, J. H., Jaffe, R., Halicioglu, T., and Koumoutsakos, P. (2001a). Carbon nanotubes in water: Structure characteristics and energies. *J. Phys. Chem. B*, 105:9980–9987.

Walther, J. H., Jaffe, R., Halicioglu, T., and Koumoutsakos, P. (2001b). Hydrophobic hydration of carbon nanotubes. Technical report, ETH, Zurich, Switzerland.

Washburn, E. W. (1921). The dynamics of capillary flow. *Phys. Rev.*, 17(3):273–283.

Wasserman, S. R., Whitesides, G. M., Tidswell, I. M., Ocko, B. M., Pershan, P. S., and Axe, J. D. (1989). The structure of self-assembled monolayers of alkylsiloxanes on silicon: A comparison of results from ellipsometry and low-angle x-ray reflectivity. *J. Am. Chem. Soc.*, 111(15):5852–5861.

Weighardt, K. E. (1953). On the resistance of screens. *The Aeronautical Quarterly*, 4:186–192.

Weigl, B. H., Bardell, R. L., and Cabrera, C. R. (2003). Lab-on-a-chip for drug development. *Advanced Drug Delivery Reviews*, 55:349–377.

Welder, T. W., Chapman, D. R., and MacCormack, R. W. (1993). Evaluation of various forms of the Burnett equations. In *AIAA Fluid Dynamics Conference July 6-9, 1993, Orlando FL, AIAA 93-3094.*

Wen, S. B. and Lai, C. L. (2003). Theoretical analysis of flow passing a single sphere moving in a micro-tube. *Proc. R. Soc. London, A*, 459:495–526.

Wendl, M. C. and Agarwal, R. K. (2002). Viscous dissipation in finite thin-gap Couette devices. *J. Heat Transfer*, 124:200–203.

Werder, T., Walther, J. H., Jaffe, R. L., Halicioglu, T., and Koumoutsakos, P. (2003). On the water-carbon interaction for use in molecular dynamics simulations of graphite and carbon nanotubes. *J. Phys. Chem. B*, 107:1345–1352.

Werder, T., Walther, J. H., Jaffe, R. L., Halicioglu, T., Noca, F., and Koumoutsakos, P. (2001). Molecular dynamics simulation of contact angle of water droplet in carbon nanotubes. *Nano Lett.*, 1:697–702.

Westermeier, R. (1990). *Electrophoresis in Practice.* VCH Publishers Inc, New York.

White, F. M. (1991). *Viscous Fluid Flow.* McGraw-Hill International Editions, Mechanical Engineering Series.

Whitesides, G. M. and Grzybowski, B. (2002). Self-assembly at all scales. *Science*, 295:2418–2421.

Wiedmann, G. (1852). Ueber die Bewegung von Flüssigkeiten im Kreise der geschlossenen galvanischen Säule. *Pogg. Ann.*, 87:321.

Wiggins, S. and Ottino, J. M. (2004). Foundations of chaotic mixing. *Phil. Trans. R. Soc. London, Ser. A*, 362:937–970.

Willemsen, S. M., Hoefsloot, H. C., and Iedema, P. D. (2000). No-slip boundary condition in dissipative particle dynamics. *Int. J. Mod. Phys.*, 11(5):881–890.

Williams, M. B. and Davis, S. H. (1982). Nonlinear theory of film rupture. *J. Colloid Interface Sci.*, 90:220–228.

Woelki, S. and Kohler, H. H. (2000). A modified Poisson-Boltzmann equation: I. basic relations. *Chem. Phys.*, 261:411–419.

Wolff, A., Larsen, U. D., Blankenstein, G., Philip, J., and Telleman, P. (1998). Rare event cell sorting in a microfluidic system for application in prenatal diagnosis. In *Micro Total Analysis Systems '98, eds. Harrison and van den Berg, Kluwer*, pages 77–80.

Woodhouse, G., King, L., Wieczorek, L., Osman, P., and Cornell, B. (1999). The ion channel switch biosensor. *Journal of Molecular Recognition*, 12(5):328–334.

Woods, L. C. (1993). *An Introduction to the Kinetic Theory of Gases and Magnetoplasmas.* Oxford Science Publications.

Woodson, H. H. and Melcher, J. R. (1968). *Electromechanical Dynamics.* Wiley, New York.

Wu, J. and Carley, R. L. (2001). Table-based numerical macromodeling for MEMS devices. In *Modeling and Simulation of Microsystems*, pages 68–71.

Xiu, D. and Karniadakis, G. E. (2001). A semi-Lagrangian high-order method for Navier-Stokes equations. *J. Comp. Phys.*, 172:658.

Xu, K. (2003). Super-Burnett solutions for Poiseuille flow. *Phys. Fluids*, 15(7):2077–2080.

Xue, H. and Chen, S. (2003). DSMC simulation of microscale backward-facing step flow. *Microscale Thermophysical Engineering*, 7:69–86.

Xue, H., Ji, H. M., and Shu, C. (2003). Prediction of flow and heat transfer characteristics in micro-Couette flow. *Microscale Thermophys. Eng.*, 7(1):51–68.

Yager, P., Bell, D., Brody, J. P., Qin, D., Cabrera, C., Kamholz, A., and Weigl, B. (1998). Applying microfluidic chemical analytical systems to imperfect samples. In Harrison and van den Berg, editors, *Micro Total Analysis Systems '98*, pages 207–212. Kluwer.

Yang, C. and Li, D. (1998). Analysis of electrokinetic effects on the liquid flow in rectangular microchannels. *J. Colloids Surfaces*, 143:339–353.

Yang, J., Huang, Y., Wang, X. B., Becker, F. F., and Gascoyne, P. R. (1999a). Cell separation on microfabricated electrodes using dielectrophoretic/gravitational field-flow fractionation. *Anal. Chem.*, 71:5:911–918.

Yang, J., Huang, Y., Wang, X. B., Becker, F. F., and Gascoyne, P. R. (2000). Differential analysis of human leukocytes by dielectrophoretic field-flow fractionation. *Biophys. J.*, 78:2680–2689.

Yang, J. M., Yang, X., Ho, C. M., and Tai, Y. C. (2001). Micromachined particle filter with low power dissipation. *J. Fluids Eng.*, 123:899–908.

Yang, X., Yang, J. M., Tai, Y. C., and Ho, C. M. (1999b). Micromachined membrane particle filters. *Sens. Actuators, A: Physical*, 73:184–191.

Ye, H., Naguib, N., Gogotsi, Y., Yazicioglu, A. G., and Megaridis, C. M. (2004). Wall structure and surface chemistry of hydrothermal carbon nanofibers. *Nanotechnology*, 15:232–236.

Ye, W., Kanapka, J., and White, J. (1999). A fast 3D solver for unsteady Stokes flow with applications to micro-electro-mechanical systems. In *Proceedings of the Second International Conference on Modeling and Simulation of Microsystems*, pages 518–521.

Yeh, I. and Berkowitz, M. L. (2000). Effects of the polarizability and water density constraints on the structure of water near charged surfaces: Molecular dynamics simulations. *J. Chem. Phys.*, 112:10491–10495.

Yong, W.-A. and Luo, L.-S. (2003). Nonexistence of H theorems for the athermal lattice Boltzmann models with polynomial equilibria. *Phys. Rev. E*, 67:051105.

Yu, S. and Ameel, T. A. (2002). Slip flow convection in isoflux rectangular microchannels. *Microscale Thermophysical Engineering*, 1:303–320.

Yuste, S. B. and Santos, A. (1994). A model for the structure of square-well fluids. *J. Chem. Phys.*, 101:2355–2364.

Zhang, L., Balasundaram, R., Jiang, S., and Gehrke, S. H. (2001). Nonequilibrium molecular dynamics simulation of confined liquids in contact with the bulk. *J. Chem. Phys.*, 114(15):6869–6877.

Zhang, Q., Zheng, J., Shevade, A., Zhang, L., Gehrke, S., Heffelfinger, G. S., and Jiang, S. (2002). Transport diffusion of liquid water and methanol through membranes. *J. Chem. Phys.*, 117:808–817.

Zhang, Y., Henson, M. A., and Kevrekidis, I. G. (2003). Non-linear reduction for dynamic analysis of cell population models. *Chemical Engineering Science*, 58:429–445.

Zhao, B., Moore, J. S., and Beebe, D. J. (2001). Surface-directed liquid flow inside microchannels. *Science*, 291:1023–1026.

Zheng, Y., Garcia, A. L., and Alder, B. J. (2002). Comparison of kinetic theory and hydrodynamics for Poiseuille flow. *J. Stat. Phys.*, 109:495–505.

Zheng, Z., Hansford, D. J., and Conlisk, A. T. (2003). Effect of multivalent ions on electroosmotic flow in micro- and nanochannels. *Electrophoresis*, 24.

Zholkovskij, E. K. and Masliyah, J. H. (2004). Hydrodynamic dispersion due to combined pressure driven and electroosmotic flow through micro-channels with a thin double layer. *Anal. Chem.*, 76:2708–2718.

Zholkovskij, E. K., Masliyah, J. H., and Czarnecki, J. (2003). Electroosmotic dispersion in microchannels with a thin double layer. *Anal. Chem.*, 75 (4):901–909.

Zhong, X. (1993). On numerical solutions of Burnett equations for hypersonic flow past 2-D circular blunt leading edges in continuum transition regime. In *AIAA 42. Fluid Dynamics Conference July 6-9, 1993, Orlando FL, AIAA 93-3092.*

Zhou, J. D., Cui, S. T., and Cochran, H. D. (2003). Molecular simulation of aqueous electrolytes in model silica nanochannels. *Mol. Phys.*, 101:1089–1094.

Zhou, N., Brown, S., and Pister, K. S. (1998). Nodal analysis of MEMS simulation and design. In *Modeling and Simulation of Microsystems*, pages 308–313.

Zhu, Y. and Granick, S. (2001). Rate-dependent slip of Newtonian liquid at smooth surfaces. *Phys. Rev. Lett.*, 87(9):096105.

Zhu, Y. and Granick, S. (2002). Limits of the hydrodynamic no-slip boundary condition. *Phys. Rev. Lett.*, 88(10):106102.

Index

Interdisciplinary Applied Mathematics

1. *Gutzwiller:* Chaos in Classical and Quantum Mechanics
2. *Wiggins:* Chaotic Transport in Dynamical Systems
3. *Joseph/Renardy:* Fundamentals of Two-Fluid Dynamics:
   Part I: Mathematical Theory and Applications
4. *Joseph/Renardy:* Fundamentals of Two-Fluid Dynamics:
   Part II: Lubricated Transport, Drops and Miscible Liquids
5. *Seydel:* Practical Bifurcation and Stability Analysis:
   From Equilibrium to Chaos
6. *Hornung:* Homogenization and Porous Media
7. *Simo/Hughes:* Computational Inelasticity
8. *Keener/Sneyd:* Mathematical Physiology
9. *Han/Reddy:* Plasticity: Mathematical Theory and Numerical Analysis
10. *Sastry:* Nonlinear Systems: Analysis, Stability, and Control
11. *McCarthy:* Geometric Design of Linkages
12. *Winfree:* The Geometry of Biological Time (Second Edition)
13. *Bleistein/Cohen/Stockwell:* Mathematics of Multidimensional
    Seismic Imaging, Migration, and Inversion
14. *Okubo/Levin:* Diffusion and Ecological Problems: Modern Perspectives
    (Second Edition)
15. *Logan:* Transport Modeling in Hydrogeochemical Systems
16. *Torquato:* Random Heterogeneous Materials: Microstructure and
    Macroscopic Properties
17. *Murray:* Mathematical Biology I: An Introduction (Third Edition)
18. *Murray:* Mathematical Biology II: Spatial Models and Biomedical
    Applications (Third Edition)
19. *Kimmel/Axelrod:* Branching Processes in Biology
20. *Fall/Marland/Wagner/Tyson (Editors):* Computational Cell Biology
21. *Schlick:* Molecular Modeling and Simulation: An Interdisciplinary Guide
22. *Sahimi:* Heterogeneous Materials I: Linear Transport and Optical Properties
23. *Sahimi:* Heterogeneous Materials II: Nonlinear and Breakdown Properties
    and Atomistic Modeling
24. *Bloch:* Nonholonomic Mechanics and Control
25. *Beuter/Glass/Mackey/Titcombe (Editors):* Nonlinear Dynamics in
    Physiology and Medicine
26. *Ma/Soatto/Košecká/Sastry:* An Invitation to 3-D Vision: From Images to
    Geometric Models
27. *Ewens:* Mathematical Population Genetics (Second Edition)
28. *Wyatt:* Quantum Dynamics with Trajectories: Introduction to Quantum
    Hydrodynamics
29. *Karniadakis/Beskok/Aluru:* Microflows and Nanoflows: Fundamentals and
    Simulation